Moonshots and Snapshots of Project Apollo

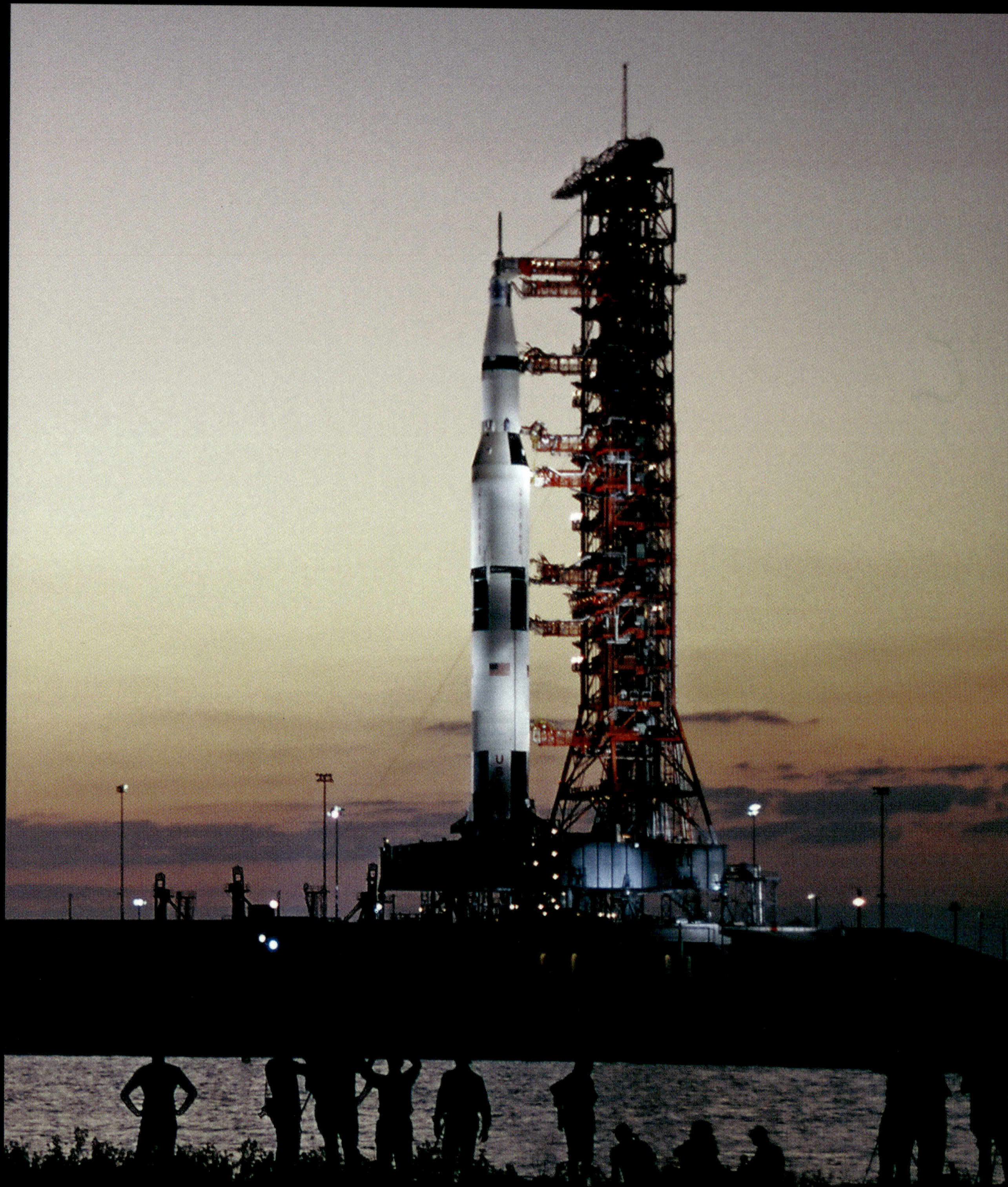

JOHN BISNEY

J. L. PICKERING

Moonshots and Snapshots of

PROJECT APOLLO

A Rare Photographic History

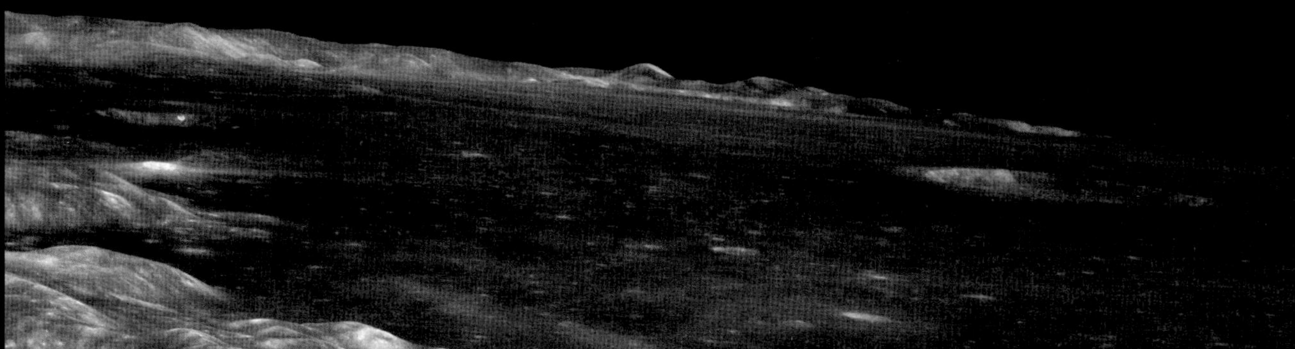

University of New Mexico Press - Albuquerque

Printed in Korea
20 19 18 17 16 15 1 2 3 4 5 6

Library of Congress Cataloging-in-Publication Data

Bisney, John, 1954–
Moonshots and snapshots of Project Apollo : a rare photographic history / by John Bisney and J. L. Pickering.
pages cm
Includes bibliographical references.
ISBN 978-0-8263-5594-2 (cloth : alk. paper)
1. Project Apollo (U.S.)—Pictorial works. 2. Project Apollo (U.S.)—History. 3. Space flight to the moon—Pictorial works. 4. Space photography—United States. I. Pickering, J. L., 1957– II. Title.
TL789.8.U6A5217 2015
629.45′4—dc23
2014038537

Cover photograph courtesy of NASA
Designed by Lila Sanchez
Composed in Adobe Garamond Pro 11/13.5
Display type is Americana and Optima

Page ii: Photographers on the Saturn Causeway capture the Apollo 16 launch vehicle on Pad 39A at sunset, looking west across Pintail Creek, on April 15. (photo by Tom Usciak)

Contents

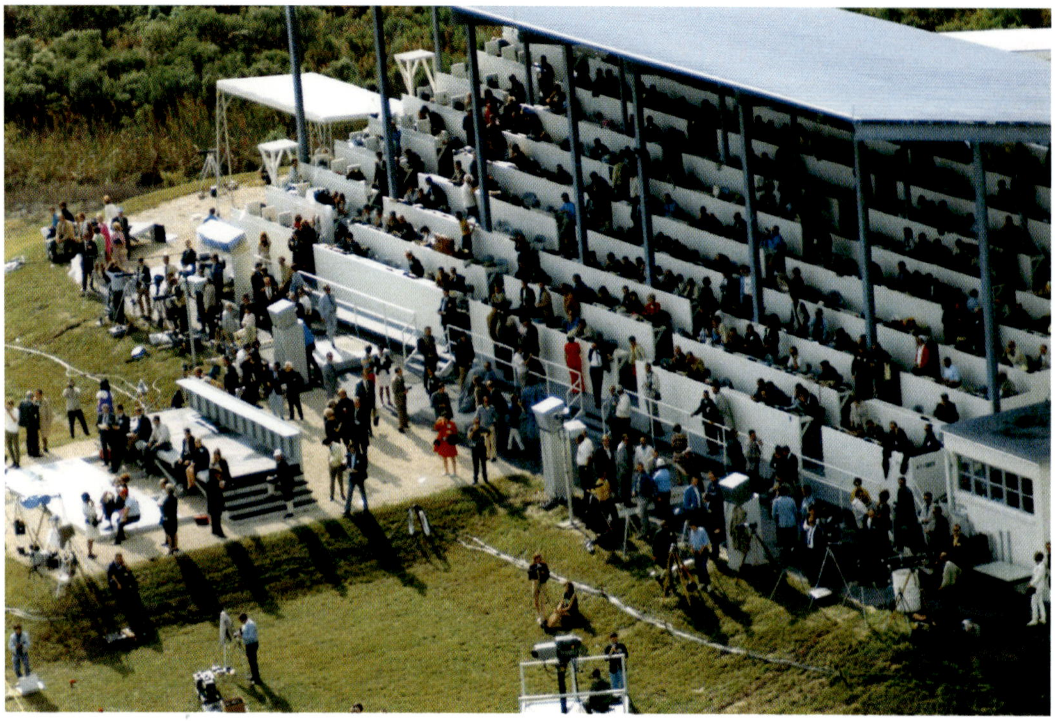

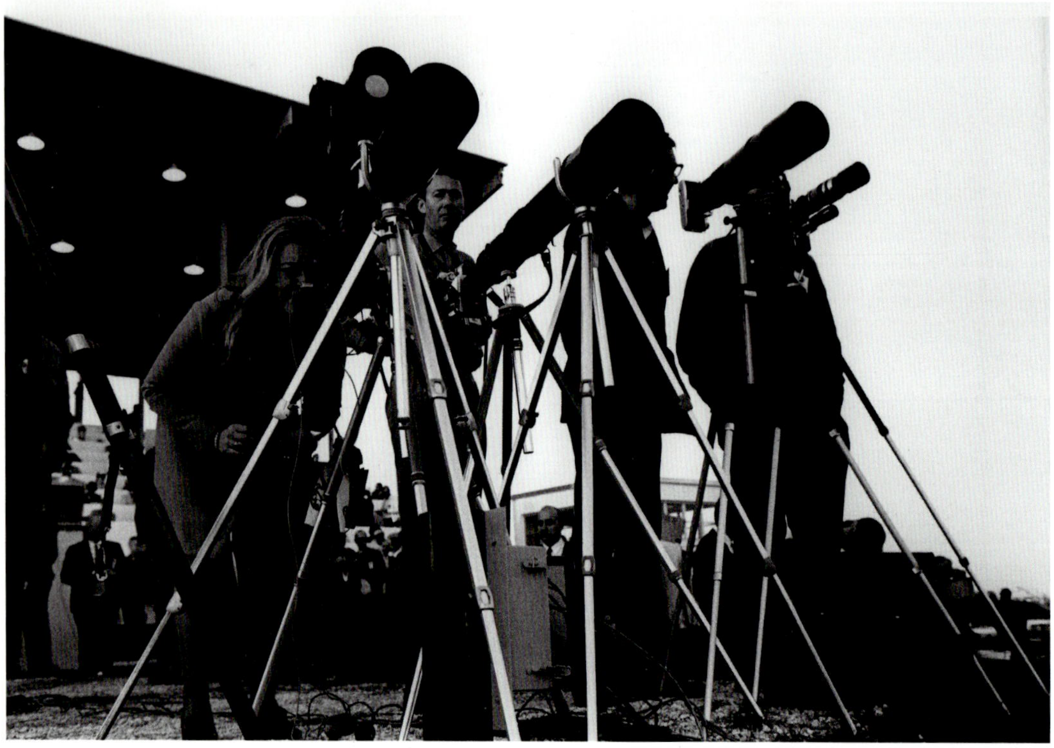

Top: Activity at the grandstand at the LC-39 Press Site picks up as the Apollo 12 launch approaches on November 14, 1969. Built in 1967, the grandstand was torn down following damage from Hurricane Frances in September 2004. The CBS-TV camera at bottom center is on the platform seen on page 212.

Bottom: A photographer sets up for the Apollo 8 launch on December 21, 1968, at the LC-39 Press Site, the first manned launch from the complex.

Acknowledgments

We wish to gratefully thank the following individuals, without whose help this effort would not have been complete.

Jay Barbree
Alan Bean
Charles Bourland
Vance Brand
Tammy Summers Bucher
John Byram
Norm Carlson
Jerry Carr
Gene Cernan
Skip Chauvin
Wes Chesser
Laura Shepard Churchley
Vic Craft
Walt Cunningham
John de Bry
Paul Donnelly
Owen Garriott
Mike Gentry
Ed Gibson
Mark Gray
Ed Harrison
Ken Havekotte
Jim Hawk
A. R. Hogan
John Johnson
Joel Kastowitz
Chris Kraft
Chuck Lewis
Sy Liebergot
Ed Mitchell
Dee O'Hara
Margaret Persinger
Don Phillips
Gwen Pittman
Bill Pogue
James Ragusa
Al Rochford
Louie Roquefort
Jody Russell
Scott Schneeweis
Rusty Schweickart
Bob Sieck
Lee Starrick
Dwight Steven-Boniecki
John Stonesifer
Larry Summers
Tip Talone
Kipp Teague
Jacques Tiziou
Jonathan Ward
Don Willis
Leigh Wilson
Ron Woods
Robert Yowell

Introduction

John Bisney

This book, which begins in 1967, chronicles the Apollo program and its two follow-ons, Skylab and the Apollo-Soyuz Test Project (ASTP). It is the companion volume to *Spaceshots and Snapshots of Projects Mercury and Gemini*. We again drew from among the more rarely seen manned spaceflight images to take you to the moon and on months-long odysseys above Earth, wrapping up with the first international manned flight in 1975.

The most important outside source for this project was our friend, French photojournalist Jacques Tiziou. I first met Jacques while covering Space Shuttle missions, and I eventually mentioned to J. L. that Jacques had been a reporter and photographer during Apollo. The two soon met; J. L. was thrilled to have hit such a gold mine. Jacques still had most of his Apollo photos as beautiful 35mm color slides stored in his basement, unseen for decades. He has generously given us permission to share some of his unpublished photos, which often provide a more artistic look than NASA's somewhat clinical photographs do. That said, government "shooters" could also capture the unusual angle or moment, and all of our contributors could be said to have taken a fair number of "snapshots" during their professional duties.

We also tried to provide a more complete understanding of all the events and processes before and after the missions than existing space-photo books have done. J. L sent the photos to me in sequence, presenting the missions in almost storyboard fashion. I then chose those I felt had exceptional visual, engineering, or human-interest appeal. We occasionally switched out a photo to maintain the flight's narrative, or in a few instances, J. L. came up with a higher-quality or more unusual image to substitute.

For the Apollo program, one group of advisors deserves special mention: the astronauts themselves. Thanks to J. L.'s reputation and to our mutual involvement in the space-history community, we have been honored to become friends with a number of the pioneering men whose exploits we illustrate. Their input has been invaluable and has helped to ensure the accuracy of this work. I would especially like to thank Walt Cunningham, Alan Bean, and Ed Gibson for their generous help.

A word about distances—we used statute (land-based) miles for distances instead of what is commonly used in space (nautical miles), simply because that's what most people are familiar with.

Each chapter again begins with that flight's crew emblem, and for these missions we use the final artwork. The detailed captions take you through the climax and conclusion of the Space Race. Now J. L. will provide you with his insights into each mission's photo pedigree.

J. L. Pickering

When I think of an iconic photo from Apollo 1, unfortunately that of the charred spacecraft comes to mind, but we instead selected the "praying" crew photo to open this book. A relatively rare image, the color version was a nice find several years ago. Chapter 1 utilizes material from the spacecraft's manufacturer, North American, showing the crew and spacecraft in Downey, California. Also included are more recent finds from the Johnson Space Center (JSC) showing the crewmen during their final month of training.

Apollo 7 turned out to be the subject of one of my favorite chapters, as there was a tremendous amount of photographic material to choose from. I was able to secure some nice "alternate" crew photos from astronaut Walt Cunningham, who was also a big help with some identification work; fellow photo historian Ed Hengeveld provided some fresh views of the recovery. My only regret with the Apollo 7 photography in general is that I wish there were more images of the crew during the mission. Nonetheless, we are pleased to give this mission, which is frequently downplayed, its due.

Apollo 8 really brings into focus the enormity of the Apollo program, from the challenges of processing and launching the first manned Saturn V, to the rewards of the first majestic images of Earth from deep space. No one would argue with the choice of the "iconic" image of Apollo 8, and we placed it in its original orientation. When sizing up the availability of pre- and postflight images from the Apollo program, Apollo 8 amazingly ranks near the bottom. For some reason, the usual "overflow" shots were not released (as was done for Apollo 9 and 11). We use some NASA images showing the news media at Kennedy Space Center (KSC); these give a sense of what it was like to cover an Apollo launch. We also include some launch photos by Jacques Tiziou and his brother, Michel.

The astronauts of Apollo 9 showed that they had a good feel for the camera, producing some of the best photos of an undocked command module (CM) and lunar module (LM). This was the only mission where the beauty of Earth served as a backdrop for the LM in flight. Apollo 9, another flight often overlooked, also provided the only photos of an astronaut (Rusty Schweickart) working outside the LM in space, and of an astronaut (David Scott) emerging from a CM hatch.

Apollo 10 finally provided photos of both Apollo spacecraft in lunar orbit. The high priority placed on color television, however—there were seventeen live and two recorded transmissions of Apollo 10—may have resulted in fewer photographs. The mission also saw the only Saturn V launch from Pad 39B.

Apollo 11 produced a treasure trove of preflight images to choose from. Not only did NASA's photo team put in extra effort, but that mission also captured the attention of photographers from all over the world. The result was one of the most documented events in history. Chapter 6 uses several creative images by the Tiziou team.

One overriding observation I made is how well the crewmen documented the flight of Apollo 11 on film. Although Buzz Aldrin took only two photos of Neil Armstrong on the moon, the lunar extravehicular activity (EVA) produced some of the most iconic space-exploration images ever. As a bonus, Michael Collins shot some of the best LM rendezvous-sequence images.

Inclement weather on Apollo 12's launch day meant poor visibility and the fewest photos taken of a Saturn V launch. And when television coverage was lost during the first lunar EVA, the photos the crew brought back became more important than ever. The black-and-white images of Pete Conrad with Surveyor III are superb. As a prank, Alan Bean had intended to set the Hasselblad camera to show himself and Conrad standing by the unmanned spacecraft, but he could not find the self-timer he had brought. The crew's fun-loving attitude is reflected in the images.

Since onboard images from Apollo 13 were somewhat limited, we concentrated on the considerable number of premission photos available. The Apollo 13 rollout shots are some of the most beautiful taken. The launch photos were great at ignition, but cloud cover eliminated decent tracking views. My only regret is that I wish there were better images of the damaged CM before reentry. (Of course, the unavailability of the surface photography by Jim Lovell and Fred Haise is another regret!)

Because it followed the aborted Apollo 13 mission, and because Alan Shepard was commander (CDR), both NASA and the media were more interested in covering Apollo 14 than they otherwise would have been. Only Apollo 11 rivals it for the sheer number of photos released. Another reason was the increase in photo opportunities with the astronauts and KSC workforce as NASA's Manned Flight Awareness (MFA) program was stepped up; this literally stressed that the crews' lives were in the workers' hands, producing a significant jump in posters, stickers, and other internal MFA safety reminders from many NASA centers. We include several examples here.

I again tapped into the collection of NASA photographer Alex Bosmeny and the Tiziou-team archives for the Apollo 14 chapter. Lunar surface photography was quite good, although it would be nice to have more color images of the crew at work.

Great available images marked Apollo 15 from start to finish. The Tiziou team shared some unique shots of NASA's Wernher von Braun at the rollout, and the Usciak brothers provided some good launch day photography. Excellent weather allowed extensive photo coverage of the Saturn V liftoff through first-stage separation.

Apollo 15's lunar surface images are breathtaking. The mission put the first four-wheeled vehicle on the lunar surface, and car lovers around the world were rewarded with images of the Lunar Roving Vehicle (LRV, or "rover") in the mountains of the moon. The only disappointments are the absence of Hasselblad shots of the rover deployment, the poor shots of the Alfred Worden EVA, and the usual lack of CM interior shots.

Apollo 16 was the only mission to have two rollouts of the Saturn V, with some creative shots by the Tiziou team. Launch day weather was probably the best of the program, providing the cleanest images. The lunar surface photography was an improvement from Apollo 15 (which was quite good), but the low point was the poor photo coverage of the Ken Mattingly deep space EVA.

Apollo 17 provided the best photography of any Apollo flight by far, producing the widest selection of images from each event before and during the mission. In addition to Tiziou and Usciak photos, we use some great crew and Saturn V shots from photographer John de Bry. The nighttime launch photography was spectacular, and all of the photographers and reporters I have met over the years refer to Apollo 17's launch as "the mother of all launches." Every phase of this mission was documented with fantastic images.

I give especially high marks to the Apollo 17 astronauts for their great camera work. They took the best photographs of Earth from cislunar space (including an iconic photo that came to be known as "The Blue Marble"), as well as the best photos of a CM's transposition and docking with an LM; lunar surface activities; the CM cabin; and a trans-Earth EVA (performed by Ron Evans). Even so, this final crew, as on the previous two missions, failed to document the deployment of the LRV.

Skylab 1 and 2 provided the only views of pads 39A and B with Saturn-launch vehicles on them simultaneously, which was as close as we ever came to early artwork that depicted Complex 39 in dual- (and triple-) pad operations. In one notable oddity, it took me thirty years to locate any nighttime views of Skylab's Saturn V at Pad 39A.

All three crews provided great photography, giving us a detailed view of life during extended stays on orbit. The second crew, Alan Bean, Owen Garriott, and Jack Lousma, did a particularly good job of documenting their EVAs and onboard activity. The final Skylab crew also did a solid job of documenting their journey, the first three-month space flight, but good EVA images are lacking. Final fly-around views of the workshop by this crew are some of the best.

ASTP provided the final views of a command/service module (CSM) in space—and the first of crews from two countries working in space together. ASTP also provided us with the first live television transmission of a crew during launch. The joint flight included the first images of US and Russian crews training on the ground; their enjoyed shared cultural experiences are quite evident.

Opposite page: Students at a Space Coast elementary school examine models of an LM and Saturn V in June 1969, weeks before the Apollo 11 launch. (photo by Tiziou News Service).

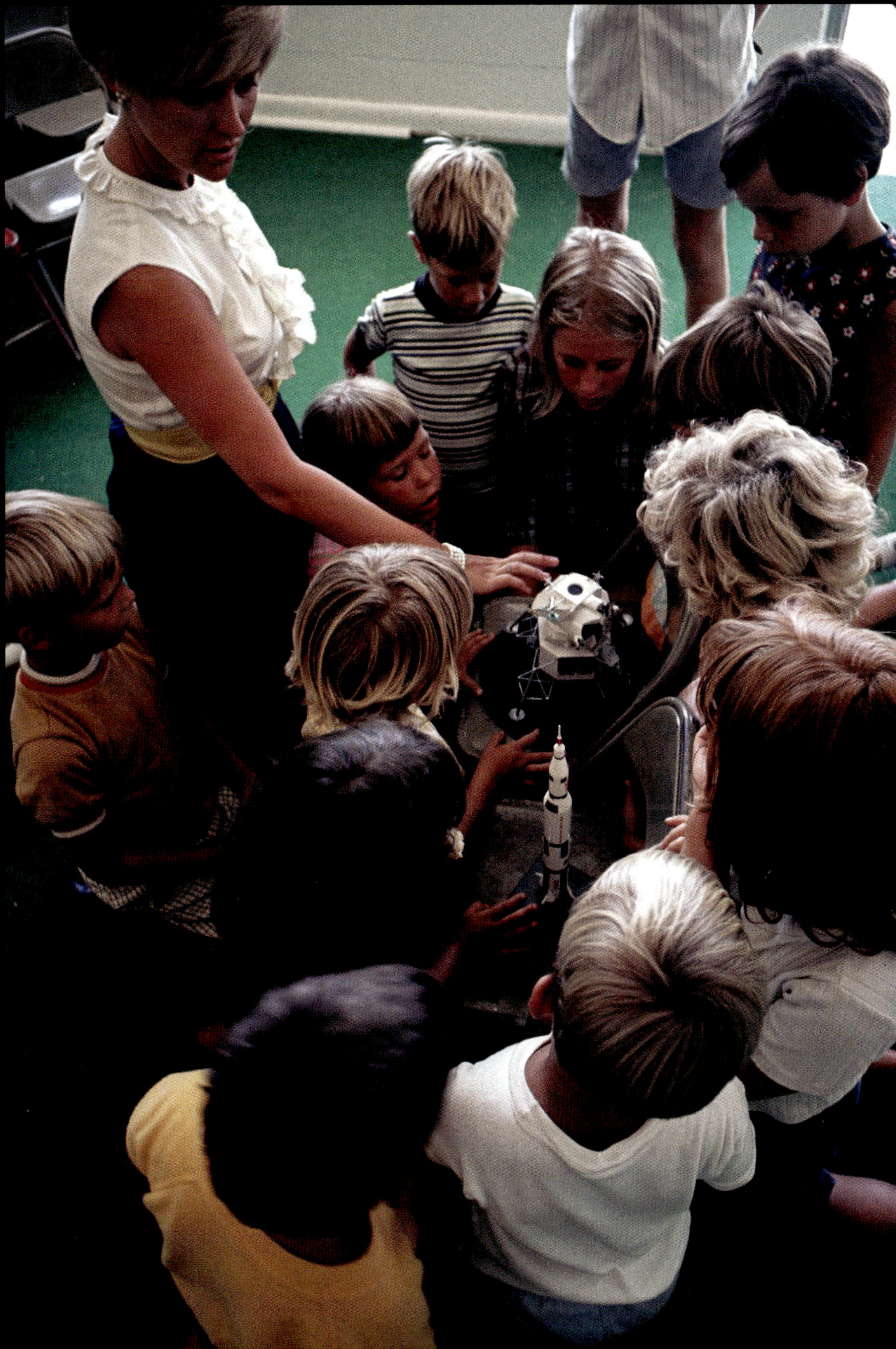

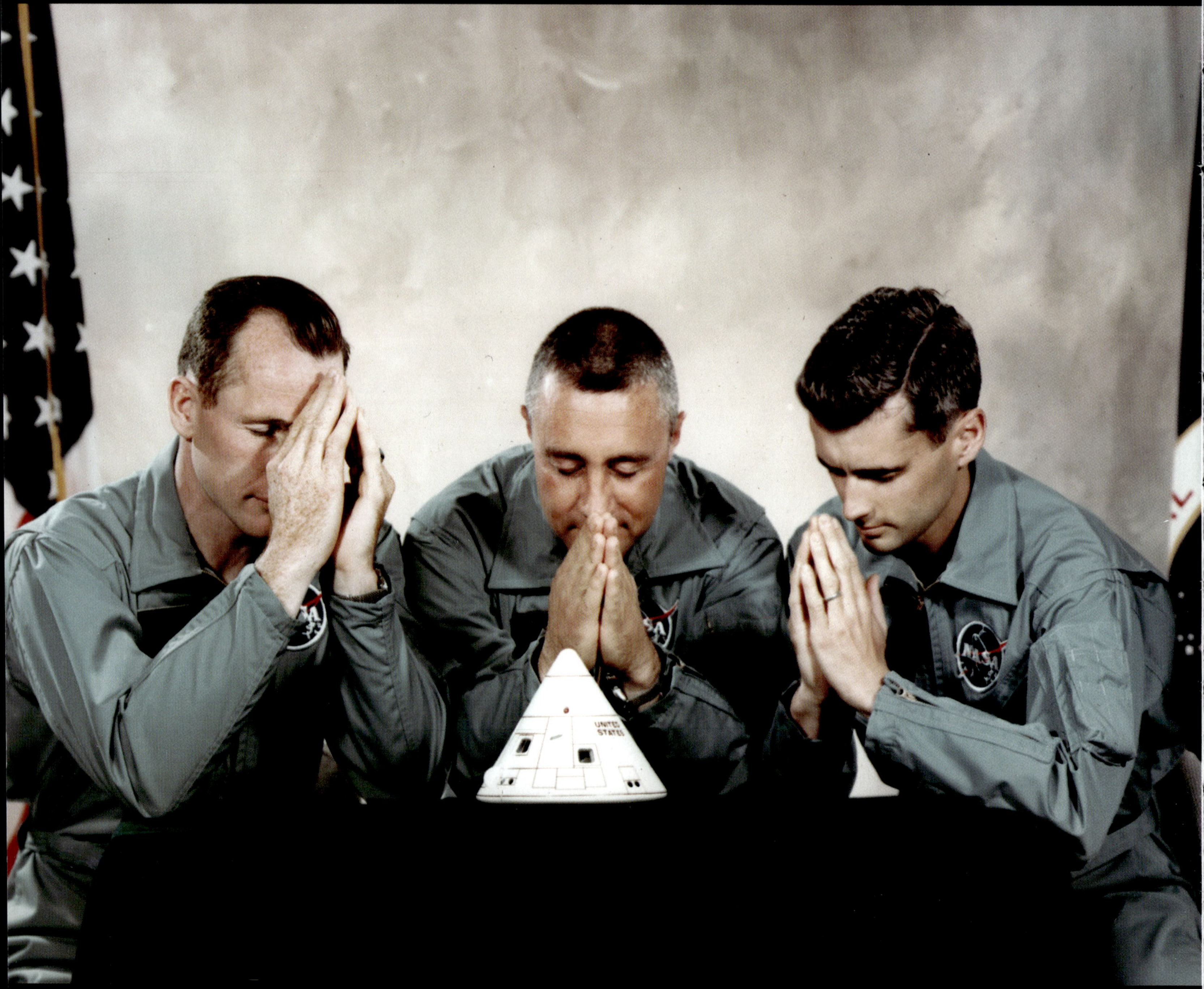

One of several gag photos taken at the Manned Spacecraft Center (MSC) on April 1, 1966, during the session for the Apollo 1 crew portrait. After a meeting during which the crewmen expressed concerns about the flammable material in the cabin, the men gave a copy of the photo to NASA's Joe Shea. The caption read, "It isn't that we don't trust you, Joe, but this time we've decided to go over your head."

CHAPTER 1

Apollo 1

January 27, 1967

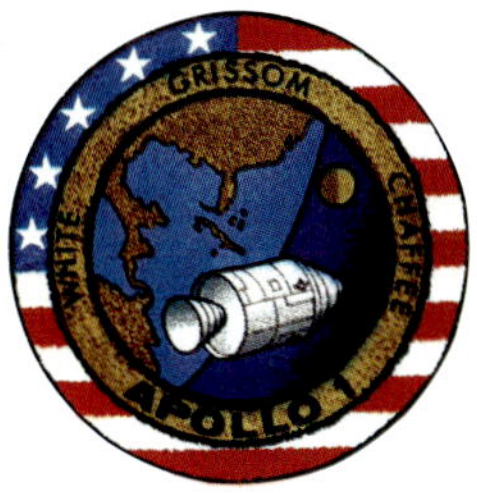

Design studies for the Apollo spacecraft had begun in 1960, so the program was well underway when Project Gemini ended in 1966. The first manned mission, Apollo Saturn-204, would test the combined CSMs. The CM was to be a "Block I" version intended for Earth orbit only, launched by a Saturn IB booster.

On March 21, 1966, NASA announced that Mercury and Gemini veteran Gus Grissom would be command pilot. The first US spacewalker, Ed White, would serve as senior pilot, while thirty-one-year-old rookie Roger B. Chaffee would be the pilot. James McDivitt, David Scott, and Rusty Schweickart were named as the backup crew. A target launch date of February 21, 1967, was set.

By December 1966, a planned second Block I flight had been canceled, so its crew—Wally Schirra, Donn Eisele, and Walt Cunningham—was reassigned as the Apollo 1 backup crew.

On August 19, 1966, in a meeting with Apollo Spacecraft Program Office Manager Joe Shea, the crew expressed concerns about the amount of flammable material (mainly nylon netting and Velcro) in the cabin used for holding tools and equipment. At the end of the meeting, the astronauts gave Shea a portrait, which depicted them with their heads bowed and hands clasped in prayer.

When CSM prime contractor North American Aviation (NAA) shipped spacecraft CM-012 to Florida from its Downey, California, factory on August 26, 1966, there were more than 100 incomplete engineering changes, with 623 more made later. Grissom was so frustrated by the training-simulator engineers' inability to keep up with the changes that he hung a lemon on the CM simulator.

On January 27, 1967, the crew entered the CM at Launch Complex (LC) 34 for a "plugs-out" test, to see whether the spacecraft would operate normally on simulated internal power while detached from all ground cables.

The CM hatch had three parts: a removable inner hatch; a hinged outer hatch, part of the spacecraft's heat shield; and an outer hatch cover. After the hatches were sealed, the cabin air was replaced with pressurized oxygen.

A number of technical problems cropped up, including poor voice communications. Grissom remarked, "How are we going to get to the moon if we can't talk between two or three buildings?" A simulated countdown was conducted at 5:40 p.m. (EST) to try to fix the problem.

The crewmen were running through checklists when a voltage transient was recorded at 6:30 p.m. Ten seconds later, after Chaffee said, "Hey," which

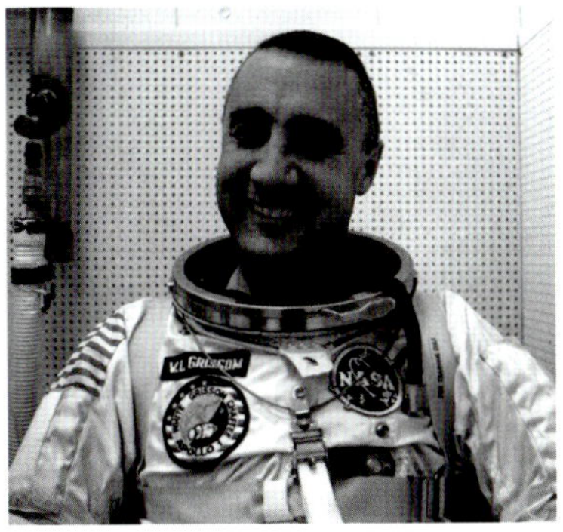

Gus Grissom, forty (January 1967)

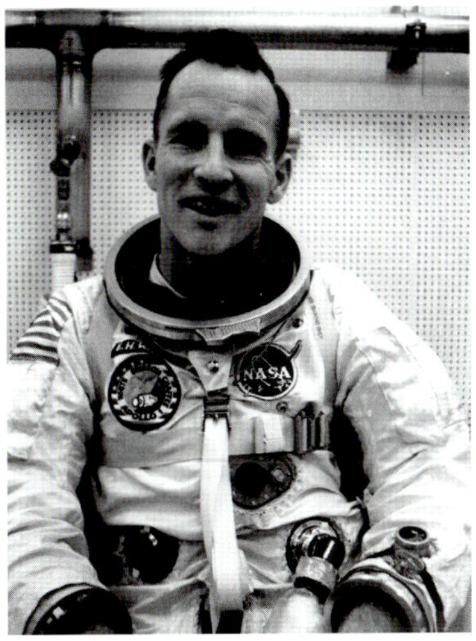

Ed White, thirty-six (January 1967)

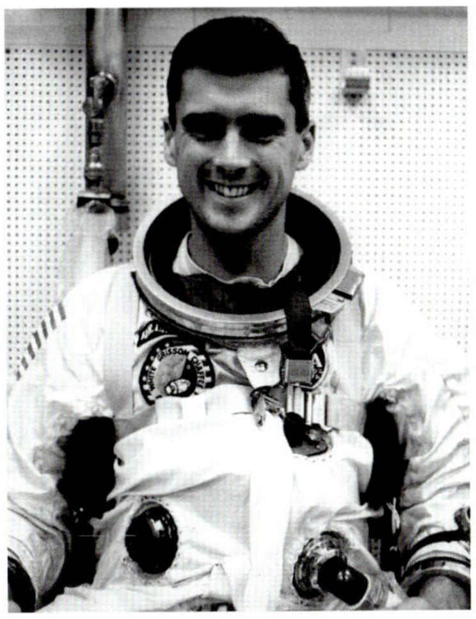

Roger B. Chaffee, thirty-one (January 1967)

was followed by scuffling sounds, Grissom reported a fire. Chaffee added, "We've got a fire in the cockpit." Seventeen seconds after the first indication of a fire, the cabin ruptured. Expanding gases from the fire had overpressurized the CM.

Flames and gases rushed into two levels of the pad service structure. Intense heat, dense smoke, and the use of gas masks designed for toxic fumes instead of heavy smoke hampered rescue efforts. It took five minutes to open all three hatches.

The flammable material in the cabin's pressurized pure-oxygen environment allowed the flames to spread quickly. The fire had partly melted Grissom's and White's nylon space suits. Grissom had removed his restraints and was lying on the floor of the spacecraft. White's restraints were burned through; he was found just below the hatch. He had tried to open it but was not able to do so against the internal pressure. Chaffee was found strapped in his seat, from which he was to have maintained communications.

NASA quickly convened the Apollo 204 Review Board. Although investigators never conclusively identified the ignition source, they attributed the fire to a wide range of design and construction flaws. The most likely cause was a spark from a short circuit in a wire bundle near Grissom's seat.

The Apollo program was delayed for twenty months while safety updates were made, including the installation of a new hatch that could be operated quickly, the removal of much of the flammable material, and the switch to a nitrogen-oxygen atmosphere. A new flame-resistant, fiberglass-based fabric of woven Teflon-covered silica fiber—Beta cloth—was developed for the outer layer of the Apollo space suits, as well as for other uses.

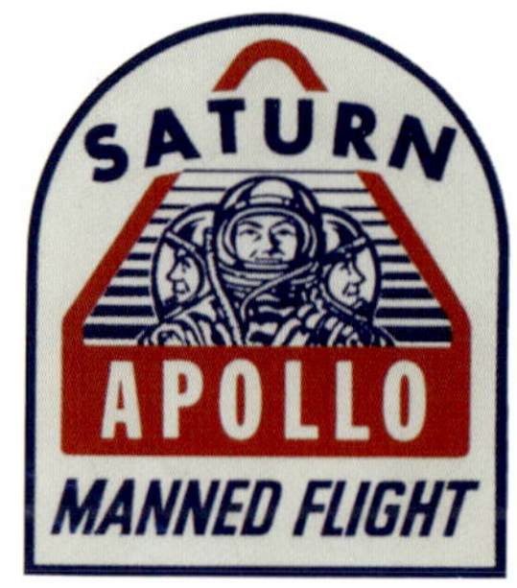

Below: The Apollo 1 prime and original backup crews at North American Aviation (NAA) in Downey, California, on August 4, 1966. *Left to right*: Paul Haney, NASA public information officer; Rusty Schweickart; Dave Scott; James McDivitt; Grissom; White; and Chaffee. A model of the Apollo CSM is at left; at right is a model of the Saturn V launch vehicle and its mobile launcher.

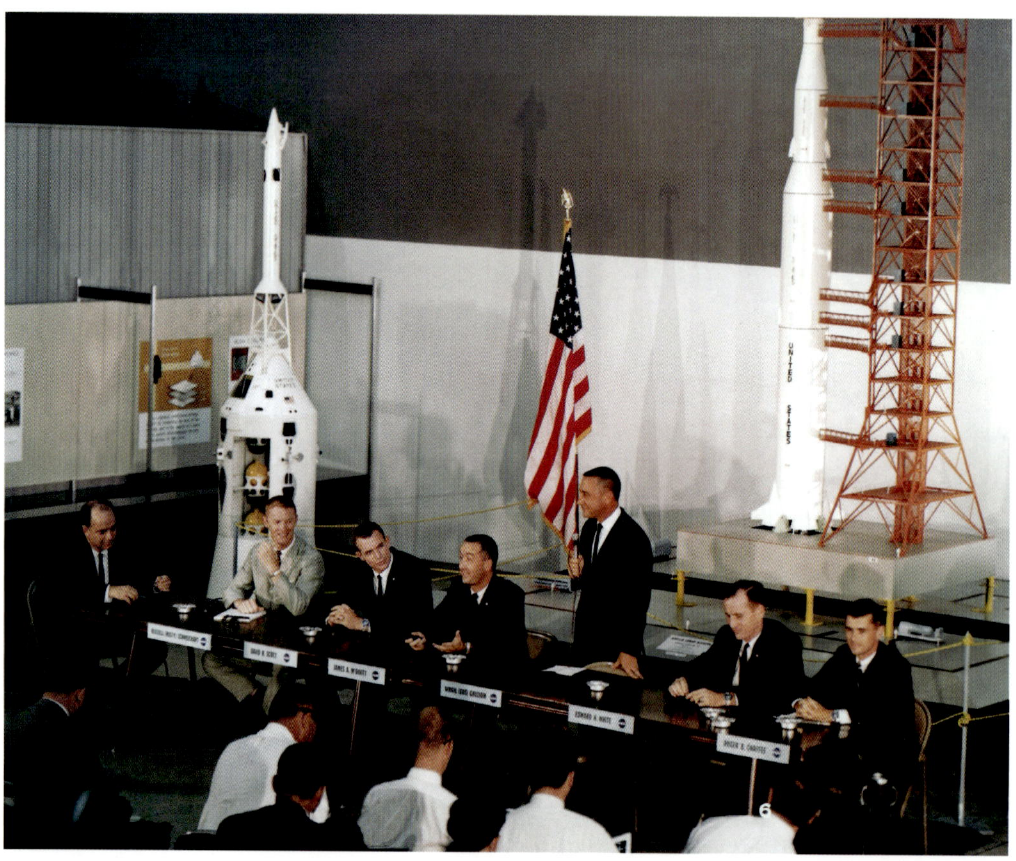

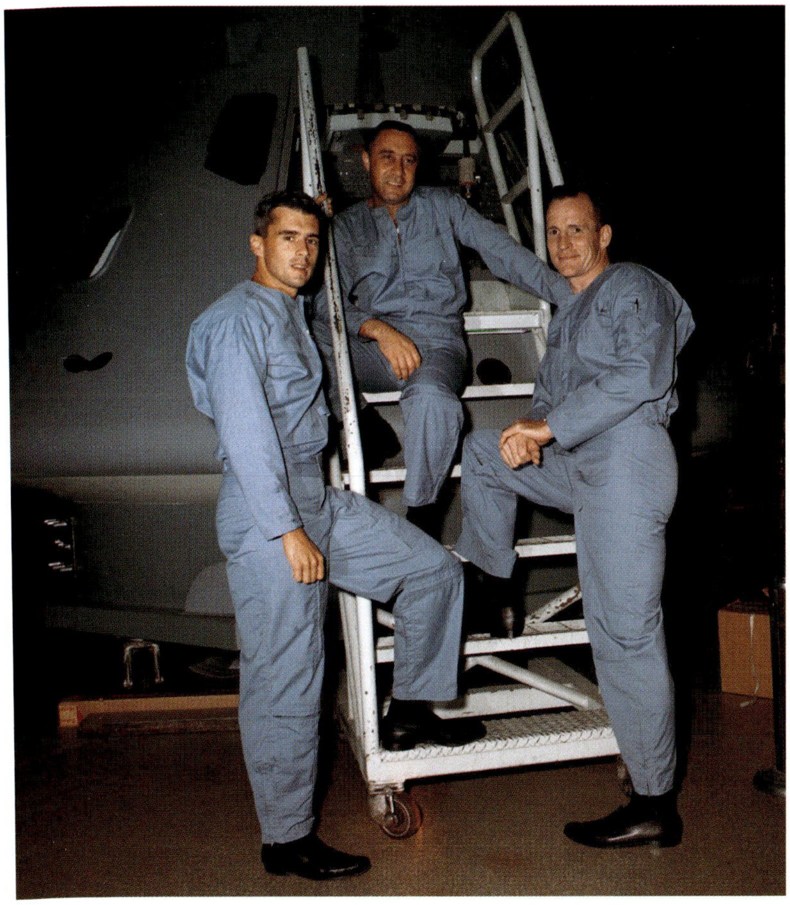

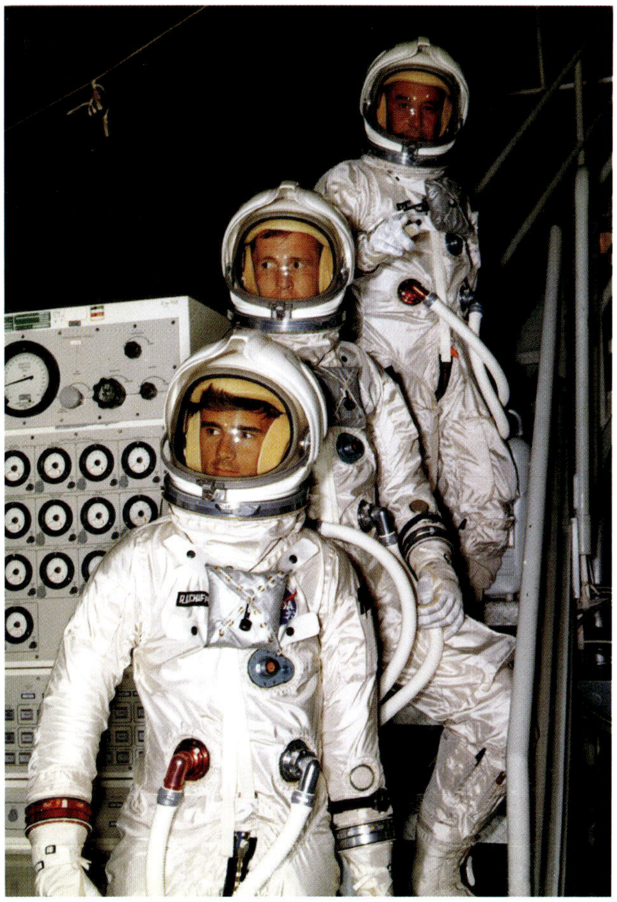

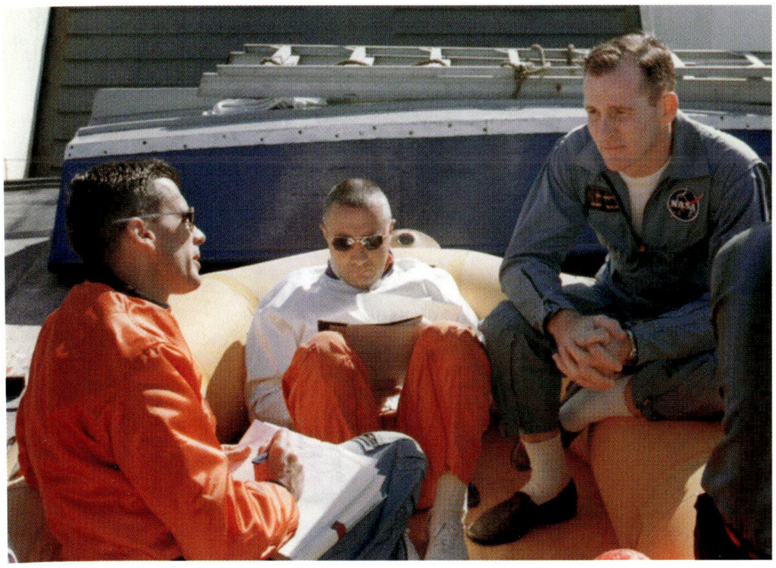

Above: The crew poses before entering the spacecraft at Downey on August 18, 1966. Because no EVAs were planned during the mission, the astronauts wore a modified Gemini G3C suit called the A1C. This version added new electrical and environmental disconnects, as well as a protective shell over the Plexiglas helmet visor. It was the final suit made by the David Clark Company, as the Government and Industrial Division of the International Latex Corporation (later ILC Dover) of Dover, Delaware, would get the contract for Apollo lunar suits. The panel at left monitored compressed gas usage in the simulator's pneumatic system, to move the spacecraft.

Top left: Chaffee, Grissom, and White with a mock-up of a Block I Apollo CM, at NAA in Downey on June 24, 1966. This version was intended for Earth-orbit missions and could not dock with an LM.

Bottom left: Chaffee, Grissom, and White go over water egress procedures aboard MV *Retriever* on October 27, 1966. The *Retriever* was a World War II tank landing craft, used to train astronauts for post-splashdown ocean recovery operations and egress from their CM. White, America's first spacewalker, was part of the second group of astronauts, chosen in 1962; Chaffee was selected as part of the third group, the following year.

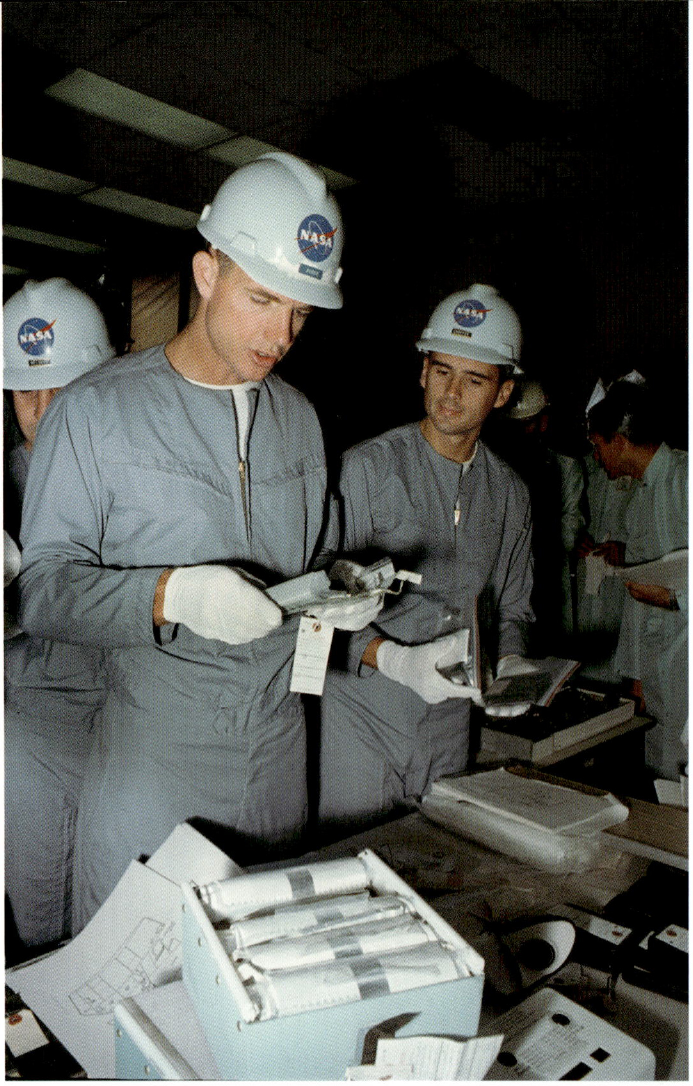

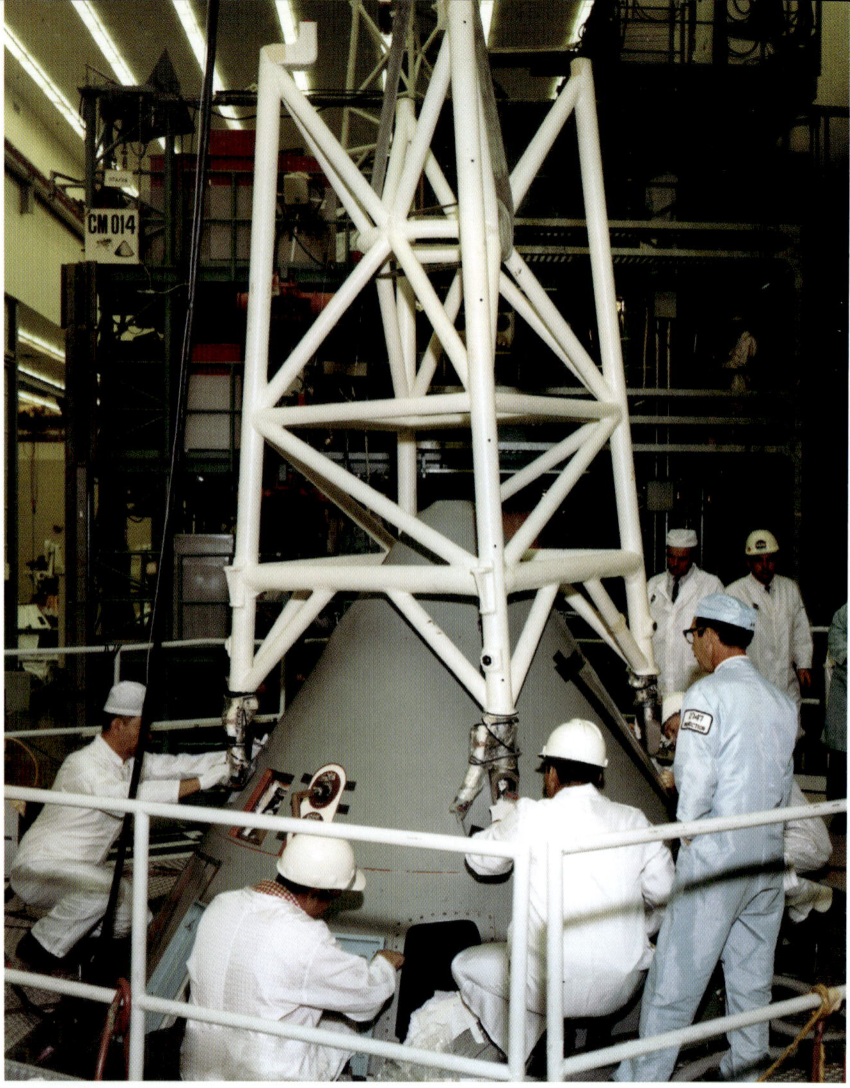

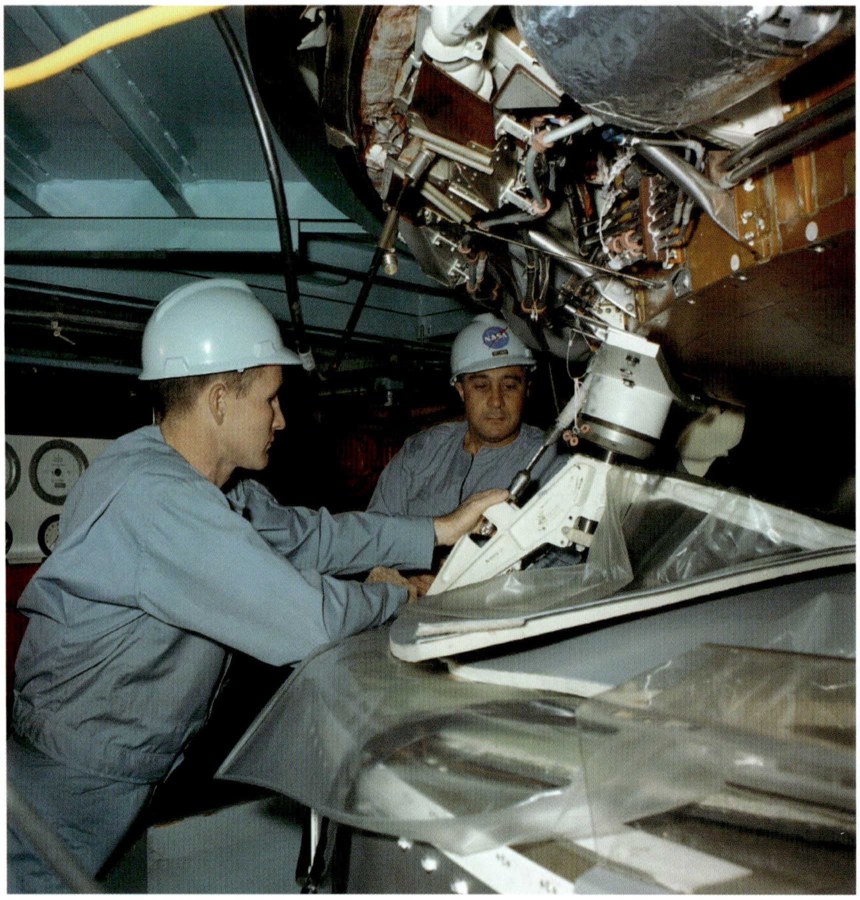

Top left: Original backup Apollo 1 senior pilot David Scott inspects spacecraft components at NAA in Downey, on June 24, 1966. Grissom is at left, behind Scott; Chaffee is at right.

Top right: At Downey in August 1966, NAA technicians check the fit of the legs of the Launch Escape System's (LES) titanium tower into openings in the forward structure of CM-204. Each leg was attached by a stud on a frangible nut, which two small detonators would break apart to separate the tower after launch. The workstand in the background contains CM-014. (CM-014 would be disassembled for study after the Apollo 1 fire.)

Left: White (*left*) and Grissom examine one of three tension ties on their CM at Downey on June 24. The ties were the structural attachment to the service module (SM), the top edge of which can be seen at the bottom of the photo. The aluminum base of the CM (*right of center*) has been detached to expose the underlying bays and components in the aft compartment.

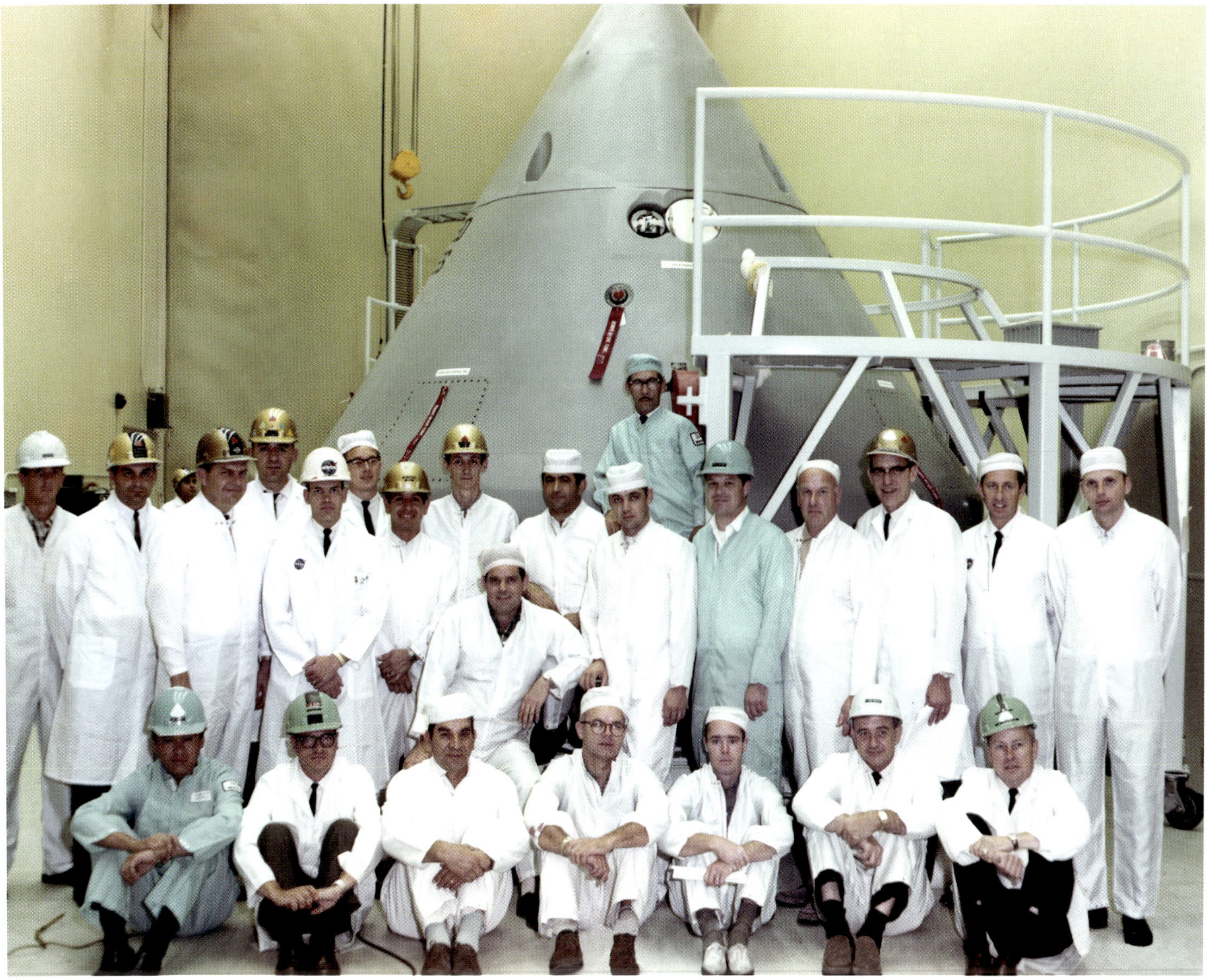

Above: NAA and NASA personnel (in clean-room garb) pose with CM-012 in the high-bay area of Bldg. 6 at Downey on August 24, 1966.

Below: NAA workers decorated their hard hats with small stickers for the spacecraft on which they had worked.

Far right: An NAA PRIDE (Personal Responsibility in Daily Effort) emblem, worn on ID badges

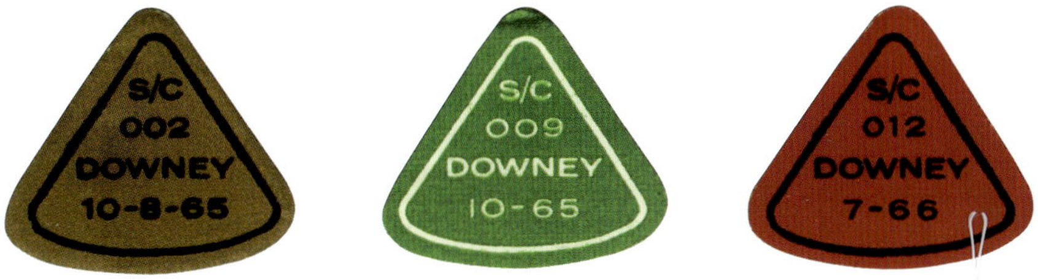

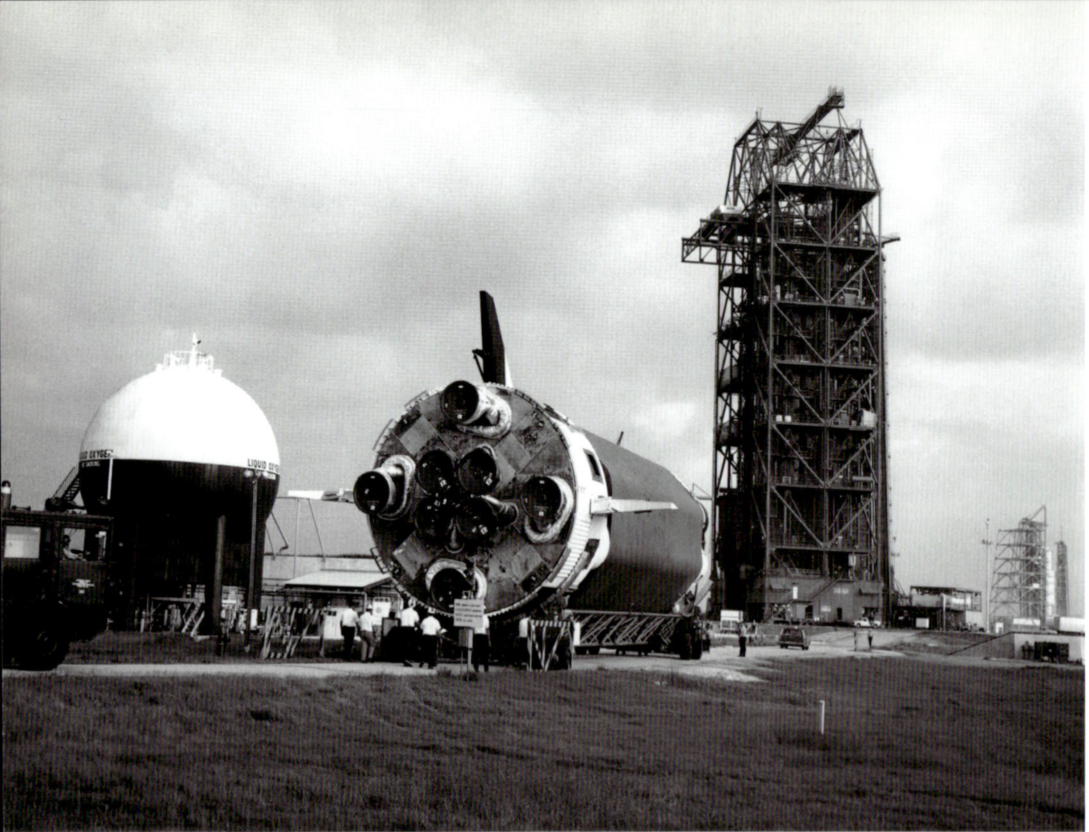

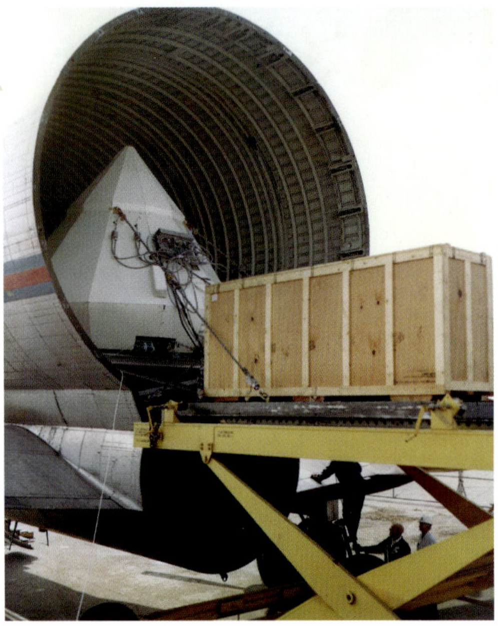

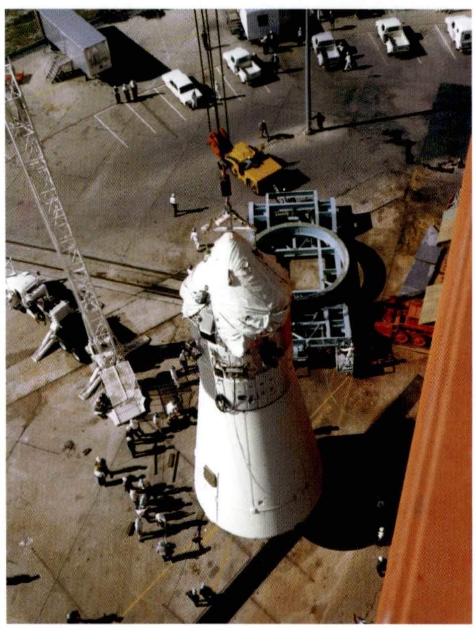

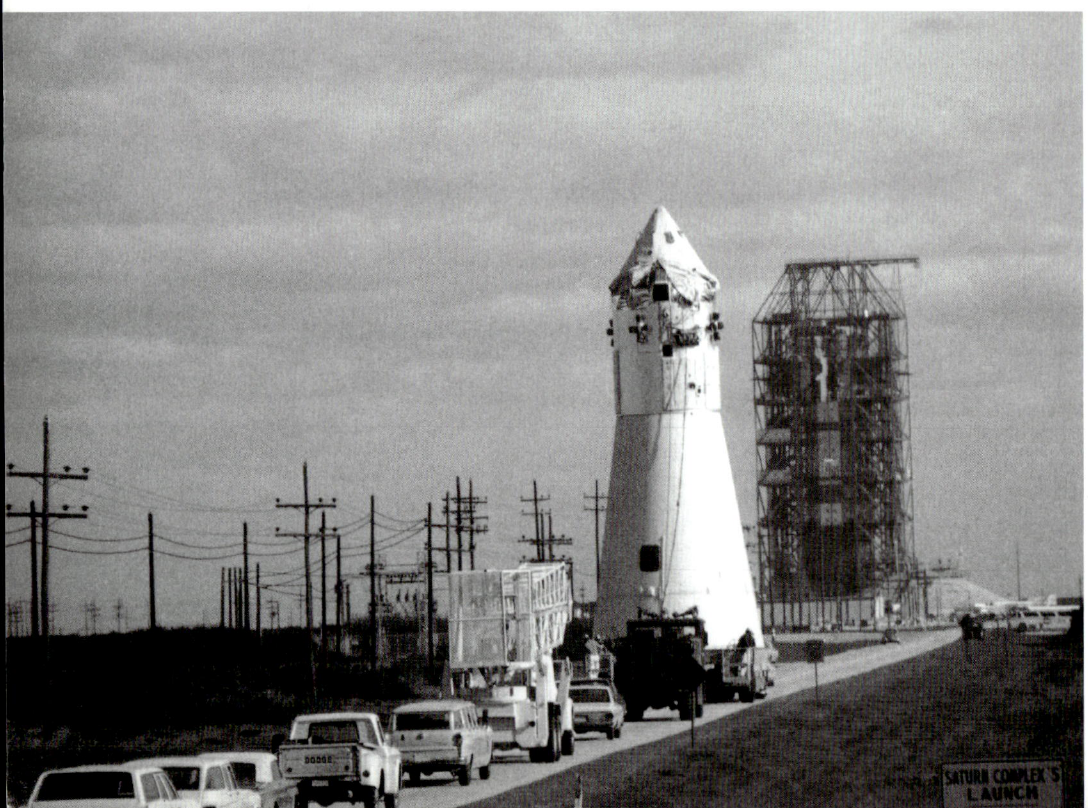

Left: The first stage of Apollo 1's Saturn IB booster, SA-204, arrives at LC-34 on August 29, 1966, showing its eight H-1 main engines. Eight Redstone tanks (four holding fuel and four holding liquid oxygen [LOX]) are clustered around a Jupiter rocket LOX tank. The four outboard engines are mounted on gimbals, allowing them to be steered to control the booster's flight. The fins surrounding the base provided aerodynamic stability and control. A 125,000-gallon LOX storage tank is at left. At right is LC-37, about 1,200 yards to the north-northwest.

Center left: CM-012 and support hardware are loaded into the *Pregnant Guppy* aircraft near Downey on August 24, 1966, on the way to Cape Kennedy Air Force Station (AFS).

Center right: The stacked spacecraft are hoisted to be mated with the Saturn booster the same day. Protective panels cover much of the SM.

Bottom: A motorcade brings the combined CSM—atop the empty Spacecraft Lunar Module Adapter (SLA)—to LC-34 on January 6, 1967. NAA also built the SLA.

Opposite page: The Saturn IB's first stage, built by the Chrysler Corporation's Space Division, is lowered toward the launch stand on August 29. Workers on the ground carefully control its sway with cables. The rail-mounted mobile service tower stood 310 feet tall. This same booster (SA-204) was used to launch Apollo 5, an unmanned test flight of the LM in Earth orbit, in 1968.

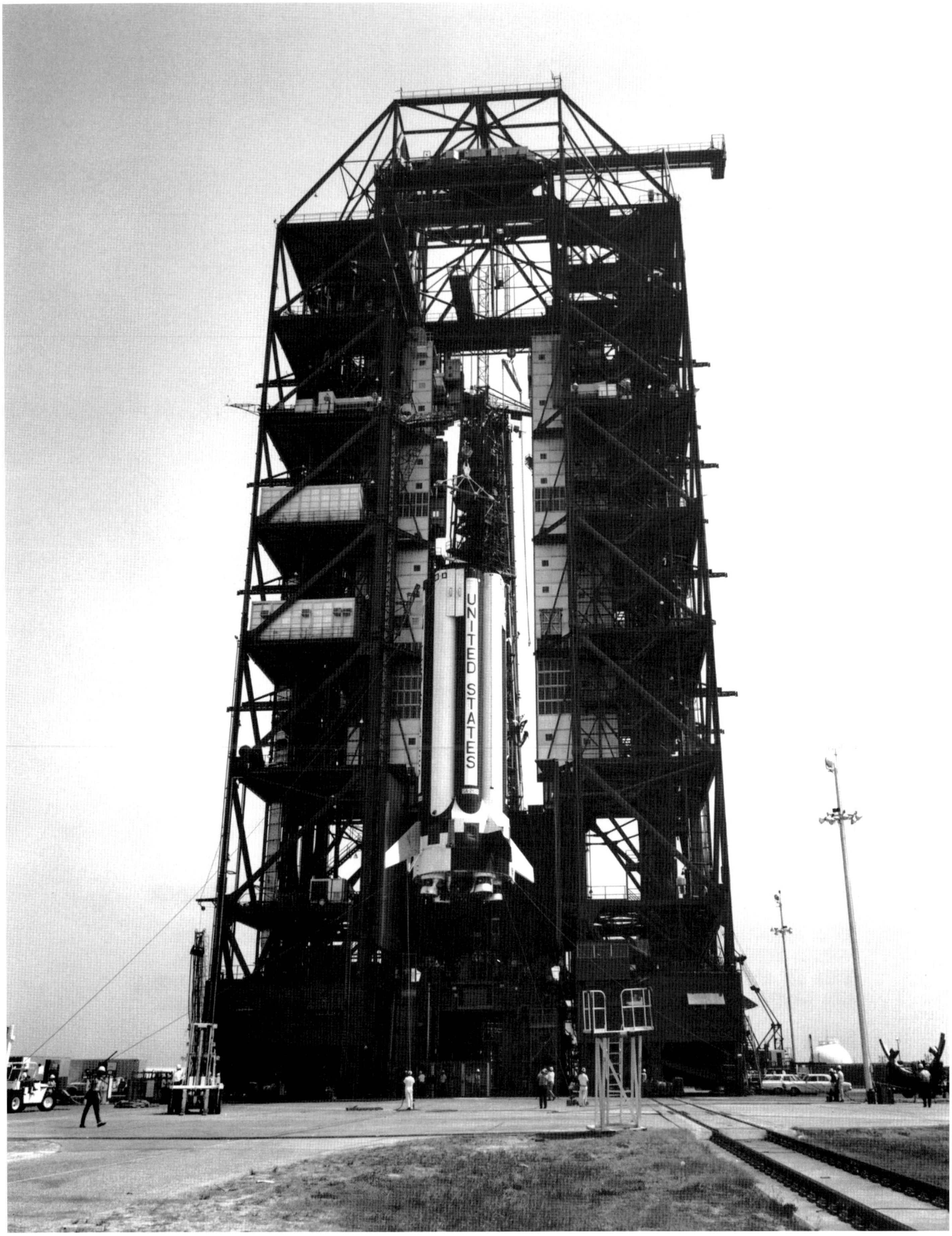
UNITED STATES

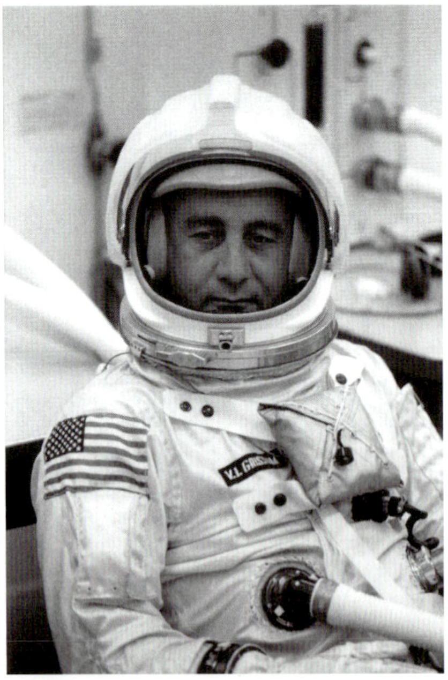

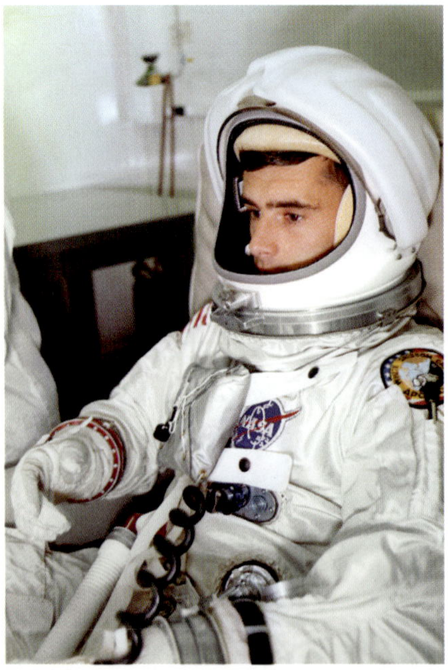

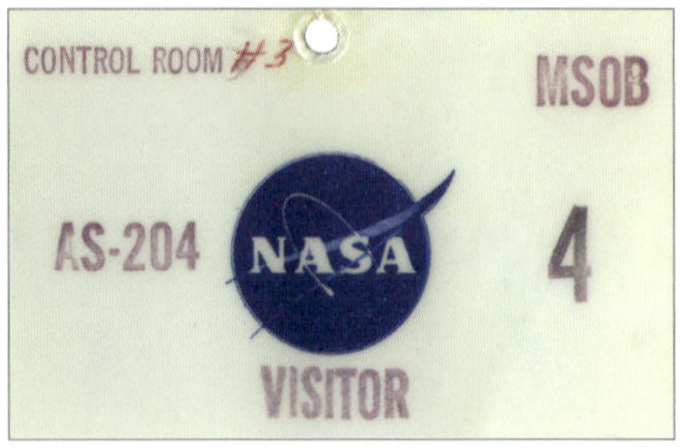

Top left: Grissom, suited for a Chamber L walk-through on October 13. Throughout the Apollo program, tests of critical spacecraft systems and astronauts took place in altitude chambers, using simulations of space's near-vacuum and extreme temperatures. Chamber L was used for Apollo CSMs, Chamber R for LMs.

Top center: Chaffee prepares for an October 18 Chamber L run. Problems during the test forced a second run the following day, when NASA announced that the last unmanned Apollo flight had been canceled. The crewmen originally wore their embroidered mission patches on the shoulders of their space suits, but these were later moved to the chest, where they were on the suits to be worn during the flight.

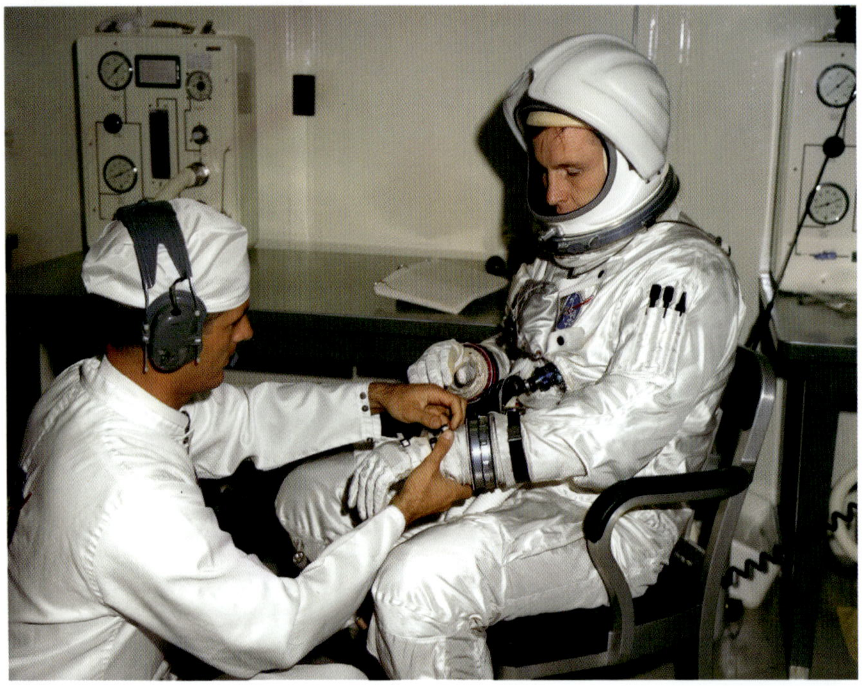

Center left: David Clark Company suit tech Joe Rebokus assists White with his glove in the Manned Spacecraft Operations Building (MSOB) on October 13, 1966, five days before a walk-through of the first manned altitude chamber run with CM-012.

Bottom left: The crewmen pose for photographers at LC-34 on January 17, 1967. Their life vests have been moved from their chests to gray pouches under their arms, on yellow harnesses. The press bus is at right.

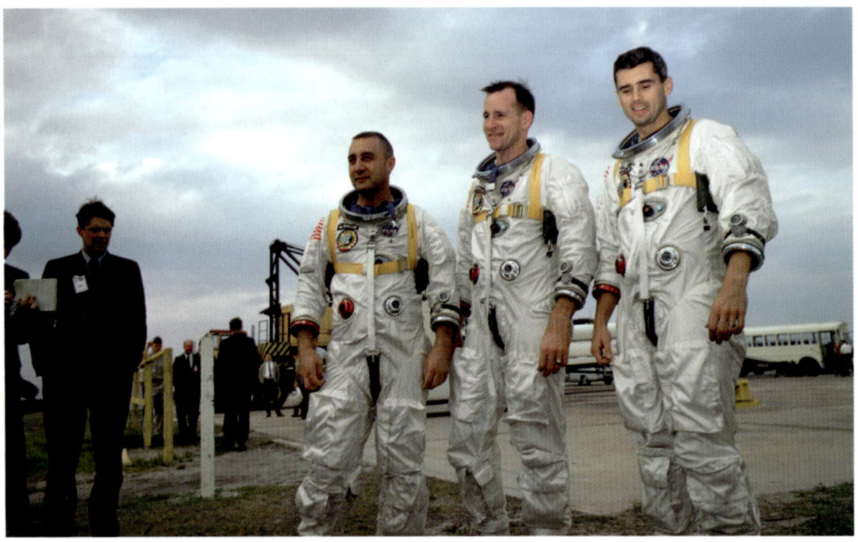

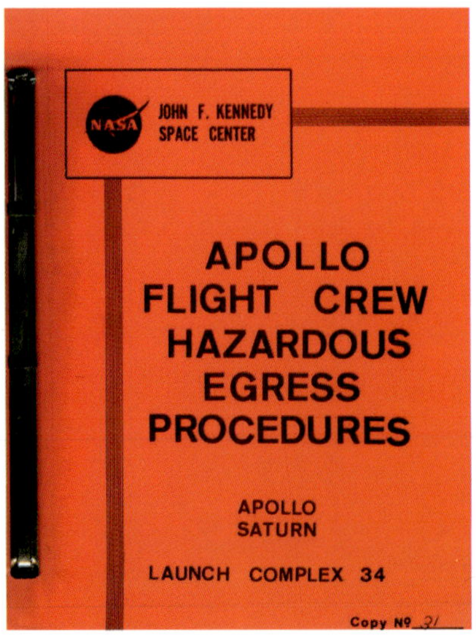

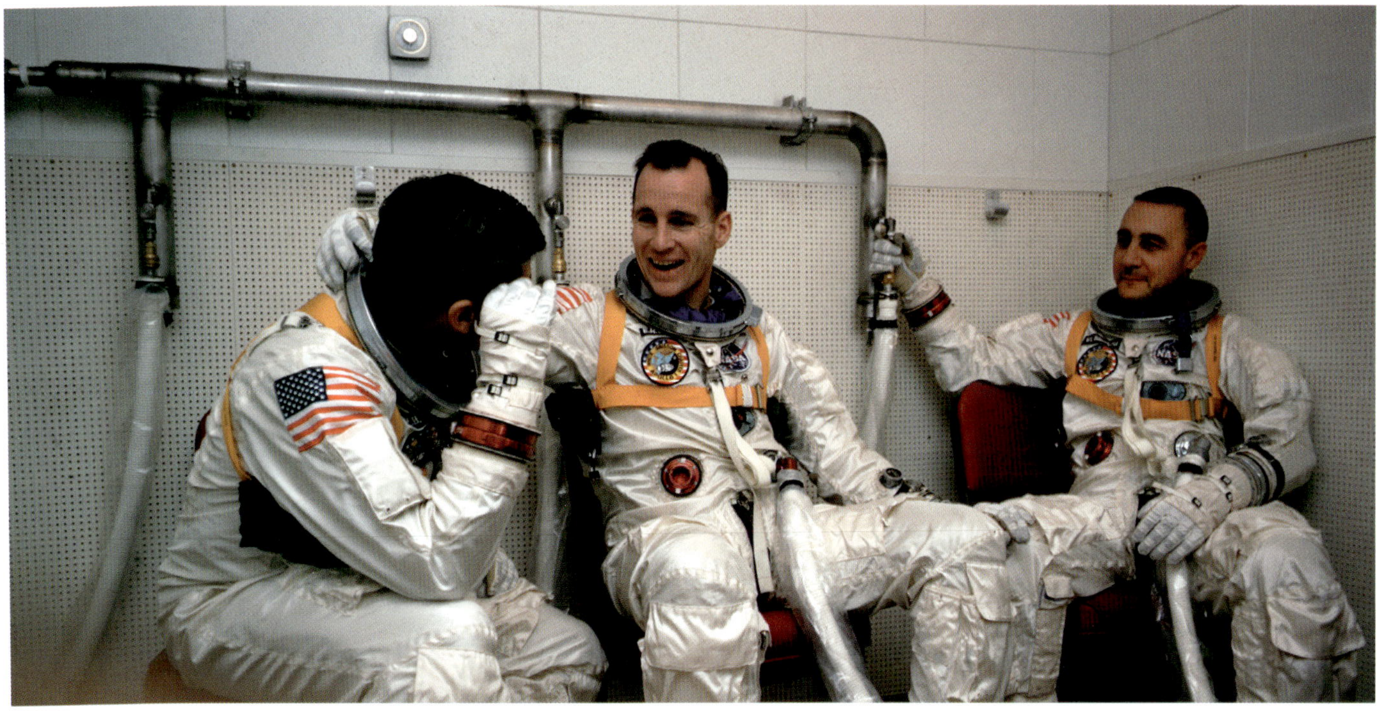

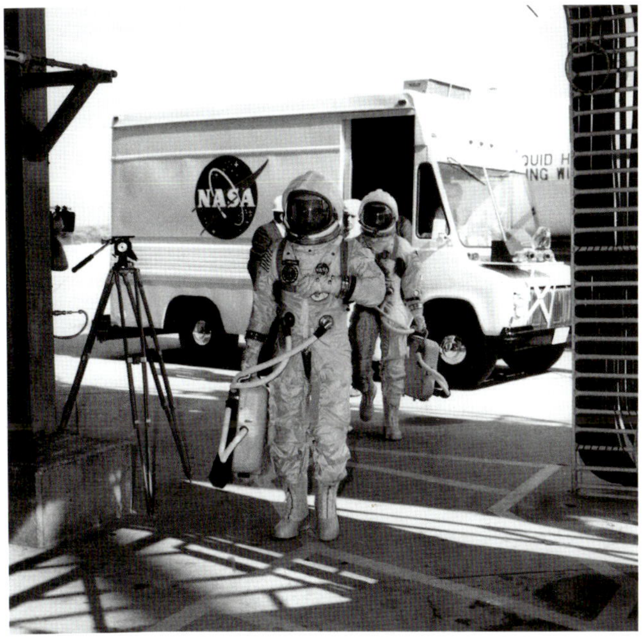

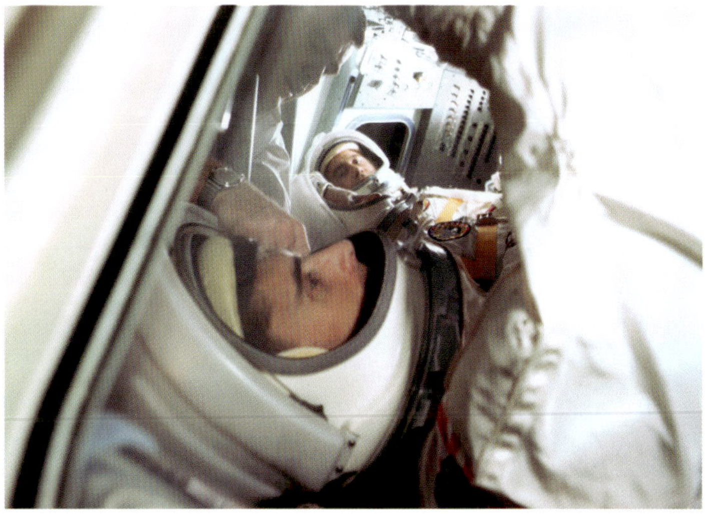

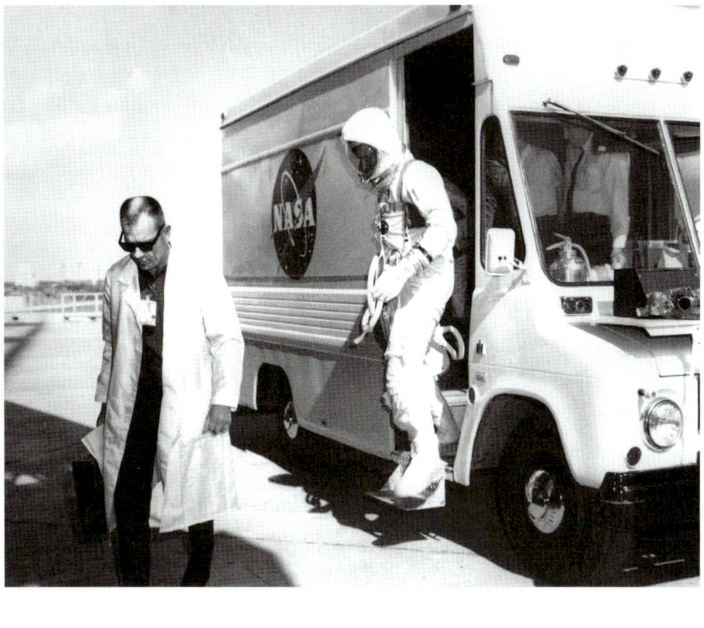

Top: Chaffee, White, and Grissom prepare for a session in the Apollo spacecraft simulator at KSC's Flight Crew Training Building on January 19, 1967. Their suit hoses are connected to air conditioning for their comfort.

Center left: Grissom (*front*) and Chaffee head toward the elevator at Pad 34 on January 27.

Center right: A NASA technician leans in through the spacecraft hatch during the January 19 flight simulation in the MSOB. *Foreground*: Chaffee. *Center*: White. *Background*: Grissom.

Bottom right: As the astronauts arrive at Pad 34, White follows Director of Flight Crew Operations Deke Slayton from the transfer van on the afternoon of January 27, 1967, the day of the fatal fire.

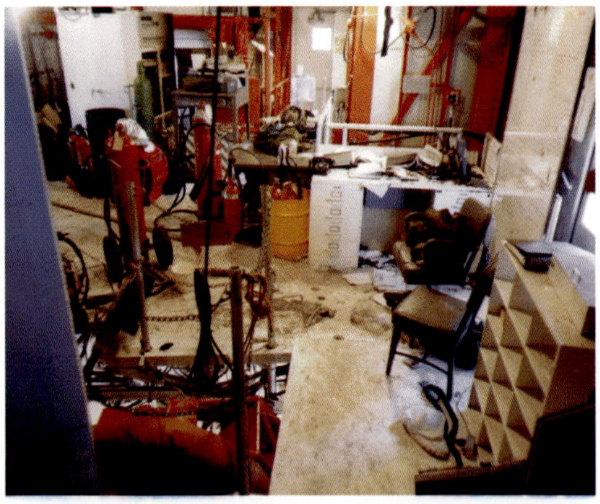

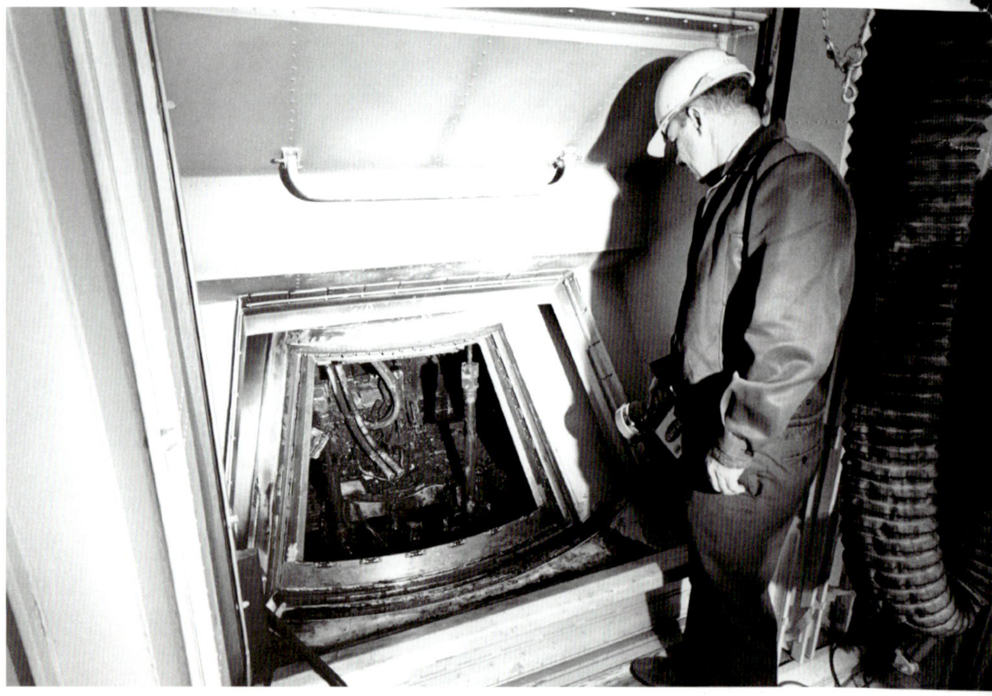

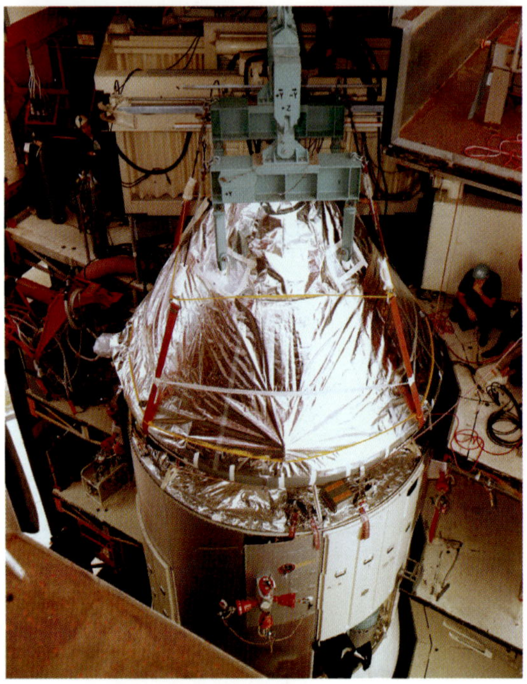

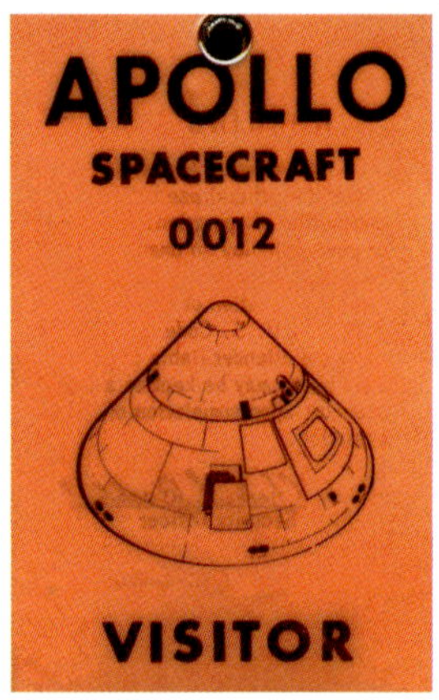

Top left: The pad leader's desk is at center in this photo of the aftermath of the fire on the A8 spacecraft–level work area in the mobile service tower. McDonnell Aircraft's Guenter Wendt, pad leader during Mercury and Gemini, had not come with the NAA contract, but the astronaut corps insisted that he be brought back after the fire.

Top right: The day after the fire, a NASA engineer inspects the charred cockpit with a flashlight. The grab bar at top was used to assist ingress and egress.

Center left: A crane lifts the wrapped CM from the SM on February 17, 1967, at Pad 34. Some interior components have been removed, and red covers protect the quad reaction control system (RCS) nozzles on the SM. The CM was taken to the Pyrotechnic Installation Building for final disassembly and inspection.

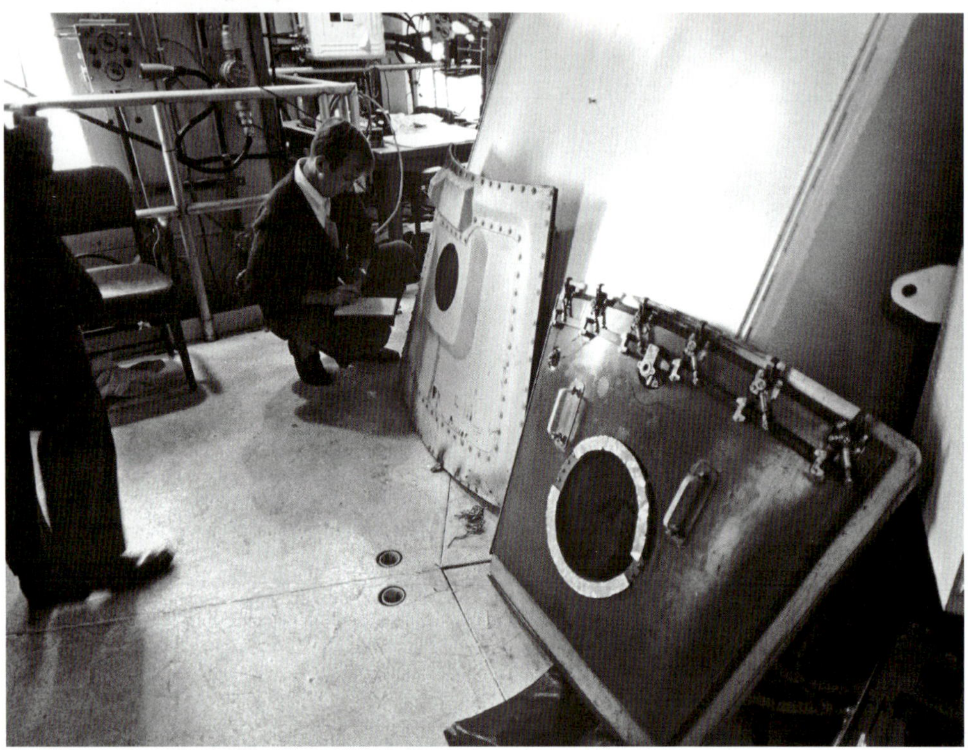

Bottom left: Pool reporter George Alexander of *Aviation Week* magazine takes notes at the CM hatch on January 30, 1967. At center is the outer hatch, with the inner hatch at right, upside down and displaying its complex latching system. Dual hatches were used, as no CM EVAs were planned for the early Apollo missions.

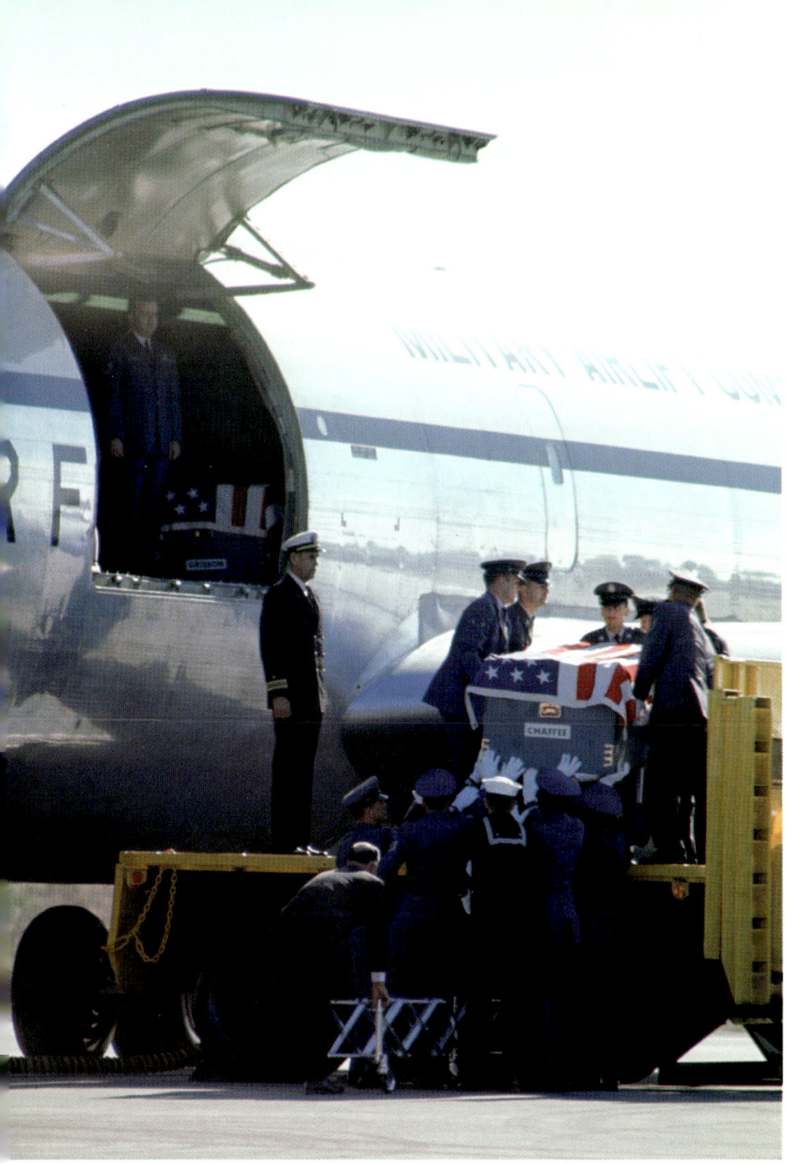

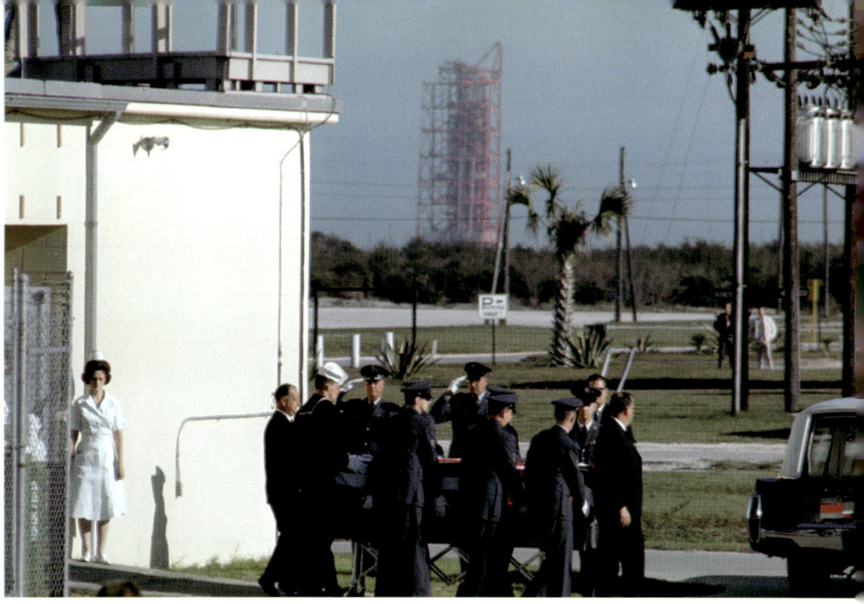

Top left: A military honor guard lifts Chaffee's casket toward the hold of a USAF Military Airlift Command C-135, heading for Andrews AFB near Washington, DC, and then to Stewart AFB near Newburgh, New York. Grissom's casket is onboard. Grissom and Chaffee were buried the next day at Arlington National Cemetery. White was buried the same day as Grissom and Chaffee at the West Point Cemetery at the US Military Academy in West Point, New York.

Top right: Grissom's casket is transferred into a hearse on January 30, 1967, following autopsies on the three astronauts at the US Air Force (USAF) Bioastronautics Operations and Support Unit Building. In the background is the gantry for LC-37, used for launching unmanned Saturn I and IB rockets.

Below: The remains of the astronauts pass by Cape Kennedy workers as the procession heads toward the Skid Strip, Cape Kennedy's airfield. The procession had driven past Mercury's Hangar S in tribute to Grissom, in the final hearse. The sign at right indicates the route for public drive-through tours on Sundays.

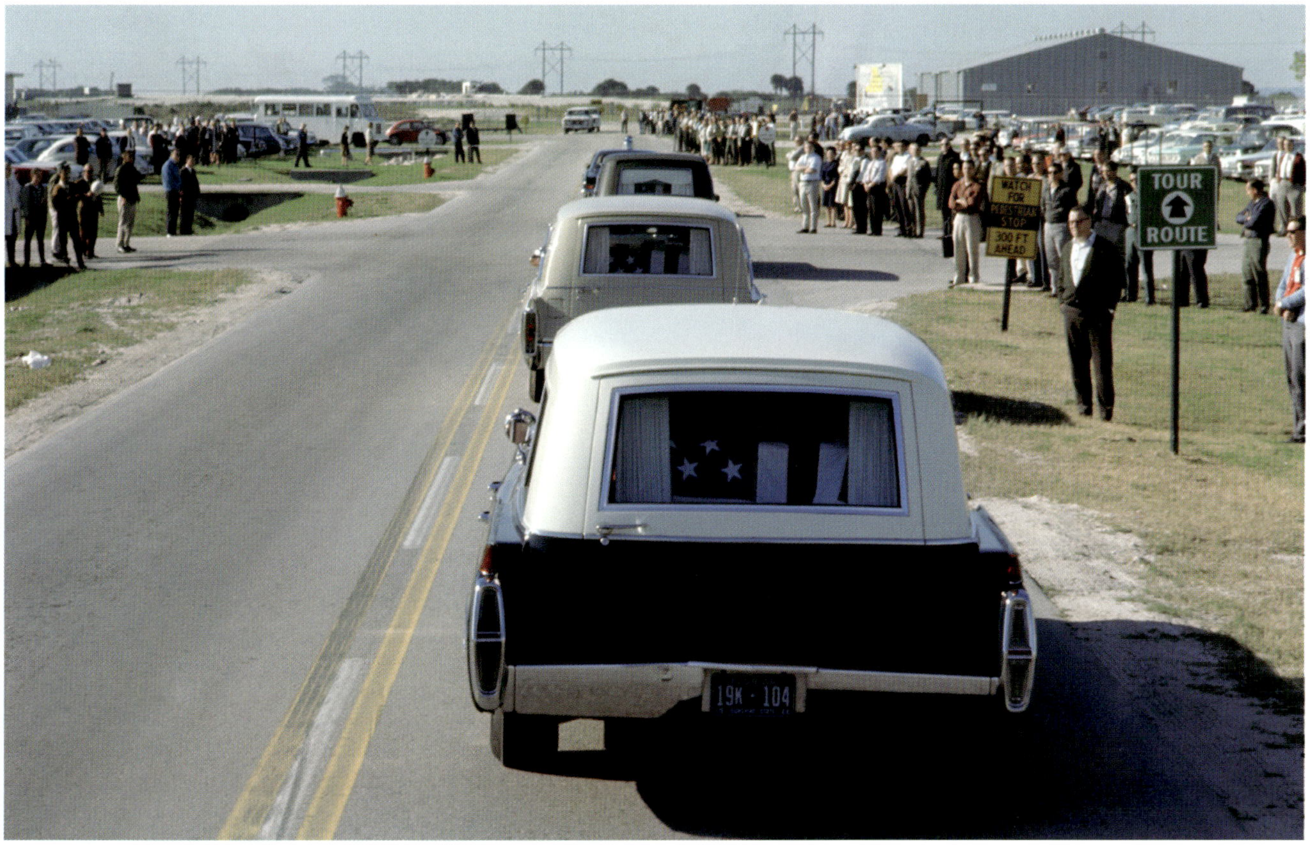

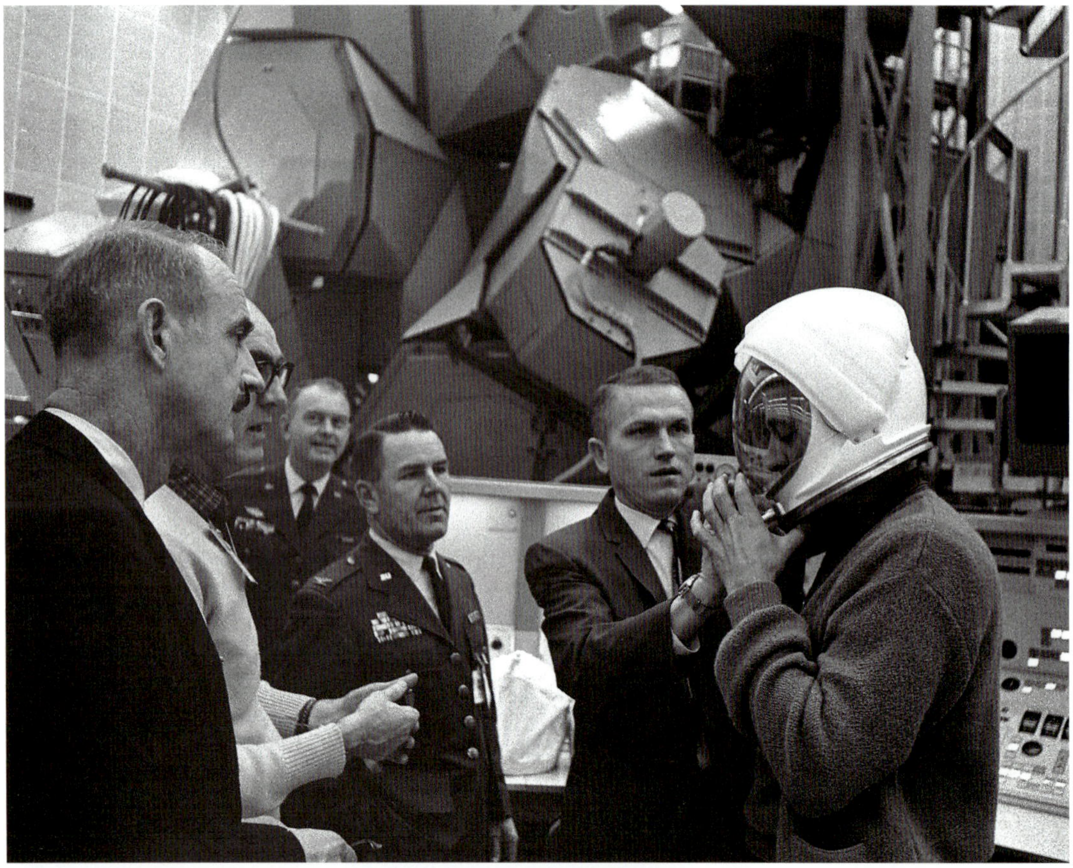

Left: Astronaut Frank Borman (*second from right*), a member of the Apollo 204 Review Board, helps astronaut Ron Evans with a helmet demonstration in the Flight Crew Training Building on February 3, 1967, as part of the board's investigation. Board member Col. Charles Strang (USAF) is next to Borman; the LM simulator is in the background.

Bottom left: This commemorative brass plaque, donated by Mark Pinchal, a private citizen, was installed on the concrete launch stand at LC-34 in 1988.

Opposite page: CM-012 in its storage container at NASA's Langley Research Center in Virginia, in 1989. The CM was transported from Florida by barge in July 1967. The spacecraft and associated hardware have been stored at Langley since, and were moved into a climate-controlled warehouse there in 2007.

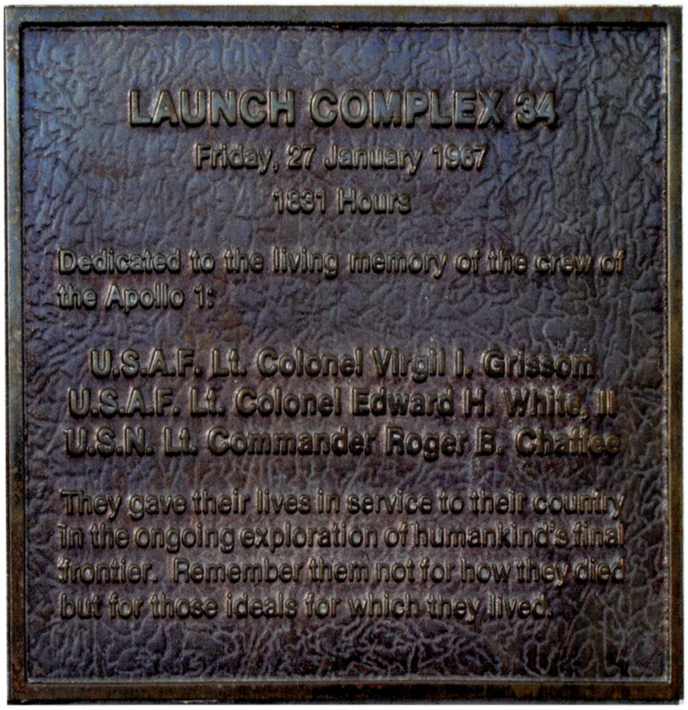

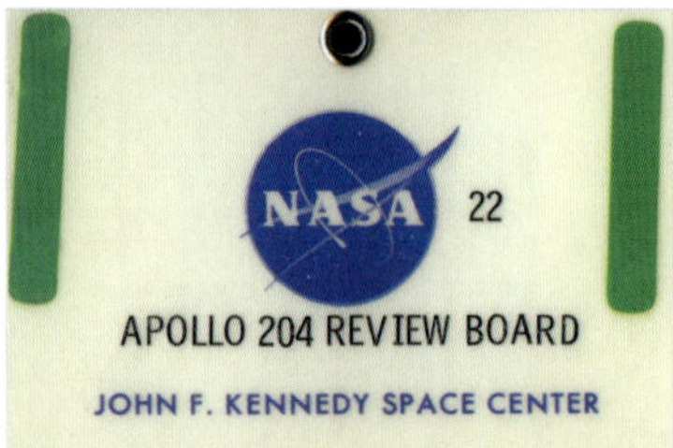

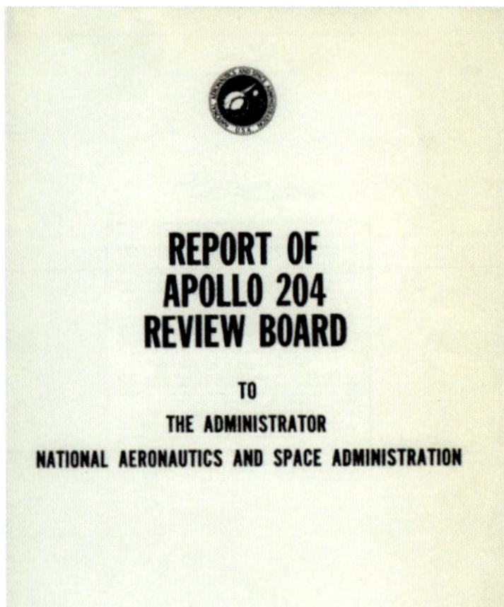

REPORT OF
APOLLO 204
REVIEW BOARD

TO

THE ADMINISTRATOR
NATIONAL AERONAUTICS AND SPACE ADMINISTRATION

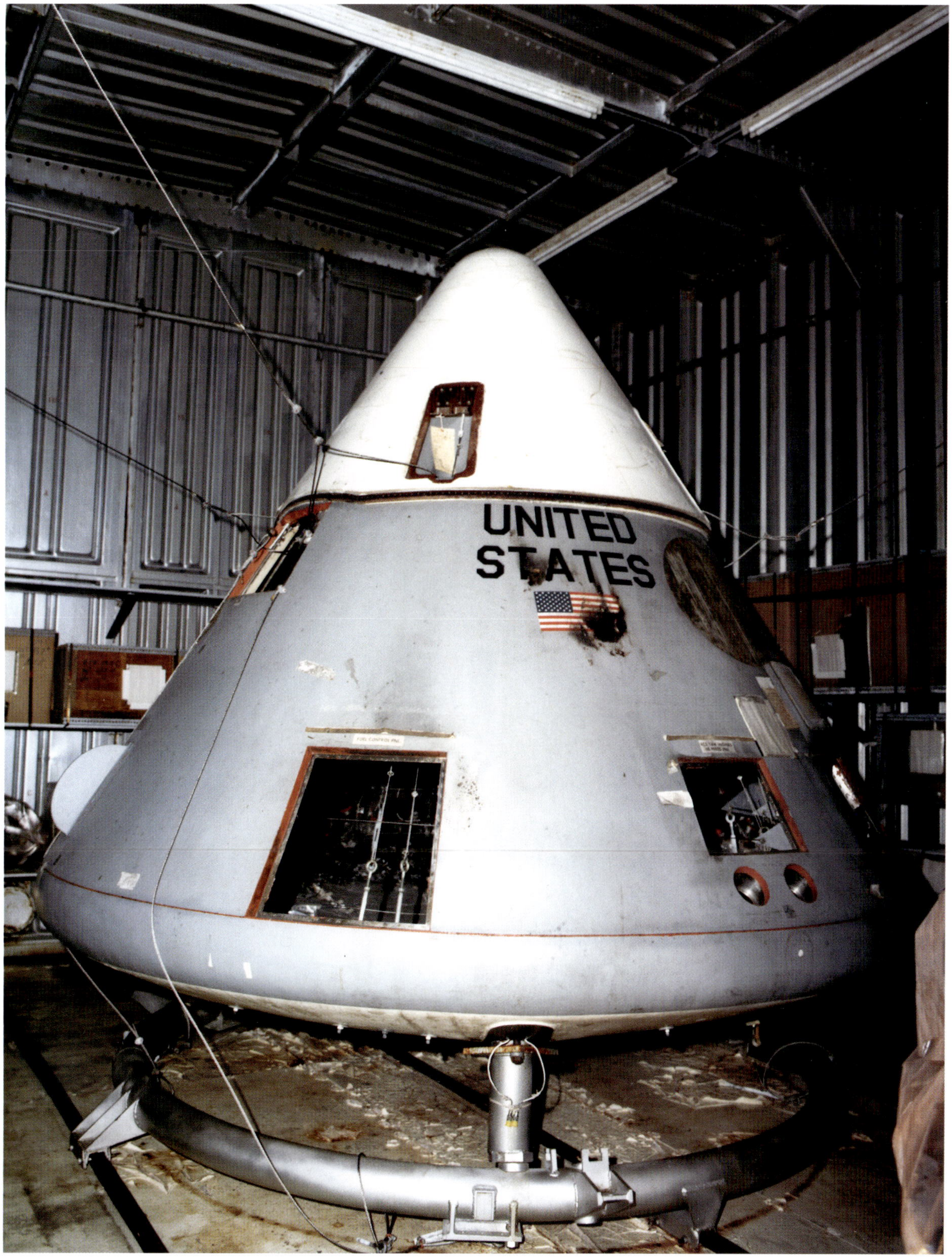
UNITED
STATES

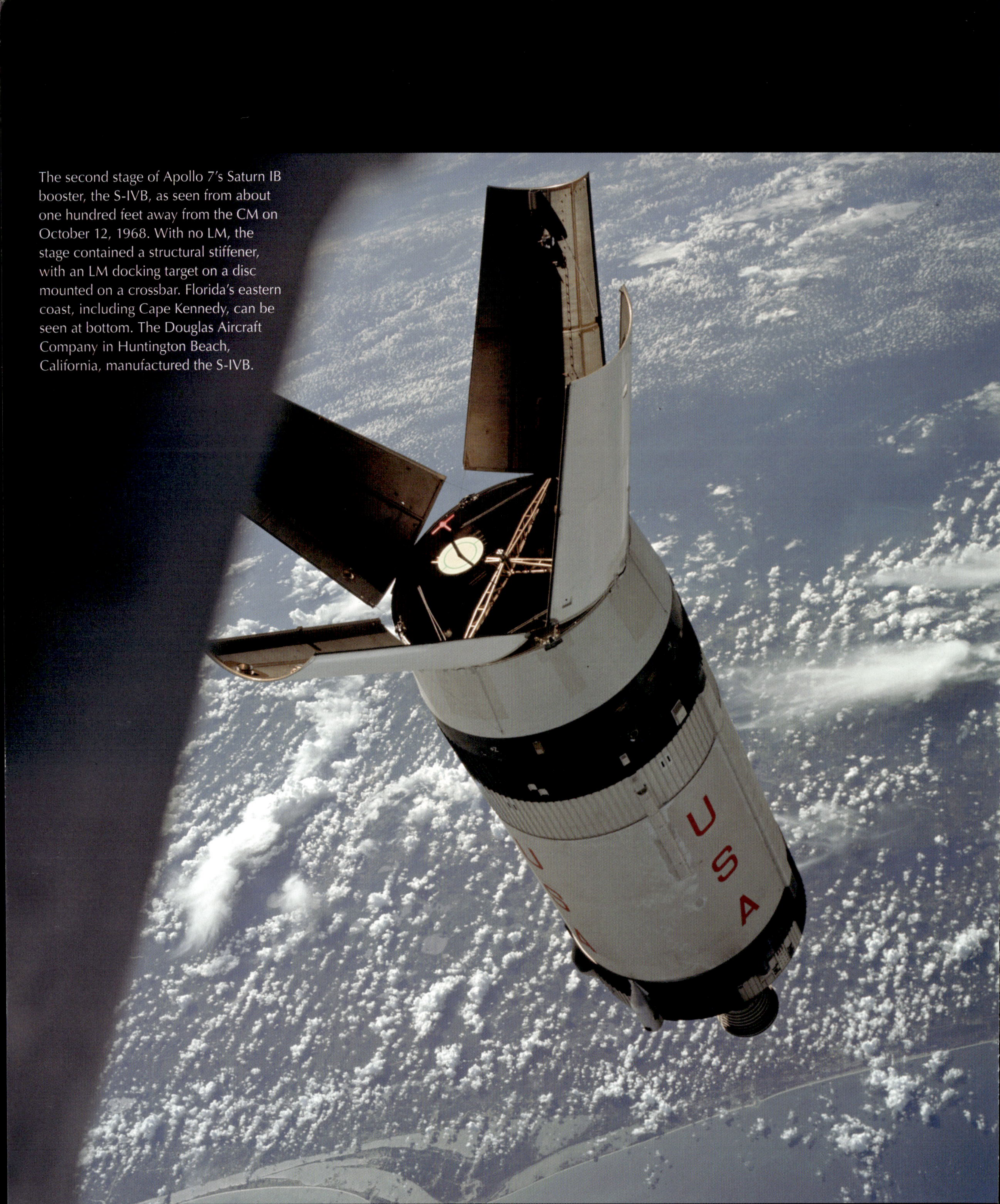

The second stage of Apollo 7's Saturn IB booster, the S-IVB, as seen from about one hundred feet away from the CM on October 12, 1968. With no LM, the stage contained a structural stiffener, with an LM docking target on a disc mounted on a crossbar. Florida's eastern coast, including Cape Kennedy, can be seen at bottom. The Douglas Aircraft Company in Huntington Beach, California, manufactured the S-IVB.

CHAPTER 2

Apollo 7

October 11–22, 1968

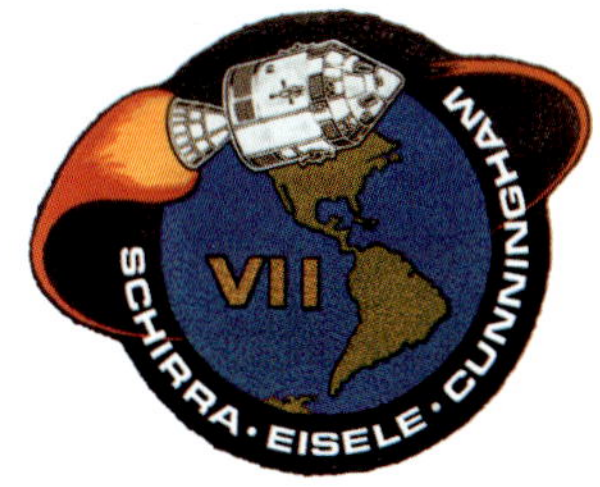

By the fall of 1968, thousands of changes had been made to all phases of the Apollo program in the wake of the Apollo 1 tragedy. On August 19, NASA announced that October 11 was the launch date for Apollo 7, a manned engineering test flight using a spacecraft that had undergone hundreds of safety and operational modifications.

To reduce fire hazards before liftoff and during initial flight, for example, the CM cabin atmosphere was changed from 100 percent oxygen to 60 percent oxygen and 40 percent nitrogen. Shortly after liftoff, the cabin atmosphere would be slowly enriched, eventually becoming pure oxygen. The original two-piece hatch was replaced by a single-piece hatch, which could quickly be opened from the inside. Materials used in the CM were replaced with less flammable substitutes, including Beta cloth.

Three unmanned Apollo test missions with CMs had flown since Apollo 1, including the first test flight of the three-stage Saturn V booster, intended to send astronauts to the moon. As an Earth-orbiting mission with no LM, however, Apollo 7 would use a Saturn IB.

The mission lifted off at 11:03 a.m. (EST) on October 11, 1968, from LC-34 at Cape Kennedy Air Force Station. With all subsequent manned flights using LC-39 at KSC, this was the last manned launch to use a pad at the US Air Force (USAF) installation.

Wally Schirra was CDR, becoming the only astronaut to fly on Mercury, Gemini, and Apollo missions. He was joined by rookies Donn F. Eisele, a USAF major, as command module pilot (CMP) and civilian Walter Cunningham as lunar module pilot (LMP). The new crew designations were based on lunar missions; in reality, Eisele became expert on the navigation system, and Cunningham, with no LM, focused on CM systems.

On the second day, the spacecraft rendezvoused with the Saturn booster's second stage. The astronauts approached to within about seventy feet, to simulate a rescue by the Apollo command ship of a stranded LM on orbit around the moon. This was also a key test of the SM's main engine, which would be needed to break out of lunar orbit on moon landing missions.

The flight is remembered for two firsts for American astronauts: the first use of live onboard television and the first time there was friction between a CDR and Mission Control. Given that the flight was nearly eleven days and largely devoted to systems checks, flight planners had filled the mission with

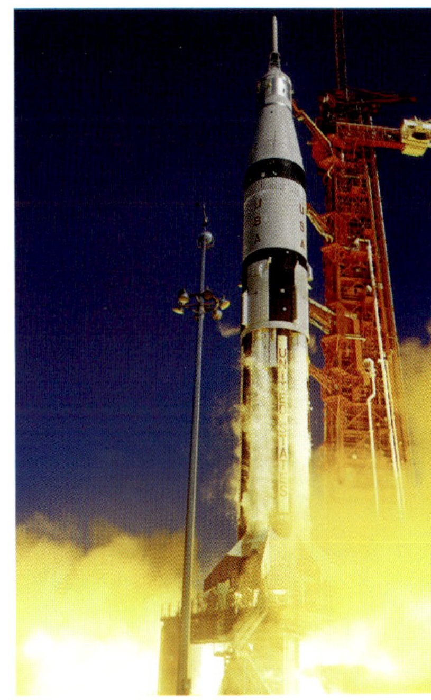

Launch: 11:03 a.m. (EDT), October 11, 1968

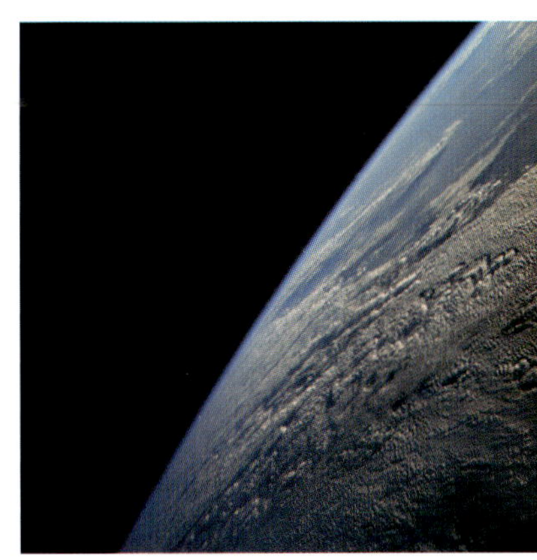

On orbit: Earth view

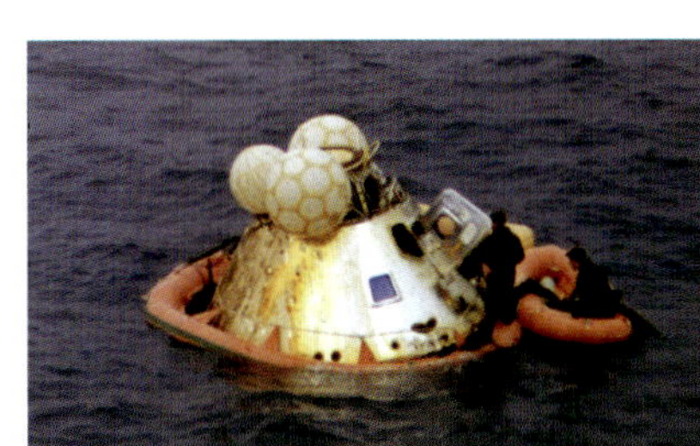

Splashdown: 7:12 a.m. (EDT), October 22

experiments, scheduling almost every minute. The tight timeline, combined with Schirra's developing a head cold, led to onboard frustration and short tempers. Schirra also once complained about ground controllers not allowing him to sleep enough. He took aspirin and decongestant tablets, but suffered for much of the mission. Hot meals and a much roomier spacecraft, however, made the flight more comfortable than Mercury and Gemini flights had been.

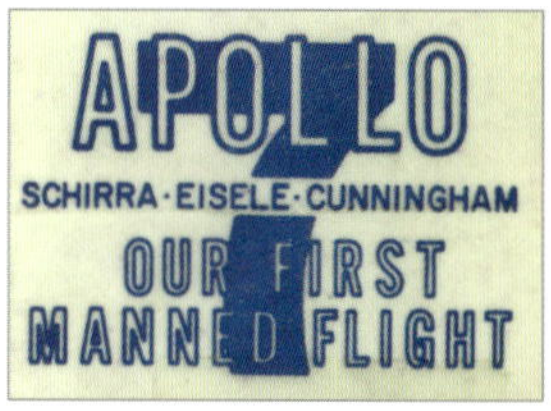

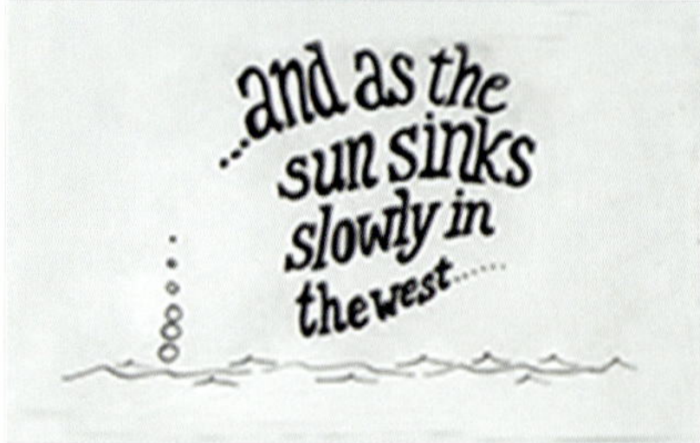

Although Gordon Cooper had experimented with a "slow-scan" TV camera during his 1963 Mercury flight, the picture quality was very poor, and the images were shown on a delayed basis. For Apollo, the RCA Corporation had developed a 4.5-pound black-and-white camera that, while still a slow-scan unit, provided much better images (which had to be converted on the ground for use by broadcasters). Live television was transmitted to ground stations on seven occasions, tied together by the crew with hand-lettered cards carrying such messages as "From the lovely Apollo Room high atop everything" and "Keep those cards and letters coming in folks." The latter prompted entertainer Dean Martin to send a telegram to Houston, reminding the crew that he used the slogan on his weekly TV show. Martin added a postscript, "Like all Americans I am proud of you," and a PPS., "I was higher last night than you are now."

Though Schirra originally criticized the camera as being an "inferior product," adding that he was embarrassed by the picture quality, the telecasts became a mission highlight.

In the last few days of the flight, the crewmen became concerned about wearing their helmets during reentry, which would prevent them from blowing their noses to relieve pressure buildup on their eardrums. Mission Control tried to persuade them to wear the helmets anyway, but Schirra refused. They took decongestant pills before reentry and landed without any problems with their ears.

Apollo 7 spent more time in space than all of the Soviet space flights up to that time combined, and it set the stage for a much more ambitious move. According to NASA's original plan, Apollo 7 would have been followed by testing of the complete Apollo spacecraft, including the LM, in Earth orbit. Since LM development was behind schedule, however, officials began considering a bold move: if Apollo 7 experienced no significant problems, they would send Apollo 8, also without a landing craft, around the moon. Apollo 7's success prompted NASA to publicly give Apollo 8's moonflight the green light, just three weeks after 7's splashdown.

Below: With the 16mm Maurer sequence film camera in front of him, Walter Cunningham takes notes. He wears a cotton constant wear garment; his lightweight headset, manufactured by Pacific Plantronics's Space Environmental Communications Division, is around his neck. The blue connector on his chest carried biomedical and communication cables.

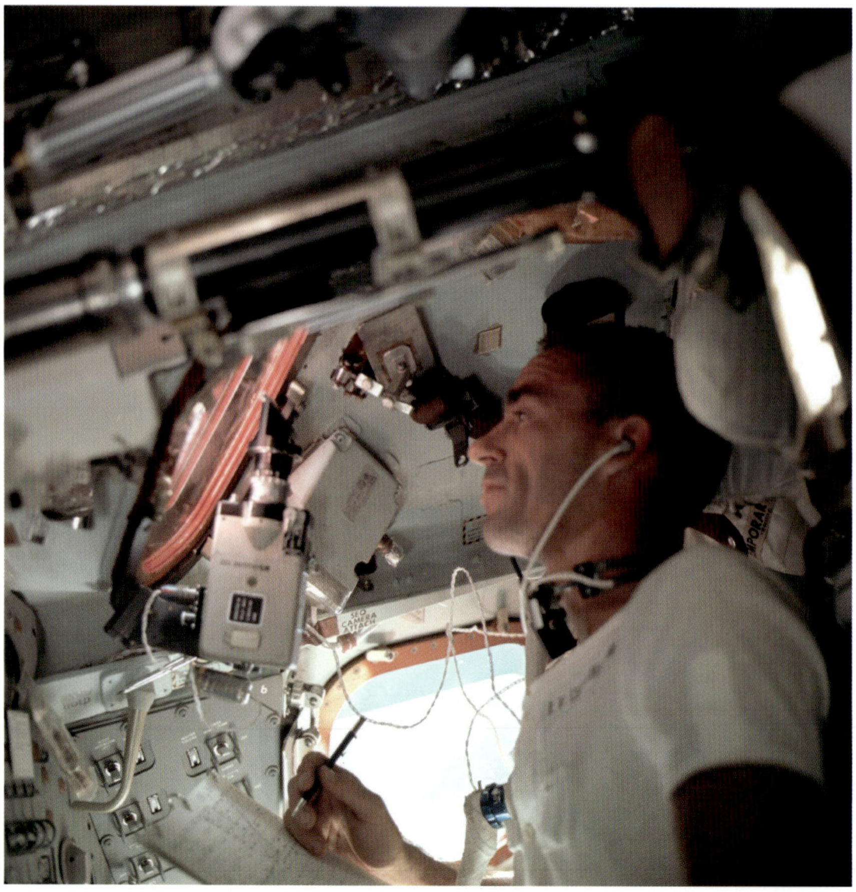

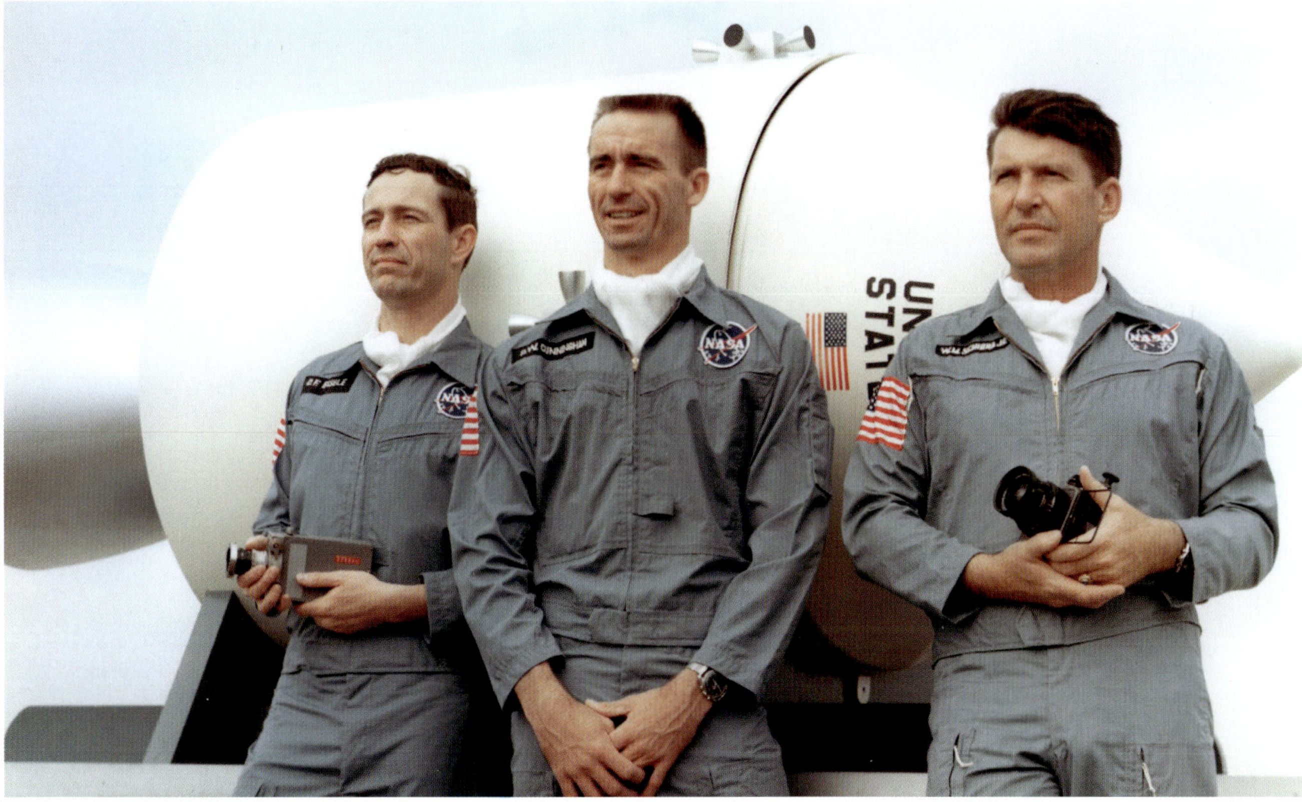

Above: Donn Eisele, Cunningham, and Wally Schirra pose with a mock-up Apollo CSM at MSC on February 13, 1968. Eisele holds a 16mm Maurer sequence camera, and Schirra holds a modified 70mm Hasselblad 500C camera with an 80mm lens. Both cameras were used for photographing Earth features, such as oceans and weather. Targets included Hurricane Gladys and Typhoon Gloria.

Bottom left: Schirra, Eisele, and Cunningham at their preflight news conference in Bldg. 2 at MSC on September 20, 1968. The model shows a Saturn I in its SA-6/SA-7 configuration, the first two launches of an Apollo CM in 1964.

Bottom right: Cunningham aboard MV *Retriever* in Galveston, Texas, before departing for spacecraft egress training in the Gulf of Mexico on August 5, 1968. A former Marine Corps fighter pilot, he was the second civilian to fly aboard an American spacecraft; Neil Armstrong had been the first.

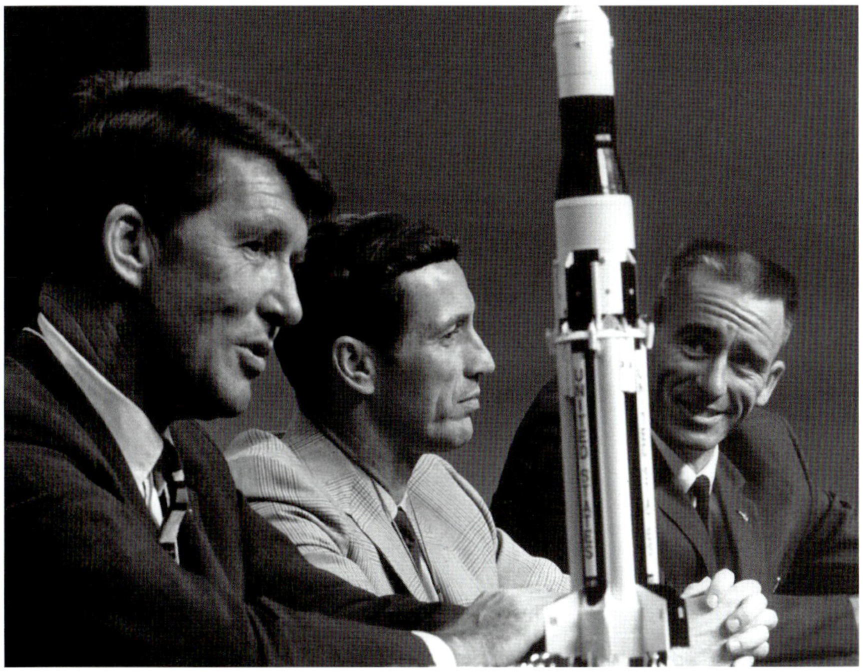

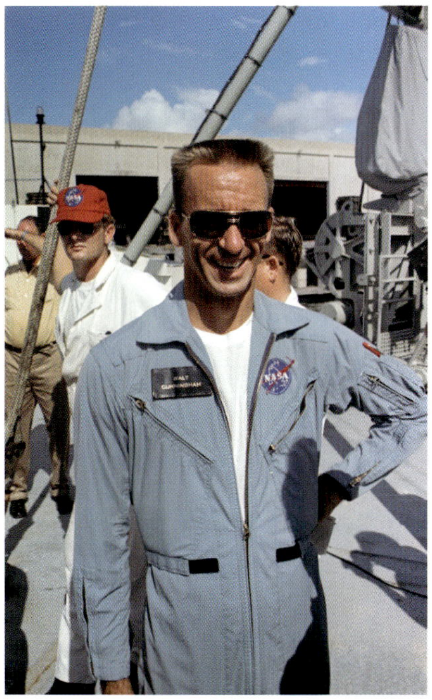

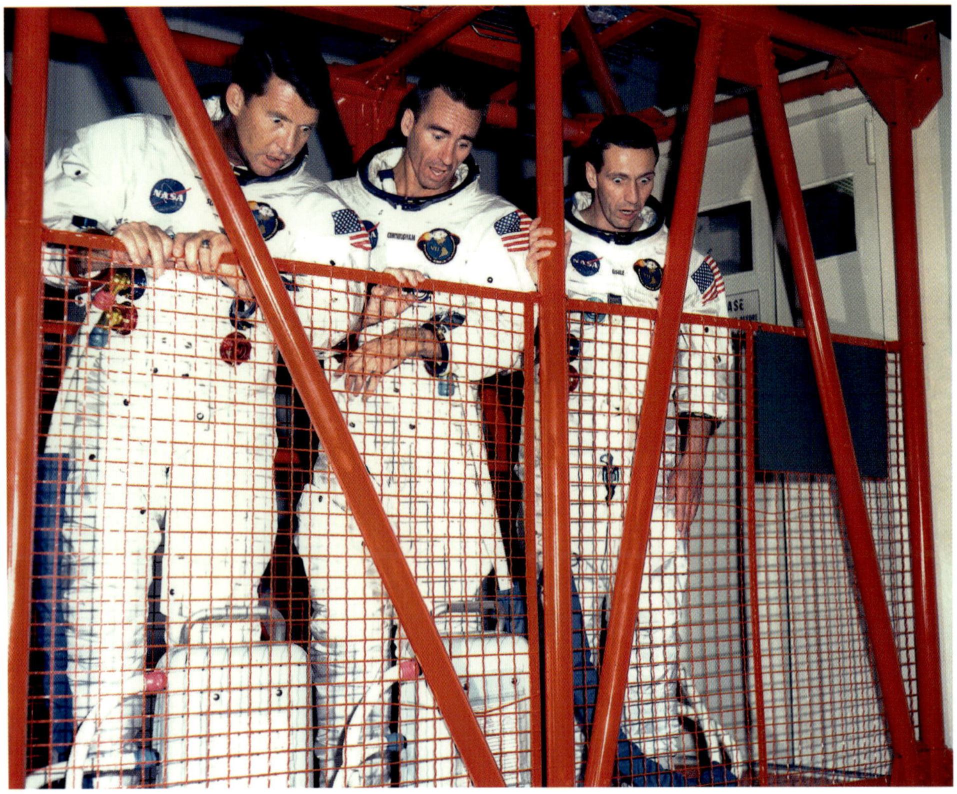

Above: In the Flight Crew Training Building at KSC on July 7, 1968, Schirra, Cunningham, and Eisele feign fear from a mock-up of the spacecraft access swing arm.

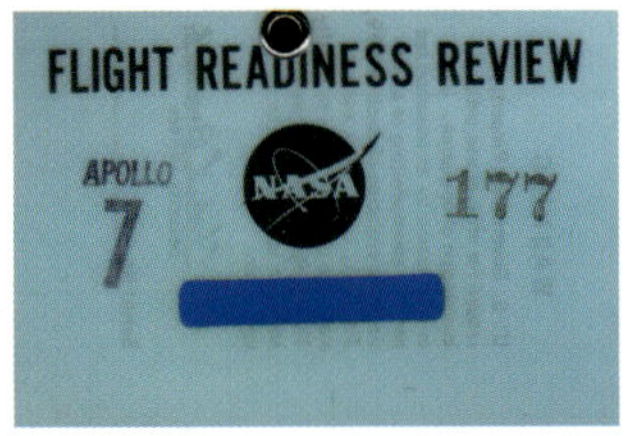

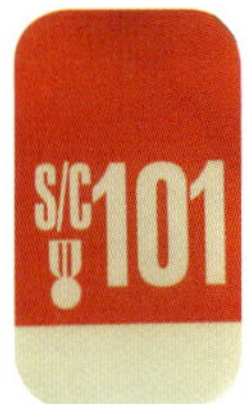

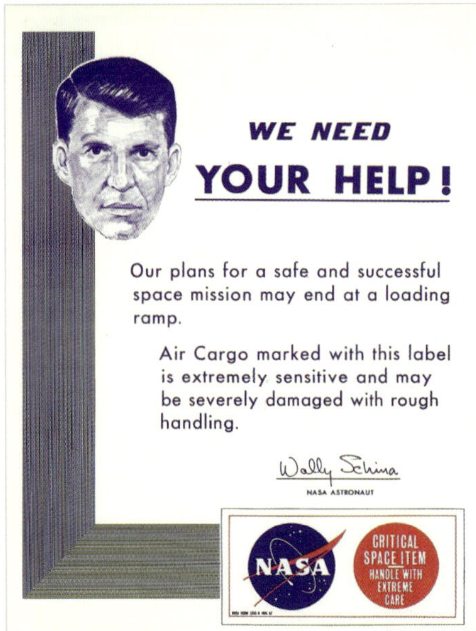

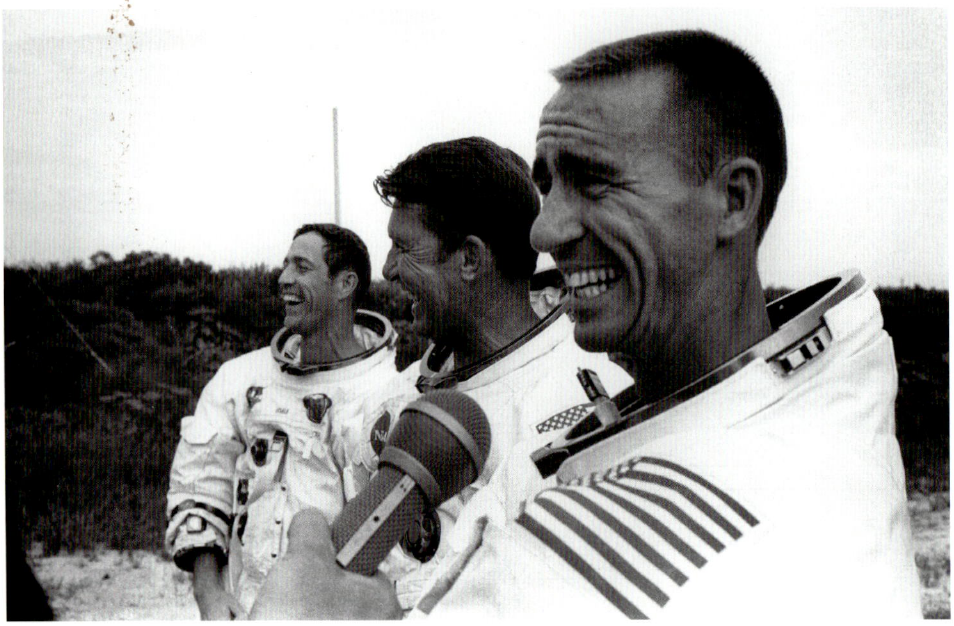

Above: Eisele, Schirra, and Cunningham speak to reporters at LC-34 after an emergency pad-escape exercise on September 9, 1968. Four days earlier, they had completed a mock countdown similar to the one that led to the deaths of the Apollo 1 crewmen nineteen months earlier.

Opposite page: Apollo 7 and its Saturn IB on the launch stand at LC-34, surrounded by the service structure, during the countdown demonstration test (CDDT) on September 15, 1968. The CM hatch was sealed for the test, in contrast to the September 9 rehearsal.

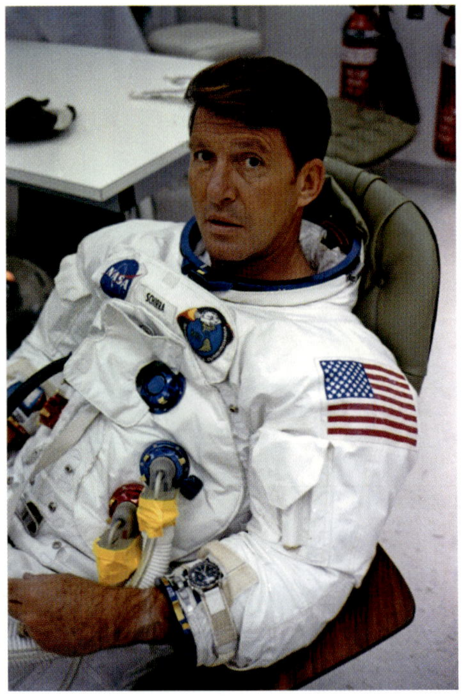

Above: Schirra during a suit test on July 26, 1968, in the MSOB. The astronauts of Apollos 7 through 14 wore the new A7L space suit, which included a cover layer of Beta cloth with a much higher melting point than the Nomex used for the Apollo 1 A1C suit covers—one of the many modifications made after the Apollo 1 fire.

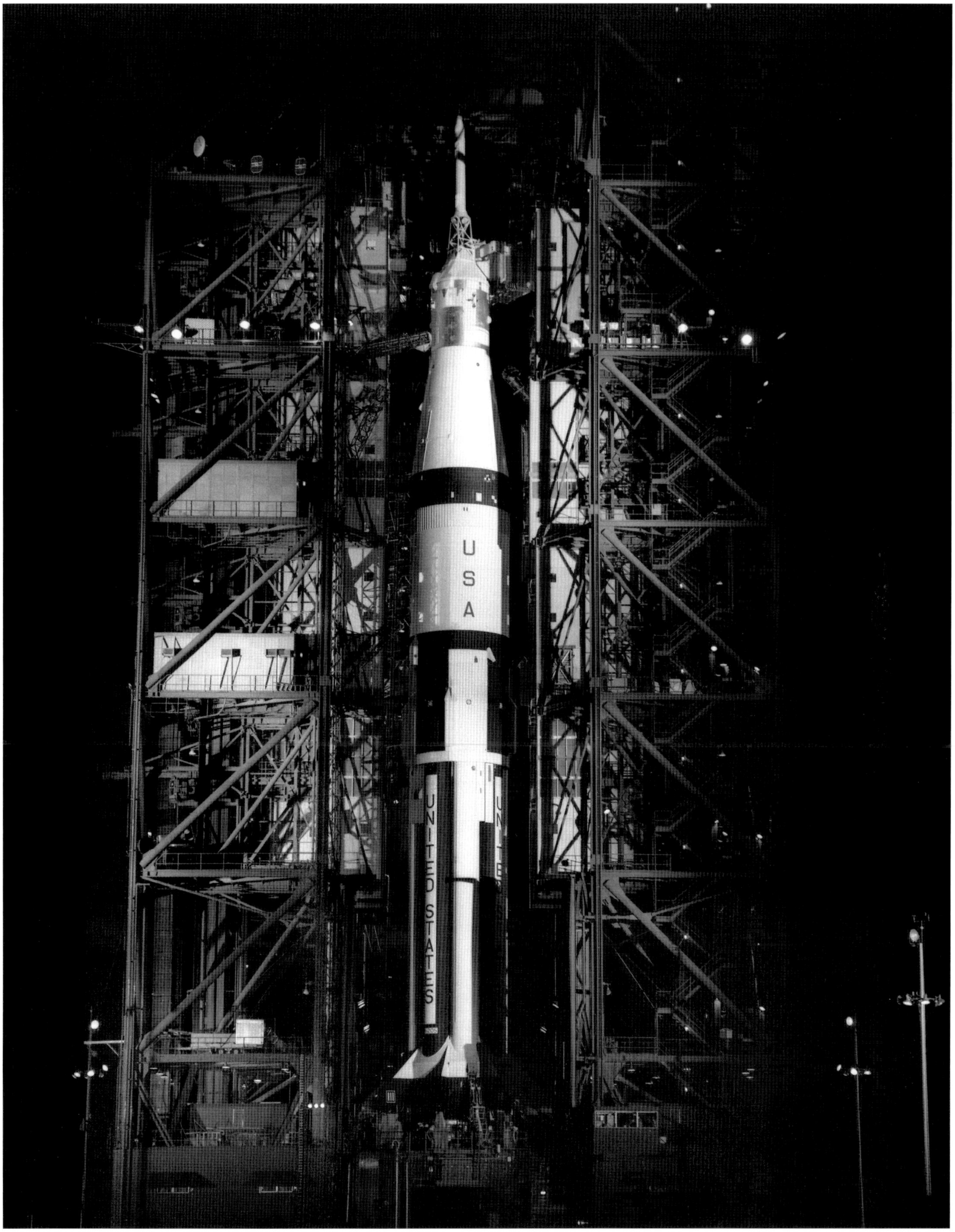
USA
UNITED STATES

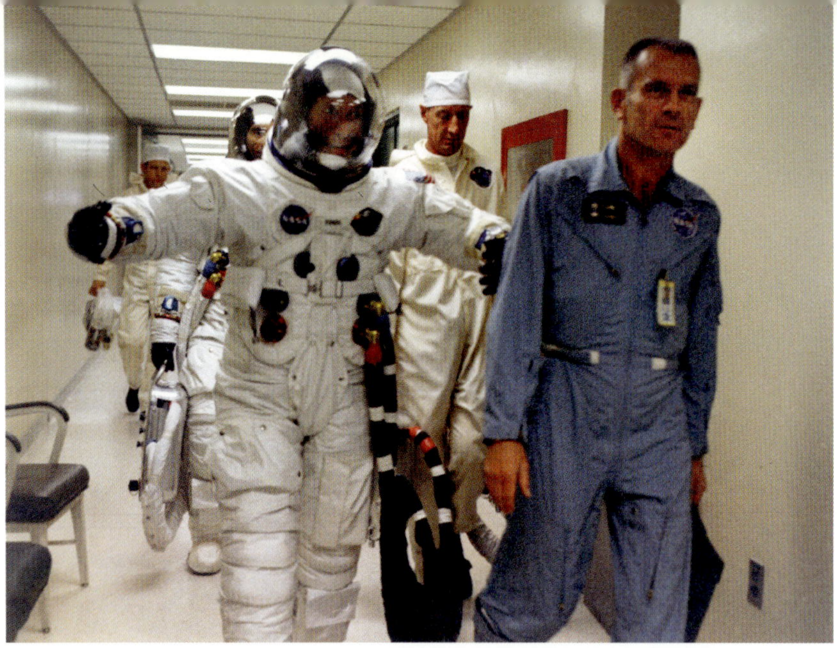

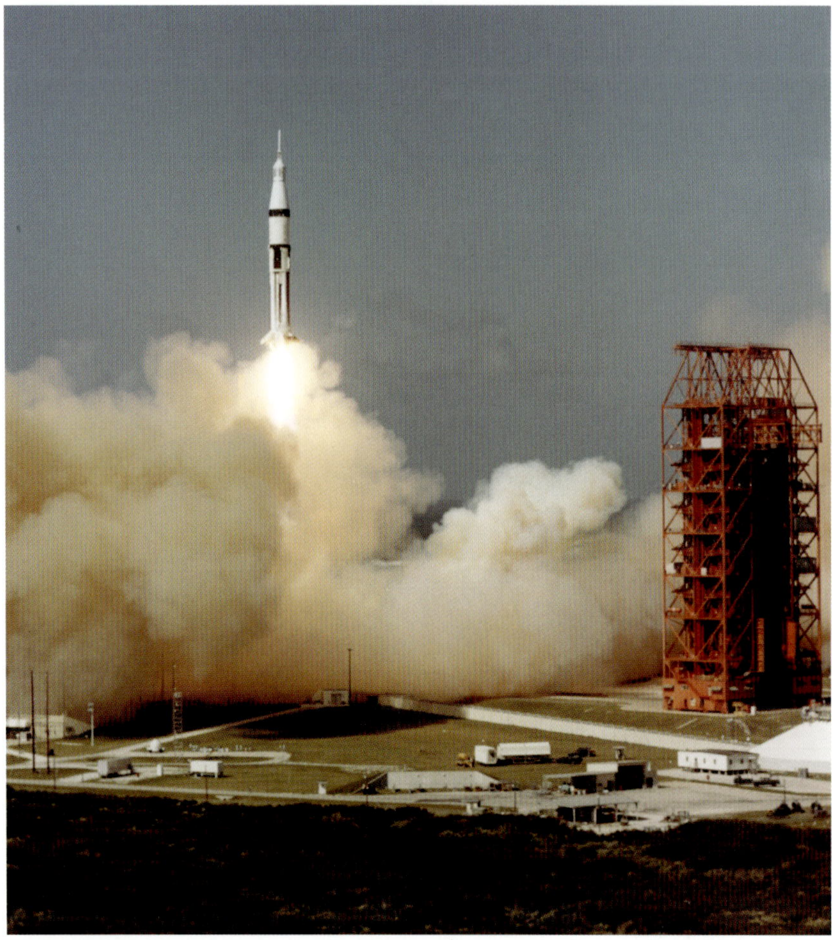

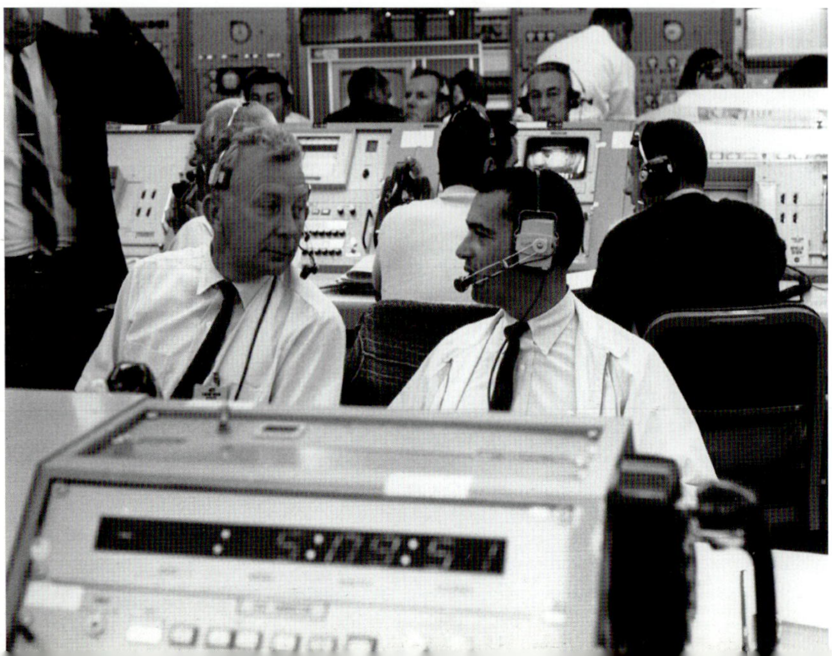

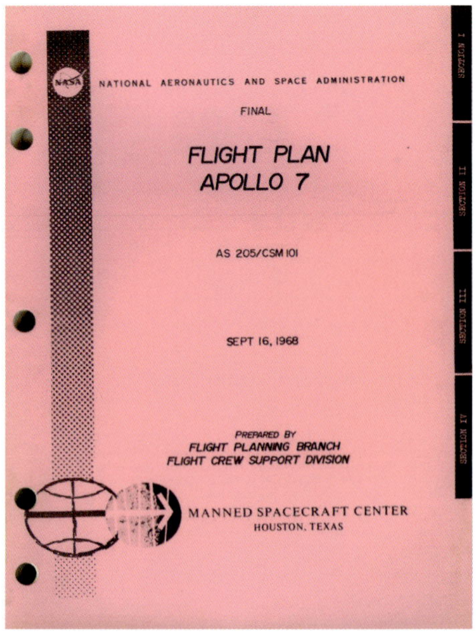

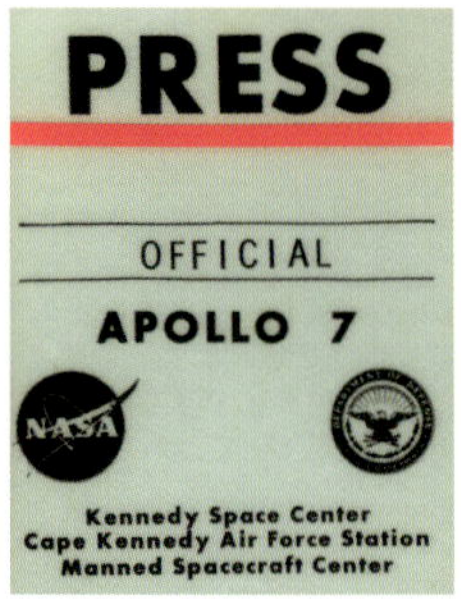

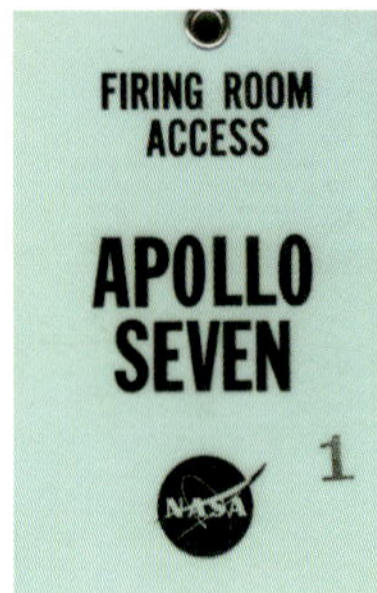

Top left: With Cunningham and Eisele (not visible) behind him, Schirra follows Director of Flight Crew Operations Deke Slayton as they leave the MSOB at 8:00 a.m. (EDT) on launch morning, October 11, 1968. Suit tech Clyde Teague is seen between Schirra and Slayton.

Center left: The red service tower has been rolled back on rails to its parked position (*right*), and Apollo 7's Saturn IB has cleared the umbilical tower, obscured by steam. The rocket went into orbit at 11:03 a.m. (EDT) on October 11. The blockhouse Launch Control Center (LCC) is at lower right.

Bottom left: KSC Launch Operations Manager Paul Donnelly and Test Supervisor Don Phillips in the LC-34 blockhouse during the September 15 CDDT

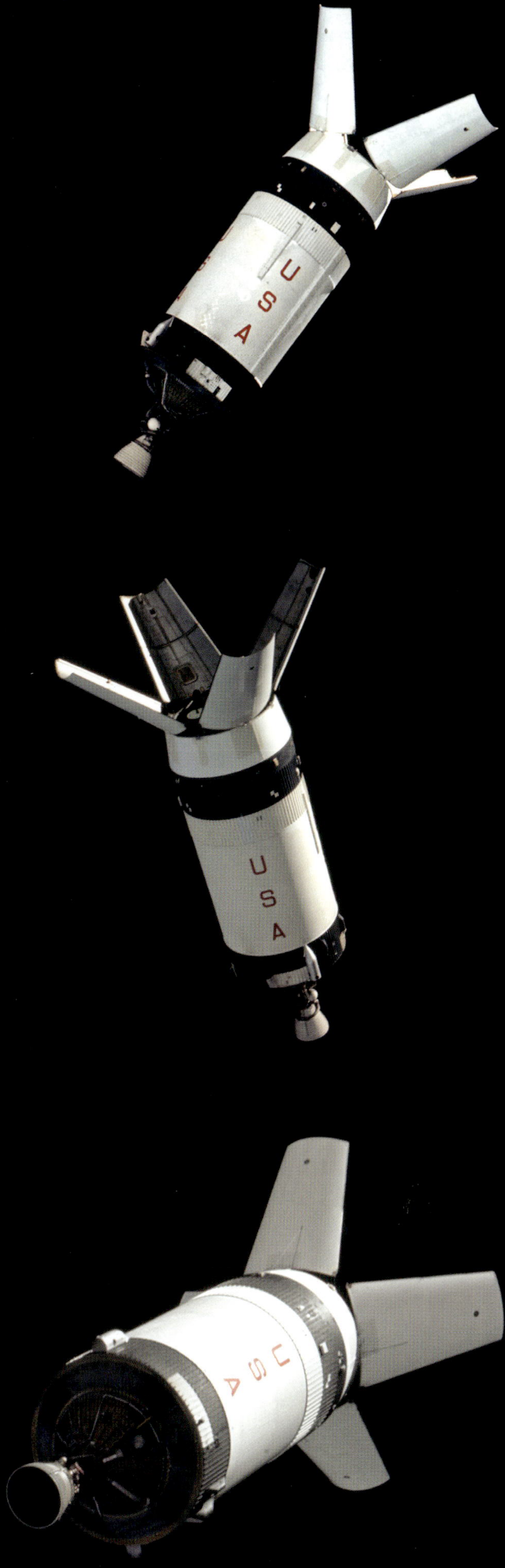

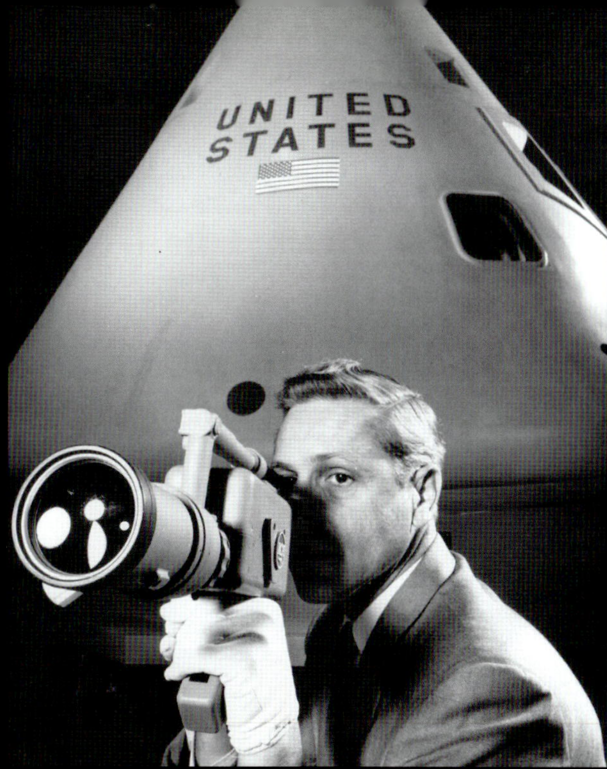

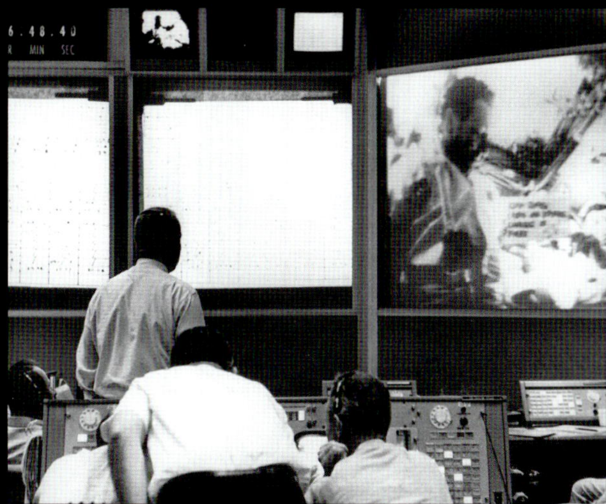

Sequence at left: The Saturn's tumbling empty second (S-IVB) stage is seen from Apollo 7, following the crew's rendezvous with it on the afternoon of October 12. The astronauts approached to within seventy feet as part of a simulated rescue by the Apollo CM of a stranded LM in lunar orbit, which was also a key test of their main engine. They hovered nearby for about thirty minutes. The stage's single J-2 engine is on the aft end. The four open adapter panels would be jettisoned on missions carrying an LM. The S-IVB decayed in Earth orbit on October 18.

Top right: Dick Dunphy, who headed the RCA team that developed the 4.5-pound TV camera, aims the camera in this RCA photo released on May 20, 1968. An identical camera would provide live video from Apollo 8 of the moon from lunar orbit, in December 1968.

Center right: Flight controllers in Mission Operations Control Room (MOCR) 1 watch as Schirra (*right*) holds up a card during the first TV transmission, on October 14, 1968. The card read, "Keep those cards and letters coming in folks"; the request generated three thousand letters to the astronauts. Eisele is at left on the projection screen. This was the first use of MOCR 1, on the second floor of Bldg. 30 of the Mission Control Center (MCC).

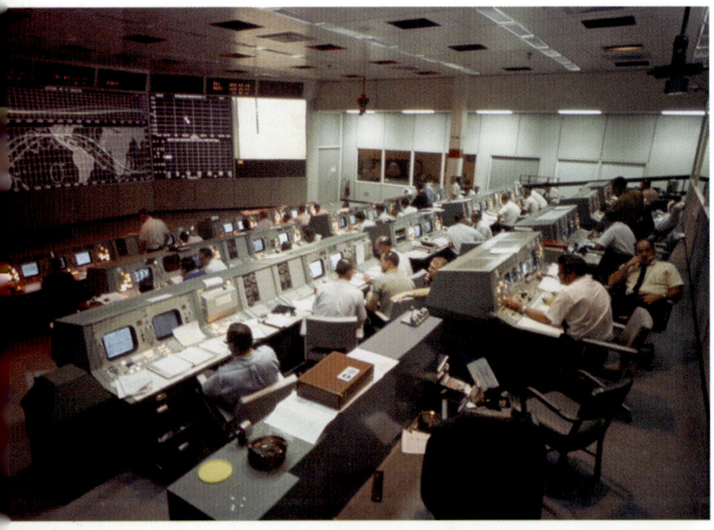

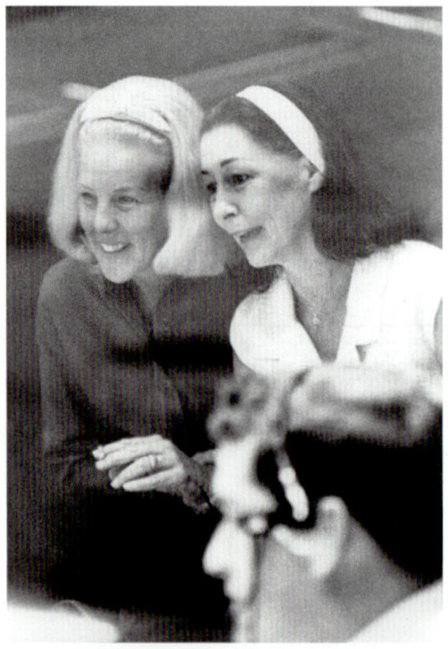

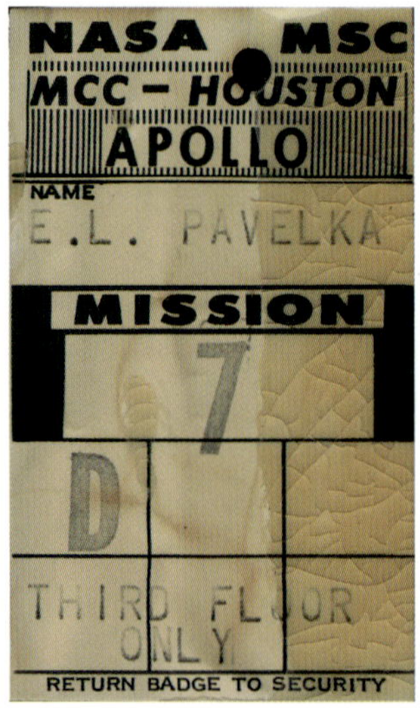

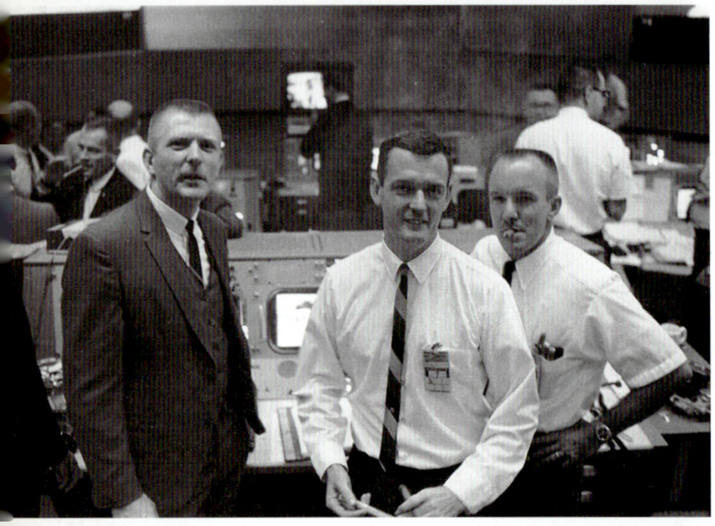

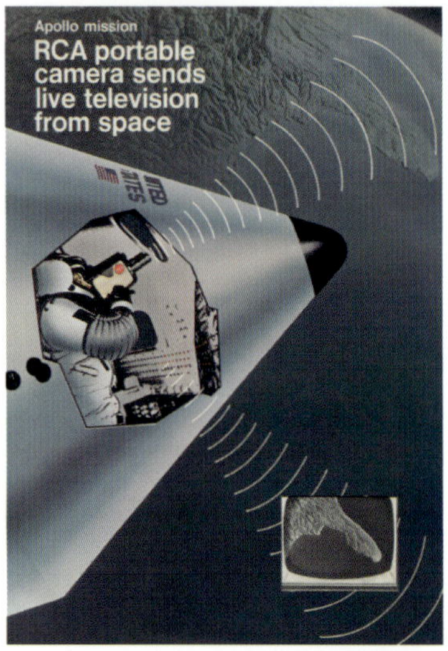

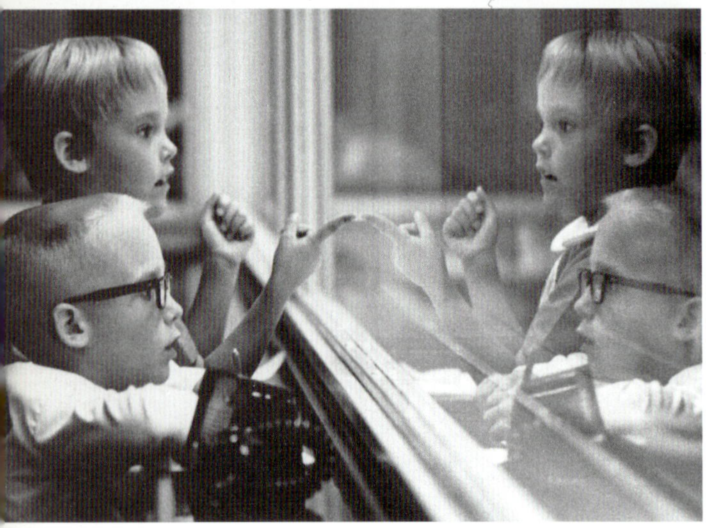

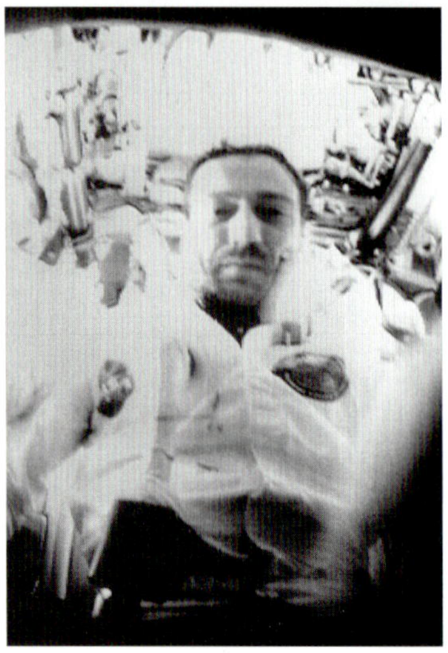

Top left: MOCR 1 after the launch on October 11. Backup CDR Tom Stafford is at the capsule communicator (CAPCOM) console in the third row from left, with Jack Swigert and Rusty Schweickart (on phone) behind him.

Top center: Josephine Schirra (*left*), smoking, and Marjorie Slayton watch the first TV transmission from Apollo 7, beginning at 10:44 a.m. (EDT) on October 14, 1968. The telecasts won the crew a special Emmy Award.

Top right: Ed Pavelka was the flight dynamics officer (FIDO).

Center left: Flight directors Gene Kranz (shift 1, White Team), Glynn Lunney (shift 2, Black Team), and Gerald (Gerry) Griffin (shift 3, Gold Team) on October 22. (Colors were used to designate MOCR teams.) Lunney, 31, was lead flight director.

Bottom left: Cunningham children Brian, seven, and Kimberly, five, are reflected in the glass window of the visitors' area at the rear of the MOCR on October 12. They visited MSC with their mother, Lo Ella Cunningham, from 9:30 to 10:30 a.m. (CDT), along with a houseguest, presidential daughter Lynda Johnson.

Bottom right: An unshaven Eisele with the hatch behind him during the October 19 telecast from the CM, 150 miles up. The live transmission cut into Saturday morning cartoon shows on the television networks. Astronauts first used razors on Apollo 10.

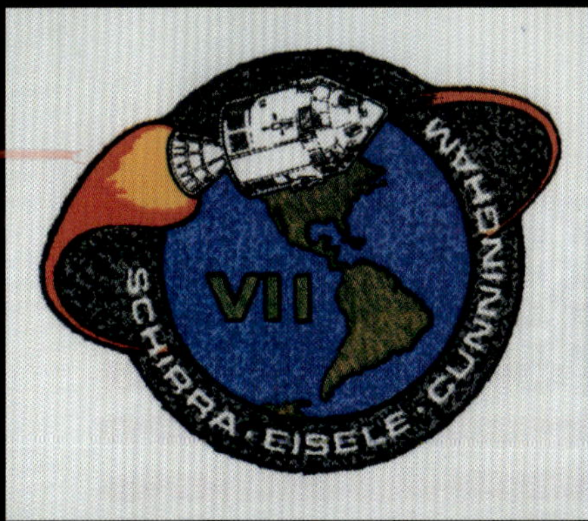

Above: The outer layer of each astronaut's space suit was made from Beta cloth, a Teflon-covered silica fiber specially developed by NASA for such use.

Sequence at right: The crewmen's name tags and mission emblem were silk-screened onto Beta cloth. The emblem was designed by Cunningham and drawn by NAR artist Allen Stevens. The crew had considered a "phoenix rising from the ashes" theme, but rejected it out of respect for the Apollo 1 crew.

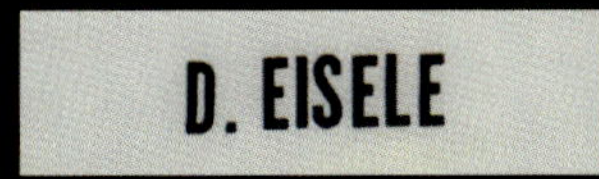

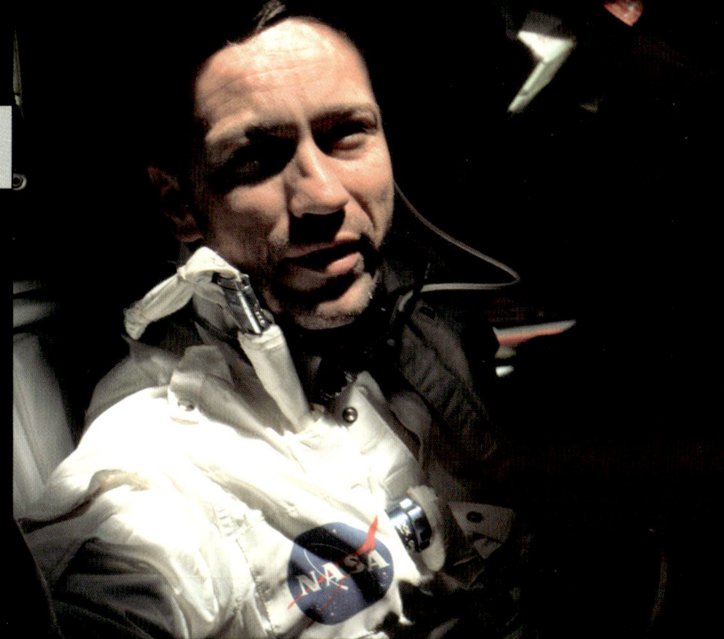

Eisele in the CM's center couch. He wears the jacket to his inflight coverall, made of Teflon fabric. The connector near his cheek is inserted into the jacket's collar and was used for headset assembly.

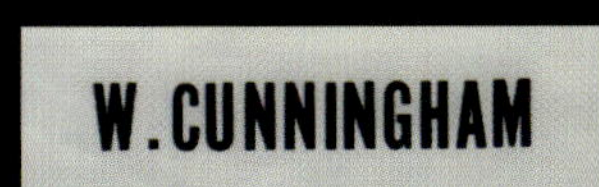

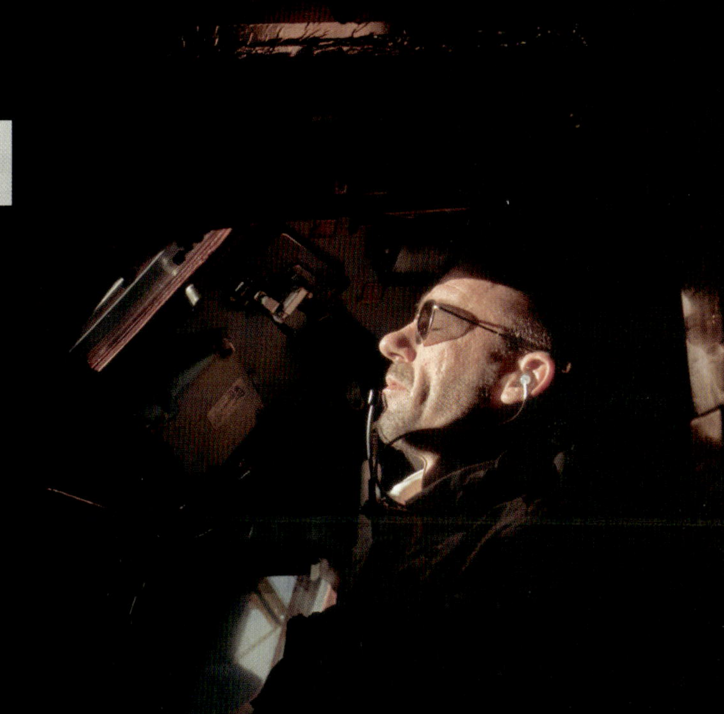

Cunningham rests in his right-hand couch, illuminated by sunlight through his rendezvous window.

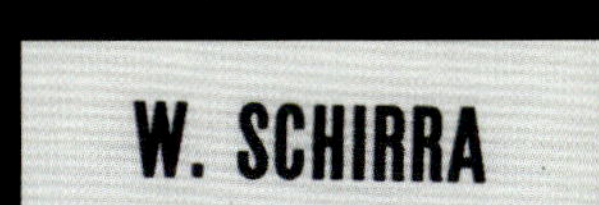

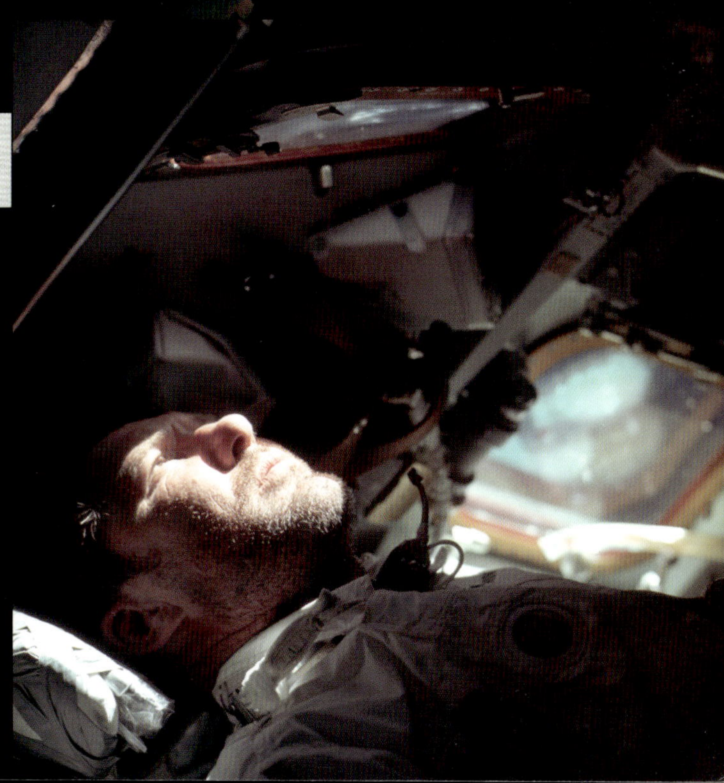

Schirra in the CDR's (left-hand) seat of the CM on October 20, with a rendezvous window above him and a side window at right.

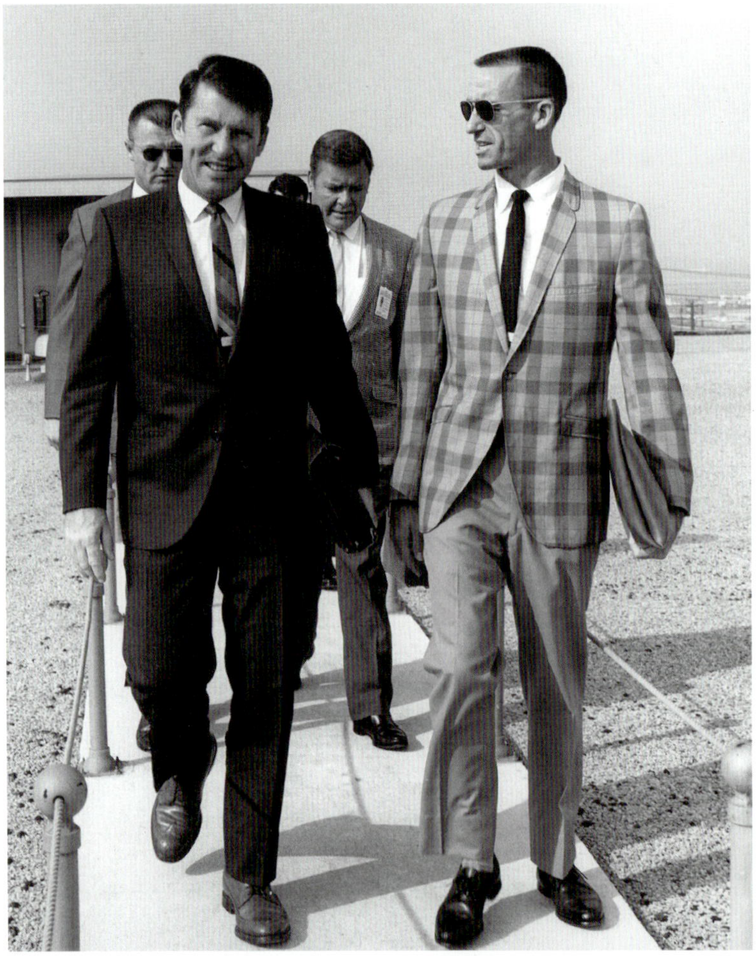

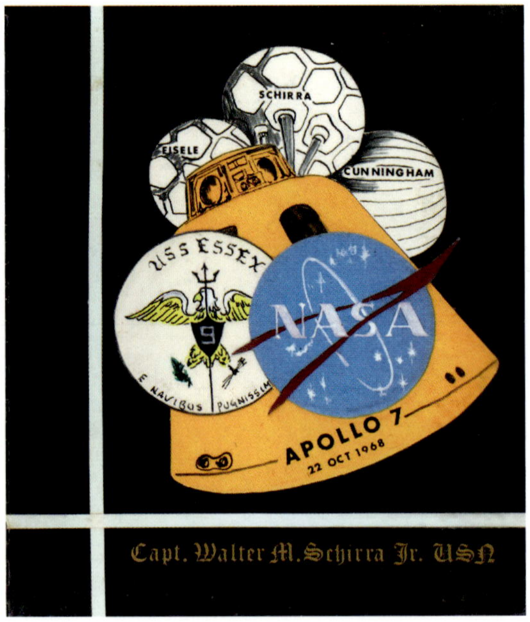

Above: Photo album presented to Schirra by the ship's crew

Left: Schirra (*left*) and Cunningham during their visit to NAR's Space Division in Downey on November 6, 1968. NAA had merged with Rockwell Standard in September 1967, creating the new company.

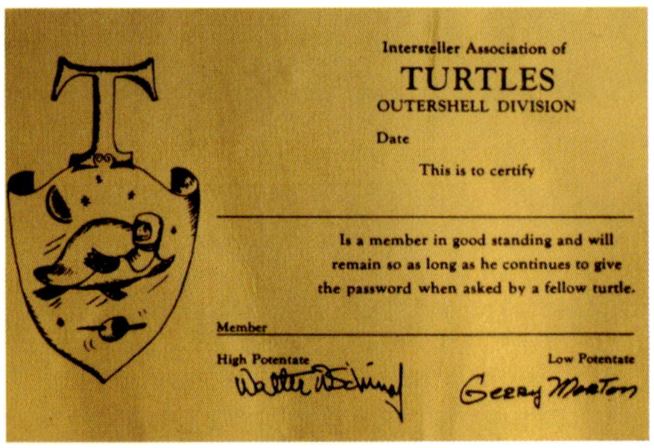

Interstellar Association of
TURTLES
OUTERSHELL DIVISION

Date

This is to certify

Is a member in good standing and will remain so as long as he continues to give the password when asked by a fellow turtle.

Member

High Potentate — Walter Schirra

Low Potentate — Gerry Morton

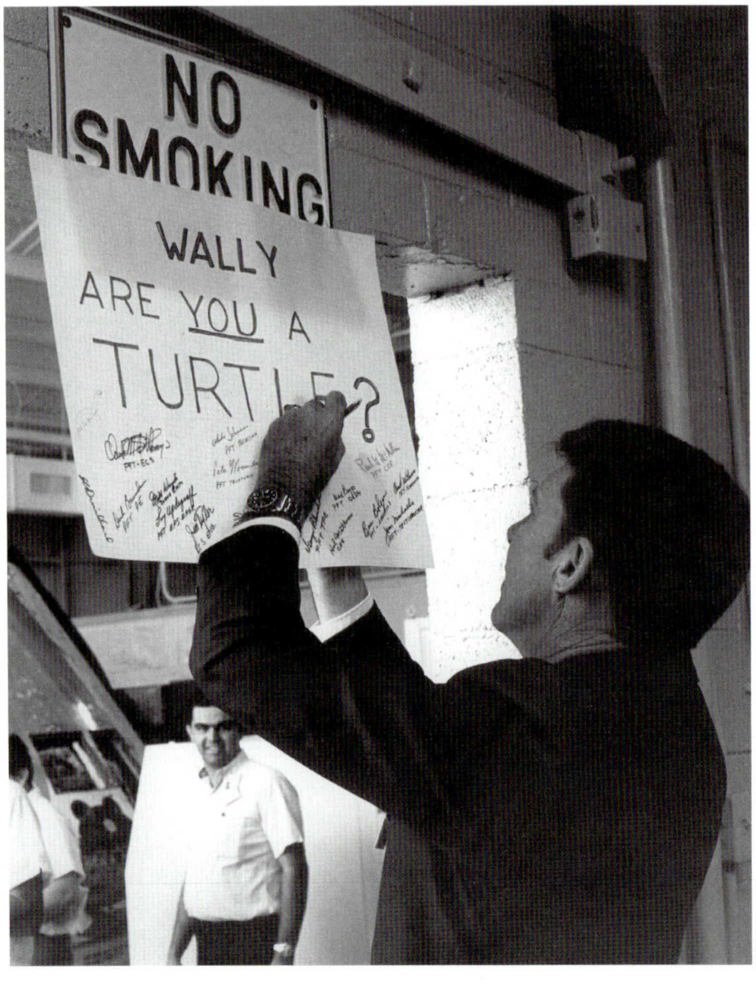

Left: Schirra autographs a sign at Downey with the same question he posed to Slayton on a card during a telecast from orbit. Failure to respond to a fellow Turtle with "You bet your sweet ass I am" required the second member to buy the first a drink. Slayton recorded his answer off the public voice circuit. The joke originated among World War II pilots; Schirra founded the "interstellar" version of the Association of Turtles.

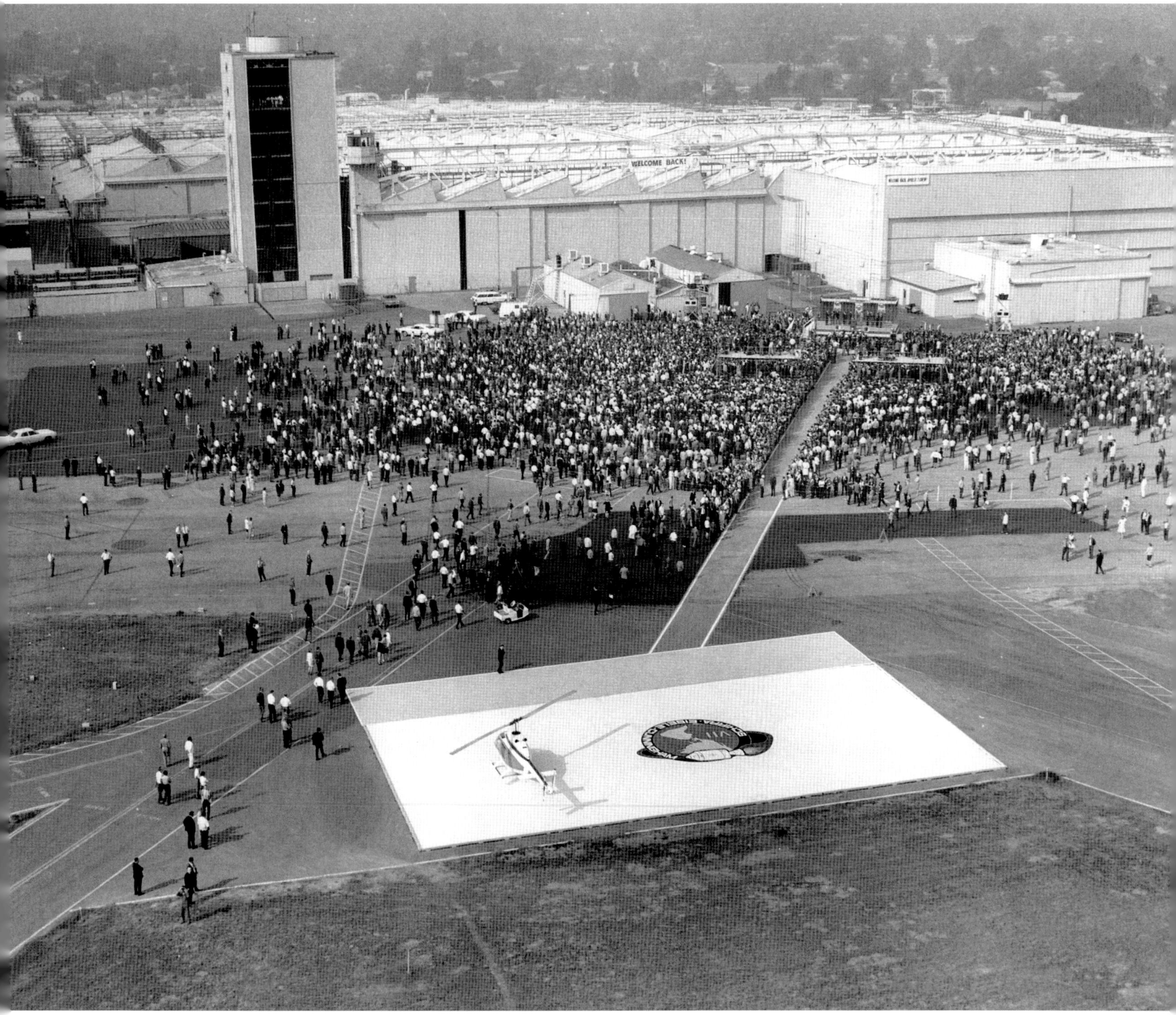

Employees assemble behind NAR's Downey plant on November 6, to hear Schirra and Cunningham thank them for their work building the Apollo spacecraft. The mission emblem is to the right of the helicopter that brought them to the plant. From a high of twenty-five thousand plant employees in 1965, the impending end of the Apollo program meant that there would be six hundred employees by 1970.

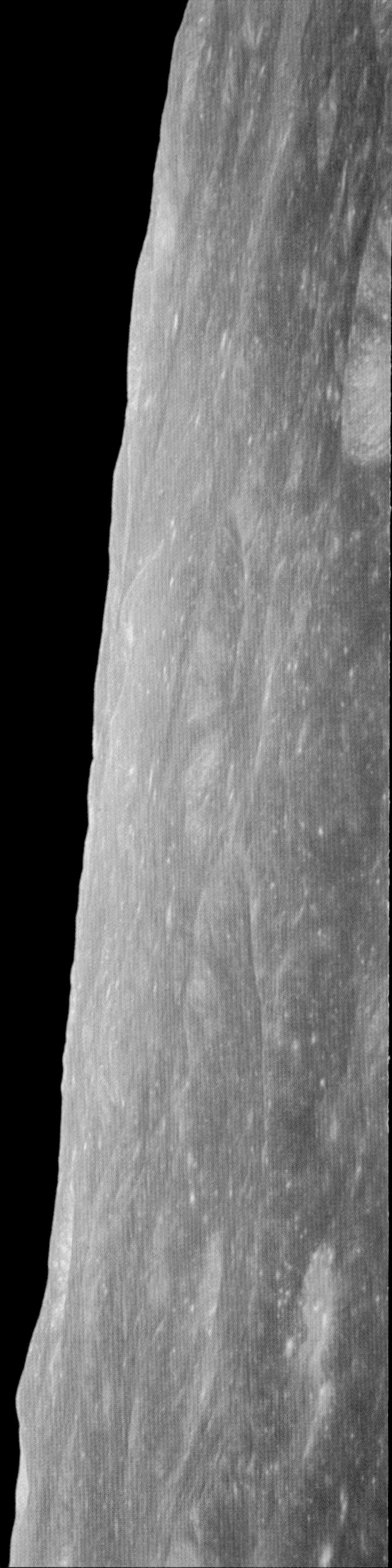

CHAPTER 3

Apollo 8

December 21–27, 1968

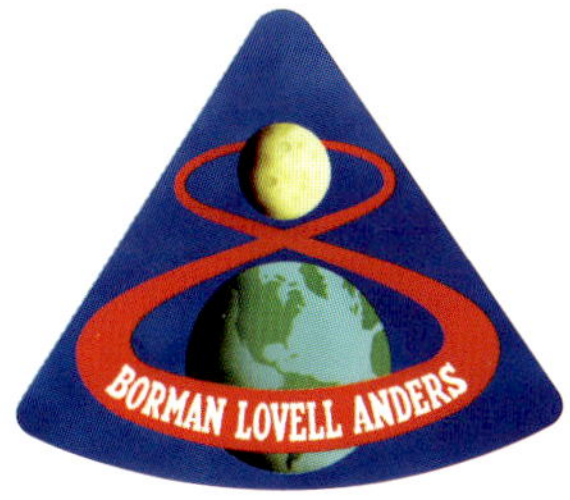

Apollo 8, the first mission to take humans to the moon and back, had been planned as the first flight of the combined CSM and LM in Earth orbit. When the LM manufacturer, Grumman Aircraft Engineering Corporation, was unable to deliver an operational spacecraft before the end of 1968, however, NASA changed the flight to be a CSM-only mission around the moon. The flight would also test the systems, trajectories, and operations for getting to the moon and back, critical for a manned landing. Rumors had circulated that the Soviet Union was preparing to attempt a manned orbit of the moon as well, adding to the pressure to get there first.

Apollo 8 was the first manned mission launched on a Saturn V, and the first from NASA's LC-39 on Merritt Island at KSC. It was also the first manned flight to escape Earth's gravity. Reaching 24,593 mph, Apollo 8 set both manned speed and distance-from-Earth records.

USAF Col. Frank Borman was CDR, while US Navy Capt. James A. Lovell Jr. served as CMP and USAF Maj. William A. Anders was the rookie LMP (though there was no LM). Apollo 8 launched at 7:51 a.m. (EST) on December 21, 1968, into a low-Earth parking orbit to check out its systems. Two hours and fifty minutes later, during the second orbit, the S-IVB stage was reignited for translunar injection.

Apollo 8 then separated from the S-IVB, and the crew rotated the CSM to photograph the spent stage. They then practiced flying in formation. The stage was sent into a solar orbit as Apollo 8 headed toward the moon.

As they sped away, the crewmen could view and photograph all of Earth for the first time. Special ground receivers near Madrid, Spain, and Goldstone, California, allowed live television transmissions to, from, and during lunar orbit from the CM, using a 4.5-pound RCA black-and-white camera identical to Apollo 7's.

Two television transmissions were made en route to the moon; two were made in lunar orbit; and two were made during the return flight to Earth. The first transmission took place on December 22, approximately halfway between Earth and the moon. The second transmission, on December 23, showed Earth from a distance of 180,000 miles. The crew was able to send back the first television pictures of Earth in its entirety.

Apollo 8's lunar trajectory required only minor midcourse correction, and on December 24, the spacecraft passed about 70 miles in front of the moon at

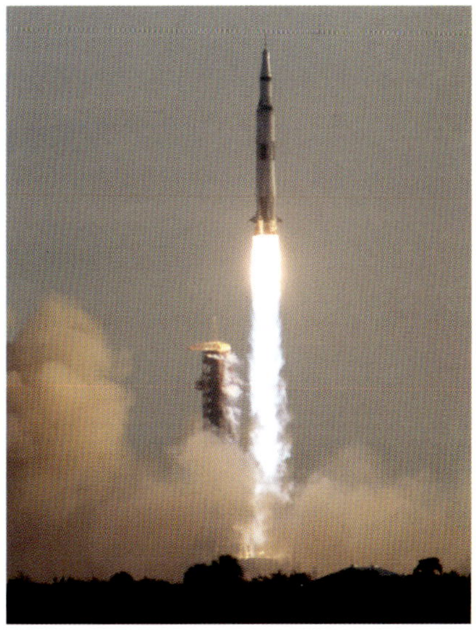

Launch: 7:51 a.m. (EST), December 21, 1968; view from the LC-39 Press Site (photo by Tiziou News Service)

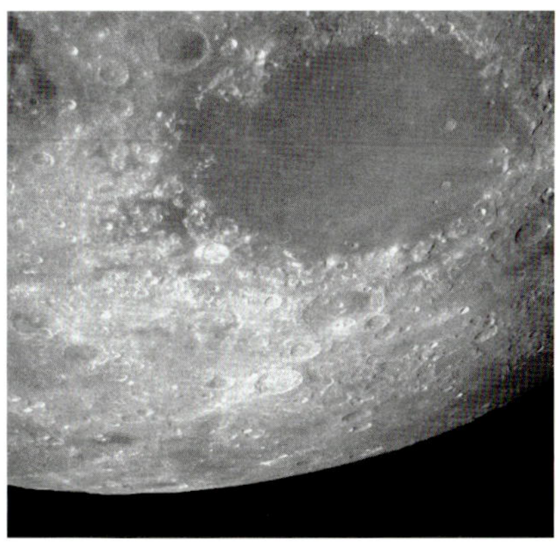

Leaving lunar orbit: December 25

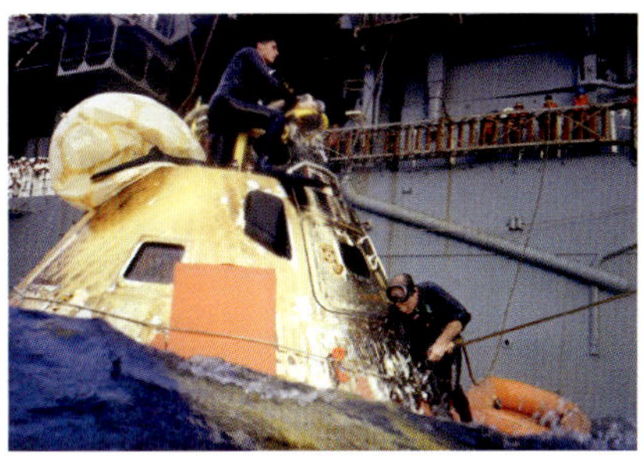

CM recovery: 6:45 a.m. local time (HST), December 27

a speed of 5,720 mph. On the far side of the moon, the astronauts fired the SM's Service Propulsion System (SPS) engine at 4:59 a.m. (EST) to brake the spacecraft into lunar orbit. After flying two elliptical orbits of 193 by 70 miles, the spacecraft was placed in a nearly circular orbit of about 69 miles for eight more orbits—for a total of twenty hours.

When Apollo 8 began its fourth pass across the near side of the moon, the crew saw an earthrise for the first time in human history (a sight rarely visible from the lunar surface due to the synchronous rotation of the moon around Earth).

Borman first saw Earth emerge from behind the lunar horizon. He called to Lovell and Anders, taking a black-and-white photo as he did. Anders then snapped the more famous color image, dubbed *Earthrise* and widely regarded as one of the most historic photos ever taken.

The first of two TV transmissions on Christmas Eve gave audiences their first closeup views of the moon. The second transmission ended with the three astronauts in turn reciting verses one through ten of chapter 1 of the book of Genesis, from the King James Version of the Bible. At the time, the broadcast was the most-watched program in television history. Borman concluded the transmission with "And from the crew of Apollo 8, we close with good night, good luck, a Merry Christmas, and God bless all of you, all of you on the good Earth."

Just two and a half hours later, at 1:10 a.m. (EST) on December 25, 1968, on the far side of the moon, Apollo 8's SPS engine was ignited for trans-Earth insertion (TEI), to accelerate out of lunar orbit. It was critical—an engine failure would have stranded the men in lunar orbit with no hope of rescue. "Please be informed . . . there is a Santa Claus," radioed Lovell when the spacecraft regained contact with Earth. Only one midcourse correction was required during the return flight.

At about 10,357 miles from Earth, the CM separated from the SM. Fifteen minutes later, Apollo 8 reentered Earth's atmosphere at 24,243 mph. The spacecraft splashed down on December 27 in the Pacific Ocean, 1,100 miles south of Hawaii and 3 miles from the aircraft carrier USS *Yorktown*. It was the first splashdown to occur in darkness (at 4:51 a.m. local time), and the crew was not recovered until dawn, forty-three minutes later.

A group of ten Soviet cosmonauts sent a telegram of good wishes to the astronauts, and Soviet Chairman Nikolai Podgorny sent a congratulatory cable to President Lyndon Johnson. Johnson called the astronauts to say that he remembered "when we saw Sputnik racing through the skies, and we realized that America had a big job ahead of it. It gave me so much pleasure to know that you men have done a large part of that job."

Although Apollo 8 was a landmark achievement in the Space Race, by 1968 NASA's budget was $3.9 billion, down from 1966's high point of $5.93 billion. Apollo was quickly nearing its zenith.

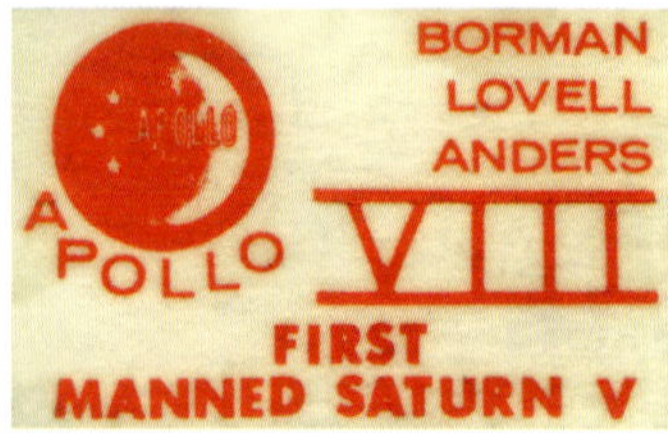

Badge sticker

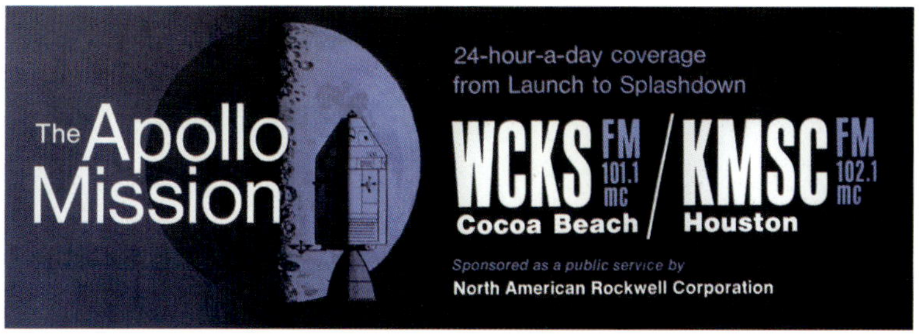

Above: WCKS-FM broadcast the NASA Public Affairs air-to-ground audio circuit, as did KMSC-FM, co-owned by John "Shorty" Powers, former Mercury public affairs officer.

Opposite page, top: Lovell locates the Sea of Serenity (near Apollo 15's eventual landing site) on a lunar chart in the MSOB on December 19, 1968. *Left to right*: NASA Chief Test Supervisor Bill Schick, Launch Operations Manager Paul Donnelly, Lovell, Borman, Launch Vehicle Test Conductor Ray Roberts, and Anders.

Opposite page, bottom left: An industrial water system test-sprays tens of thousands of gallons of water from the Launch Umbilical Tower (LUT) onto the tower's swing arms and the Saturn V launch vehicle at Pad 39A on November 12, 1968. The Mobile Service Structure (MSS) is at right. The Firex system, operated by Bendix, not only provided fire control, but also protected the pad and launch vehicle from damage caused by the sound and heat of liftoff.

Opposite page, bottom right: Anders exits the three-man gondola of the centrifuge in the Flight Acceleration Facility in MSC's Bldg. 29 on November 1, 1968. The control room is at rear. The centrifuge could create twenty gees continuously, or up to thirty gees for short durations.

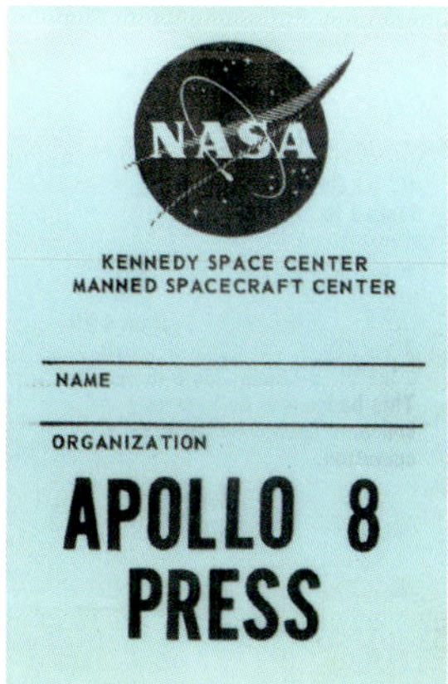

Top right: CBS News anchor Walter Cronkite pauses during the network's coverage of the Apollo 8 launch, from the second floor of the CBS building at the LC-39 Press Site. Producer Jeff Gralnick is at right.

Second right: NBC News anchor David Brinkley in the NBC trailers, next door to CBS's. The camera is from WFGA-TV in Jacksonville. Gulf Oil sponsored the NBC broadcasts.

Third right: ABC News Science Editor Jules Bergman works on a script in that network's building, on the other side of the CBS building. A technician adjusts a light at right.

Bottom right: A journalist from the Cocoa Beach, Florida, radio station WKKO-AM in the station's mobile unit. All photos were taken in the hours before sunrise on December 21.

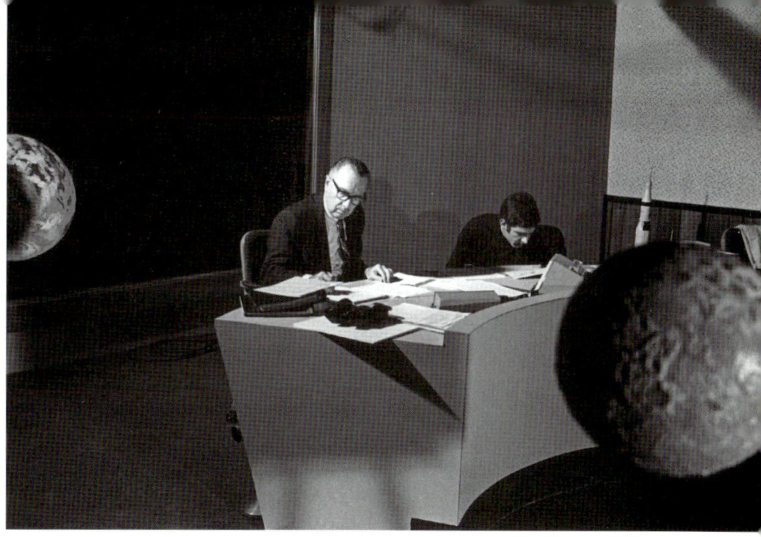

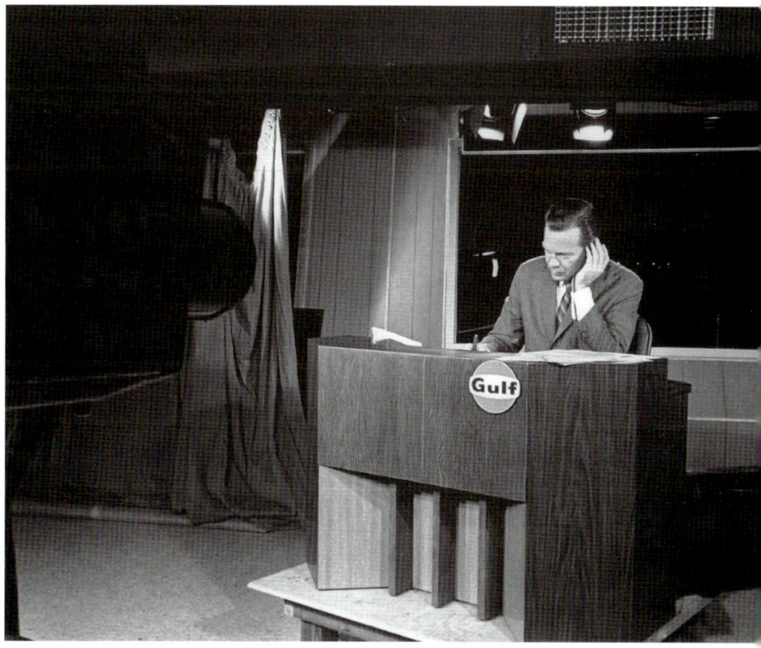

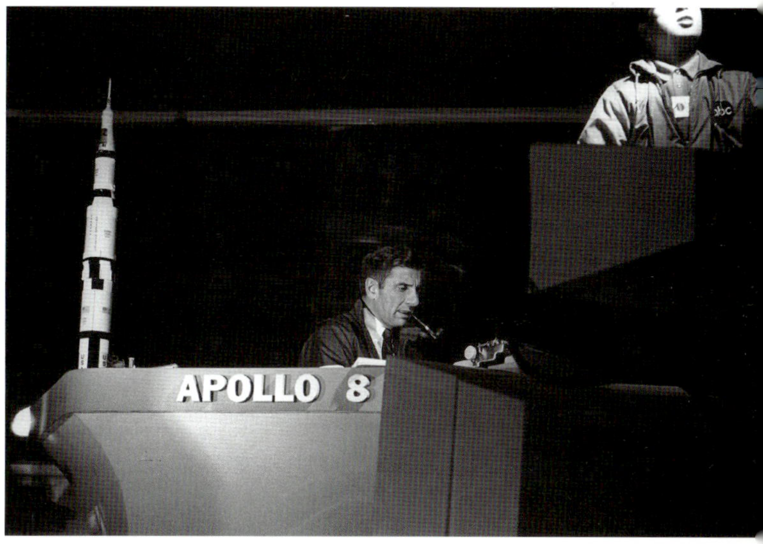

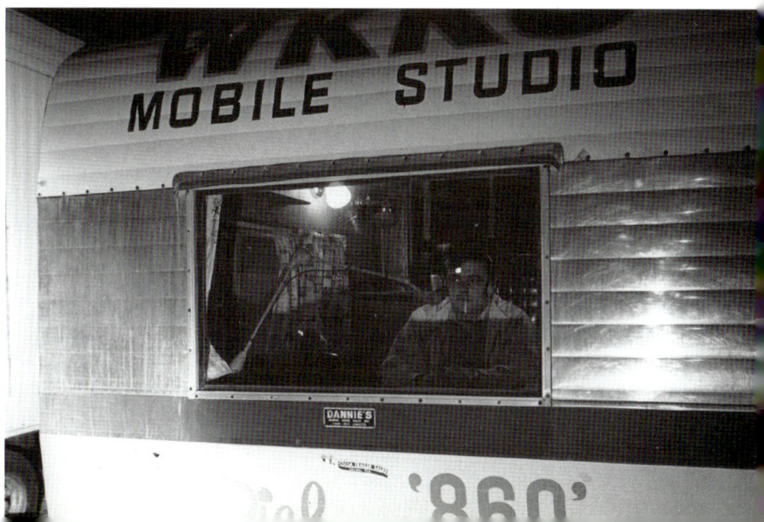

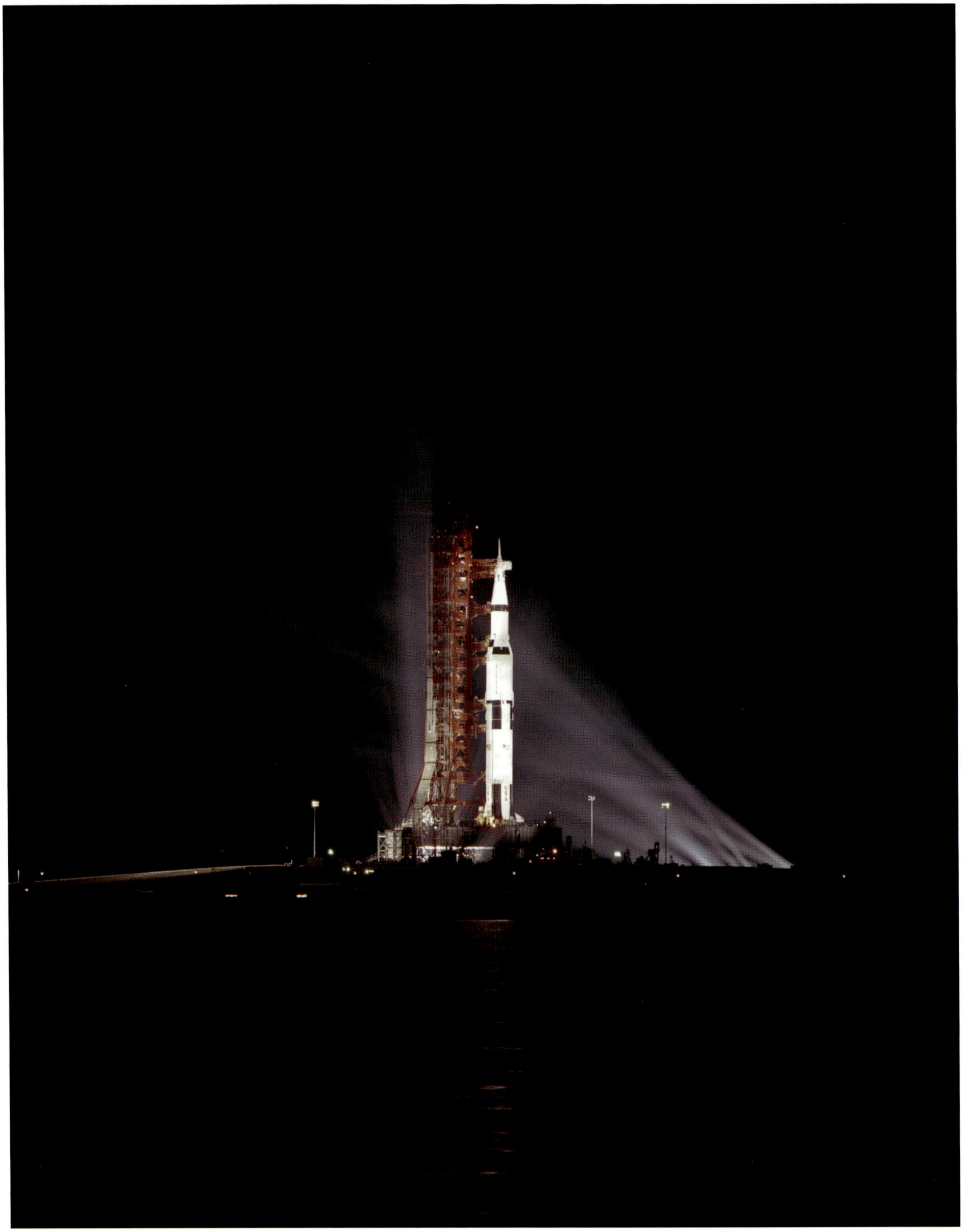

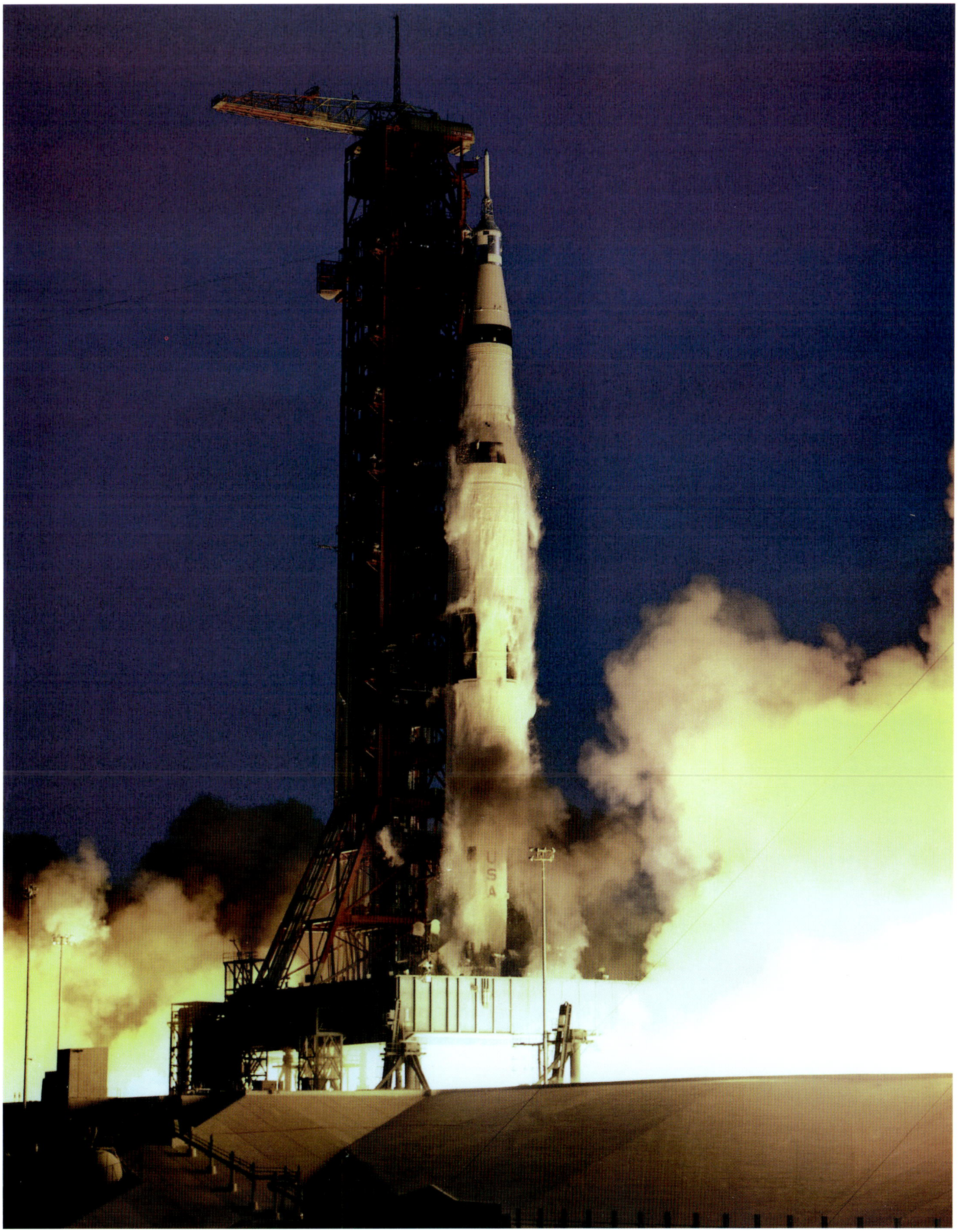
USA

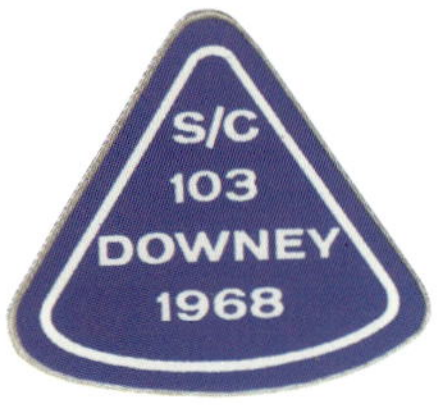

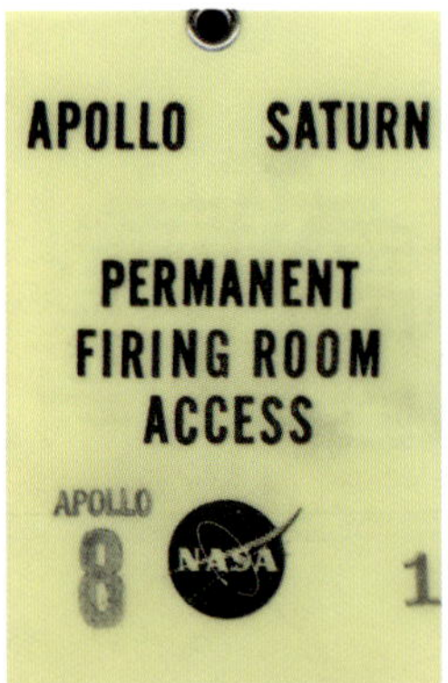

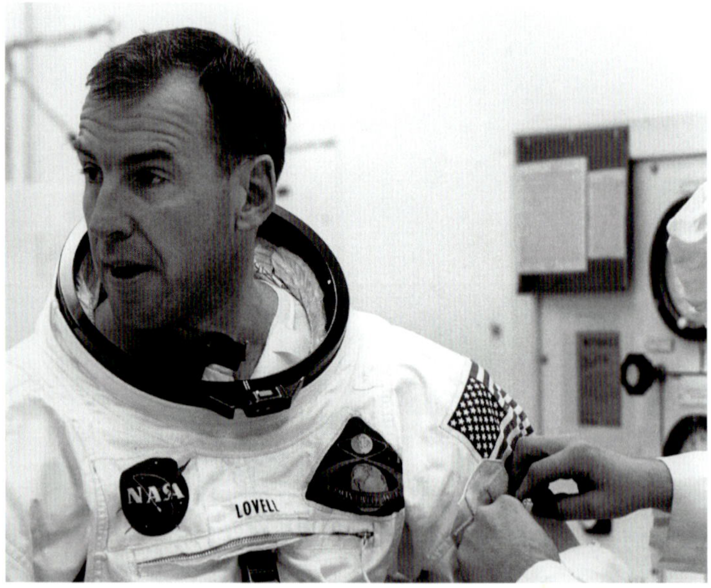

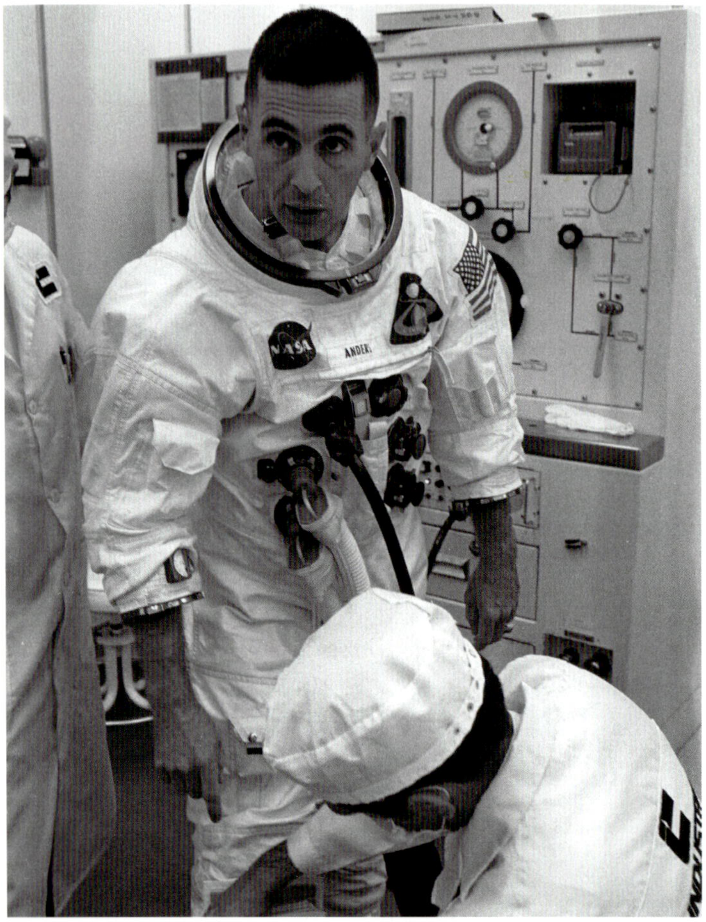

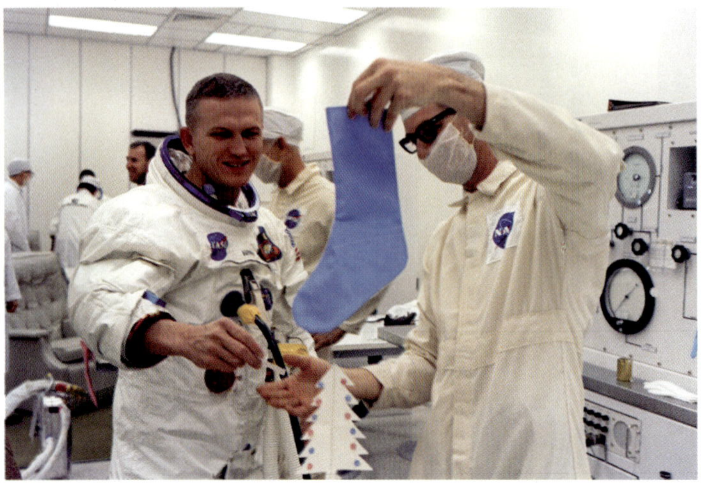

Page 38: Apollo 8's Saturn V sits on its ML at Pad 39A, with only hours to go before launch in the early morning of December 21, 1968. This view looks northeast; the Atlantic lies beyond.

Page 39: Apollo 8's Saturn V, captured at the moment of liftoff at 7:51 a.m. (EST) by a remote camera at Pad 39A. Weather was high cirrus clouds and the temperature was fifty-nine degrees Fahrenheit. US teletype operators for the hot line between Washington and Moscow provided updates during Apollo 8's mission to their Soviet counterparts, in response to a congratulatory message a Soviet operator had sent after the Apollo 7 mission. Curiously, in mid-December, a KGB agent in the United States mailed a letter to KSC Security, claiming that the launch vehicle had been sabotaged.

Top right: Lovell in the MSOB preparing for the CDDT on December 11, 1968. As the mission's navigator, he used the CM's built-in sextant to determine the spacecraft's location by measuring star positions. That information was then used to calculate midcourse corrections.

Center left: Anders undergoes suiting on launch morning, December 21, 1968, with the help of an ILC technician. Anders was born in Hong Kong after his father, Lt. Arthur F. Anders, was assigned to USS *Panay*, which provided protection for US merchant vessels. On December 12, 1937, Lt. Anders was severely injured but took command of the *Panay* after renegade Japanese bombers attacked it on China's Yangtze River.

Bottom left: NASA suit tech Dick Sandridge presents Borman with a Christmas stocking in the MSOB suiting room on launch morning. No prior manned space flight had been conducted during Christmas.

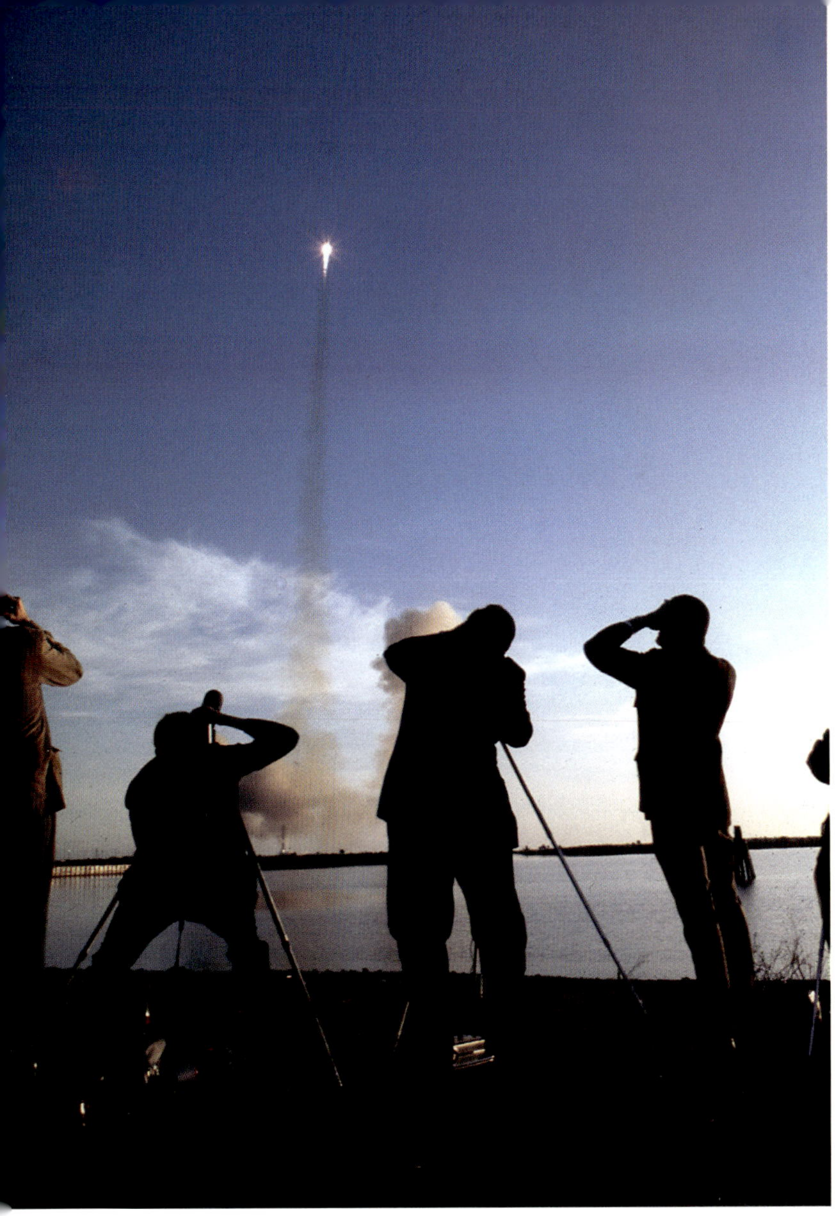

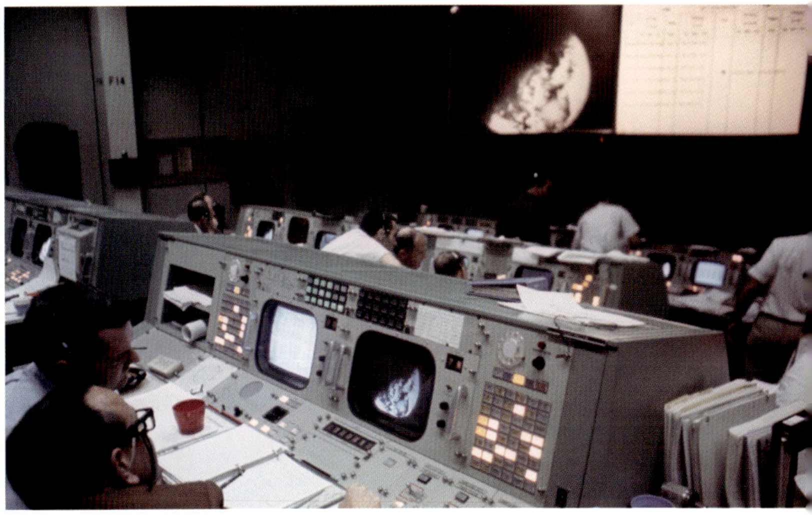

Above: Chris Kraft, director of flight operations at MSC, and Glynn Lunney, shift 2 flight director, watch the December 26 television transmission from Apollo 8 at Lunney's console in MOCR 2.

Left: News photographers record the launch of Apollo 8 from the LC-39 Press Site. (photo by Tiziou News Service)

Below: Ice particles surround the top of the S-IVB third stage, as seen from the CM after translunar injection on December 21, 1968. The top of the liquid hydrogen tank can be seen below a structural spacer. The S-IVB's two clusters of three small auxiliary propulsion system motors were fired to place it into solar orbit.

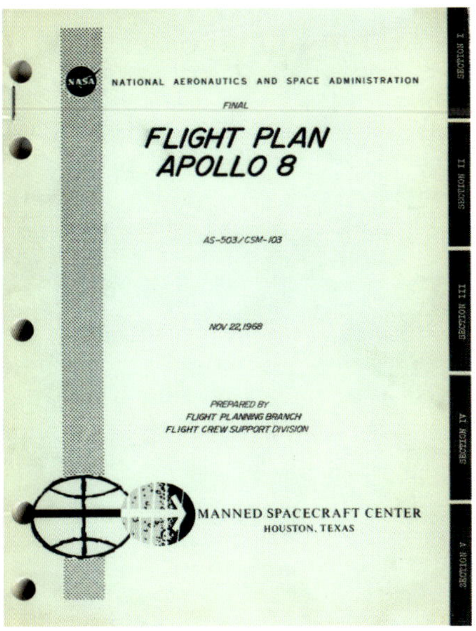

NATIONAL AERONAUTICS AND SPACE ADMINISTRATION

FINAL

FLIGHT PLAN
APOLLO 8

AS-503/CSM-103

NOV 22,1968

PREPARED BY
FLIGHT PLANNING BRANCH
FLIGHT CREW SUPPORT DIVISION

MANNED SPACECRAFT CENTER
HOUSTON, TEXAS

SECTION I
SECTION II
SECTION III
SECTION IV
SECTION V

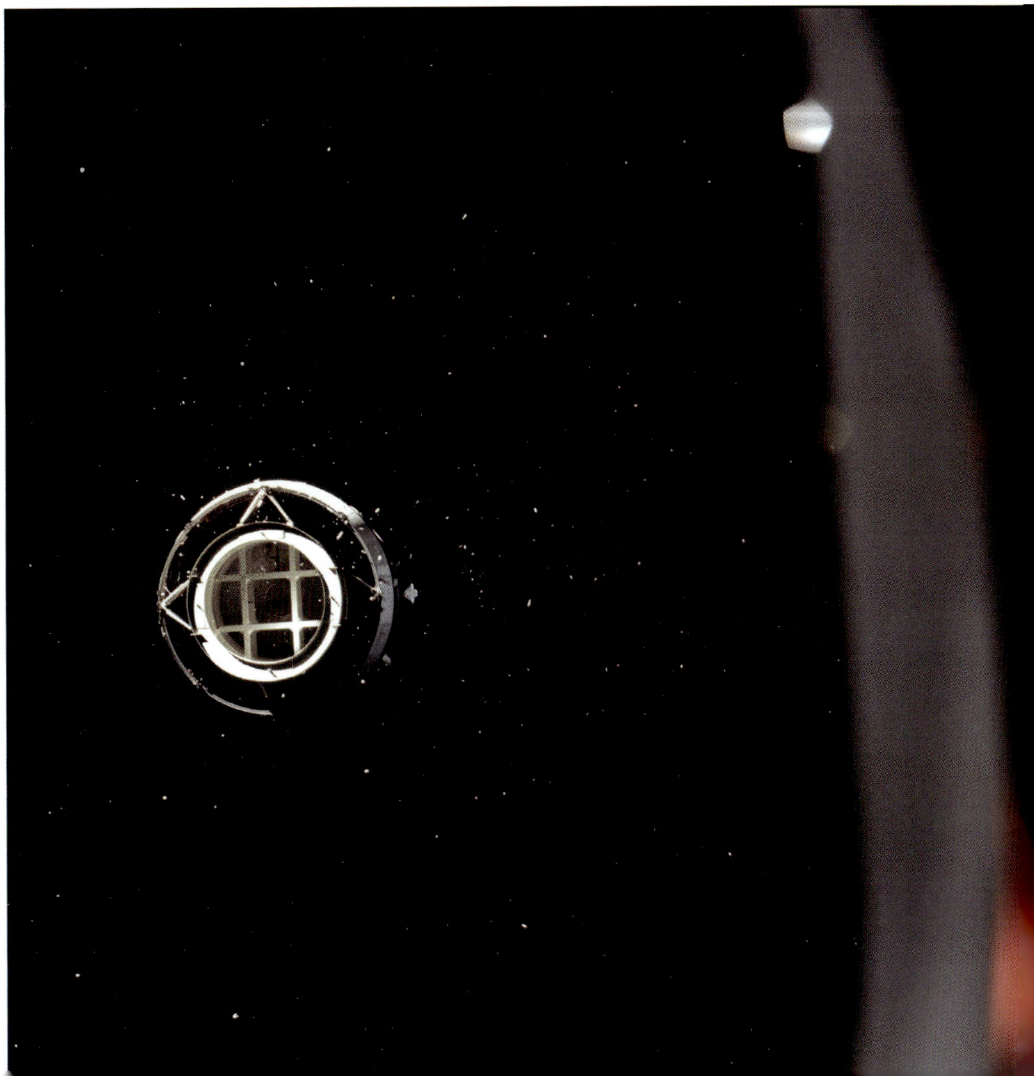

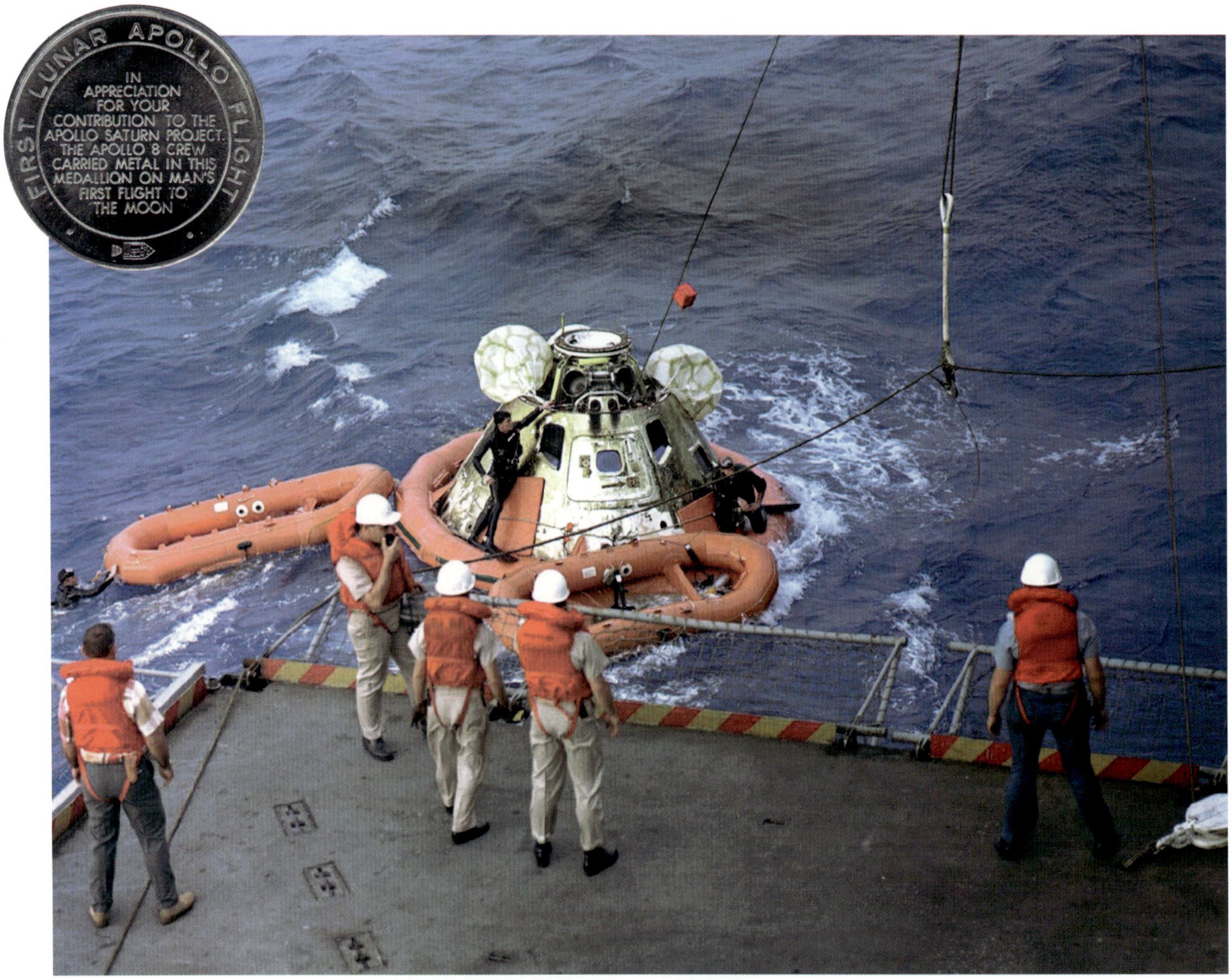

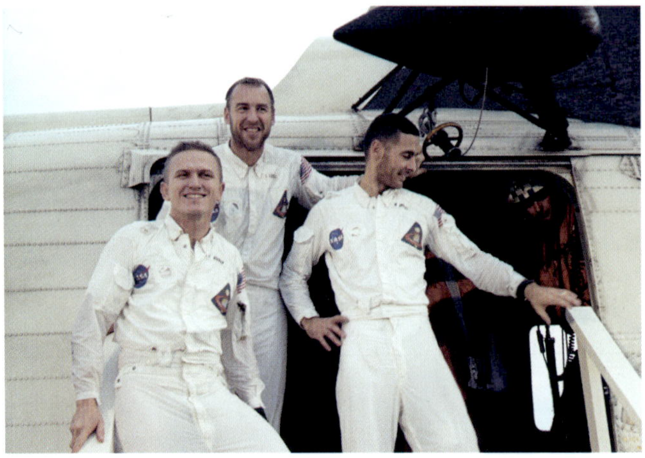

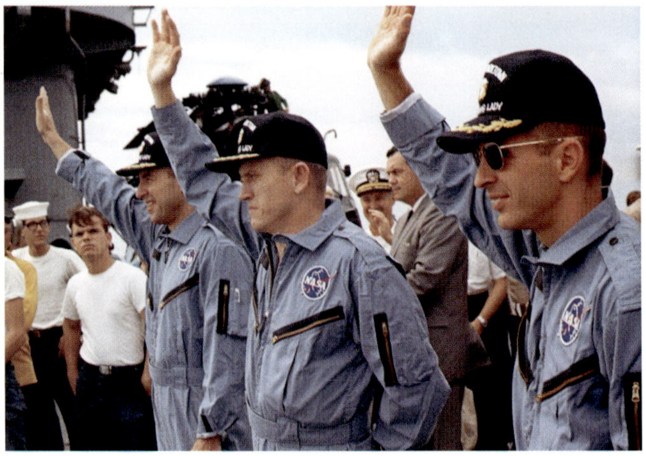

Top: Navy sailors aboard USS *Yorktown* prepare to recover the five-and-a-half-ton Apollo 8 CM from the Pacific Ocean south of Hawaii, at about 7:00 a.m. local time. Splashdown, two hours earlier, had occurred within 2.6 miles of the *Yorktown*. The spacecraft was off-loaded from the aircraft carrier on December 29 at Ford Island, Hawaii, and then flown to Long Beach, California. President Richard Nixon would later spend the night on the *Yorktown*, before watching the Apollo 11 recovery from USS *Hornet* on July 24, 1969. Today the *Yorktown* is on display near Charleston, South Carolina.

Top-right inset: Aluminum MFA medallion

Bottom left: Borman, Lovell, and Anders arrive on the deck of the *Yorktown* at 6:19 a.m. local time. Borman had borrowed a razor aboard the helicopter and shaved during the brief flight.

Bottom right: Lovell, Borman, and Anders wave to crewmen aboard the communications relay ship USS *Arlington* as it passes near Hawaii on December 27.

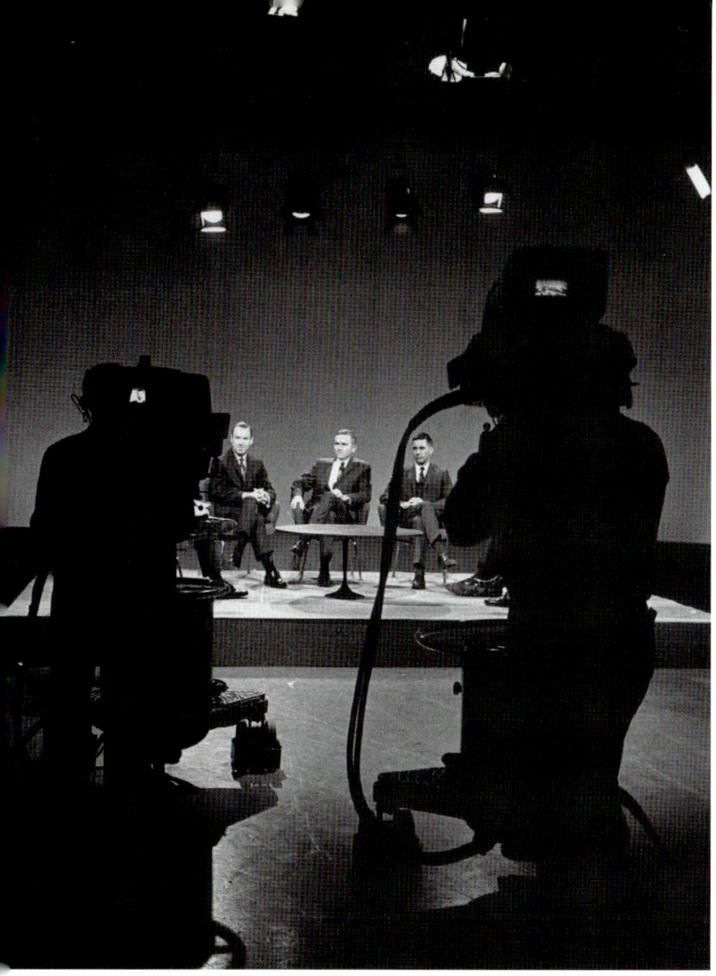

Above: The astronauts tape an interview on January 11 at the CBS News Washington, DC, bureau. The interview aired the next morning on *Face the Nation*.

Top right: Anders signs an autograph after the crewmen are welcomed home at Houston's Ellington Field on December 29, 1968. Although it was past 2:00 a.m. (CST), more than two thousand people were on hand as the astronauts arrived from Hickam AFB, Hawaii.

Center right: At the White House, President Nixon sends the crew on a European goodwill tour on January 30, 1969.

Bottom right: Lovell, Borman, and Anders are honored at a luncheon in Chicago on January 14, 1969, in the International Room of the Hilton. The city proclaimed it Astronaut Day and made the crewmen honorary citizens of Chicago. The crew was honored with a ticker-tape parade through downtown; Mayor Richard Daley gave them gold medals.

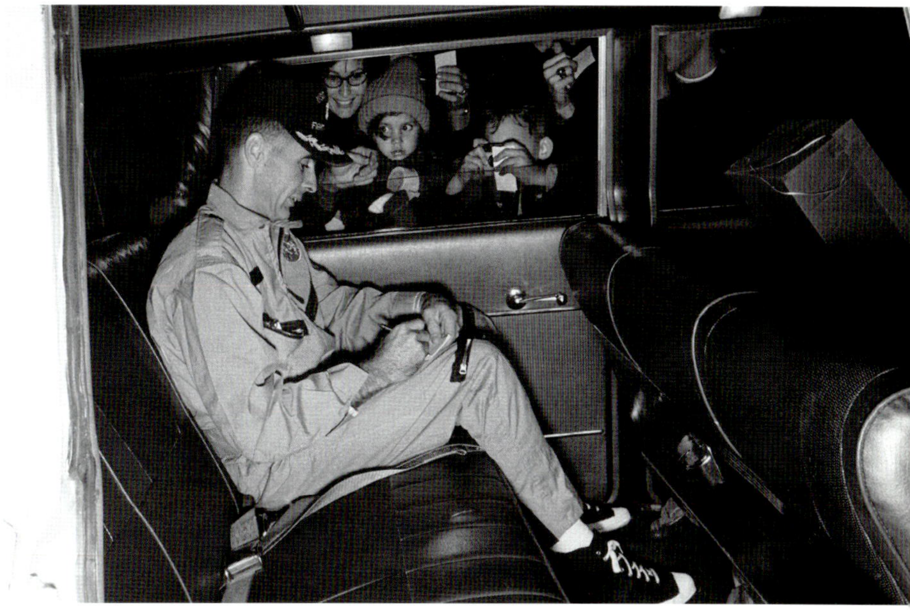

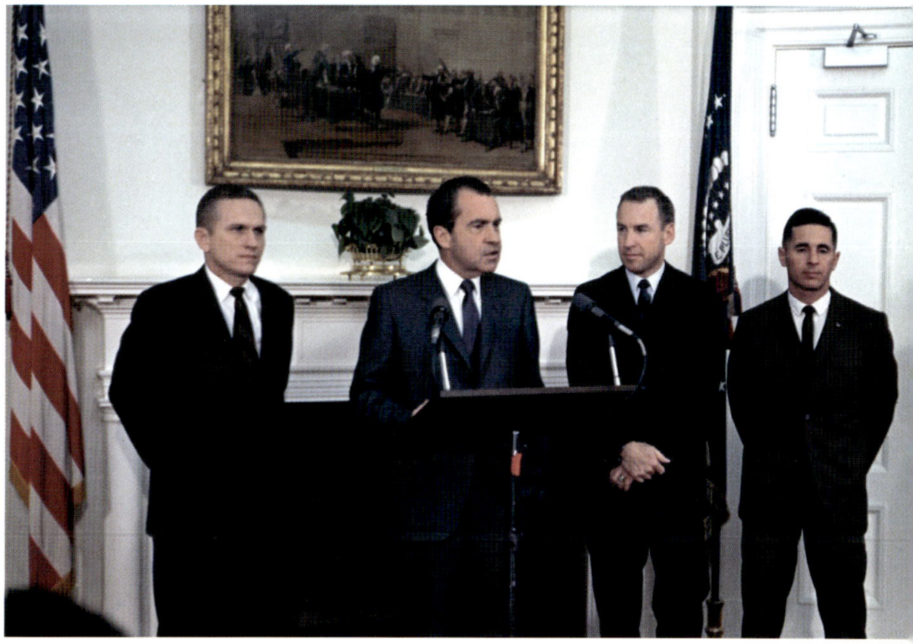

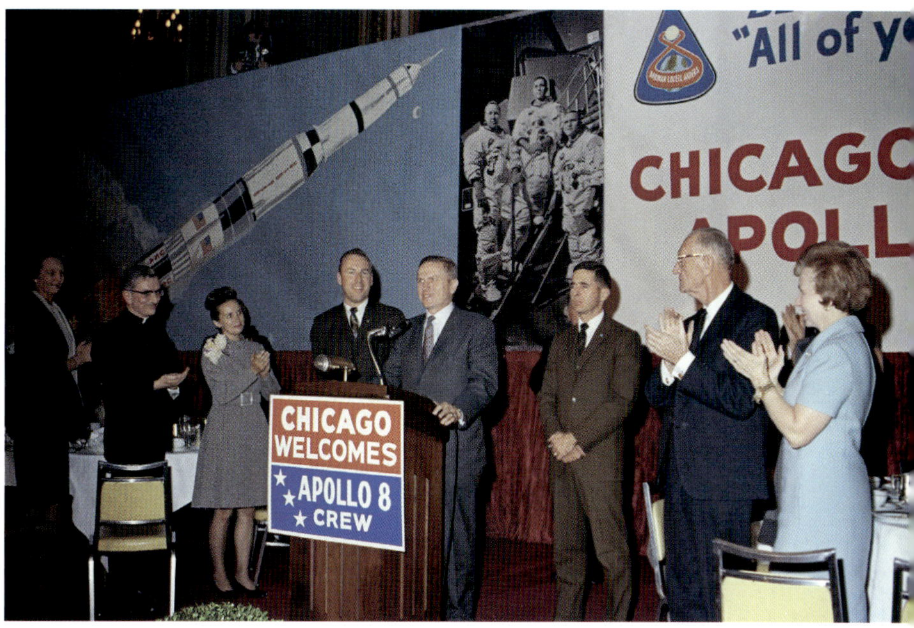

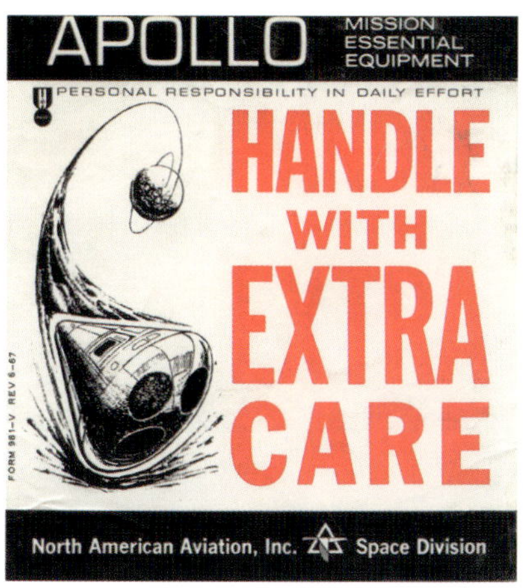

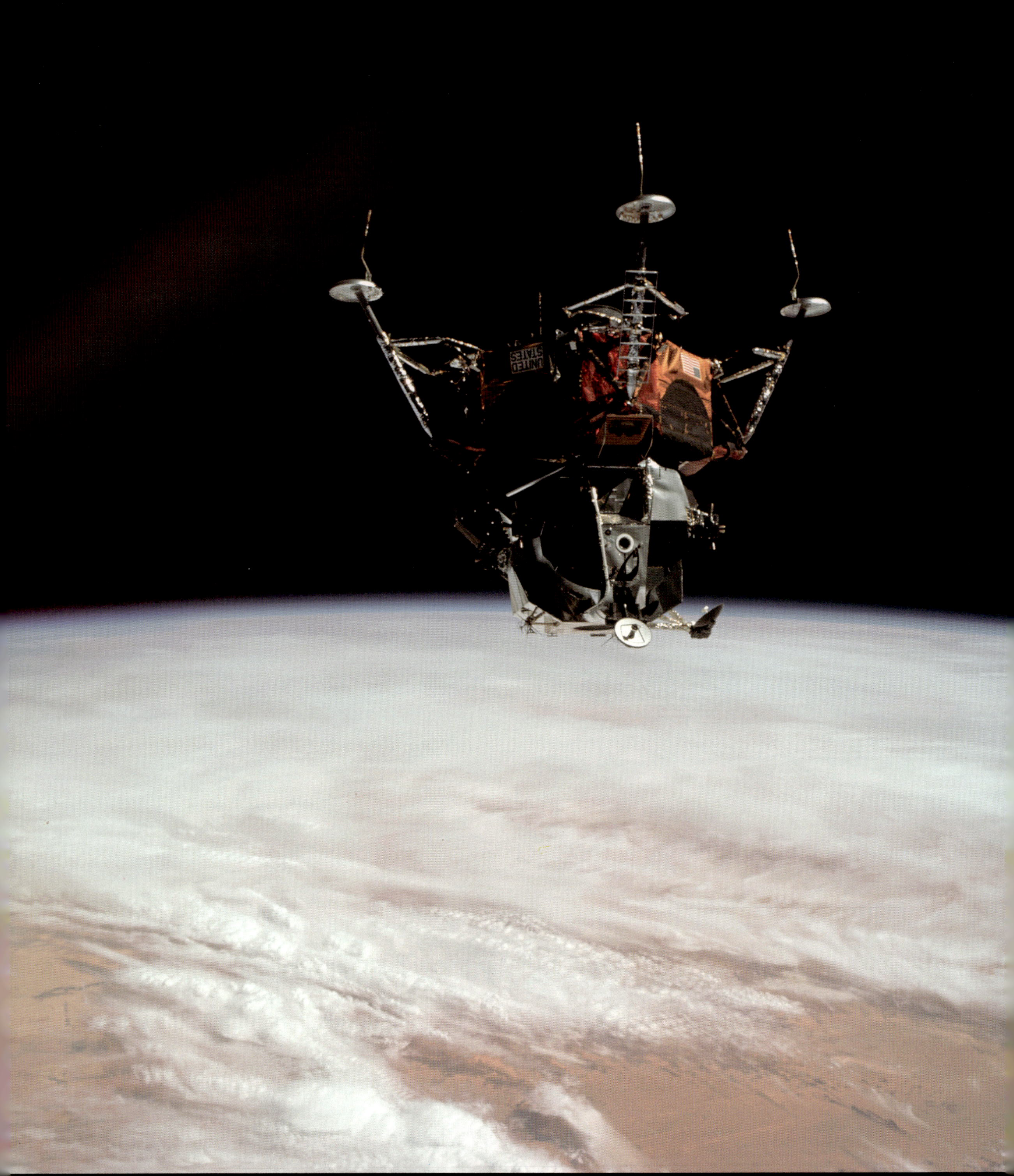

Jim McDivitt and Rusty Schweickart fly *Spider* in Earth orbit (145 miles high), as seen from *Gumdrop* on March 7, 1969. The LM's landing gear has been deployed, with lunar surface sensing probes extending from its footpads. The CM's side window edge is at top left.

CHAPTER 4

Apollo 9

March 3–13, 1969

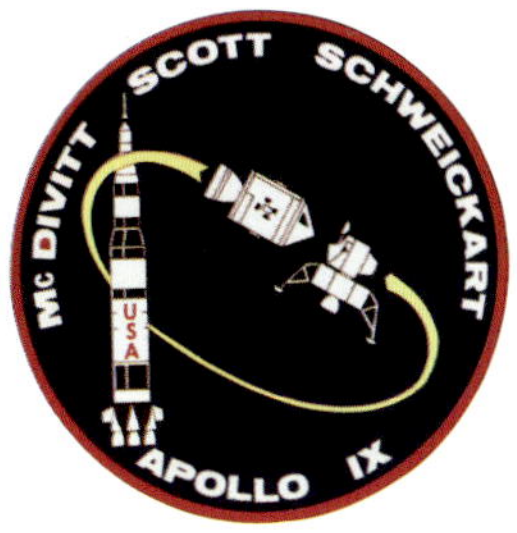

Because it came between the first two moon missions—Apollos 8 and 10—the Earth-orbiting Apollo 9 flight has often received less attention than it deserves. In addition to being a necessary step in NASA's deliberate but steady progression toward a landing, the mission was, at the time, the riskiest that astronauts had ever undertaken.

It was to be the first manned test of the LM in space. Since the LM had not been built to return to Earth, if anything went significantly wrong, the two astronauts inside could have been stranded on orbit. They were also planning a two-hour EVA, including a simulated LM crew rescue by the CSM. Add the fact that this would be just the second manned launch of a Saturn V, and the stakes went up again.

Apollo 9 CDR Jim McDivitt and CMP David (Dave) Scott, both Gemini veterans, were joined by rookie Rusty Schweickart, a civilian, as LMP. Their goals were not only to check out the LM in space, but also to determine the types of space suit and backpack needed for use on the lunar surface. The two vehicles required separate radio call signs, so the crew chose the descriptive names *Gumdrop* for the CM and *Spider* for the LM.

The flight was delayed from February 28 to give the astronauts time to recover from a mild respiratory illness. It launched on time from Pad 39A at 11:00 a.m. (EST) on March 3, 1969, into a low-Earth orbit, about 118 miles up.

After the CSM separated from the SLA at 3:08 p.m. (EST) and the SLA's protective panels were jettisoned, the CSM turned around and docked with the exposed LM. *Spider* was connected to the SLA at four points around the lower, stationary panels. The astronauts docked *Gumdrop* to the LM, and then blew charges to separate the connections. A guillotine severed the LM from its Instrument Unit umbilical. The charges fired, and springs pushed the LM away from the S-IVB.

The docked spacecraft then moved two thousand feet away. The S-IVB engine was restarted twice to send it into solar orbit—simulating a translunar injection of the stage, planned for subsequent lunar missions.

The flight's goals included conducting spacecraft-engine tests as well. The LM descent and ascent engines were fired to change orbits, simulating lunar orbit rendezvous and abort procedures. The CSM SPS fired five times, including a simulation of an active rendezvous to rescue a stranded LM.

Astronauts had used one vehicle to push another into a higher orbit, but

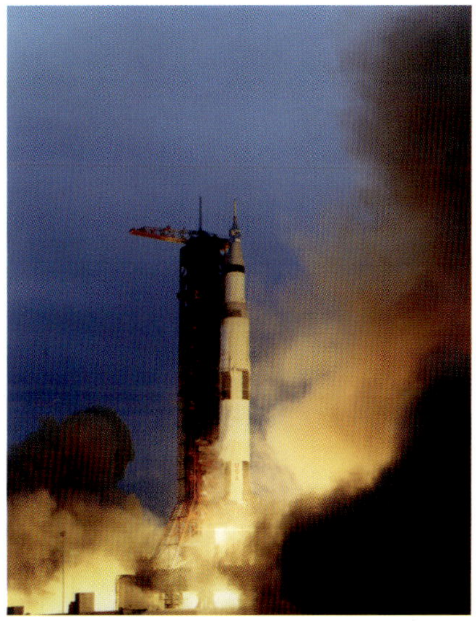

Launch: 11:00 a.m. (EST), March 3, 1969

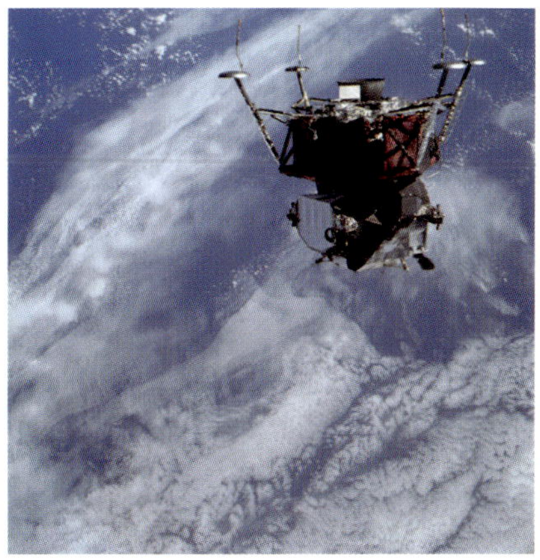

LM undocking: March 7

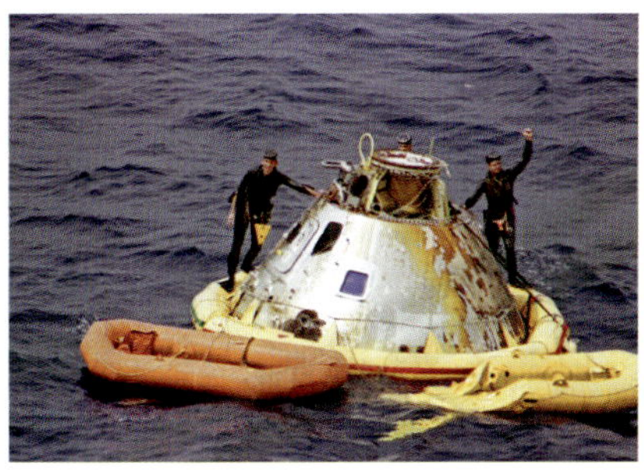

Splashdown: 12:01 p.m. (EST), March 13

they had never used a spacecraft as big as the LM. Six hours into the mission, they made the first test burn, which lasted five seconds. Mission Control considered this the most critical SPS burn; it went smoothly. The remaining four burns, late the next morning, also went well.

After breakfast on March 5, McDivitt and Schweickart put on their pressure suits, entered *Spider* at about 6:30 a.m. (EST), and prepared for the planned EVA. But after Schweickart became sick to his stomach twice, the EVA was rescheduled for the following day. The two did test *Spider*'s descent engine and radar.

The next day, Schweickart felt better than expected. He wore the new Apollo space suit outfitted with a Portable Life Support System (PLSS) backpack, which provided communications and oxygen and circulated water through the suit to keep him cool, eliminating the need for an umbilical connection to the spacecraft. Schweickart was attached to the LM by only a twenty-five-foot nylon rope. He worked his way out of the hatch around noon and put his boots into a pair of foot restraints—dubbed the "golden slippers"—on the LM's porch, as Apollo 9 prepared to cross the United States.

Scott, in the meantime, had opened *Gumdrop*'s hatch and stuck his head out. Schweickart was originally to transfer to the CM for the rescue simulation, but he had stayed with *Spider*, gradually making his way along a vertical handrail on the front face of the LM to its radar rendezvous antenna. Scott and Schweickart filmed each other, snapped photos, and retrieved small patches of material samples from the exterior of each craft. The EVA, scheduled to last more than two hours, ended after about fifty minutes, in part to avoid stressing Schweickart after his illness.

On March 7, the crew prepared for the key events of the mission: the separation of *Gumdrop* and *Spider* and their subsequent rendezvous and docking. With McDivitt and Schweickart in the LM, Scott separated from *Spider* at 7:40 a.m. (EST), firing the RCS thrusters to move about three and a half miles away. After McDivitt fired the LM's descent engine, the stage was jettisoned at 11:16 a.m. (EST), and its engine was fired until the fuel had been depleted.

McDivitt and Schweickart then fired *Spider*'s ascent stage to rendezvous with *Gumdrop*. The LM successfully docked with the CSM after being up to 114 miles away during the six-and-a-half-hour separation. McDivitt and Schweickart returned to *Gumdrop*, and *Spider*'s ascent stage was jettisoned at 4:22 p.m. (EST).

During the rest of the flight, the astronauts tracked *Pegasus III*, a meteoroid-detection satellite launched in 1965; took multispectral photos of Earth; and tested the CSM's systems.

Gumdrop splashed down in the Atlantic Ocean 180 miles northeast of Puerto Rico on March 13 at 12:01 p.m. (EST). The crew was picked up by helicopter within an hour and flown to the recovery ship USS *Guadalcanal*.

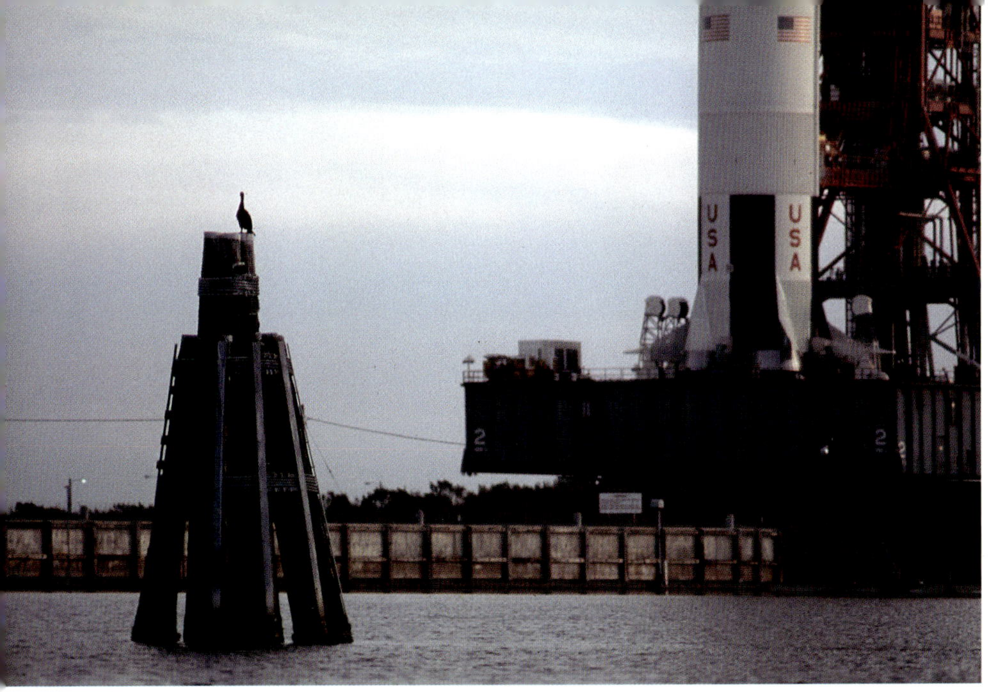

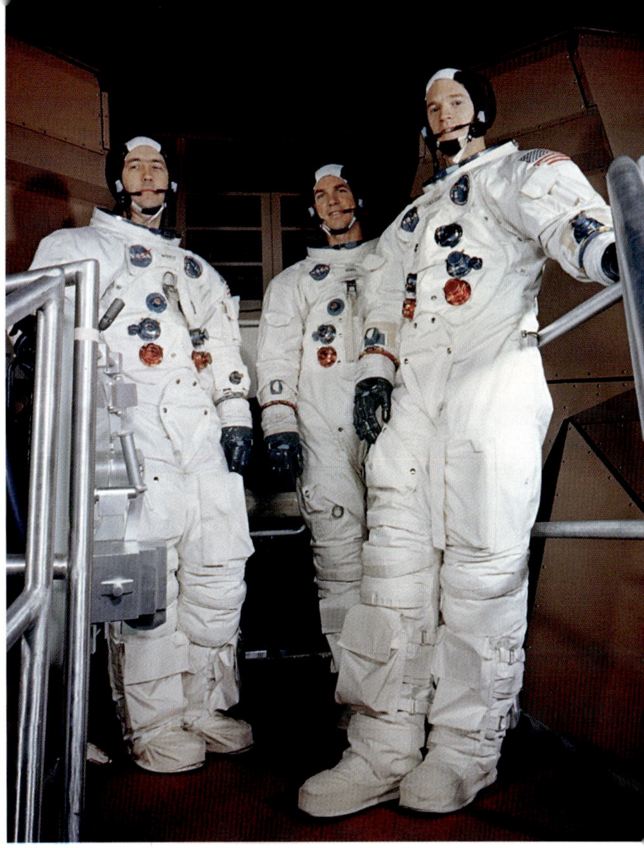

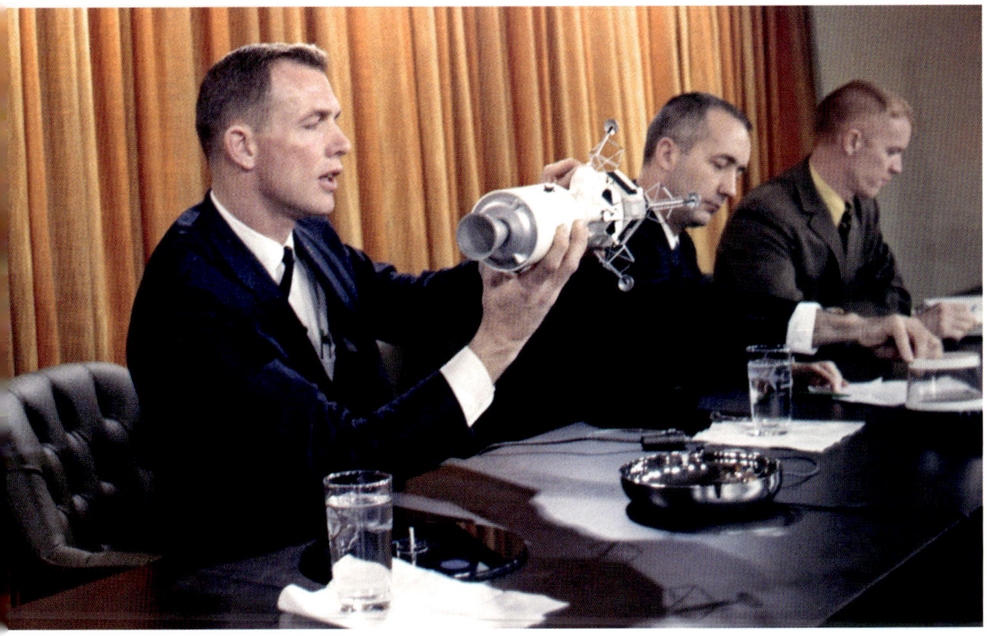

Top left: Apollo 9's Saturn V rolls to Pad 39A atop ML-2 on the morning of January 3, 1969. A cormorant sits on a pile mooring for Saturn barges in the Turning Basin. The crew was in Houston that day for a briefing from the Apollo 8 astronauts. (photo by Tiziou News Service)

Top right: McDivitt, Scott, and Schweickart pause in front of the Apollo Mission Simulator in the Flight Crew Training Building at KSC on February 13, 1969.

Center left: Scott holds docked models of the CM and LM during a two-hour news conference with the crew on January 25, 1969. The news conference was held at the Grumman Aircraft Engineering Corporation plant in Bethpage, New York, on Long Island.

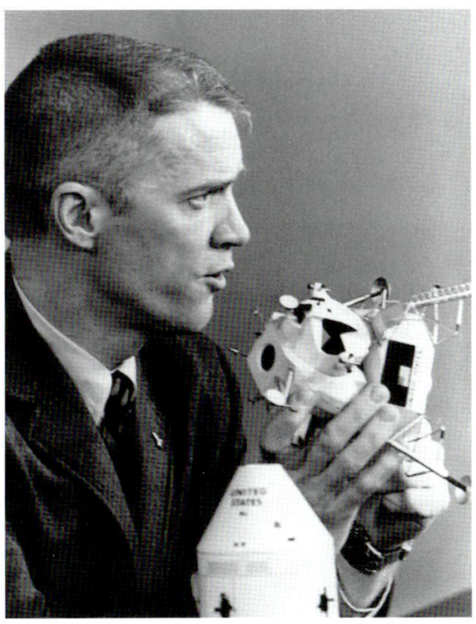

LMP Russell L. "Rusty" Schweickart, thirty-three, was a fighter pilot–turned–research scientist at MIT when he was chosen as an astronaut in 1963.

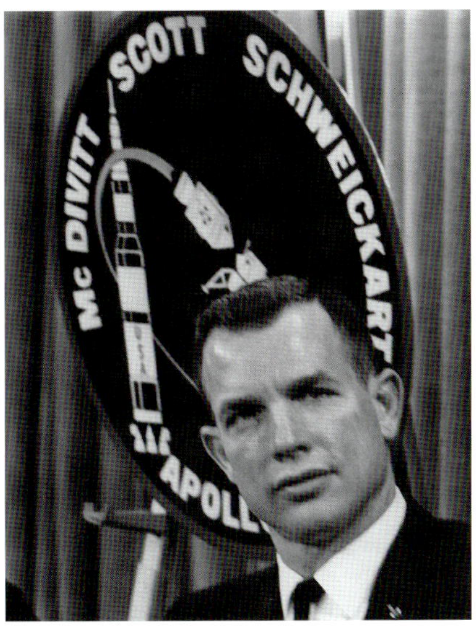

CMP David Scott, thirty-six, would become the last American to fly a spacecraft in Earth orbit solo.

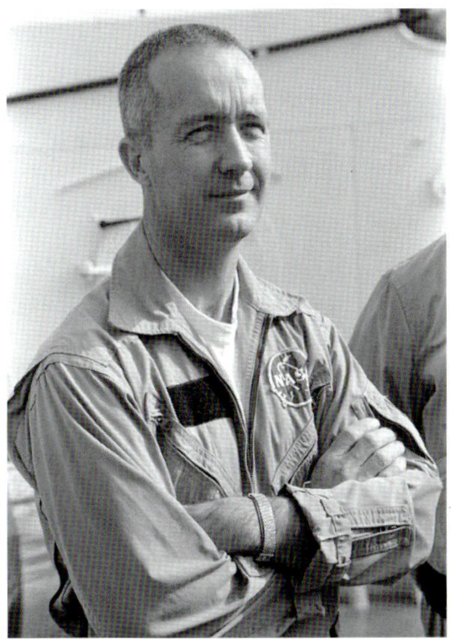

CDR Jim McDivitt, thirty-nine, aboard MV *Retriever* on November 5, 1968. Promoted to brigadier general in 1971, McDivitt served as Apollo program manager for flights 12 through 16.

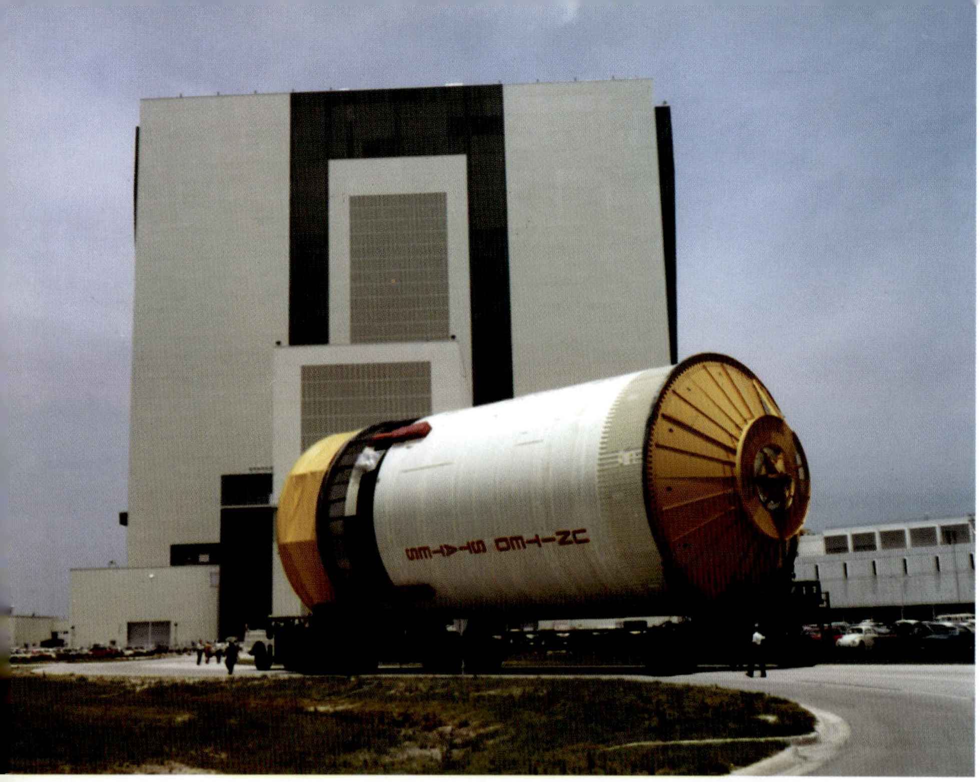

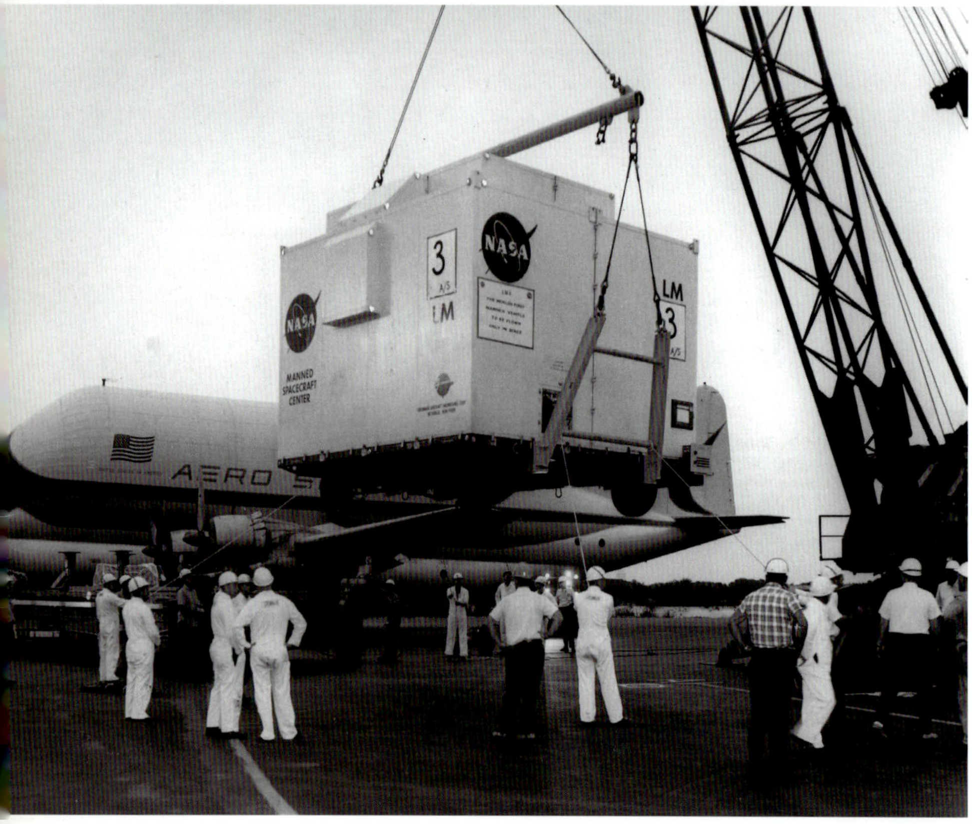

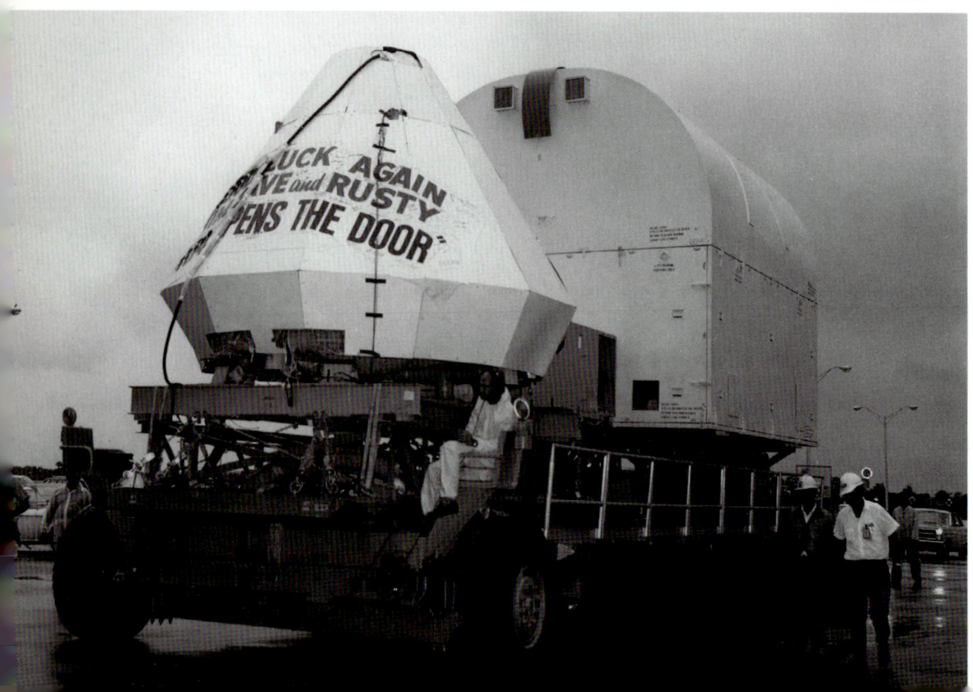

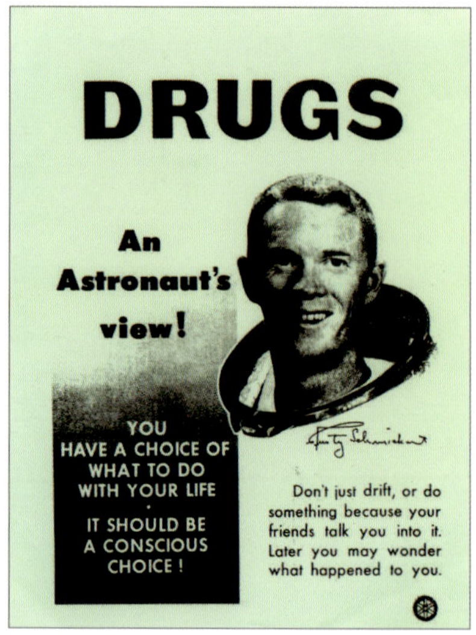

Above: Schweickart served as honorary chairman of the Harris County, Texas, Committee on Drug Abuse after his flight.

Top left: The Saturn's second stage (S-II-4) arrives at the VAB on May 15, 1968, with the back of the lower LCC at right. The forward bulkhead of the liquid hydrogen tank is protected by the yellow handling cap at right, also the VAB-crane attach point. The stage, designed and built by NAR's Space Division, was at that time the largest hydrogen-fueled rocket stage ever built. It was shipped by barge from California through the Panama Canal to KSC.

Center left: Grumman workers steady the wheeled shipping container for LM-3's ascent stage, which is being moved by crane from its trailer (*left*) after arriving with the descent stage on the *Pregnant Guppy*. The *Guppy* brought both stages to the Cape Kennedy AFS Skid Strip from Grumman's Bethpage, New York, airfield on June 13, 1968.

Bottom left: CM-104 and its SM wait in shipping containers on a trailer outside the MSOB after arriving at the Skid Strip—also via *Guppy*—on October 7, 1968, from NAR in Downey, California. The banner reads, "Good luck again / Jim, Dave and Rusty / 104 opens the door." The previous CM, originally intended for this crew, bore a similar message; that CM, 103, was used for Apollo 8.

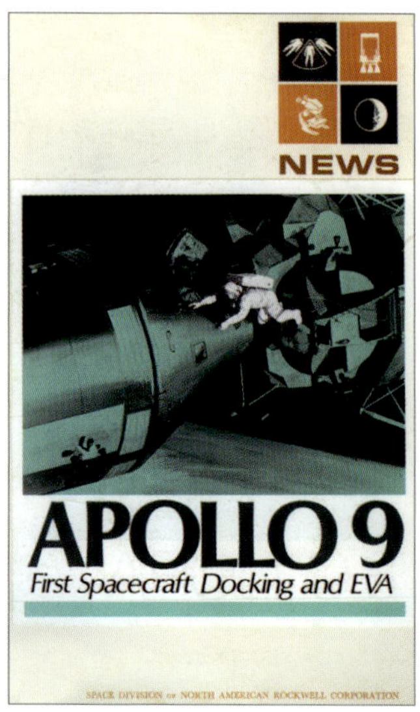

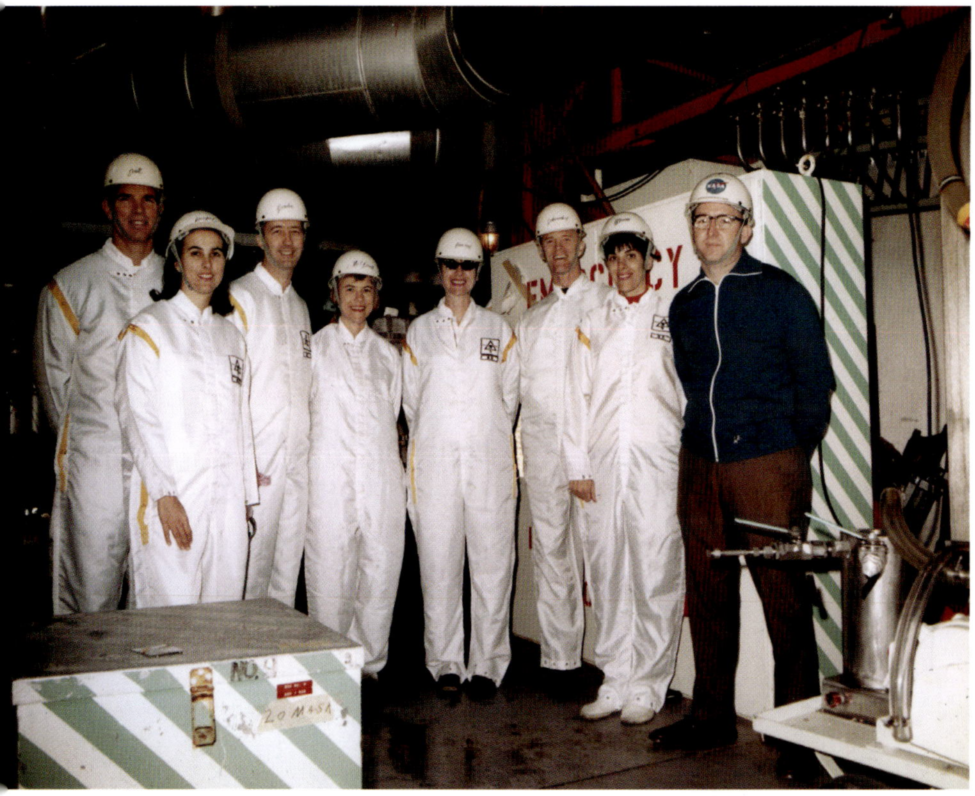

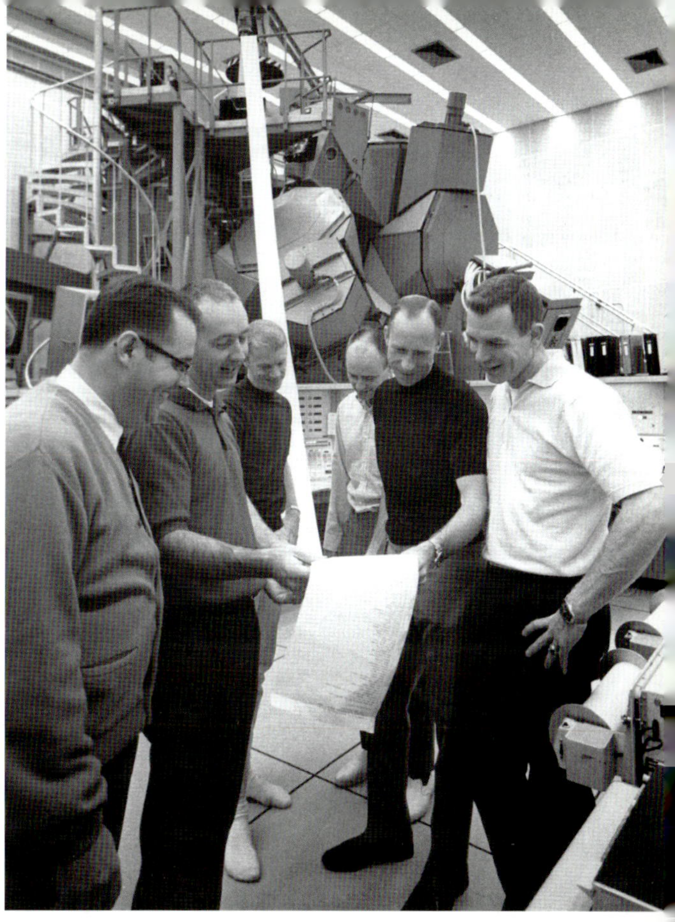

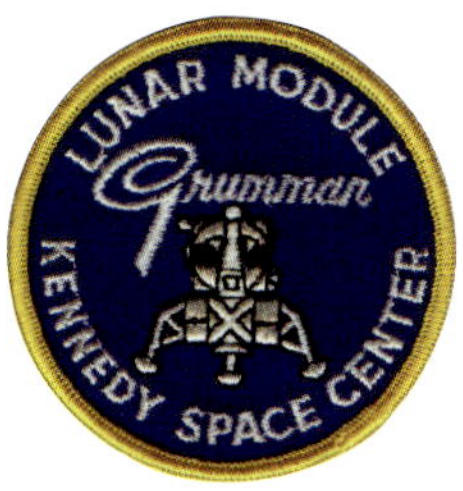

Top left: The astronauts and their wives tour the base of ML-3. *Left to right*: Dave and Lurton Scott; Jim and Pat McDivitt; NASA nurse Dee O'Hara; Rusty and Clare Schweickart; and Jim Harrington, KSC senior test supervisor. Their hard hats read (*left to right*) Scott, Cooper, Gordon, McDivitt, Conrad, Schweickart, and Bean.

Top right, left to right: Riley D. McCafferty (chief of the Flight Crew Operations Branch at MSC), McDivitt, Schweickart, Alan Bean (backup LMP), Al Worden (on the support crew), and Scott. They read a lengthy "best wishes" message from the residents of Jackson, Michigan, in the Flight Crew Training Building on February 22, 1969, near the mission simulator. Worden was a Jackson native; McDivitt graduated from junior college there.

Bottom right: Grumman and NAR technicians run checks on the CM (*bottom*) and LM (*top, facing right*) in an MSOB altitude chamber before a docking test on October 30, 1968. The NAR tech wears the NAA logo above a NAR patch. The two had merged in September 1967.

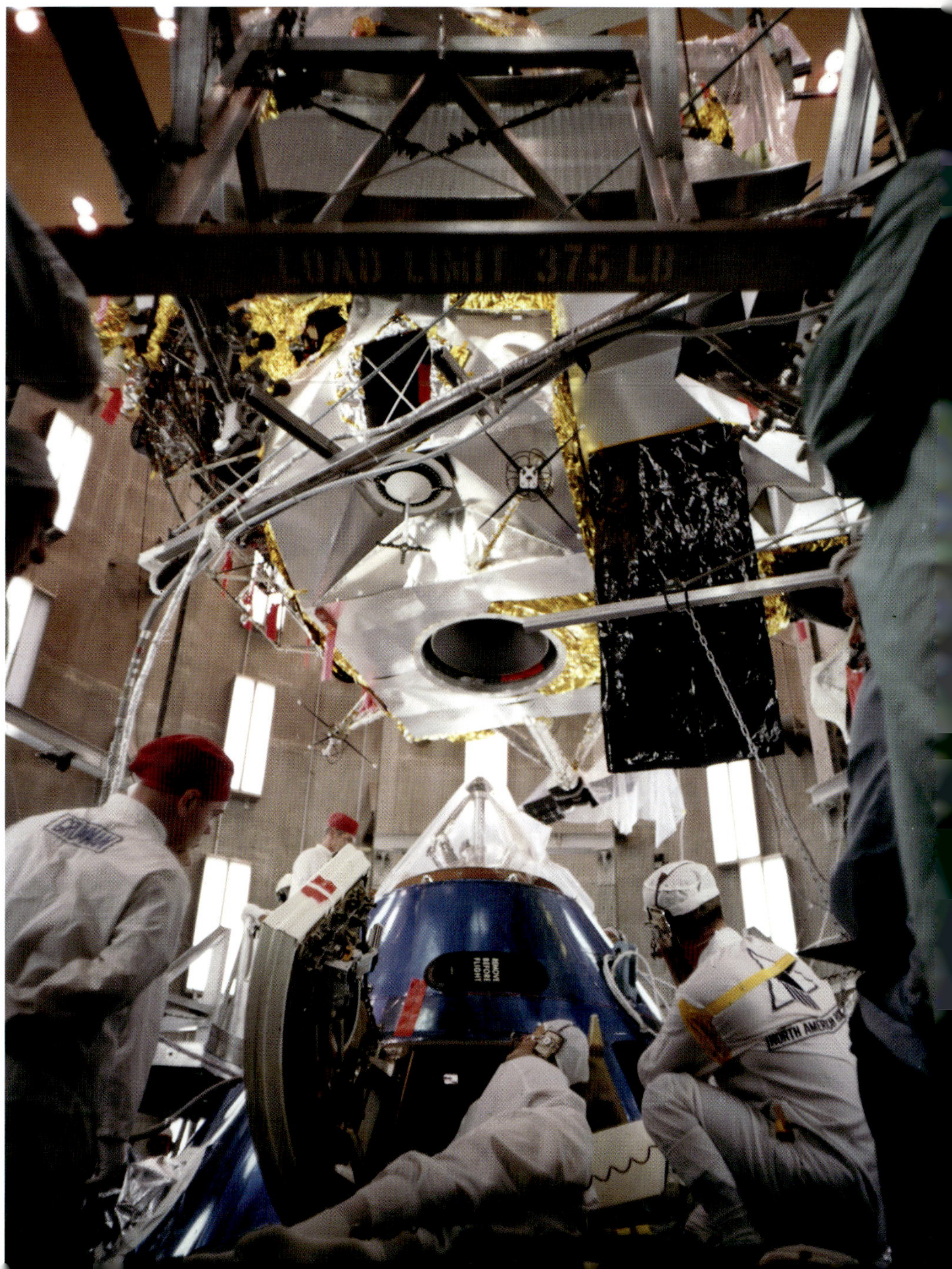

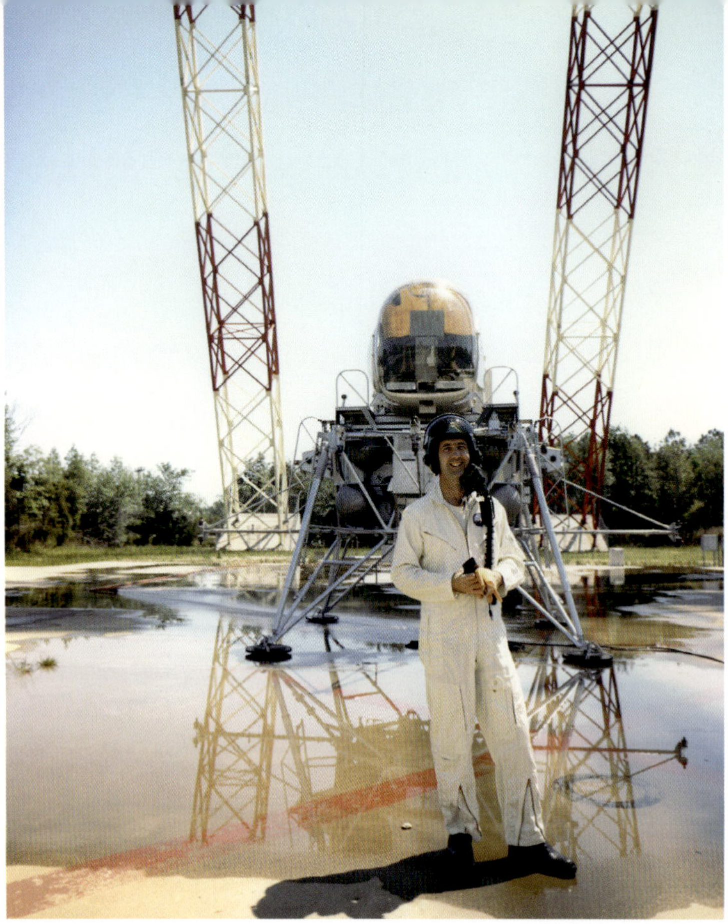

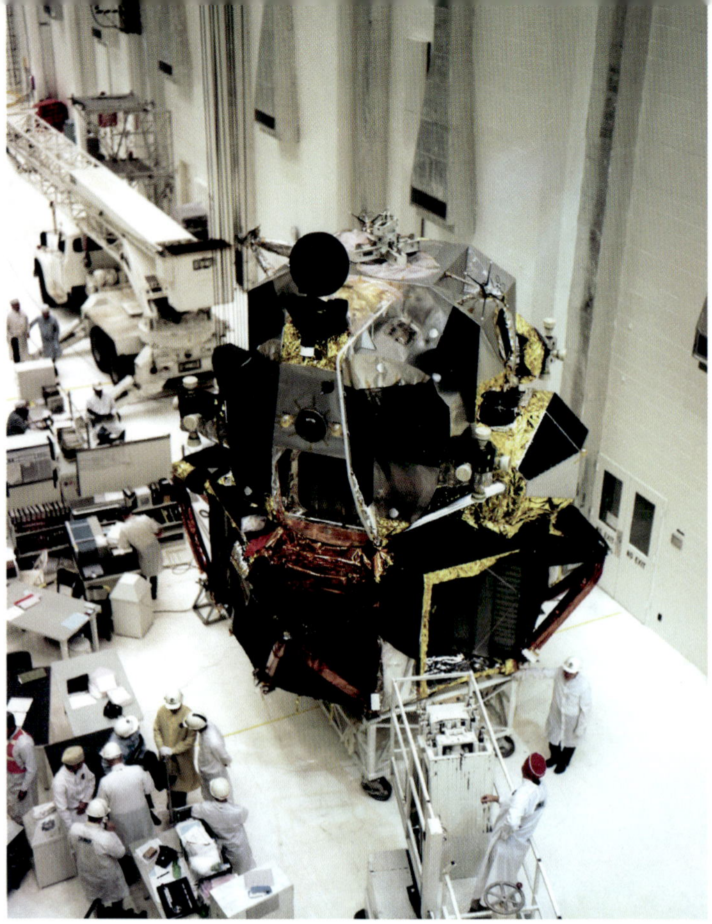

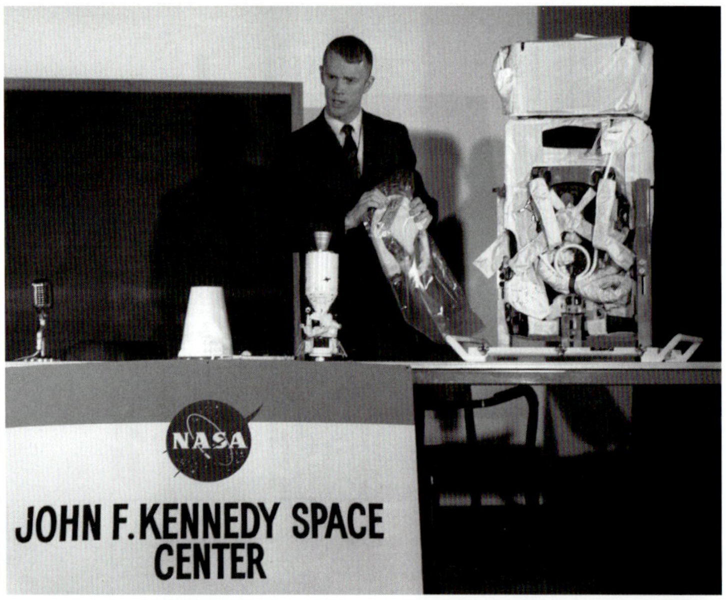

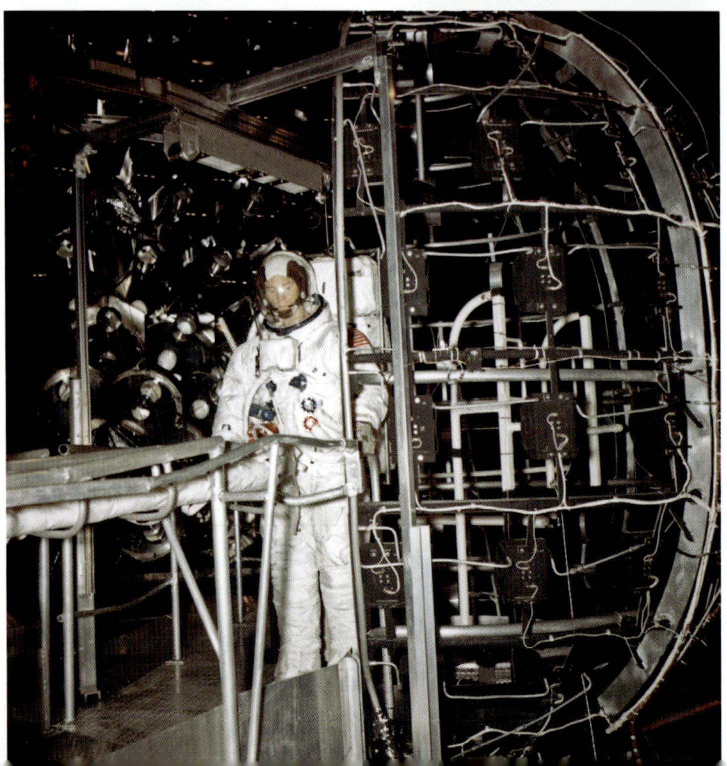

Top left: McDivitt with the Lunar Excursion Module Simulator (LEMS) at the Lunar Landing Research Facility of Langley Research Center, Virginia, on May 7, 1968. The rocket-powered LEMS was suspended from a four-hundred-foot-tall A-frame gantry. It could simulate three minutes of the final 150 feet of a lunar landing. The cabin was later replaced with a stand-up cabin, which offered better visibility for the pilot. Many tests were conducted at night, to allow simulated lunar surface lighting conditions.

Top right: A legless LM-3 during checkout in the MSOB on August 24, 1968. The quadrant 1 bay is exposed at right, showing one of the descent stage's aluminum-alloy support beams inside. This quadrant would later contain the tank of compressed supercritical helium, used to pressurize the stage's fuel systems.

Center left: At a news conference on February 14, 1969, Schweickart discusses the lunar backpack he plans to test during his EVA. He holds the chest pack by its air circulation and audio volume control unit. The entire crew also trained in the KSC CM simulator that day.

Bottom left: Schweickart, with the backpack, inside Chamber A at the Space Environment Simulation Laboratory in MSC's Bldg. 32 on December 11, 1968. The chamber allowed thermal-vacuum testing; carbon arc lights, used to mimic lunar surface temperatures, surround Schweickart. Built in 1965, the chamber remains the largest of its type in the world. This was a dry run before a simulation of Schweickart's planned EVA, which would demonstrate emergency transfer capability between the CM and LM. It was the last EVA before the Apollo 11 landing.

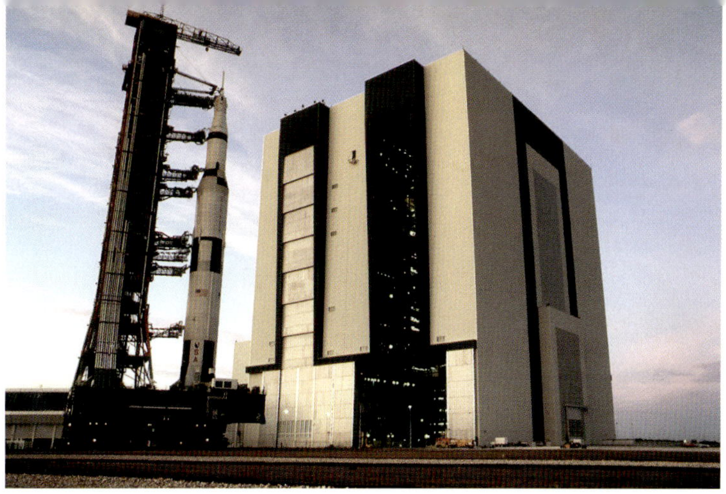

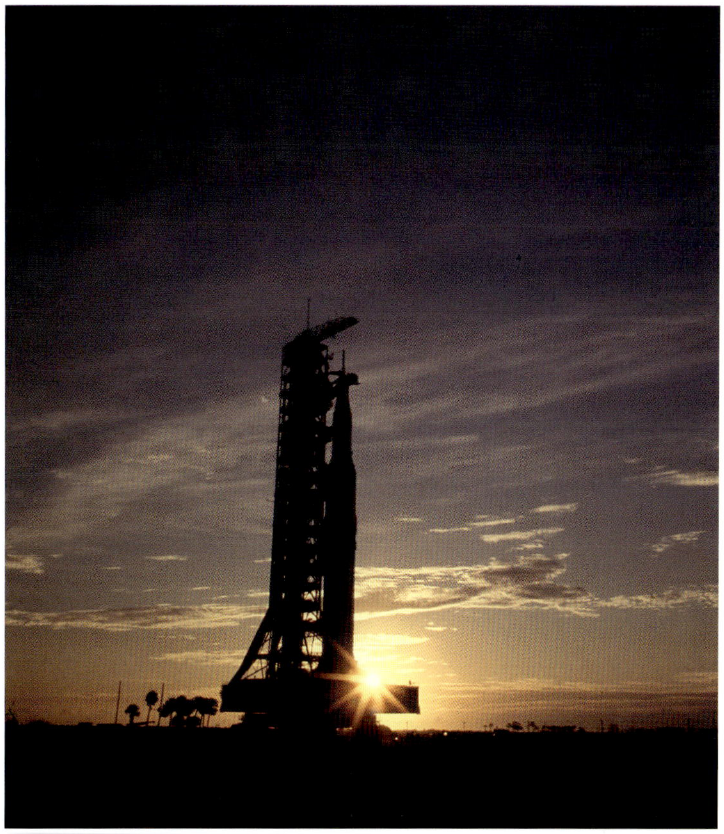

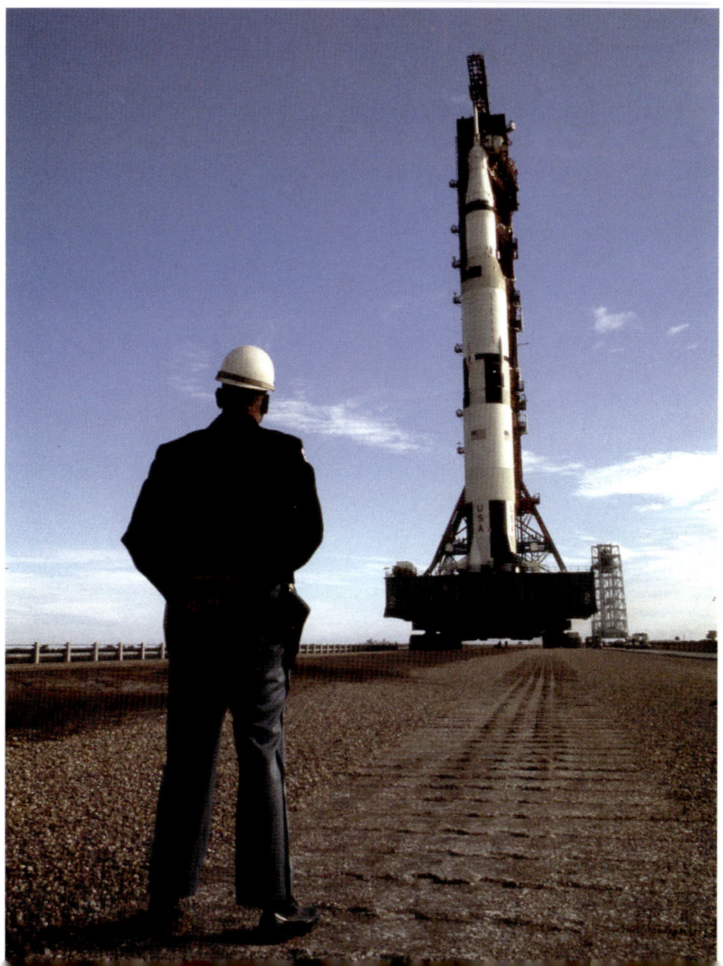

Top: Apollo 9's Saturn V moves from the VAB's high bay 3, headed for Pad 39A on January 3, 1969. With the Apollo 9 crew at MSC for an all-day debriefing with the Apollo 8 crew for the astronaut corps, the only VIPs at the rollout were actor Ken Curtis, who played Festus Haggen on the CBS television series *Gunsmoke*, and his wife, Torrie Ahern Connelly Curtis. (photo by Tiziou News Service)

Center: The launch vehicle inches its way along the crawlerway as the sun rises over Merritt Island on a cool January morning. This view is from north of the crawlerway, looking east-southeast. (photo by Tiziou News Service)

Bottom: A KSC Security patrolman stands in the crawler-transporter's tracks along the rock-covered crawlerway near the VAB. The crawler-transporter carries Apollo 9's Saturn V toward Pad A. In the distance is the MSS (*right*). The vehicle left the VAB at 6:25 a.m. (EST) and arrived at the pad at 1:20 p.m. (photo by Tiziou News Service)

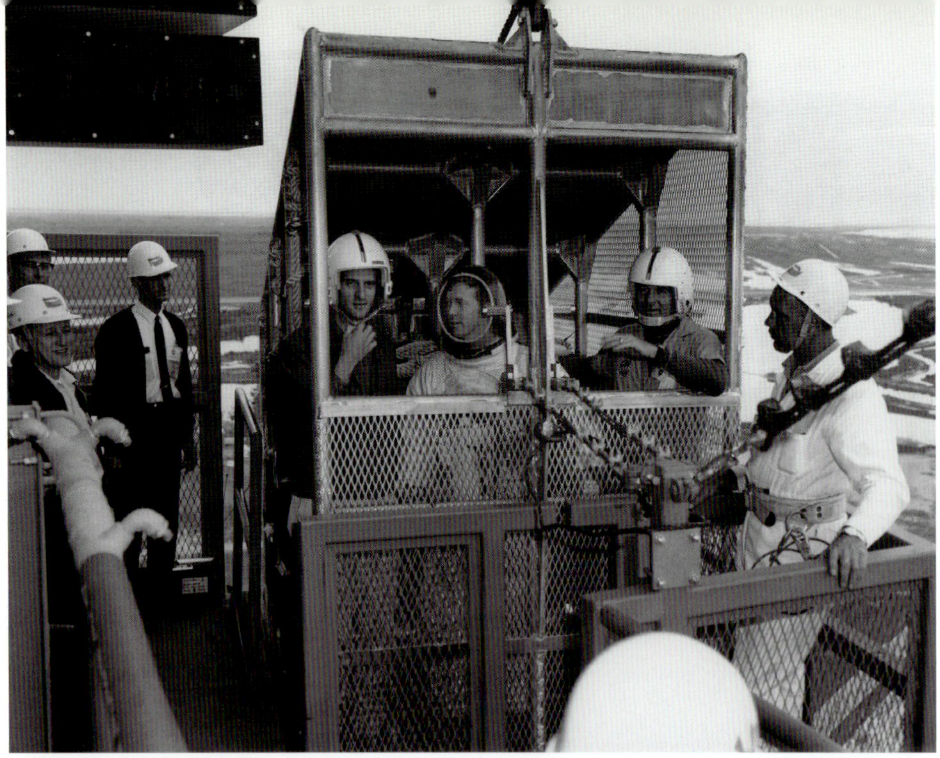

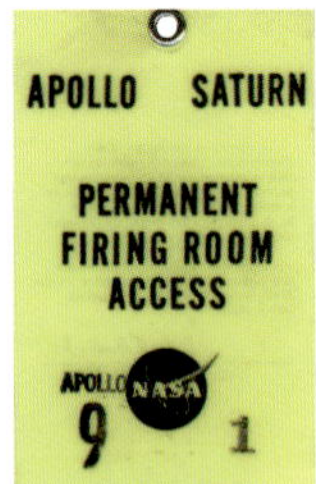

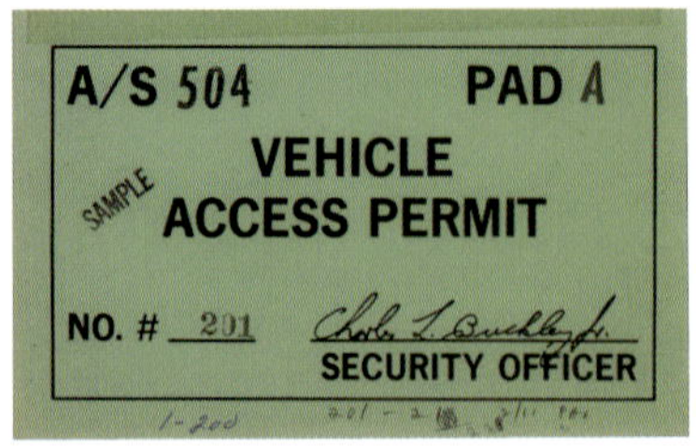

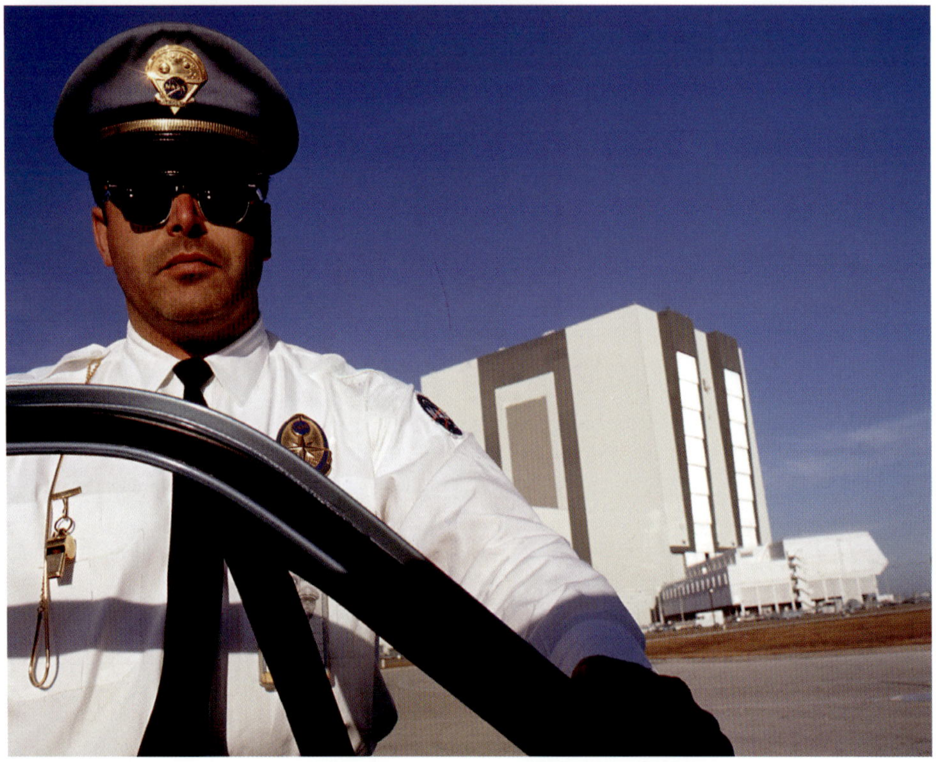

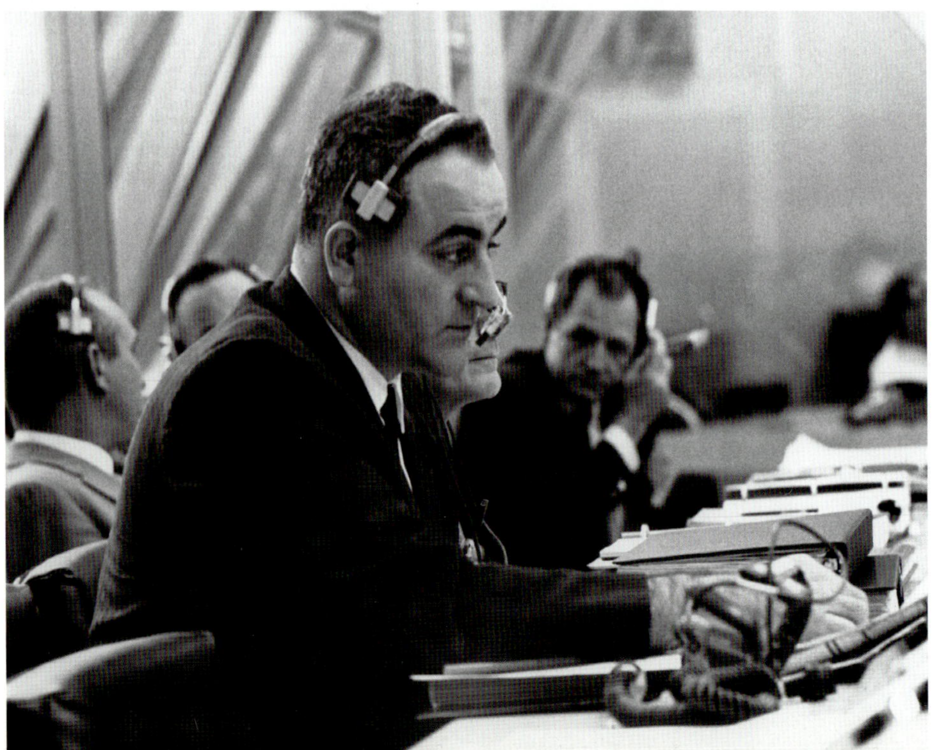

Top left: On January 26, 1969, Chuck Billings of the KSC Safety Office (*left*) joins astronaut Stu Roosa and Art Porcher of the KSC Design Engineering Office in the emergency escape slide-wire gondola at the 320-foot level of the LUT. Bendix personnel prepare to send them down the 2,500-foot wire west of Pad 39A, as the first live test subjects. Six dummies occupied the other two rows. A group of program officials at the other end watched as the gondola came to a stop. "It was a good ride," said Billings.

Center left: In 1964 Wackenhut Services won a $44 million, seven-year contract to provide security at the space center as a subcontractor to Trans World Airlines (TWA), the prime KSC contractor for base support activities. (photo by Tiziou News Service)

Bottom left: KSC Launch Operations Director Rocco Petrone in Firing Room 2 in the LCC during the CDDT on February 19, 1969. In September of that year, he moved to Washington, DC, to become Apollo program director. A former football star at West Point, Petrone was an imposing and demanding figure. He succeeded Wernher von Braun and Eberhard Rees as Marshall Space Flight Center (MSFC) director in 1973.

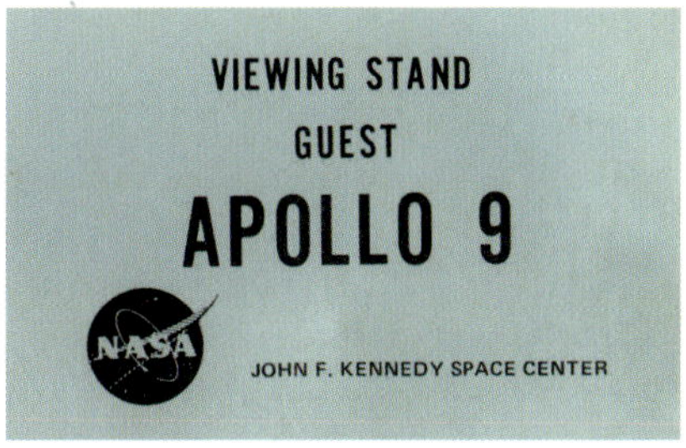

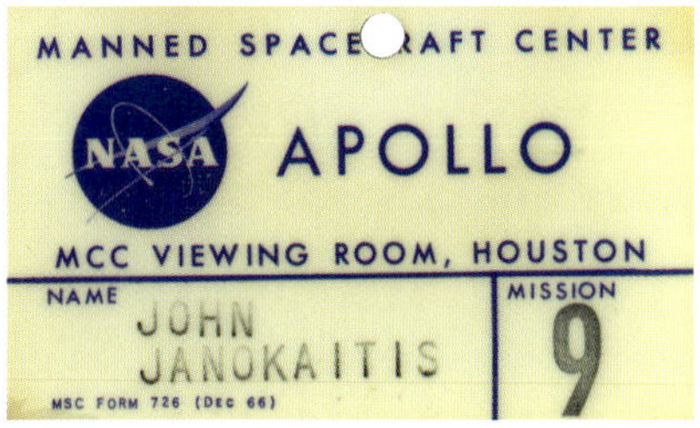

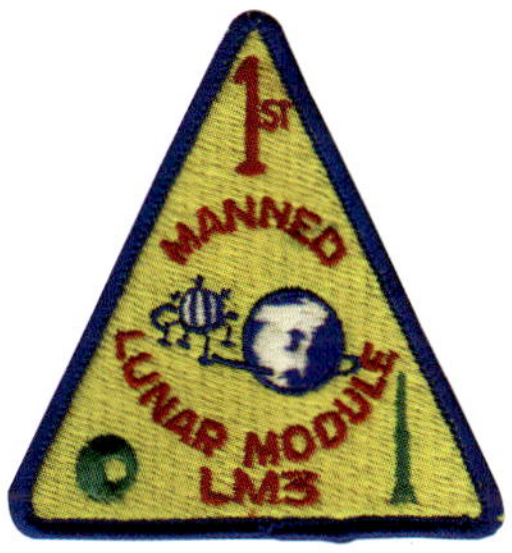

Top right: Vice President Spiro Agnew gives remarks after landing at the Cape Kennedy AFS Skid Strip at 9:30 a.m. (EST) on March 3. Behind him are (*left to right*) USAF Secretary Robert Seamans, who flew with Agnew from Washington, DC; Apollo 10 crewmen Gene Cernan, Tom Stafford, and John Young; and Apollo 8 CDR Frank Borman. The group watched the launch from the firing room, and then Agnew briefly toured LC-39, with Borman and Stafford as guides, after lunch. Agnew, chairman of the National Aeronautics and Space Council (NASC), also attended the remaining Apollo launches (except for that of Apollo 15).

Second right: Borman and Agnew, with Seamans, Governor Claude Kirk of Florida, and Kirk's wife, Erika, in the background

Third right: Comedian Jack Benny (*center*) and his brother-in-law, TV producer Hilliard Marks (*left*), were among one thousand guests at the VIP site near the VAB.

Bottom right, left to right: Seamans; KSC Deputy Director Mike Ross (*in back*); Agnew; Eberhard Rees, MSFC's deputy director for Research and Development (*in back*); acting NASA administrator Tom Paine (*in back*); MSFC Director Wernher von Braun; and George E. Mueller, NASA associate administrator for Manned Space Flight, in Firing Room 2. Paine was officially named administrator two days later, succeeding Jim Webb, who had resigned at the end of the Johnson administration.

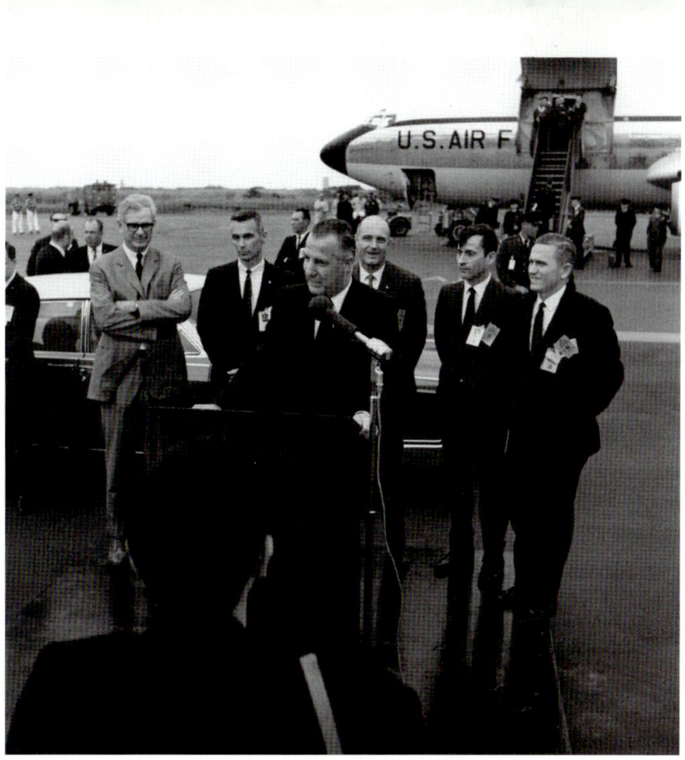

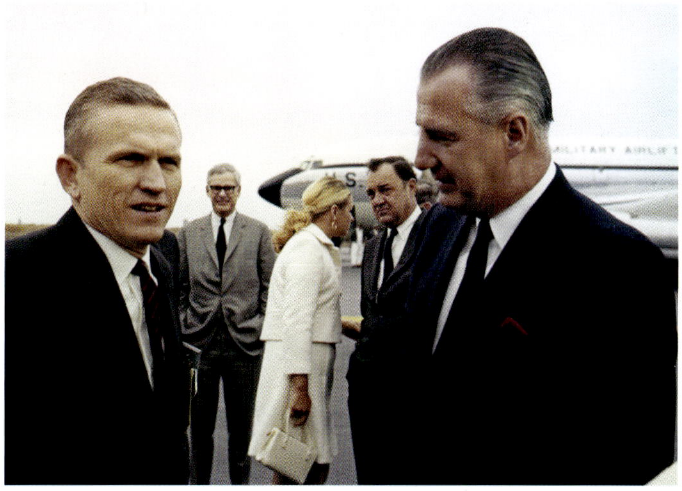

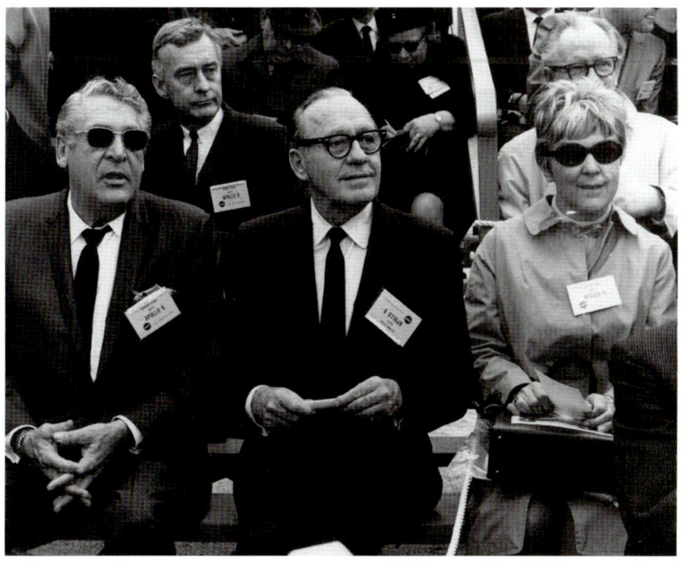

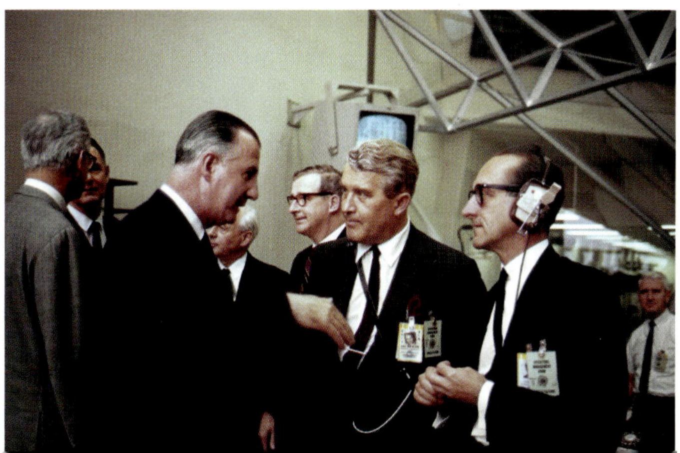

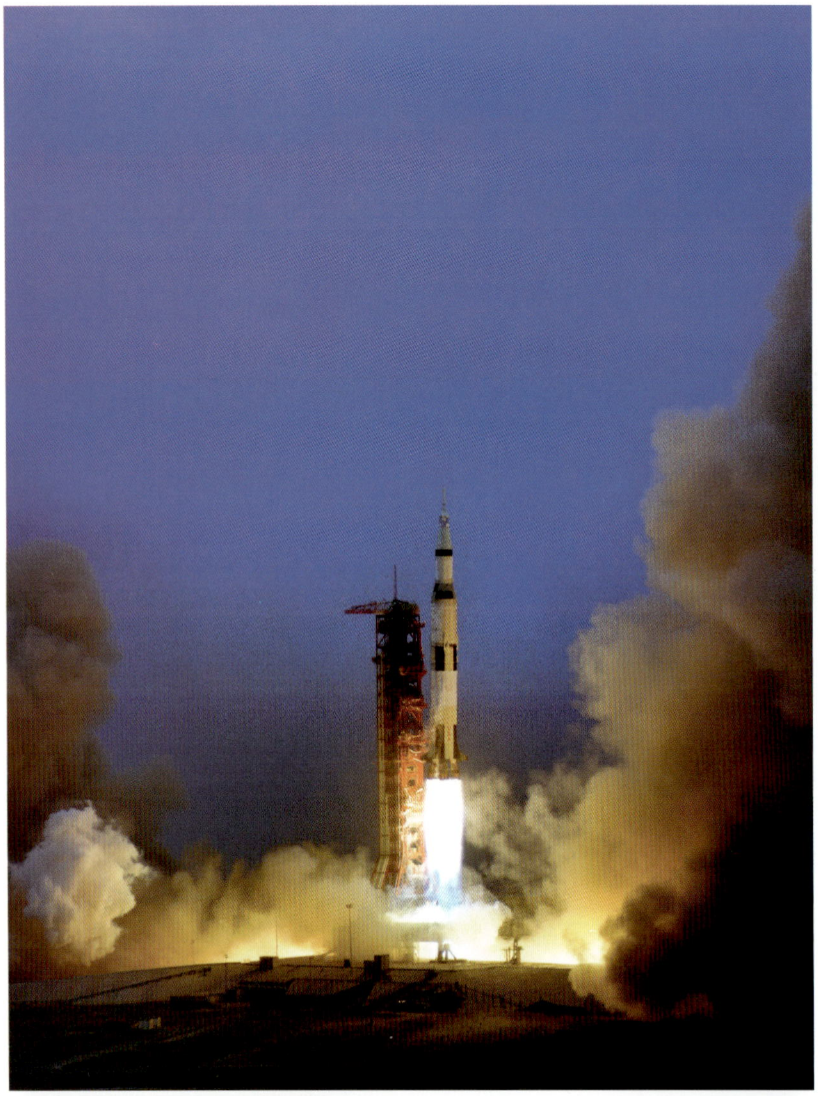

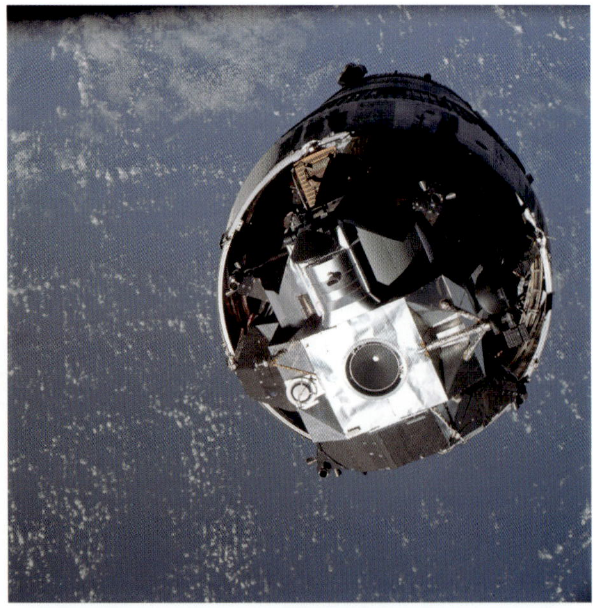

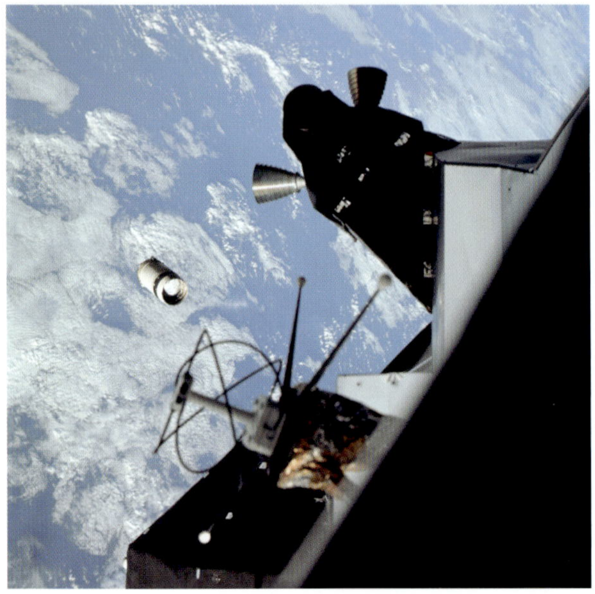

Top left: The five F-1 engines powering the Saturn V's first stage billow yellow and brown smoke from their RP-1 kerosene fuel seconds after liftoff at 11:00 a.m. (EST) in this view of Pad 39A looking southwest. The stage will take the vehicle thirty-six miles high in two minutes and forty-three seconds.

Top right: LM *Spider*, still attached to the S-IVB, as seen from the CM *Gumdrop* in Earth orbit, 119 miles up, at about 1:50 p.m. (EST), following CSM/LM-S-IVB separation but before the CM docked and extracted the LM from the S-IVB. The SLA panels have already been jettisoned. The LM was secured to the S-IVB with explosive bolts at the tops of its four landing struts.

Second right: The S-IVB stage drifts away from the docked CSM/LM at about 3:10 p.m. (EST) in this view from the CM, showing the forward fuel tank bulkhead. One of the LM's two very high frequency (VHF) antennas, used for inflight communications with the CM, is at center, with RCS cluster No. 1 at top. The stage was sent into solar orbit.

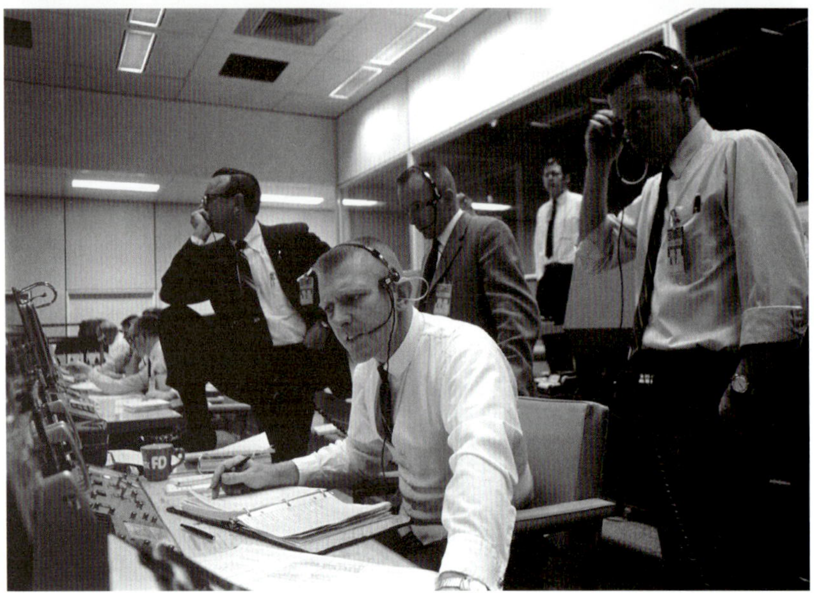

Above: With MSC Director of Flight Operations Chris Kraft at left, Apollo 9 flight directors Gene Kranz (White Team; *foreground*), Gerry Griffin (Gold Team; *left with headset*), and M. P. "Pete" Frank (Orange Team; *right*) monitor events during the first day of the mission in MOCR 2.

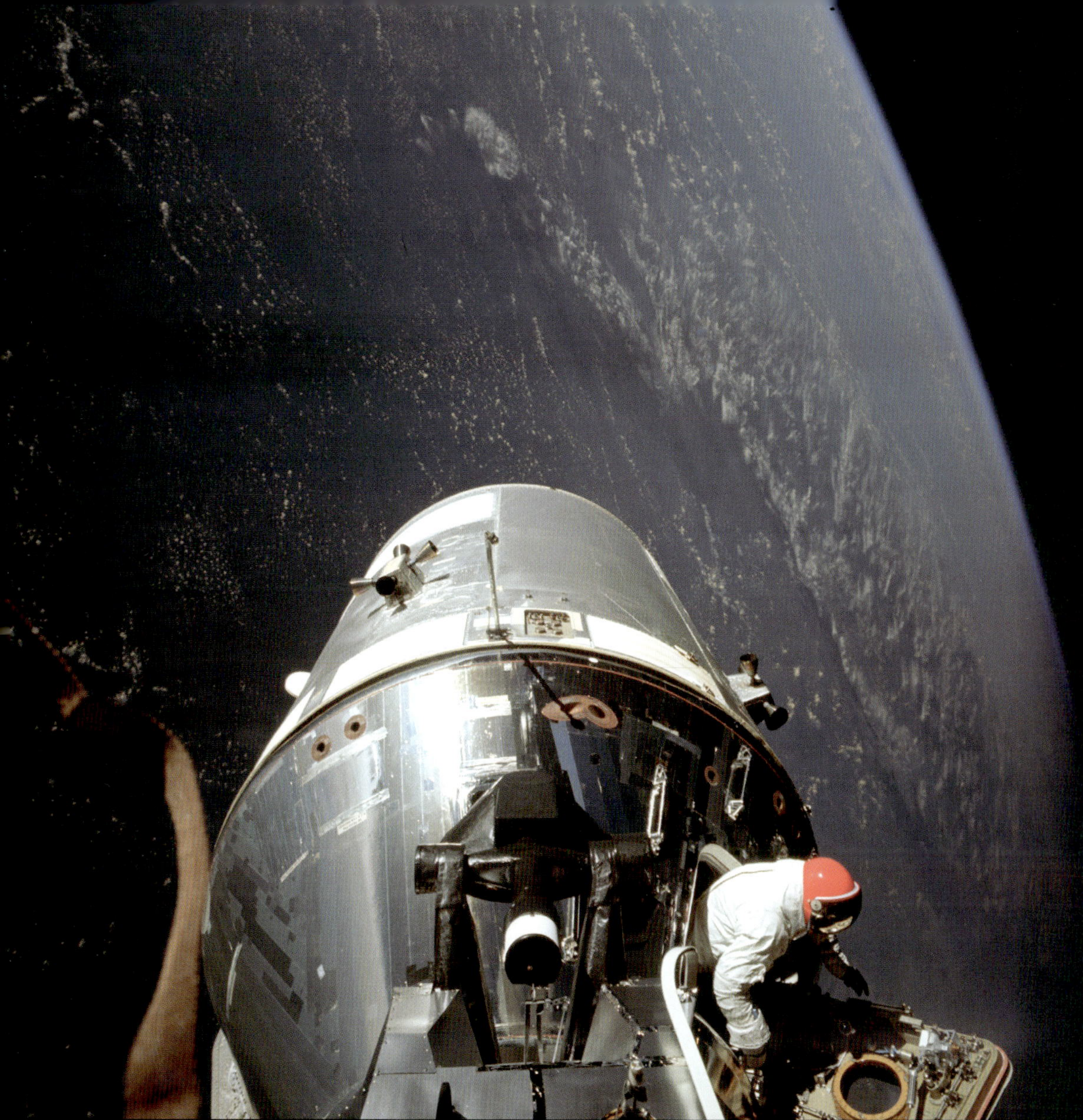

Left: Spider's rendezvous radar antenna, lower center, points away from Schweickart as he takes this photo of Scott and the CSM. A portion of his tether is at left. Both astronauts retrieved thermal samples from their spacecrafts' hulls. The entire EVA lasted about fifty minutes.

Bottom left: McDivitt's photo of Schweickart through *Spider*'s right-hand front window, as Schweickart transfers along a handrail on the LM's face during his EVA about 12:40 p.m. EST.

Bottom right: Scott, photographed by Schweickart, emerges through *Gumdrop*'s open hatch about 12:20 p.m. EST as the two docked spacecraft pass over the US. The hatch was operated by pumping the striped handle above the hatch window, which drove a ratchet mechanism to open or close fifteen latches. The 16 mm Maurer camera mounted left of the window recorded about eight minutes of Schweickart's EVA. *Spider*'s VHF in-flight antenna is at upper right.

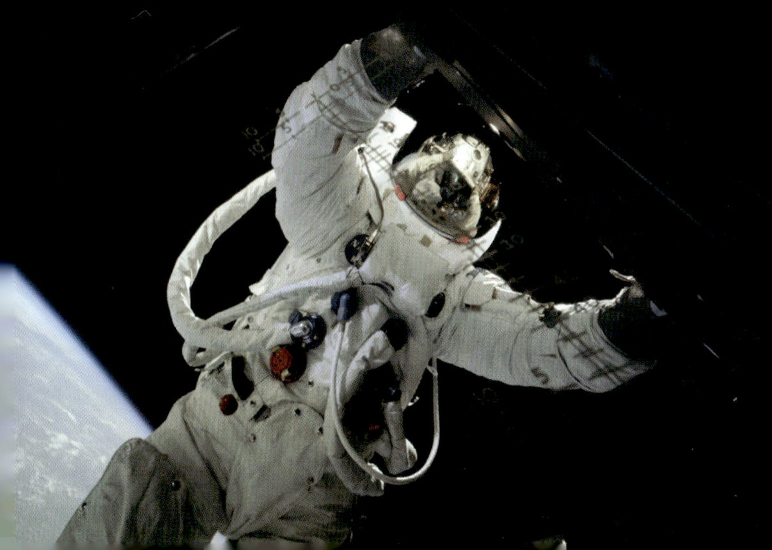

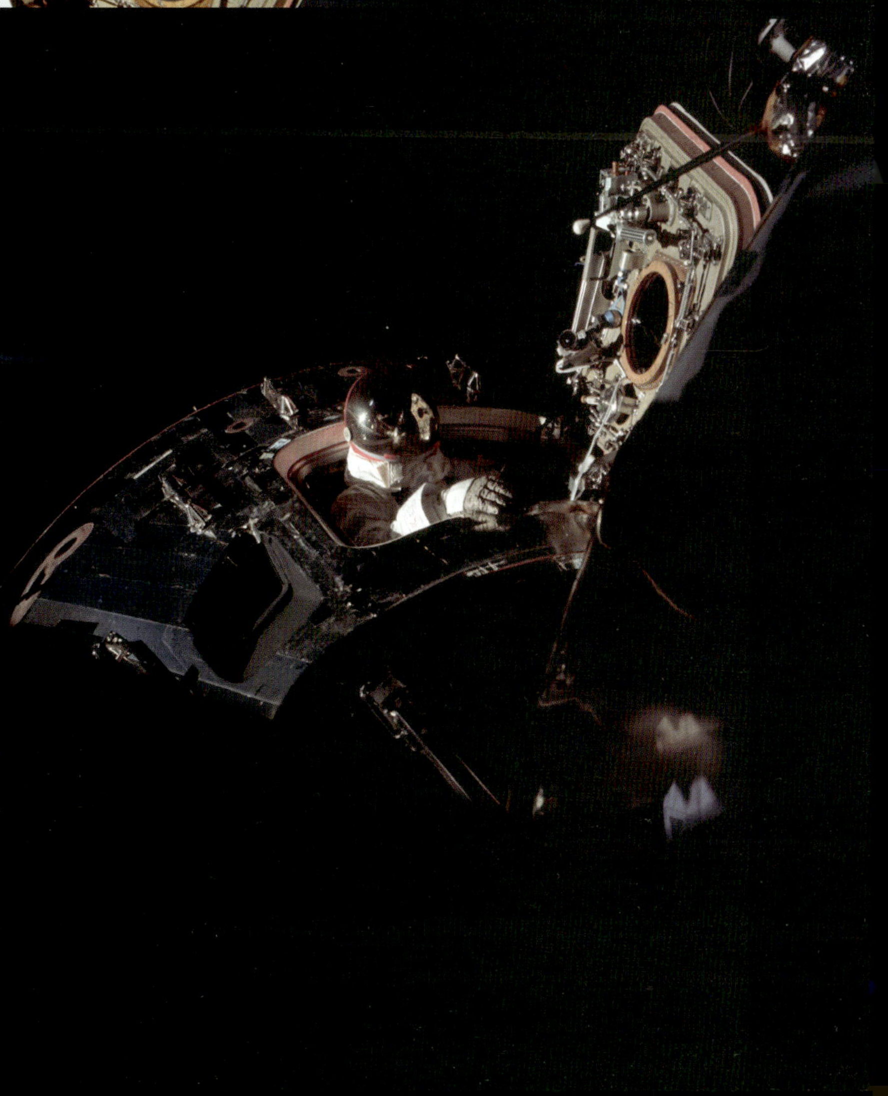

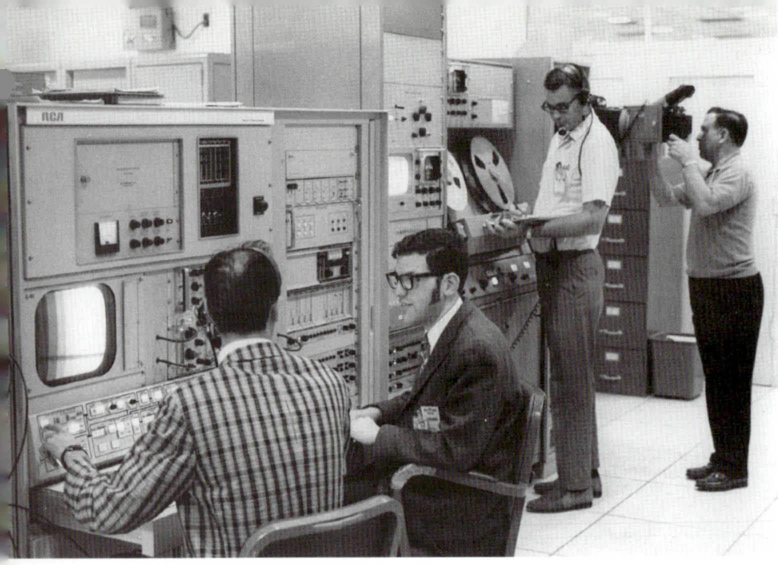

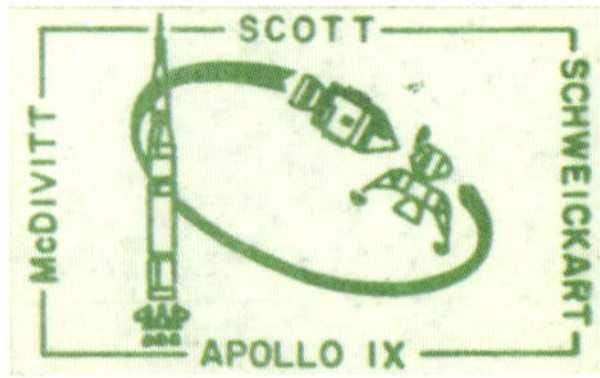

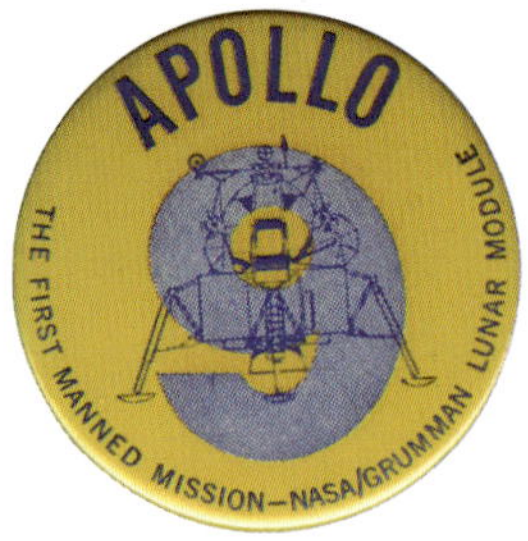

Above: Badge sticker and button

Top left: Much of the television from Apollo 9 was received at the Merritt Island Launch Area (MILA) Spaceflight Tracking and Data Network station at KSC, using a thirty-foot S-band dish. The station's controls are shown in this photograph. Goldstone, California, handled the rest.

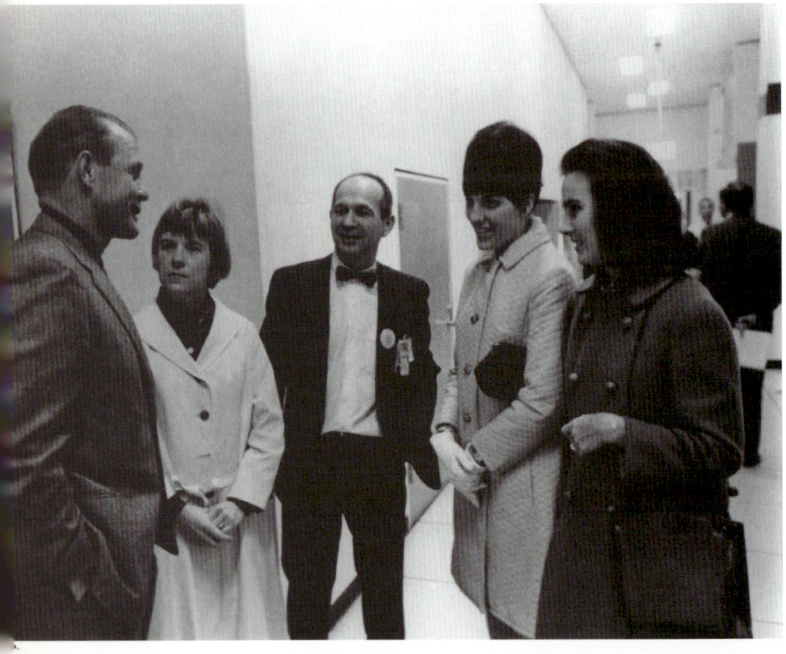

Center left: John Glenn (*left*) chats with (*left to right*) Pat McDivitt; MSC Director of Engineering and Development Max Faget; Clare Schweickart; and Lurton Scott during a visit to the MCC on March 5, 1969. Glenn, a Royal Crown Cola executive, would soon move from New York back to Ohio to make a second US Senate run in 1970.

Bottom left: Apollo 9's splashdown, as seen from the open side hatch of the helicopter *Recovery Three*. The splashdown occurred 3 miles from the amphibious assault ship USS *Guadalcanal* (near Grand Turk Island in the Atlantic Ocean) at 12:01 p.m. (EST) on March 13, 1969. Because of a storm—and because the seas were calm and visibility was good—the target was moved 480 miles south the day before. Live color-TV coverage, which was transmitted to Europe, was relayed for the first time via satellite.

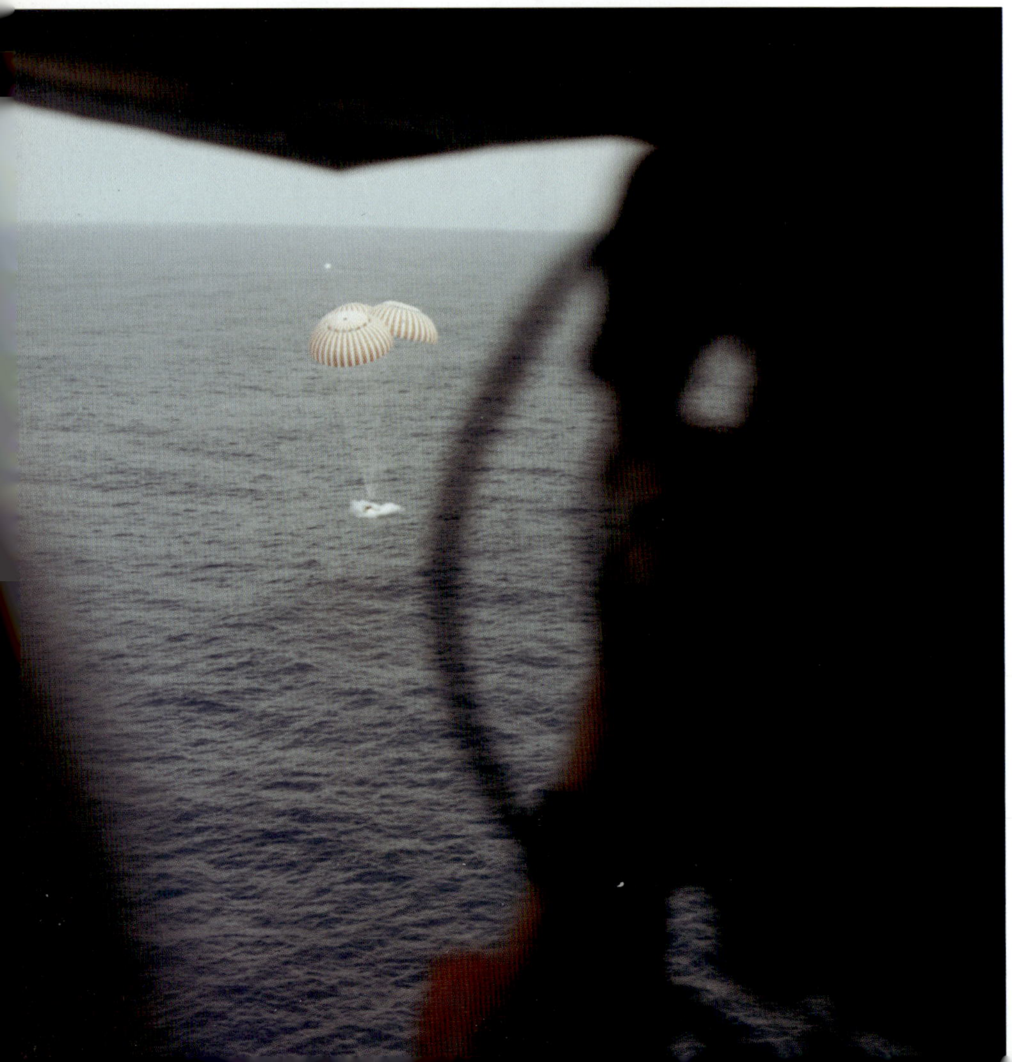

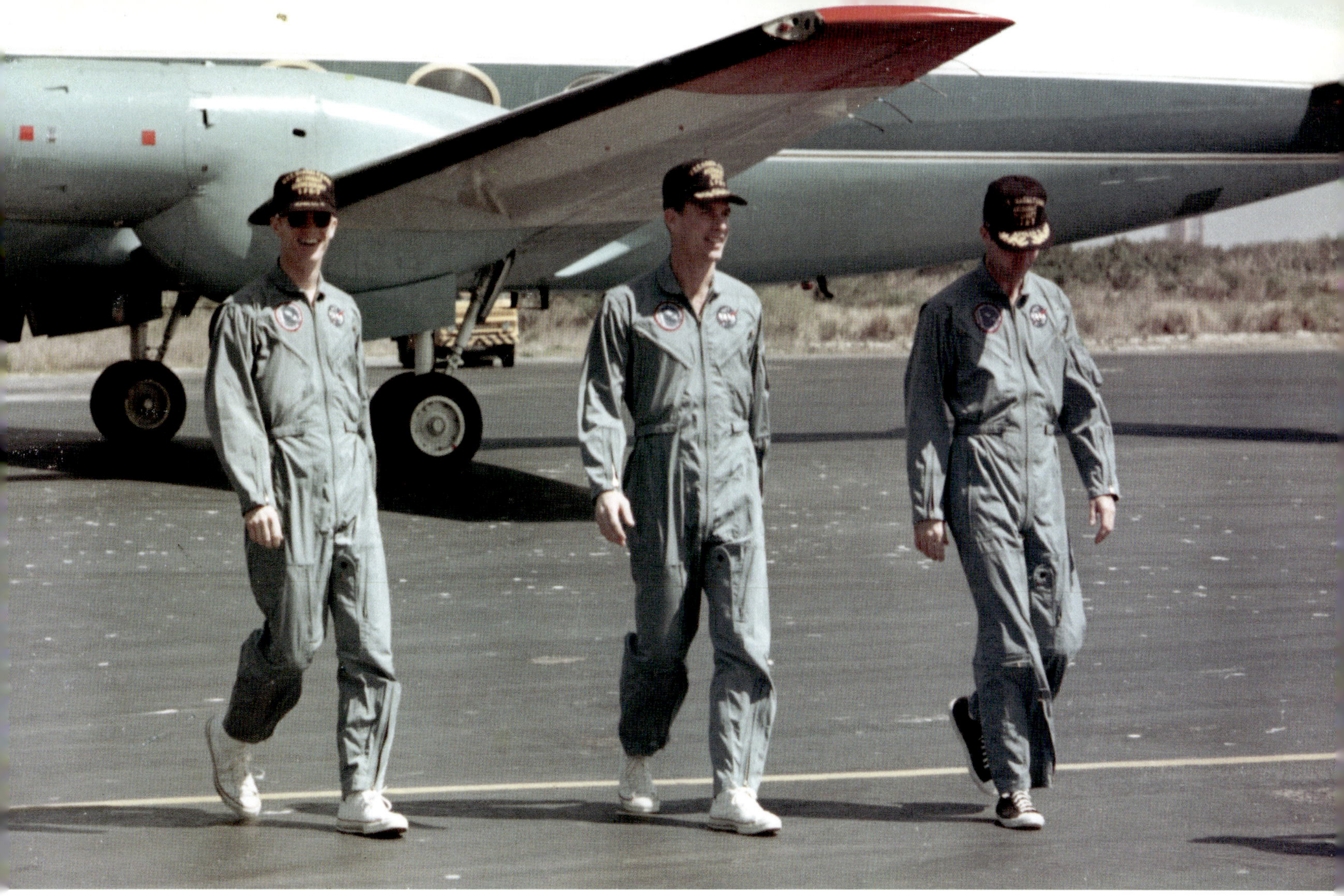

Schweickart, Scott, and McDivitt stretch their legs during a forty-one-minute refueling stop at the Cape Kennedy Skid Strip, en route to Ellington Field on March 14, 1969. The NASA Grumman Gulfstream I twin turboprop behind them brought them from Eleuthera Auxiliary Air Base in the Bahamas, where they had landed that morning. They had flown into Eleuthera from the *Guadalcanal* on helicopters.

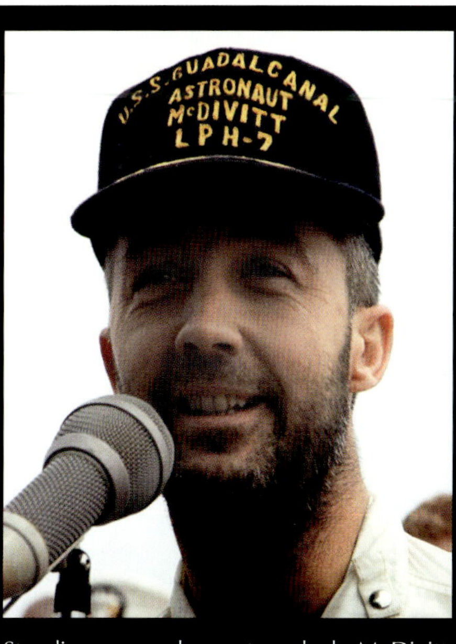

Standing on a red carpet on deck, McDivitt tells cheering sailors, "It's sure good to be back. Matter of fact, I think you guys kept us, here on the *Guadalcanal*, kept us going the last few days. We heard you had a big cake here."

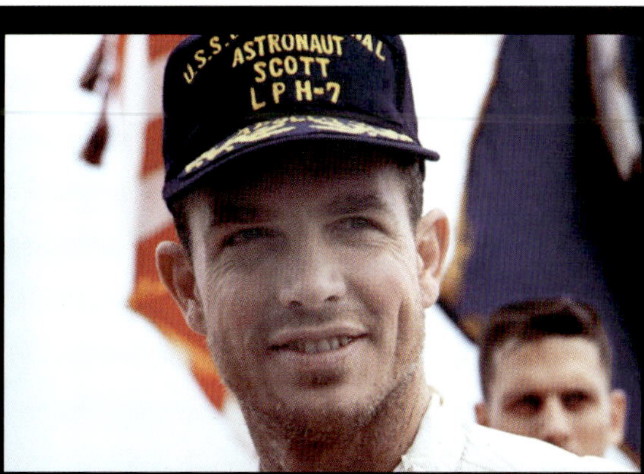

Downdrafts from the helicopter rotor created waves, making egress from the CM tricky. Scott (*shown*) made it on deck relatively dry; McDivitt slipped out of a raft into the Atlantic and, like Schweickart, was inadvertently dunked into the ocean as he was being raised to the recovery helicopter.

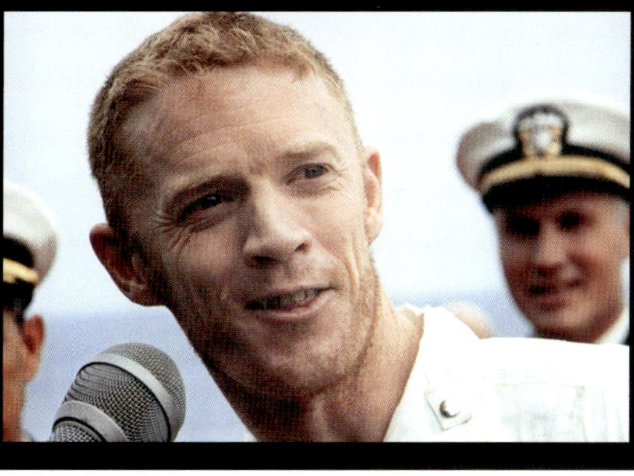

Schweickart told the sailors that the recovery was "the thing that was prettiest" about the flight. Neither he nor McDivitt would fly in space again. The *Guadalcanal* was decommissioned in 1994 and used as a target; it sunk in the Virginia Capes area of the Atlantic in 2005.

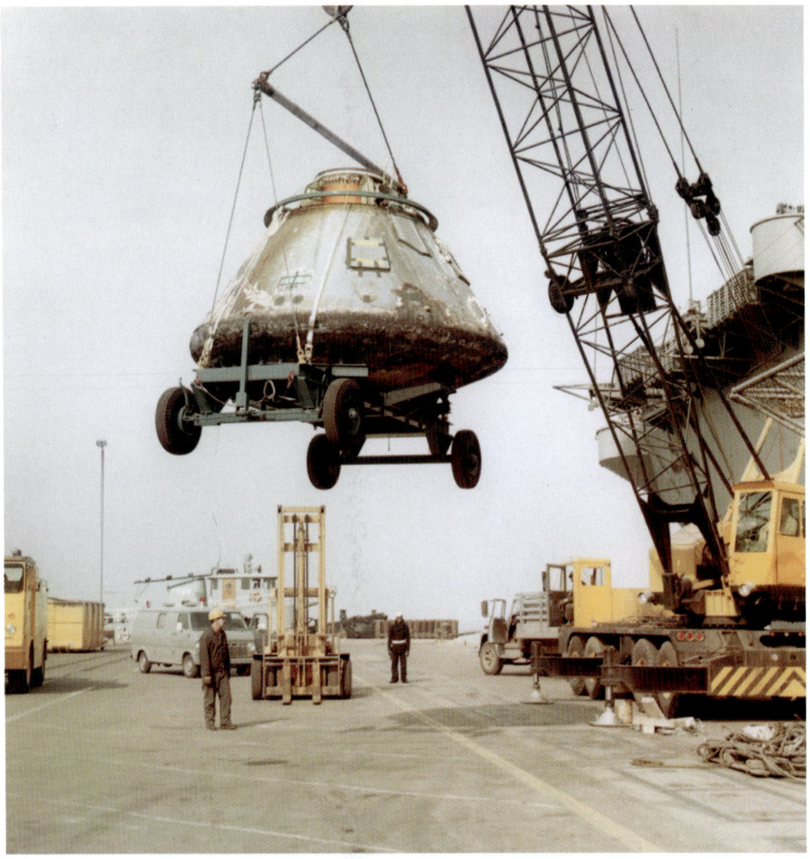

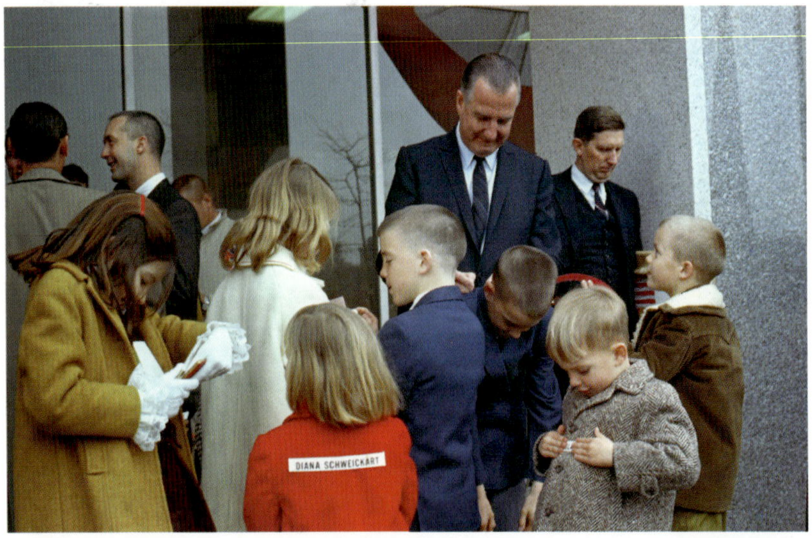

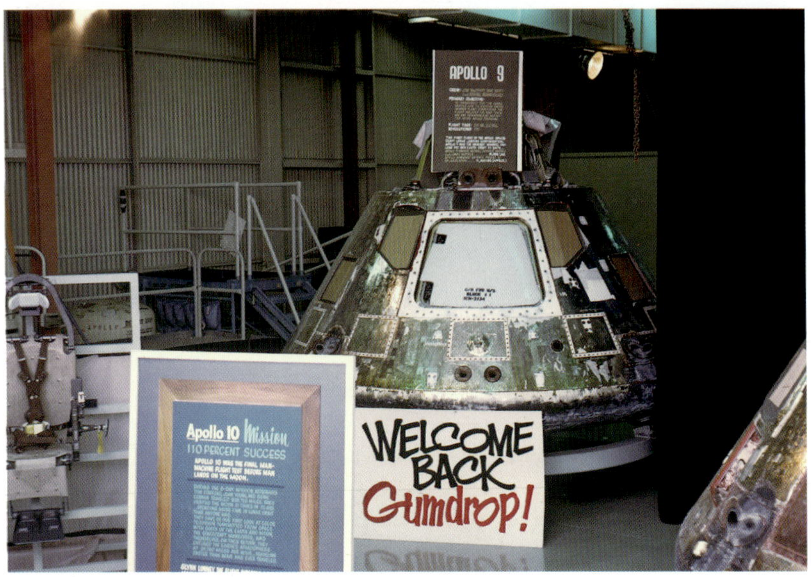

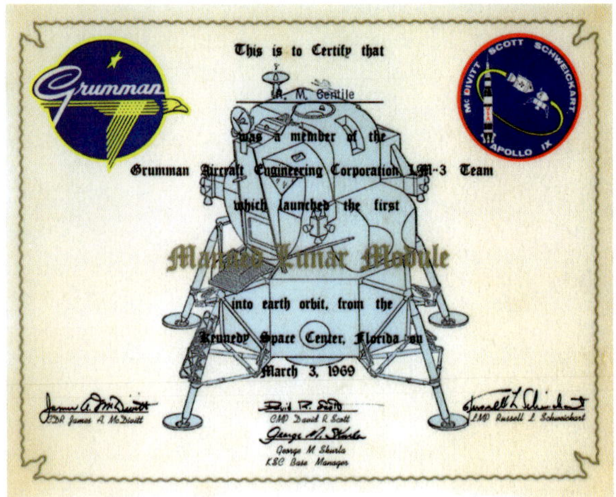

Above: Grumman certificate

Top left: The CM is off-loaded from *Guadalcanal* on March 16 at the Norfolk Naval Air Station in Norfolk, Virginia. After deactivation, the spacecraft was flown to Long Beach, California, on March 19 and trucked to NAR's Space Division in Downey, California, for postflight analysis, arriving there March 21.

Center left: Agnew signs autographs for the crewmen's children after presenting each astronaut with the NASA Distinguished Service Medal. The ceremony took place in front of NASA Headquarters (HQ) in Washington, DC, on March 26, 1969. *Left to right*: Vicki Schweickart, nine; Ann McDivitt, ten; Diana Schweickart, four; Mike McDivitt, eleven; Patrick McDivitt, eight; Doug Scott, five; and Rusty Schweickart, eight. McDivitt chats with Scott and others at left; the Agnew aide at right has a US flag, which the crew presented to Agnew. They had carried the flag on their mission.

Bottom left: Apollo 9's CM on display at NAR in Downey, California, in June 1969. A corner of the Apollo 10 CM is at lower right; at left is a CM couch.

Opposite page: The grandstand is at the center of this aerial view of the LC-39 Press Site in May 1969, with the broadcast network and wire service trailers at left. The large countdown clock has not yet been installed near the edge of the Turning Basin at right.

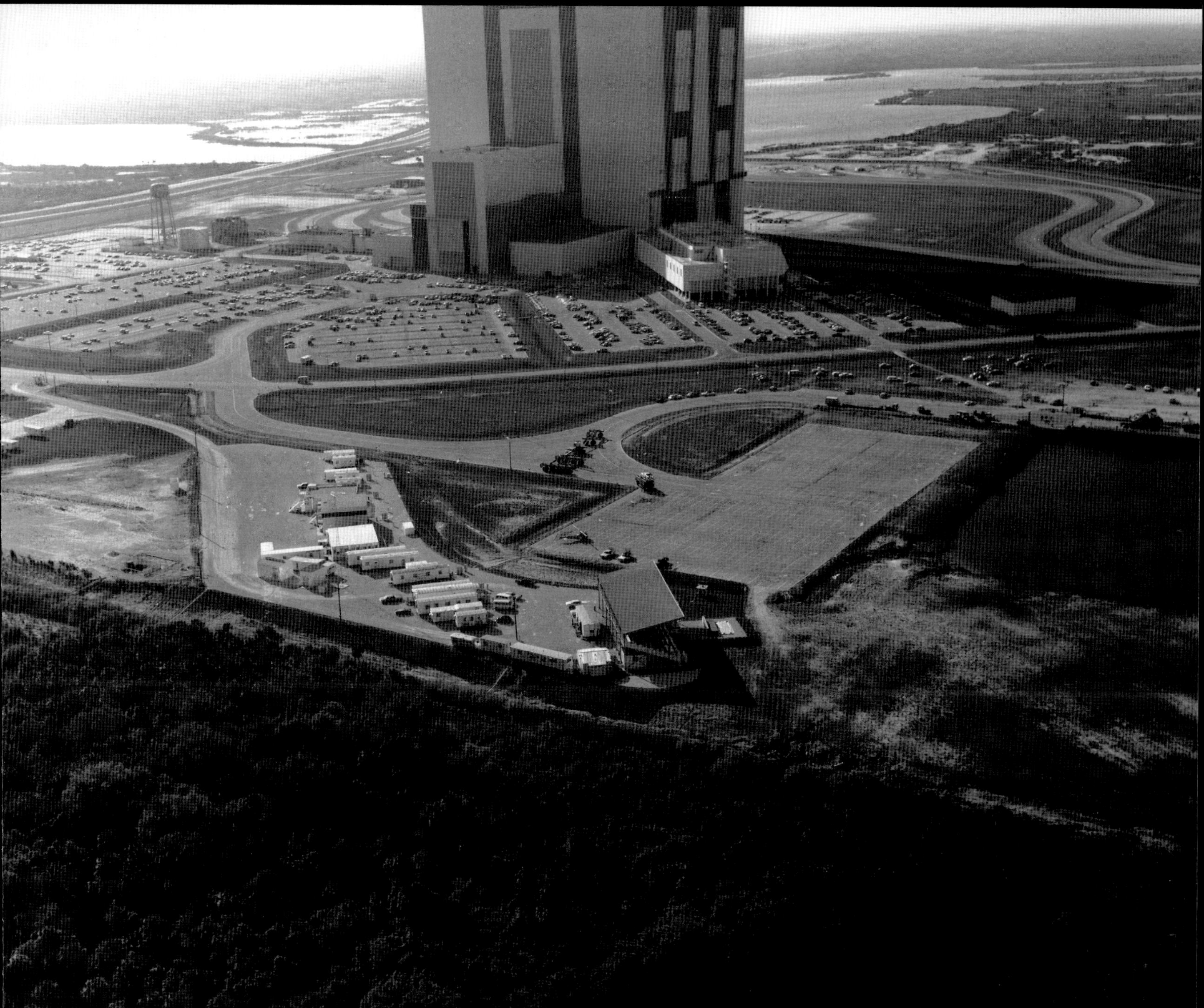

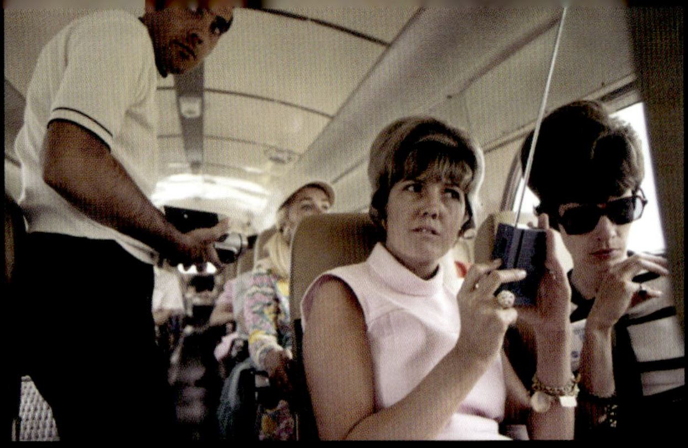

Barbara Cernan monitors the mission on WCKS-FM. She is aboard a NASA bus heading to NASA's launch guest center in Cocoa Beach, after having viewed her husband's launch near the VAB. She was the only one of the three astronauts' wives to attend. Gene Cernan had first seen Barbara, then a stewardess, while buying an airplane ticket. After overhearing her name, he called her and asked her for a date.

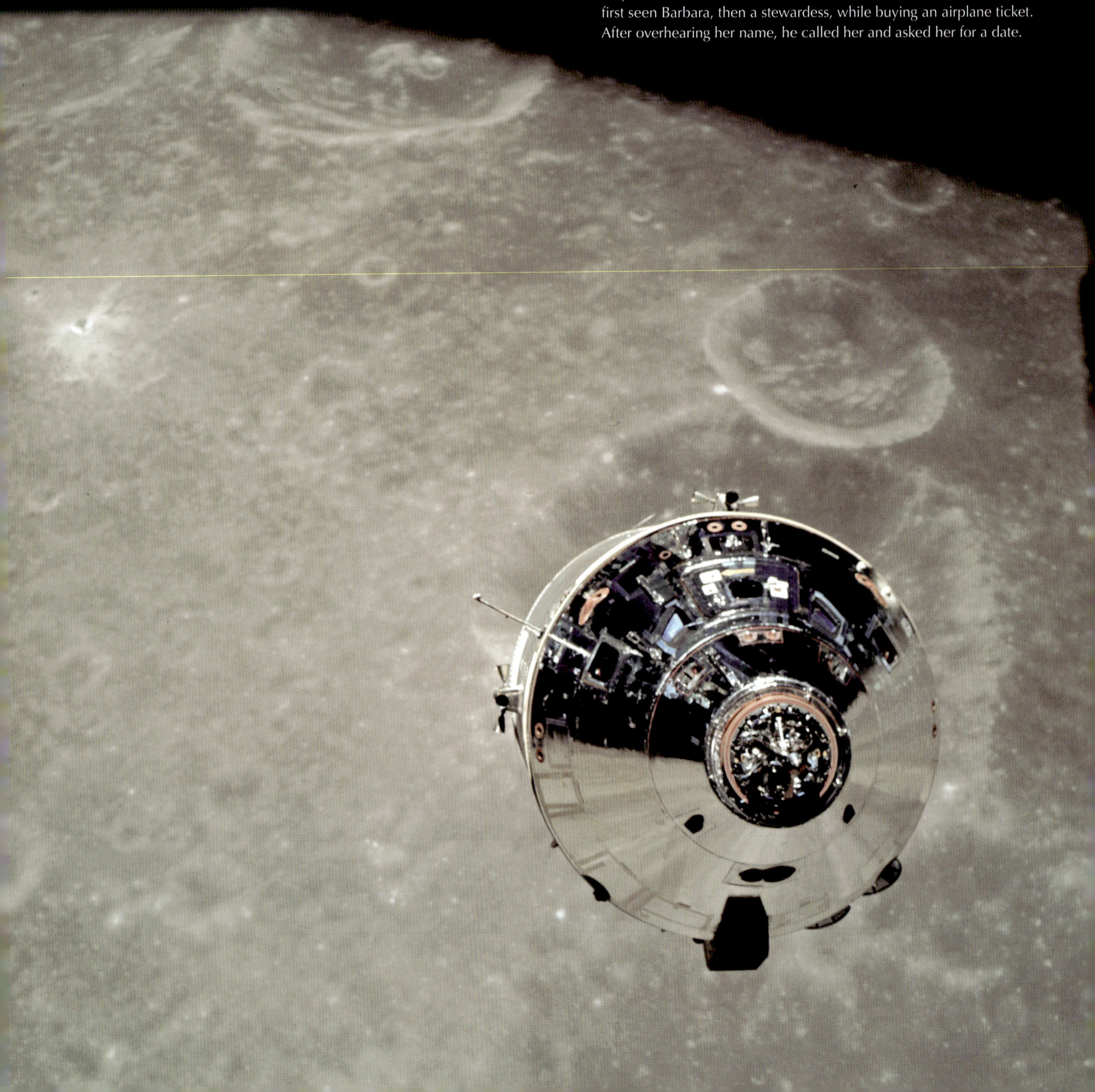

CM *Charlie Brown*, with John Young aboard, is photographed from LM *Snoopy* on May 22, 1969. The two spacecraft orbited sixty miles above the northwest corner of Smyth's Sea on the far side of the moon. The twelve small engines spaced around the CM provided three-axis attitude and rotational control to orient it for reentry. The docking probe is at center, and the EVA floodlight is extended from the SM at left. The umbilical fairing, which covered the wiring and plumbing between the CM and SM, is in shadow at bottom, with two of the SM's RCS clusters visible (*top and left*). The SM's high-gain S-band antenna is partially visible at bottom right.

CHAPTER 5

Apollo 10

May 18–26, 1969

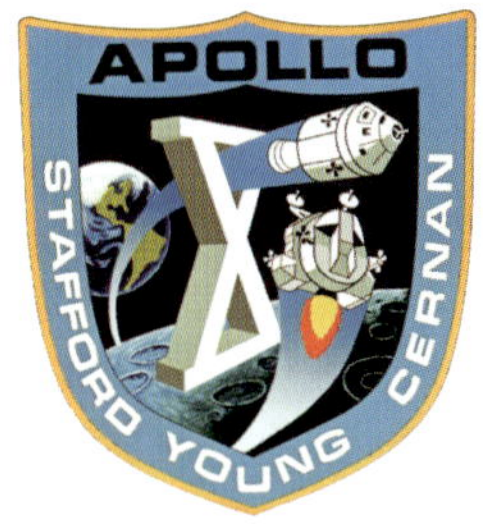

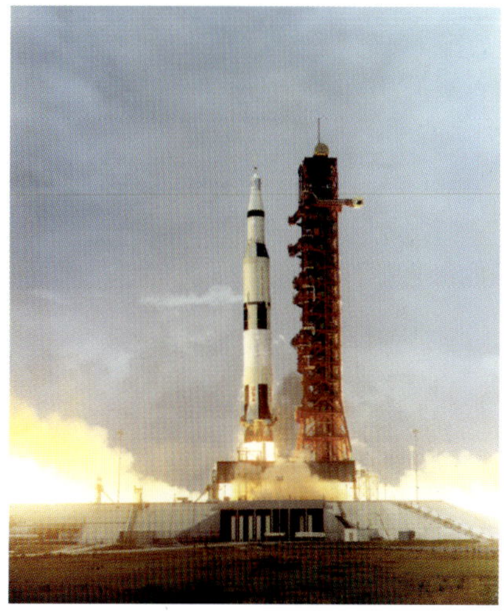
Launch: 12:49 p.m. (EDT), May 18, 1969

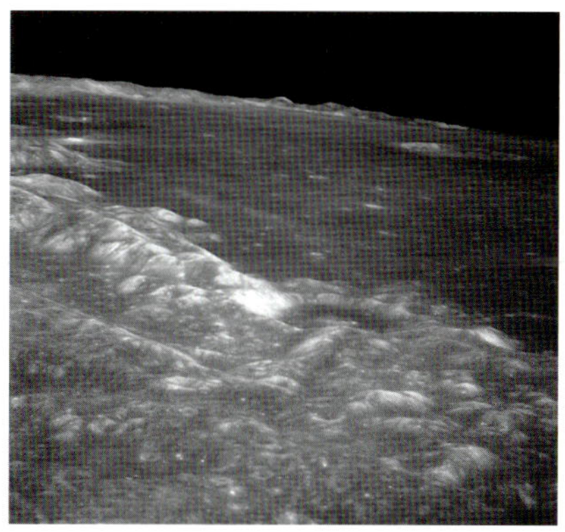
Lunar orbit: May 22 (Sea of Crises)

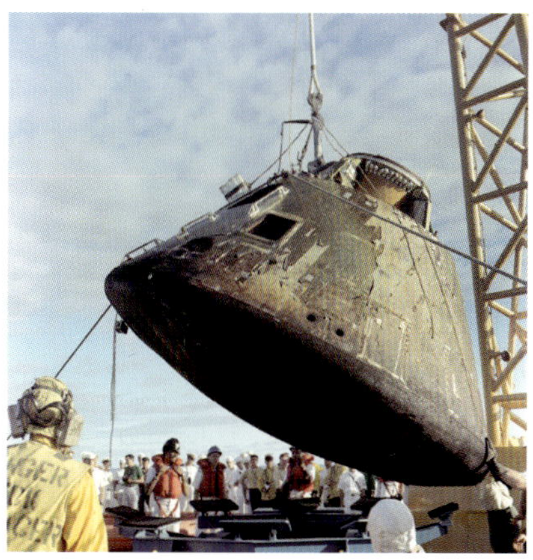
Splashdown: 12:52 p.m. (EDT), May 26

LC-39 was a busy place in the late spring of 1969, as the deadline for President John F. Kennedy's goal approached. Three Saturn Vs were being stacked in the VAB for the first time, including that of Apollo 10, the first to be stacked in the building's high bay 2. As Apollo 9 prepared to splash down, Apollo 10 was rolled to the pad; it was also the first to use Pad 39B. Apollos 11 and 12 were coming together in two other high bays.

Apollo 10 was the bridge to the first lunar landing—it was essentially an eight-day version of Apollo 9, except this time, the spacecraft would be tested in lunar orbit. CDR Tom Stafford was joined by his Gemini IX pilot, Gene Cernan, as LMP. CMP John Young completed the first all-veteran crew of space travelers. (Apollo 11 would be the second.)

The flight's primary goals were to demonstrate rendezvous and docking in the lunar gravitational field and to evaluate navigation between the two spacecraft. The LM—dubbed *Snoopy* in honor of the cartoon character's role as the MFA program's mascot—would undock from the CM, *Charlie Brown*, for eight hours. With Stafford and Cernan at the controls, *Snoopy* would descend to about nine miles above the moon, before ascending for rendezvous and docking with Young in the CSM. The mission would also test the Manned Space Flight Network and its Unified S-Band System for tracking two vehicles at lunar distances, and would check out the LM's programmed trajectories, radar, and flight control systems.

The crew awoke at about 10:00 a.m. (EDT) on May 18, 1969, for their 12:49 p.m. on-time launch into a 115-mile circular Earth parking orbit. One and a half orbits later, the S-IVB fired to push the spacecraft out of orbit and toward the moon.

The first live color-TV transmissions to Earth began three hours after launch, when Apollo 10 was 3,570 miles from Earth. They showed *Charlie Brown* docking with *Snoopy*, as well as the interior of the CM. About four hours after launch, the crew docked with the LM. The LM separated from the S-IVB stage, which was sent into solar orbit.

Stafford was an enthusiastic booster of the new color TV, and nine more transmissions—seven live and two recorded—on the way to the moon gave viewers scenes of the receding Earth, the approaching moon, and onboard activities.

During meals, the crew was the first to use spoons. Young reported that he enjoyed a chicken salad sandwich made from a spread and two slices of Houston grocery store bread chosen to resist crumbling.

On May 21, the SPS fired to send the spacecraft into lunar orbit. A second firing circularized the orbit to sixty-nine miles. The crew sent back the first color-TV transmissions of the moon's surface.

The following afternoon, Stafford and Cernan entered the LM and prepared for undocking. The two vehicles separated and briefly flew in a station-keeping orbit. To simulate the Apollo 11 landing, *Snoopy*'s descent engine fired for twenty-seven seconds, sending it into a new orbit of 9.7 by 70.5 miles.

Two and a half hours after undocking, *Snoopy* flew over Apollo 11's planned landing site in the moon's Sea of Tranquility. The descent engine then fired in two bursts to put the LM into an orbit of 13.7 by 219 miles.

When it was time to drop the spacecraft's descent stage, Stafford had *Snoopy* in the right attitude (12.7 miles above the surface) ten minutes early. Suddenly, the LM began to swing wildly. Cernan exclaimed, "Son of a bitch!" Stafford told Cernan to thrust forward, and then hit the switch to release the descent stage. But the LM continued its gyrations, and a warning light indicated that its inertial measuring unit was about to reach its limit and go into "gimbal lock," which happens when the axes of two of the three gimbals are driven into a parallel configuration. Stafford took over in manual control, made a pitch maneuver, and started firing the RCS thrusters. *Snoopy* was back under control within three minutes. It turned out that earlier, the astronauts had accidentally switched the guidance system to the automatic mode. *Snoopy*'s descent stage remained in orbit, but it eventually crashed onto the lunar surface.

Shortly after, the ascent engine fired for fifteen seconds to rendezvous with *Charlie Brown*, and, with Young at the controls in the CM, the spacecraft were redocked at 9:11 p.m. (EDT). *Snoopy*'s ascent stage was jettisoned and sent into orbit around the sun, becoming the only remaining intact LM ascent stage of the ten sent into space. (The others were left in lunar orbit. They either crashed—having been intentionally steered into the moon to obtain readings from seismometers on the surface—or, as was the case with Apollo 9's LM, they burned up in Earth's atmosphere.)

May 23 was spent on lunar landmark tracking and photography. Early the next morning, the SPS was fired to push the CSM out of orbit and back toward Earth.

After a midcourse correction and CM and SM separation, Apollo 10 reentered Earth's atmosphere on May 26. *Charlie Brown* splashed down in the Pacific Ocean 460 miles east of American Samoa at 12:52 p.m. (EDT), within television range of its primary recovery ship, USS *Princeton*.

When the crewmen stepped out of the recovery helicopter, it was apparent they had shaved—a first. The men used tube shaving cream and wiped it off with swabs, alleviating concerns about whiskers escaping into the cabin and causing problems.

The first manned moon landing was now less than two months away.

Top left: The Apollo 10 crew discusses lunar surface features with scientist-astronaut Jack Schmitt (*second from left*), a geologist, on May 15, 1969, in the astronaut crew quarters in the MSOB at KSC. Apollo 10 had the world's first all-veteran space crew—and all three men would fly in space again. The maps, prepared by the USAF's Aeronautical Chart and Information Center, were based in part on images returned from five unmanned Lunar Orbiter missions. A CSM model by the Walter J. Hyatt Company is at left.

Bottom left: Grumman workers pause near the hatch while transferring LM-4's ascent stage from a workstand to Chamber R (where they will conduct altitude testing) in the MSOB on December 2, 1968. The rendezvous antenna is not yet installed.

Second right: USAF Col. Tom Stafford, thirty-eight, was Apollo 10 CDR. A respected, plainspoken Oklahoman, Stafford was a former fighter pilot who wrote flight manuals. When he was accepted among the second group of astronauts in 1962, the Naval Academy graduate left studies at Harvard Business School. He flew on two Gemini missions.

Third right: LMP, US Navy Cdr. Gene Cernan, thirty-five. Stafford's Gemini IX pilot, the personable and articulate Cernan received an electrical engineering degree from Purdue University and became a navy officer through ROTC there. After flight school, he was assigned to attack squadrons at Miramar, California. In 1963, he received a master's degree in aeronautical engineering from the Naval Postgraduate School and was selected as an astronaut.

Bottom right: CMP, US Navy Cdr. John Young, thirty-eight, remained reserved and succinct, a cool customer who could come off as detached, aloof, or even cocksure. He was known for his deadpan wit. These three pictures of the crewmen were taken in April 1969, for their MSC photo ID badges.

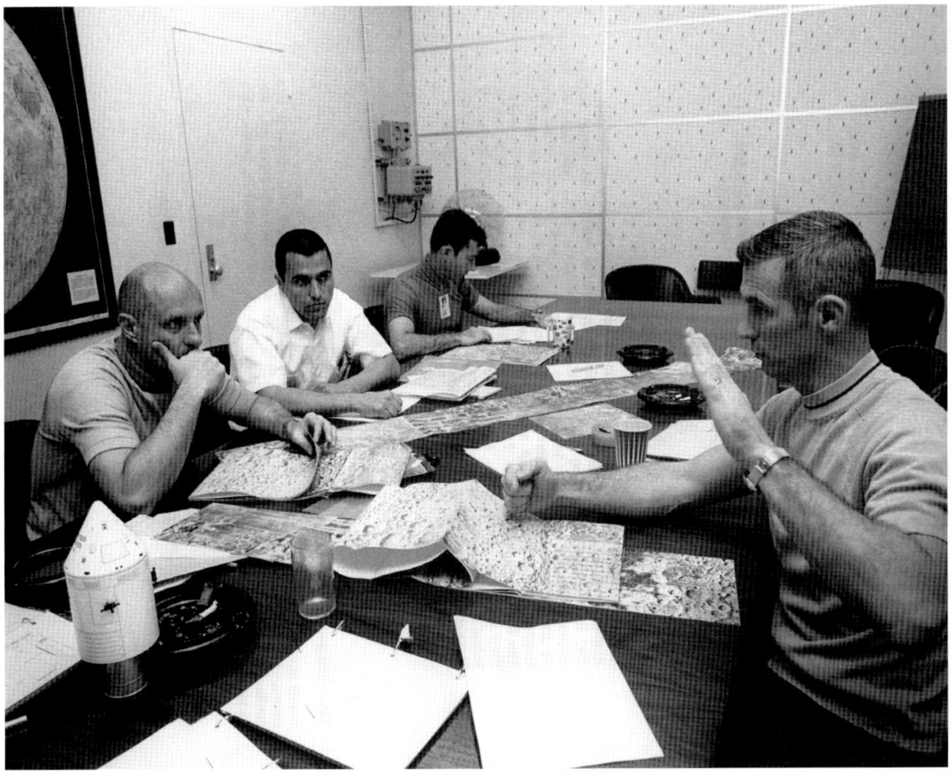

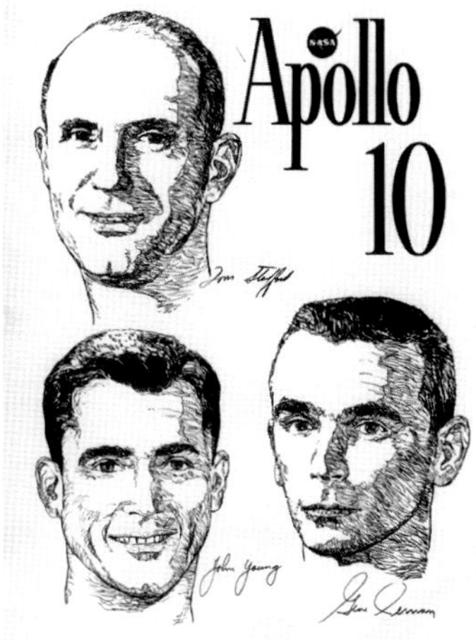
NASA
Apollo
10

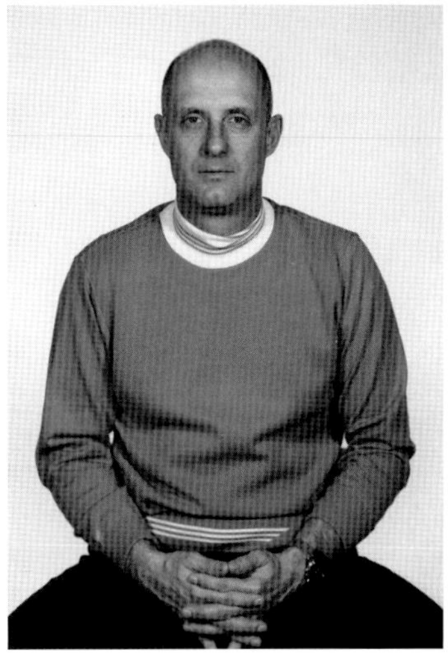

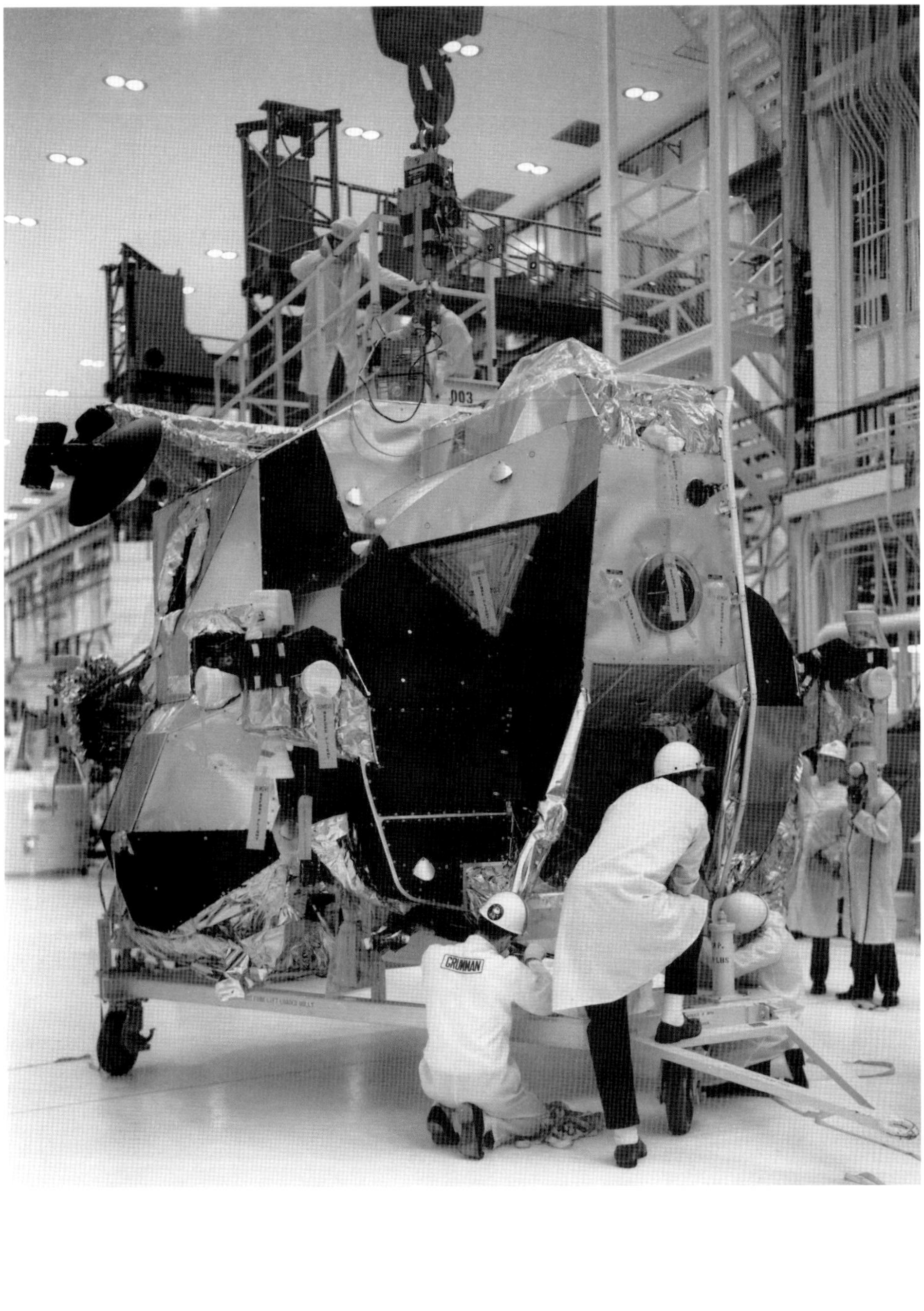
003
GRUMMAN

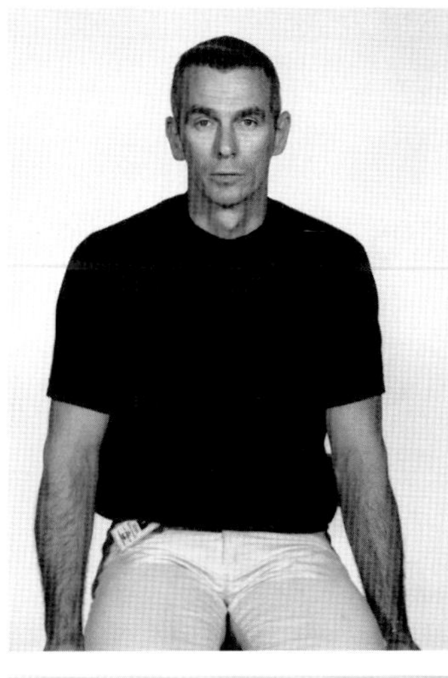

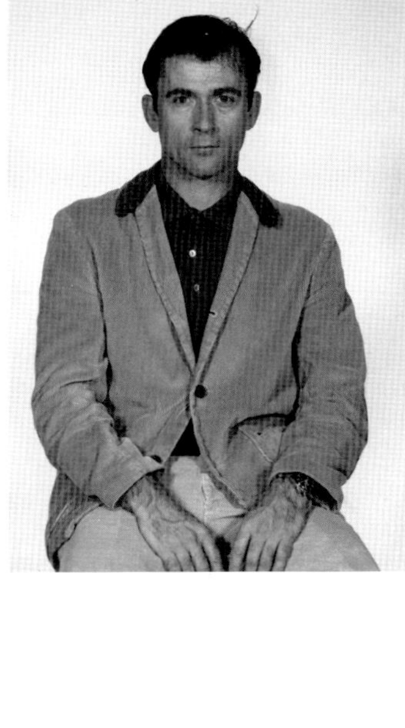

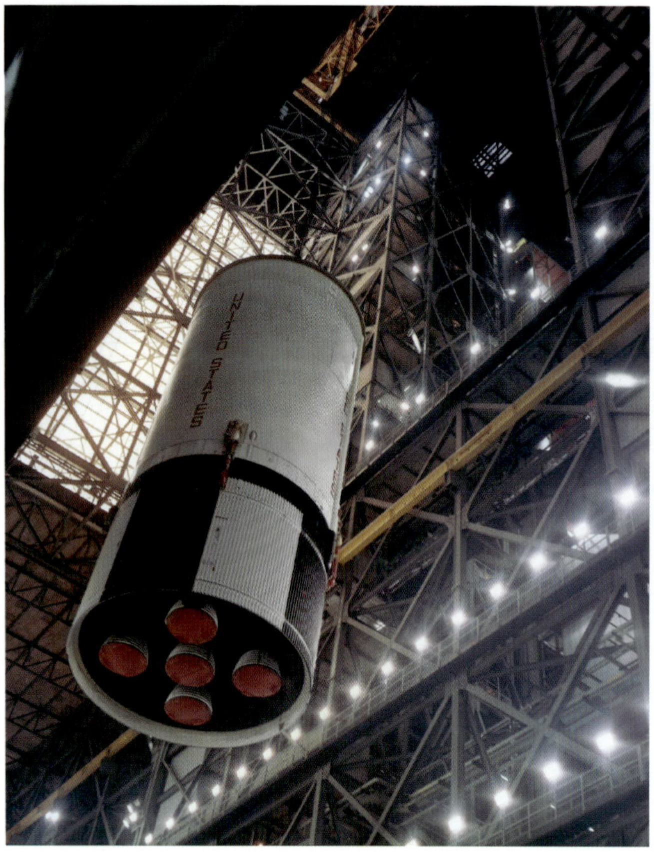

Above: The Saturn's S-II second stage is hoisted toward high bay 2 for stacking in the VAB on December 30, 1968. It was the first Saturn to use the bay. This photo, taken from the transfer aisle, looks up to the open bay, with the S-IC stage visible through the gridwork at lower right on LUT-3. Near the stage's center is a liquid hydrogen feed line, with a red ground support equipment protective cover. The white systems tunnel runs down the right side. The S-II stage ignited at 38 miles in altitude and boosted the third stage and Apollo spacecraft by 144 miles.

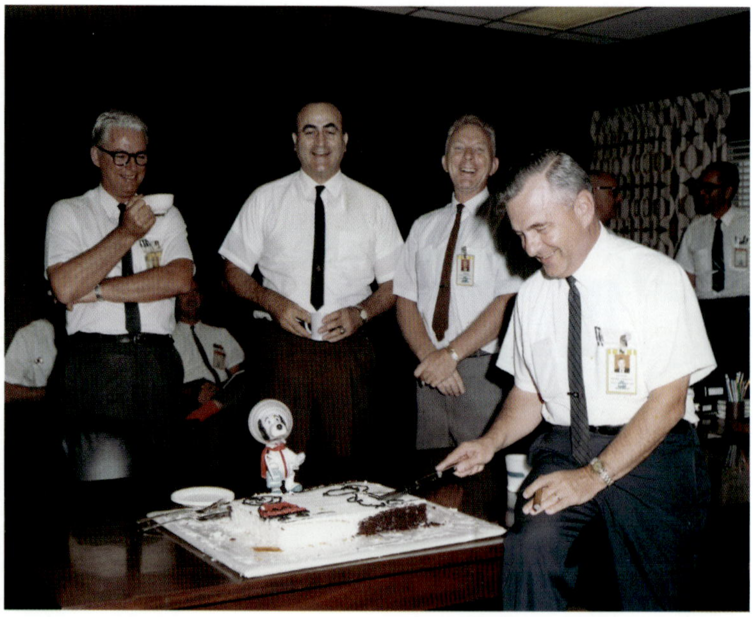

Above: Apollo 10 Deputy Launch Director Walt "Kappy" Kapryan (*right*) celebrates his forty-ninth birthday on May 16, 1969, in his office at KSC HQ. His cake features *Peanuts* characters Charlie Brown and Snoopy, the mission's call signs; a Snoopy doll made by Determined Productions is behind it. From left are John J. Williams, spacecraft operations director; Rocco Petrone; and Launch Operations Manager Paul Donnelly. Kapryan would succeed Petrone as KSC director of launch operations in August.

Bottom right: Personnel from NAR, Grumman, and NASA inspect the edge of the fixed-panel section of the SLA, with LM-4 in place, in the MSOB on January 17, 1969. Its steerable S-band antenna is folded against the ascent stage (*center*). After the four deployable panels were lowered and attached as a unit, and the CSM mated above, the unit was towed to the VAB for stacking with the Saturn's third stage.

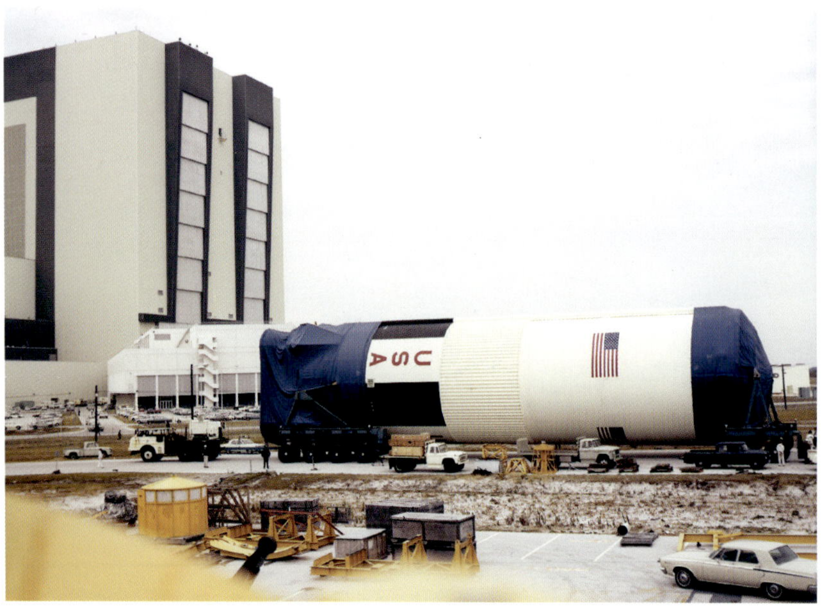

Above: After being off-loaded from a barge at the LC-39 Turning Basin dock on November 27, 1968, Apollo 10's Saturn S-IC stage is prepared for towing into the VAB for checkout. The stage weighed three hundred thousand pounds when dry.

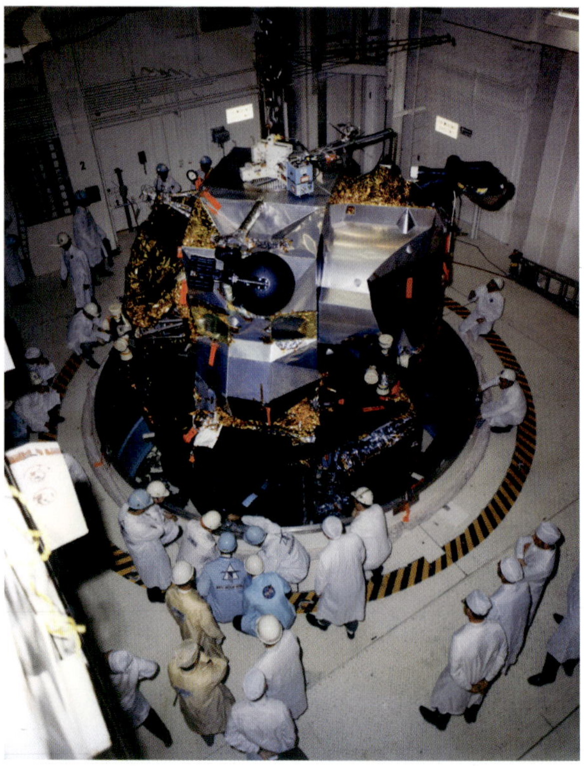

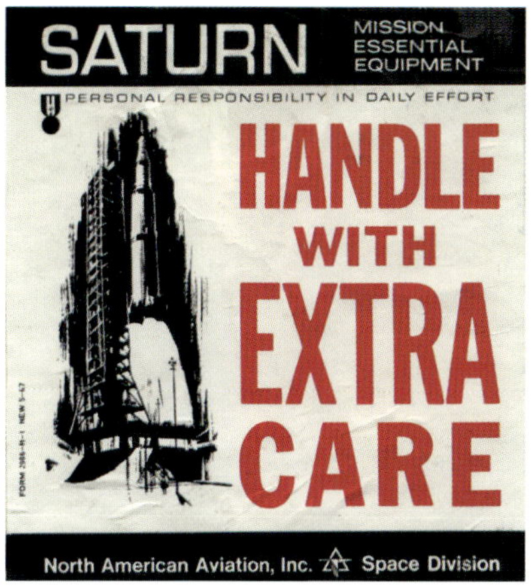

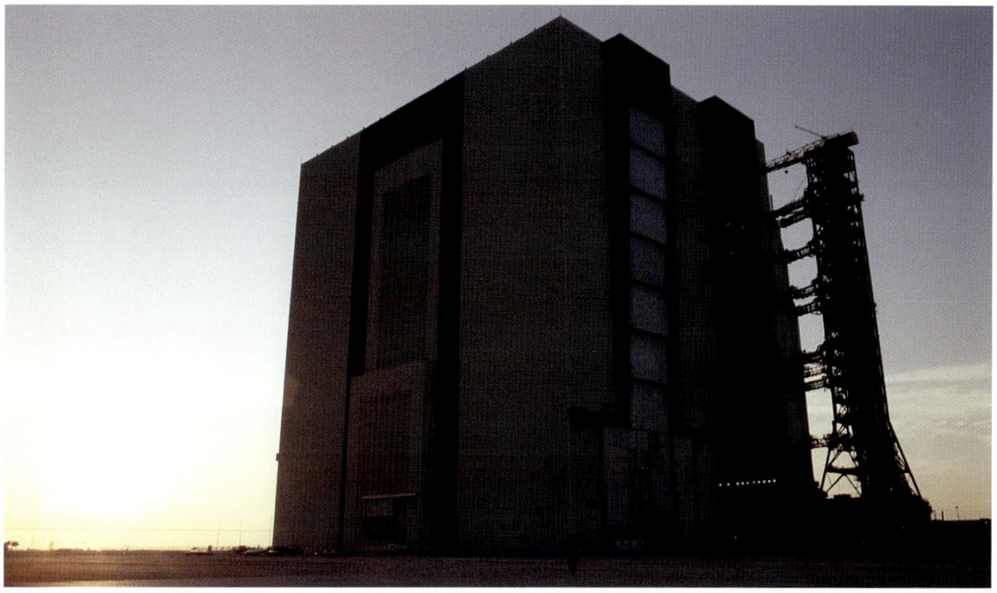

Right: A lone figure approaches the VAB at 6:30 a.m. (EST) on March 11, 1969, as Apollo 10's Saturn V rolls out of high bay 2 (on the west side of the building) on ML-3. This was the first of only two Saturn rollouts from that side of the VAB; Skylab 1 was the other. (Apollo 11 stacking was underway in high bay 1 and Apollo 12 in high bay 3.) The vehicle was heading to a first launch from Pad 39B—another indication of how busy the LC had become. Apollo 9 was two days from splashdown. (photo by Tiziou News Service)

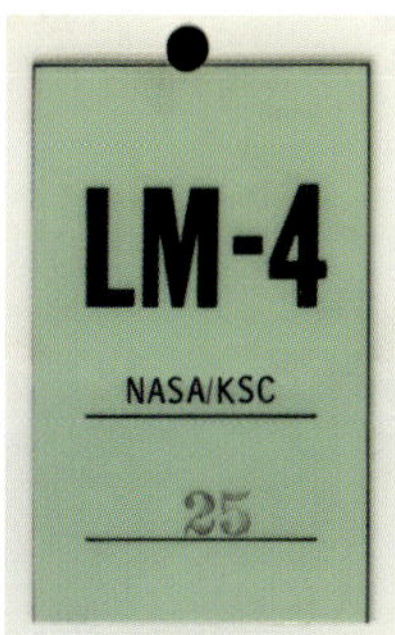

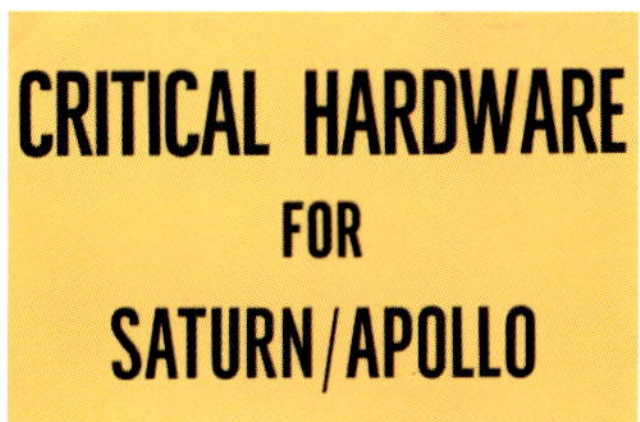

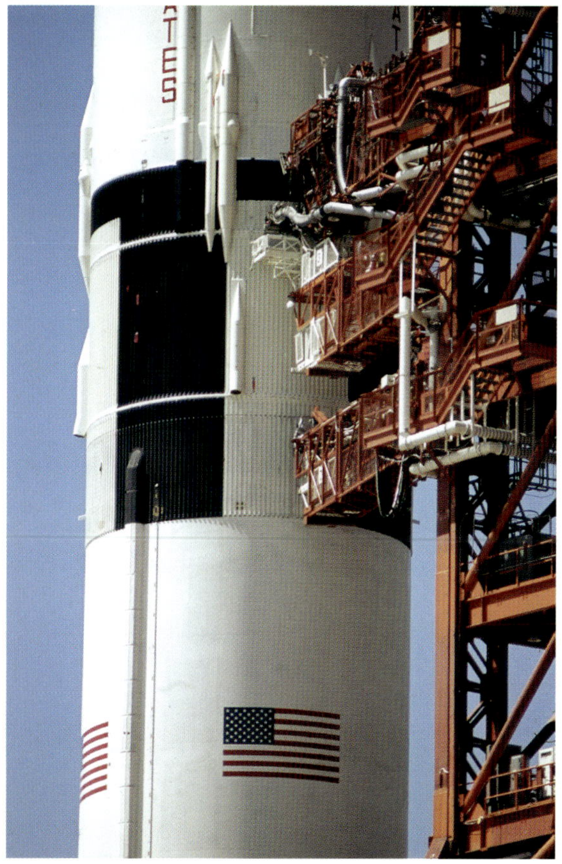

Above: Three service arms extend to the vehicle from the LUT. The S-II intermediate arm at top provided fueling and was retracted at launch; the S-II aft arm (*center*), which provided access, and the S-IC forward arm (*below*, providing pneumatic, electrical, and air-conditioning interfaces) were both retracted before launch. Two of the eight ullage motors are visible on the Saturn's interstage, to the left of the center arm. These motors fired to briefly nudge the stage forward, to settle its fuel prior to ignition of its five J-2 main engines. The two fairings to the right of the service tunnel above the interstage housed liquid hydrogen feed lines. (photo by Tiziou News Service)

Below: The crawler-transporter keeps the mobile launcher level as it begins to climb the 5 percent–grade ramp from the crawlerway to the hardstand at Pad 39B. The Apollo 8 crew returned to Houston on March 11 for the biggest parade in the city's history to that time. (photo by Tiziou News Service)

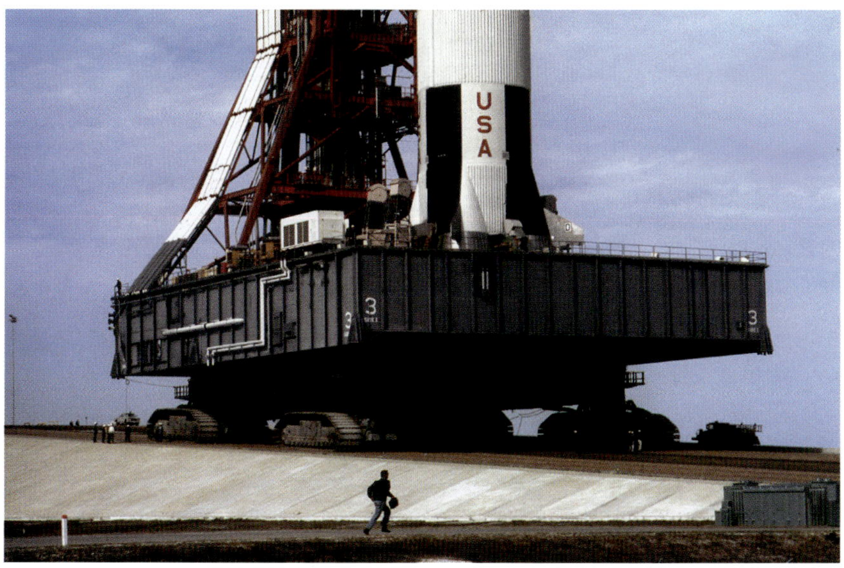

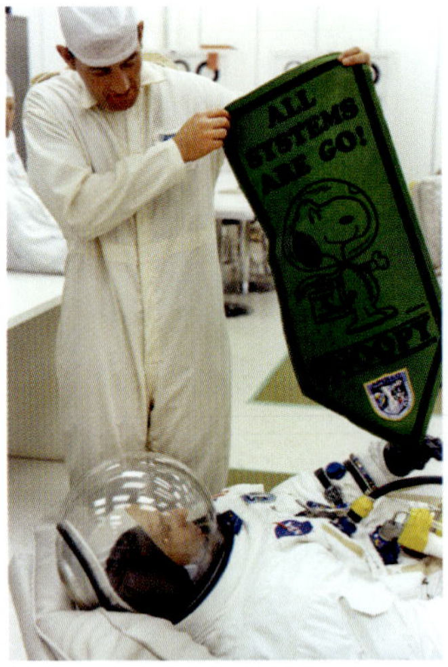

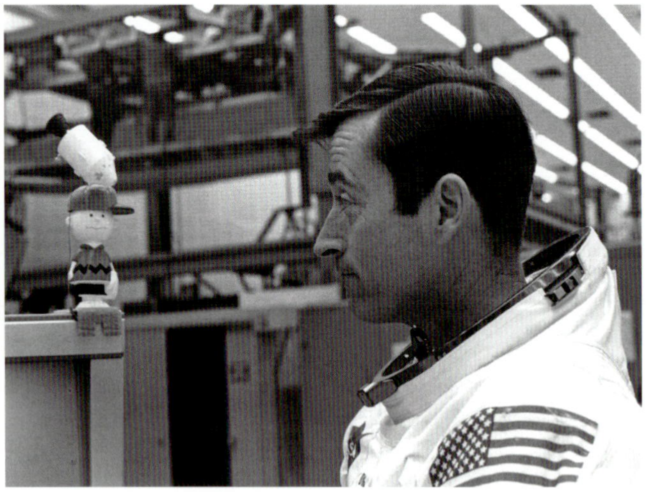

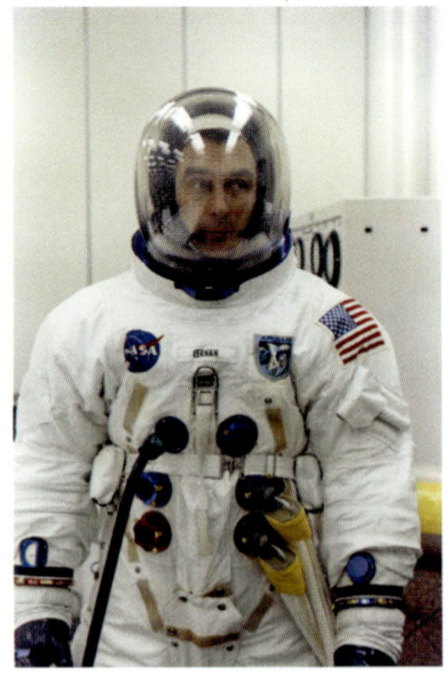

Top left: USAF suit technician Barry Lewis holds up a cloth Snoopy banner for Stafford in the MSOB's suiting room, before an abbreviated launch dry run on May 6, 1969. NASA had chosen the cartoon beagle in 1968 as an icon to "emphasize mission success and act as a 'watchdog' for flight safety." The CDDT had wrapped up the previous day.

Top center: Young eyes Charlie Brown, which was the radio call sign for the CM (and an apparent CSM crash victim) on April 15 in the Flight Crew Training Building. The crew obtained permission from *Peanuts* cartoonist Charles Schulz to use his characters' names for the flight.

Top right: Cernan undergoes suit checks on May 6. After the mock launch, he was the primary cameraman when the crew ran a twenty-five-minute test transmission to MSC, using two new color-TV cameras inside the CM. This was the final step in qualifying a television camera for the mission. A black-and-white camera carried onto the flight was never used. Cernan and Stafford wore cloth headsets officially known as communications carrier assemblies, but dubbed the "Snoopy caps." The units were manufactured by Air-Lock, Inc., a subsidiary of the David Clark Company, the Gemini space-suit maker.

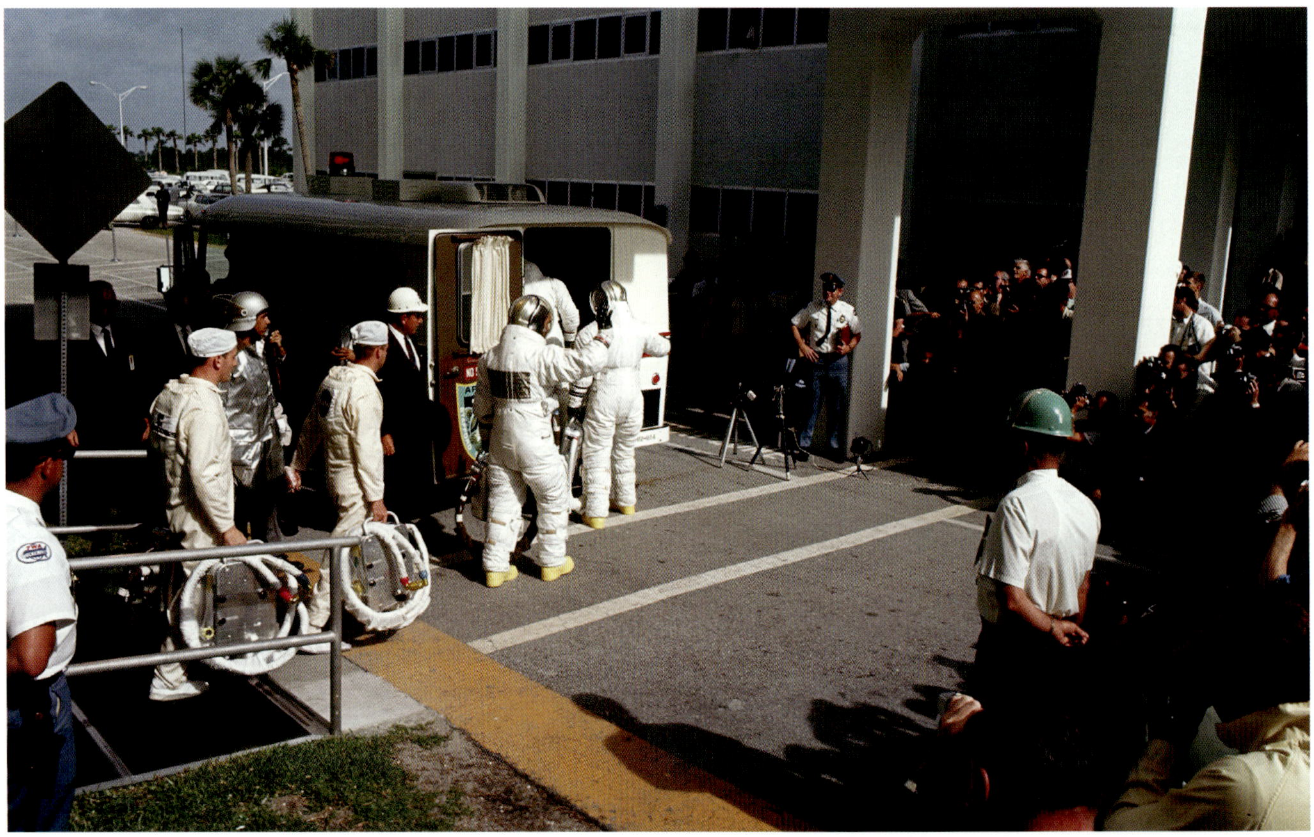

With news photographers and reporters at right, Young enters the transfer van, followed by Stafford and Cernan, on May 18 at the MSOB. Suit techs Al Rochford and ILC's Jim McNair follow with three additional suit ventilators, which they would swap out before the crewmen left the van at the pad. This ensured that the astronauts had sufficient pure-oxygen supplies during "prebreathe" before CM ingress, necessary to prevent the bends. (photo by Tiziou News Service)

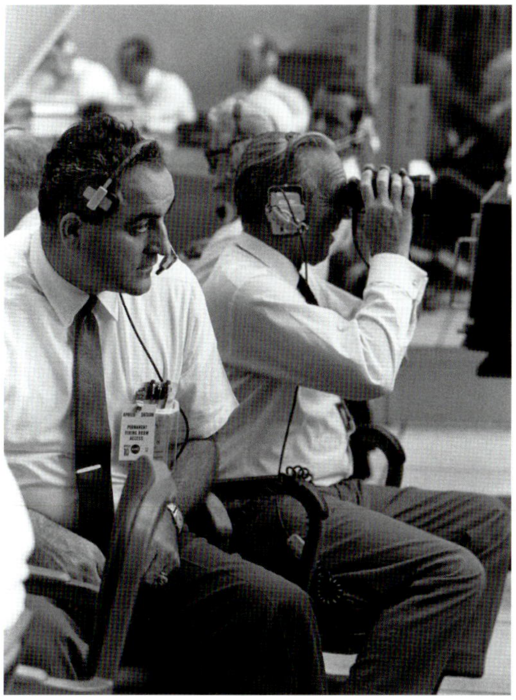

Petrone and KSC Director Kurt Debus follow the final minutes of the countdown from Firing Room 3, its first use. Apollo 10 used ML-3; each ML was controlled by the same-numbered firing room.

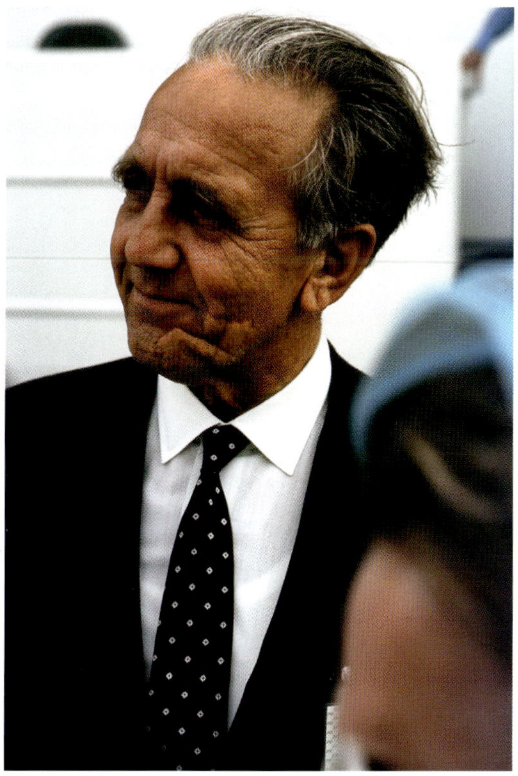

Debus waits to speak to reporters at the LC-39 Press Site after launch. Before turning his attention in 1962 to building complexes 34, 37, and 39, Debus had supervised the design and construction of several early launch facilities at Cape Canaveral, beginning in 1952. (photo by Tiziou News Service)

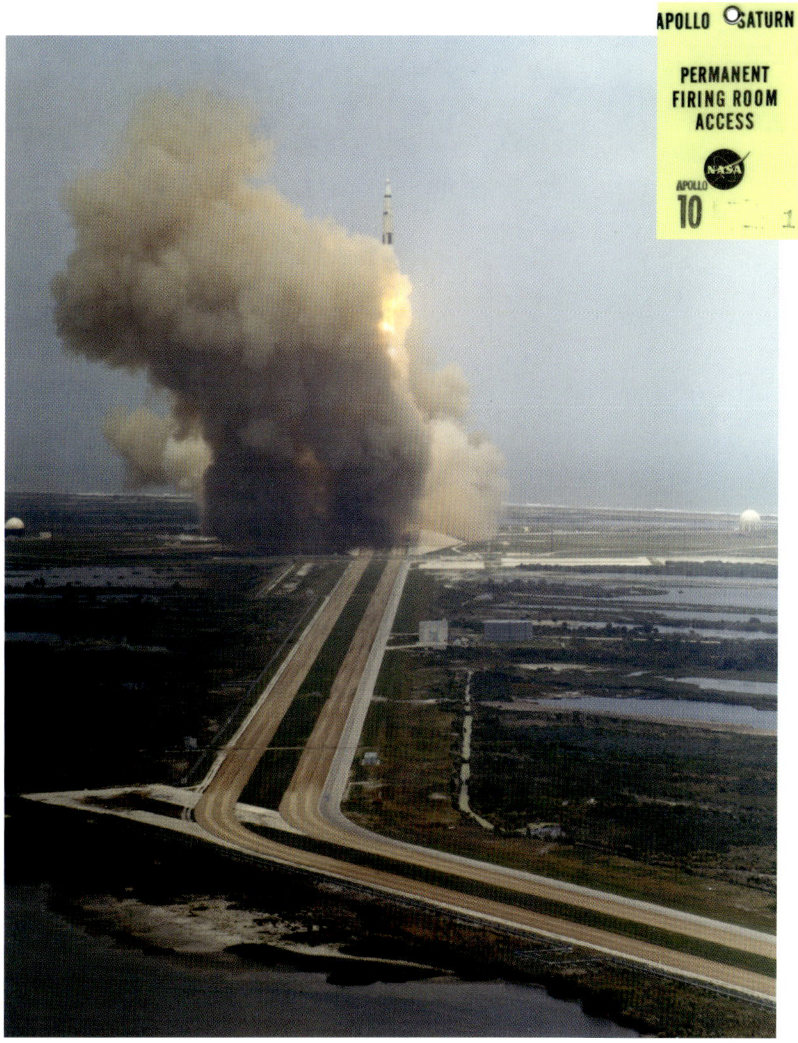

Waves crash on the Atlantic beach in the background as Apollo 10 begins to pitch and roll. It reached an Earth parking orbit in less than twelve minutes, just after launch at 12:49 p.m. (EST) on May 18, from Pad 39B. This photo is from the MSS. At bottom left is what would have been the branch of the crawlerway to the planned, but never built, Pad 39C.

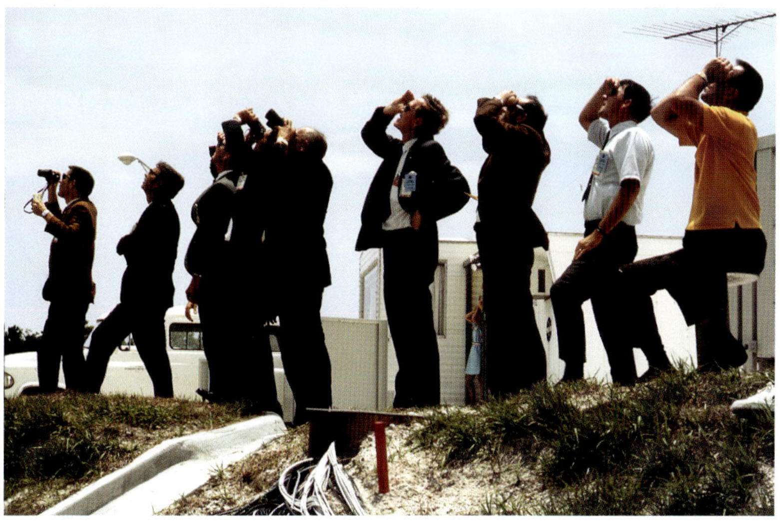

NASA managers follow the launch from the mound at the LC-39 Press Site, with a cable run at bottom center. (photo by Tiziou News Service)

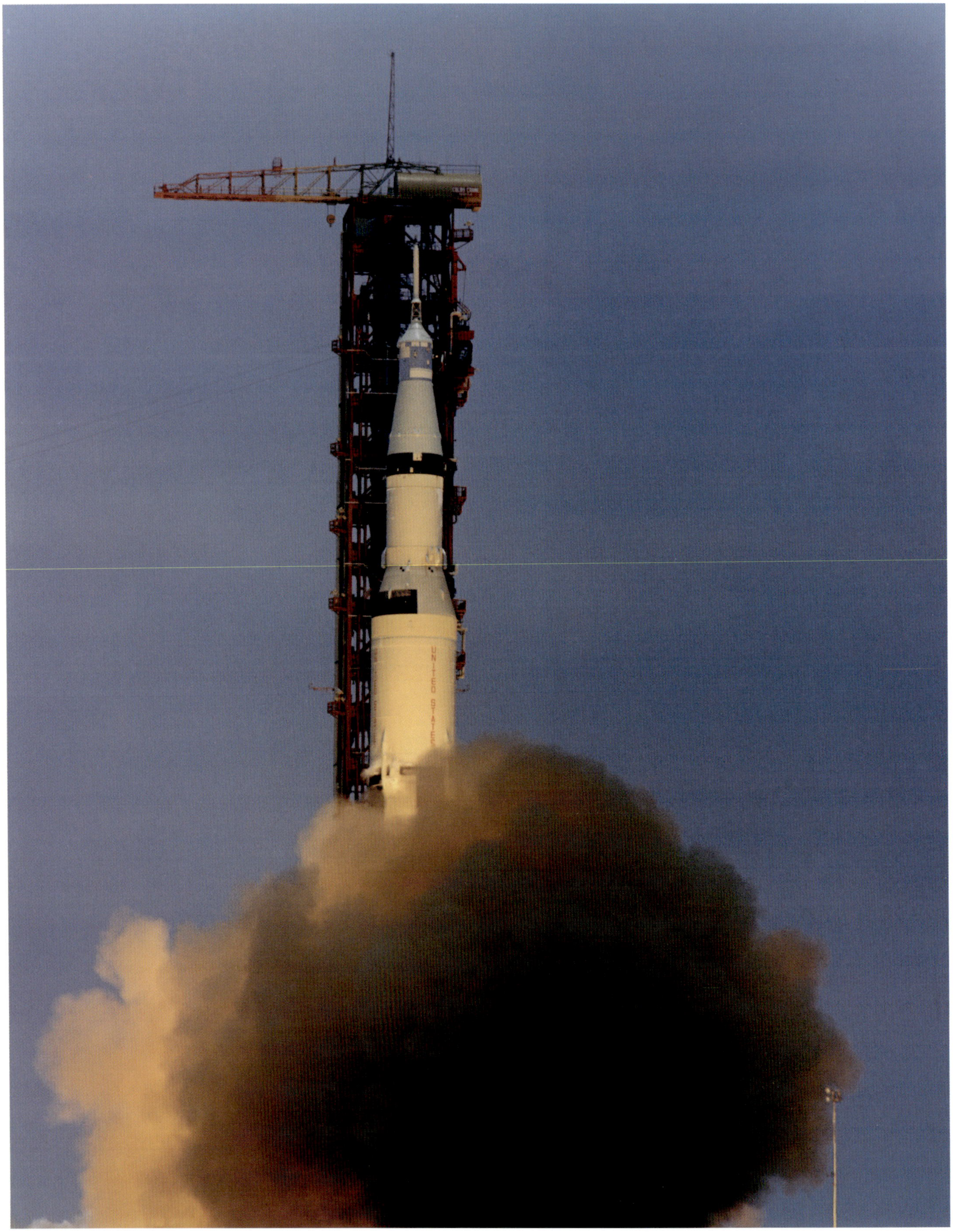

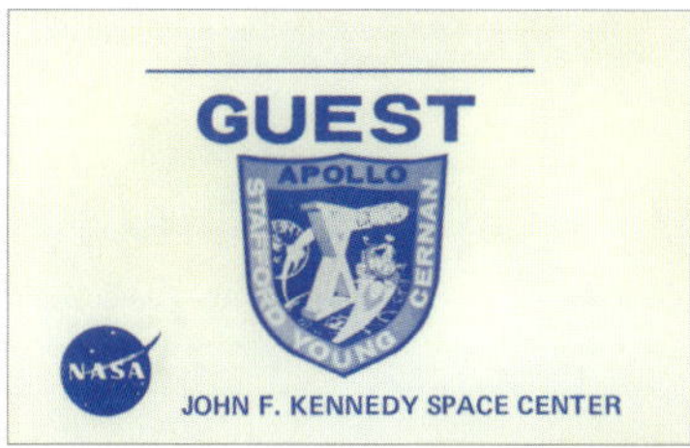

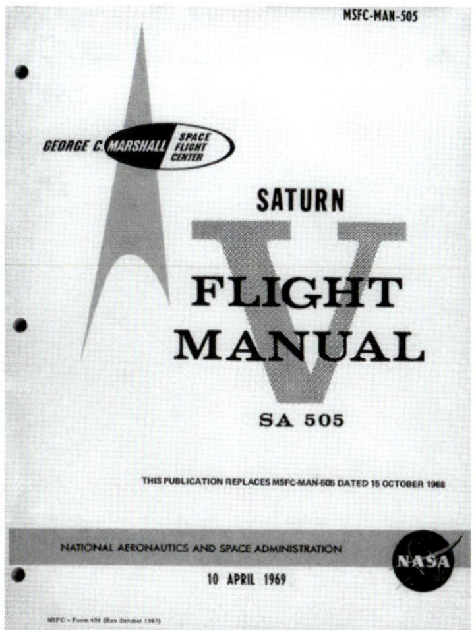
MSFC-MAN-505

GEORGE C. MARSHALL SPACE FLIGHT CENTER

SATURN V FLIGHT MANUAL

SA 505

THIS PUBLICATION REPLACES MSFC-MAN-505 DATED 15 OCTOBER 1968

NATIONAL AERONAUTICS AND SPACE ADMINISTRATION

10 APRIL 1969

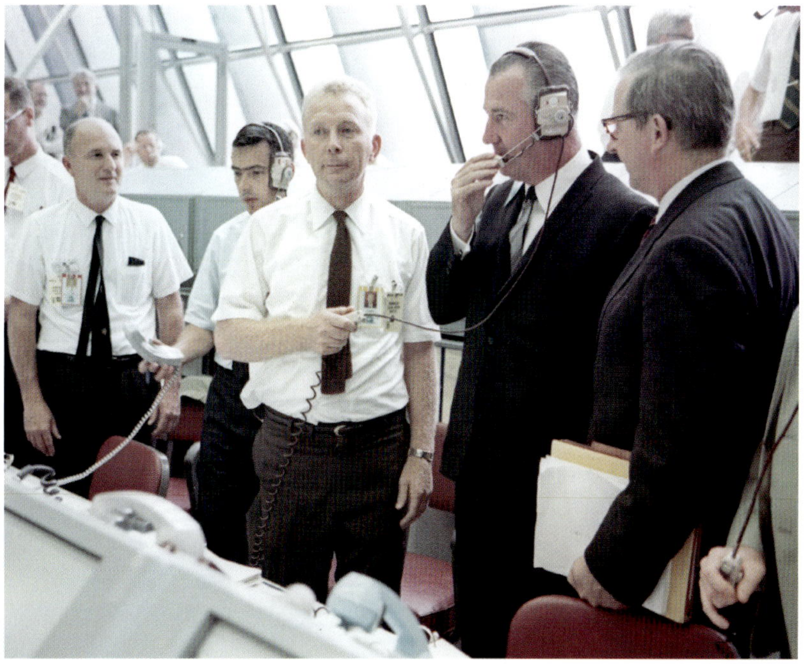

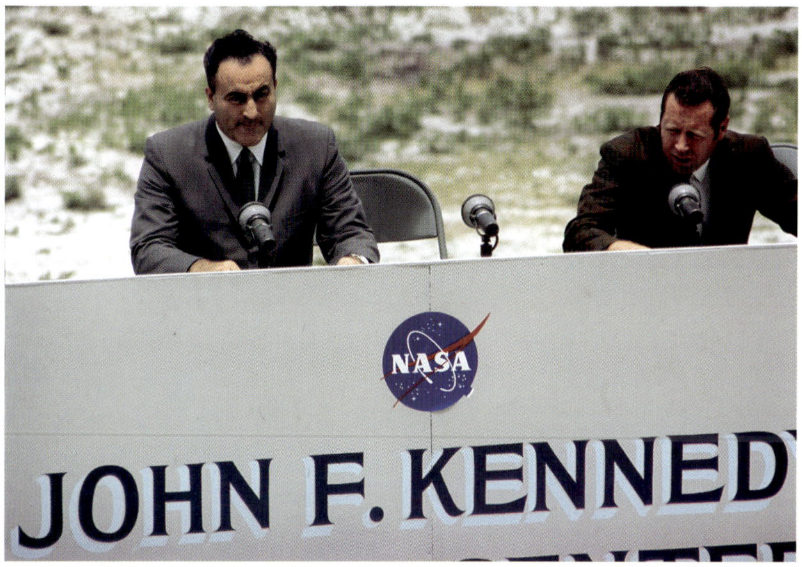

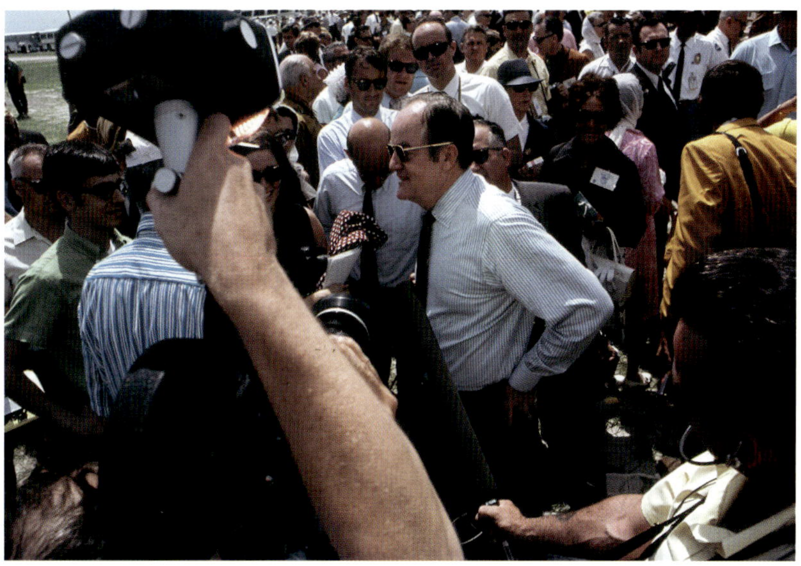

Opposite page: The Saturn V's five first-stage F-1 engines build up thrust, moments before liftoff from Pad 39B. Apollo 10 would leave Earth orbit for the moon two and a half hours later.

Top right: Following launch, Vice President Spiro Agnew congratulates the NASA/contractor team in Firing Room 3 over their headsets. *Left to right*: Bob Moser, chief test planning officer (*in glasses*); Karl Striby, test conductor; Don Phillips, chief space vehicle test supervisor; Donnelly; Agnew; and NASA Administrator Tom Paine. KSC Public Affairs Chief Jack King sits at his console behind Striby.

Center right: Petrone (*left*) takes reporters' questions in front of the Press Site grandstand after launch. At right is King, who was also the launch commentator. (photo by Tiziou News Service)

Bottom right: Former US vice president Hubert Humphrey, who had been defeated in his presidential campaign by Richard Nixon six months earlier, speaks with a film crew at the VIP launch viewing site near the VAB. A strong supporter of the space program, Humphrey had served as chairman of the NASC during the Johnson administration. (photo by Tiziou News Service)

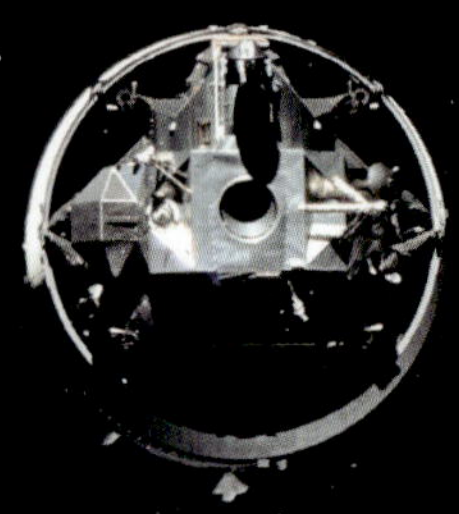

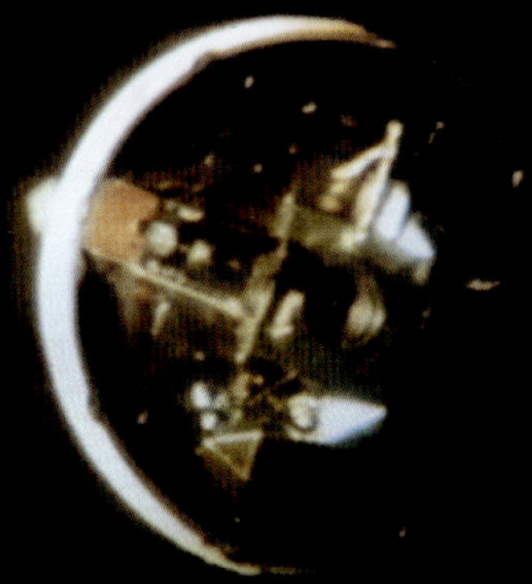

Above left: Busy with the color-TV camera, the crew took this lone 70mm photo, which looks down on LM *Snoopy*, still secured in the S-IVB. The ascent stage is exposed; the four SLA deployable panels were jettisoned following translunar injection at 3:28 p.m. (EST) on May 18. Ice crystals from the stage sparkle in the sunlight.

Above center: This view of *Snoopy* and the S-IVB stage is from the first color-TV transmission from space. The crew, aboard *Charlie Brown*, approached for docking at about 4:00 p.m. (EST) on May 18. CAPCOMs during the transmission were backup CDR Gordon Cooper and Charlie Duke.

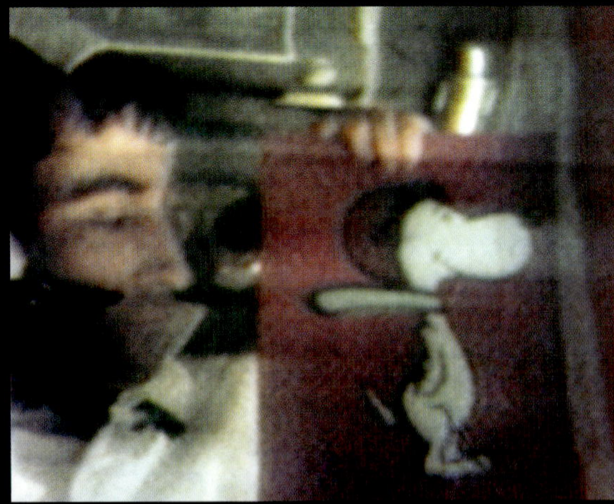

Young holds a sketch of Snoopy as a World War I flying ace during a TV transmission from the CM on May 19, 1969. Cernan remarked of Snoopy, "He's been quiet for two days. He's gonna get a chance to do a little woofin' here in the next couple days." The spacecraft were 132,000 miles from Earth and about 100,000 miles from the moon. Stafford was a passionate advocate for the color-TV transmissions as a means to better share the experience with the public. The crew made sixteen transmissions totaling almost six hours.

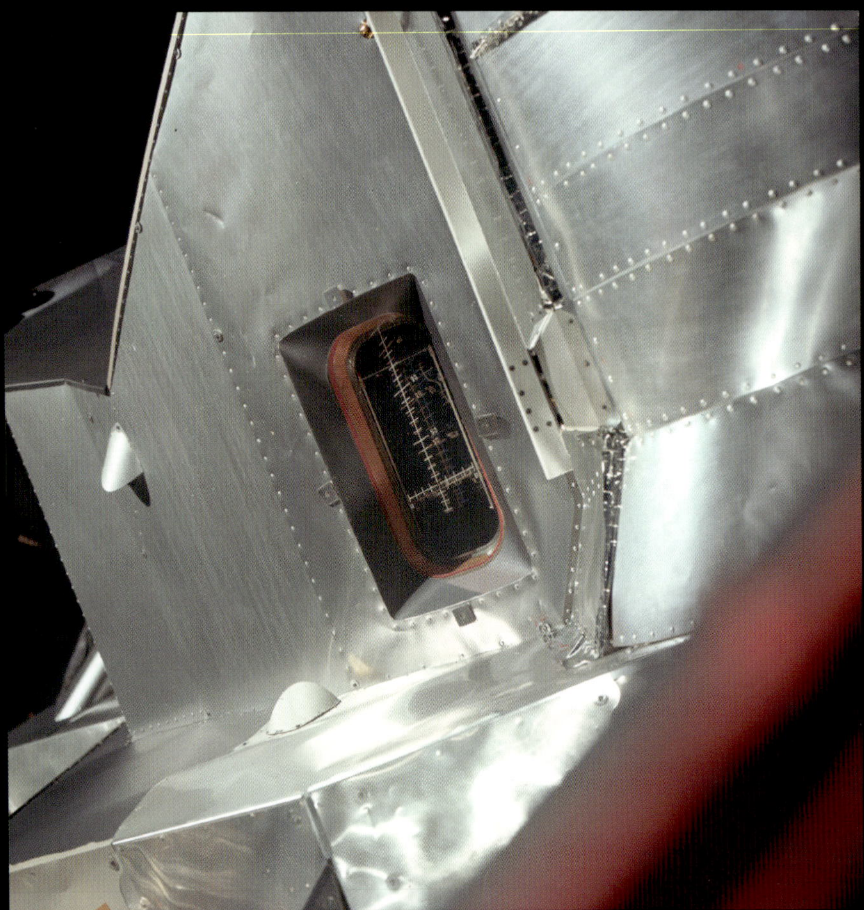

Snoopy's docking window above the CDR's station, seen from *Charlie Brown*'s right-hand rendezvous window (No. 4, Cernan's side) as the two docked spacecraft coast toward the moon on May 19. This docking window and both front LM windows had electrical heaters in case of fogging.

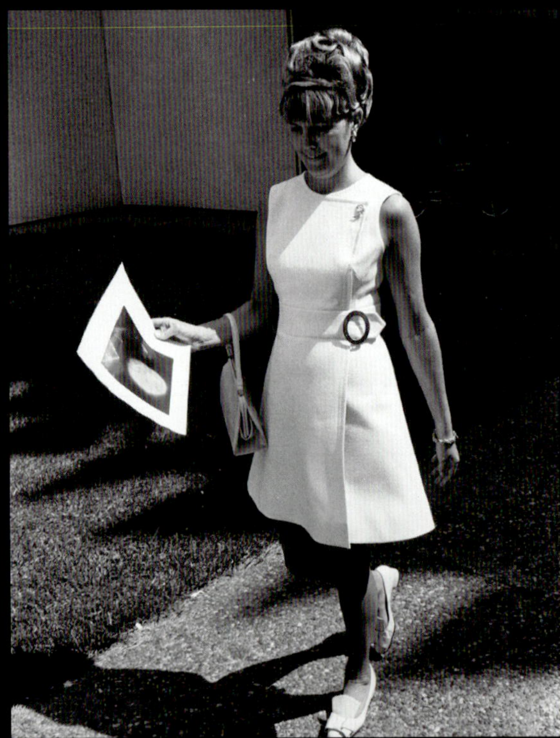

Barbara Cernan leaves Bldg. 30 at MSC after a visit to the MCC on May 19, carrying a color print of Earth from the crew's TV transmission.

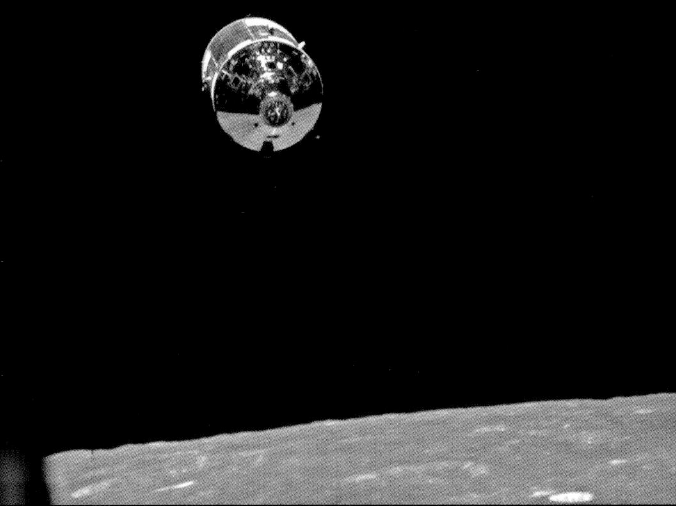

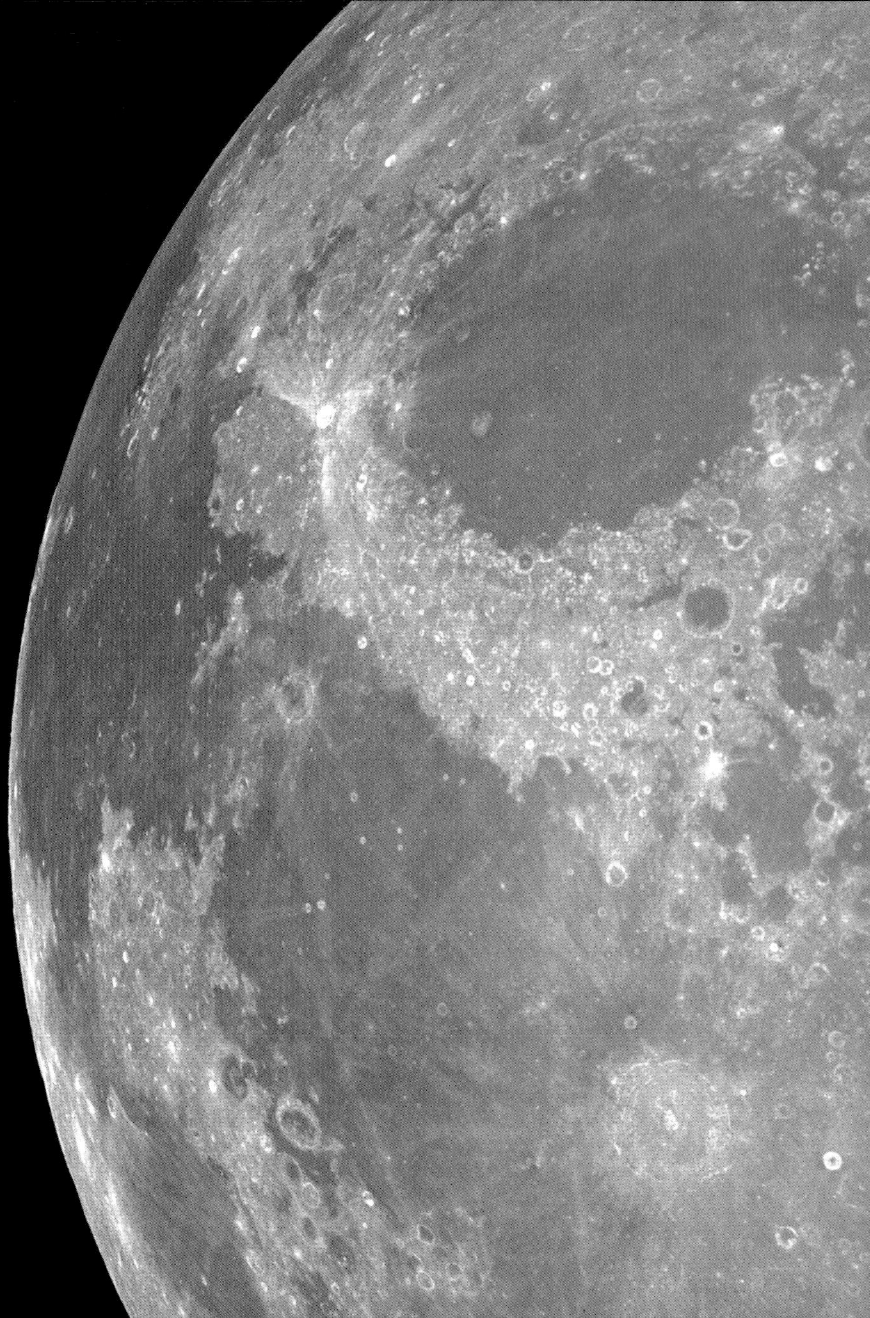

Above: *Charlie Brown* in lunar orbit, as seen from Cernan's right-forward window aboard *Snoopy* on May 22, shortly after the two spacecraft separated. They are 69 miles above the moon's far-side highlands, somewhere east of 110 degrees east longitude. *Charlie Brown* spent two and a half days in lunar orbit, with each orbit taking two hours and nine minutes. *Snoopy* was never more than 340 miles away.

Right: The Sea of Tranquility, Apollo 11's landing site, is at left center, with the Sea of Crises at top right and the Sea of Fertility below.

Below: CAPCOMs Charlie Duke, Bruce McCandless, and Joe Engle in MOCR 2 on May 22, with the flight's mascots taped to the top of their console. Duke and Engle were also part of the Apollo 10 astronaut support crew. To the left of Duke is Gold Team Flight Director Gerry Griffin; Flight Surgeon Dr. Charles Berry is in the background at far right. A chart of Apollo 10's one and a half Earth orbits after launch is at left; CSM and LM models are on the console at right; and the VIP viewing area is behind the windows. Three closed-circuit TV cameras hang from the ceiling.

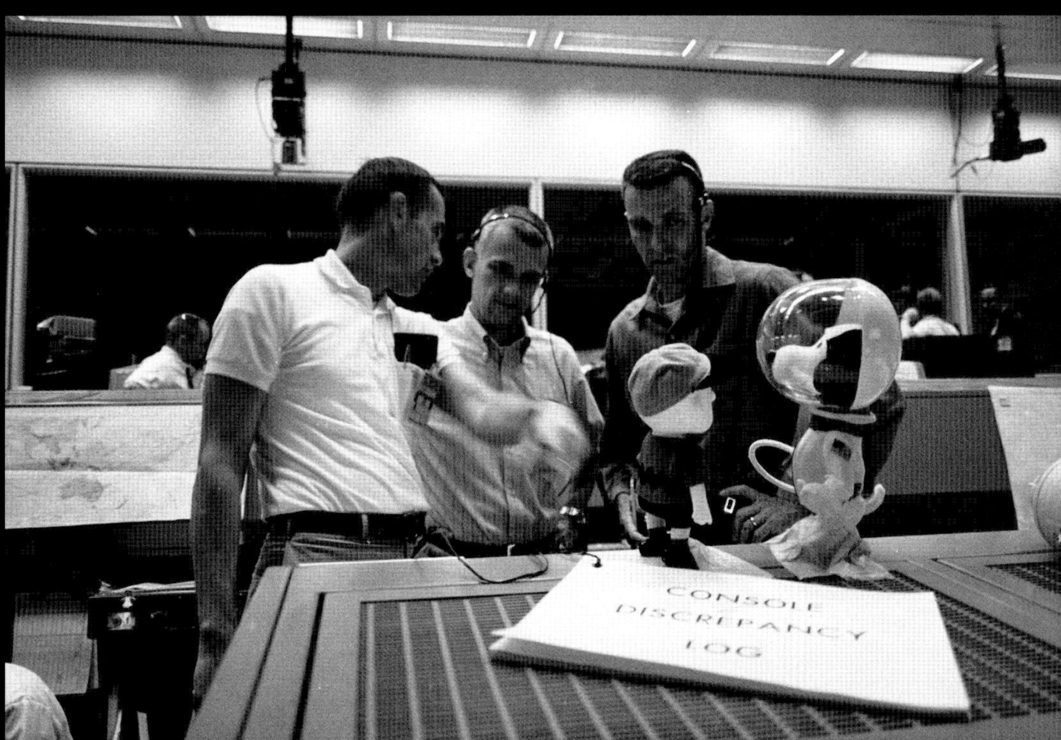

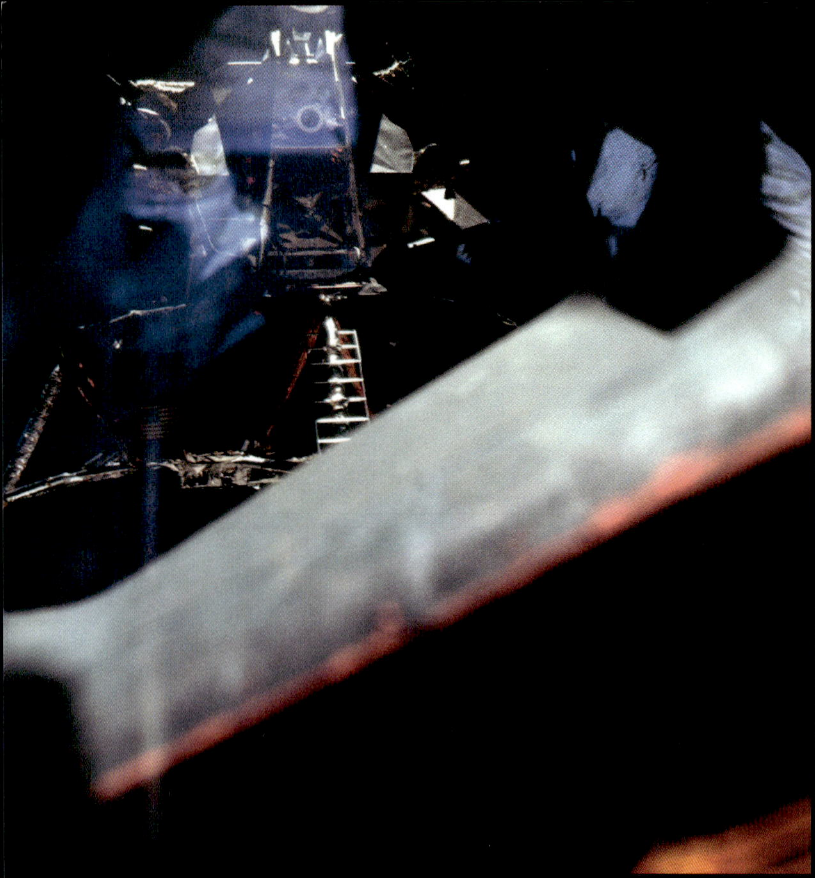

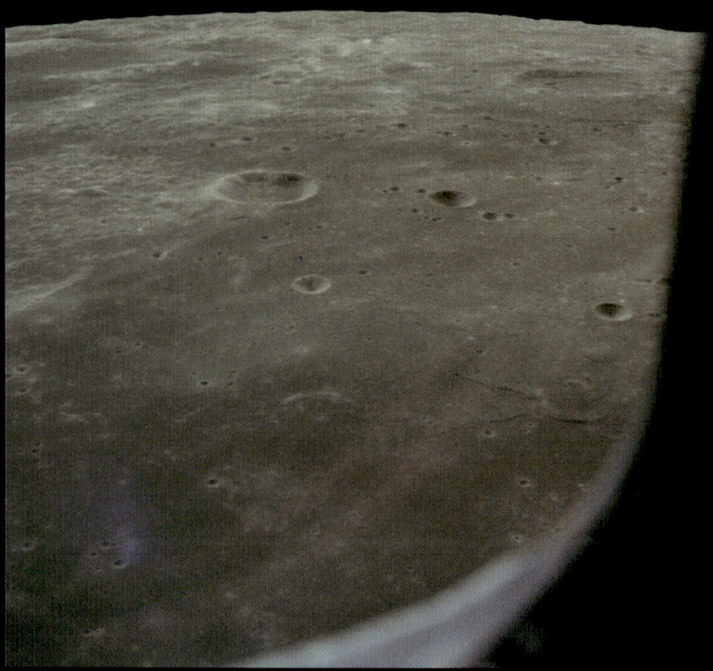

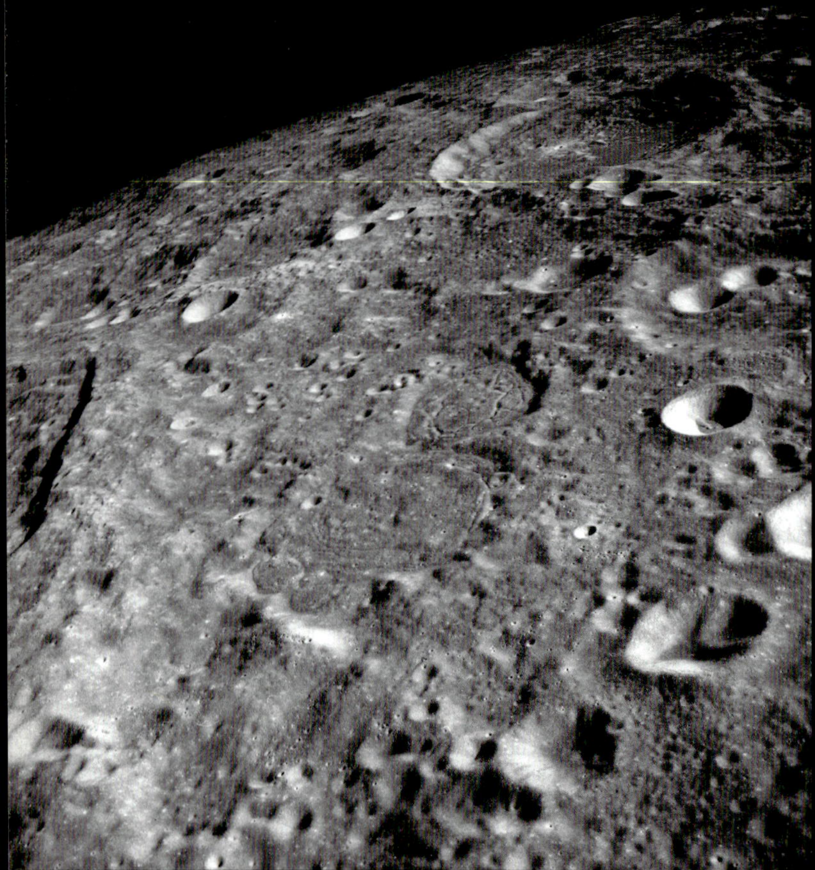

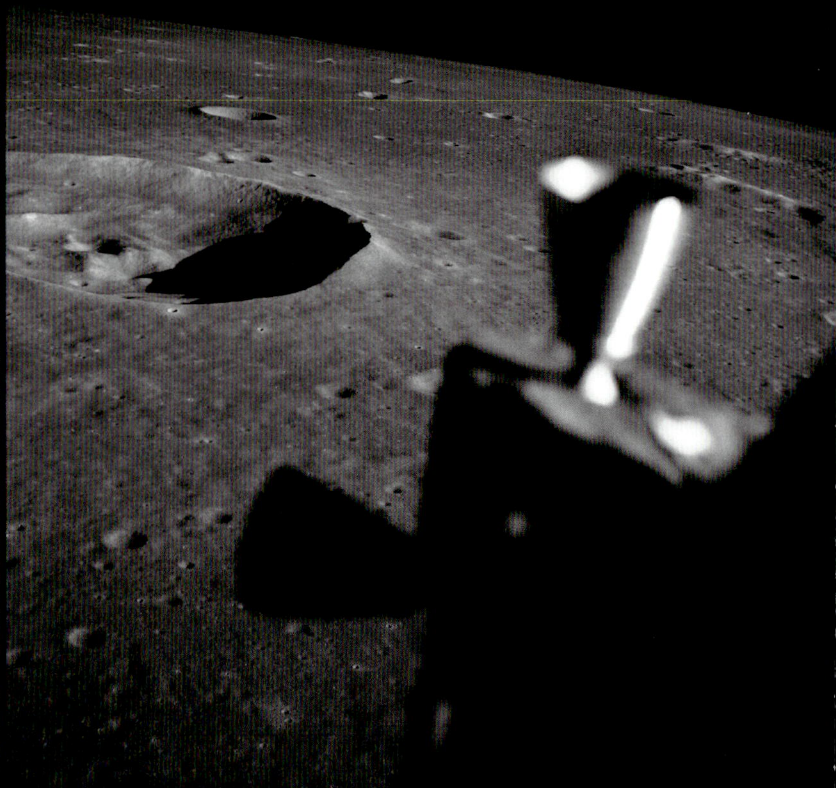

Top left: Young took this photo of *Snoopy* from one of *Charlie Brown*'s rendezvous windows, shortly after undocking on May 22.

Top right: Young's view from orbit—through a CM side window—shows crater Webb B (*center*) and three satellite craters on the eastern edge of the Sea of Fertility. This view looks east toward the Foaming Sea, and is about five hundred miles southeast of Apollo 11's landing site.

Bottom left: Looking south from the CM's orbit, this view on the far side shows photographic Target of Opportunity No. 29, craters produced by volcanic uprisings. At left is the west rim of Chaplygin, an impact crater. The large impact crater at top right, Marconi, is sixty-five miles in diameter.

Bottom right: Crater Maskelyne in the southern Sea of Tranquility dominates this view looking west from Cernan's window aboard *Snoopy*, with Apollo 11's landing site about 125 miles to the west-southwest. Two RCS nozzles are in the foreground. *Snoopy* and *Charlie Brown* flew independently for eight and a half hours, with *Snoopy*'s ascent and descent stages separating almost four hours into its flight.

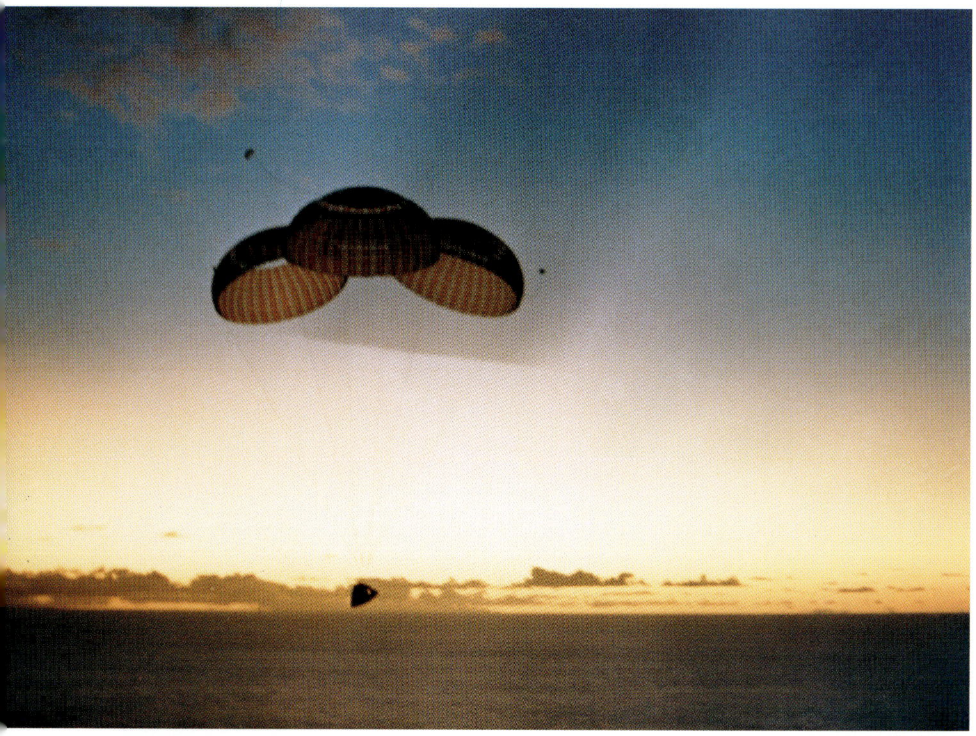

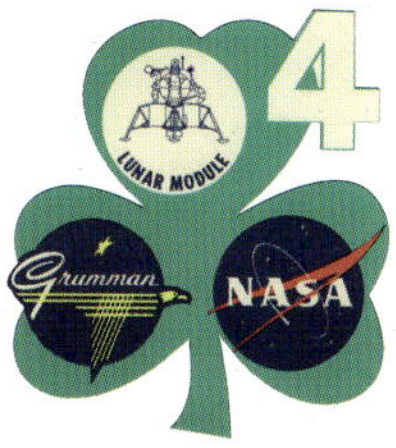

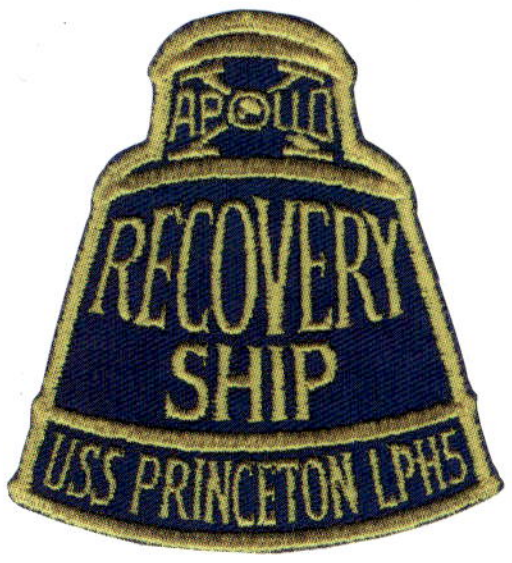

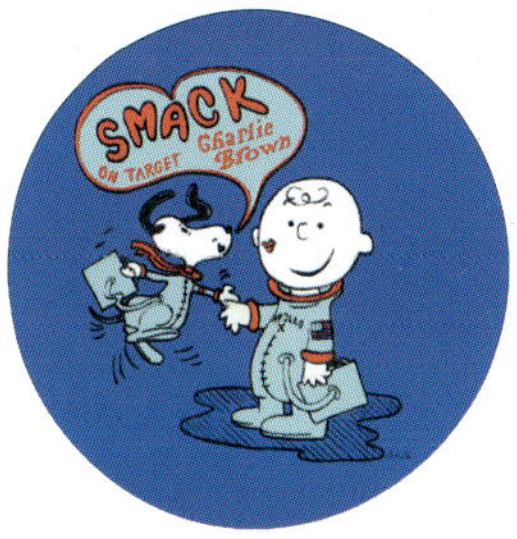

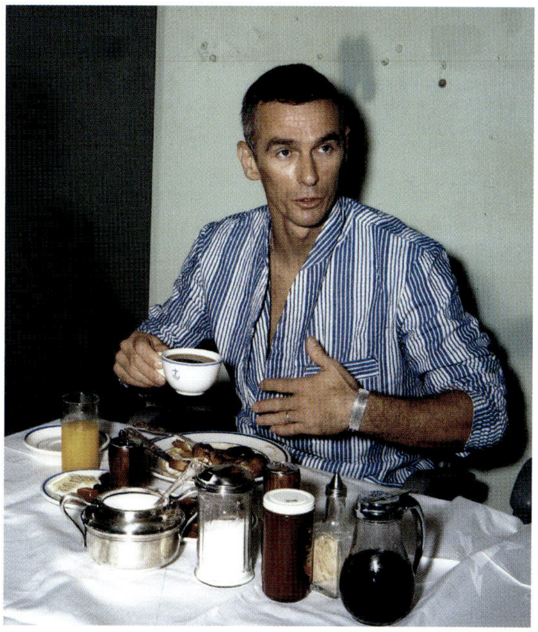

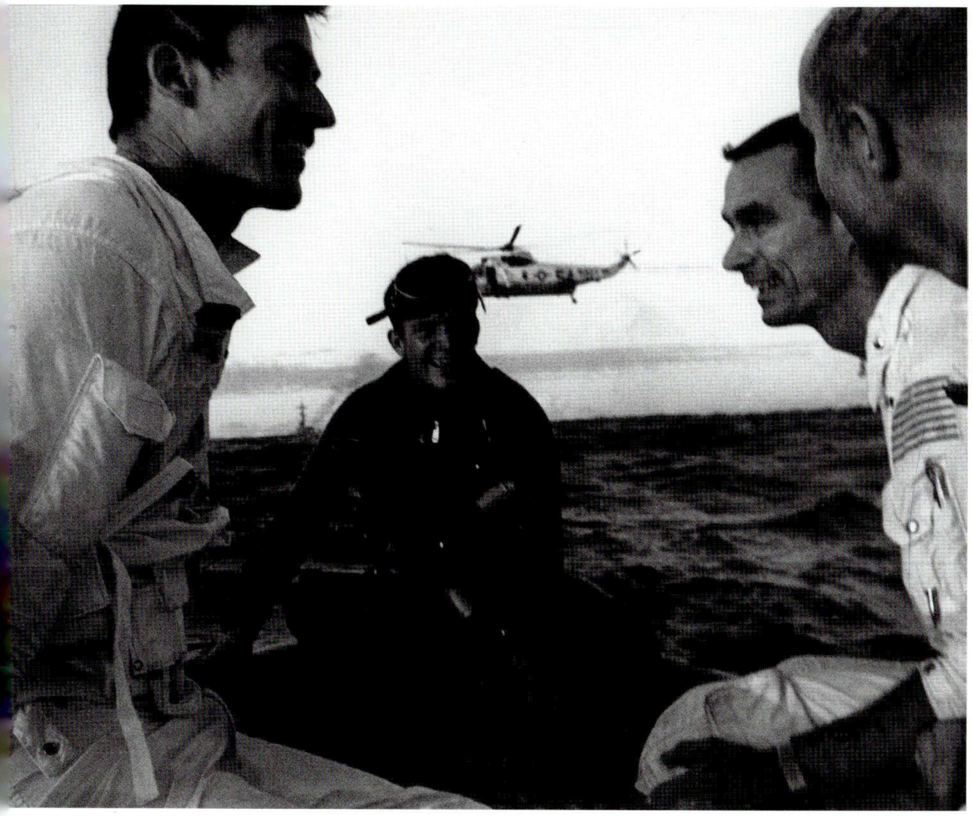

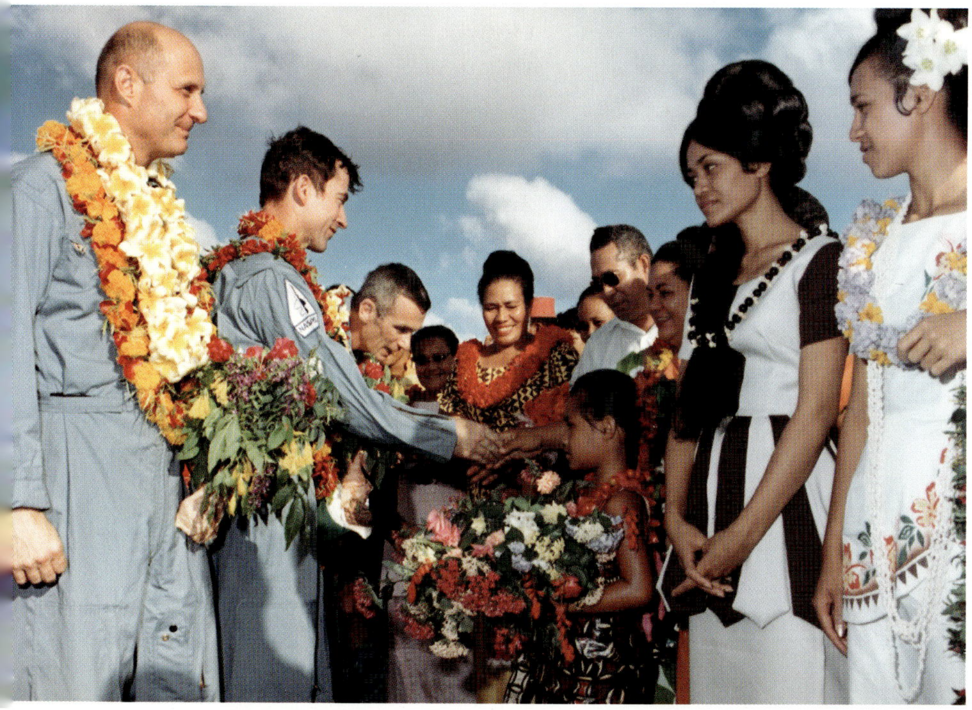

Above: Cernan enjoys a breakfast of steak and eggs aboard USS *Princeton*. The Apollo 10 crew was the first to shave in space—using safety razors and brushless shaving cream, which retained the whiskers—so they could clean up on the way home.

Top left: Sunrise greets Apollo 10, descending on its main parachutes toward a splashdown in the South Pacific Ocean at 5:52 a.m. local time (HST) on May 26, about four hundred miles east of American Samoa and within three miles of the prime recovery ship, the amphibious assault carrier *Princeton*. The CM is tilted to mitigate water impact.

Center left: Young; navy diver Wes Chesser of Underwater Demolition Team (UDT) 11, lieutenant junior grade; Cernan; and Stafford wait in a recovery raft, with the *Princeton* on the horizon to the left of Chesser. They were aboard the ship thirty-nine minutes after splashdown.

Bottom left: The astronauts, after medical tests and a meal aboard *Princeton*, were flown by helicopter to Pago Pago International Airport, where they were greeted by island dignitaries and beauty queens. It was then a thirteen-hour flight back to Ellington Field.

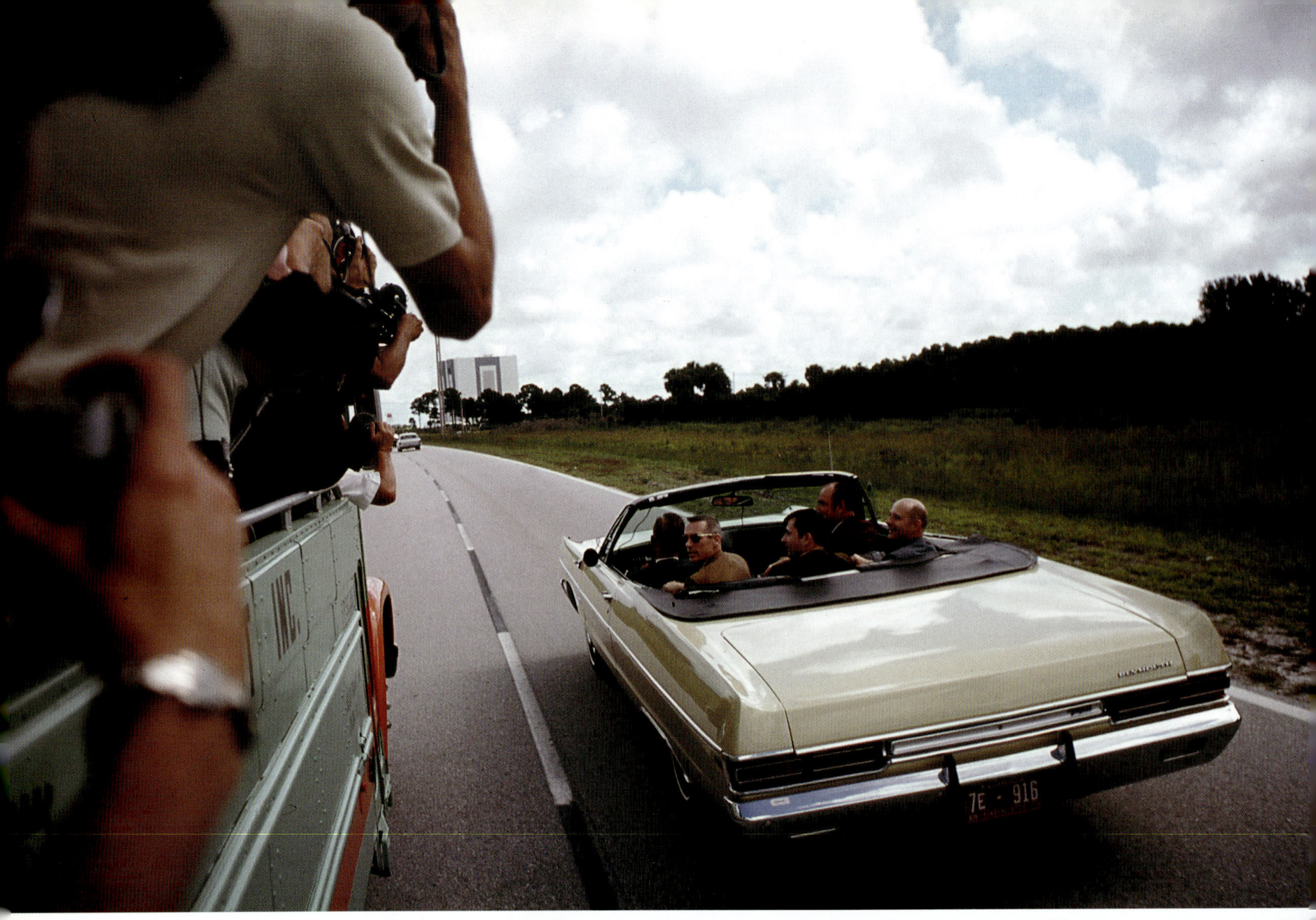
INC.
7E - 916

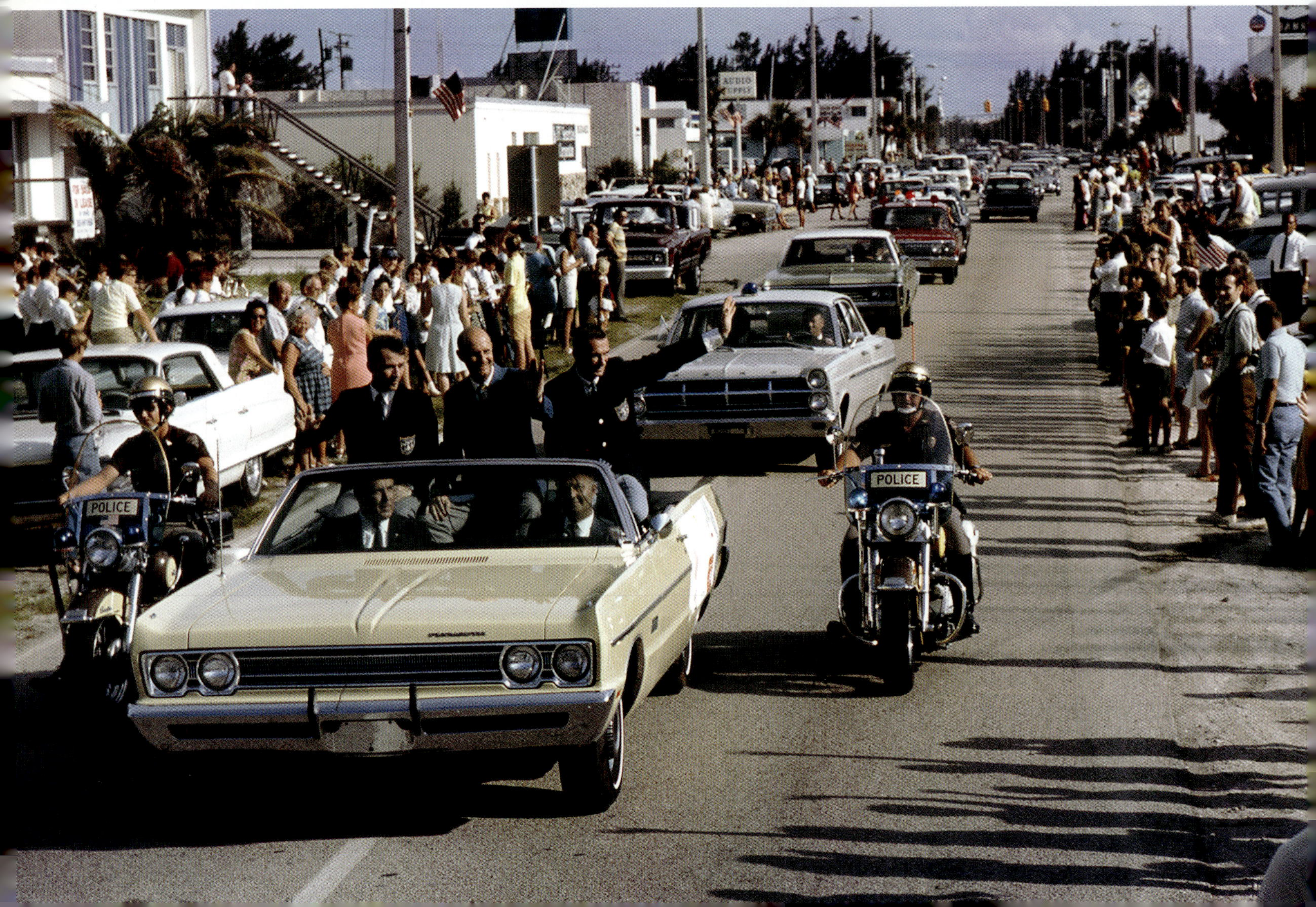
POLICE
POLICE

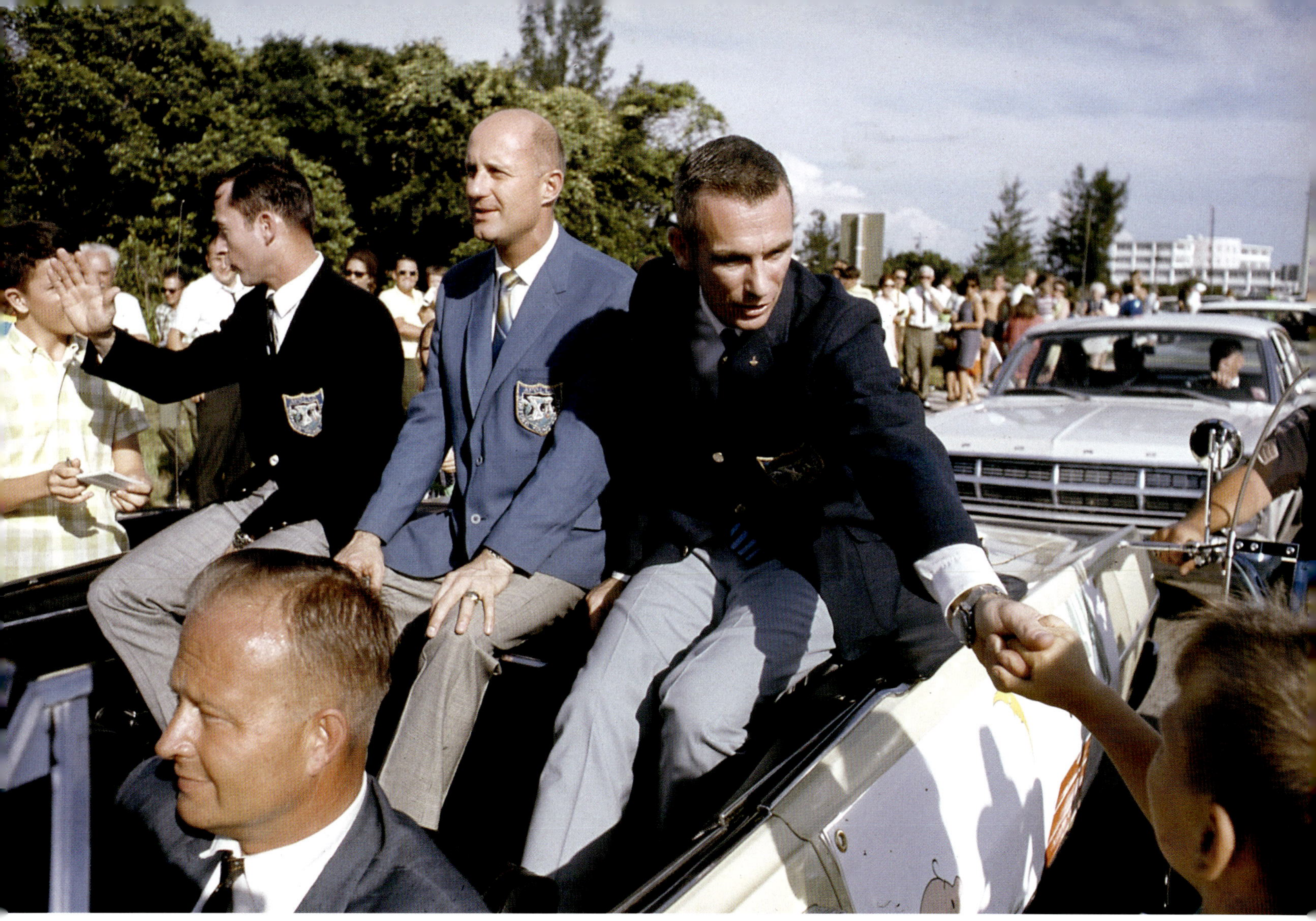

Opposite page, top: News photographers jockey for position to photograph the crew as they head northbound on the Kennedy Parkway on the morning of June 11, 1969, toward the VAB and a reunion with the KSC workforce. Some ten thousand workers turned out for the ceremony in the VAB. (photo by Tiziou News Service)

Opposite page, bottom: Late that afternoon, a motorcade with the crewmen heads north on Florida State Road A1A in Cocoa Beach, near North 1st Street, with KSC Security Chief Charles Buckley in the front passenger seat. It was the first motorcade there for returning astronauts since Gemini 3 in 1965. John Glenn, Scott Carpenter, and Gordon Cooper had received the first three Spacecoast motorcades during Mercury; the sixth and final one was for the Skylab 2 astronauts in 1973. (photo by Tiziou News Service)

Above: A youngster hopes for an autograph from Young (*left*), as Cernan, wearing his gold astronaut lapel pin, reaches out to a boy along A1A. The next day, NASA set July 16 as Apollo 11's launch date. (photo by Tiziou News Service)

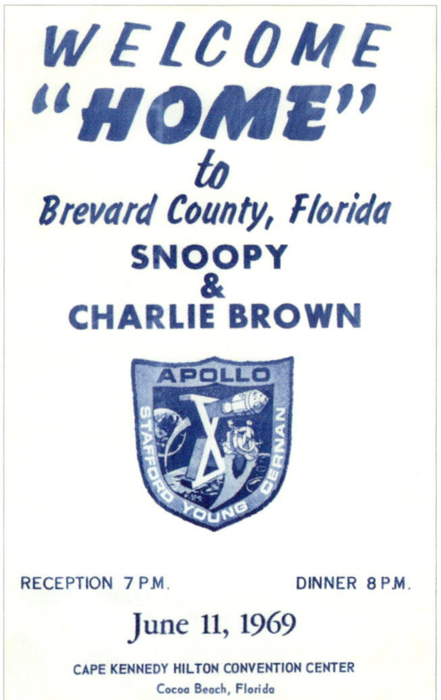

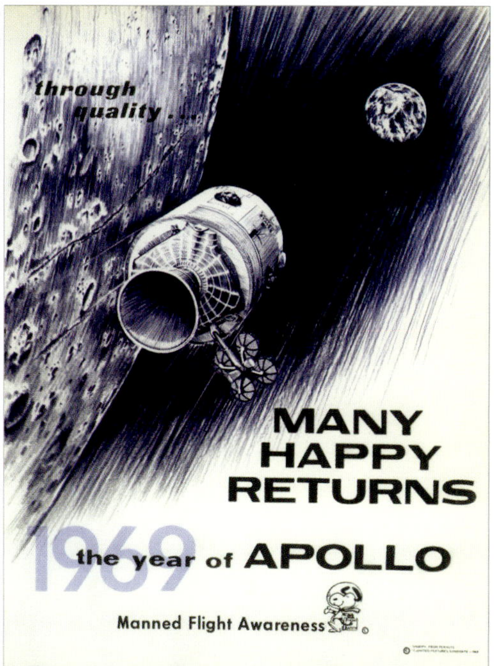

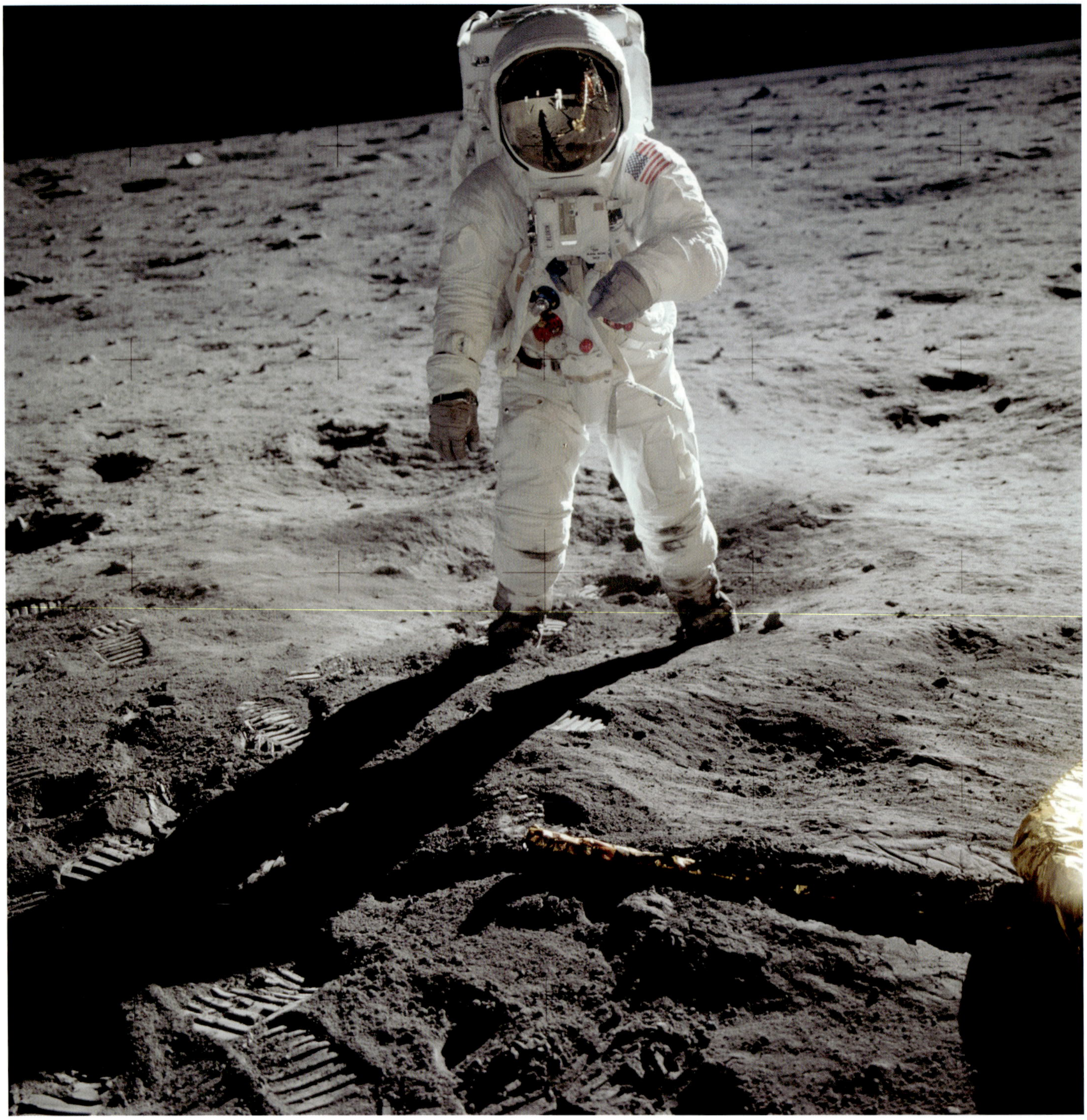

Buzz Aldrin stands near *Eagle*'s right footpad—with its sixty-eight-inch lunar surface sensing probe extended—in NASA's iconic image, No. AS11-40-5903. This photo, taken early in the EVA, looks west-southwest. Aldrin may be consulting the checklist on the wrist cover of his left glove. Neil Armstrong and the LM are reflected in his EVA visor, and modern photo analysis has shown that the tiny white dot near the top of his visor is very likely a reflection of Earth. The red knob near Aldrin's navel, called the red apple, would start a thirty-minute emergency oxygen flow when pulled. Armstrong's framing cut off the white VHF communication antenna, which extended from the top of Aldrin's emergency oxygen purge supply unit behind his helmet and is usually not evident in photographs. A black background is commonly added to the top of the photo to provide better visual balance.

CHAPTER 6

Apollo 11

July 16–24, 1969

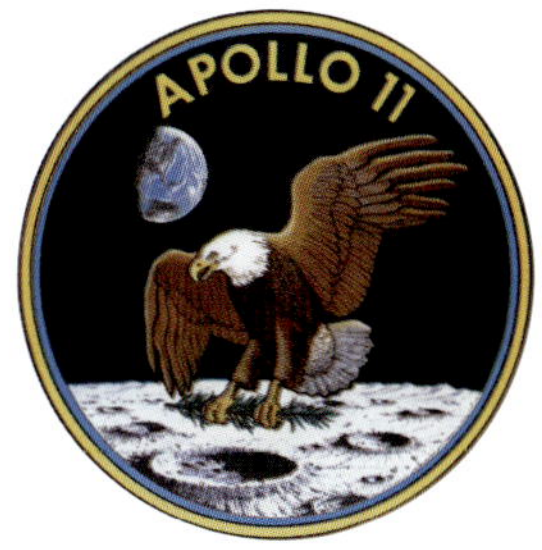

The Apollo 11 mission represented several major accomplishments. It was the historic first visit by humans to the surface of another heavenly body; the fulfillment of President Kennedy's bold 1961 commitment that those humans would be Americans and would succeed within eight years; a US victory in the Space Race with the Soviet Union; and the gaining of scientific knowledge that the experiments and 47.5 pounds of lunar samples would bring.

Less appreciated at the time were the important advances in science, engineering, and data processing that the rush to the moon produced. For the NASA industry team responsible, the flight's success was the expected result of a long series of well-planned incremental steps, which had begun with Alan Shepard's fifteen-minute suborbital trip in 1961.

It was only the second time NASA had assigned an all-veteran crew to a flight; Apollo 10 had been the first. CDR Neil Armstrong, a survivor of Gemini VIII's near-death experience and a Lunar Landing Training Vehicle (LLTV) crash, was joined by LMP Buzz Aldrin Jr. and CMP Michael Collins. Each had a Gemini flight to his credit.

Just three days before launch, the Soviets announced that an unmanned probe, Luna 15, was heading toward the moon. Although Moscow maintained silence on its mission, this was the third Russian attempt to land on the surface and return a lunar sample.

Apollo 11 got underway on a beautiful morning on July 16, 1969, from Pad 39A. It launched right on time at 9:32 a.m. (EDT), before a crowd estimated to have approached a million spectators—the largest ever to witness a rocket launch. Dignitaries included Vice President Spiro Agnew and Army Chief of Staff Gen. William Westmoreland, commander of US forces in Vietnam until March 1968. Jan Armstrong was the only crewman's wife to attend. Postlaunch traffic was backed up for ten miles.

After the standard Earth-orbit checkout, the Saturn's S-IVB was restarted to send Apollo 11 toward the moon. As the astronauts neared the midpoint of the outbound trip, Luna 15 went into lunar orbit on July 17.

On July 20, LM *Eagle* separated from CM *Columbia*, and the LM descent engine fired to put the craft into a lower orbit. The engine fired again to drop *Eagle* toward the surface. The LM landed at 4:17 p.m. (EDT) in the Sea of Tranquility, after Armstrong had steered around a boulder field. He reported, "Houston, *Tranquility Base* here. The *Eagle* has landed."

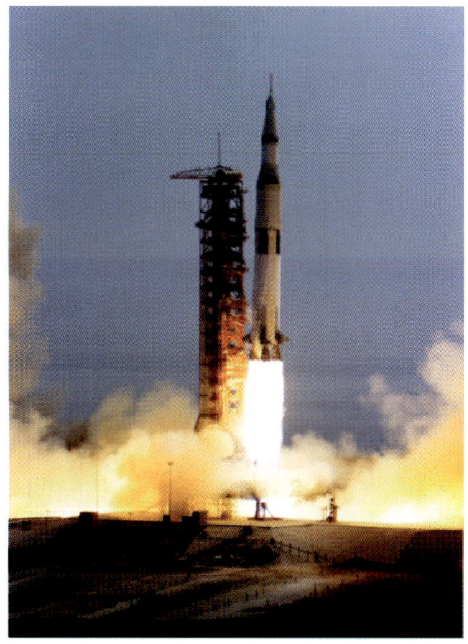

Launch: 9:32 a.m. (EDT), July 16, 1969

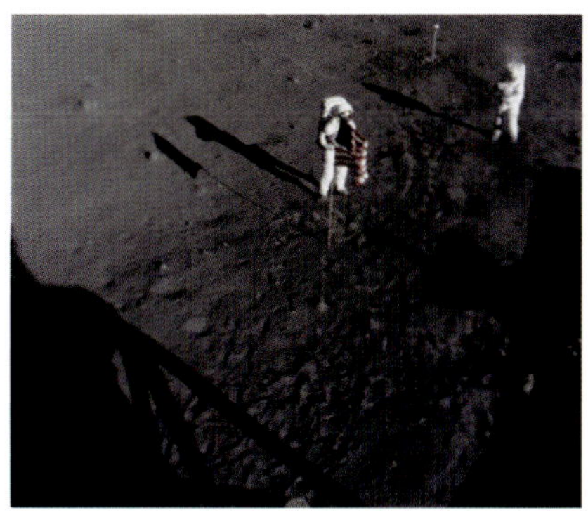

Lunar landing: 4:17 p.m. (EDT), July 20.
Ascent: 1:54 p.m. (EDT), July 21

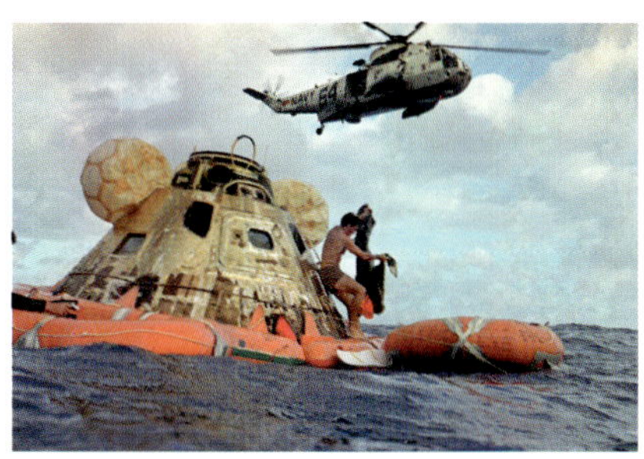

Splashdown: 12:50 p.m. (EDT), July 24

Armstrong stepped onto the surface at 10:56 p.m., stating, "That's one small step for [a] man, one giant leap for mankind." He collected a contingency lunar sample. Aldrin joined him nineteen minutes later. They unveiled a plaque on the strut behind the ladder and read the inscription: "Here men from the planet Earth first set foot upon the moon July 1969, A.D. We came in peace for all mankind." They put up an American flag. President Richard Nixon called with congratulations.

The astronauts then deployed the Solar Wind Composition Experiment (SWCE), and Aldrin set up the two experiments comprising the Early Apollo Surface Experiment Package (EASEP): a Laser Ranging Retro Reflector (LRRR) and a Passive Seismic Experiment (PSE) package. The seismometer began operating almost immediately; it was sensitive enough to detect the men's movements on the surface. The astronauts also took panoramic photos and collected lunar rocks and soil, each venturing about three hundred feet from *Eagle*. They took two core tube soil samples and packed them into sample boxes, along with the other samples and foil collector from the solar wind experiment.

Aldrin returned to the LM after an hour and forty-one minutes outside. Armstrong followed about twelve minutes later, after transferring the sample boxes up to Aldrin with a cable. The two spent the next seven hours resting and checking out systems. During that time, the USSR's unmanned Luna 15 attempted to land about seven hundred miles northeast of the Apollo 11 landing site, in the Sea of Crises, but it crashed.

Eagle's ascent from the moon, its redocking with *Columbia*, and the crew's return voyage were without incident. In the early morning of July 24, just more than eight days after launch, *Columbia* splashed down in the North Pacific Ocean, about 230 miles south of Johnston Island. Divers from the aircraft carrier USS *Hornet* arrived quickly and tossed three biological isolation garments (BIGs) into the spacecraft, to guard against any spaceborne pathogens. An hour after splashdown, the astronauts were inside the Mobile Quarantine Facility (MQF) aboard *Hornet*, where they participated in a brief ceremony with President Nixon. Later that day, Apollo Program Director Sam Phillips announced from Houston that the Apollo 12 crew would launch on November 14, and that it would try to land near an unmanned US probe, Surveyor 3.

Below left: Michael Collins had become part of the Apollo 11 crew by happenstance: his flight assignments had been changed by the Apollo 1 fire, the decision to send Apollo 8 around the moon, and a neck surgery he had in 1968.

Below right: Aldrin, with an LM model by Precise Models, was a Gemini veteran who, like Collins, had EVA experience. Both photos are from a January 10, 1969, news conference at MSC, the day after the crew was officially named.

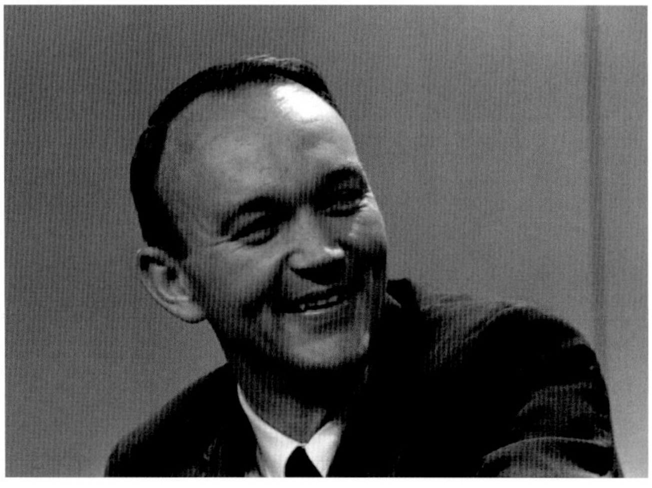

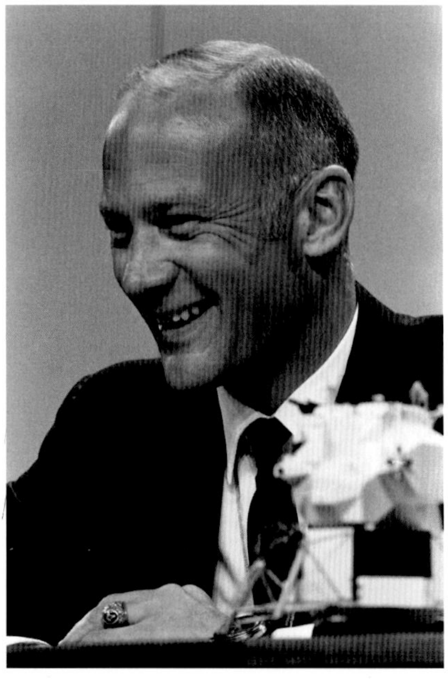

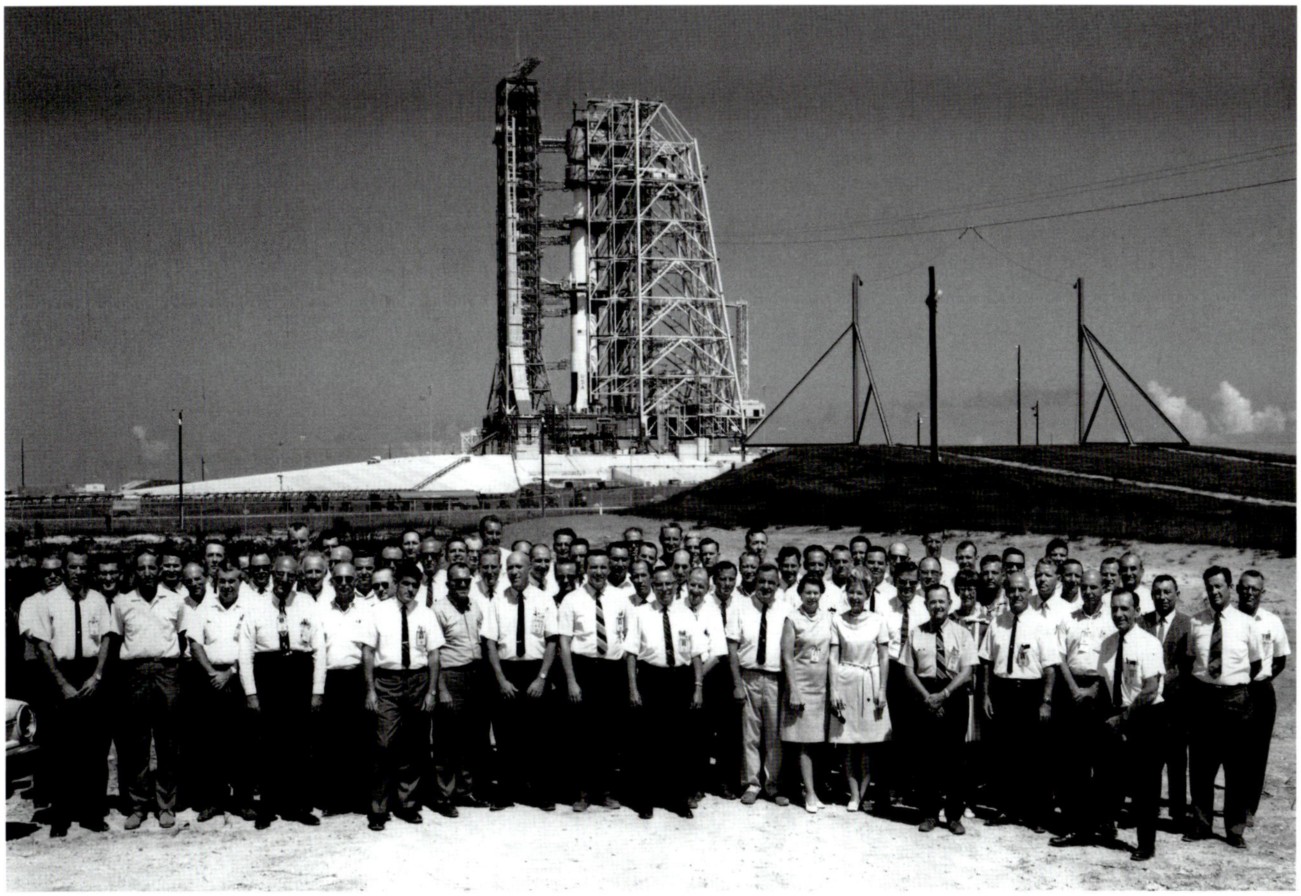

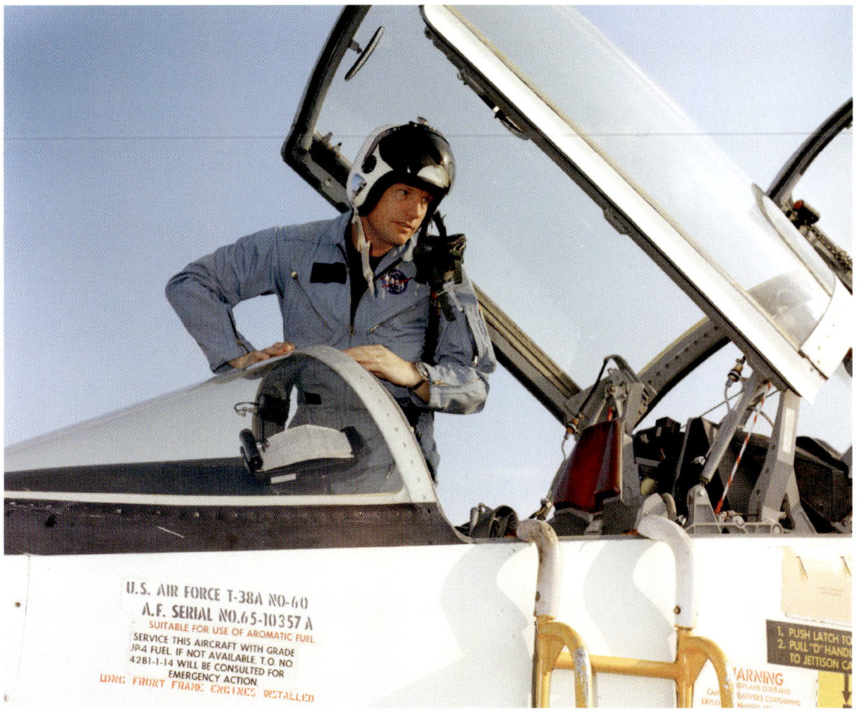

Above: KSC employees assemble at LC-39A on June 26, 1969. Apollo 11 is surrounded by the MSS. The emergency slide-wires run from the white room (320-foot) level, heading out of frame to the right.

Right: Armstrong arrives at Patrick AFB on June 17, 1969, from Ellington AFB. Deke Slayton was in the second seat on the flight from Houston. Armstrong had flown the Lunar Landing Training Vehicle (LLTV) twice the day before at Ellington, bringing it to three hundred feet before practicing powered descents to the moon. All three crewmen spent time in the KSC simulators on June 17. That evening, after the day's flight readiness review, Apollo Program Director Sam Phillips announced that the launch was on for July 16.

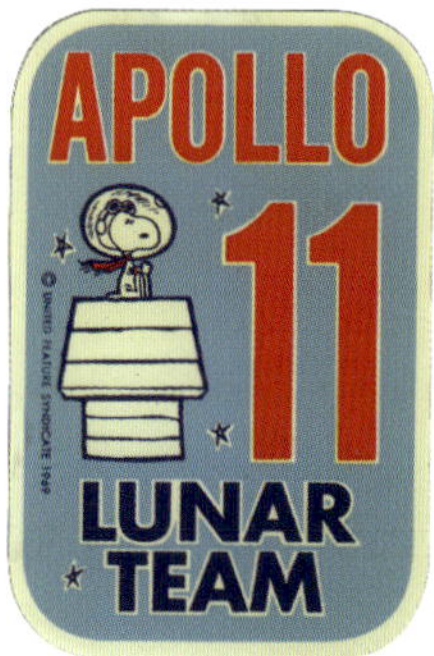

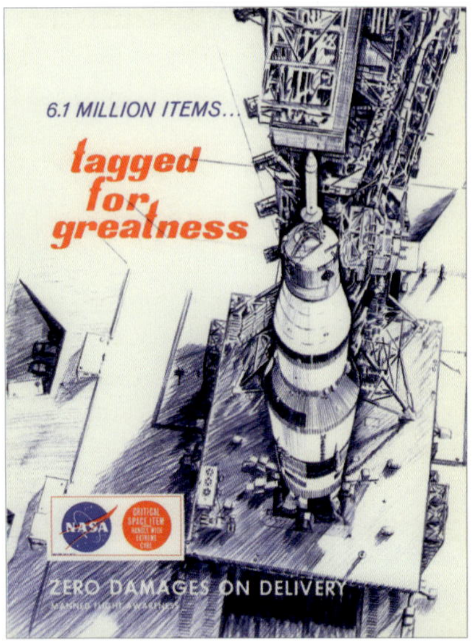

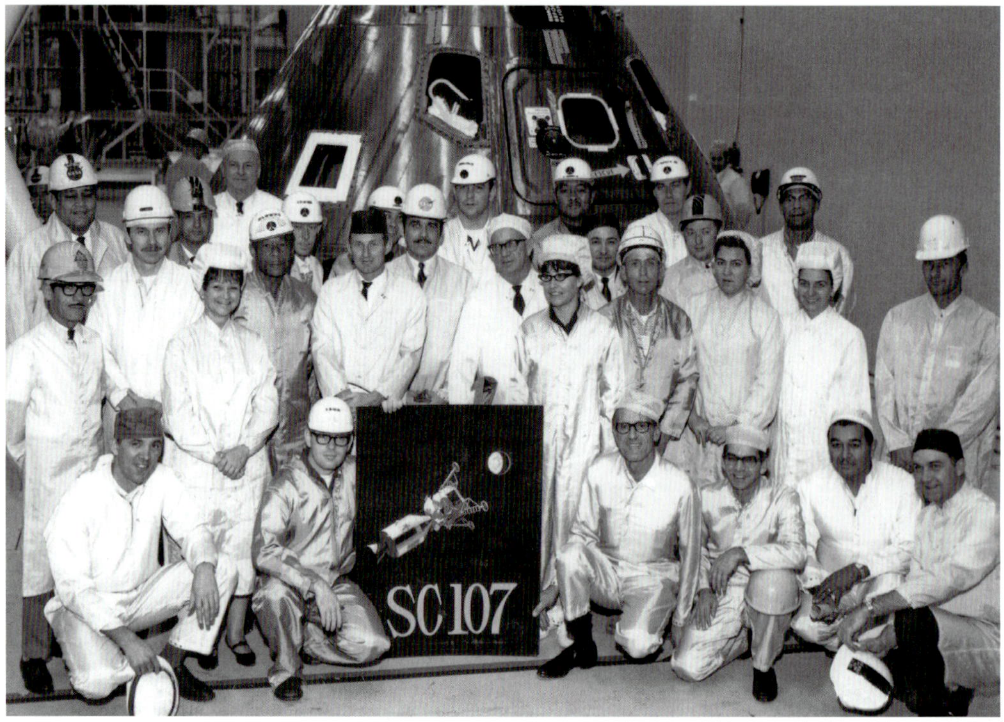

Above: NAR and NASA personnel pose with spacecraft 107, to be dubbed *Columbia* by its crew, at NAR's Downey, California, plant on January 17, 1969.

Right: Apollo 11's S-IVB third stage is raised by a crane in the VAB's transfer aisle on March 5, 1969. It would be stacked with the first and second stages in high bay 1. The single J-2 engine had to burn twice: first to help get Apollo 11 into Earth orbit, and then to push it out toward the moon, two hours and forty-four minutes after launch.

Below: Apollo 11's CM and SM in the MSOB on January 24, 1969. The spacecraft had arrived the day before, aboard the *Super Guppy* aircraft, at Cape Kennedy's Skid Strip.

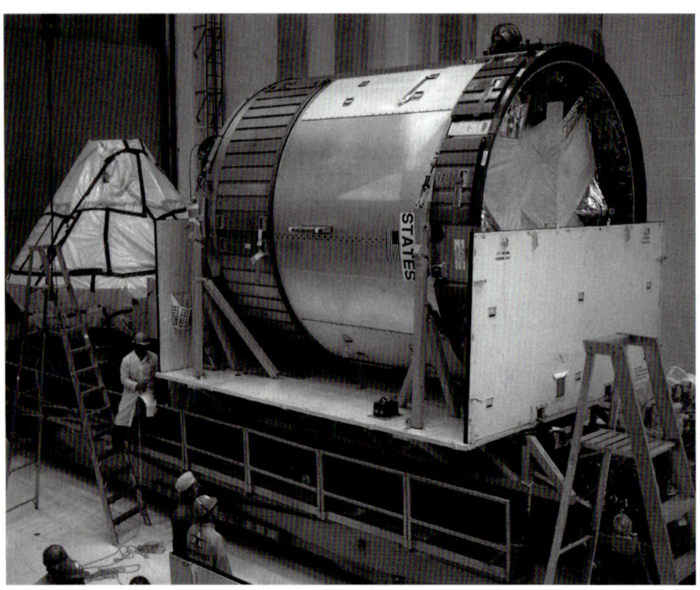

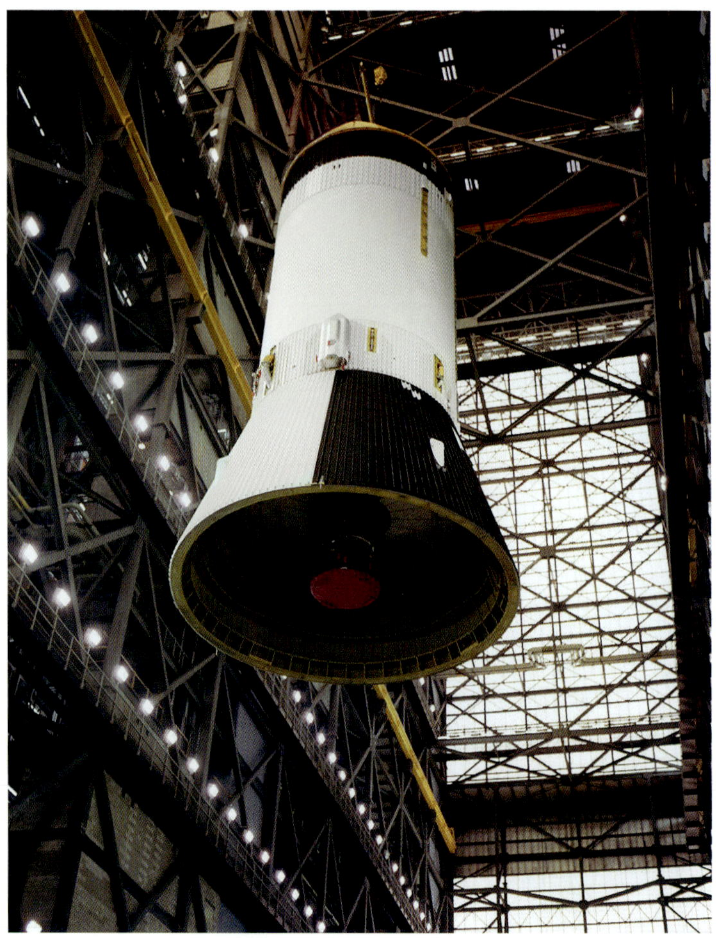

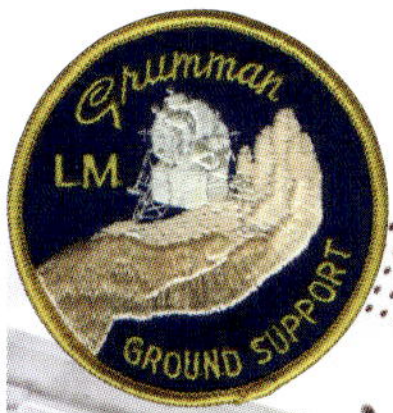

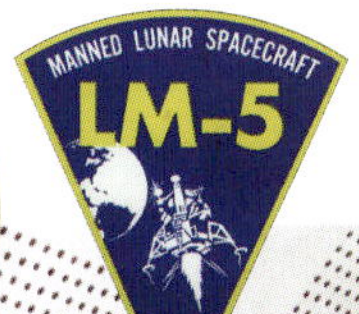

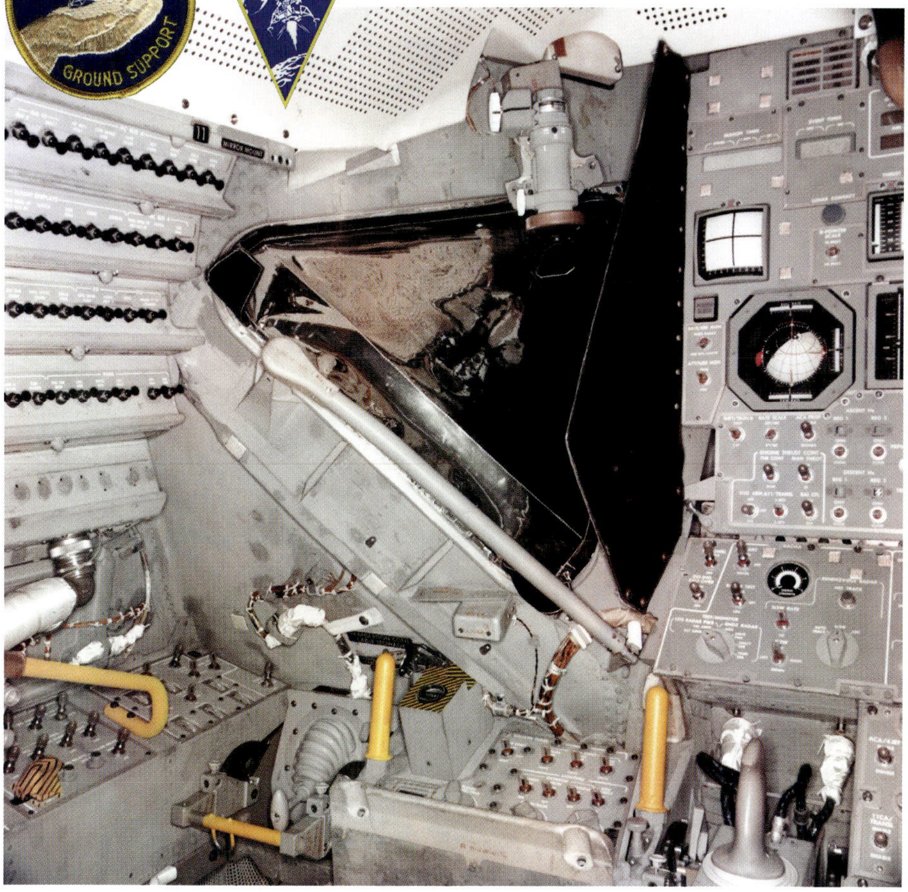

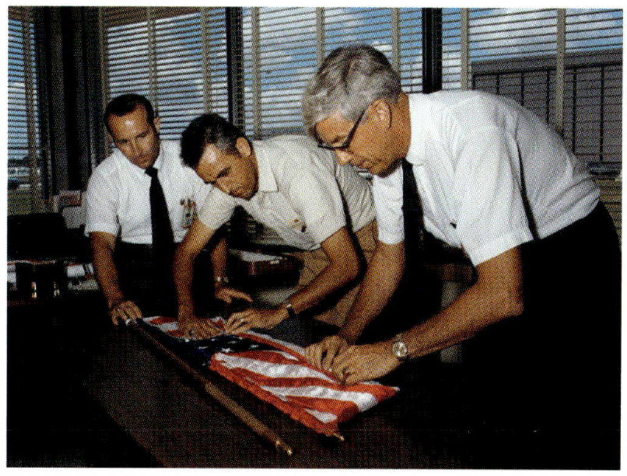

Above: MSC personnel test-fold the US flag that Armstrong and Aldrin would deploy on the moon. The bottom half of the gold-anodized pole is at left. The flag, made by the Annin Flag Company of Roseland, New Jersey, is believed to have been one of three purchased for use on the mission. MSC secretaries bought them at separate Sears department stores in Houston.

Bottom left: A Grumman propulsion technician (*wearing a red hat*) examines the LM's descent engine nozzle as it sits in a test stand at the MSOB at KSC on April 11, 1969. The thin lunar surface sensing probes are hinged back against both visible footpads at left and right.

Bottom right: LM-5's descent stage, under construction at Grumman Aerospace in Bethpage, New York, in late 1968. The gold Mylar–wrapped tank at left is for water; the large white tank at right contained supercritical helium, used to pressurize the LM's fuel lines and tanks.

Above: The CDR's flight station inside LM-5, before a vacuum chamber run at KSC on March 18, 1969. The black glareshield reduced reflections on the front window. The grey handle at lower right is the attitude hand controller; handholds are yellow. At the window's top right is the crew optical alignment sight for docking. The start button for the ascent engine is surrounded by yellow-and-black striping at lower center.

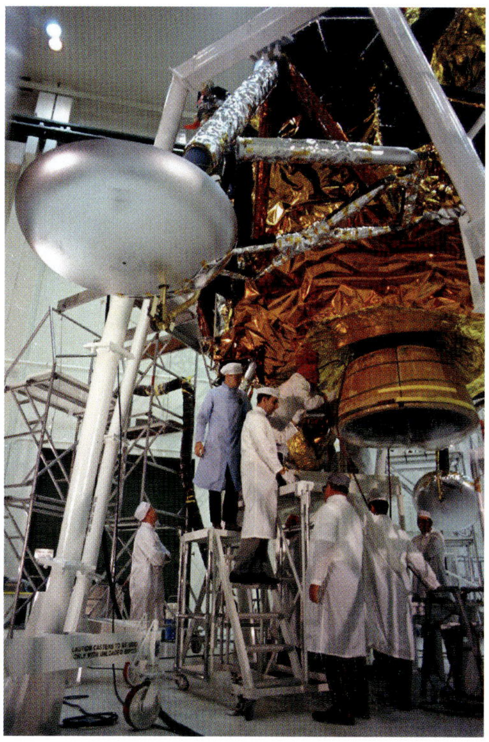

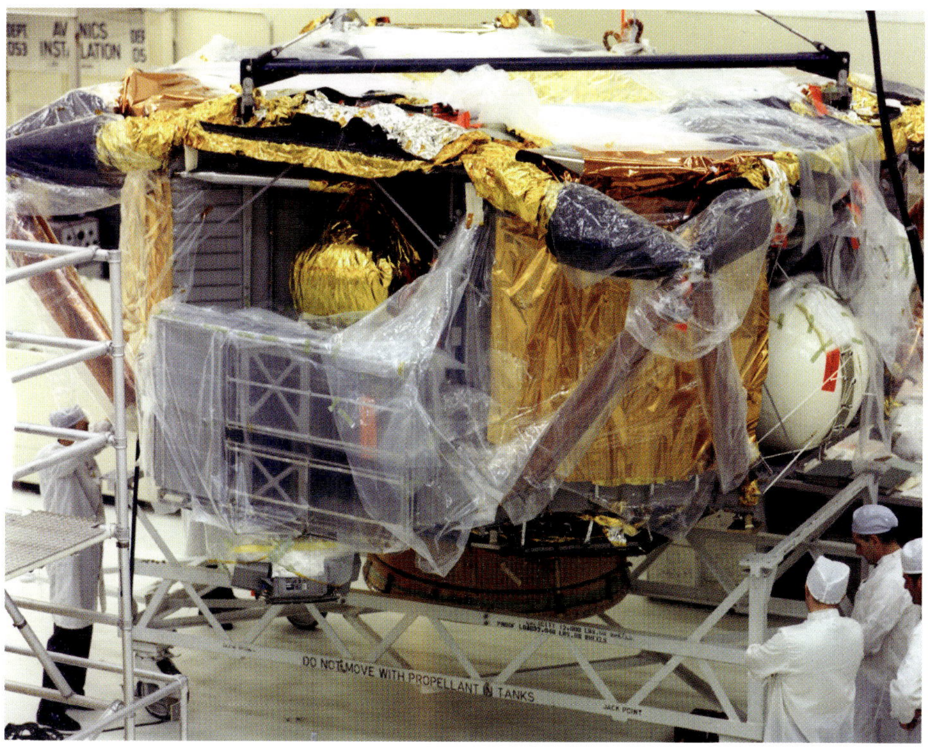

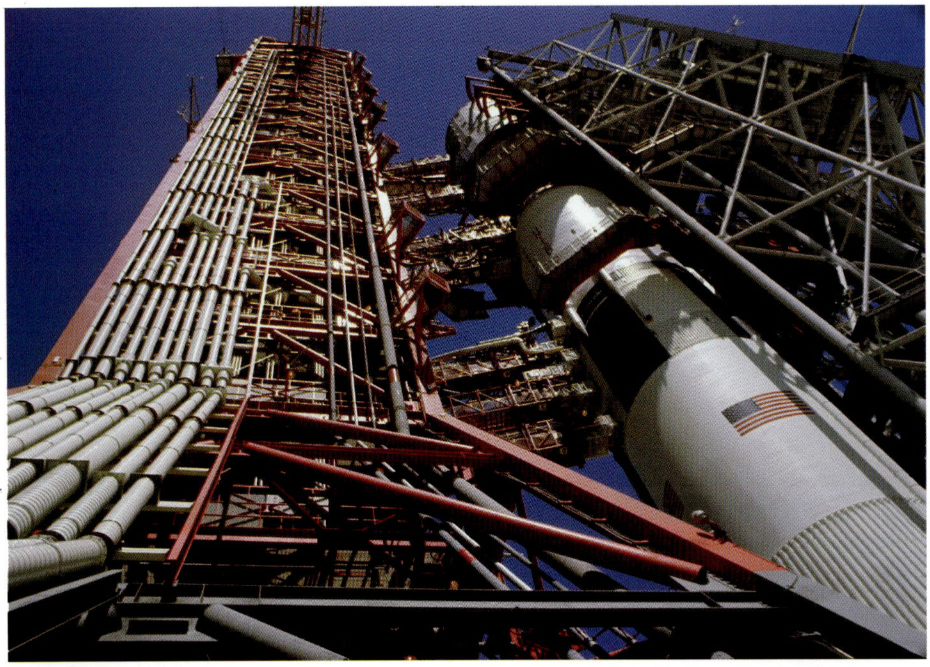

Left: With its LUT to the left, Apollo 11 is enclosed in the three-level service compartment near the top of the MSS (*right*) at Pad 39A. The series of white pipes on the tower carried water, used to dampen powerful sound waves at launch and protect against the heat. (photo by Tiziou News Service)

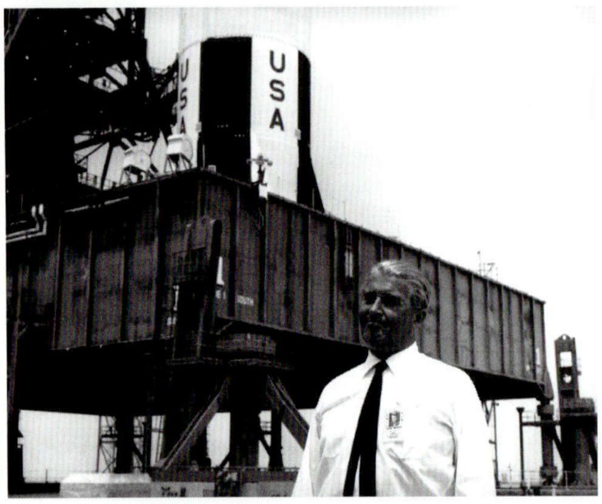

Center left: MSFC Director Wernher von Braun with ML-1 and Apollo 11's Saturn V resting on the hardstand, after the crawler-transporter has departed on May 20, 1969. The Saturn V was von Braun's crowning achievement.

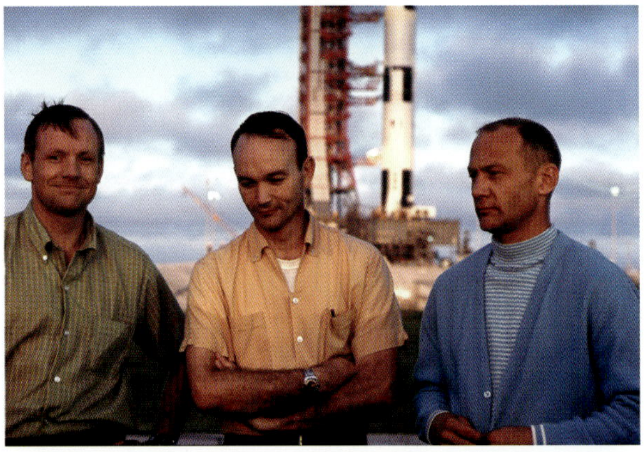

Center right: Armstrong, Collins, and Aldrin speak with reporters on the afternoon of May 20. Pad 39A and the Atlantic Ocean are in the background. The previous day, at the dedication of the Goddard Library at Clark College in Worcester, Massachusetts, Aldrin had heard Senator Edward Kennedy call for redirecting space funding to social needs at the end of the Apollo program. Vietnam War spending had just peaked. (photo by Tiziou News Service)

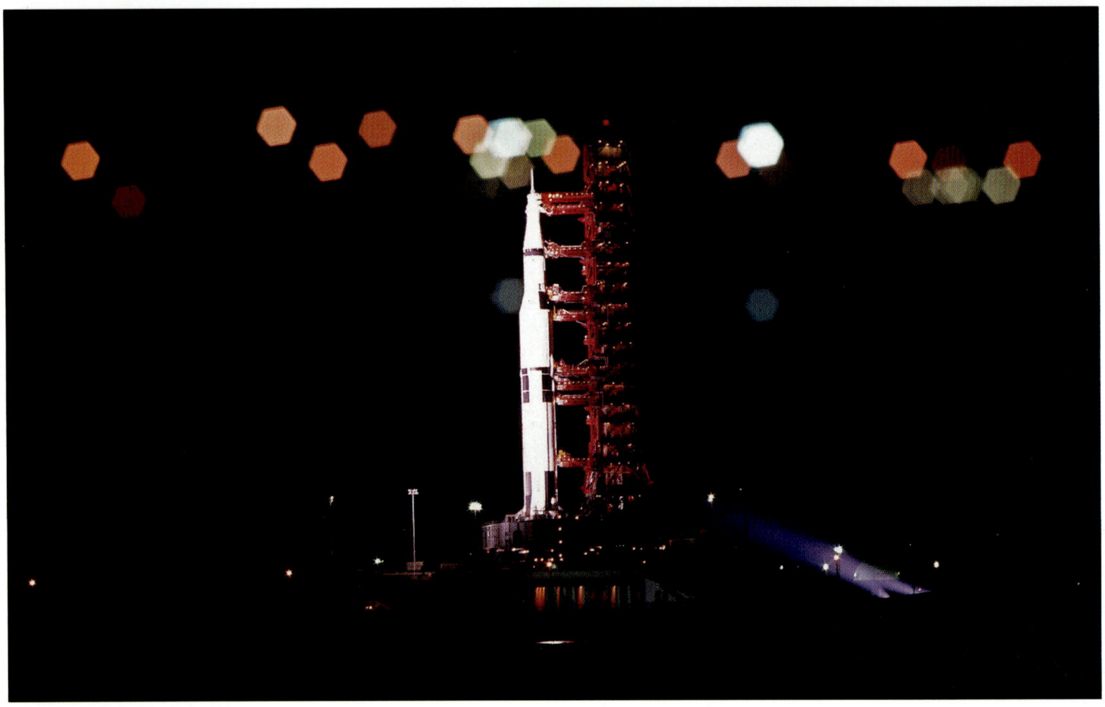

Bottom left: Apollo 11 and its Saturn V booster on Pad 39A on the night of May 20. Apollo 10, meanwhile, was only hours away from entering lunar orbit. (photo by Tiziou News Service)

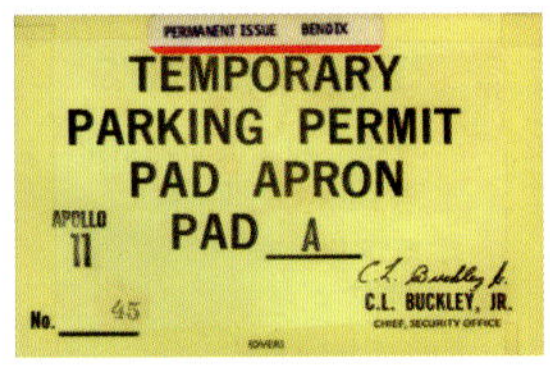

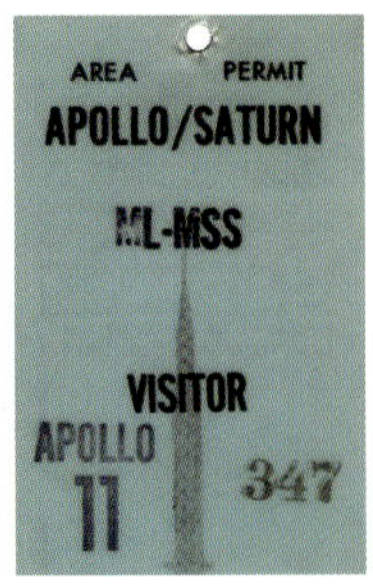

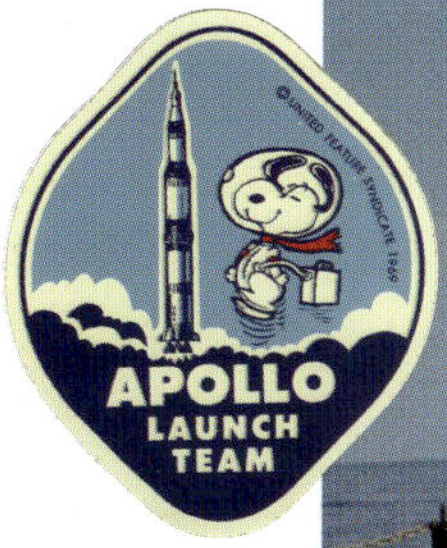

Top right: Aldrin, Armstrong, and Collins before recovery training on the deck of MV *Retriever* in the Gulf of Mexico (off Galveston, Texas) on May 24, 1969. They wear Beta cloth inflight coveralls. Aldrin would overcome depression and alcoholism in the years after the mission to become a relentless promoter of manned space exploration and his accomplishments. Armstrong taught and became reclusive, while Collins became an official with the US State Department, and later director of the National Air and Space Museum in Dulles, Virginia.

Center left, left to right: KSC Launch Operations Manager Paul Donnelly, Launch Operations Director Rocco Petrone, and Director of Design Engineering Donald "Buck" Buchanan on the hardstand at Pad 39A on May 20, 1969. Buchanan designed the ML and crawler-transporters for LC-39.

Center right: Armstrong (*left*) with MSC pilot Harold E. "Bud" Ream on June 14, 1969, at Ellington Field, with one of three LLTVs, LLTV B2. The previous May, Armstrong had ejected from a modified version of one of its two predecessors, the Lunar Landing Research Vehicle, just before it crashed. This LLTV crashed in 1971. Built by Bell Aerosystems of Niagra Falls, New York, each LLTV used a single General Electric turbofan jet engine to offset Earth's gravity and help simulate a moon landing.

Bottom right: A KSC bucket truck stops at the perimeter-fence guard gate on May 20. The Saturn is at the pad. (photo by Tiziou News Service)

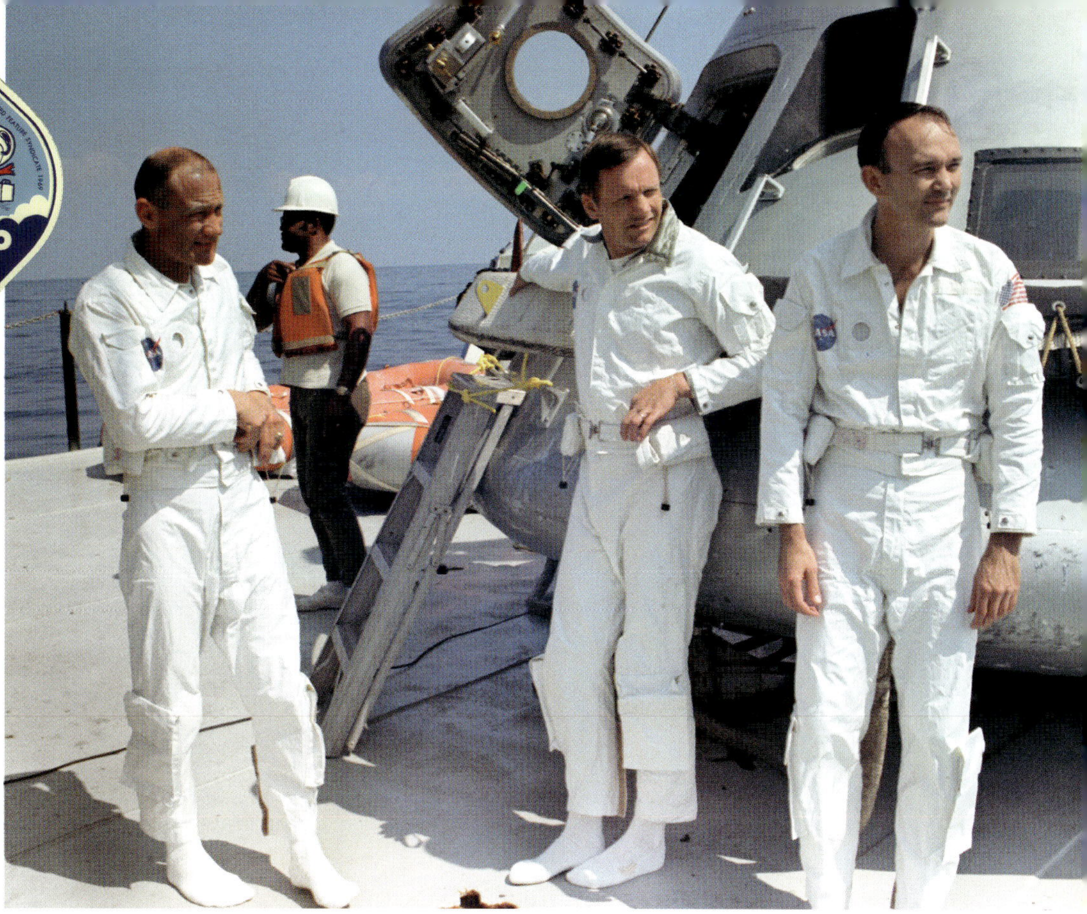

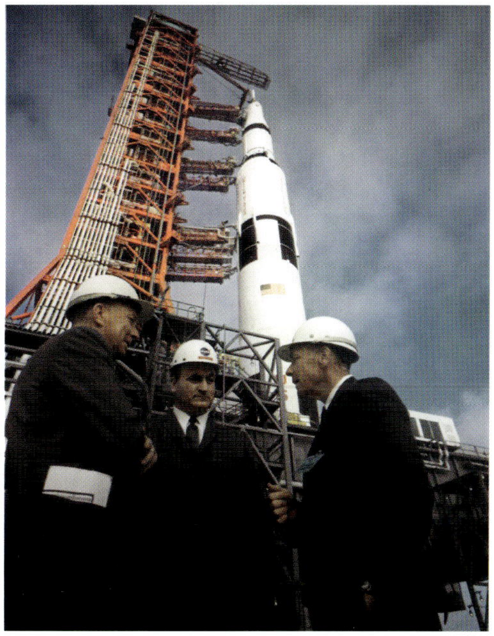

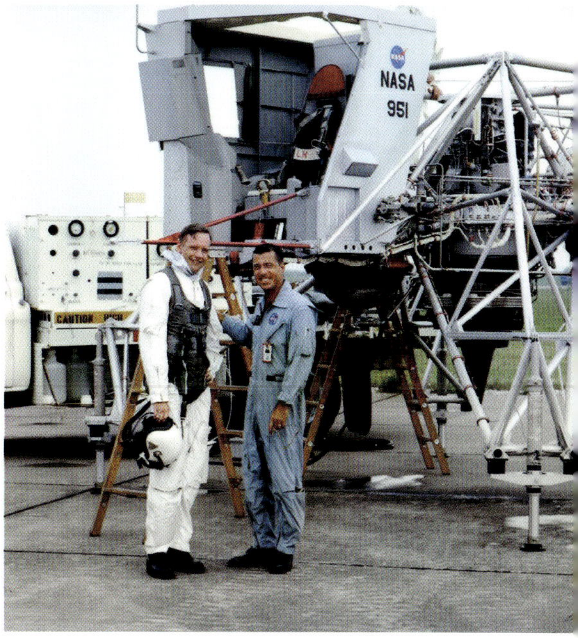

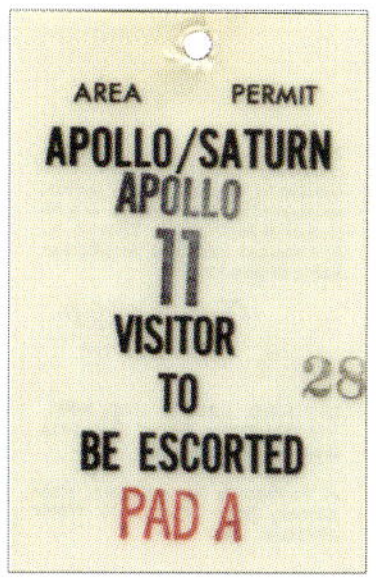
AREA PERMIT
APOLLO/SATURN
APOLLO
11
VISITOR
TO
BE ESCORTED
PAD A
28

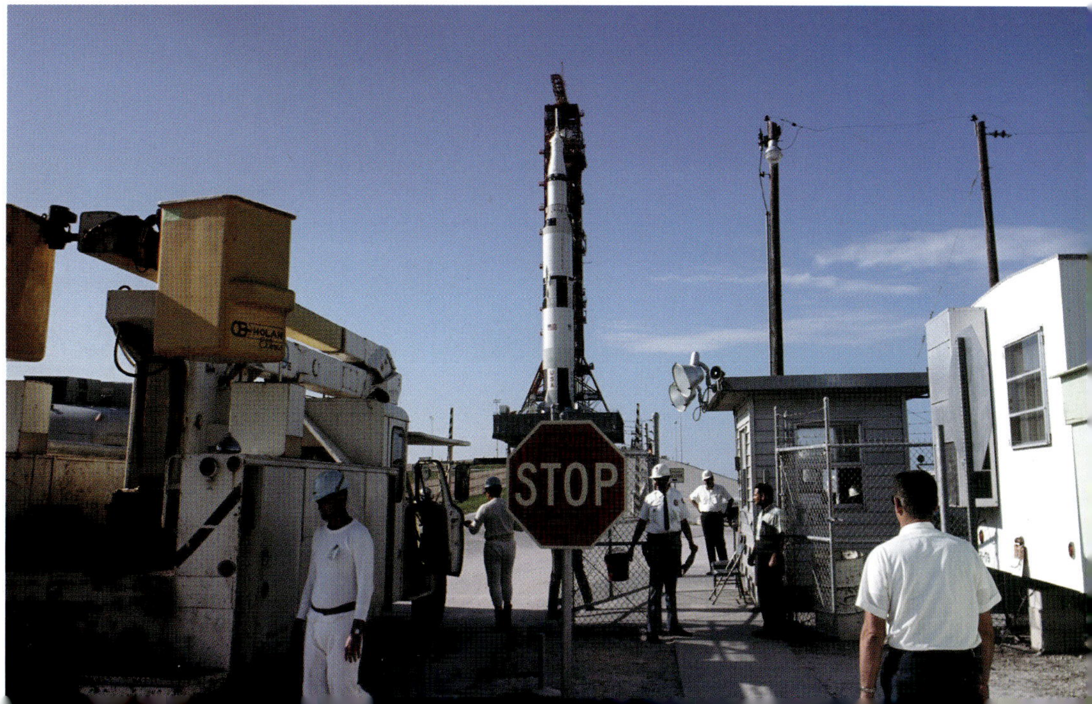

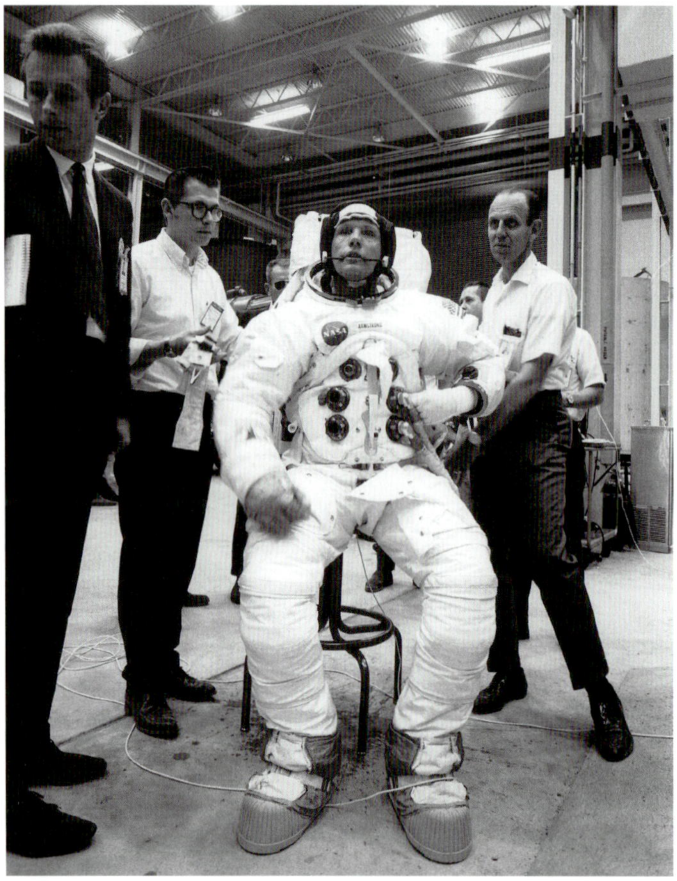

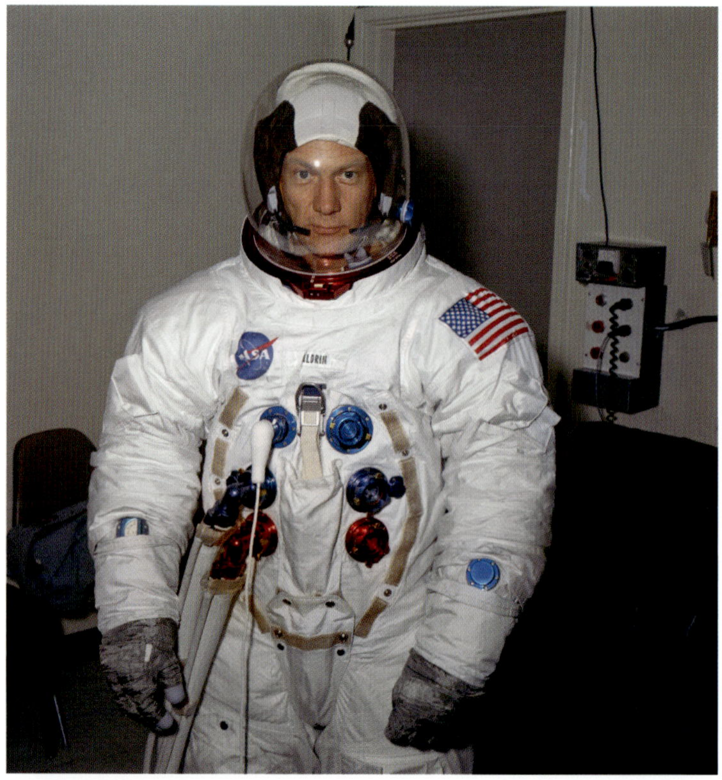

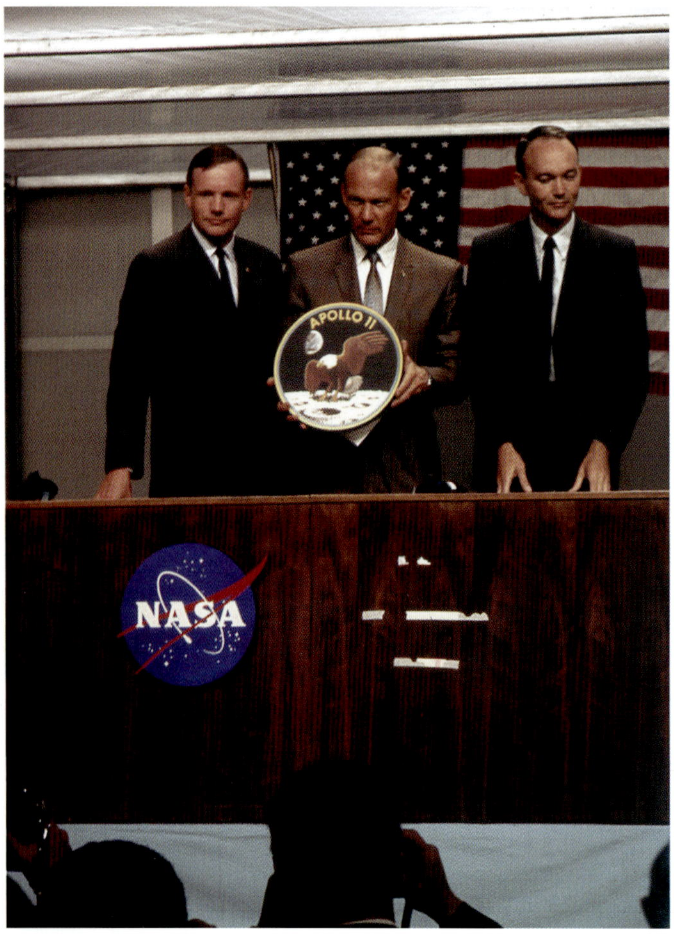
APOLLO 11
NASA

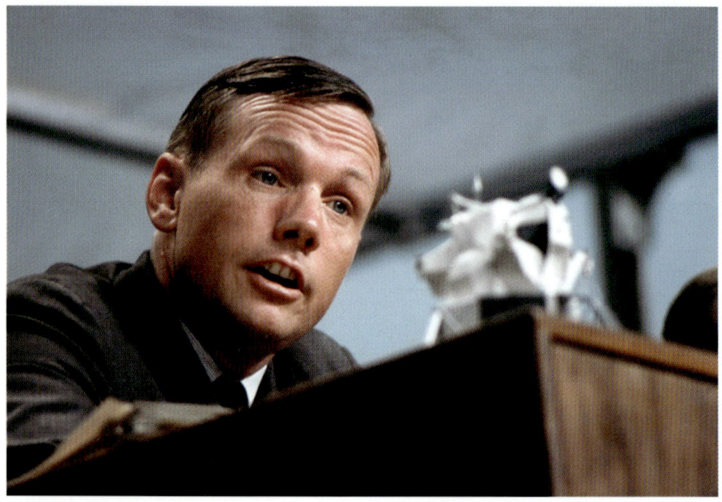

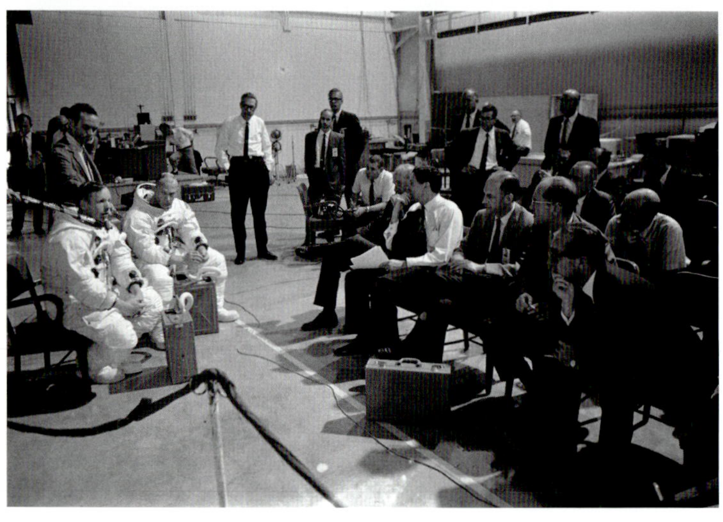

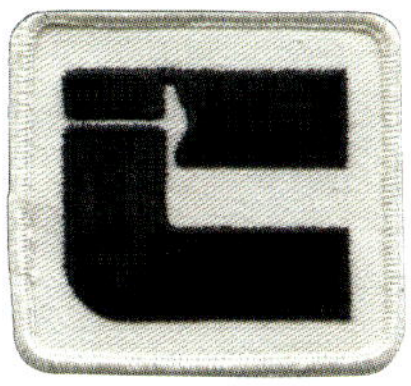

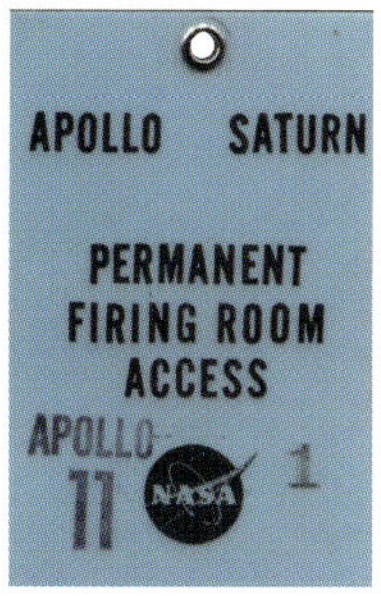

Opposite page, top left: Armstrong prepares for EVA training in Bldg. 9 at MSC on April 18, 1969. For lunar surface EVAs, the space suit, PLSS backpack, and associated systems were called the Extravehicular Mobility Unit (EMU). Hamilton Standard of Windsor Locks, Connecticut—a division of United Aircraft—manufactured the PLSS.

Opposite page, top right: Aldrin prepares for a vacuum chamber test of the backpacks (each known as a PLSS) he and Armstrong will wear on the lunar surface, in Bldg. 9 at MSC on April 4, 1969. He wears the A7L torso-limb suit used for Apollos 7 through 14, and his gloves are covered in the same Chromel-R steel cloth that had been used for Gene Cernan's pants during Gemini IX-A.

Opposite page, center right: Armstrong told reporters that the spacecraft radio call signs would be *Eagle* and *Columbia*, and that he had no involvement in the decision to have him be the first to step onto the moon. A Grumman LM model is on the desk. (photo by Tiziou News Service)

Opposite page, bottom right: Armstrong and Aldrin brief NASA officials after their training on April 18. *Seated front row, left to right*: unidentified; Mission Director George Hage; Apollo Spacecraft Program Office Manager George Low; Apollo Program Director Lt. Gen. Sam Phillips; and Associate Administrator for Manned Space Flight George Mueller. *Houston Chronicle* space reporter Art Hill is seated at center, taking notes.

Opposite page, bottom left: Aldrin holds a mission emblem—which has been hastily removed from the front of the desk—for photographers, after the crew's news conference in Bldg. 2 at MSC on July 5, 1969. To prevent crew exposure to pathogens, the media was kept fifty feet away, and fans blew air toward the audience from a three-sided plastic enclosure. (photo by Tiziou News Service)

Direct left: Donnelly's badge

Top right: ILC suit tech Ron Woods watches as Aldrin dons a glove in the MSOB on launch morning.

Center right: Two members of the Chi Alpha Christian Fellowship, a college ministry, sell souvenir Apollo 11 patches on launch day, along the Indian River in Titusville, Florida, twelve miles west of LC-39. NASA did not authorize commercial reproductions of the actual Apollo mission emblems until Apollo 13.

Bottom right, left to right: Apollo 12 crewmen Alan Bean and Pete Conrad, with Ed Mitchell—who would be named as Apollo 14 LMP within three weeks—chat with Donnelly and Phillips in the VIP viewing room in the LCC's Firing Room 1 on launch morning, July 16.

Below: The patch worn by the astronauts on their recovery ship was produced by Texas Art Embroidery of Houston.

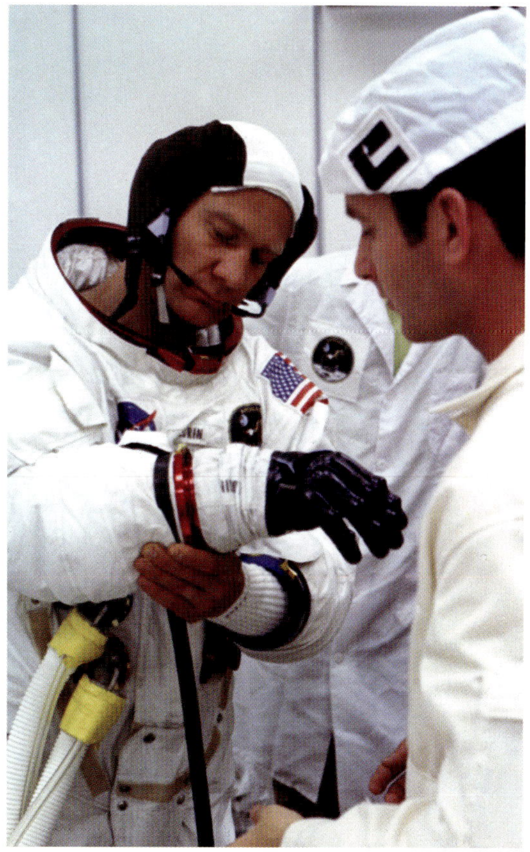

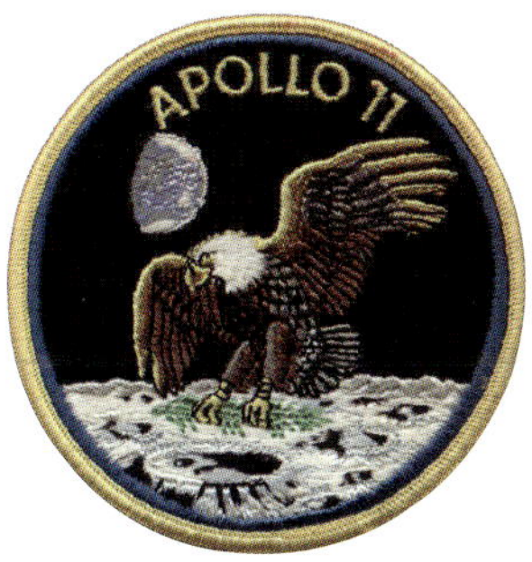

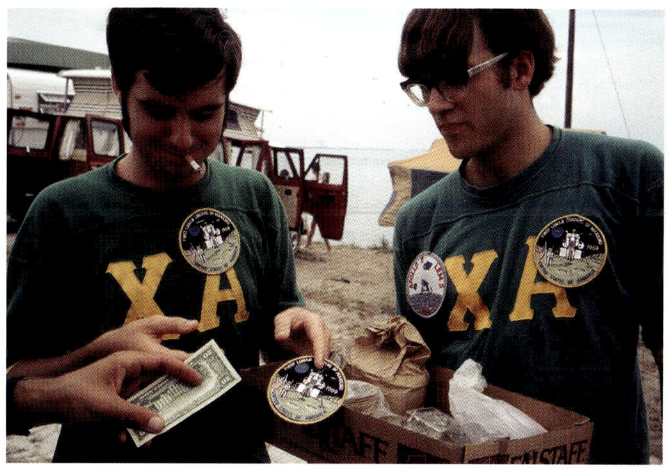

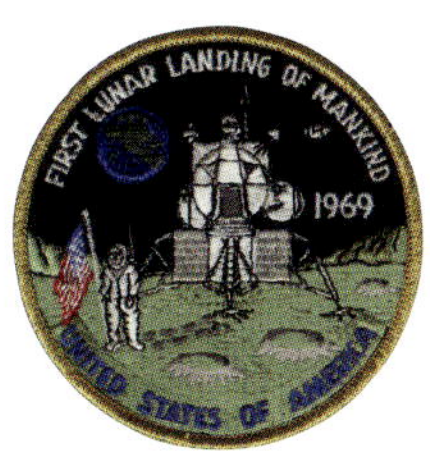

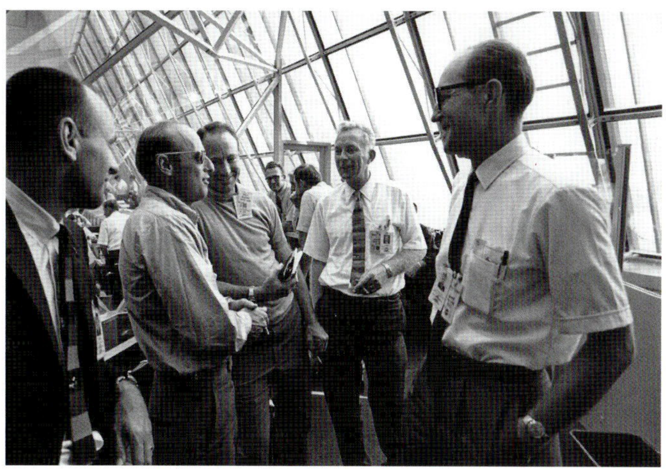

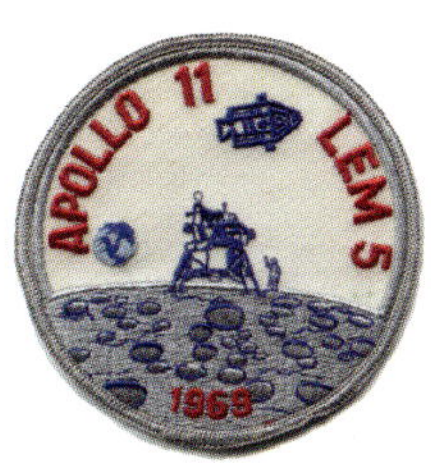

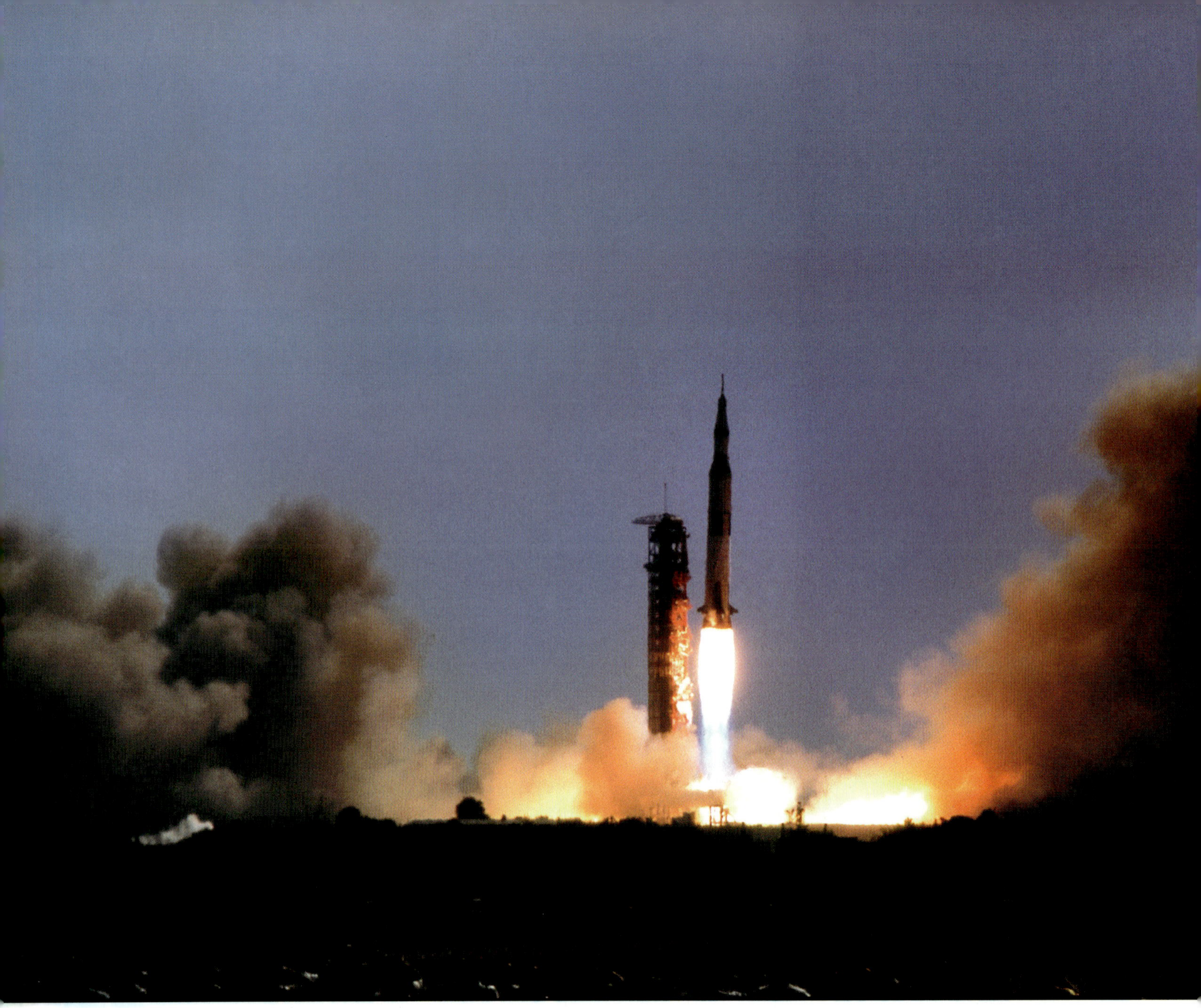

Above: Apollo 11 launches from Pad 39A at 9:32 a.m. (EDT). The Saturn V is already executing a roll and yaw maneuver, tilting slightly from vertical to avoid striking the LUT in case of a wind gust or other problem.

Opposite page, center left: Apollo 11 slowly picks up speed. In this view from near the VAB looking east-northeast, it is some eight hundred feet off Pad 39A, traveling eighty-three mph just fourteen seconds after liftoff. The Saturn's automatic engine shutdown function was inhibited during the first thirty seconds to prevent the vehicle from falling back onto the pad during a launch failure, seen as the least preferable option at that stage of the flight.

Opposite page, center right: Armstrong leads Collins and Aldrin from the MSOB at 6:25 a.m. (EDT) on launch day, to board the transfer van for the eight-and-a-half-mile trip to Pad 39A. The crewmen wear yellow protective covers on their boots.

Opposite page, bottom left: NASA gave forty members of the Poor People's Campaign against poverty, led by Rev. Ralph D. Abernathy, VIP seats at the last minute. Though his demonstration was against the amount of money spent on space, Abernathy was thrilled by the launch. (photo by Tiziou News Service)

Opposite page, bottom right: Mrs. Gay Debus (Kurt Debus's wife), Lady Bird Johnson, former president Lyndon Johnson, and KSC Deputy Director Al Siepert in the VIP viewing stands before launch. Apollo 10 CDR Tom Stafford is at left, behind Mrs. Debus. Johnson was the first former US president to attend a rocket launch. (John Kennedy had watched a Polaris missile launch from the submarine USS *Andrew Jackson*, thirty miles east of Port Canaveral, on November 16, 1963.)

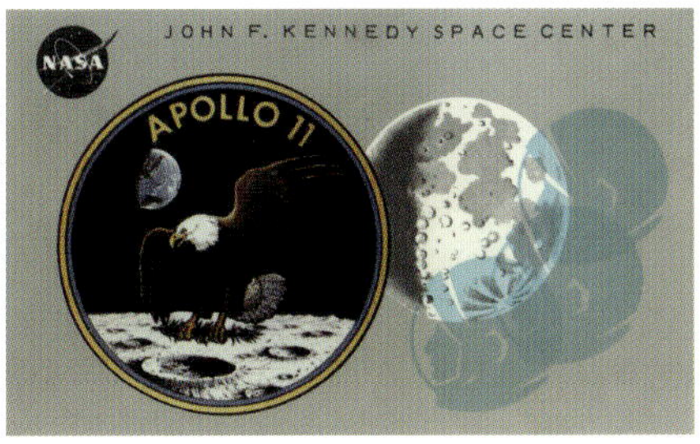
JOHN F. KENNEDY SPACE CENTER
NASA
APOLLO 11

VIP badge

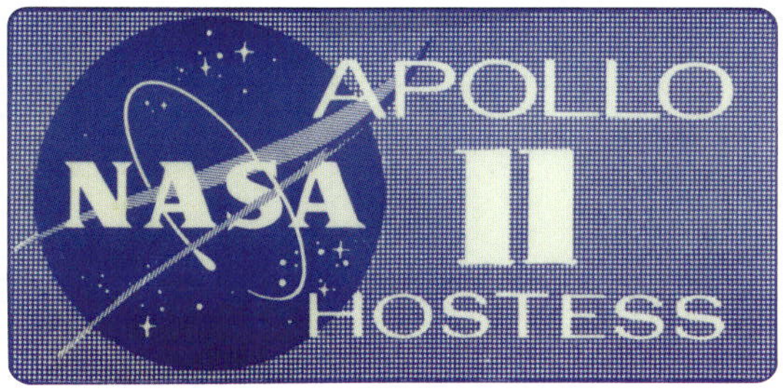
NASA
APOLLO
II
HOSTESS

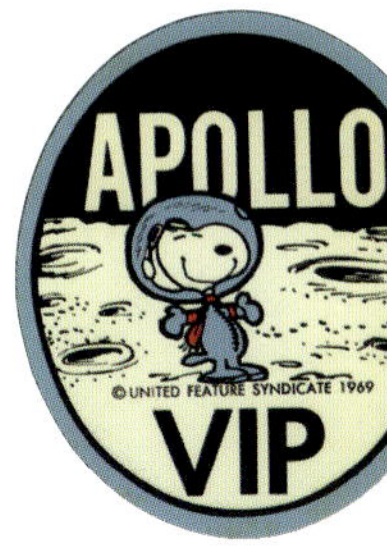
APOLLO
©UNITED FEATURE SYNDICATE 1969
VIP

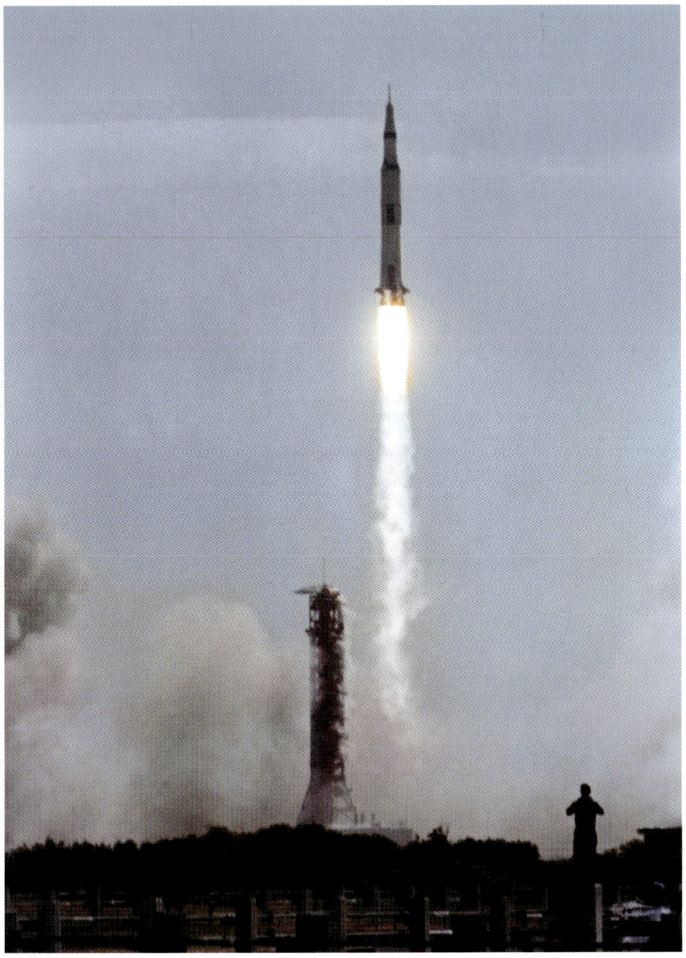

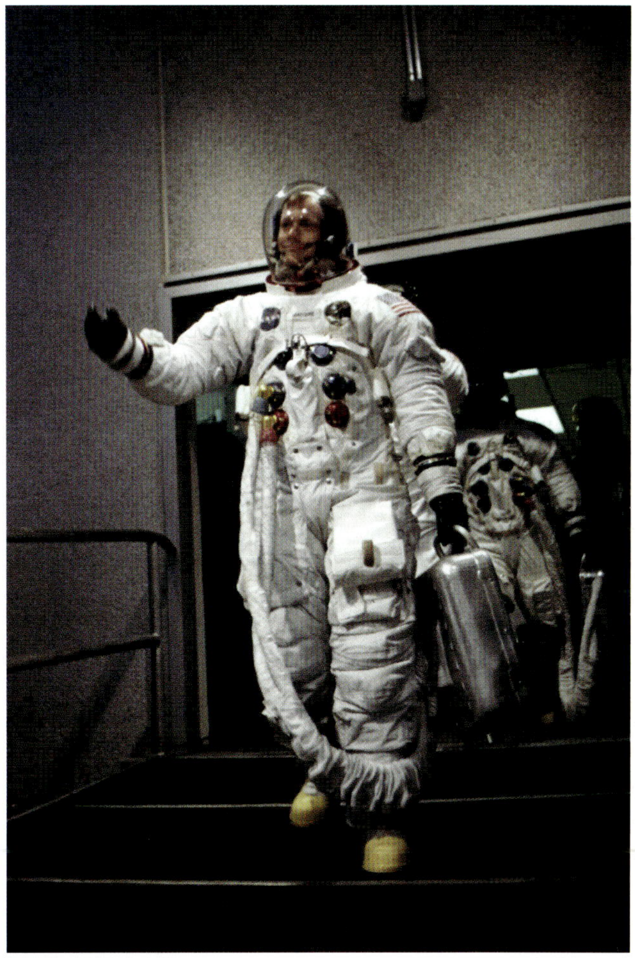

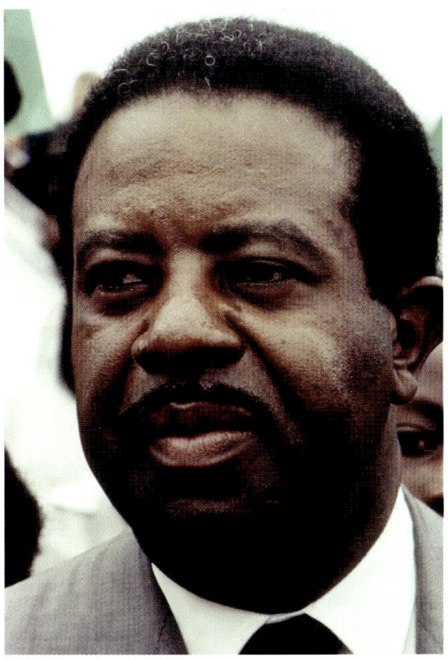

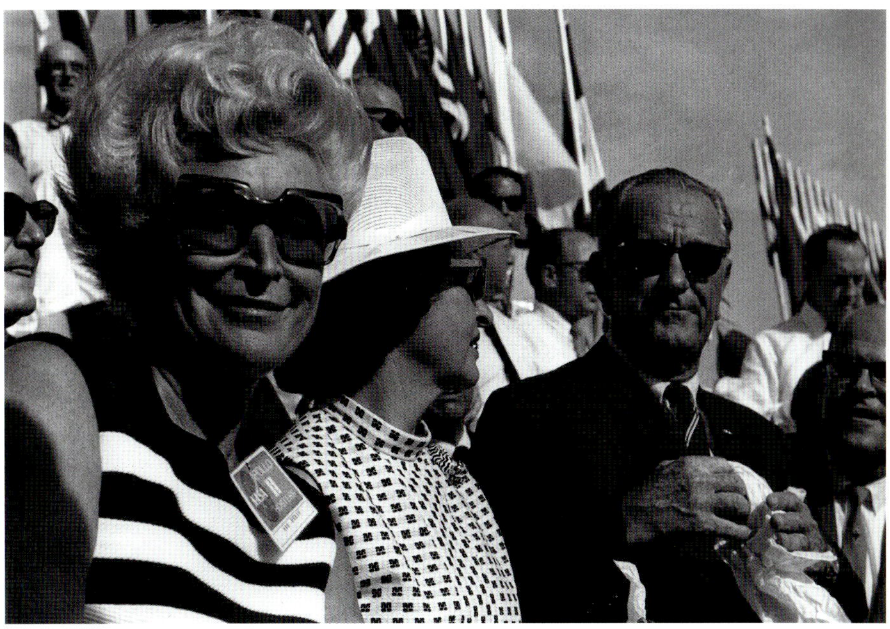

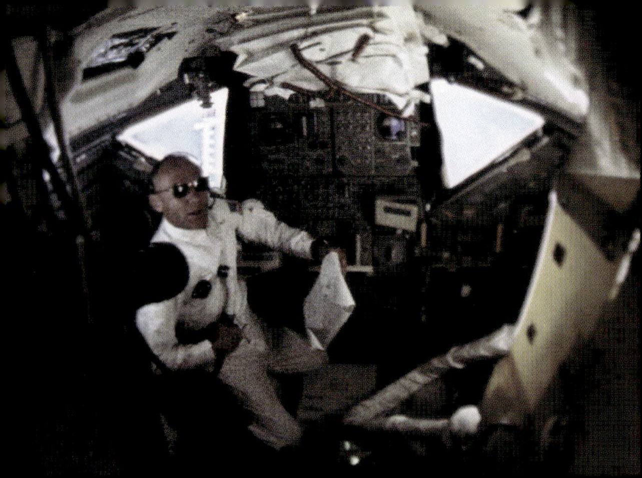

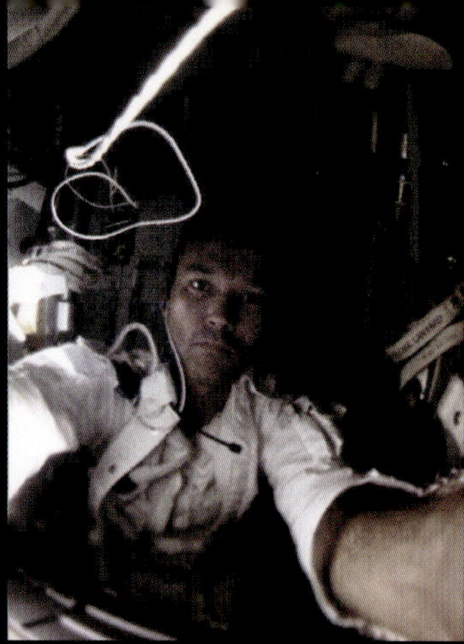

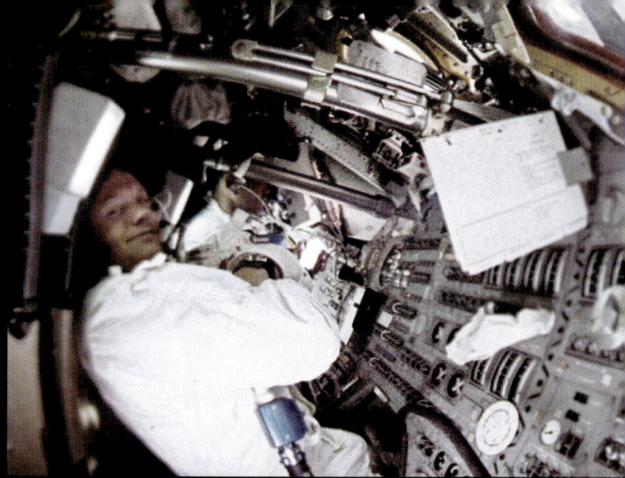

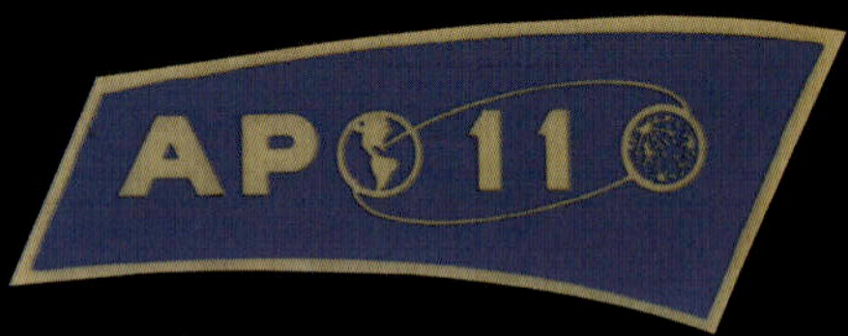

Top left: In this 16mm frame, Aldrin holds a Beta cloth pouch containing flight data (as the mission paperwork was known) at the CDR's flight station aboard *Eagle*, en route to the moon. The docking window is above his head.

Top center: Collins stands in the lower equipment bay aboard *Columbia*, with the guidance and navigation station behind him, in this 16mm frame taken during the outbound leg. The center couch seat has been folded down to provide more space. When Collins likened the CM to a cathedral, he called the navigation station the altar. His sightings using its scanning telescope and sextant were critical to determining precisely the spacecraft's position.

Top right: Armstrong in the center of *Columbia*'s main display console, with the No. 4 rendezvous window at upper right. Collins is behind him in this 16mm frame outbound; two couch struts stretch above them. Several flight-data pages are in a clip attached between panel 2F (containing controls and displays for the environmental control subsystem) and panel 3A (*far right*), which controls the service propulsion subsystem and a VHF antenna.

Direct right: Earth is framed in this view from *Columbia*, looking down on the top of *Eagle* as the two docked spacecraft near the moon. Three of the four nozzles on the left-rear RCS thruster cluster (No. 2) are at left, with the docking target at center right. The Marquardt Corporation of Van Nuys, California, manufactured the one-hundred-pound thrusters, which could be fired or pulsed.

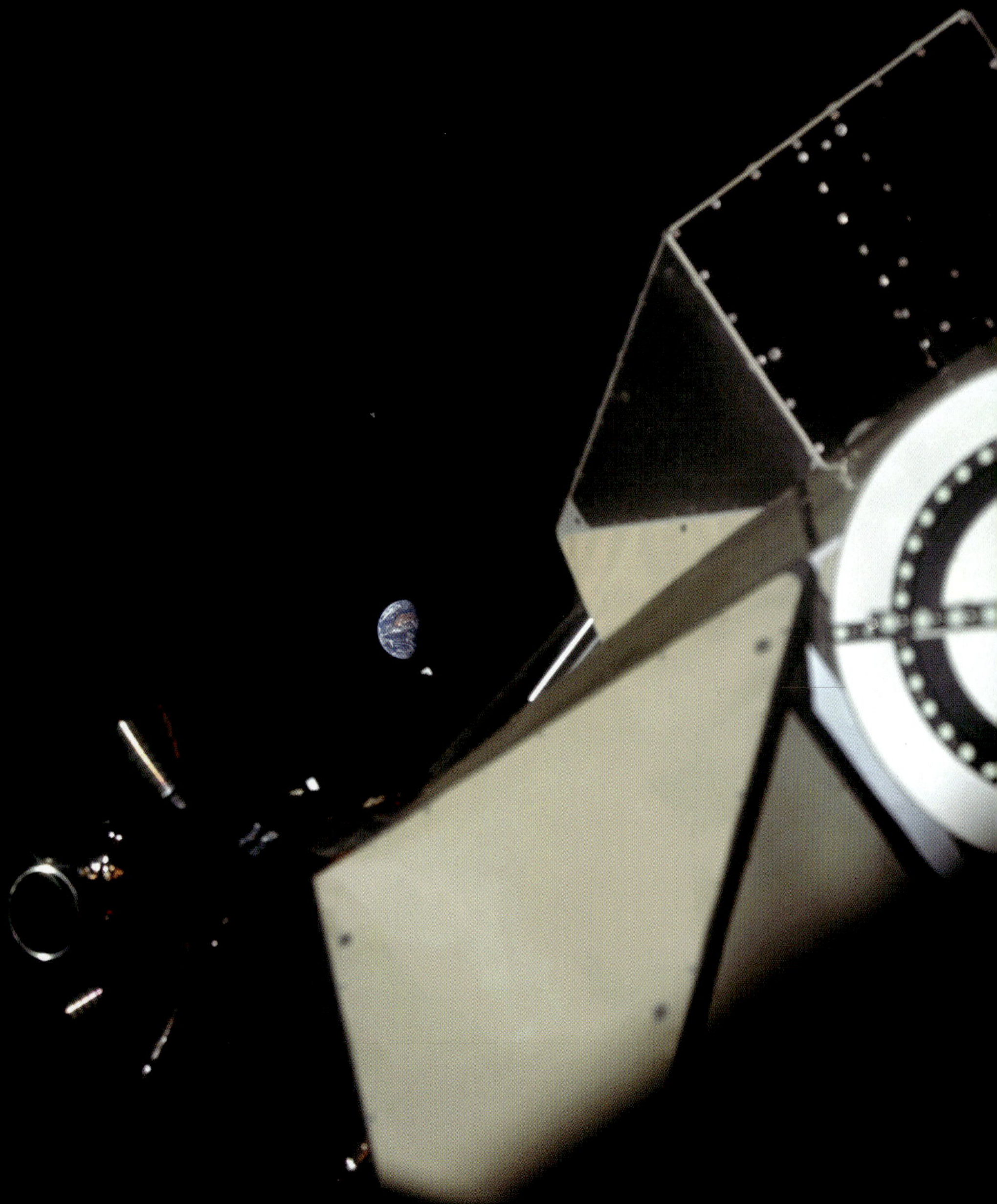

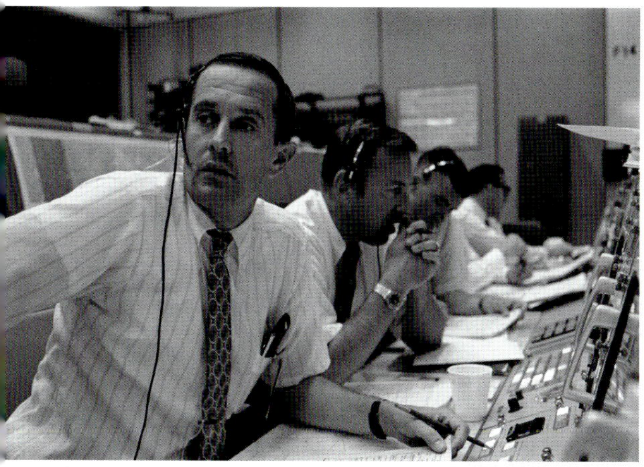

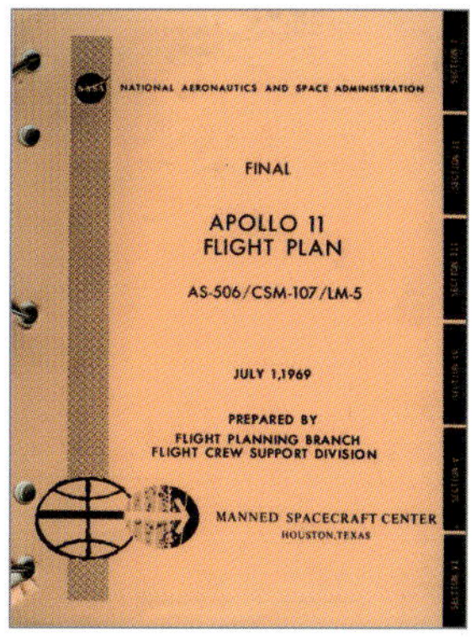

NATIONAL AERONAUTICS AND SPACE ADMINISTRATION

FINAL

APOLLO 11
FLIGHT PLAN

AS-506/CSM-107/LM-5

JULY 1,1969

PREPARED BY
FLIGHT PLANNING BRANCH
FLIGHT CREW SUPPORT DIVISION

MANNED SPACECRAFT CENTER
HOUSTON, TEXAS

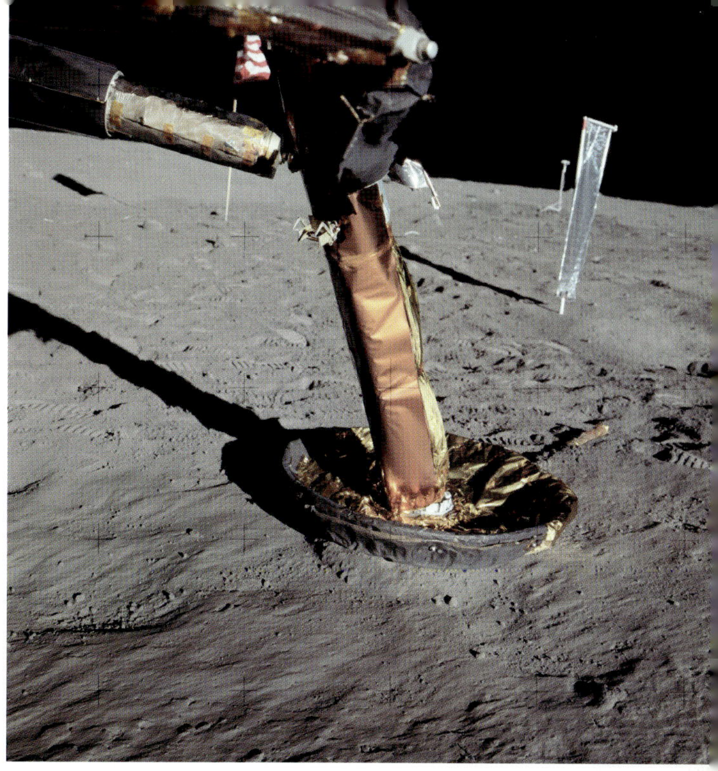

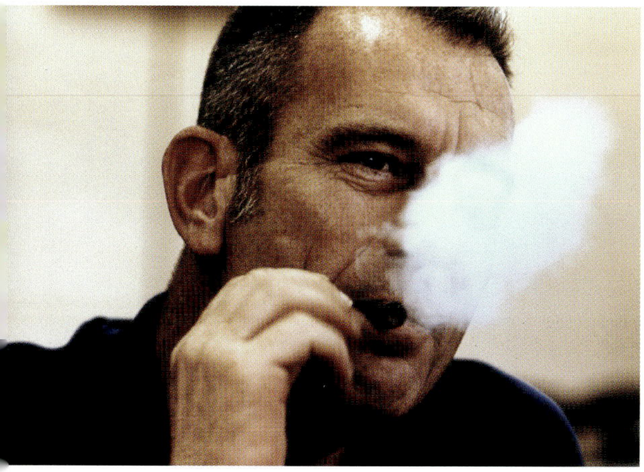

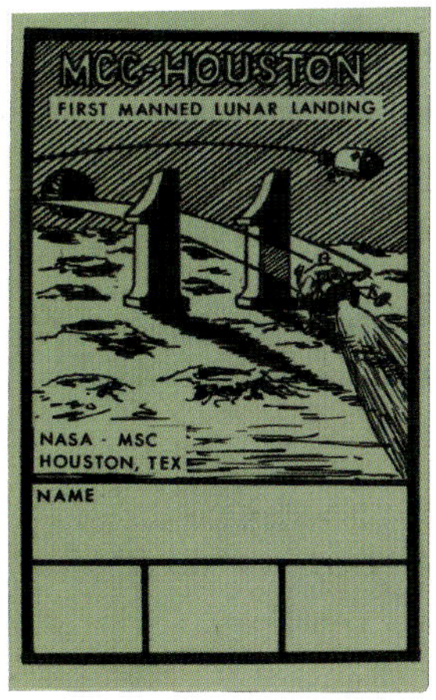

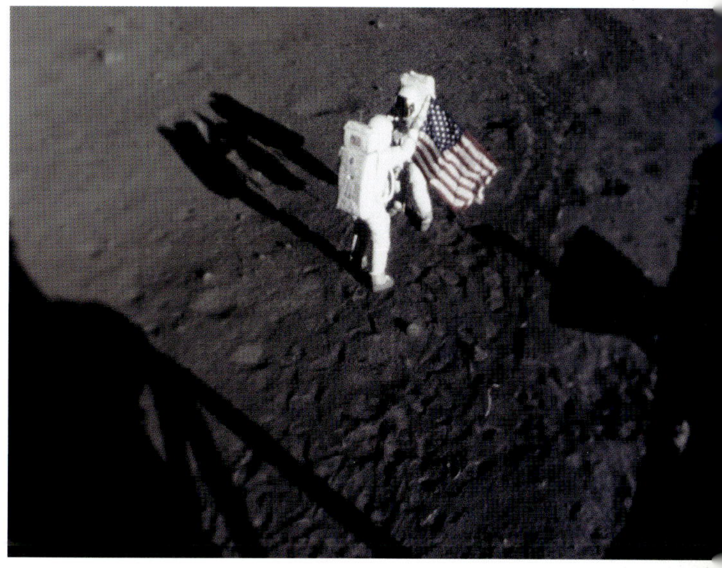

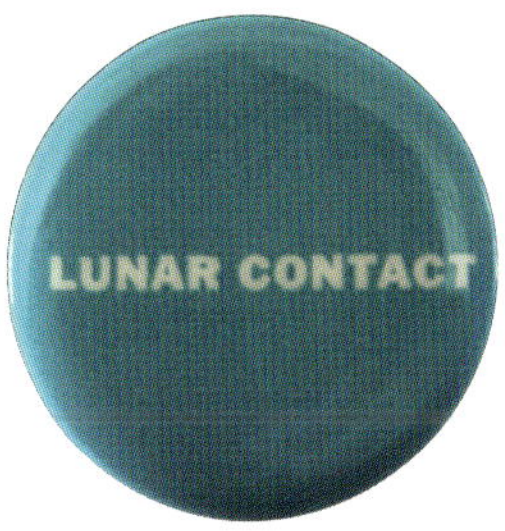

Grumman button

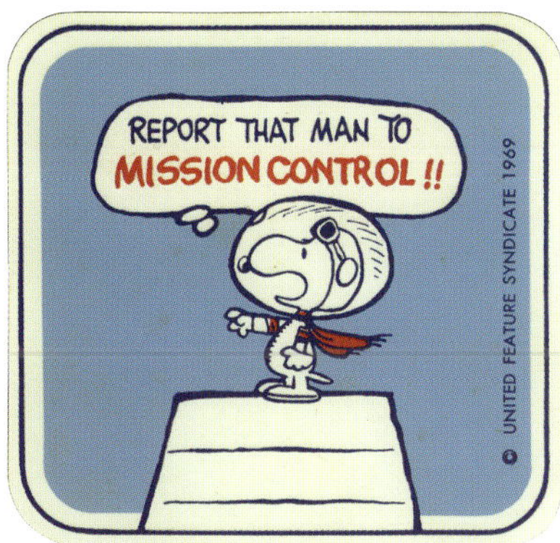

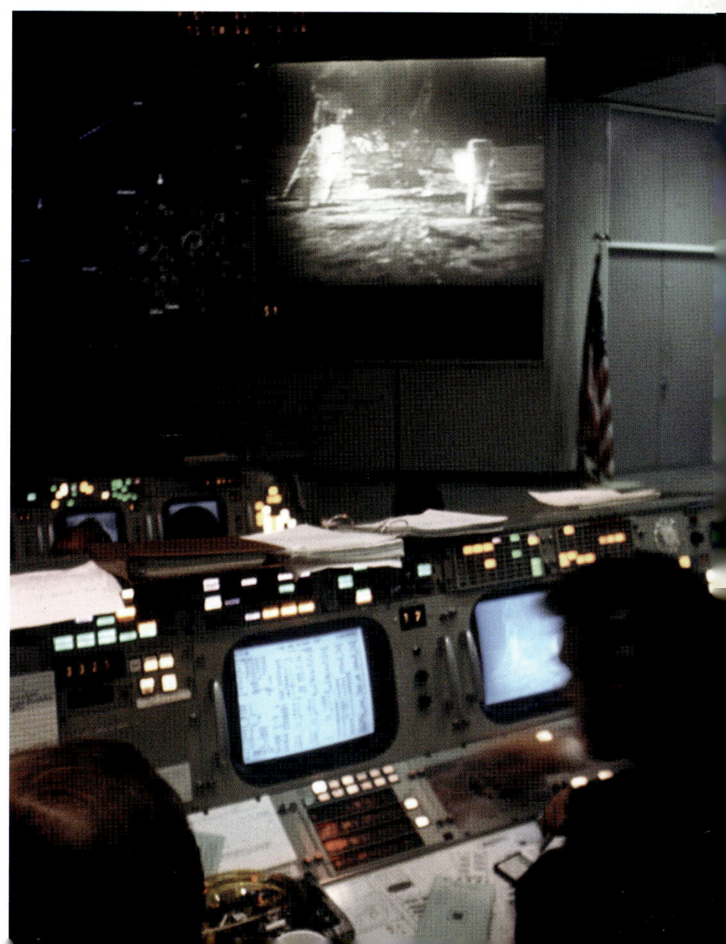

Top left: CAPCOM Charles Duke at his console in MOCR 2 as *Eagle* prepares to land. To his right are astronauts Jim Lovell and Fred Haise. Dr. Charles Berry, MSC medical director, is at the flight surgeon console (*far right*).

Top right: Close-up of the right footpad looking west-northwest, with the Solar Wind Collector (SWC) to the right; to its left is the TV camera. The US flag is to the left of the strut, partially blocked by the strut support. The SWC foil was returned to Earth with the crew.

Center left: Slayton puffs on a trademark cigar in his office at MSC on July 25, 1969. (photo by Jacques Tiziou)

Center right: Armstrong and Aldrin set up the US flag northwest of *Eagle*, with the TV camera at upper right, in this frame shot by the Maurer 16mm camera through Aldrin's window. The camera catches an RCS nozzle on cluster No. 4 at right, and the shadow of *Eagle*'s right leg is to the left.

Bottom right: Flight controllers during the mission's EVA, shown on the projection TV screen. The EVA lasted two hours and thirty-six minutes. The plot board at left depicts the landing site on a lunar chart.

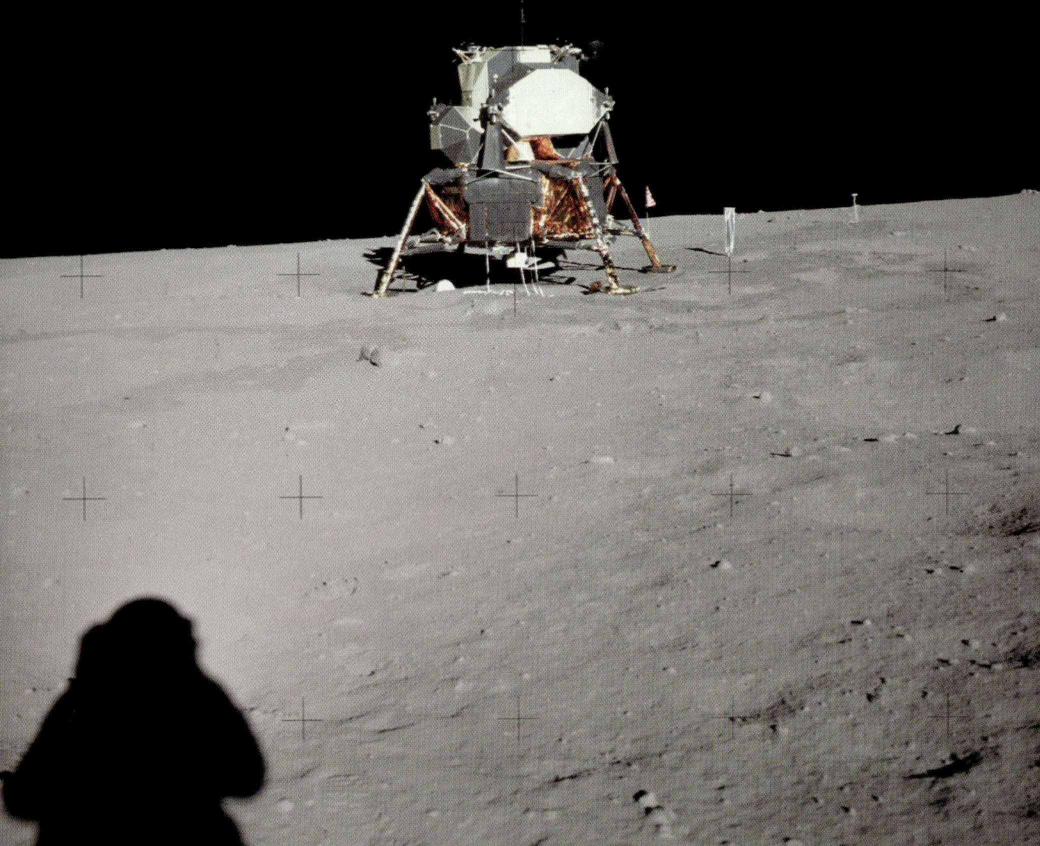

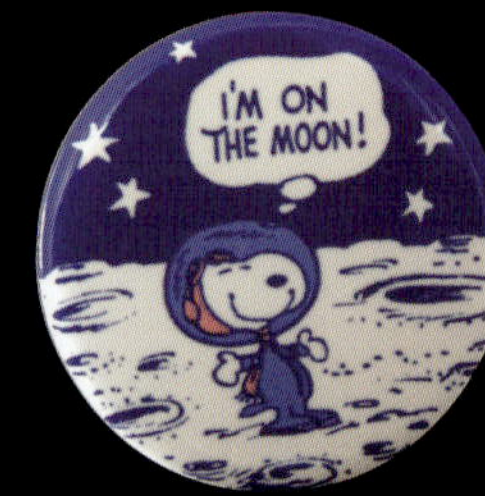

Top left: Armstrong's photo of the landing site during his return from the rim of Little West Crater about 160 feet to the east, with the flag, SWC, and TV camera to the right of *Eagle*. Aldrin is behind the LM.

Center right: Aldrin at his flight station aboard *Eagle* on July 21, with the lunar surface and horizon visible through his front window. The 16mm motion picture camera used to take the image on the center right of the previous page is in the upper-right corner.

Bottom left: At about 12:30 a.m. (EDT) on July 21, Aldrin drops off the LRRR before carrying the PSE to its location about twenty-five feet to the south. The LRRR bounced laser beams from Earth back to their origin for precise distance measurements; it remains in use. The PSE measured moonquakes and meteoroid impacts, but failed after three weeks. It was powered by solar panels and heated by 1.2 ounces of plutonium-238.

Bottom right: Aldrin sets up the PSE about 65 feet south of *Eagle*. He uses its built-in handle to level the experiment. The deployed LRRR is to his left, with the TV camera beyond, about 150 feet north. The eastern rim of Double Crater is at the left edge.

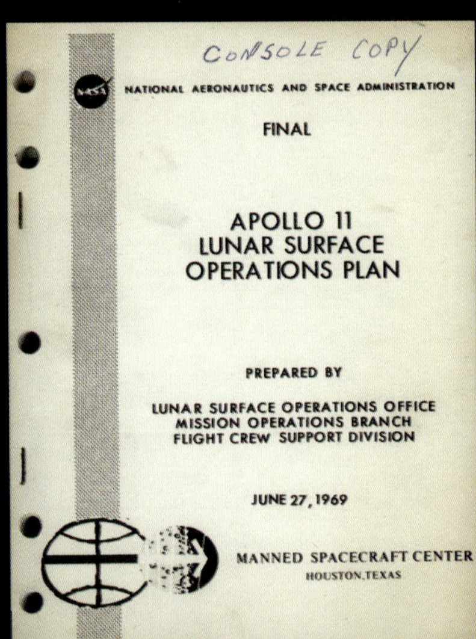
CONSOLE COPY

NATIONAL AERONAUTICS AND SPACE ADMINISTRATION

FINAL

APOLLO 11
LUNAR SURFACE
OPERATIONS PLAN

PREPARED BY

LUNAR SURFACE OPERATIONS OFFICE
MISSION OPERATIONS BRANCH
FLIGHT CREW SUPPORT DIVISION

JUNE 27, 1969

MANNED SPACECRAFT CENTER
HOUSTON, TEXAS

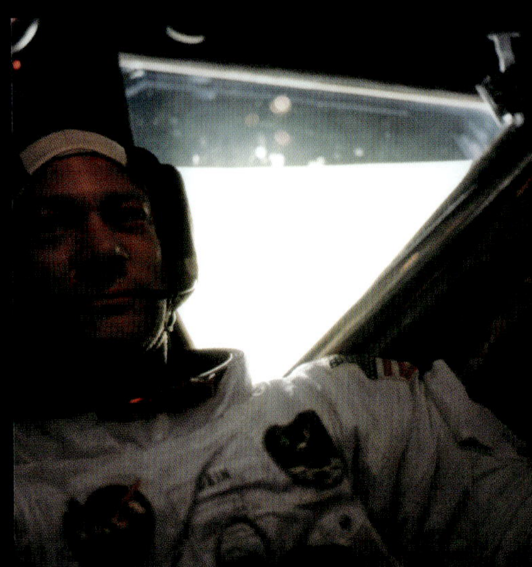

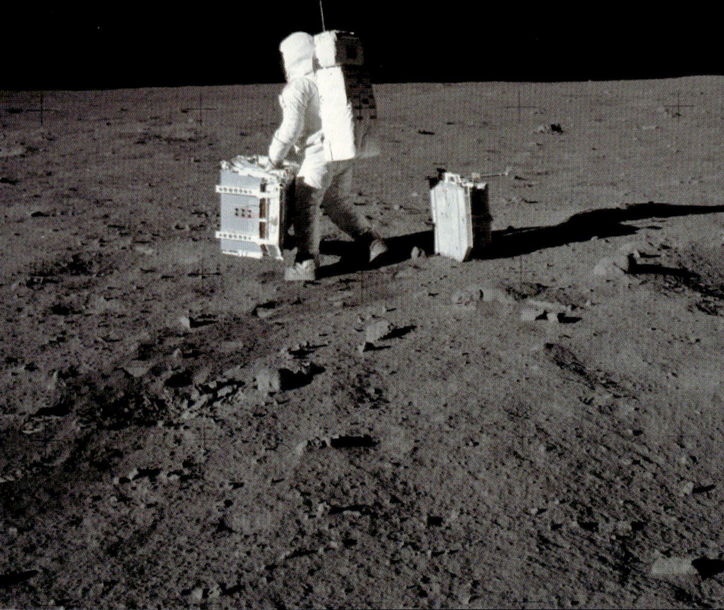

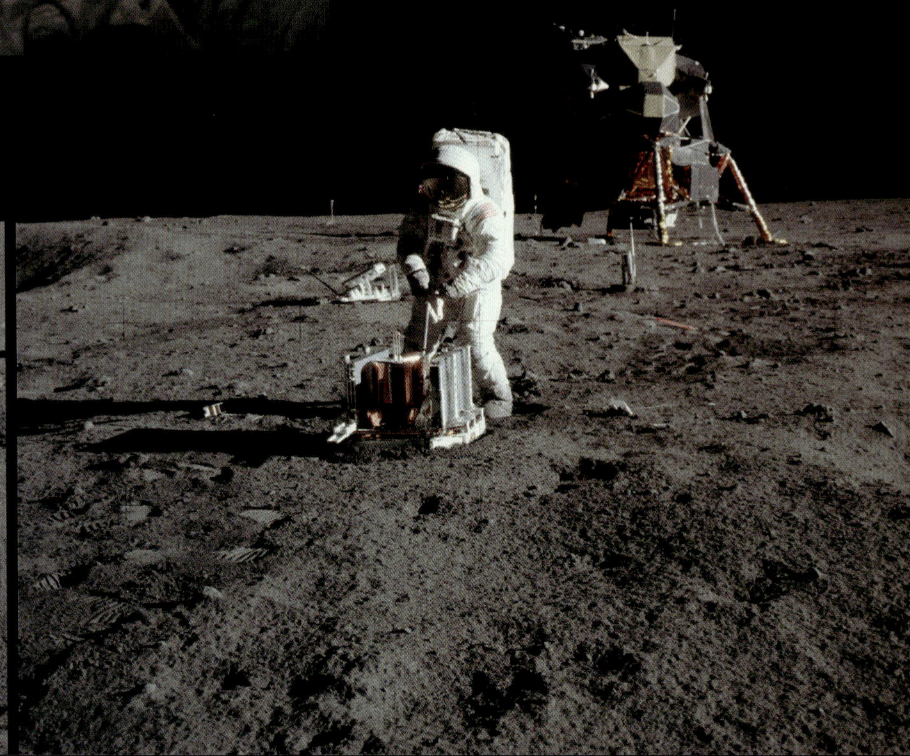

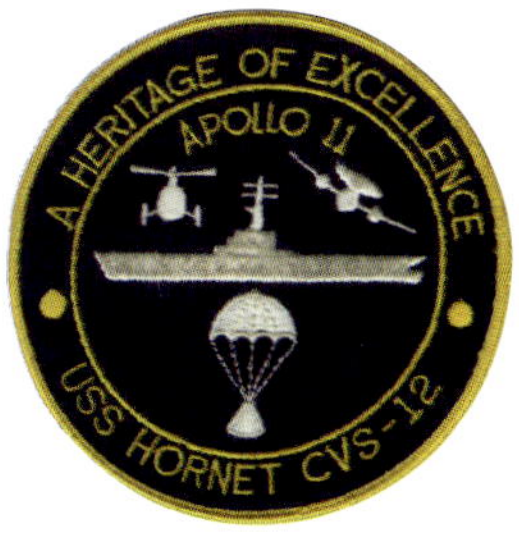

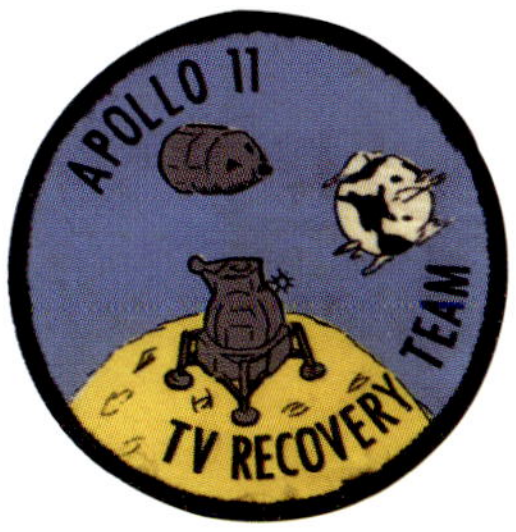

Top right: First Lt. Clancy Hatleberg, a US Navy SEAL (*right*), assists Aldrin before he is raised to the Sea King recovery helicopter in a Billy Pugh Company net, about an hour after splashdown. Both wear BIGs to prevent releasing any unknown moon pathogens. As decontamination officer, Hatleberg then sank all the equipment he had used, including this raft. The CM's flotation collar is at left.

Center left: The gold Mylar foil on *Columbia*'s hatch (aboard USS *Hornet*) is peeling, revealing the honeycomb structure of the heat-protective ablative material covering the CM. Just below the cabin pressurization valve at left is a urine dump valve. An EVA handhold is at upper right. In an emergency, a large hex-key wrench opened the hatch, using the fitting at lower right.

Center right: Robert Gilruth stops to sign NBC News correspondent Jay Barbree's Apollo 11 flight plan as he leaves a news conference in Bldg. 2 at MSC. The news conference, with other senior program officials, followed the crew's ascent from the moon on July 21. (photo by Tiziou News Service)

Bottom right: In MOCR 2 after recovery, NASA officials celebrate at the flight director's console. *Left to right*: the mission's three flight directors, Cliff Charlesworth (Green Team, launch and EVA), Glynn Lunney (Black Team, lunar ascent), and Gene Kranz (White Team, lunar landing); MSC Director of Flight Operations Chris Kraft; Associate Administrator for Manned Space Flight George Mueller; MSC Director Robert Gilruth; Deputy Administrator George Low; and Deputy Associate Administrator for Manned Flight Charles Mathews.

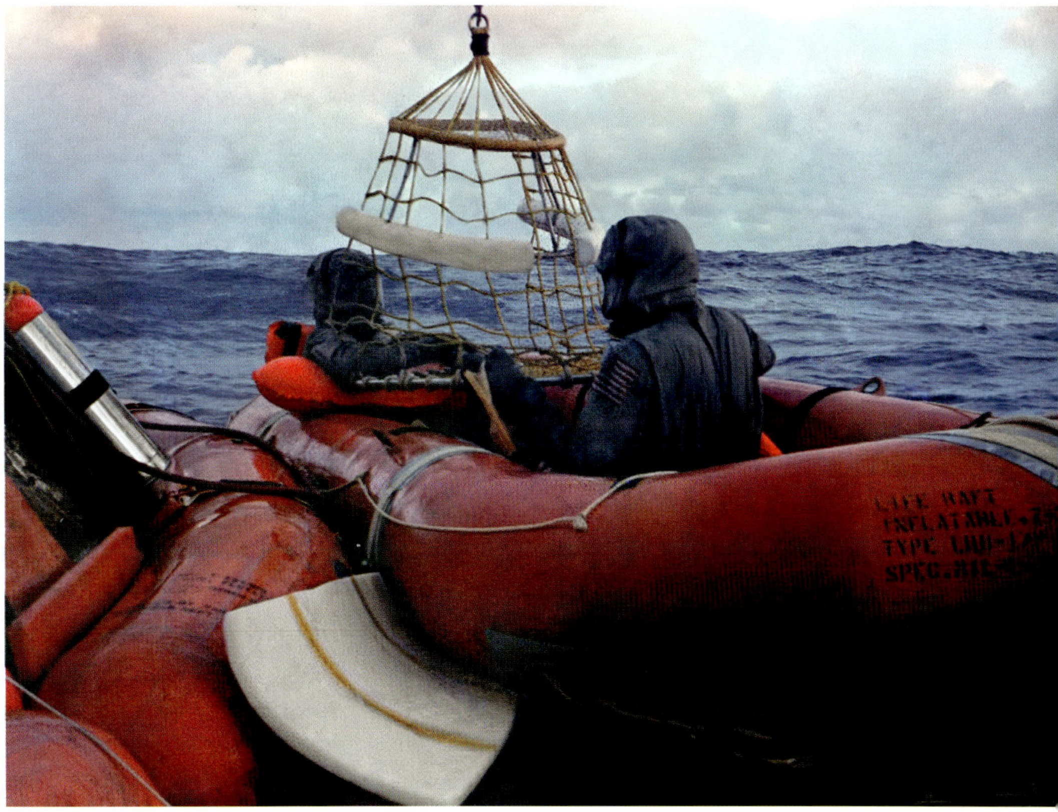

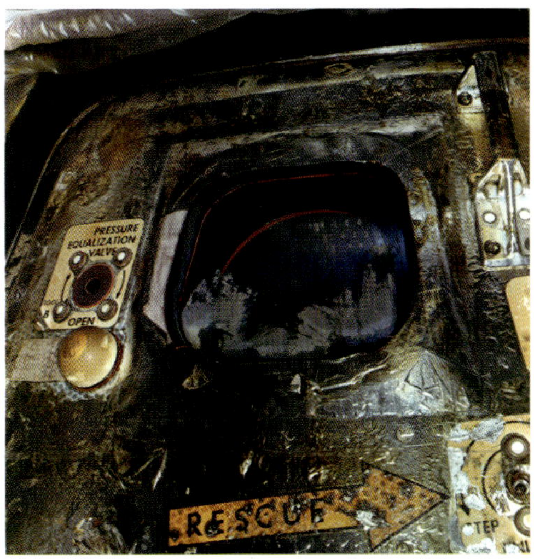

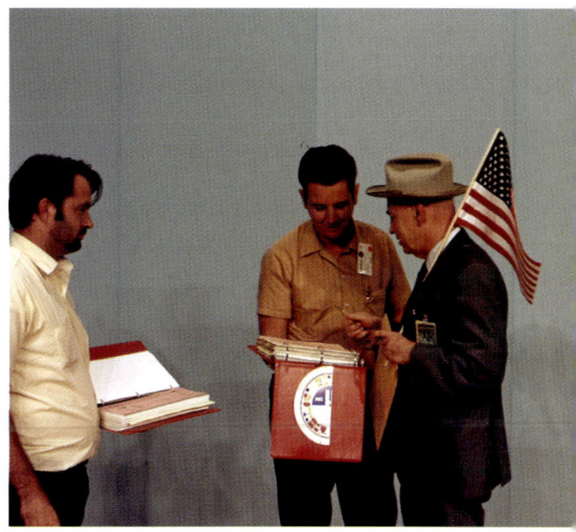

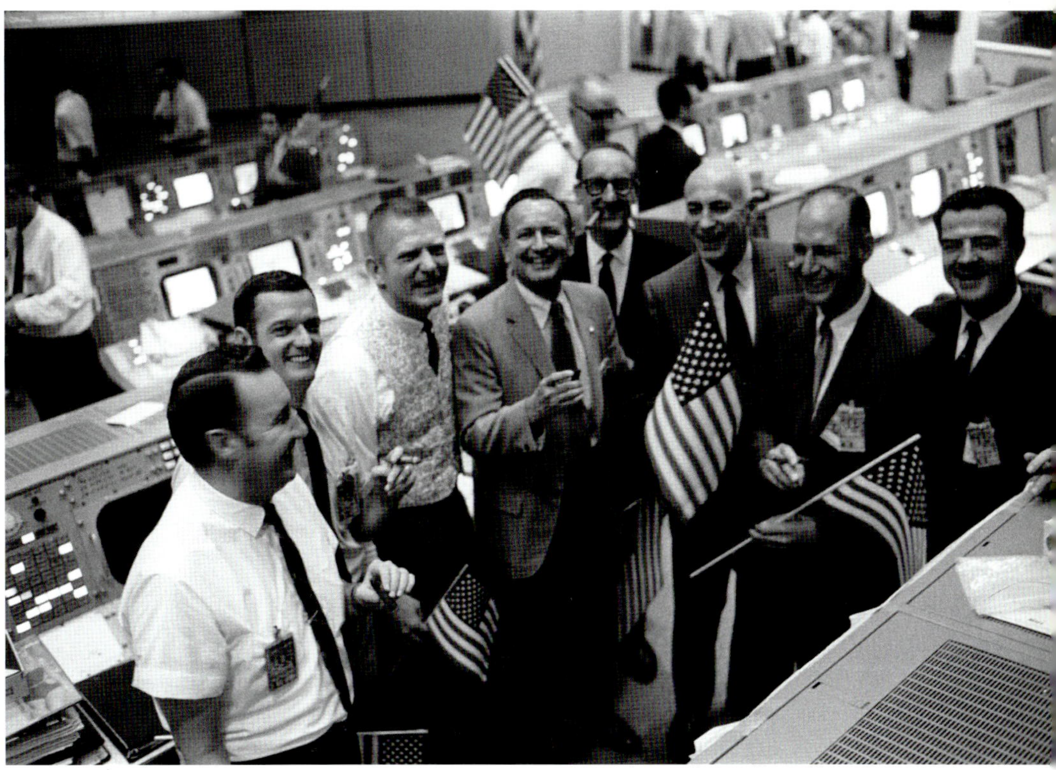

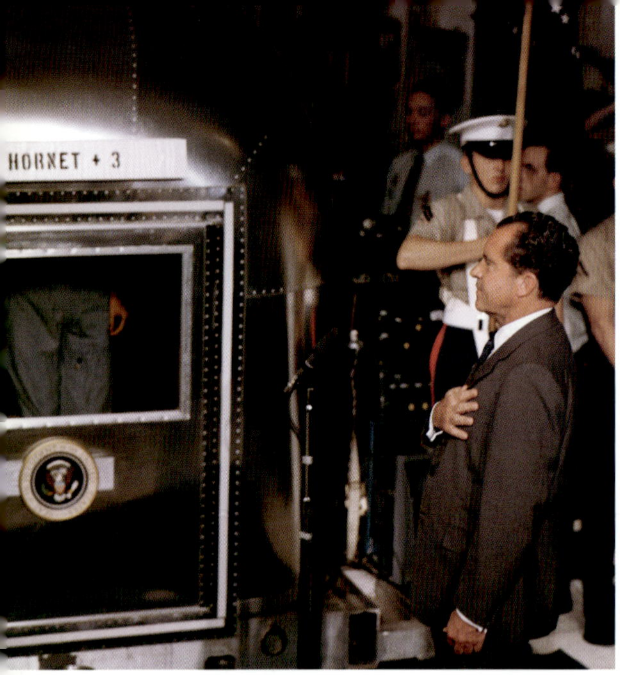

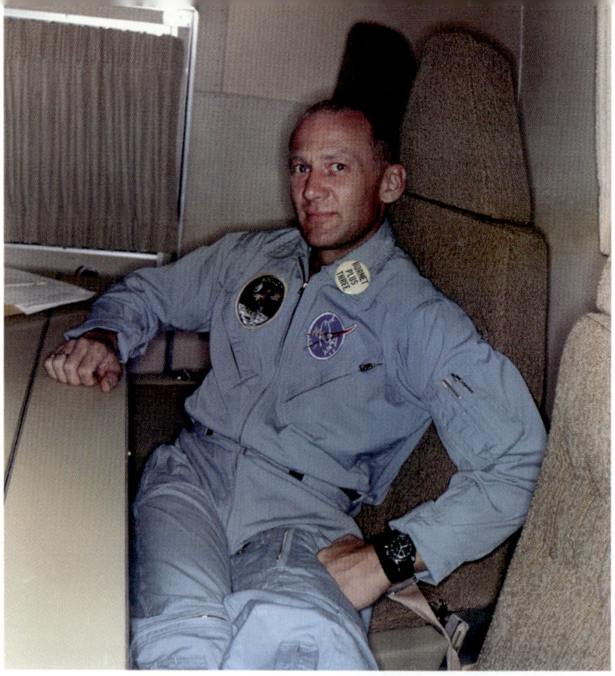

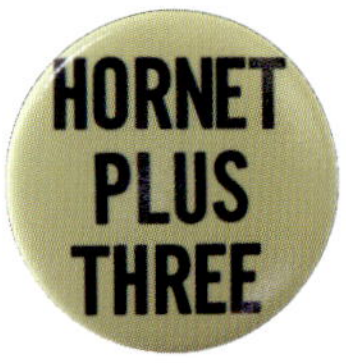

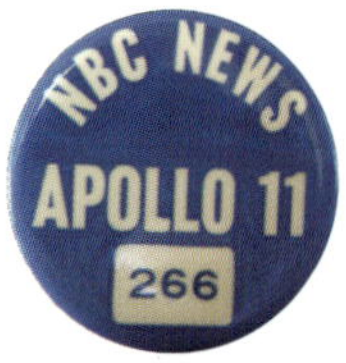

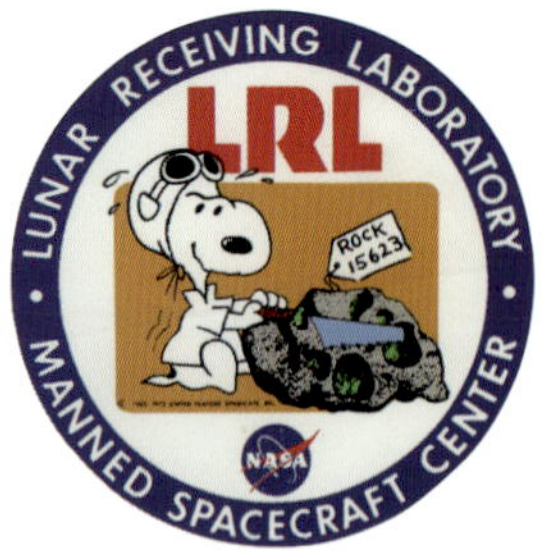

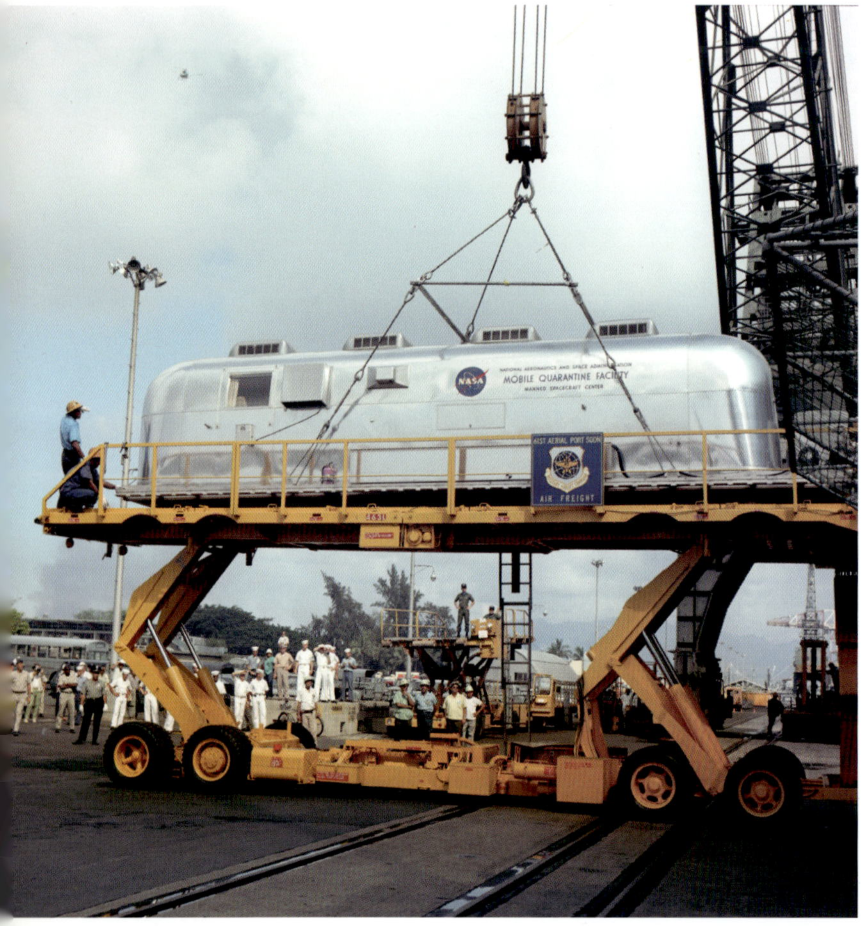

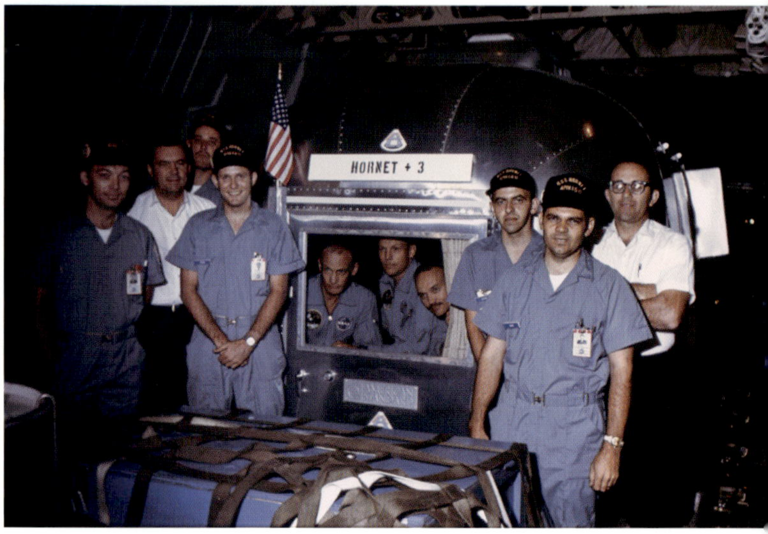

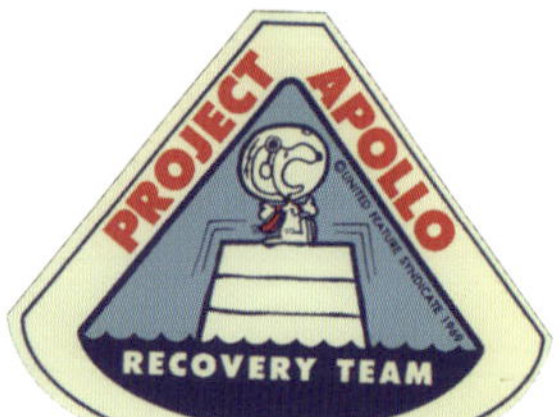

Top left: Aldrin (*left*) stands for the national anthem in the MQF, before President Nixon delivers remarks on *Hornet*'s hangar deck on July 24. Navy regulations did not allow the first lady aboard; she stayed in Honolulu. The presidential seal covers the American Standard plate on the MQF door (American Standard supplied plumbing for the MQF). When Nixon left Hawaii, he embarked on a tour of Asia, including South Vietnam.

Top right: Aldrin relaxes later in the MQF, a converted thirty-two-foot Airstream trailer, one of four built for Apollo. It is on display at the National Air and Space Museum.

Center left: The MQF is off-loaded from *Hornet* at Pearl Harbor, Hawaii, on July 26. The trailer was trucked to Hawaii's Hickam Field and loaded aboard a C-141 cargo aircraft.

Center right: MSC personnel with the MQF and crew aboard the C-141, en route to Ellington AFB. NASA Public Affairs Officer Ben James is second from left; John Stonesifer, chief of MSC's Recovery Systems Branch, Landing and Recovery Division is at right.

Bottom left: The MQF remains at the loading dock as *Columbia* arrives at the LRL at MSC's Bldg. 37 on July 30. The spacecraft had been deactivated on Ford Island, Hawaii, and then flown to Ellington on a C-133 aircraft. (photo by Tiziou News Service)

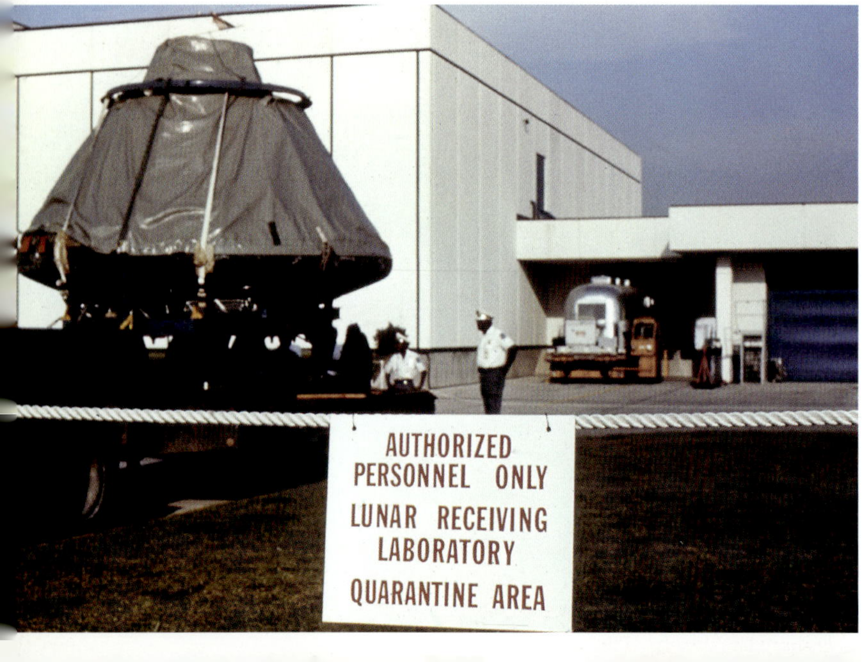

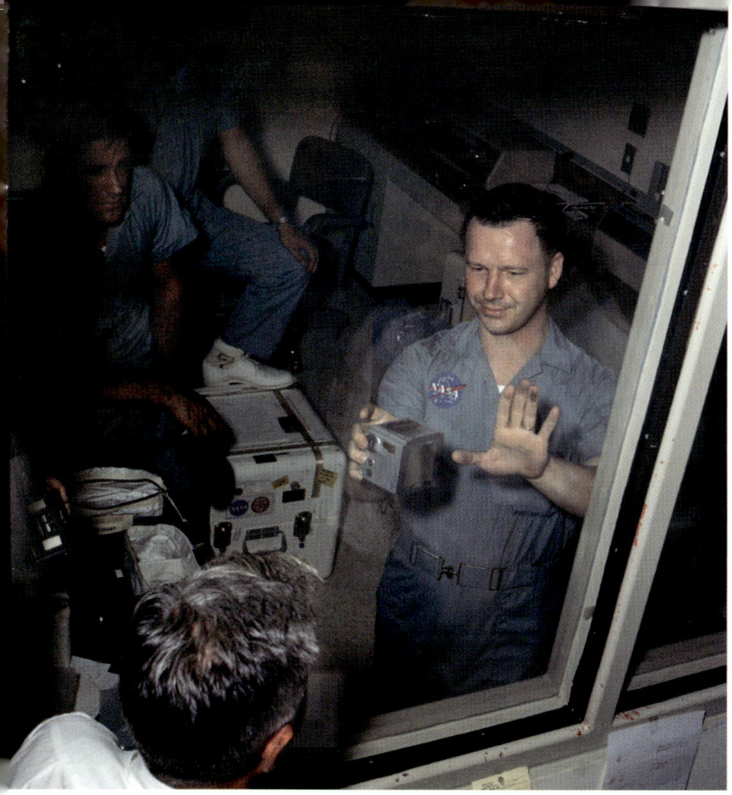

NASA photographer Terry N. Slezak, inside the Lunar Receiving Laboratory (LRL) before the crew has arrived, shows fingers smudged with moon dust on July 25. The 70mm film magazine he is holding had fallen to the lunar surface. He and four other technicians were required to shower. Slezak kept a one-inch piece of cellophane tape carrying the dust, which he sold in 2001 for $25,000 to a German dealer, who then resold sections. One section was returned to NASA in 2011, after authorities noticed it at a public auction.

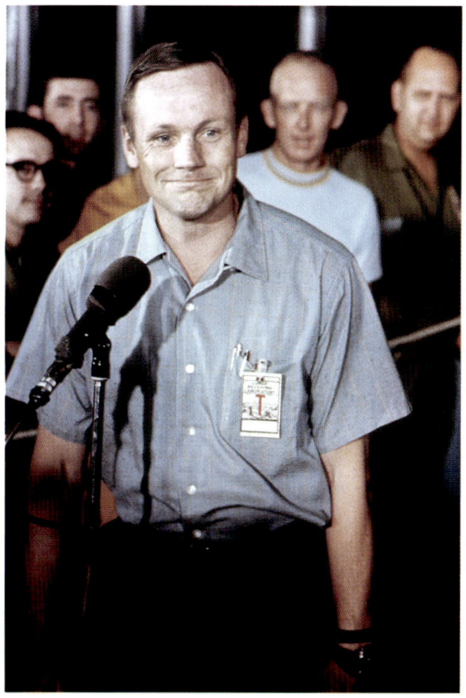

Armstrong speaks to a crowd of 1,500 at MSC after being released from the LRL on August 10, 1969. He, Aldrin, and Collins were in the LRL for three weeks. The crewmen went home and had the next day off. (photo by Tiziou News Service)

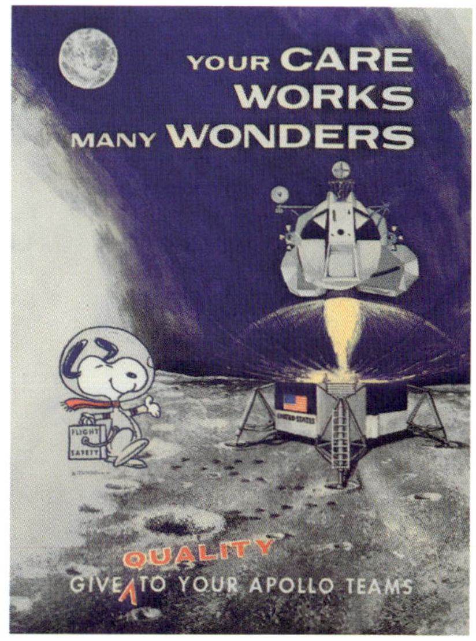

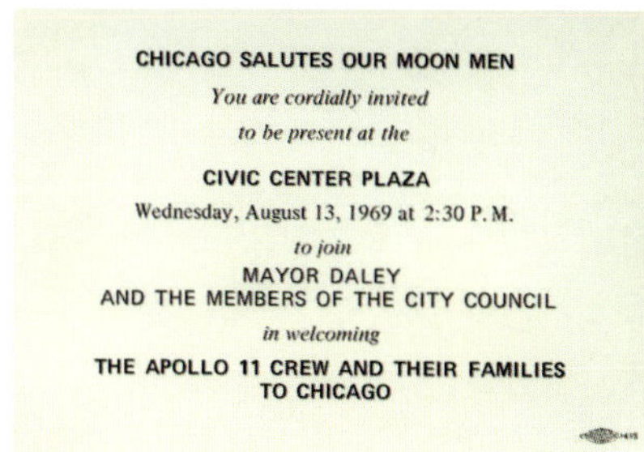

CHICAGO SALUTES OUR MOON MEN

You are cordially invited
to be present at the

CIVIC CENTER PLAZA

Wednesday, August 13, 1969 at 2:30 P. M.

to join

MAYOR DALEY
AND THE MEMBERS OF THE CITY COUNCIL

in welcoming

THE APOLLO 11 CREW AND THEIR FAMILIES
TO CHICAGO

Chicago motor police form a cordon on either side of the crew's car on August 13, as a parade along State Street passes underneath the city's elevated train system and into the Theater District. A ceremony was then held at the Civic Center Plaza. The day had begun in Houston, then included a ticker-tape parade in Manhattan and an appearance at the United Nations. From Chicago, it was on to a state dinner with President Nixon that night in Los Angeles, the first held outside Washington, DC. The NASA party was given the use of Air Force One for the day.

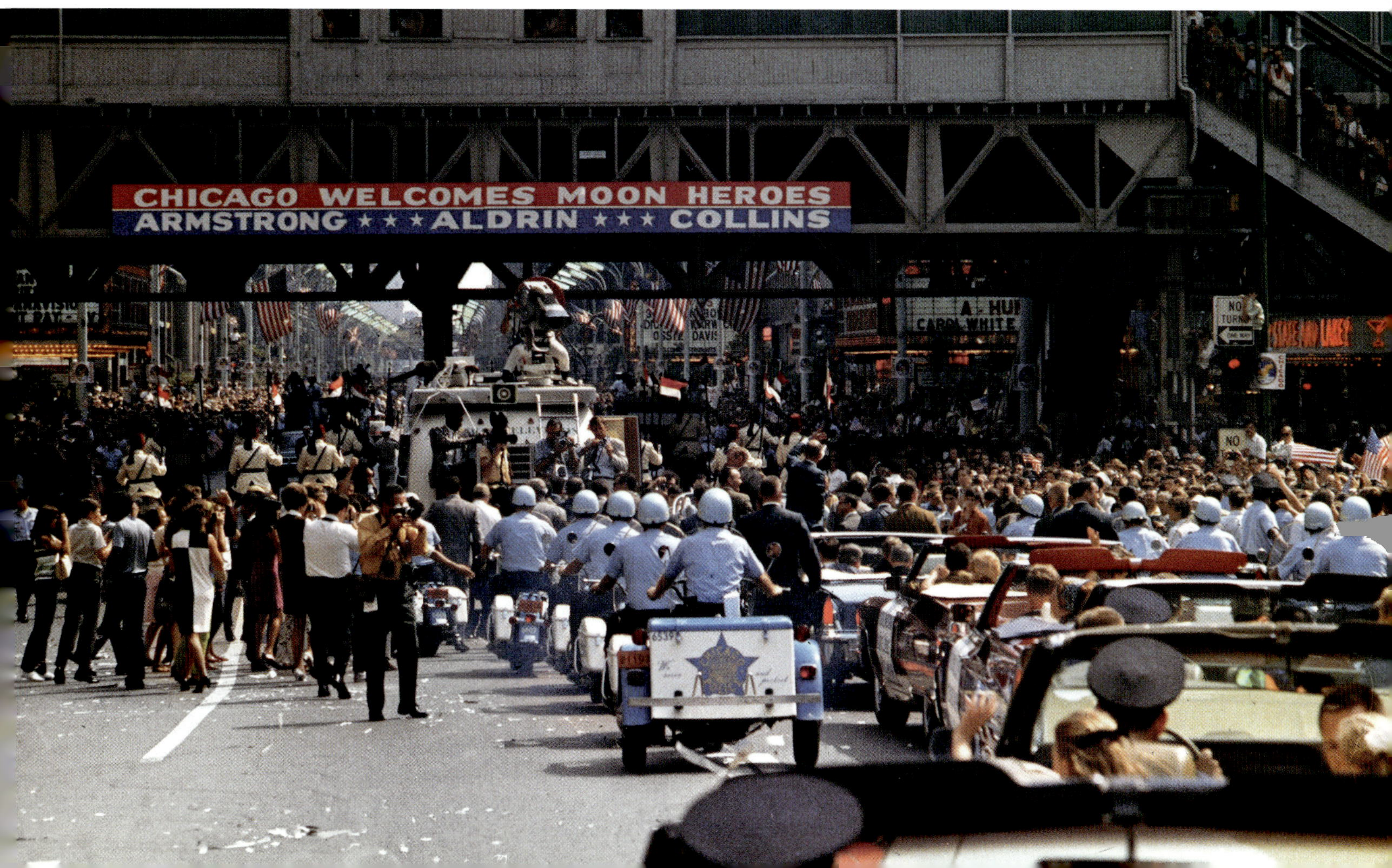

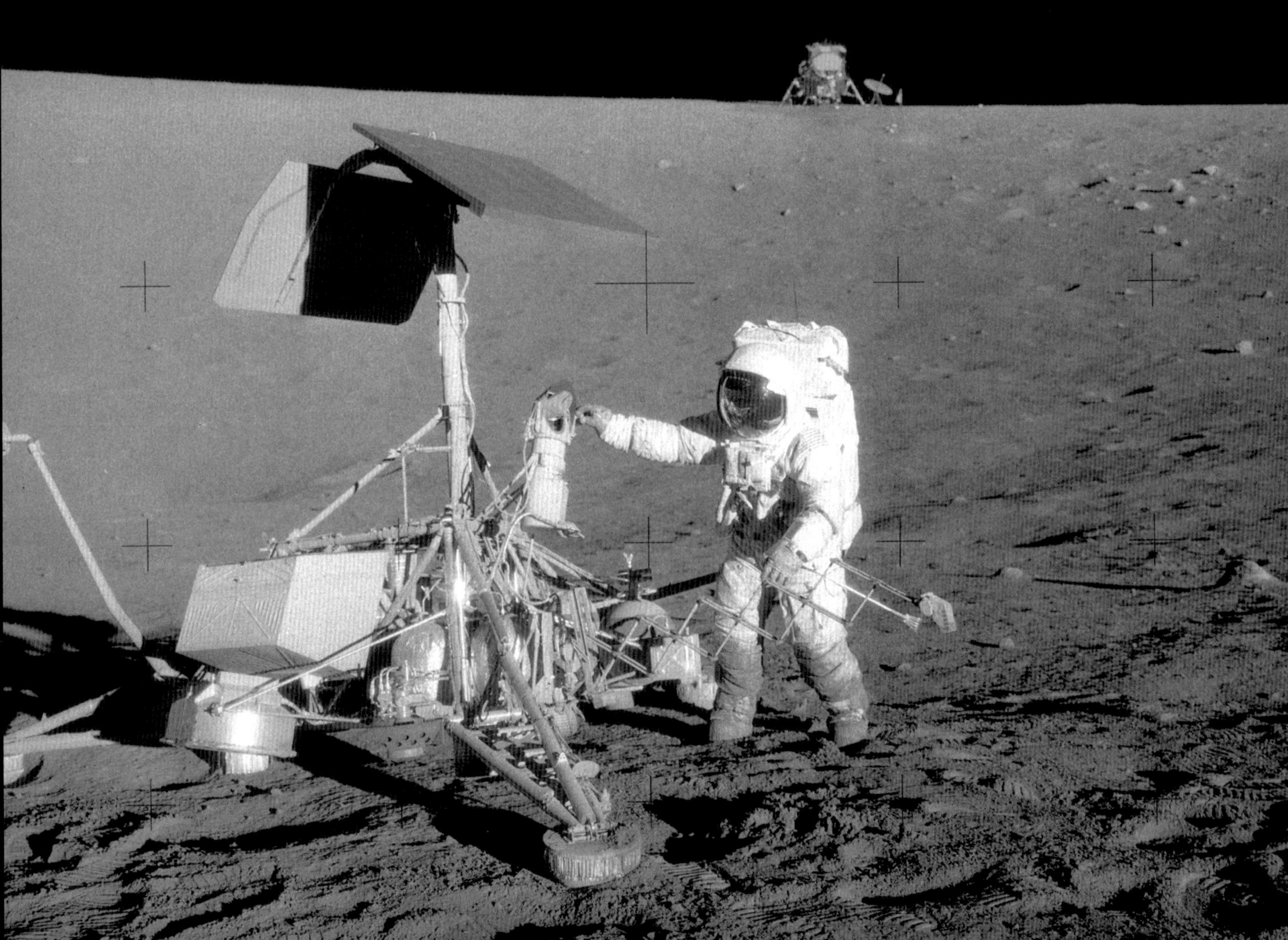

Pete Conrad with the housing of Surveyor 3's TV camera in Surveyor Crater during EVA 2 on November 20, 1969. His right hand holds the arm for the Surveyor's soil-sampling scoop, the first on the moon. Surveyor 3 landed two and a half years earlier, on April 20, 1967. *Intrepid* and the portable S-band antenna are two hundred yards to the northwest.

CHAPTER 7

Apollo 12

November 14–24, 1969

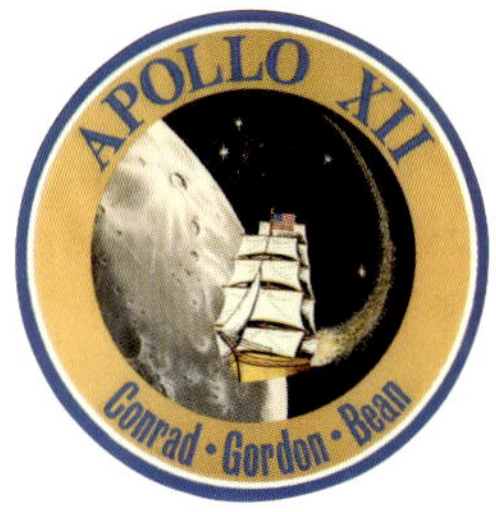

With less attention (and pressure) on Apollo 12, the second lunar landing, the flight took on a more casual feel, which was exemplified by its crew. Two-time Gemini astronaut Pete Conrad, already legendary for his sense of humor, was CDR; he was joined by his Gemini XI comrade Dick Gordon as CMP and rookie Alan Bean as LMP. The trio enjoyed driving matching gold-and-black 1969 Corvettes at Cape Canaveral and were known as an easygoing, fun crew. Bean had joined Conrad and Gordon on the backup crew for Apollo 9 after rookie astronaut Clifton Williams was killed in a T-38 crash in 1967. When they became the prime crew for Apollo 12, Bean, on his first space flight, suddenly became the fourth person scheduled to walk on the moon.

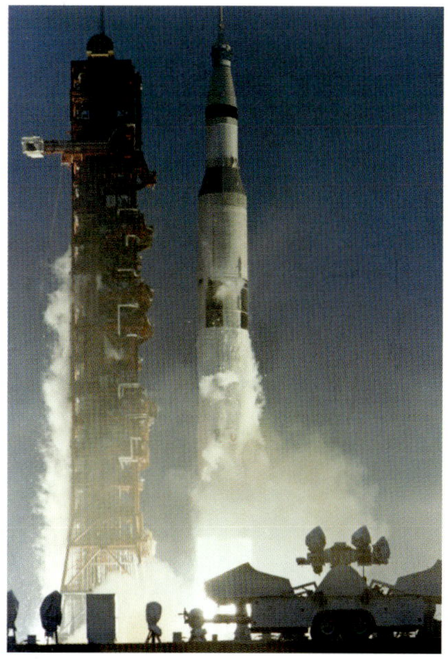
Launch: 11:22 a.m. (EST), November 14, 1969

Apollo 12 was the first of the H missions, which called for precision landings, two-day stays on the moon, and two lunar EVAs. Its goals included a pinpoint landing on the Ocean of Storms, about 930 miles west of Apollo 11's landing site, and the recovery of equipment from an unmanned Surveyor spacecraft, which had landed two and a half years earlier. The all–US Navy crew dubbed their CM *Yankee Clipper* and their LM *Intrepid. Yankee Clipper* had been chosen in a contest among NAR employees.

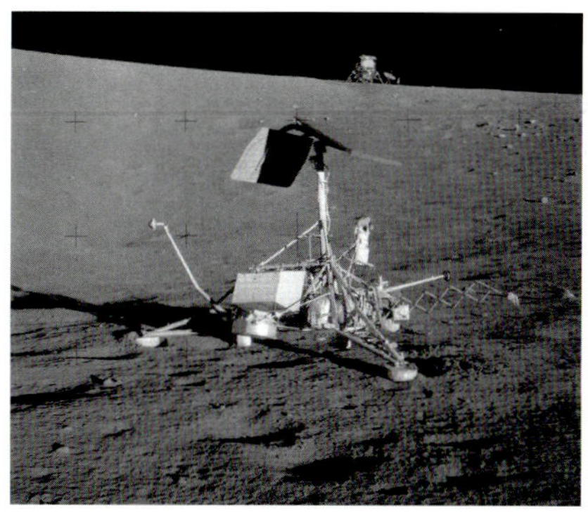
Lunar Landing: 1:54 a.m. (EST), November 19. Ascent: 9:25 a.m. (EST), November 20

No matter how interesting the mission itself may have seemed, however, the launch was even more electrifying, quite literally. The Saturn V was cleared for liftoff at 11:22 a.m. (EST) on November 14, 1969, amid a lightning storm and heavy rain—the worst weather for any manned US launch. Among the guests were President Richard Nixon, First Lady Pat Nixon, and their daughter Tricia; Vice President Spiro Agnew; golfer Arnold Palmer; actor Jimmy Stewart; and Aga Khan IV, spiritual leader of a branch of Shia Islam.

The rocket disappeared into the gray rain clouds. Because the Saturn's exhaust plume acted as a giant lightning rod, the vehicle was struck twice by lightning during the first minute of flight. Protective circuits took all three fuel cells and much of the CM's instrumentation offline. The Saturn V's Instrument Unit, however, was not affected, and the booster performed normally.

At Mission Control in Houston, Electrical, Environmental, and Communications Manager (EECOM) John Aaron remembered the telemetry failure pattern from an earlier test, when a power supply had malfunctioned in the CSM Signal Conditioning Equipment (SCE). (The SCE converted raw signals from instrumentation to standard voltages for spacecraft instrument

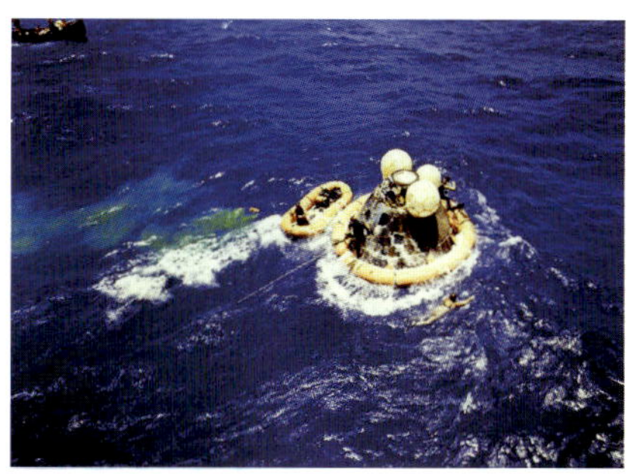
Splashdown: 3:58 p.m. (EST), November 24

displays and telemetry.) Aaron's call, "Try SCE to aux," was relayed to the crew. Neither Flight Director Gerry Griffin, the CAPCOM, nor Conrad recognized the obscure switch. Bean, as the CSM systems engineer, remembered the switch from training a year earlier, when the same failure had been simulated. Aaron's quick thinking and Bean's memory saved what could have been an aborted mission. Bean switched the SCE to a backup power supply to put the fuel cells back online. With telemetry restored, the flight continued successfully.

After a checkout in Earth orbit showed all systems operating normally, Apollo 12 was declared "go for TLI," or translunar insertion.

After LM separation, the empty S-IVB third stage was to be sent into solar orbit. A small error in the Saturn's guidance system, however, left it in Earth orbit until 1971. The stage has been in solar orbit since, although it was briefly recaptured in Earth orbit for about a year, after it was "discovered" by an amateur astronomer in 2002.

Gordon remained in lunar orbit as *Intrepid* landed on the northwest rim of Surveyor Crater, in the southeastern portion of the Ocean of Storms, on November 19, just 553 feet from Surveyor 3, which had landed in April 1967. When Conrad stepped from the last rung of the LM's ladder onto *Intrepid*'s footpad, he said, "Man, that may have been a small one for Neil, but that's a long one for me!"

That initial descent was relayed back to Earth by a Westinghouse TV camera, the first time an astronaut's descent to the surface was sent back in color. Not long after, as Bean was moving the camera to show the entire scene, he inadvertently pointed the lens toward the sun, destroying its low-light capability and rendering the camera useless. Bean finally struck the camera twice with a hammer, which briefly improved the picture, but the TV coverage was over.

During their two EVAs over almost eight hours, Conrad and Bean deployed the first Apollo Lunar Surface Experiments Package (ALSEP), collected about seventy-six pounds of lunar samples, and retrieved parts from the nearby Surveyor.

After a surface stay of a little more than one day and seven hours, the two launched from the moon to dock with *Yankee Clipper*. *Intrepid*'s ascent stage was commanded to crash on the moon, and the resulting moonquake was registered by the ALSEP. The astronauts then performed their trans-Earth injection after forty-five lunar orbits, lasting eighty-six hours and fifty-six minutes.

On November 24, after just more than ten days, the astronauts splashed down in the Pacific Ocean and were recovered by the aircraft carrier USS *Hornet*.

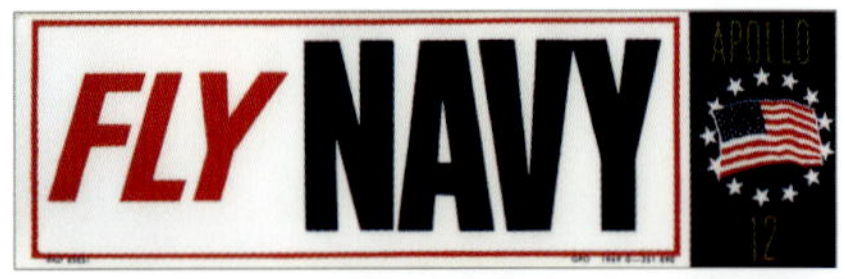

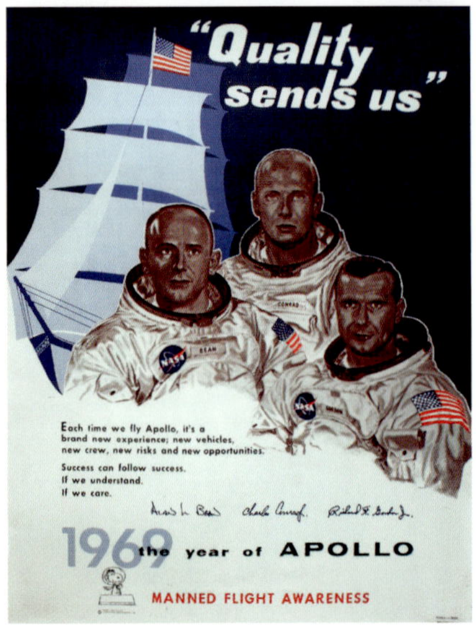

Opposite page, top left: Three days before launch, and after Apollo mission simulator runs, Pete Conrad, Dick Gordon, and Alan Bean pose with Bean's 1969 Chevrolet Corvette in the parking lot of the Flight Crew Training Building at KSC on November 11, 1969. The three drove matching Corvettes, thanks to favorable lease deals for astronauts from Chevrolet dealer Jim Rathman, a Melbourne, Florida–based former race-car driver. (Rathman won the Indianapolis 500 in 1960.) Bean, an artist, liked the car's appearance more than its performance. It is now owned by a collector in Texas, who has kept it original, including by maintaining the "LMP" on the fenders.

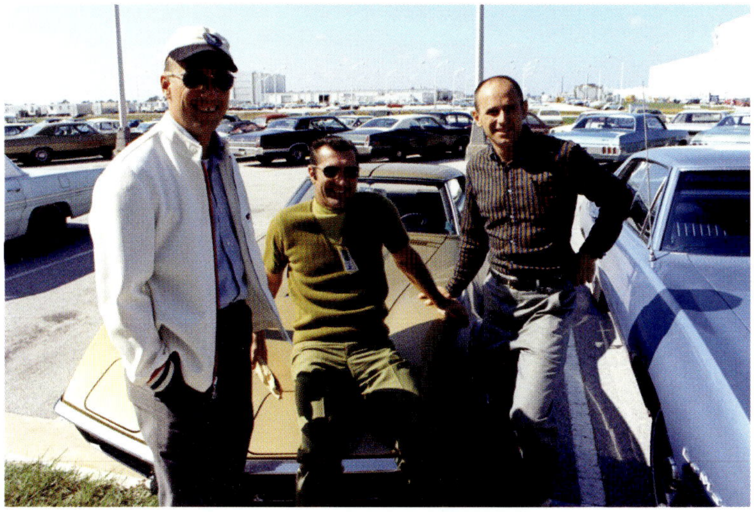

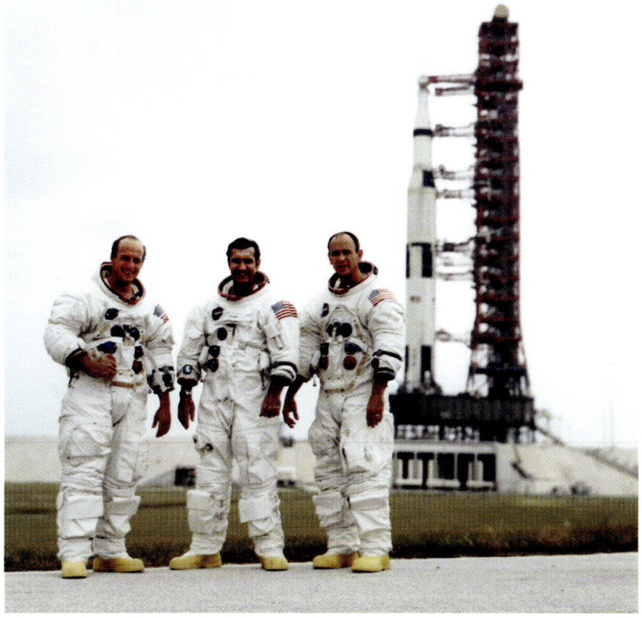

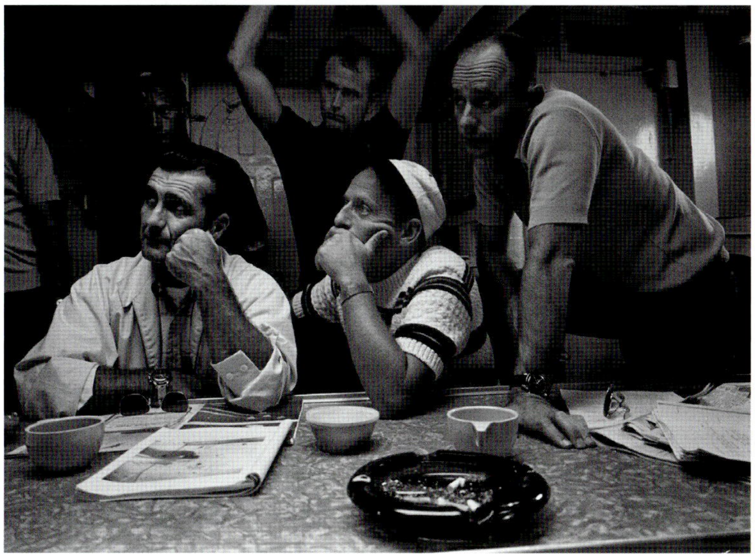

Above: Gordon, Conrad, and Bean aboard MV *Retriever* before spacecraft egress and recovery training in the Gulf of Mexico near Galveston, Texas, on September 20, 1969. The Gordon and Bean photos below are also from this meeting.

Above: Conrad, Gordon, and Bean at Pad 39A on October 29, 1969, after the CDDT. As CMP, Gordon had a suit with fewer oxygen hose connectors; Conrad's and Bean's suits also connected to their lunar surface PLSS backpacks.

Bottom center: US Navy Cmdr. Pete Conrad during geology classroom training in Flagstaff, Arizona, on October 14, 1969. Conrad, thirty-nine, was considered the best pilot in the second group of astronauts. Apollo 12 was his third space flight, and his second as CDR. In 1973, he would lead a flight to rescue the Skylab space station.

Bottom left: US Navy Cmdr. Dick Gordon, thirty-nine, had flown with Conrad on Gemini XI. Apollo 12 was his final space flight. In 1972, he became executive vice president of the New Orleans Saints, owned by John Mecom (whom he had met racing boats in 1965).

Bottom right: US Navy Cmdr. Alan Bean, thirty-seven, had applied for the second astronaut group, which included Conrad. He was accepted the following year, with Gordon.

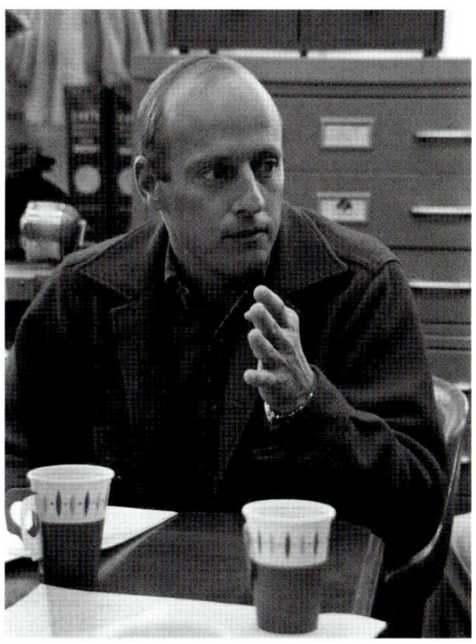

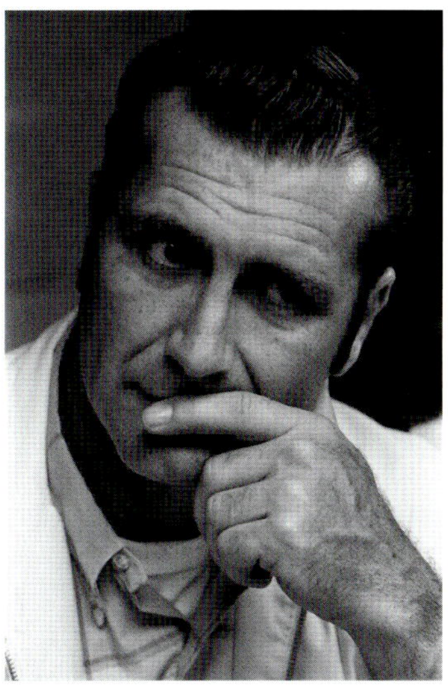

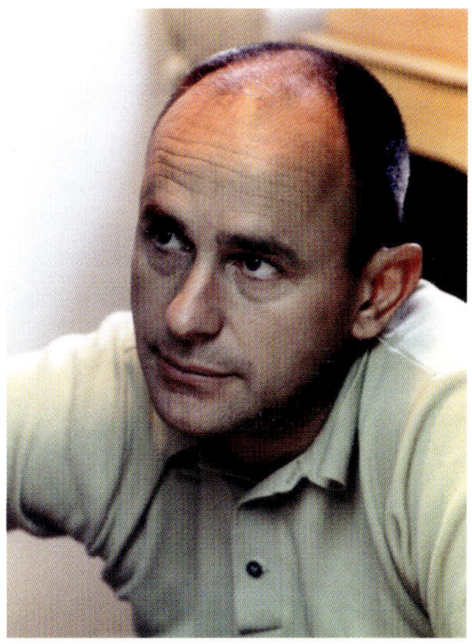

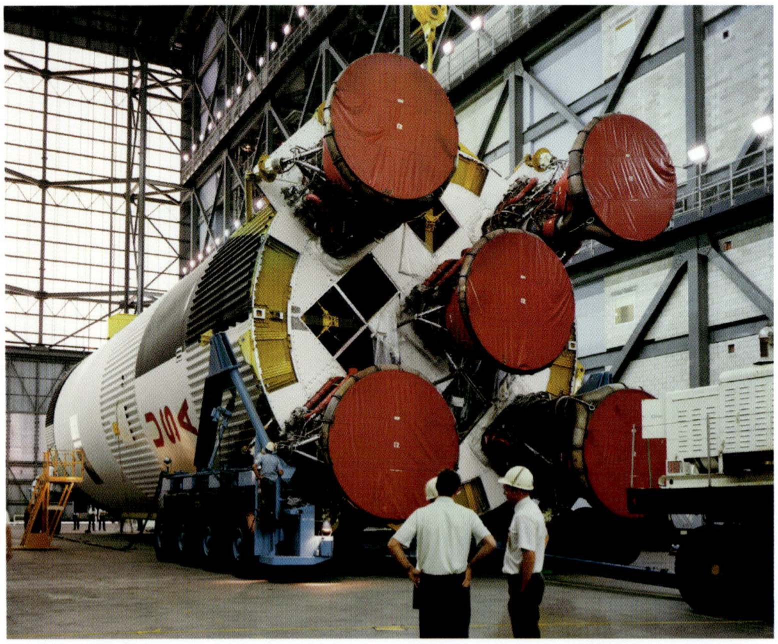

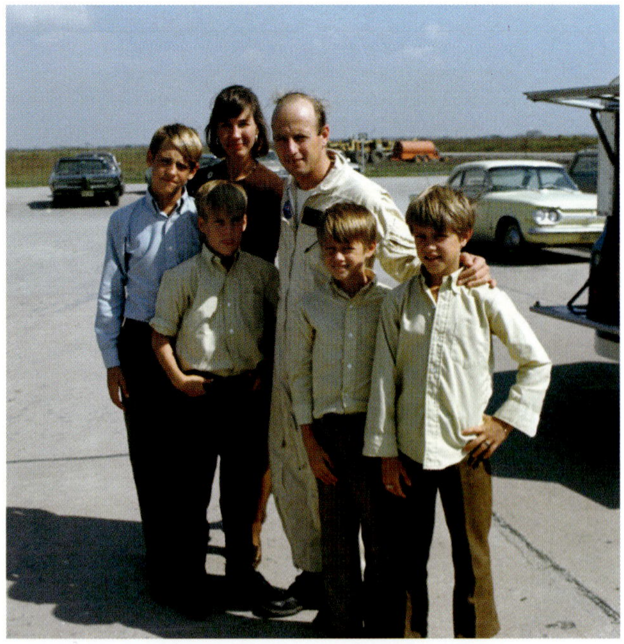

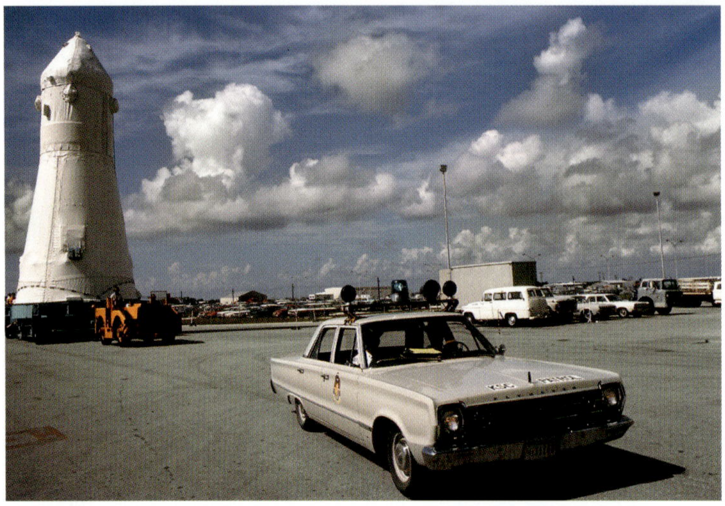

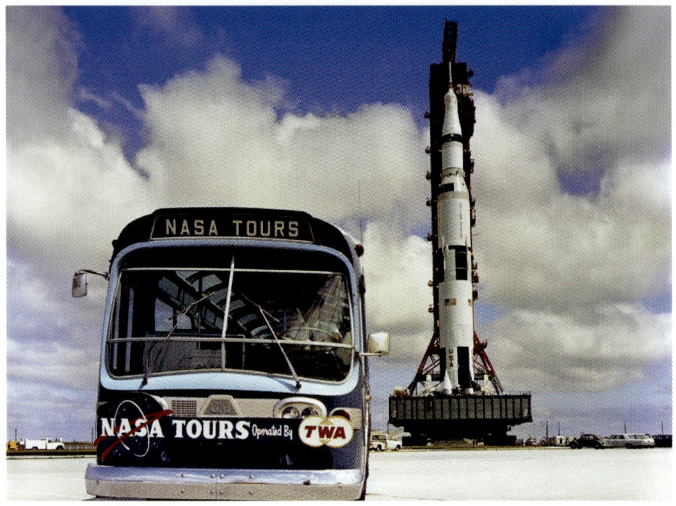

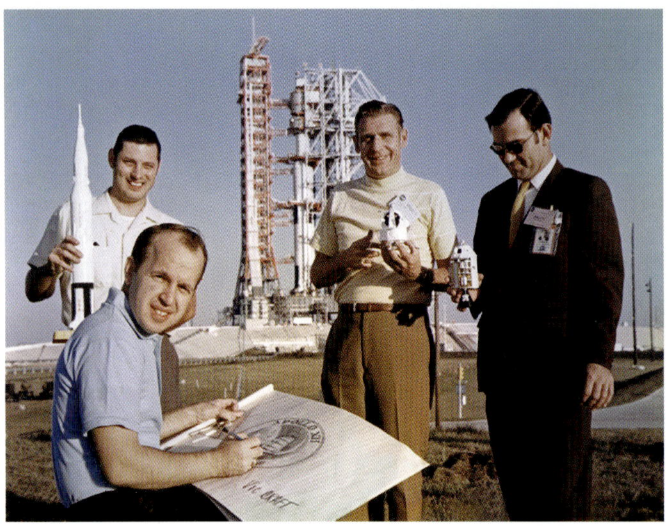

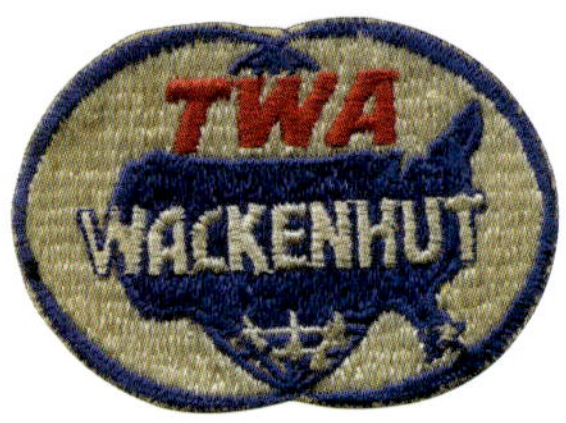

Top left: The Saturn V's S-IC first stage in the VAB's transfer aisle, prior to stacking on May 7, 1969. Red protective covers are over its five F-1 Rocketdyne engines.

Top right: Jane and Pete Conrad with sons Peter, Andrew, Christopher, and Thomas at Ellington AFB after Conrad's training run in the LLTV on October 25, 1969.

Center left: A KSC Security Patrol 1966 Plymouth Belvedere escorts the Apollo 12 spacecraft on June 30, 1969. The spacecraft is being towed toward the VAB from the MSOB. (photo by Tiziou News Service)

Center right: A TWA NASA Tours bus parks on the Saturn Causeway, approaching the perimeter of Pad 39A on September 8, 1969. Apollo 12 is being rolled toward the hardstand on ML-2. Public tours were not permitted this close to a launch vehicle.

Above: The astronauts chose the mission emblem design—by Victor R. Craft of the RCA Service Company in nearby Melbourne, Florida—from among two hundred submitted. Behind Craft are (*left to right*) Walt Mack, editor of RCA's magazine; Howard Enders of RCA corporate publicity; and a NASA tour guide. They hold Revell kit models. The MSS is in place to the right of the Saturn V at Pad 39A.

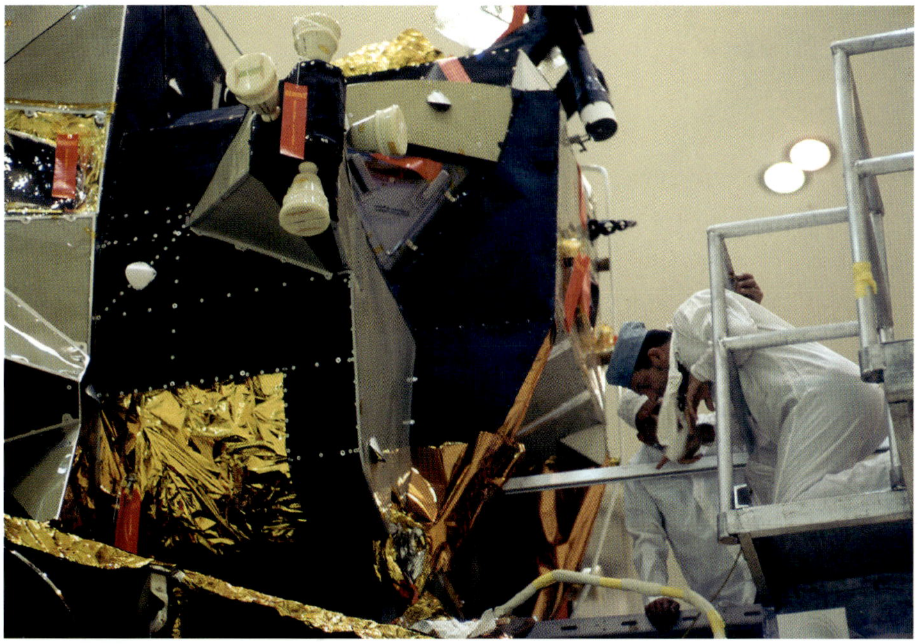

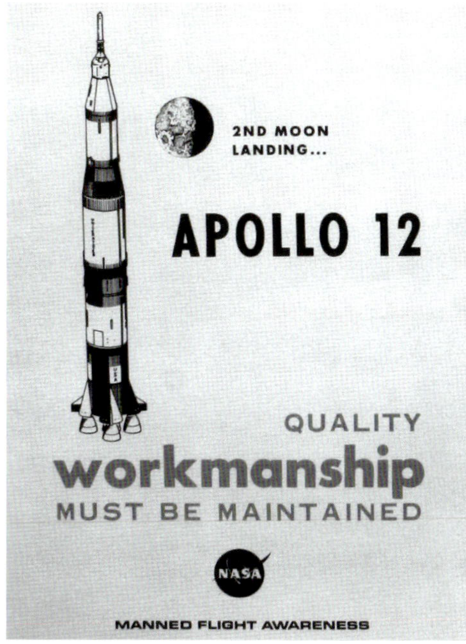

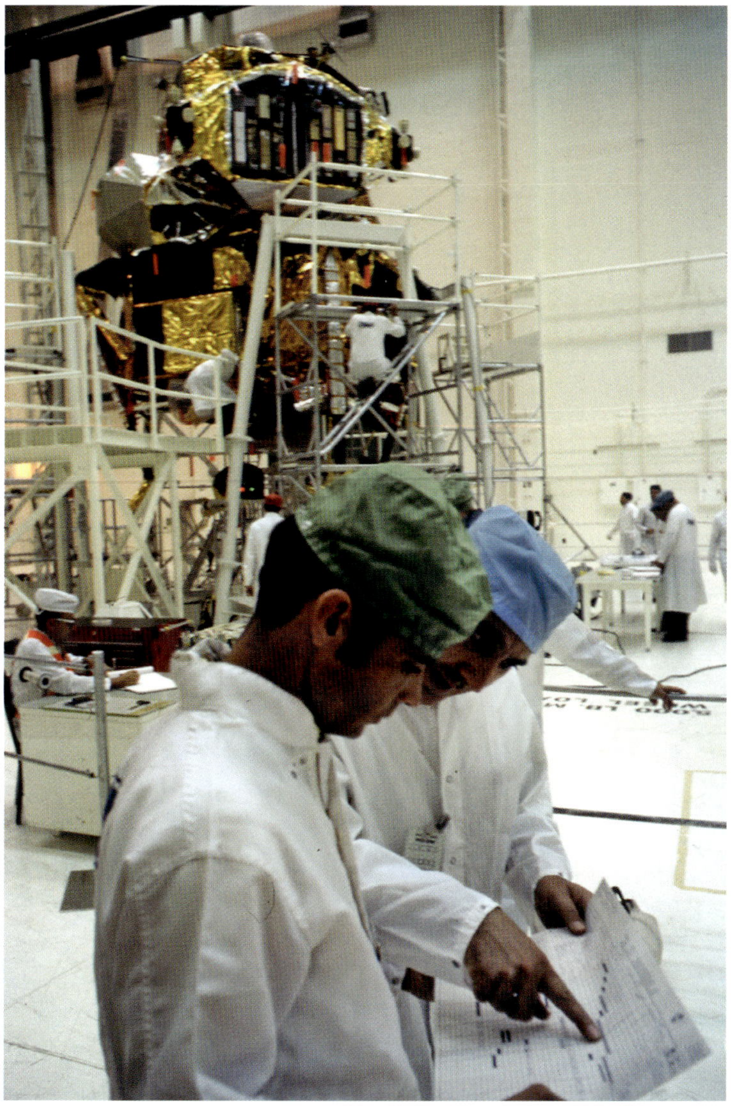

Above: Technicians look over a processing schedule during closeout procedures on LM-6, in a test stand in the MSOB on June 26, 1969. An access control desk is at left. Metal alloy foil panels have not yet been installed on the rear of the ascent stage. (photo by Tiziou News Service)

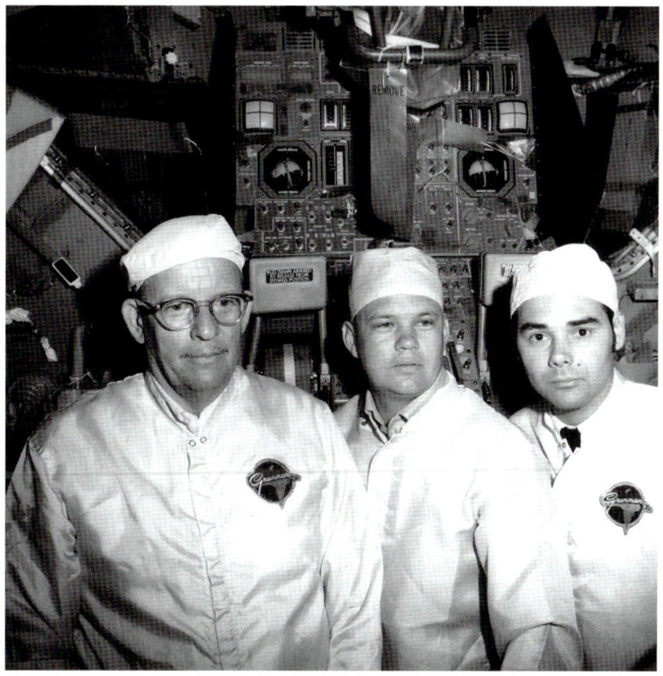

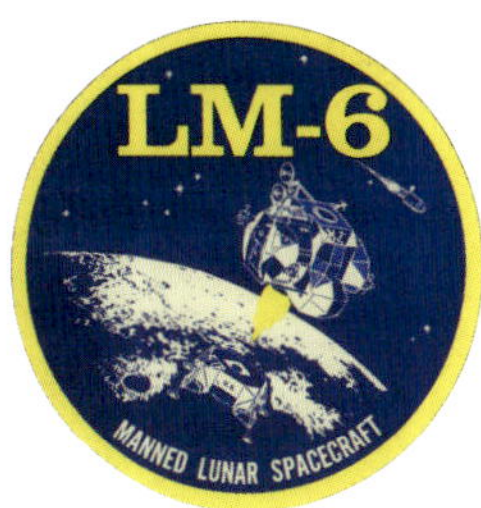

Top left: Technicians inspect the porch of the Apollo 12 LM ascent stage during closeout procedures in the MSOB on June 26, 1969. Black areas are foil sheets of the superalloy Inconel and nickel. (photo by Tiziou News Service)

Center right: Grumman Aerospace technicians in the crew compartment of LM-6 on August 27, 1969, during closeout operations inside the SLA near the top of the Saturn V stack. They would leave with the "Remove Before Flight" ribbons on the LM's windows and instrument panels (*behind them*). Rollout to Pad 39A occurred twelve days later.

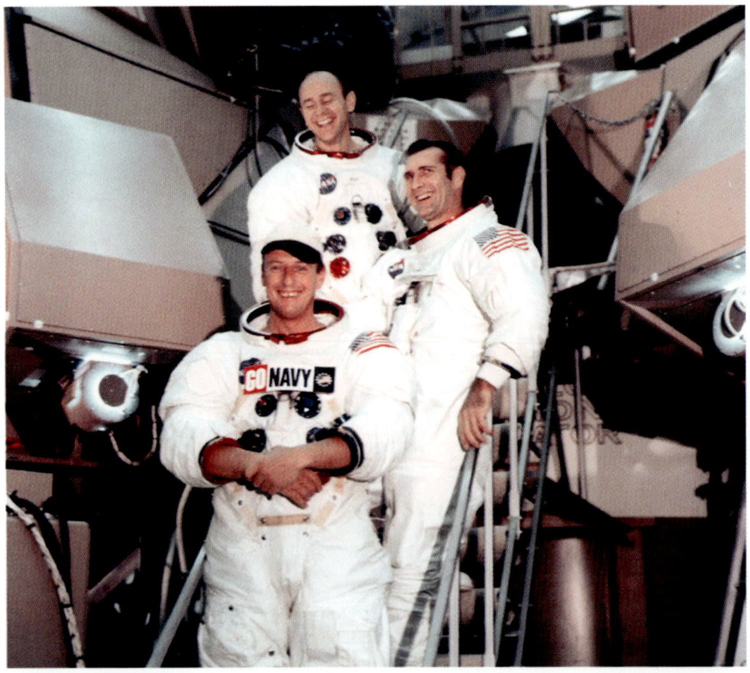

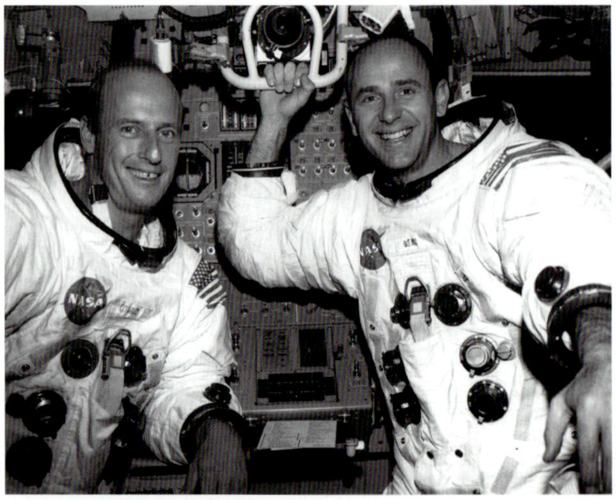

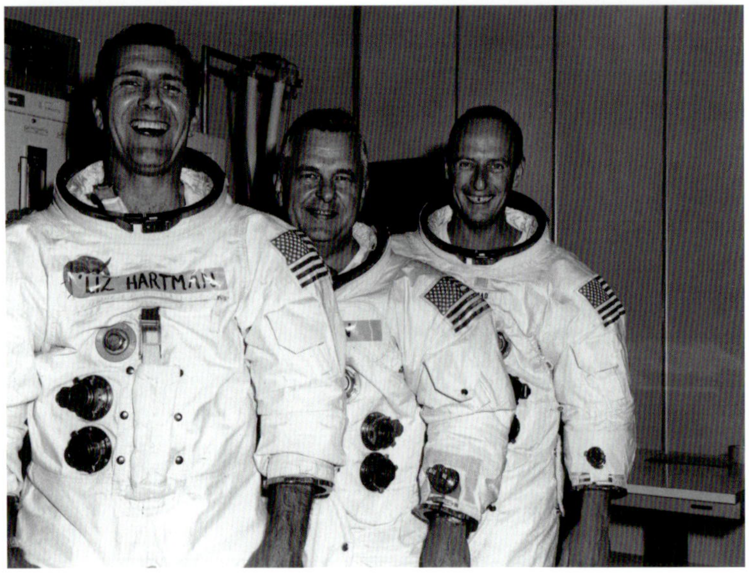

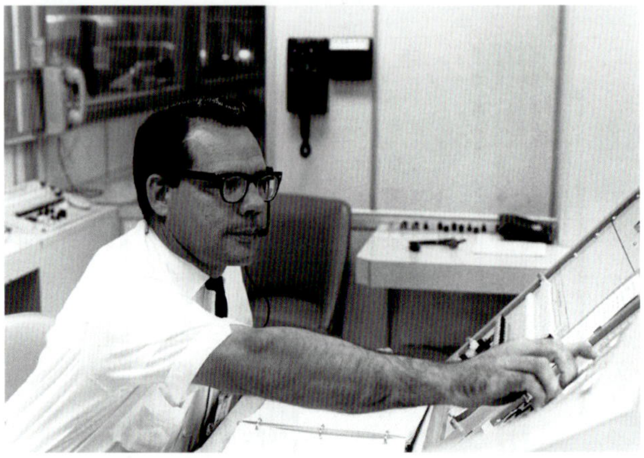

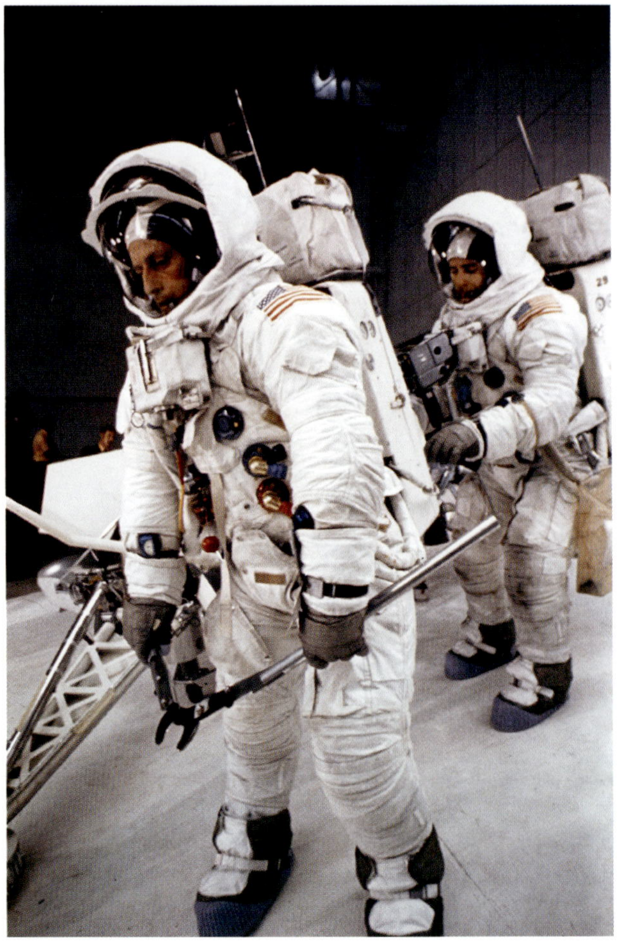

Top left, top to bottom: Bean, Gordon, and Conrad (displaying his crew's US Navy pride) on the steps of the Command Module Simulator at the Flight Crew Training Building at KSC on October 22, 1969.

Top right: Conrad and Bean at their flight stations inside the Lunar Module Simulator at the Flight Crew Training Building on October 22, 1969. Above Bean's hand is the Alignment Optical Telescope, used to determine the LM's position based on stars and crew input. The lower panel between them contains the push buttons and data displays of the LM guidance computer.

Bottom left: Gordon and Conrad pose in space suits with Launch Director Walter Kapryan (*center*) in April 1969, as part of a twenty-seventh birthday joke on Elizabeth K. Hartman, who worked in Kapryan's office at KSC.

Center right: Lead Command Module Test Conductor Clarence "Skip" Chauvin mans a console in an Acceptance Checkout Equipment (ACE) station during prelaunch activities in the MSOB at KSC on November 7, 1969. The ACE system monitored 3,500 different spacecraft measurements.

Bottom right: Conrad and Bean during lunar surface training with a mock-up of Surveyor 3, in the Flight Crew Training Building at KSC on September 30, 1969. Conrad holds the cutters used to remove equipment from the spacecraft.

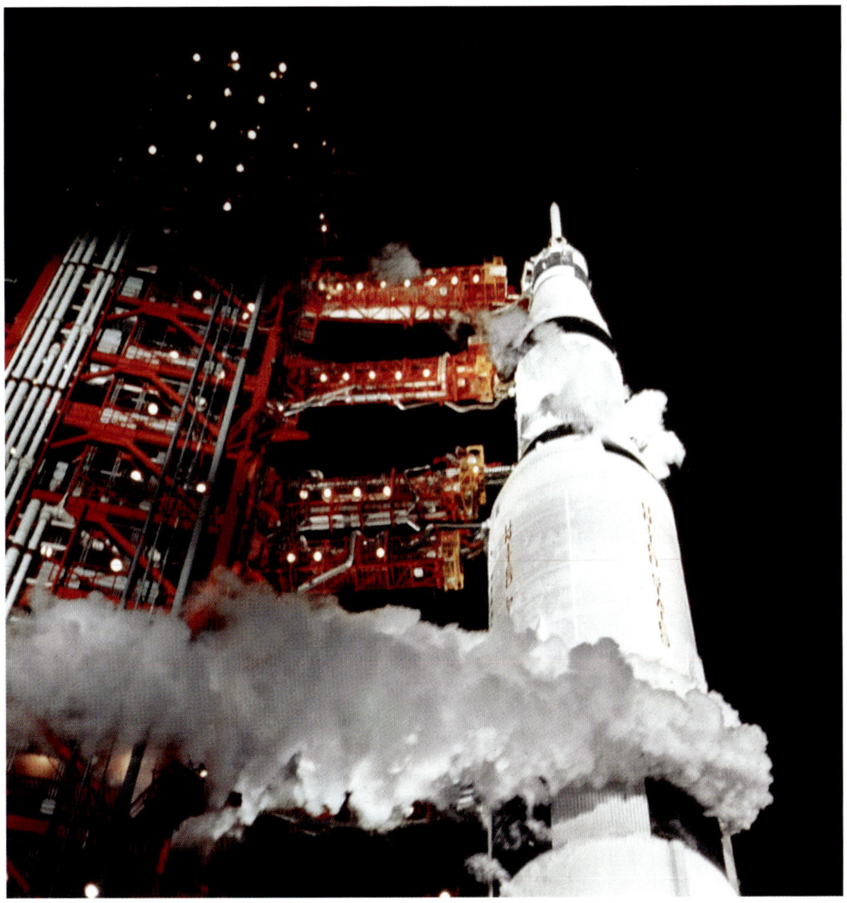

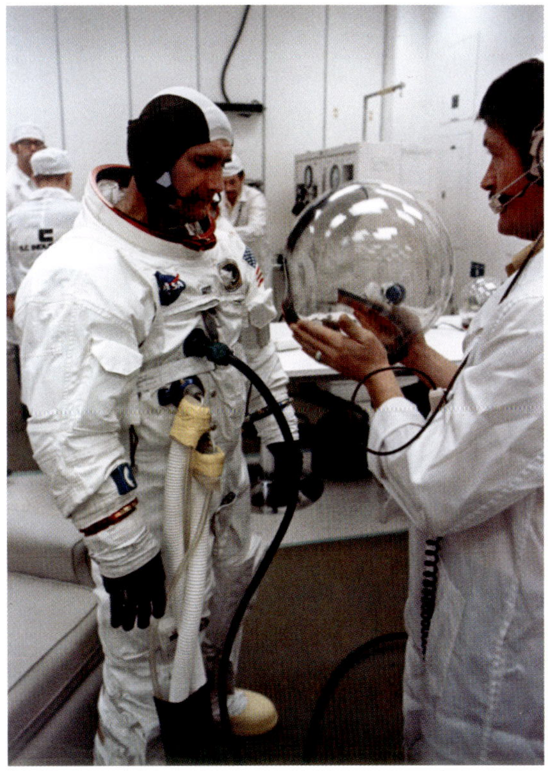

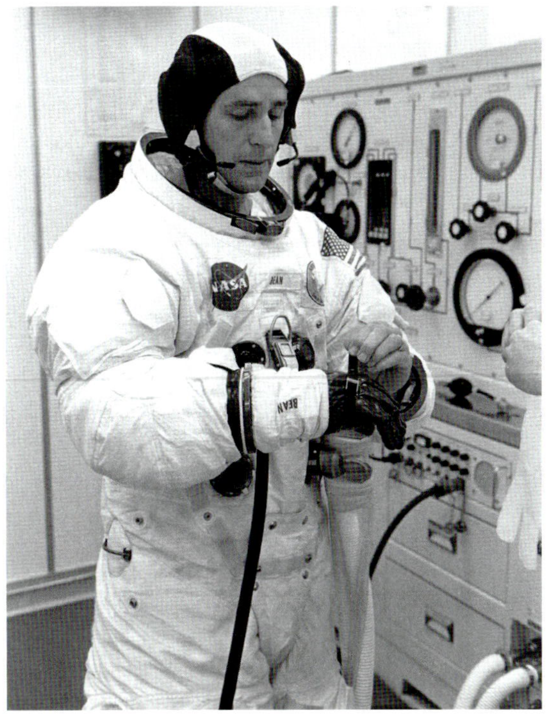

Top left: The Saturn V's three stages vent super-cold oxygen vapor—creating condensation clouds in the Florida humidity—during fueling operations at Pad 39A on November 13, 1969, the night before launch.

Top right: NASA suit tech Bill Prior assists Gordon during suiting on launch morning. Suit tech John Bolan is to the right of Gordon.

Center right: Bean adjusts a glove while suiting on launch morning.

Bottom center: Conrad during a session with the news media following an EVA training exercise, at the Flight Crew Training Building at KSC on September 30, 1969.

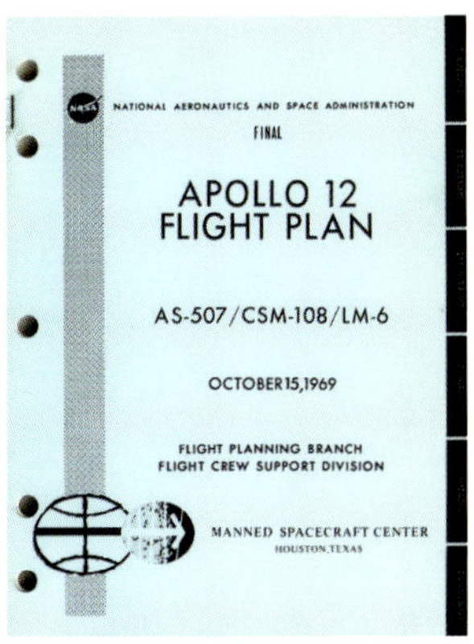

NATIONAL AERONAUTICS AND SPACE ADMINISTRATION

FINAL

APOLLO 12
FLIGHT PLAN

AS-507/CSM-108/LM-6

OCTOBER 15,1969

FLIGHT PLANNING BRANCH
FLIGHT CREW SUPPORT DIVISION

MANNED SPACECRAFT CENTER
HOUSTON, TEXAS

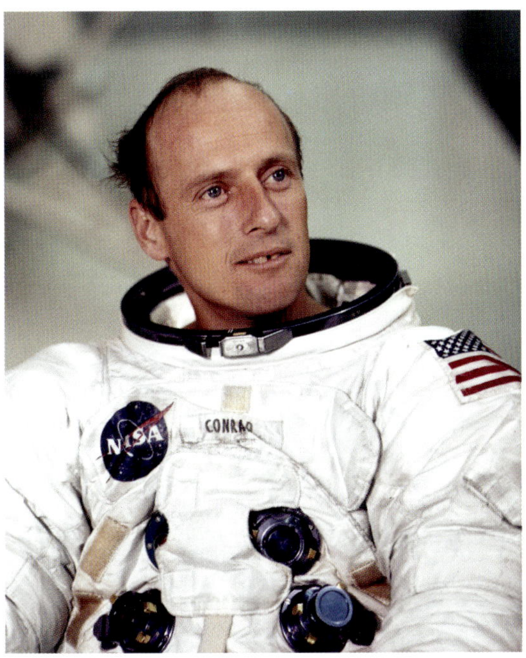

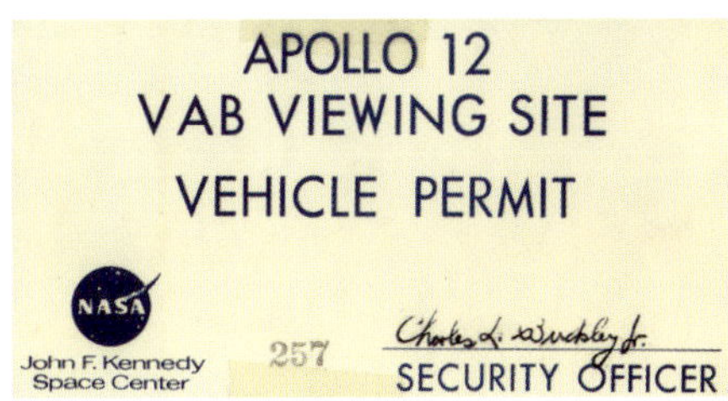

APOLLO 12
VAB VIEWING SITE
VEHICLE PERMIT

John F. Kennedy
Space Center

257

SECURITY OFFICER

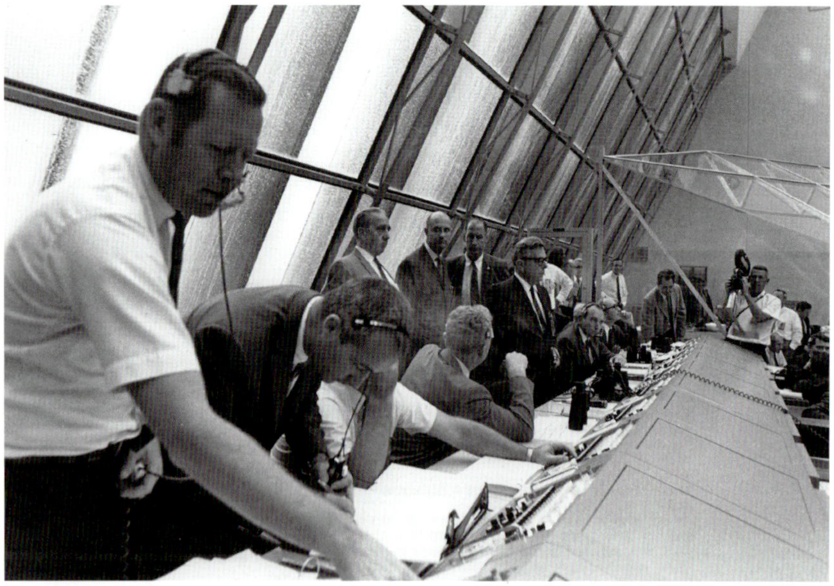

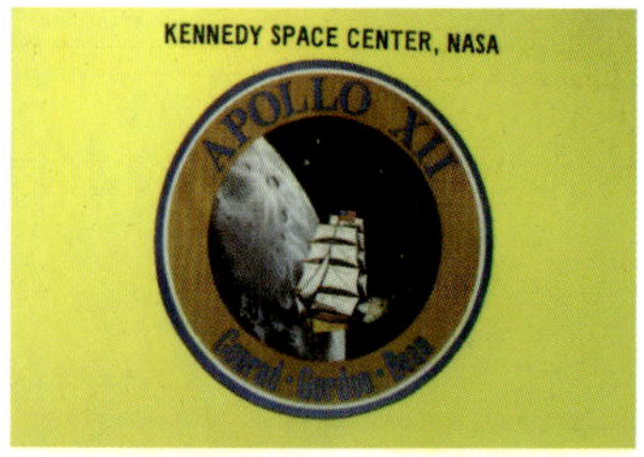

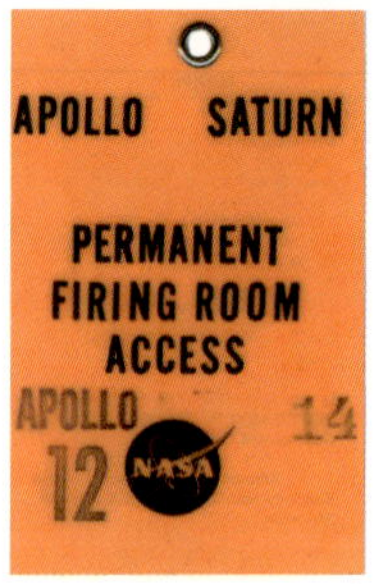

Top left: Management row in Firing Room 2 on launch morning. *Front row, left to right*: Jack King, Jim McDivitt, John Williams (hidden), Paul Donnelly, Walt Kapryan, and Andy Pickett, director of mechanical and propulsion. Deputy Director of Launch Vehicle Operations Ike Rigell stands at the end of the console. *Back row, left to right*: KSC Director Kurt Debus, Tom Stafford, NASA Deputy Administrator George Low, and Rocco Petrone (hidden behind Kapryan). Al Bransford is the film cameraman at right.

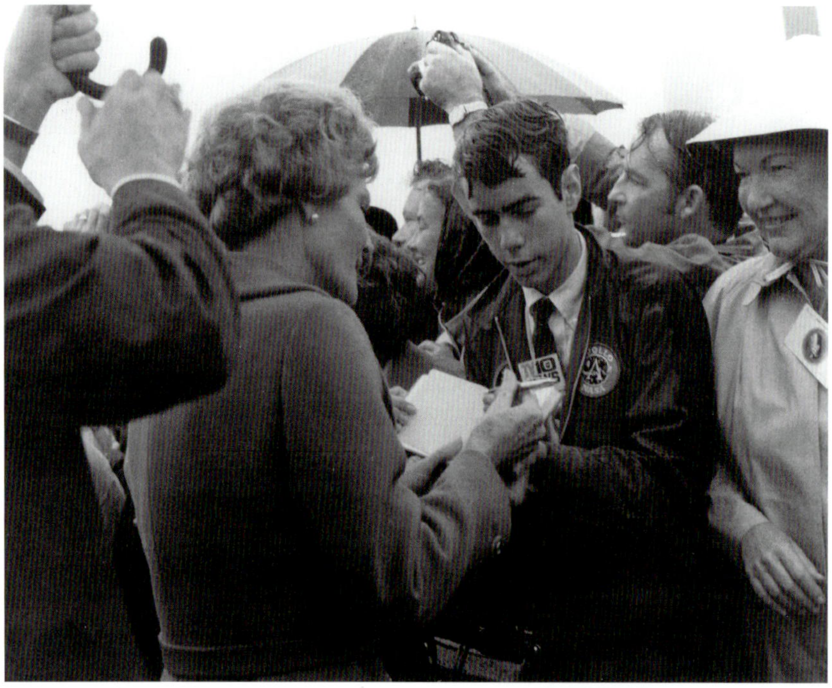

Center left: On launch morning, First Lady Pat Nixon signs an autograph for John Bisney, along the rope line leading from President Nixon's helicopter landing pad to the VIP viewing stand on the north side of the VAB. Bisney had written to Nixon requesting a launch invitation; only later did the president decide to attend.

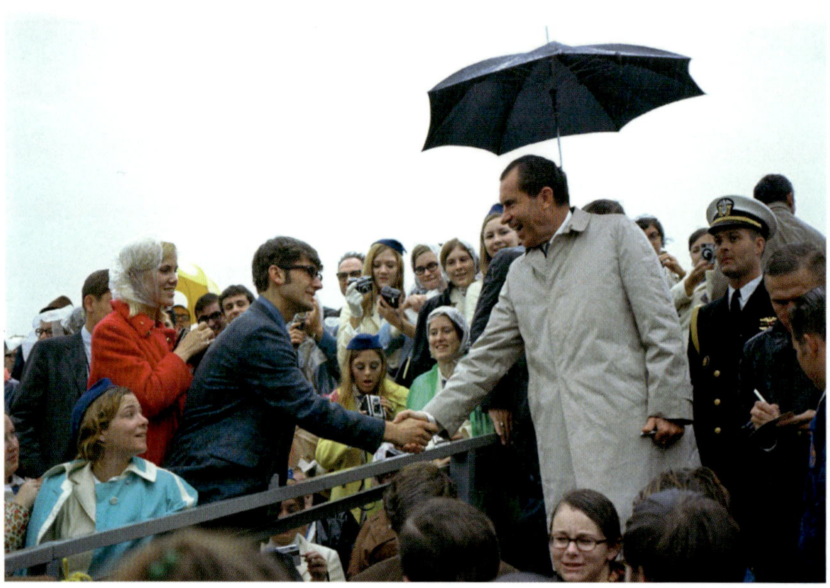

Bottom left: President Nixon greets guests at the VIP viewing stand on launch morning. His naval aide behind him, Lt. Cdr. Charles R. Larson, became a four-star admiral. Astronaut Frank Borman, who had been assigned as a liaison to the White House, signs an autograph at right. The following August, Nixon sent Borman on an unsuccessful twelve-nation tour to seek the release of US POWs held by North Vietnam. Nixon was the first incumbent president to attend a launch from KSC or Cape Canaveral. He was also the only one to do so, until then-president Bill Clinton attended the STS-95 Space Shuttle launch of John Glenn in 1998.

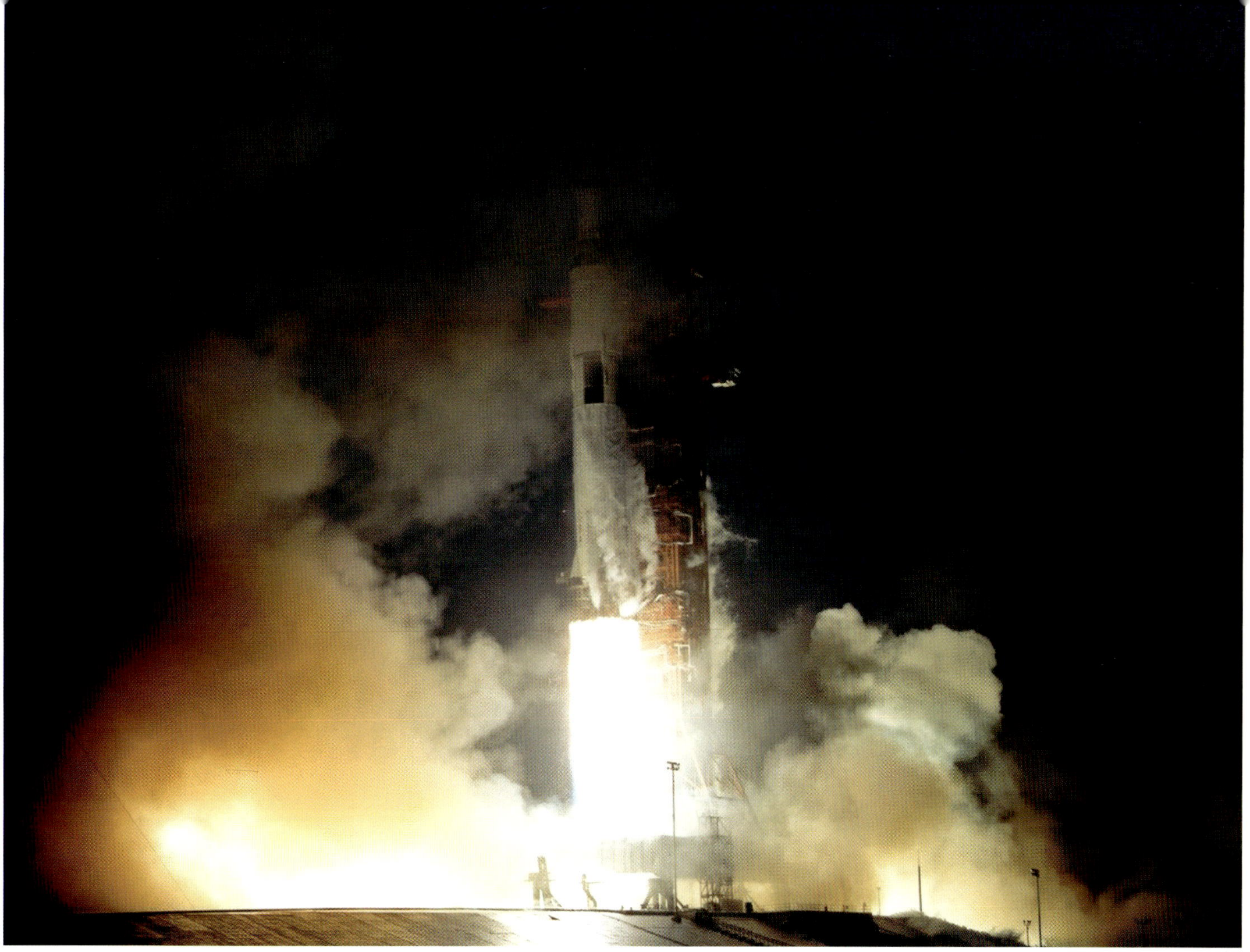

Above: Apollo 12 slowly roars away from Pad 39A at 11:22 a.m. (EST) on November 14, 1969.

Right: Raindrops dot the lens of a camera at the 360-foot level of the LUT (near the top), looking south. Lightning struck the launch vehicle twice less than one minute after liftoff.

Below: Some of the estimated three hundred thousand launch spectators line the water at Jetty Park Campground, 13.5 miles south of Pad 39A. This was the northernmost public viewing location along the Atlantic. The photograph's view looks west into Port Canaveral, with most vehicles pointing north-northeast. Viewing spots along US Route 1 and the Indian River in Titusville (to the north) were slightly closer, at 12 miles west of the pad. The VIP stands at the VAB, by comparison, were 3 miles west of the launch.

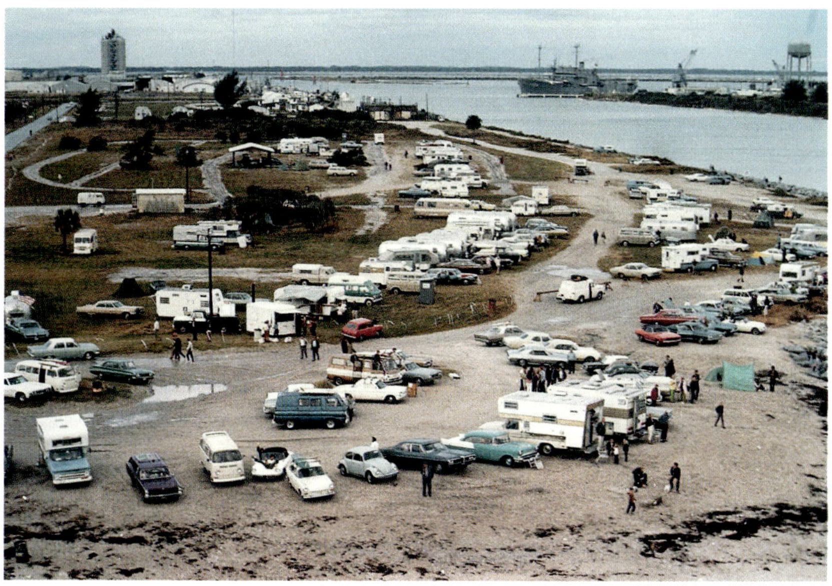

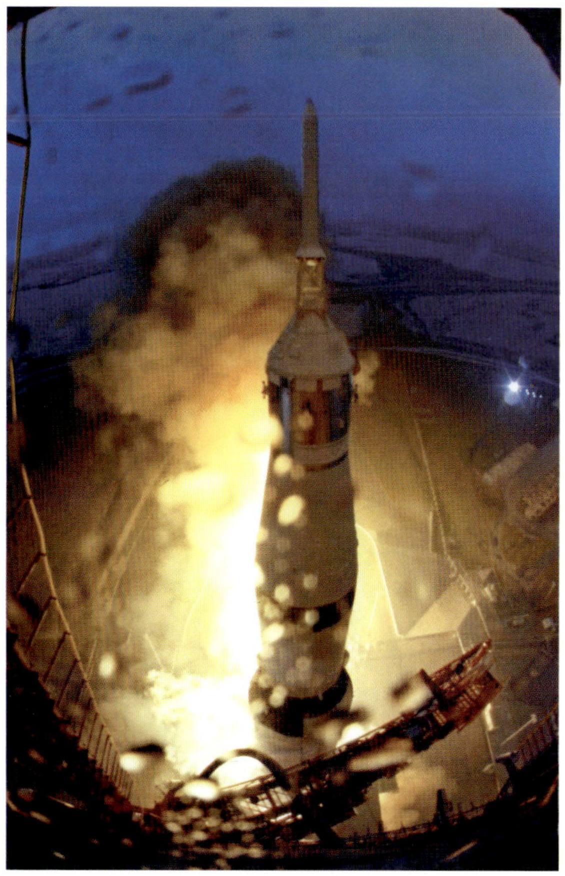

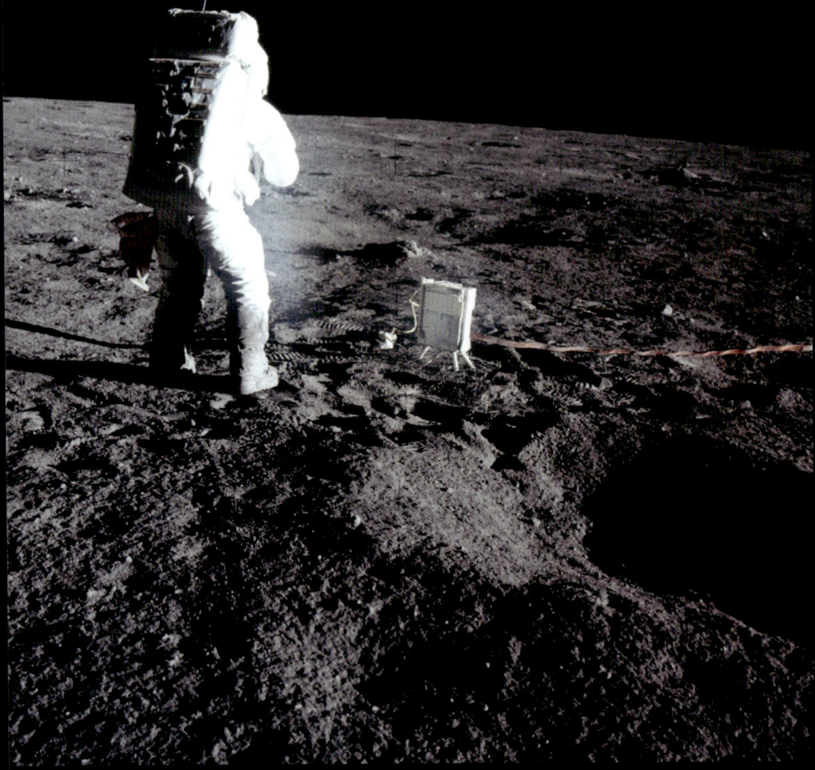

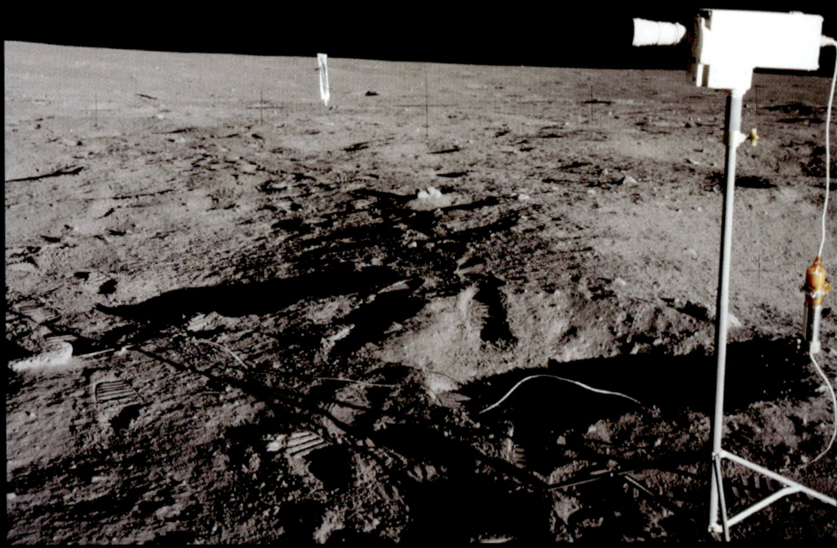

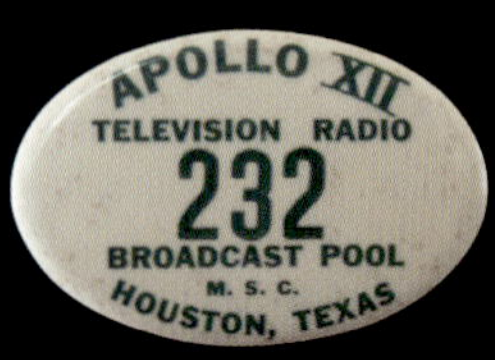

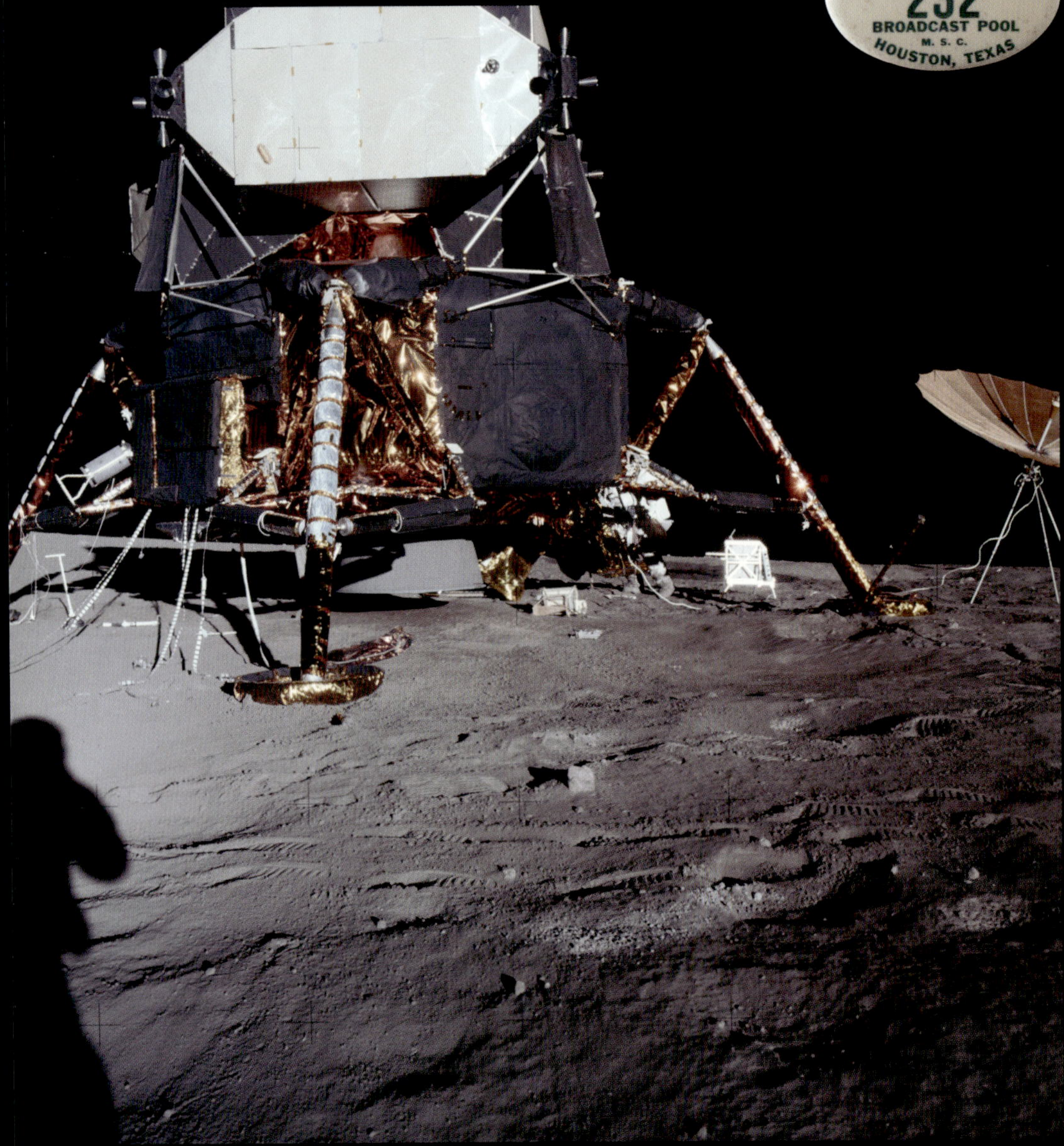

Top left: Bean takes a photo of the Suprathermal Ion Detector Experiment (SIDE) during EVA 1 on November 19. The small cold cathode ion gauge, which failed within fourteen hours, is to the left of the SIDE; the ribbon cable connecting SIDE to the ALSEP Central Station runs to the right. The blue "fog" is a dust smudge.

Top right: A view looking north of the LM during EVA 1, with the SWCE at center and the burned-out Westinghouse color TV camera at right. The camera, painted white for thermal control, was from the Apollo 10 CM. The orange area covers an adapter, which was required because the camera's connector did not match the extension cable back to the LM. The cable was designed for a black-and-white camera; the late decision to use the color camera precluded a more elegant solution.

Bottom left: The first in a series of photos by Bean, taken for a panorama during EVA 1 on November 19. The photo shows the rear of *Intrepid*, with a lunar contact probe bent upward at right, near the portable S-band antenna. Conrad works at the Modularized Equipment Stowage Assembly (MESA), just right of the LM. The gray cylinder at left is the cask for the plutonium-238 fuel capsule, used to power the ALSEP for a year. Its SNAP-27 (Systems Nuclear Auxiliary Power) generator was the first nuclear electrical power source on the moon. It remained functional when the ALSEP program was discontinued in 1977.

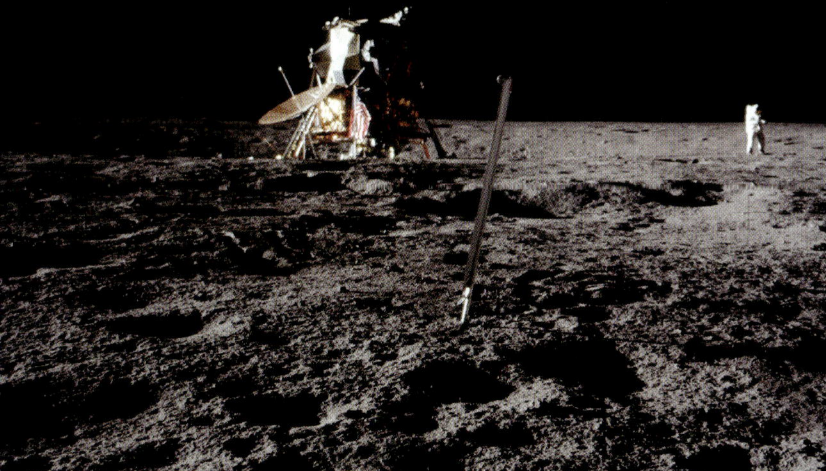

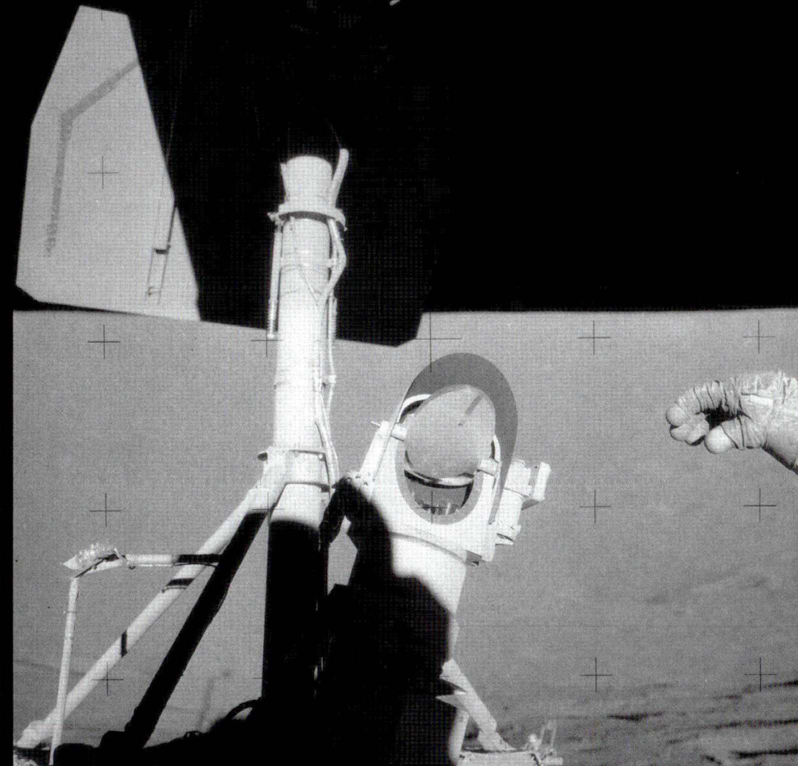

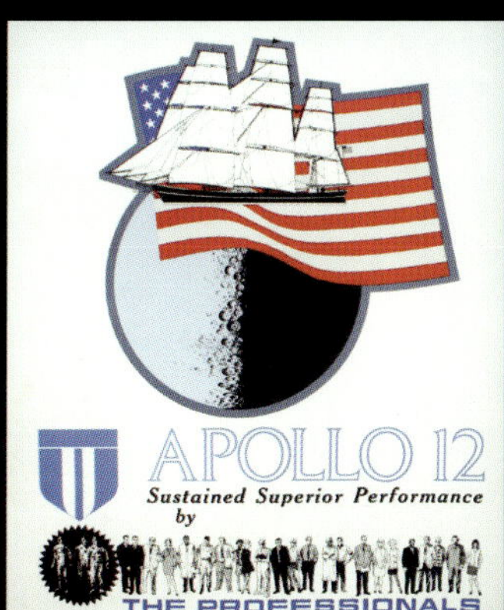

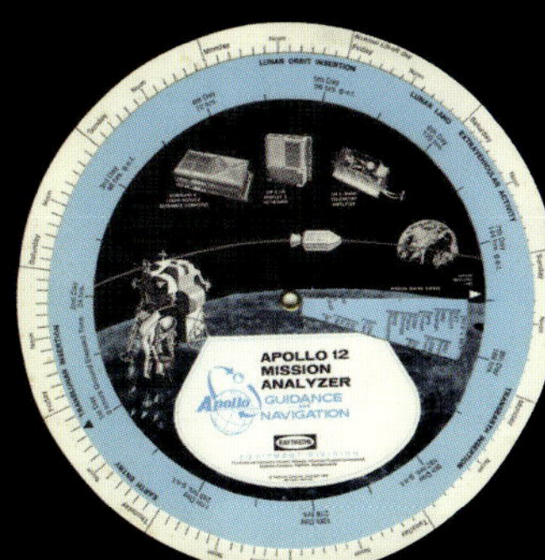

Top left: Bean's photo during EVA 1 on November 19, looking south toward *Intrepid*. The SWCE is in the foreground and the portable S-band antenna points toward Earth. Conrad, at right, is taking a series of panoramic photos, which would include Bean. Surveyor Crater is behind *Intrepid* at left.

Top right: Conrad reaches toward the Surveyor 3 television camera assembly during EVA 2 on November 20. With one finger, he has already drawn a line in the dust on the mirror, which reflected images to the camera below. Surveyor's two solar power panels are at top.

Center right: "Mission Analyzer" promotional item from the Raytheon Company, which built the Apollo guidance computer designed by MIT

Bottom right: Bean's photo, looking southwest from the LM, of the US flag, with a partial shadow from the portable S-band antenna at left. The flat object in the foreground is the S-band antenna cover.

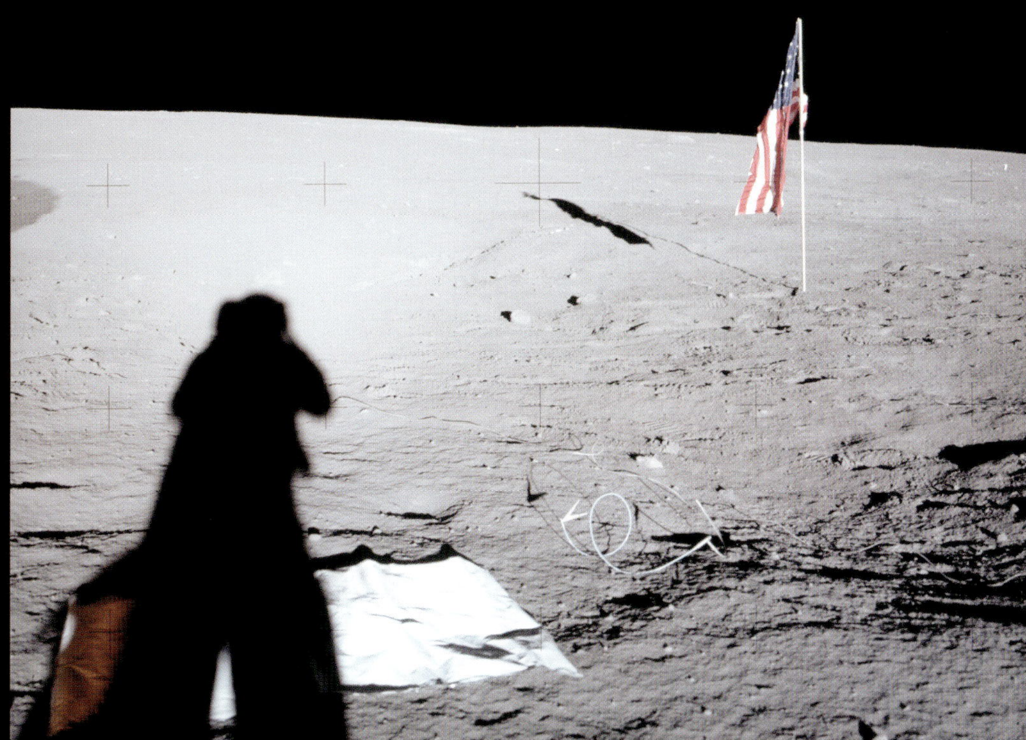

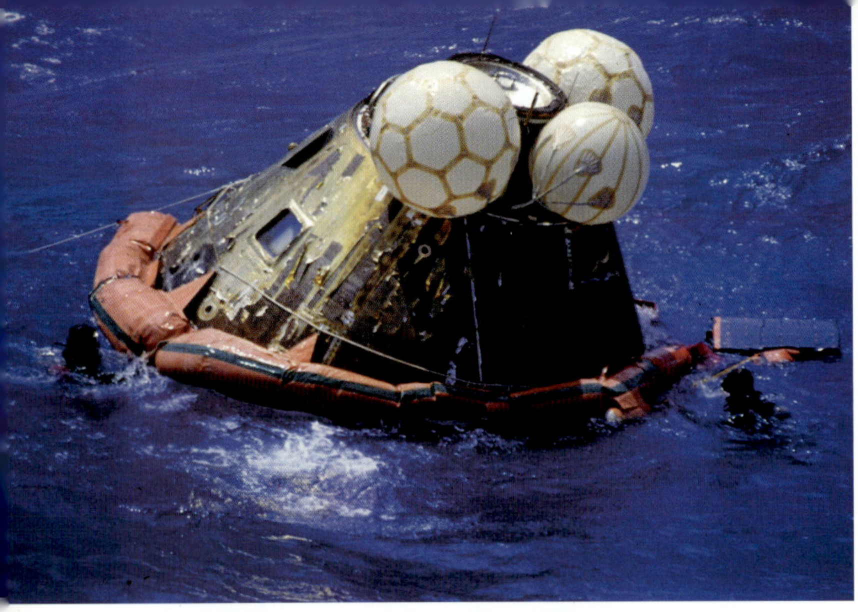

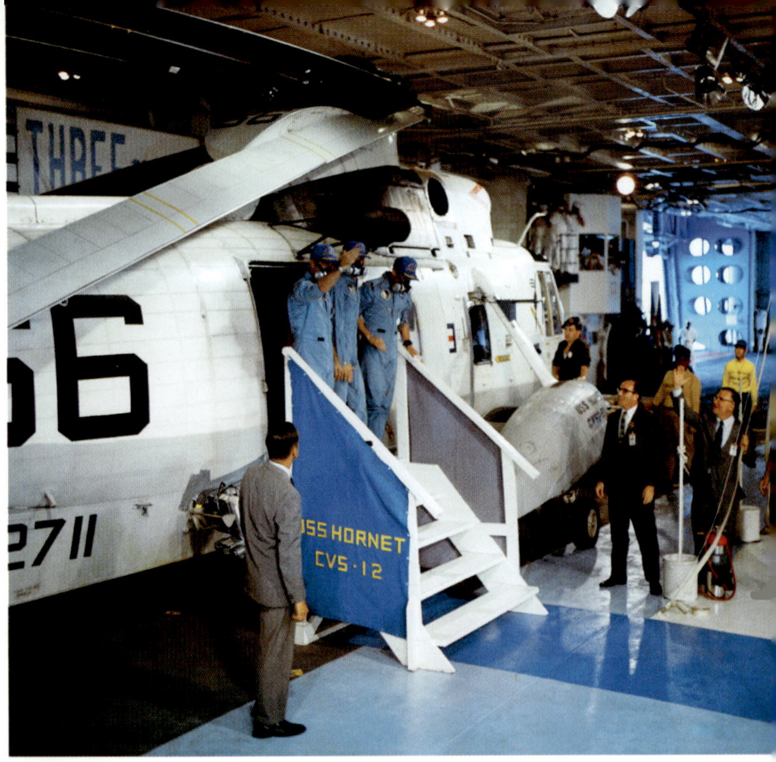

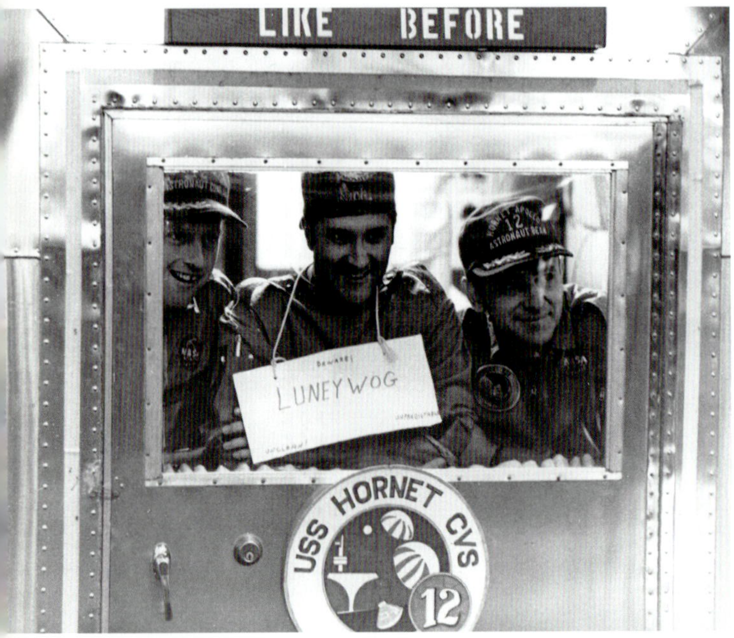

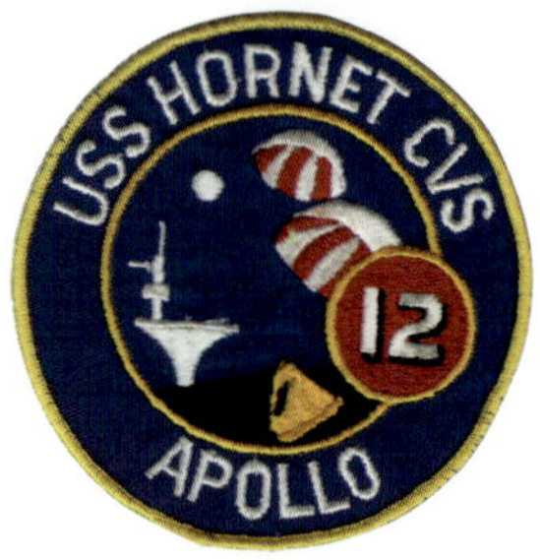

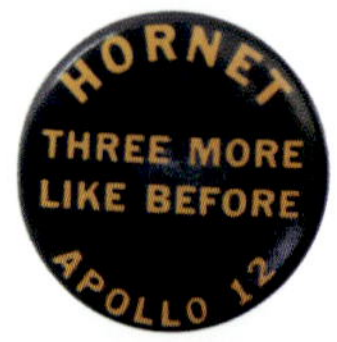

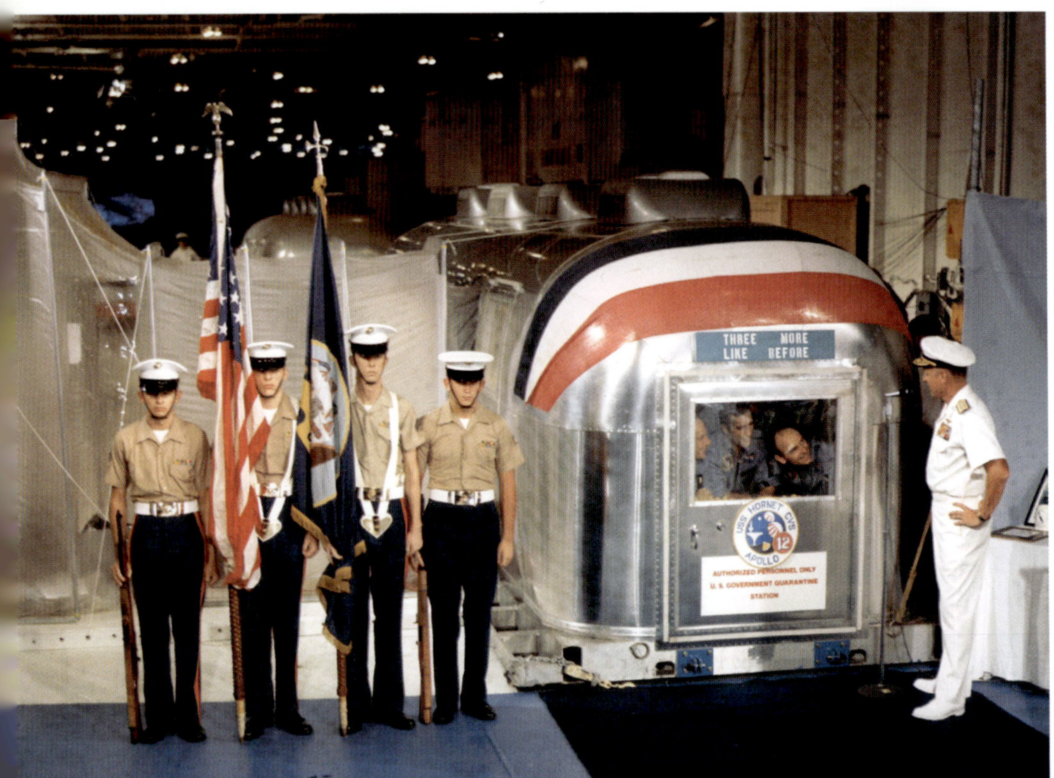

Top left: Divers from UDT-13 attach the flotation collar around *Yankee Clipper* during recovery operations, just three miles from the recovery ship, on November 24. The CM's three flotation bags are inflated. Recovery personnel often kept as souvenirs the small pieces of gold Mylar foil that peeled off the spacecraft.

Top right: Wearing respirators to prevent the release of any biohazards from the moon, Gordon, Bean, and Conrad step from their Sea King recovery helicopter onto the hangar deck of USS *Hornet*, an antisubmarine warfare support aircraft carrier, on November 24. To the right of the steps is John Stonesifer, chief of MSC's Recovery Systems Branch, Landing and Recovery Division.

Center left: Conrad and Bean label Gordon a "luneywog" in the MQF on November 27. (In a navy initiation rite, first-time equator-crossers are dubbed "pollywogs.") Gordon wears his uniform backward for the occasion.

Bottom left: The crewmen, inside their MQF, are welcomed aboard the *Hornet* by Rear Adm. Donald C. Davis, commander of the Pacific Manned Spacecraft Recovery Force, on November 24, 1969. A Marine Corps color guard is at left. The *Hornet* was decommissioned seven months later, and is now on display in Alameda, California. This MQF is displayed at the US Space & Rocket Center in Huntsville, Alabama.

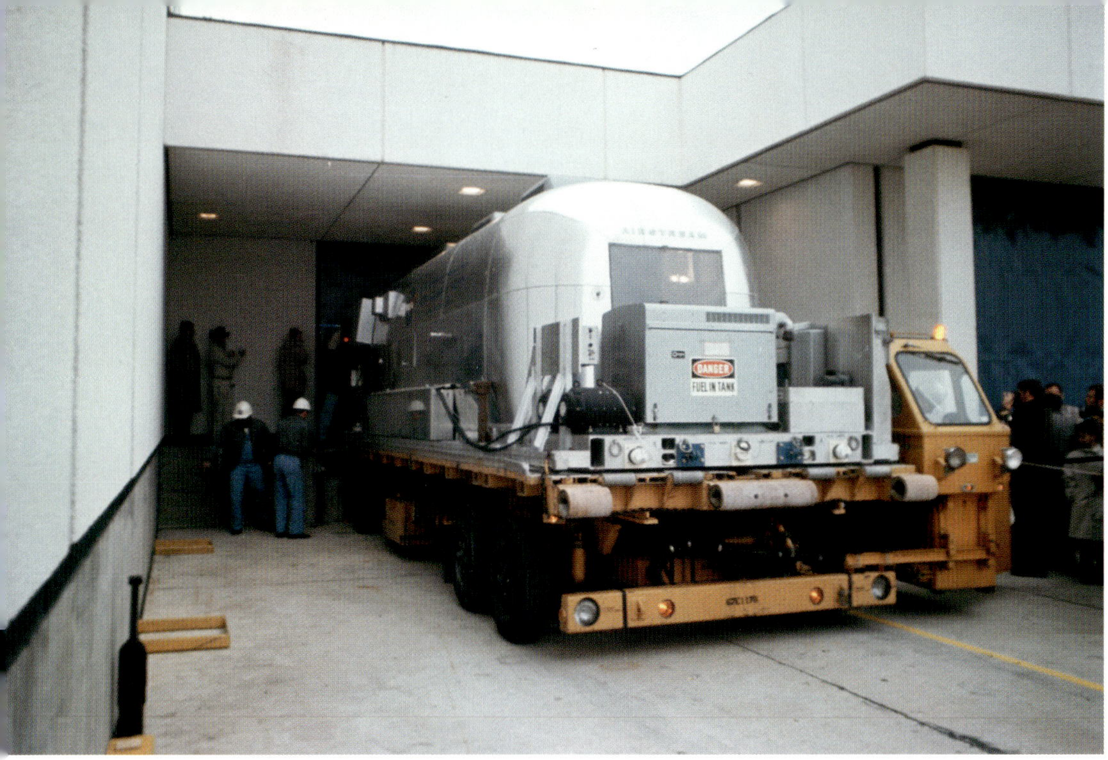

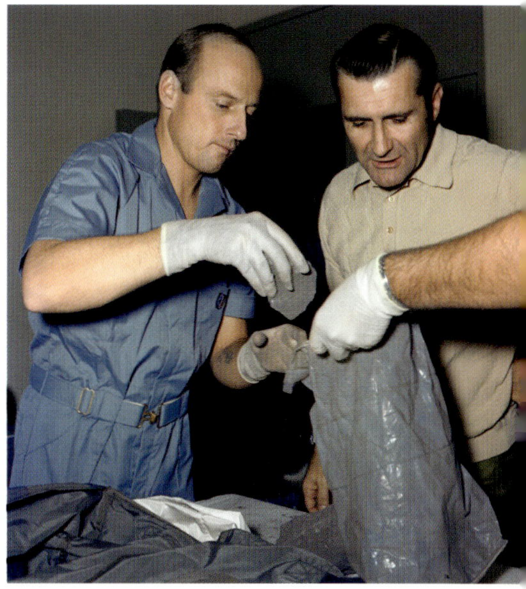

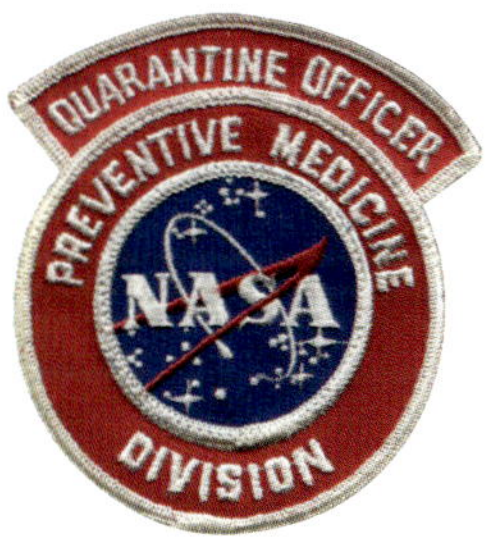

Top left: Apollo 12's MQF is backed into the loading dock of Bldg. 37 at MSC, after arriving by C-130 at Ellington AFB on November 29. Two shipments of lunar rocks had arrived four days earlier, sent via Pago Pago in American Samoa. The $7.8 million LRL contained the lunar lab, biological facilities, crew isolation area, a gas analysis lab, and a radiation-counting lab.

Top right: Conrad and Gordon examine lunar samples in the LRL on November 29. The crew returned with more than seventy-five pounds of material, three times the amount from Apollo 11.

Center right: Bean, Gordon, and Conrad pose with the LRL staff, just before being released from isolation on December 10.

Bottom right: Conrad's suit undergoes inspections in Room 2020 of Bldg. 7 at MSC. *Left to right*: T. J. Tirado and Jay Aljoe of ILC; unidentified; Gene Nitch and Dick Sandrich of NASA; and ILC's Dan McCarthy.

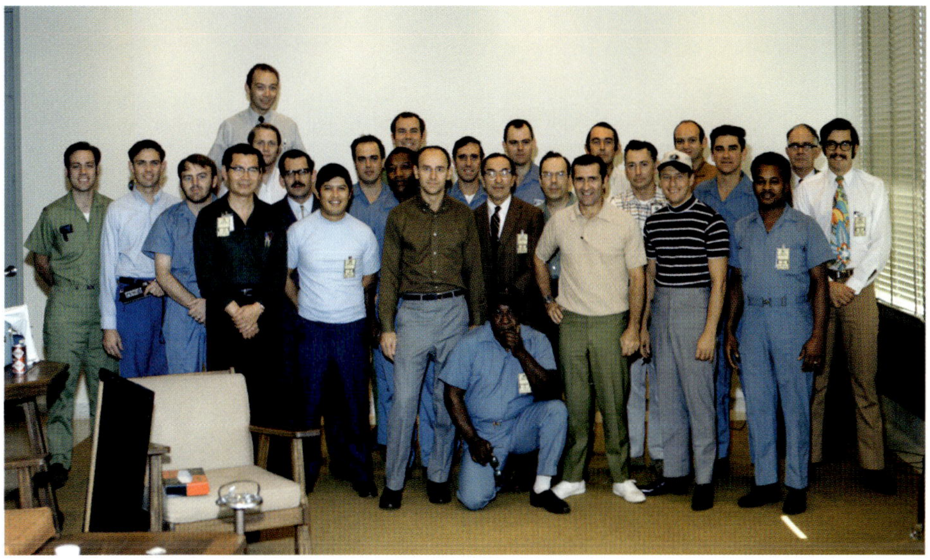

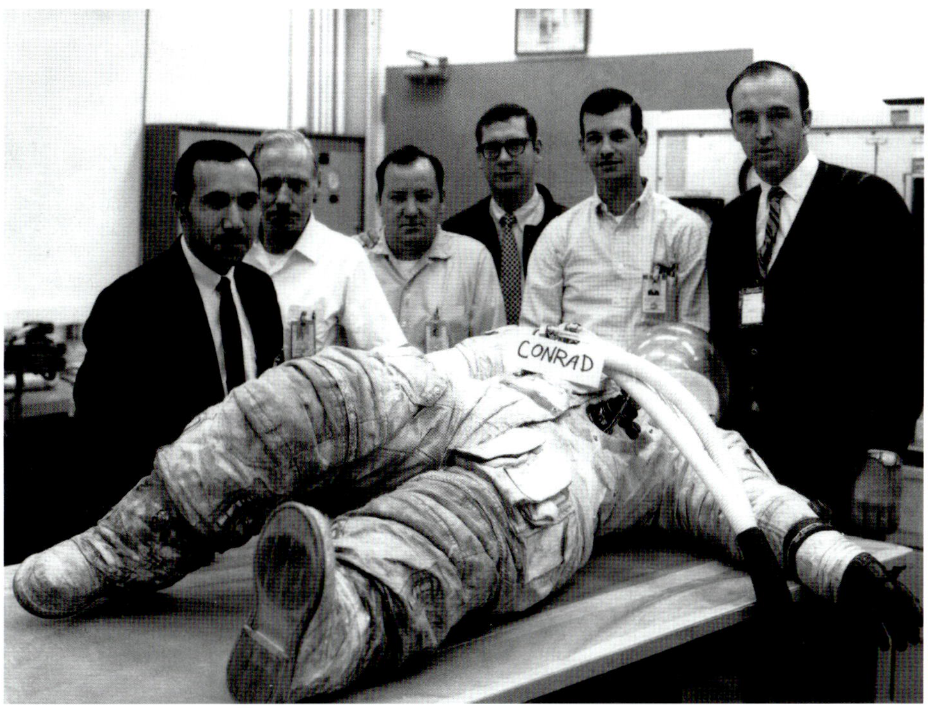

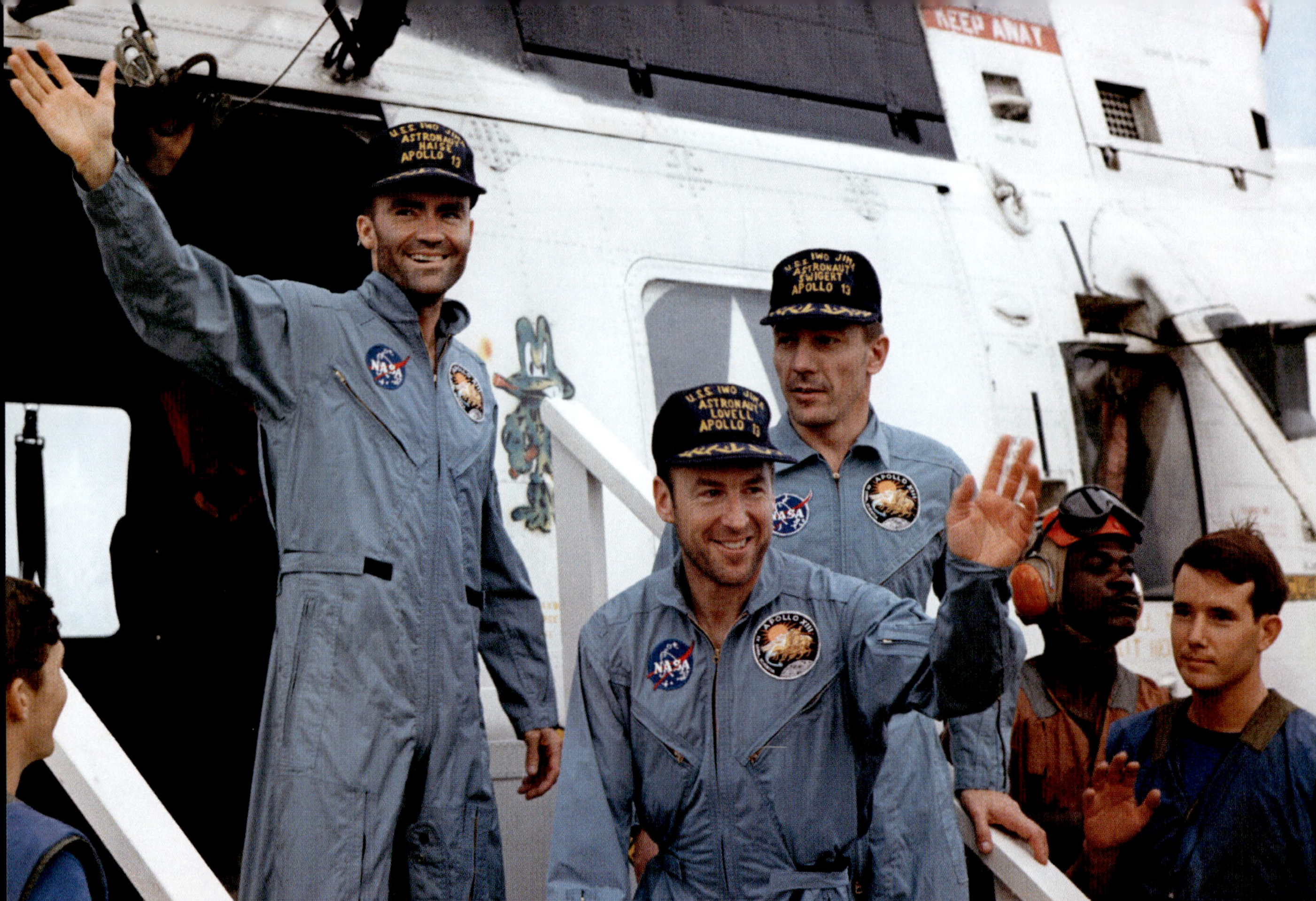

Above: Fred Haise, Jim Lovell, and Jack Swigert step from SH-3D Sea King helicopter No. 66 onto the deck of USS *Iwo Jima* on April 17, 1970. The "Freddie the Frog" logo to the right of Haise was painted by Pete Carolan, a US Navy SEAL who was a swim team leader on the UDT-13 recovery crew. This was the chopper's fifth and final Apollo crew recovery; it crashed off the California coast near San Diego in 1975, during a training flight for hunting Soviet submarines.

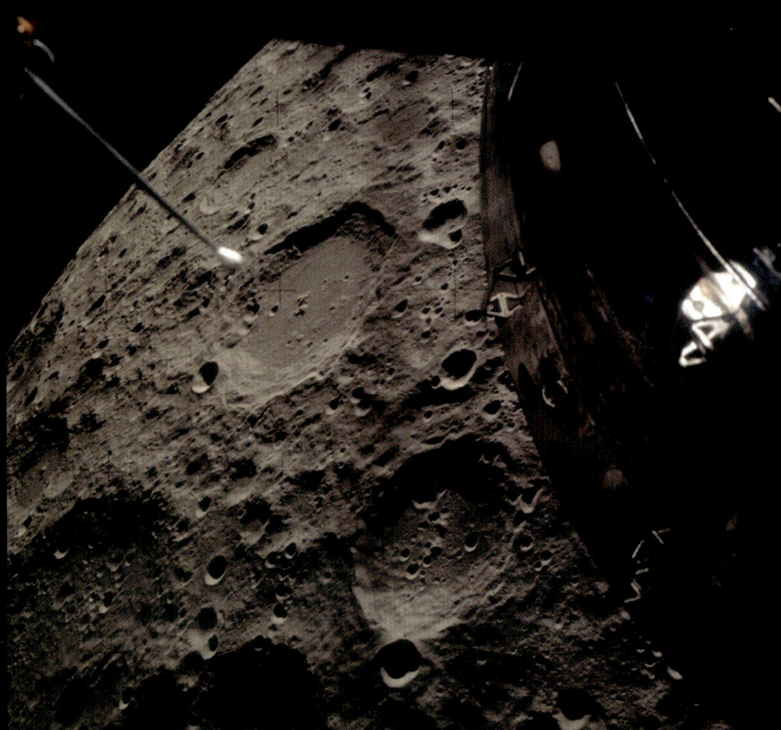

Left: CM *Odyssey*'s hatch, seen from the docking window aboard LM *Aquarius* as the linked spacecraft swing around the far side of the moon on April 15. The two major craters are Chaplygin (*left of center*) and Schliemann (*lower center*). At upper left is the VHF EVA antenna.

CHAPTER 8

Apollo 13

April 11–17, 1970

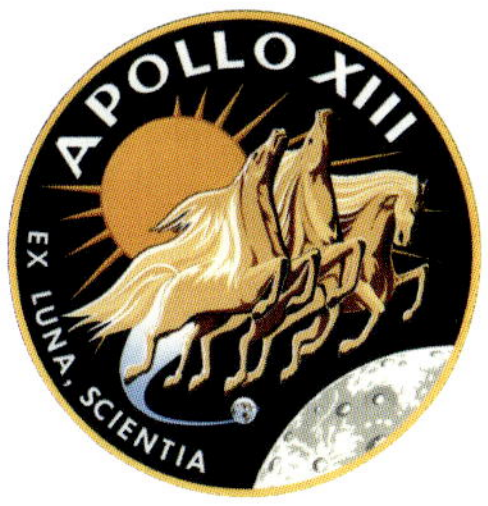

The heroic Apollo 13 mission, just more than four months after Apollo 12, might have been the flight that triggered the coming decline in public interest in the program. Instead, it is remembered for the skills and ingenuity of the crewmen and the ground personnel, who saved the astronauts' lives and perhaps Apollo itself.

The CDR was Jim Lovell, who was joined by rookies Fred W. Haise as LMP and John L. "Jack" Swigert as CMP. Swigert had been the backup to prime crewman Thomas "Ken" Mattingly; he only joined the crew two days before launch. The previous week, they had all been training with backup LMP Charles Duke. When Duke broke out in a rash caused by rubella (German measles) NASA officials realized they faced a dilemma. They could either substitute Swigert—who, like Lovell and Haise, was immune—for Mattingly forty-eight hours before launch, or they could pay a half-million-dollar penalty and delay the mission until May.

NASA Administrator Tom Paine made the final decision to proceed with the unprecedented switch, and Apollo 13 was launched on April 11, 1970, into an Earth parking orbit. At 4:54 p.m. (EST), an S-IVB burn placed the spacecraft and third stage on a translunar trajectory. The CSM then separated from the S-IVB and docked with LM *Aquarius* on live television. The booster's auxiliary propulsion system aimed the spent stage for the moon, striking the lunar surface three days later.

On April 13, when the mission was about three quarters of the way to the moon, CAPCOM Jack Lousma in Mission Control asked Swigert to turn on stirring fans in the SM's liquid hydrogen and oxygen tanks, to get a better reading of their quantities. The astronauts soon heard a "large bang." They reported an undervoltage alarm on the CSM main electrical bus B, rapid loss of pressure in one oxygen tank, and dropping power from the other two fuel cells.

One of the two liquid oxygen (LOX) tanks in the SM had exploded. Damaged insulation on the wires to the stirring fan inside the tank had short-circuited, igniting the insulation. Pressure from the fire blew out the tank. Valves on the other two fuel cells closed from the shock of the blast, which also created a leak in the remaining oxygen tank. The explosion blew off a thirteen-foot outer aluminum panel, damaging the nearby high-gain S-band antenna used for translunar communications.

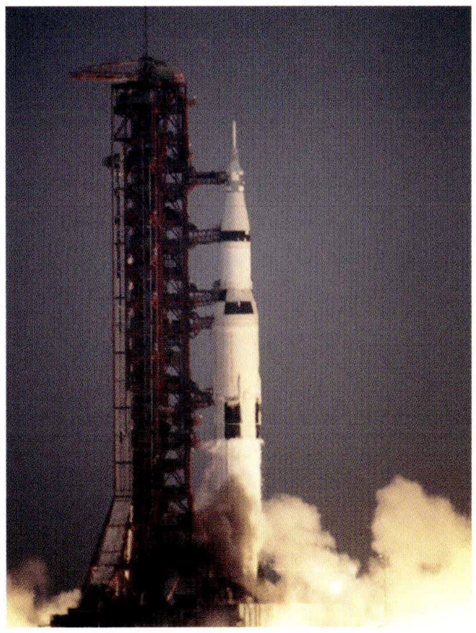

Launch: 2:13 p.m. (EST), April 11, 1970 (photo by John de Bry)

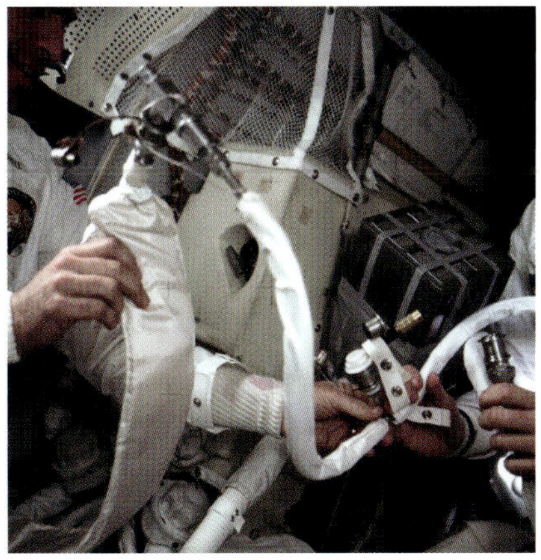

Trans-Earth coast: April 16

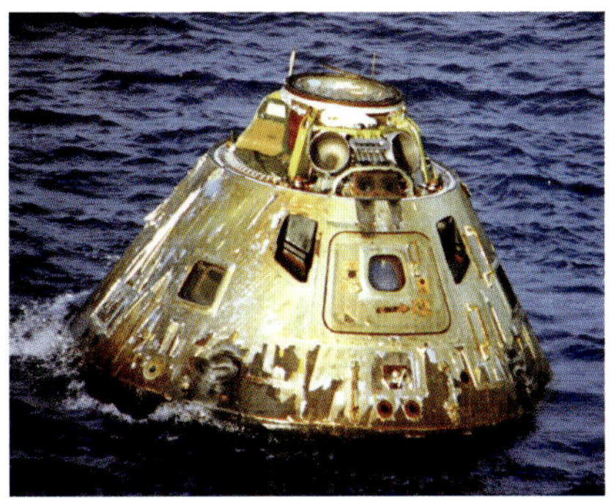

Splashdown: 1:07 p.m. (EST), April 17

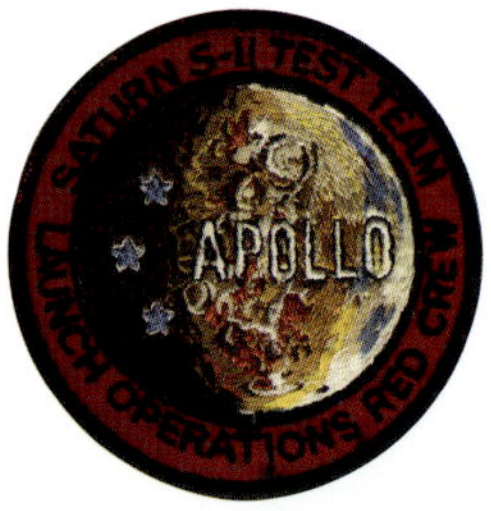

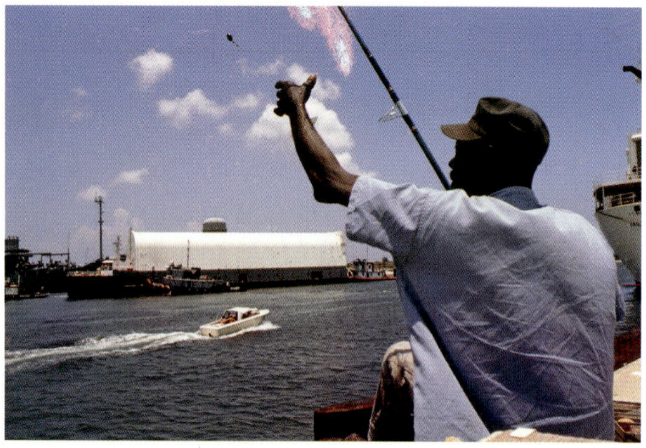

Above: The Saturn's S-II stage arrives on the barge *Orion* at Port Canaveral, south of the Cape, on June 29, 1969. At right is the stern of the tracking ship USNS *Vanguard*. (photo by Tiziou News Service)

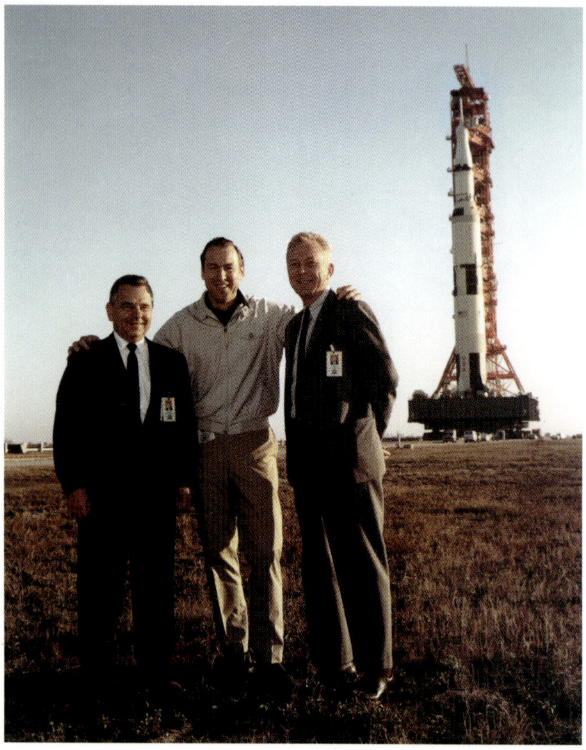

Above: Launch Director Walt Kapryan, Lovell, and Launch Operations Manager Paul Donnelly pose during the Apollo 13 rollout on the morning of December 15, 1969. In the aftermath of the lightning strike during Apollo 12's launch, Kapryan joked that he would order a rollback to the VAB after just one raindrop on Apollo 13's launch day.

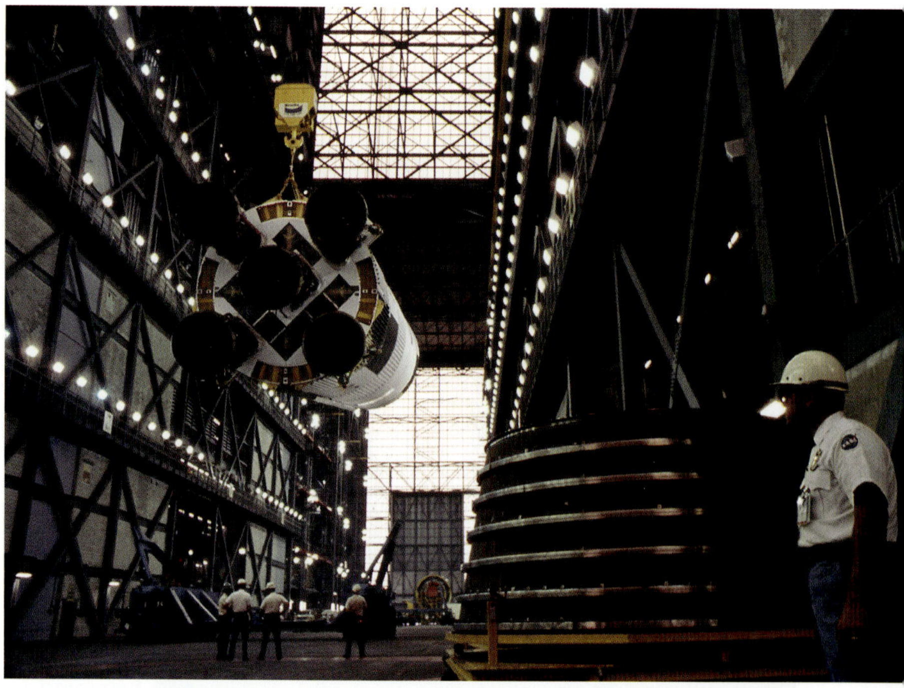

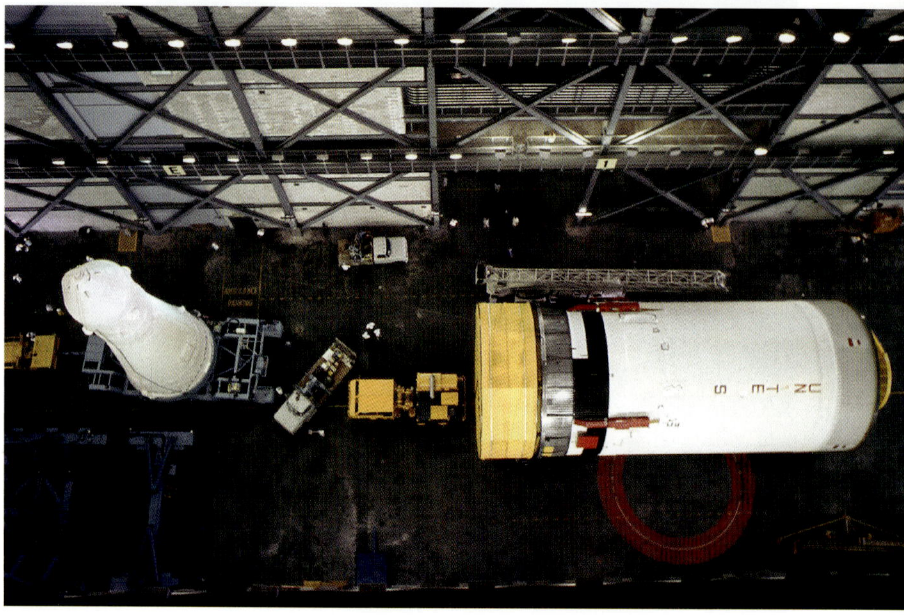

Top right: The Saturn's S-IC stage is lifted horizontally by crane in the VAB transfer aisle on June 18, 1969. At lower right is an engine skirt for an F-1 engine, next to a KSC Security patrolman. (photo by Tiziou News Service)

Bottom right: Apollo 13's S-II stage (*right*) is parked in this overhead shot of the busy VAB transfer aisle on July 1, 1969. The Apollo 12 spacecraft at left is ready for hoisting, to be mated to its Saturn V. Also at left is a blue cradle for an S-IC stage. (photo by Tiziou News Service)

Opposite page: A view of the Apollo 13 spacecraft from the MSS, as the MSS moves away from Pad A on the evening of April 10, 1970. The Atlantic beach is in the background. The clamshell sections enclosed the LES when a much sturdier three-level compartment (attached below to the MSS) surrounded the CSM and SLA.

CHAPTER 8

Apollo 13

April 11–17, 1970

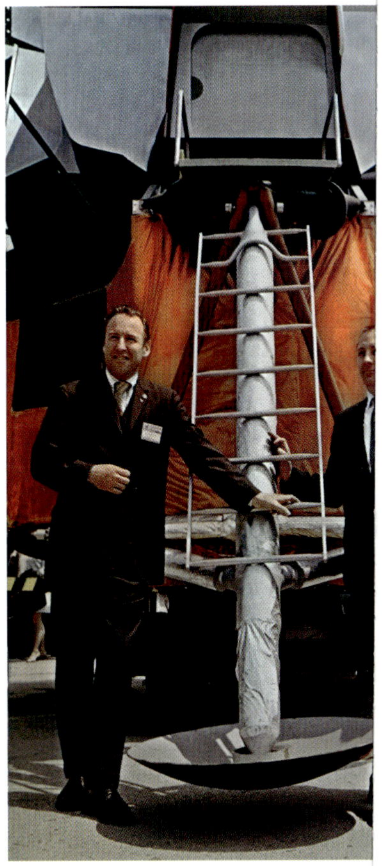

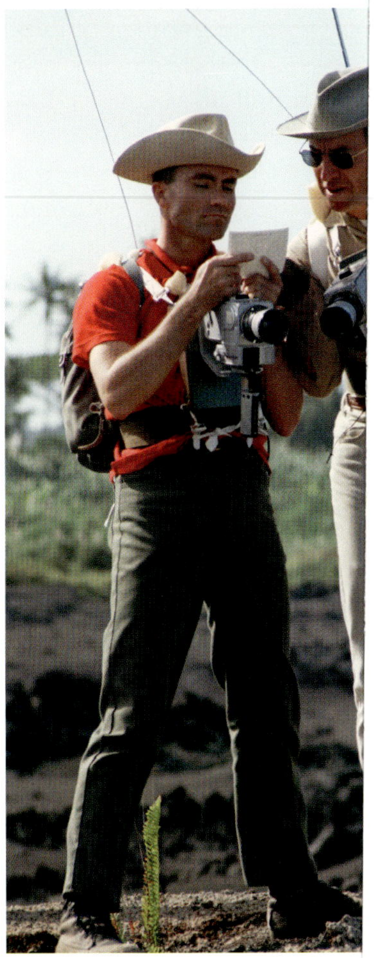

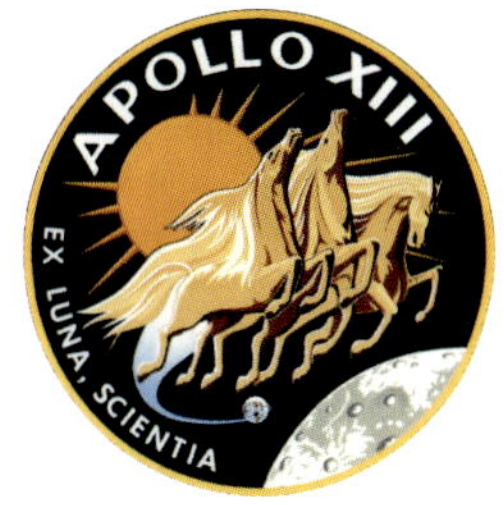

The heroic Apollo 13 mission, just more than four months after Apollo 12, might have been the flight that triggered the coming decline in public interest in the program. Instead, it is remembered for the skills and ingenuity of the crewmen and the ground personnel, who saved the astronauts' lives and perhaps Apollo itself.

The CDR was Jim Lovell, who was joined by rookies Fred W. Haise as LMP and John L. "Jack" Swigert as CMP. Swigert had been the backup to prime crewman Thomas "Ken" Mattingly; he only joined the crew two days before launch. The previous week, they had all been training with backup LMP Charles Duke. When Duke broke out in a rash caused by rubella (German measles) NASA officials realized they faced a dilemma. They could either substitute Swigert—who, like Lovell and Haise, was immune—for Mattingly forty-eight hours before launch, or they could pay a half-million-dollar penalty and delay the mission until May.

NASA Administrator Tom Paine made the final decision to proceed with the unprecedented switch, and Apollo 13 was launched on April 11, 1970, into an Earth parking orbit. At 4:54 p.m. (EST), an S-IVB burn placed the spacecraft and third stage on a translunar trajectory. The CSM then separated from the S-IVB and docked with LM *Aquarius* on live television. The booster's auxiliary propulsion system aimed the spent stage for the moon, striking the lunar surface three days later.

On April 13, when the mission was about three quarters of the way to the moon, CAPCOM Jack Lousma in Mission Control asked Swigert to turn on stirring fans in the SM's liquid hydrogen and oxygen tanks, to get a better reading of their quantities. The astronauts soon heard a "large bang." They reported an undervoltage alarm on the CSM main electrical bus B, rapid loss of pressure in one oxygen tank, and dropping power from the other two fuel cells.

One of the two liquid oxygen (LOX) tanks in the SM had exploded. Damaged insulation on the wires to the stirring fan inside the tank had short-circuited, igniting the insulation. Pressure from the fire blew out the tank. Valves on the other two fuel cells closed from the shock of the blast, which also created a leak in the remaining oxygen tank. The explosion blew off a thirteen-foot outer aluminum panel, damaging the nearby high-gain S-band antenna used for translunar communications.

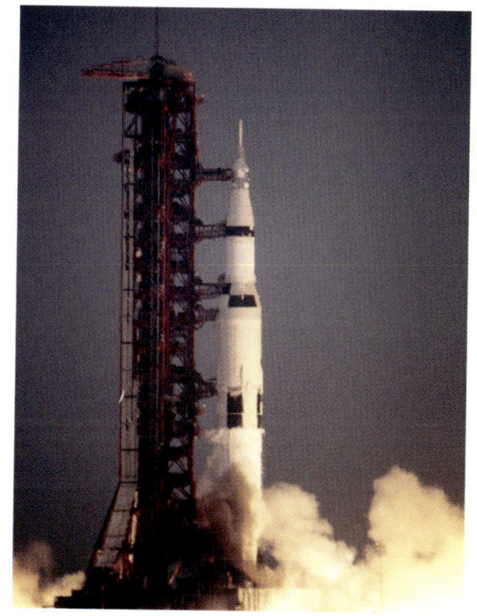

Launch: 2:13 p.m. (EST), April 11, 1970 (photo by John de Bry)

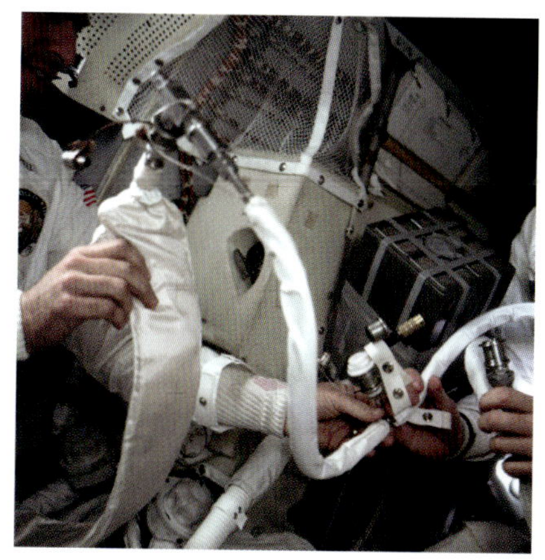

Trans-Earth coast: April 16

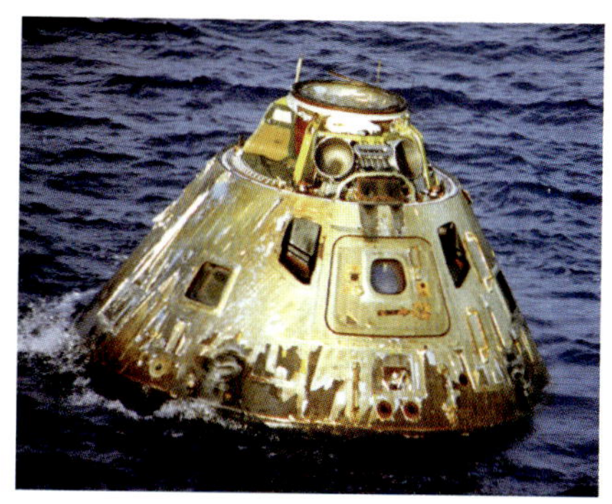

Splashdown: 1:07 p.m. (EST), April 17

The fuel cells, first te
to produce electricity ar
remaining one shut dow
crew was forced to shut
LM *Aquarius* as a "lifeb
manned lunar landing
getting the three men b
dominated television co
The LM consumable
half, not three people fo
trical power and water
cal consumables. To ke
powered down to the lo
When Apollo 13 loo
tact with Earth for alm
descent engine to speed
Removing carbon d
The LM's stock of lithiu
down, and the remaind
CM had an adequate su
didn't fit the LM's cylin
vised a solution by usin
traption the crew built
suffered through a grin
the LM caused cabin te
The final major pro
CM, something never
the support of Ken Ma
a successful sequence u
About four hours b
the crew photographed
was powered up and *A*
1:07 p.m. (EST) in the
and four miles from th
taken by helicopter to
suffering from a urinar
but the astronauts were
After an intensive i
the cause of the explos
been degraded by too
been upgraded to mat
oxygen tank heaters. T
Apollo 10's SM before
for modifications. A r
which had also left th

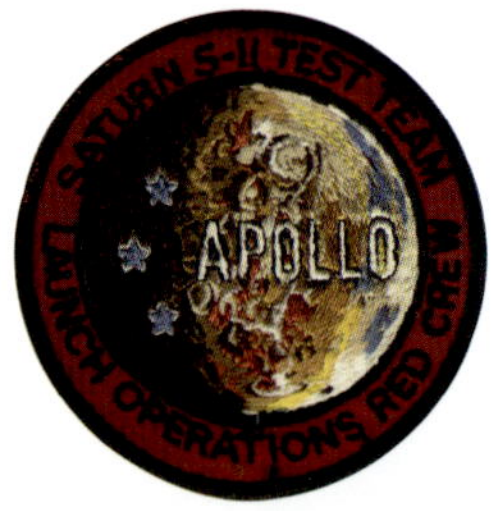

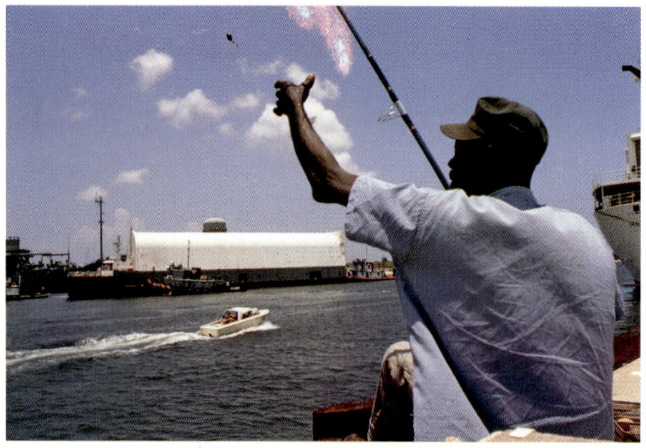

Above: The Saturn's S-II stage arrives on the barge *Orion* at Port Canaveral, south of the Cape, on June 29, 1969. At right is the stern of the tracking ship USNS *Vanguard*. (photo by Tiziou News Service)

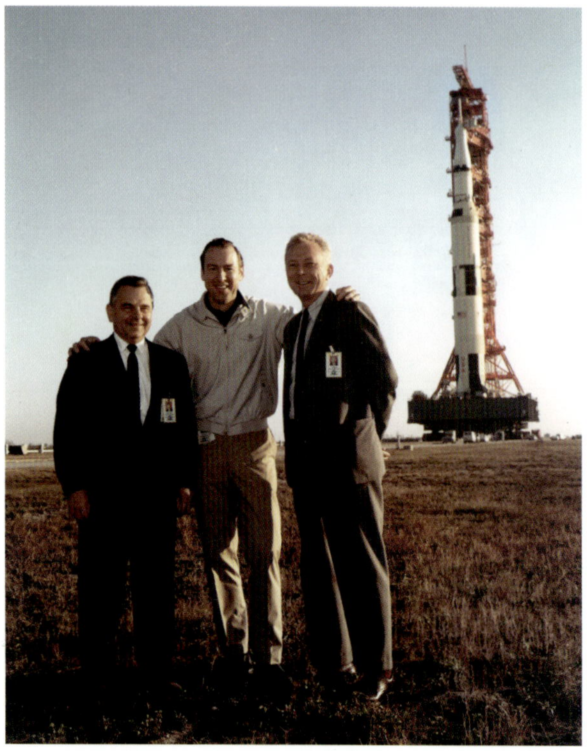

Above: Launch Director Walt Kapryan, Lovell, and Launch Operations Manager Paul Donnelly pose during the Apollo 13 rollout on the morning of December 15, 1969. In the aftermath of the lightning strike during Apollo 12's launch, Kapryan joked that he would order a rollback to the VAB after just one raindrop on Apollo 13's launch day.

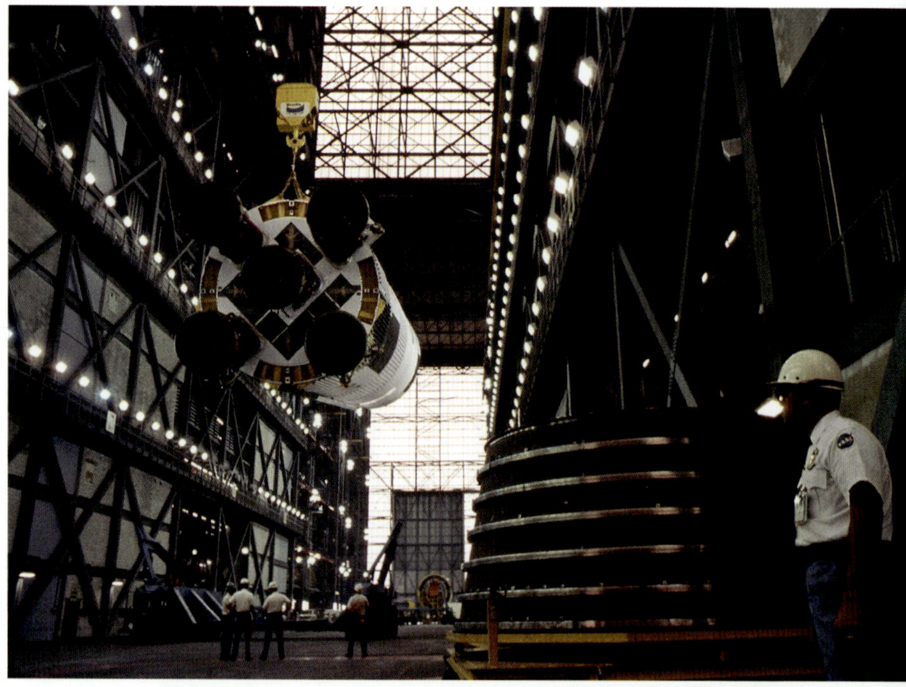

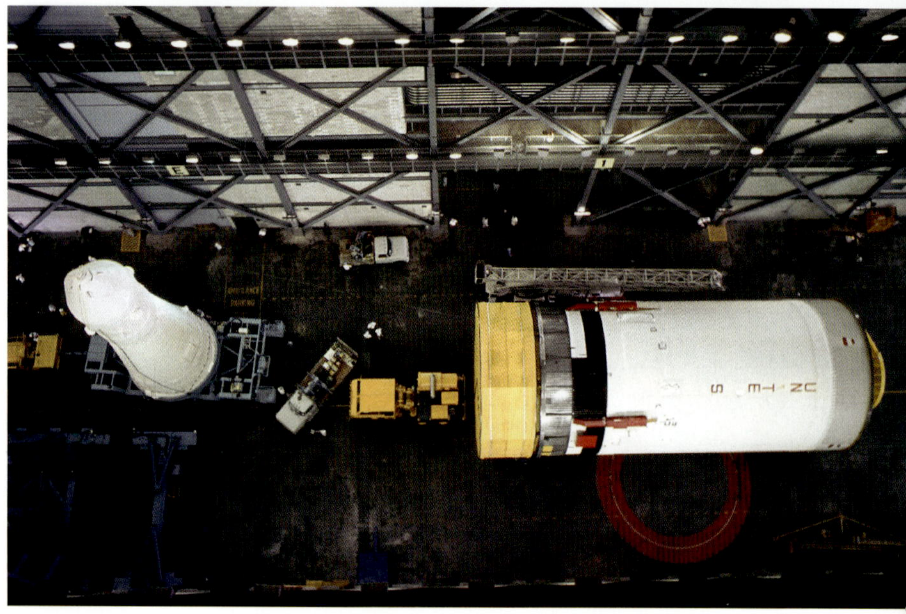

Top right: The Saturn's S-IC stage is lifted horizontally by crane in the VAB transfer aisle on June 18, 1969. At lower right is an engine skirt for an F-1 engine, next to a KSC Security patrolman. (photo by Tiziou News Service)

Bottom right: Apollo 13's S-II stage (*right*) is parked in this overhead shot of the busy VAB transfer aisle on July 1, 1969. The Apollo 12 spacecraft at left is ready for hoisting, to be mated to its Saturn V. Also at left is a blue cradle for an S-IC stage. (photo by Tiziou News Service)

Opposite page: A view of the Apollo 13 spacecraft from the MSS, as the MSS moves away from Pad A on the evening of April 10, 1970. The Atlantic beach is in the background. The clamshell sections enclosed the LES when a much sturdier three-level compartment (attached below to the MSS) surrounded the CSM and SLA.

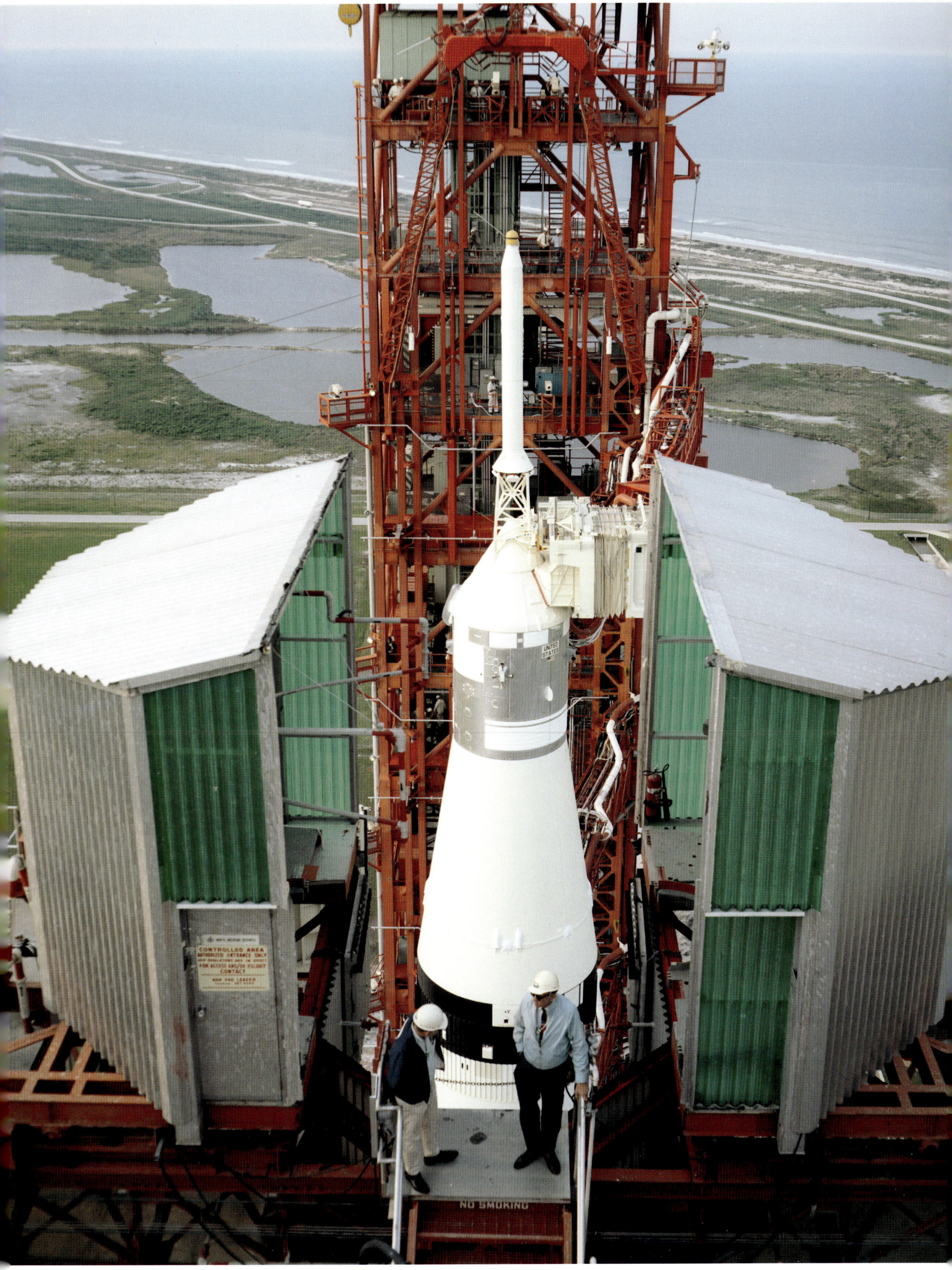
UNITED STATES
CONTROLLED AREA
CONTACT
NO SMOKING

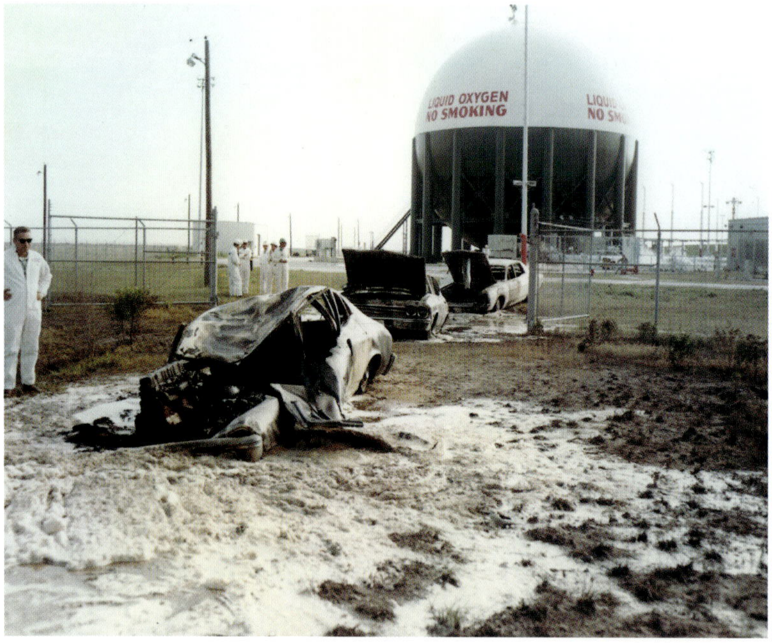

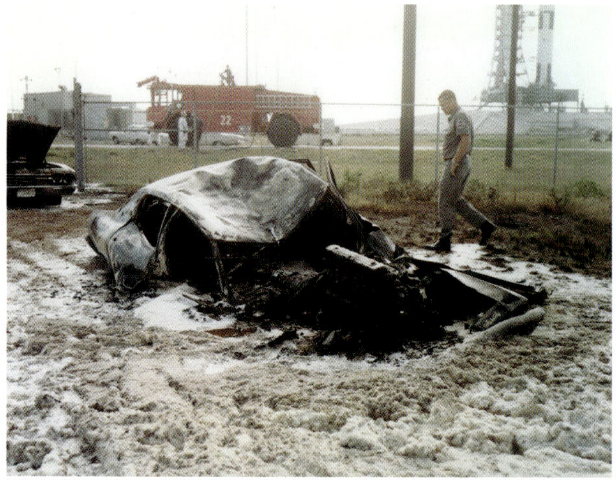

Top left and top right: A trial tanking countdown was delayed by almost four hours on the morning of March 25, 1970, after three KSC Security cars were destroyed. Foam surrounds the cars. They caught fire when they were accidentally driven through a fog-like LOX vapor concentration on the pad perimeter, 1,100 feet from the Saturn V. Officers had just finished clearing personnel as tanking for the S-IVB began. They abandoned the vehicles and no one was injured.

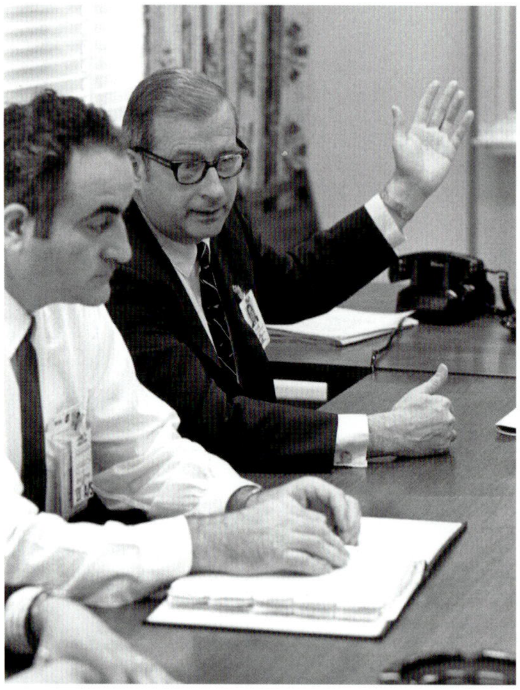

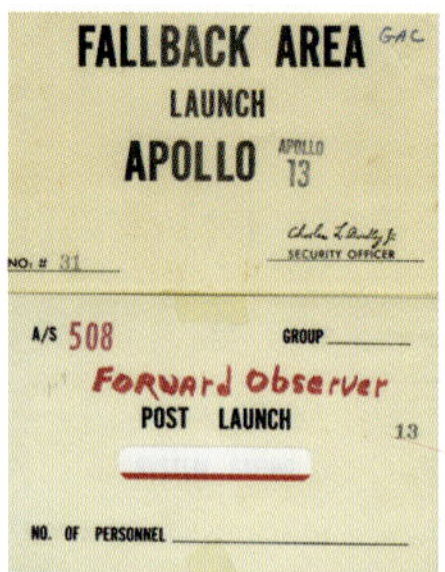

Center left: KSC Director of Launch Operations Rocco Petrone (*left*) and NASA Administrator Tom Paine during a launch status meeting at KSC on April 10, 1970. The decision to replace Mattingly with Swigert as CMP was made during this meeting.

Below: The second shift of Grumman's LM team at KSC poses with Apollo 13's Saturn V at Pad 39A in January 1970, with the MSS at right. *Front row, left to right*: LeRoy Kullman, Stewart Taylor, Joe Myer, Julio Juristo, Charles Cava, Darrell Kidd, and Bob Martin. *Back row, left to right*: Ed Moran, Dale Polk, Fred Syrett, Charles Kolonoski, Ronnie Linge, Dave Dandenaur, Julie Goldfarb, Nick Povini, Sam Feldman, Bob Sadler, Charles Sweeney, and Bruce Gray.

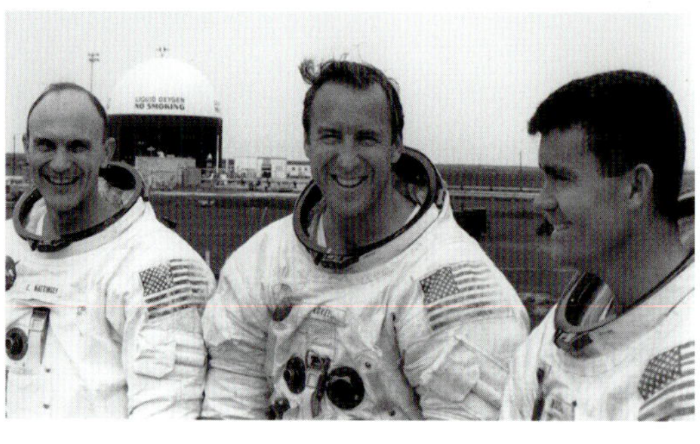

Above: Mattingly, Lovell, and Haise in training suits at Pad 39A on April 6, 1970, a day and a half after the countdown toward launch had begun.

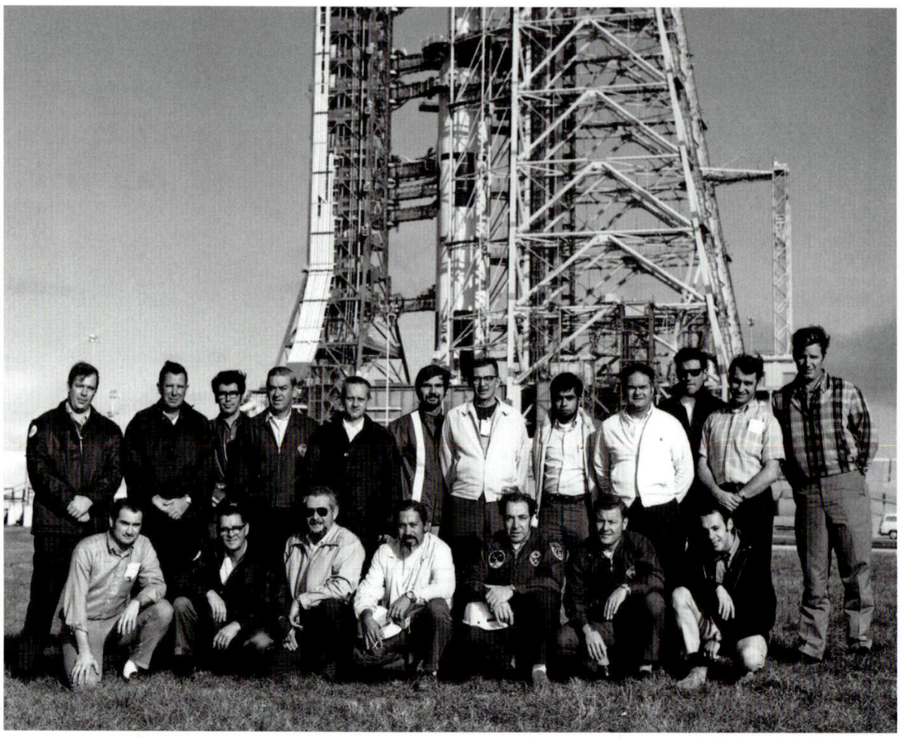

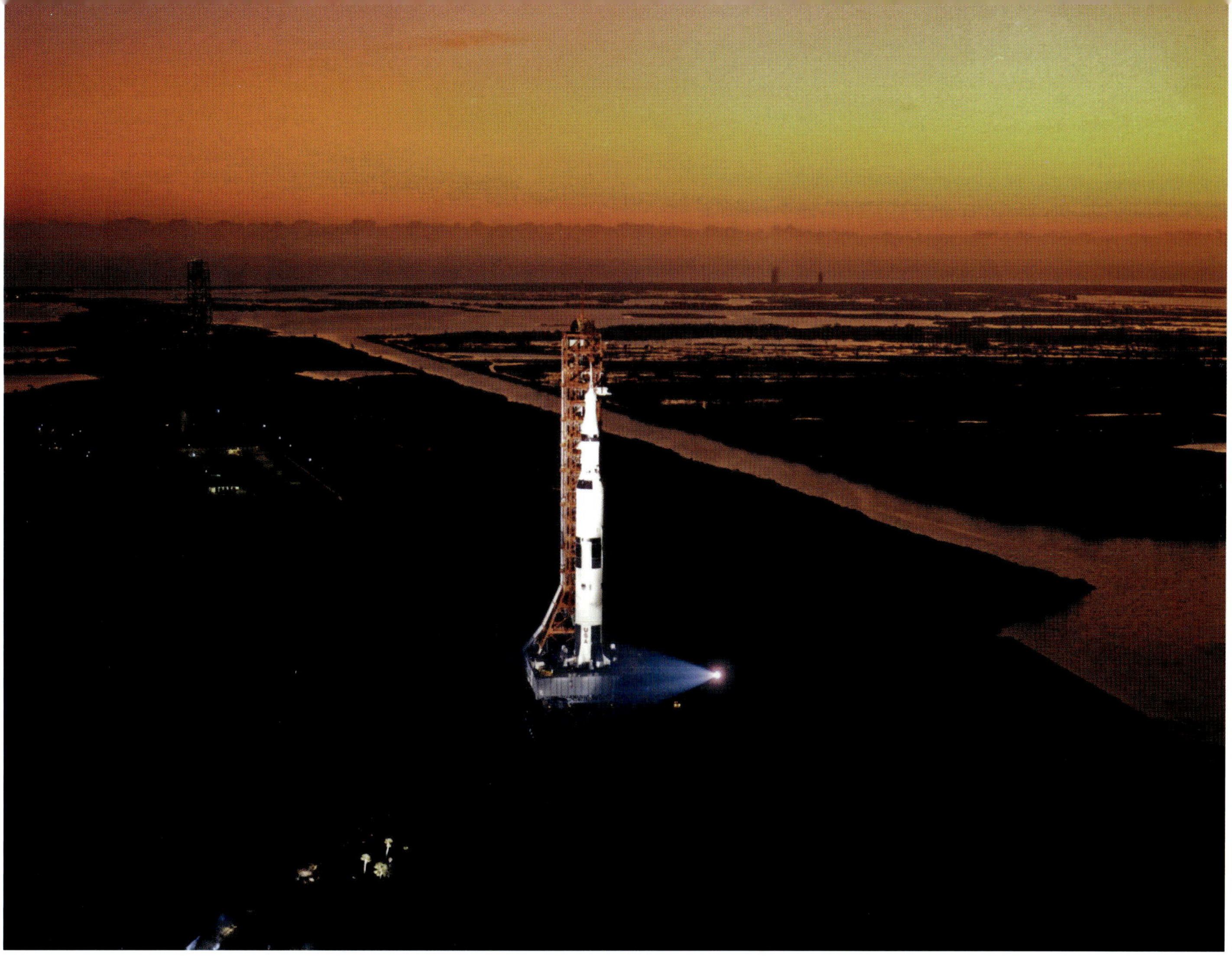

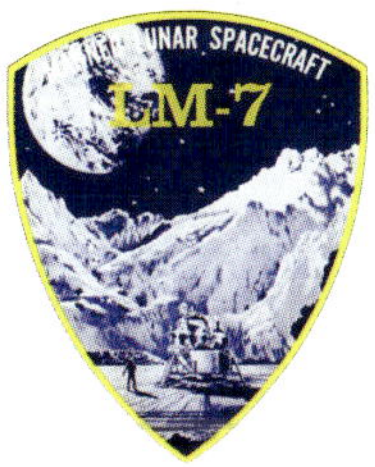

Above: Sunrise along Florida's Atlantic coast on December 15, 1969, as Apollo 13 is moved along the crawlerway to Pad 39A. The Turning Basin is at right, and the MSS is parked in the distance at left. At center right are the gantries at LC-41 and LC-40 at Cape Kennedy AFS.

Bottom right: Lacking its legs, LM-7 is moved by crane from Chamber R to a low-bay workstand in the MSOB on October 7, 1969. The descent propulsion system engine could be throttled between 10 and 60 percent of full thrust, which was 10,125 pounds.

Bottom left: Two primary struts and landing pads for the descent stage of LM-7 remain in plastic wrap in the MSOB on July 1, 1969. A gold Mylar–wrapped lunar surface sensing probe has been pulled back and is perpendicular to the strut in the foreground. The strut in the background, which lacks a probe, was likely for the LM's front leg. (photo by Tiziou News Service)

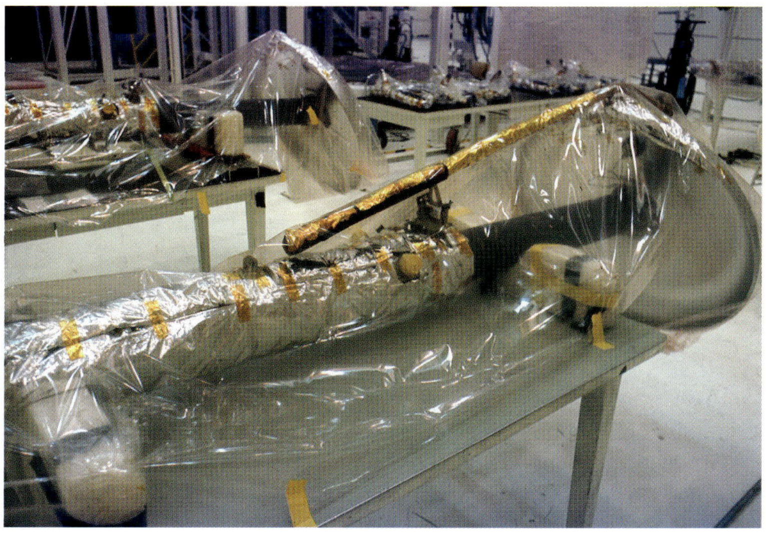

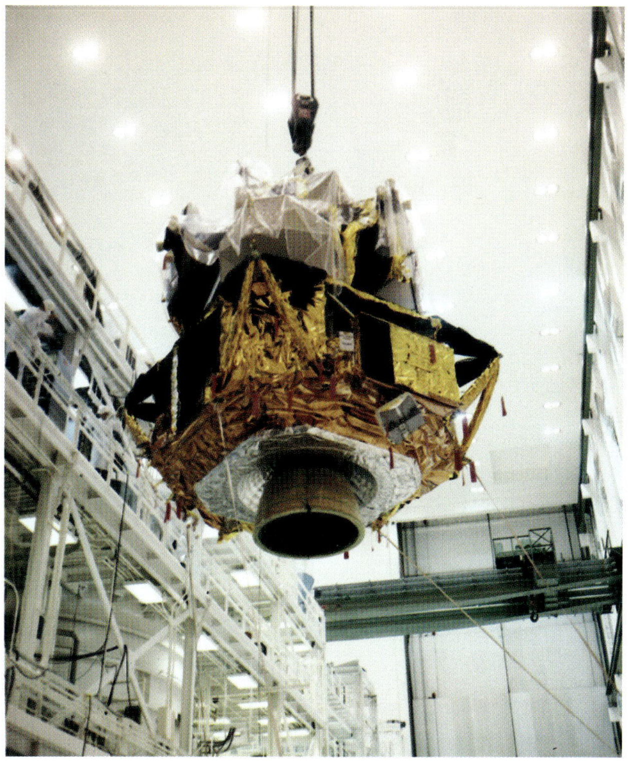

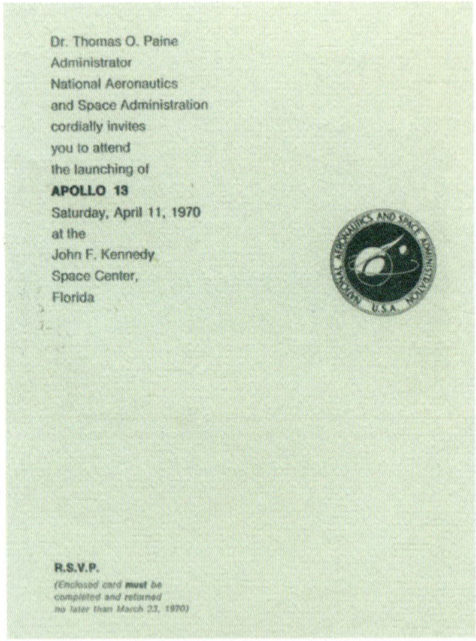
Dr. Thomas O. Paine
Administrator
National Aeronautics
and Space Administration
cordially invites
you to attend
the launching of
APOLLO 13
Saturday, April 11, 1970
at the
John F. Kennedy
Space Center,
Florida

R.S.V.P.
*(Enclosed card **must** be completed and returned no later than March 23, 1970)*

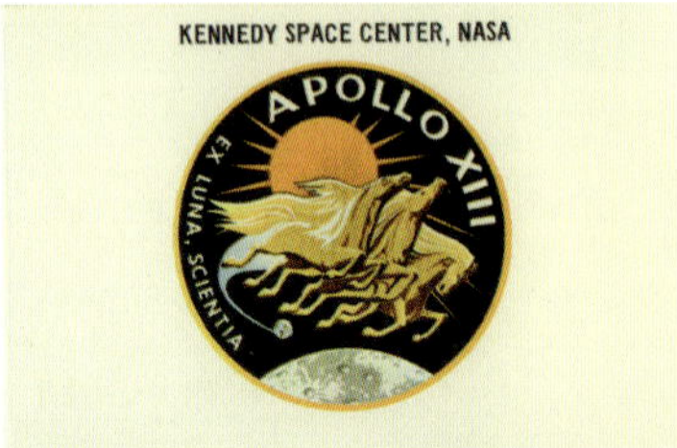

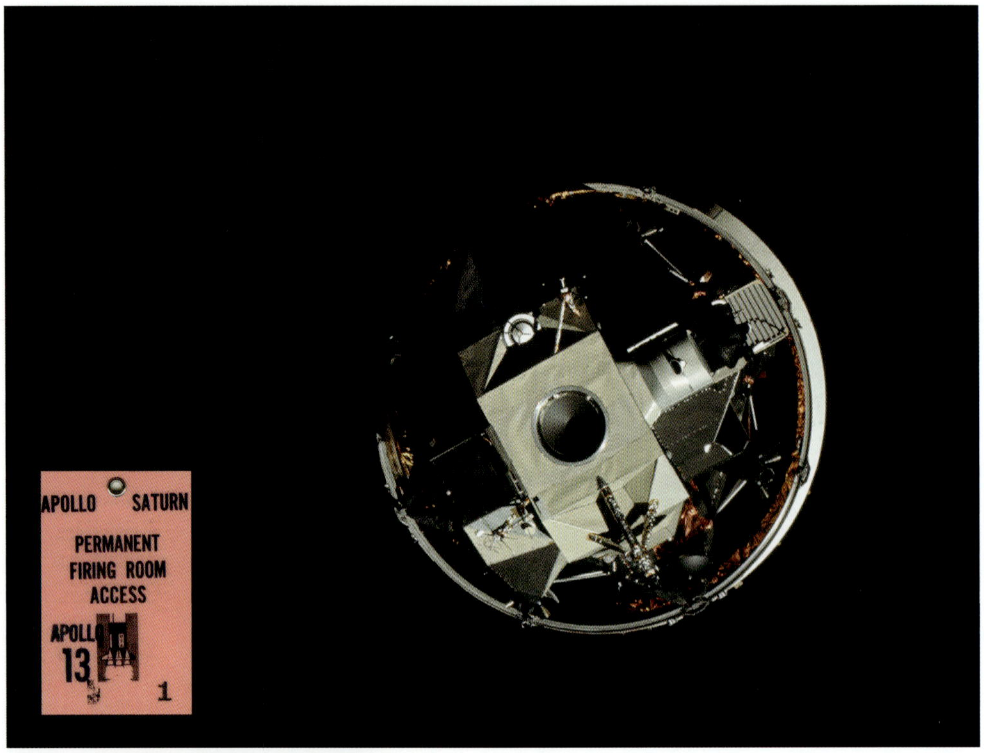

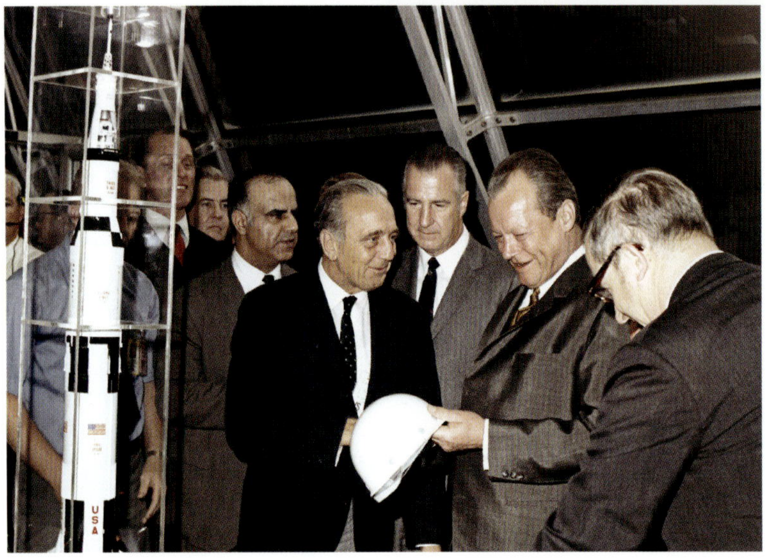

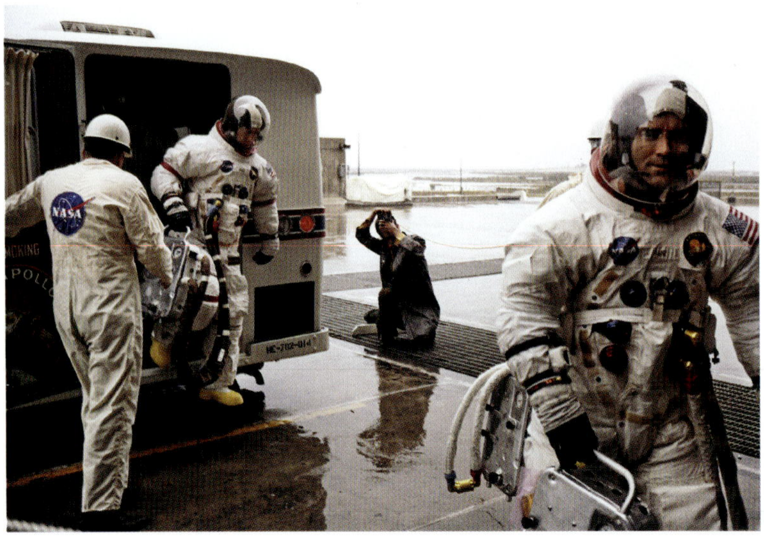

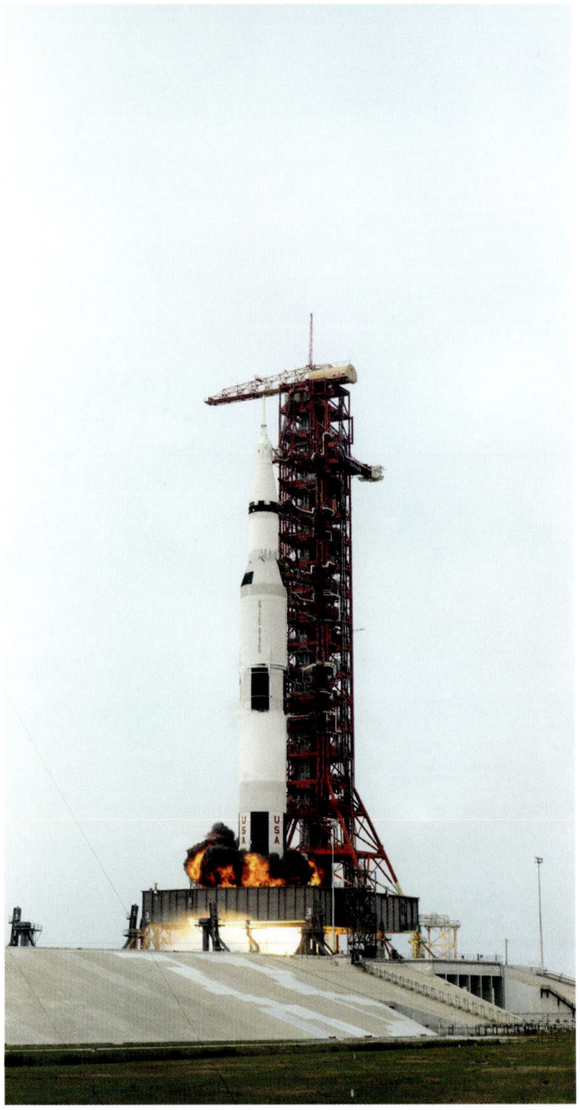

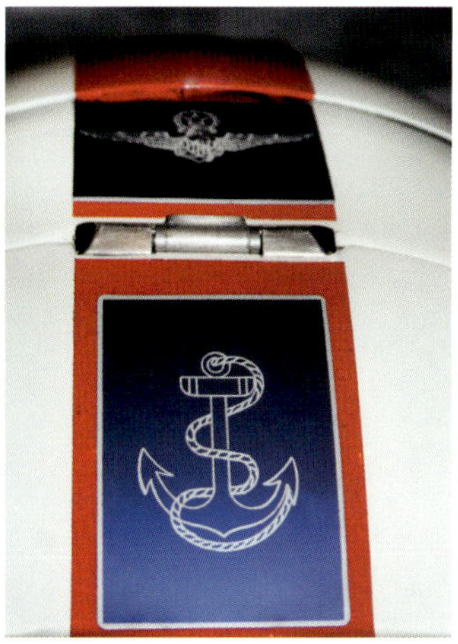

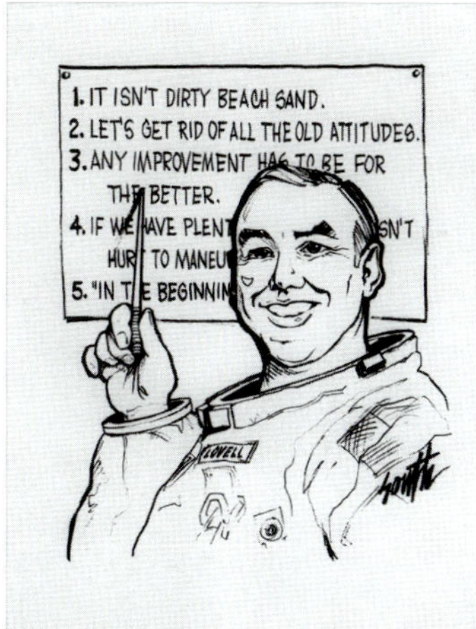

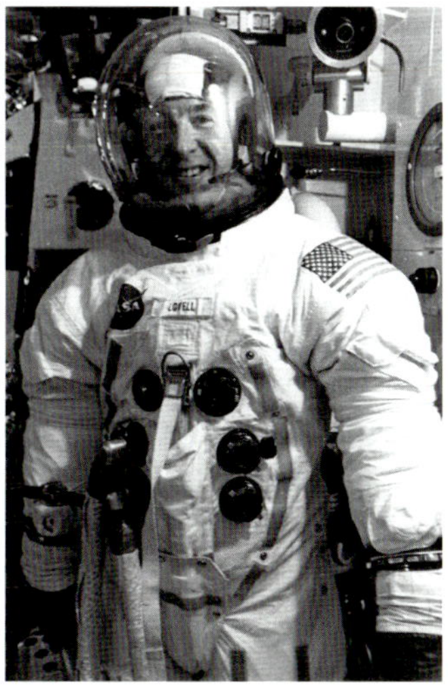

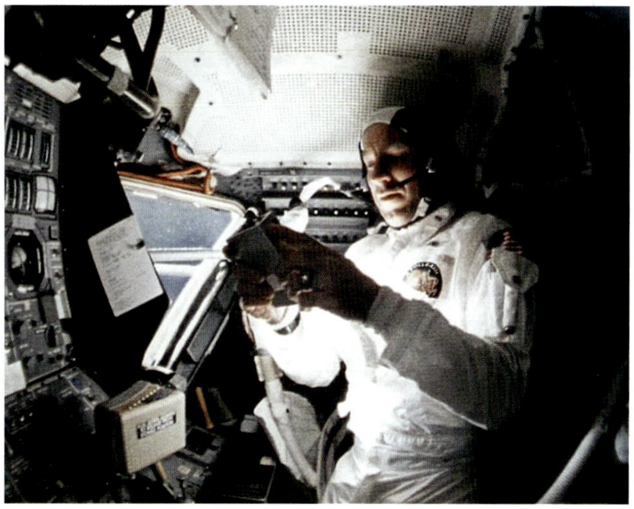

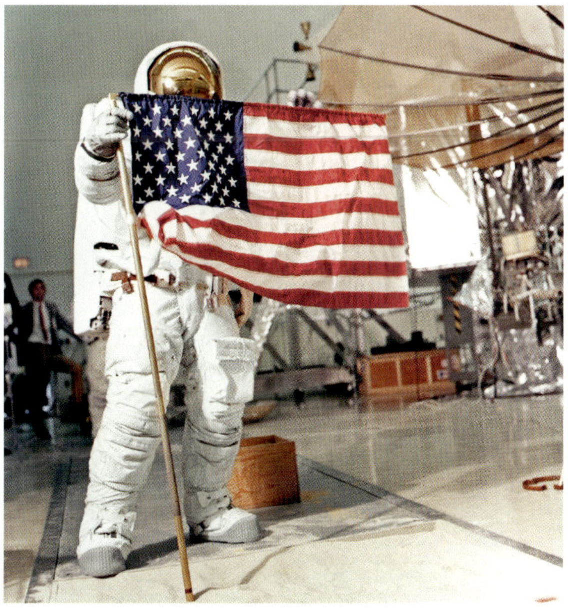

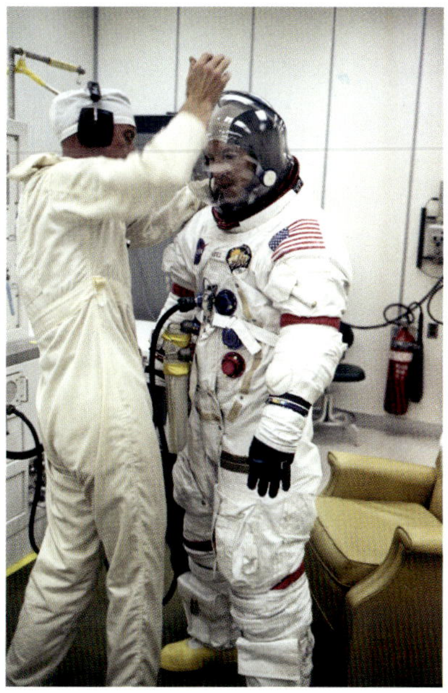

Opposite page, top right: With the four SLA petals on the S-IVB jettisoned, the top of LM *Aquarius* is exposed prior to CM docking, about three hours and twenty minutes after launch on April 11, 1970. The S-IVB would later be crashed into the moon for seismic studies.

Opposite page, third left: In Firing Room 1, KSC Director Kurt Debus (*center*) presents a hard hat and a Saturn V model from MSFC to West German Chancellor Willy Brandt following the launch. Vice President Spiro Agnew stands between them; Paine is at right. Donnelly is behind the model, in blue. Brandt received the Nobel Prize the following year for his efforts to improve West German relations with East Germany, Poland, and the Soviet Union.

Opposite page, bottom left: In the rain at Pad 39A, Lovell leaves the transfer van to follow Haise for the CDDT on March 26, 1970. KSC Security Chief Charles Buckley is at left.

Opposite page, bottom right: The base of the Saturn V is engulfed in flames as the first-stage engines ignite, seconds short of liftoff at 2:13 p.m. (EST) on April 11, 1970.

Top left: A close-up of the EVA visor Lovell was to wear on the moon, on display at the Adler Planetarium in Chicago. Lovell, a naval aviator, had received NASA's permission to add an anchor to the visor. (photo by J. L. Pickering)

Top right: Lovell had described the lunar surface as "sort of grayish beach sand" during Apollo 8.

Center left: Lovell, forty-two, prepares to enter the Apollo 13 LM for an altitude chamber test on September 18, 1969, at KSC. That week, NASA announced manpower cutbacks at the center and a launch slowdown, and President Richard Nixon revealed that Vice President Agnew's task force had recommended targeting a manned Mars landing for 1986.

Center right: On April 16, 1970, during the return trip to Earth, Lovell operates a Sony TC-50 audiocassette recorder on the pilot's side of *Aquarius*, with the craft's front window at left. The blue lunar contact light is at the bottom right of the panel at left, bounded by three white lines.

Bottom left: Lovell practices erecting the US flag during EVA training at KSC on December 4, 1969. The umbrella of the portable S-band antenna is at upper right. A smaller horizontal rod held the flag away from the pole, as if windblown. After the Apollo 11 crewmen had trouble fully extending the telescoping top rod, creating a rippled appearance, later crews intentionally left the rod partially retracted for the same effect.

Bottom right: Suit tech Clyde Teague checks Lovell's helmet in the MSOB on launch morning, April 11, 1970.

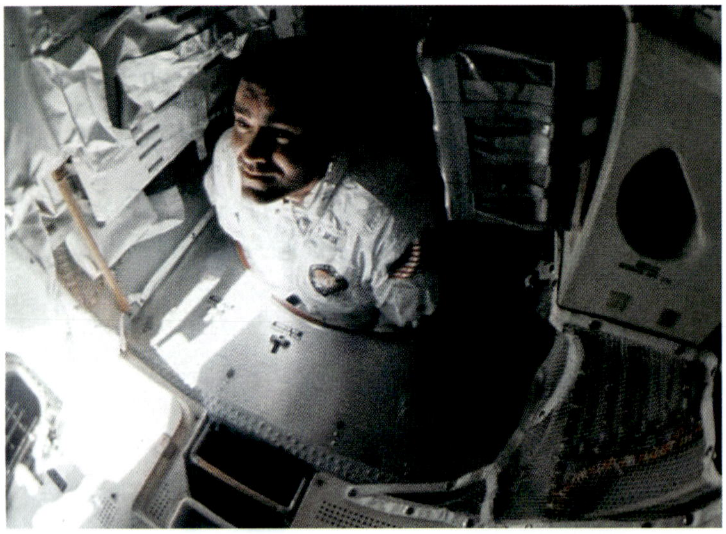

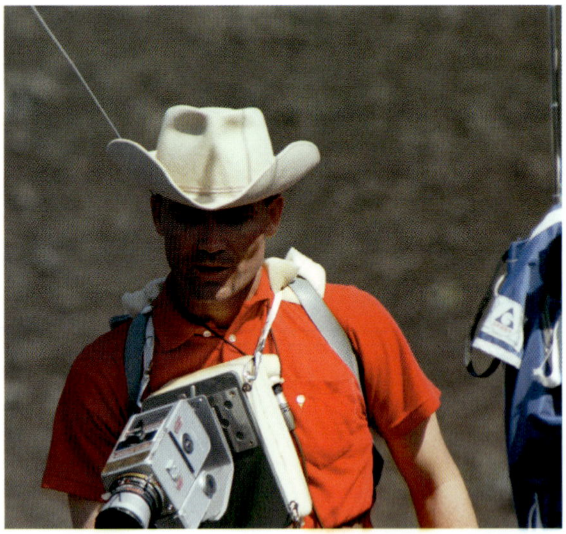

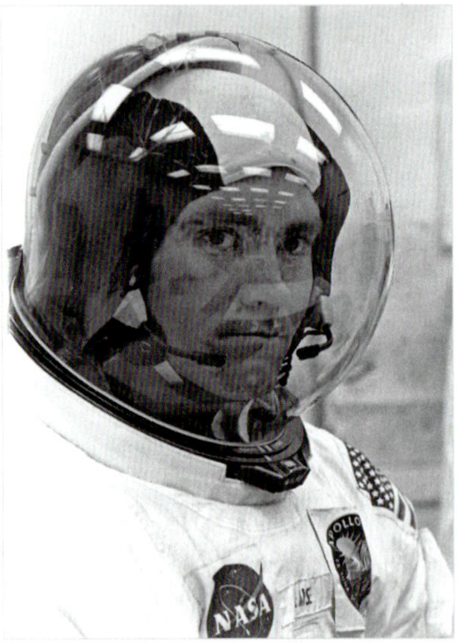

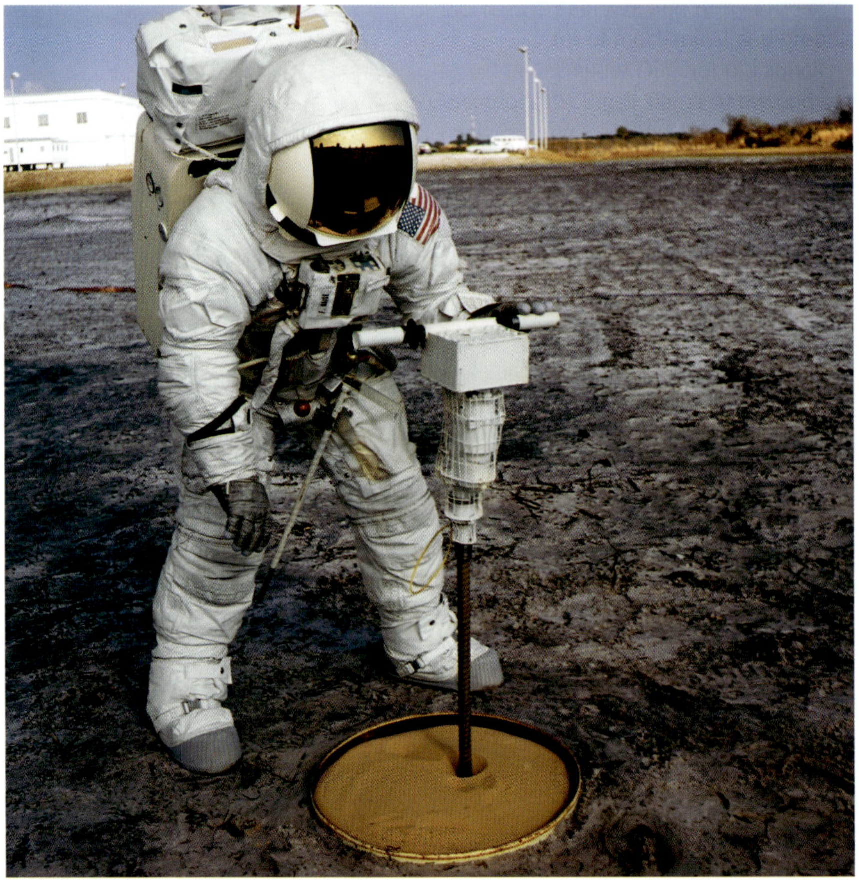

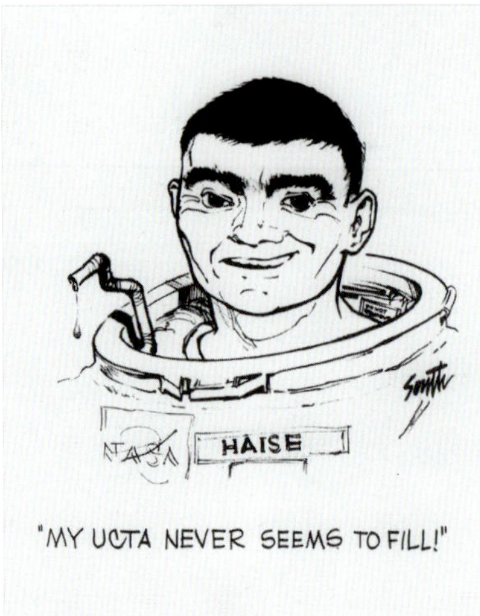

Top left: Haise in the tunnel between the CM and LM on the trip home. At right is the boxy makeshift air filter attached to the LM's environmental control subsystem housing.

Top right: Haise with a 70mm Hasselblad electric data camera (EDC) during geology training in Kilauea, Hawaii, which lasted from December 17 to 20, 1969.

Center left: Haise during the CDDT on March 26, 1970. A rookie selected by NASA in 1966, he and Swigert were the first of that group to fly. Haise, thirty-six, was a former military and NASA research pilot. He had been on the backup crews for two Apollo missions.

Bottom right: On January 28, 1970, Haise practices using the battery-powered lunar core drill during an EVA timeline walk-through near the Flight Crew Training Building. He would have been the sixth person to walk on the moon. The drills were used on Apollos 15, 16, and 17 to obtain samples nearly ten feet deep.

Bottom left: In this caricature, as his drinking straw drips, Haise is puzzled that his Urine Collection and Transfer Assembly "never seems to fill."

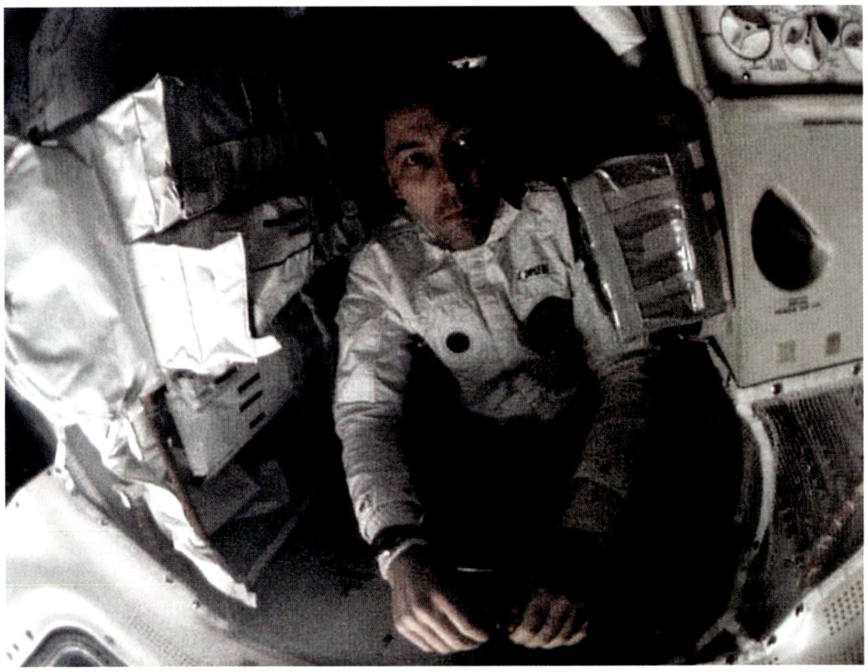

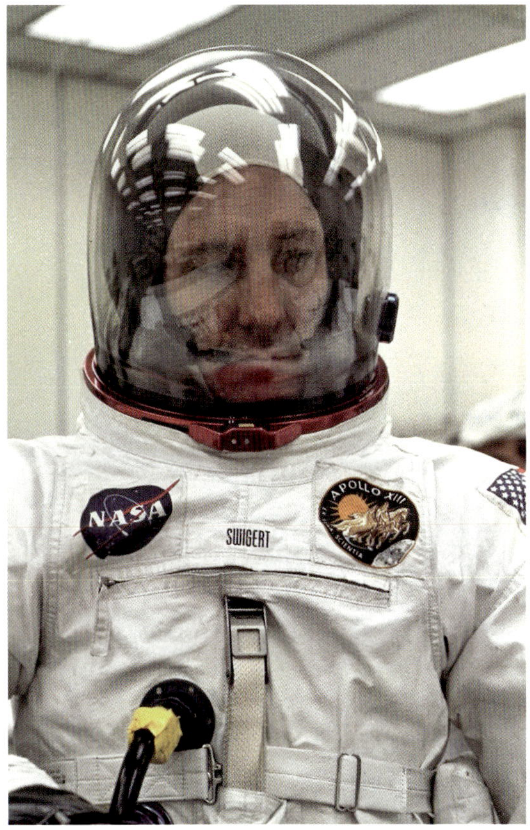

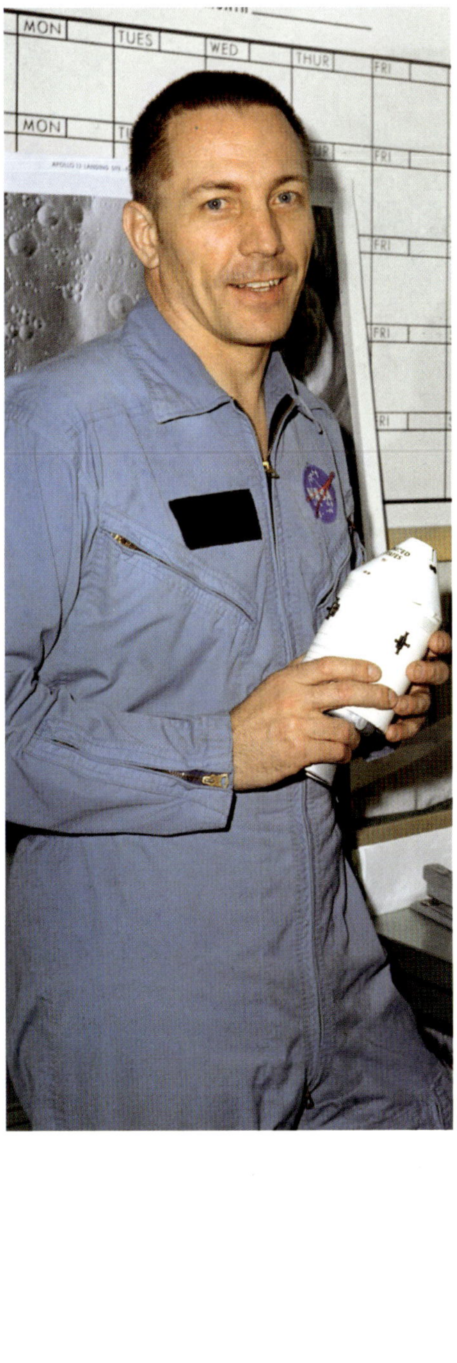

Top left: Swigert on April 16, 1970, in the tunnel between *Aquarius* below and *Odyssey* above. The duct tape–wrapped oxygen filter is at right. At left are the US flag and pole assembly, and storage pouches for the flight-data file cards.

Top right: CMP Jack Swigert, suited for launch on April 11. A lifelong bachelor with an eye for women, Swigert, thirty-eight, was turned down by NASA twice before being selected in 1966. He was elected to the US House of Representatives from Colorado in 1982, but he died from cancer before taking office.

Bottom left: Swigert, holding a Precise Apollo spacecraft model, played football at the University of Colorado and served with the USAF from 1953 to 1956. His service included a tour of duty as a fighter pilot in Japan and Korea.

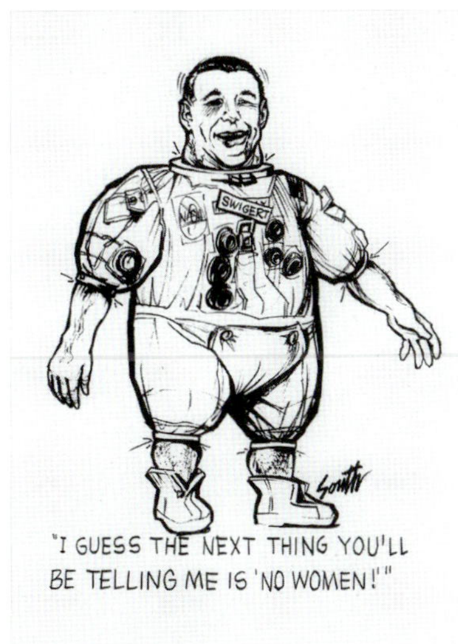

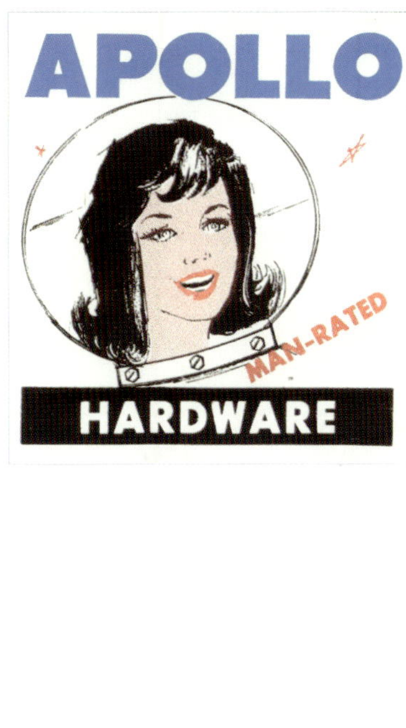

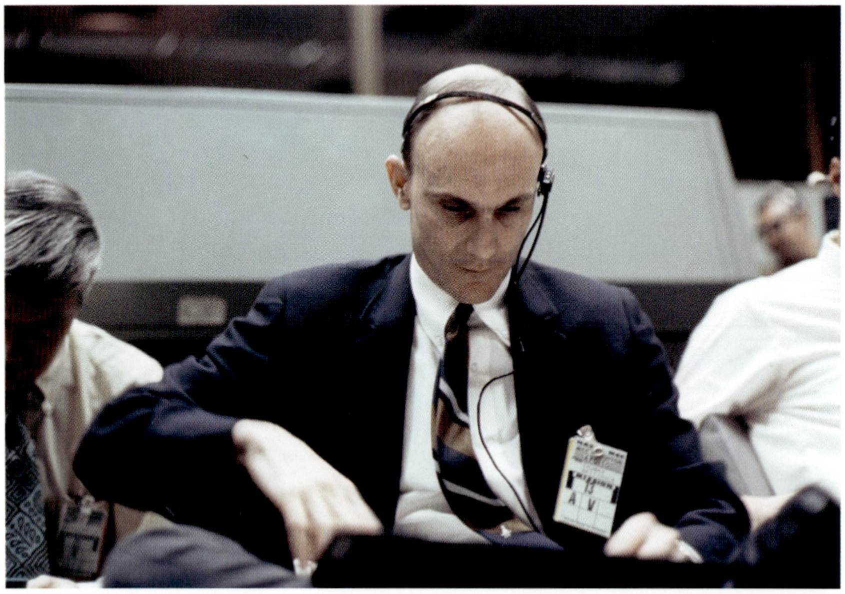

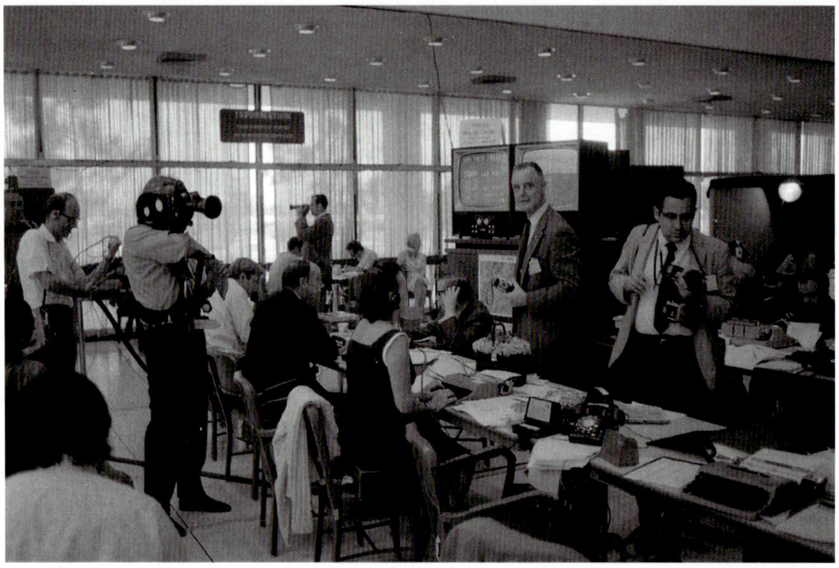

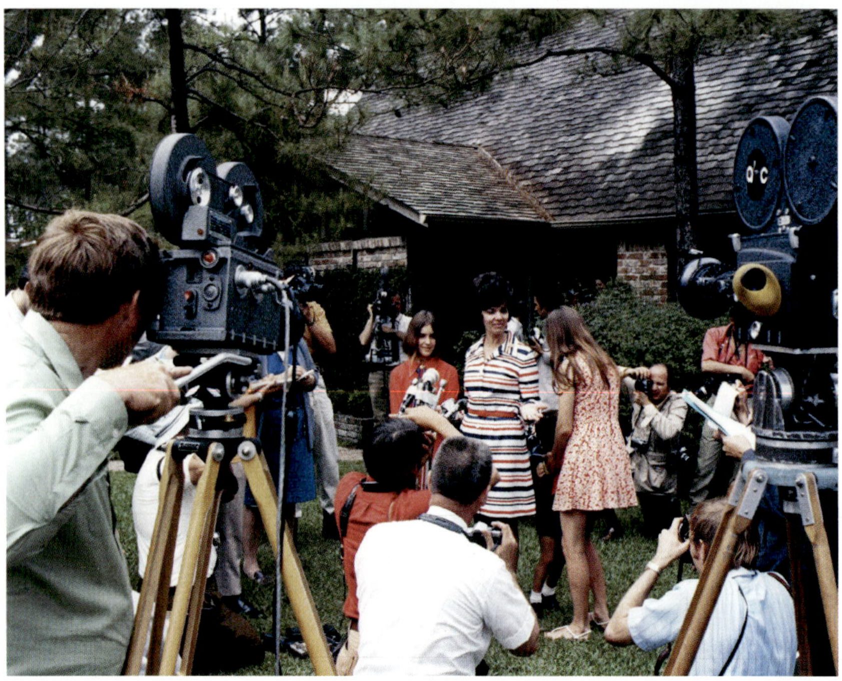

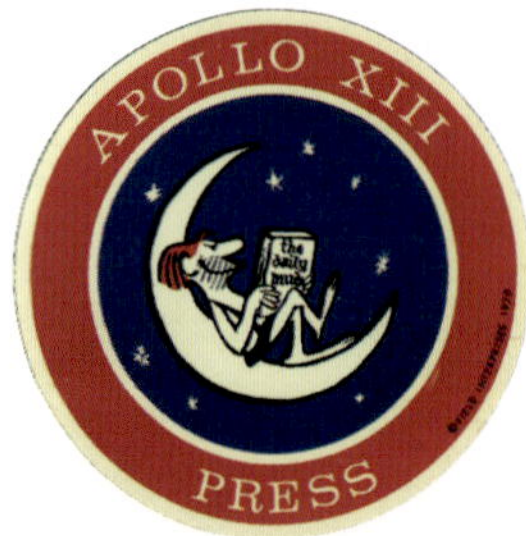

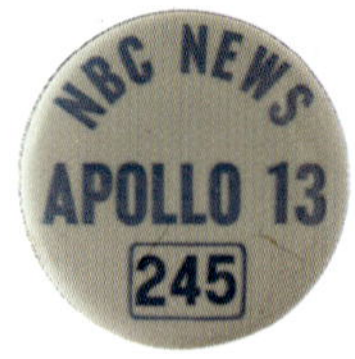

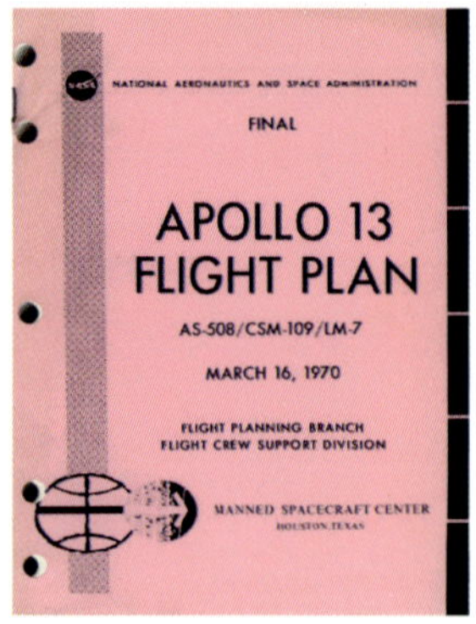
NATIONAL AERONAUTICS AND SPACE ADMINISTRATION

FINAL

APOLLO 13
FLIGHT PLAN

AS-508/CSM-109/LM-7

MARCH 16, 1970

FLIGHT PLANNING BRANCH
FLIGHT CREW SUPPORT DIVISION

MANNED SPACECRAFT CENTER
HOUSTON, TEXAS

Top left: Mattingly follows launch day procedures in MOCR 2. At left is NASA Public Affairs Officer Brian Duff.

Center left: NASA contract cameraman Jerry Bray films news media activities in the lobby of Building 2 at MSC during Apollo 13 with sound engineer Glenn Osborne. The man standing at center is John Peterson. With her back to the camera is Fairchild reporter Mary Bubb. The small "squawk boxes" on either side of the telephone provided mission audio; the left TV monitor gave the times of mission events.

Bottom left: Marilyn Lovell and daughters Susan and Barbara speak to reporters at their home in the Houston suburb of Timber Cove (near MSC), following recovery of the crew on April 17, 1970. Son Jeffrey is behind Susan. The 16mm camera at left is an Auricon Super 1200, while the ABC camera at right is an Auricon Pro 600. Both have Angénieux 10x lenses.

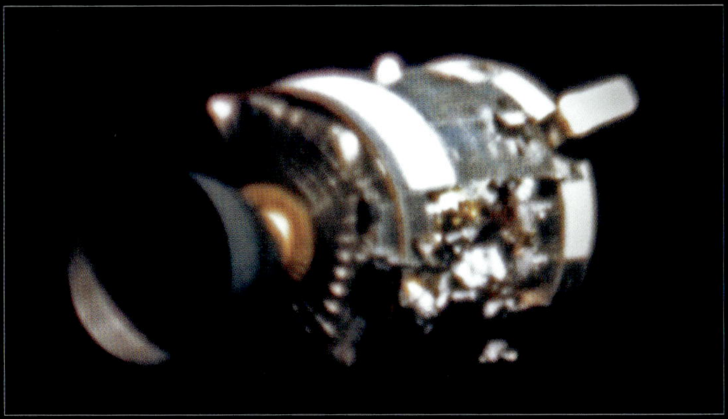

Apollo 13's damaged SM drifts away on April 17, having separated from the CM *Odyssey*. The thirteen-foot exterior aluminum panel is missing, exposing the Sector 4 fuel cell bay.

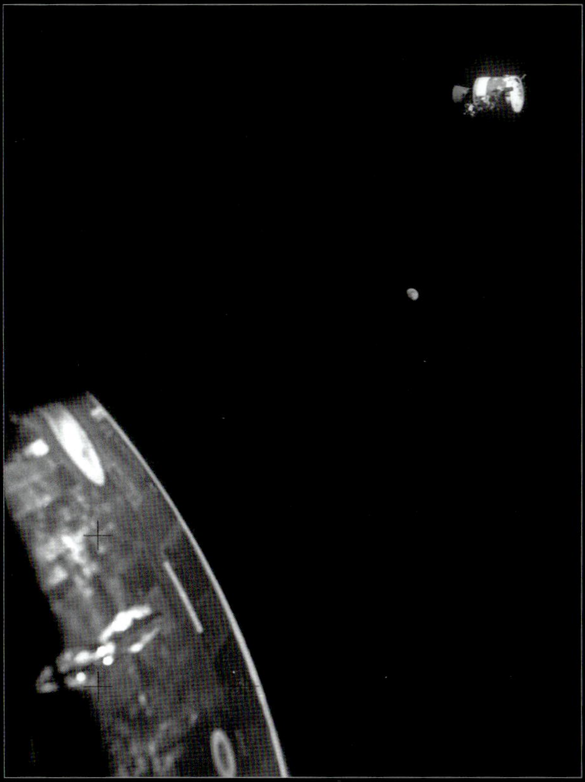

With the moon between them, the SM drifts away while *Odyssey* is at lower left in this photo from *Aquarius*.

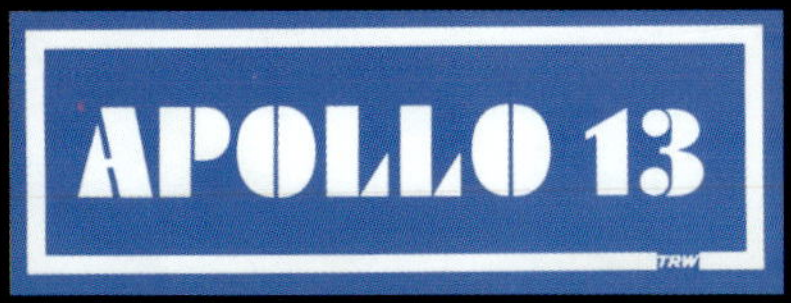

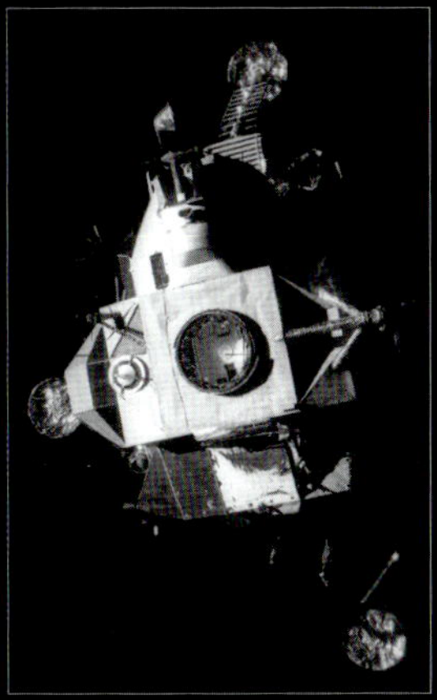

Direct right: *Aquarius* separates from *Odyssey* on April 17, just before CM reentry.

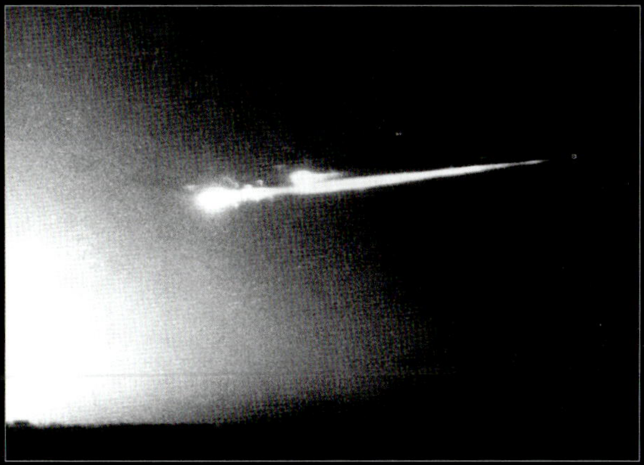

A commercial airline passenger captured this image of the SM and LM reentering the atmosphere over the Pacific Ocean on the morning of April 17.

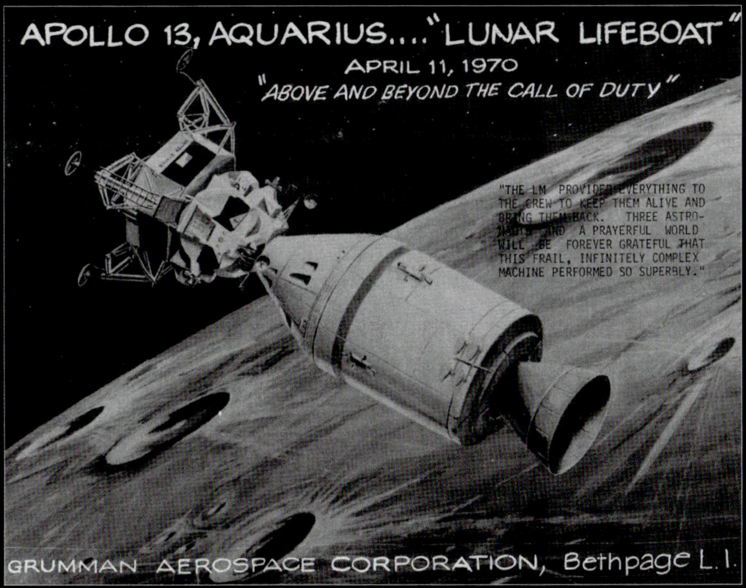

Above: Grumman Aerospace Corporation team certificate

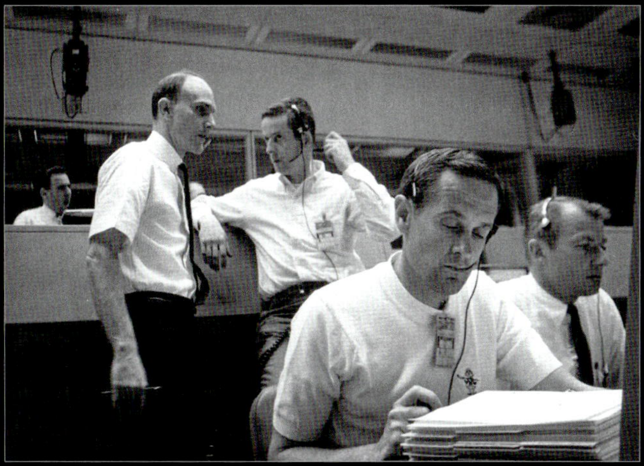

Right, left to right: Astronauts Mattingly, Joe Kerwin, Charles Duke, and Vance Brand at the CAPCOM console in MOCR 2 on April 14. While they were still in radio contact, Brand had given the crew the details for a burn of the SM's propulsion system, which would take place once they were behind the moon (hours away), to return them to Earth. Duke had first exposed the crew to German measles weeks earlier. Black Team Flight Director Milt Windler is behind Mattingly.

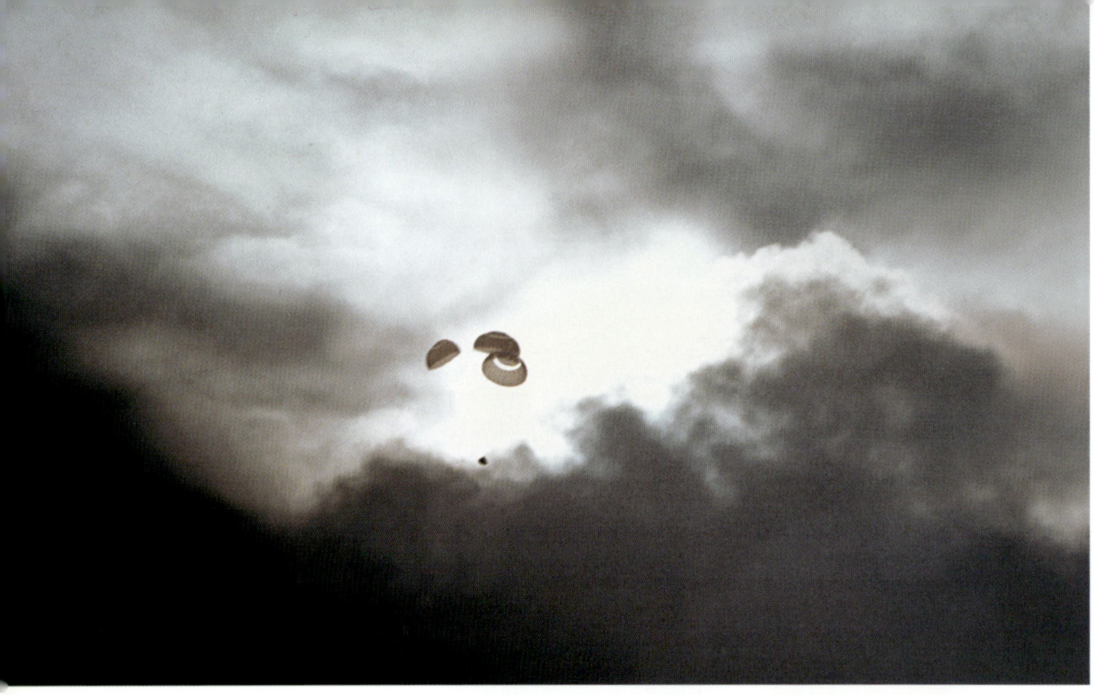

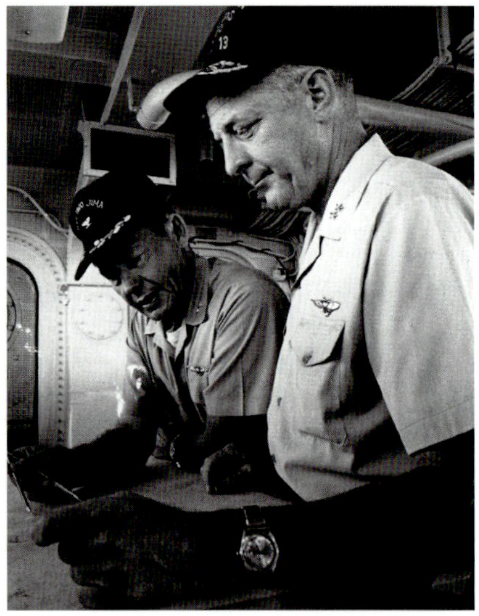

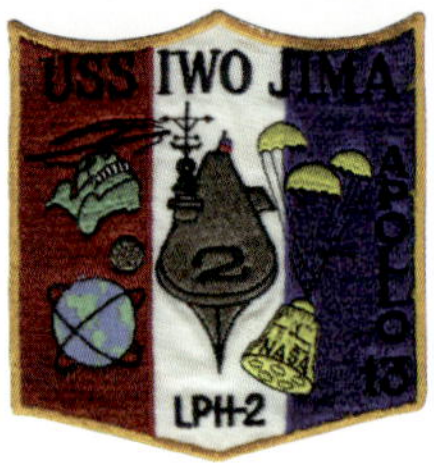

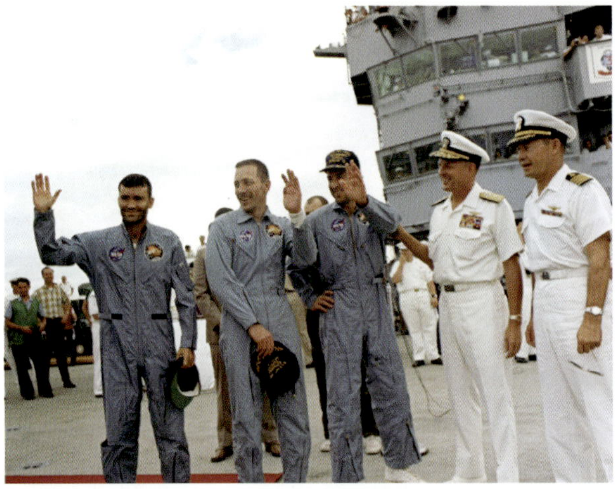

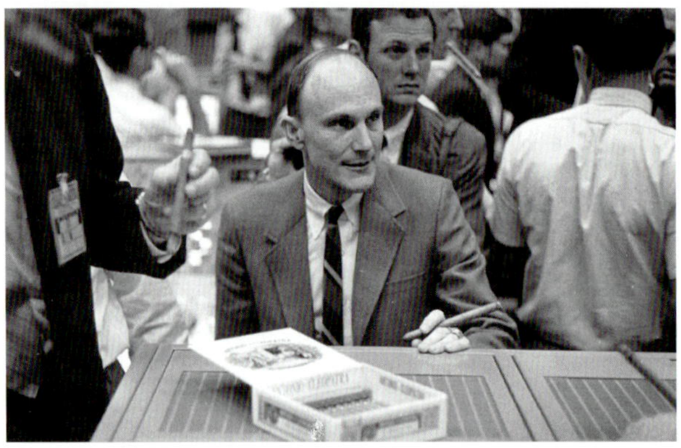

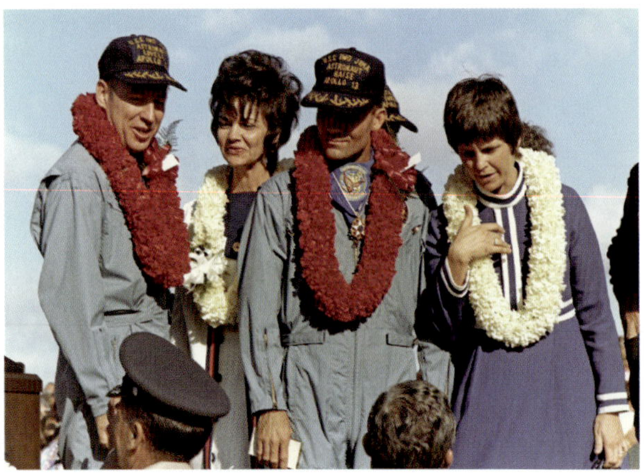

Top left: *Odyssey* descends through clouds just before splashdown at 1:07 p.m. (EST) on April 17, 1970, in the southeast Pacific Ocean, west of Australia. Local time at splashdown was 6:07 a.m.

Top right: Capt. Leland E. Kirkemo (*left*), *Iwo Jima* commanding officer, and Rear Adm. Donald C. Davis, commanding officer of Task Force 130, the US Navy's Pacific recovery force, chart a course for Pago Pago, American Samoa, from which the crew would depart for Hawaii.

Second left: Haise, Swigert, and Lovell prepare to go below deck for medical exams. Davis and Kirkemo are at right.

Third left: Mattingly smokes a cigar in MOCR 2 after splashdown on April 17. He would serve as CMP on Apollo 16.

Bottom left: Jim and Marilyn Lovell (*left*) and Fred and Mary Haise at Honolulu International Airport on April 19, 1970. The astronauts had just received the Presidential Medal of Freedom from President Nixon. Swigert is behind Haise.

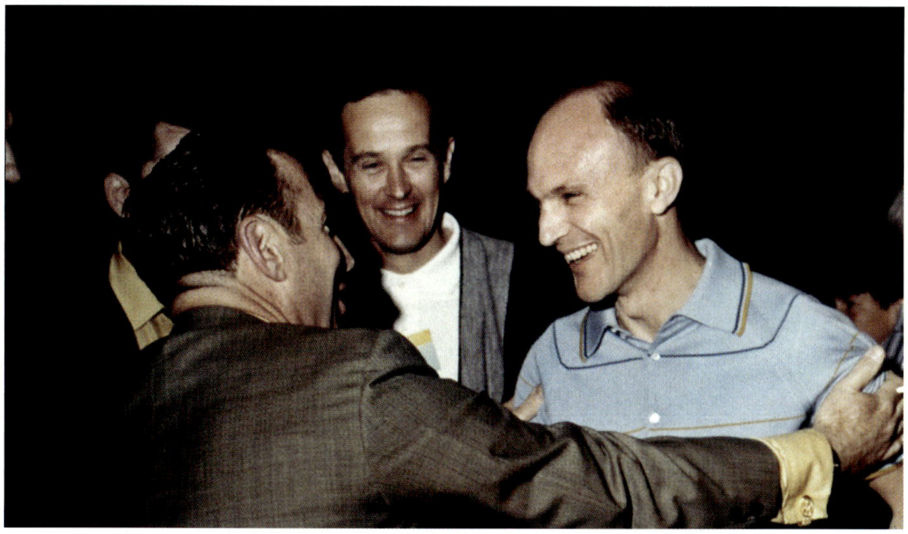

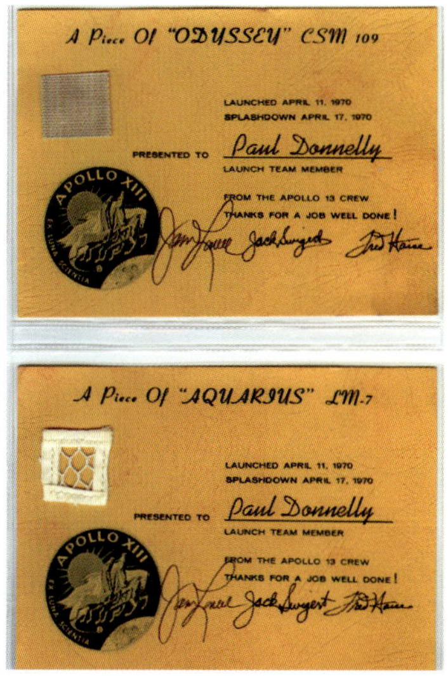

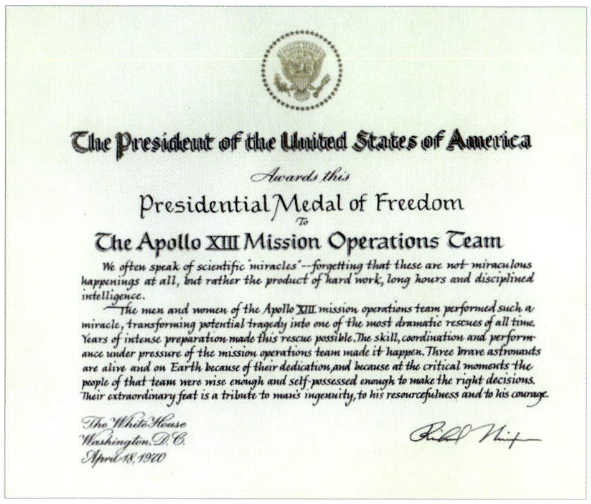

The President of the United States of America

Awards this

Presidential Medal of Freedom

To

The Apollo XIII Mission Operations Team

We often speak of scientific "miracles"--forgetting that these are not miraculous happenings at all, but rather the product of hard work, long hours and disciplined intelligence.

The men and women of the Apollo XIII mission operations team performed such a miracle, transforming potential tragedy into one of the most dramatic rescues of all time. Years of intense preparation made this rescue possible. The skill, coordination and performance under pressure of the mission operations team made it happen. Three brave astronauts are alive and on Earth because of their dedication, and because at the critical moments the people of that team were wise enough and self-possessed enough to make the right decisions. Their extraordinary feat is a tribute to man's ingenuity, to his resourcefulness and to his courage.

The White House
Washington, D.C.
April 18, 1970

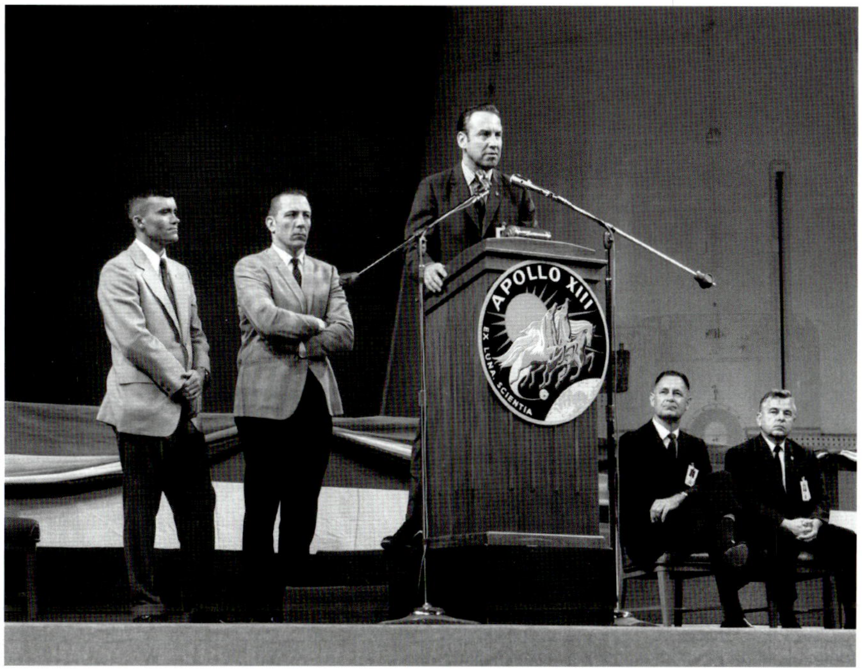

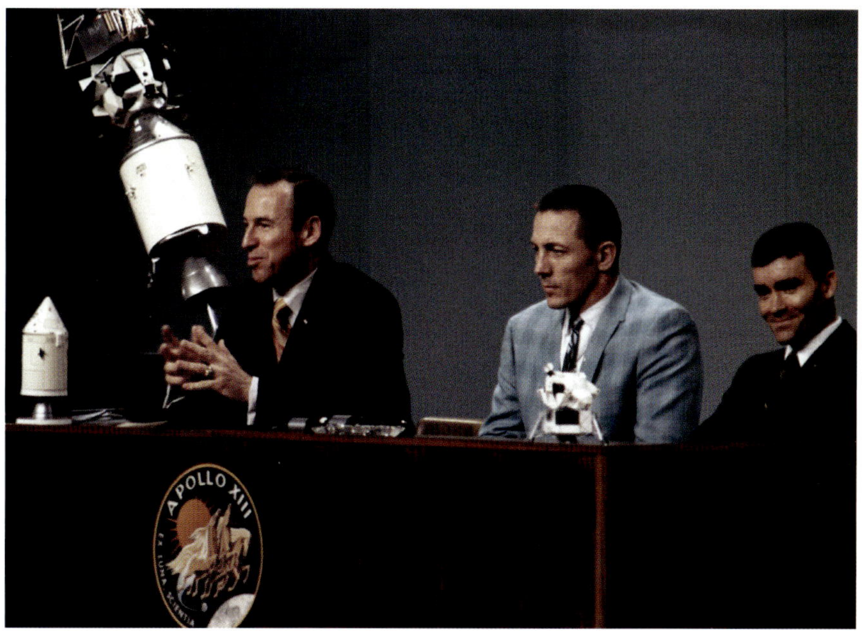

Top left: Lovell greets Mattingly at Ellington Field as the crew arrives home on April 19, 1970. Duke is at center.

Top right: Flown fabric and netting swatches given to KSC workers

Center right: With Haise and Swigert at left, Lovell addresses thousands of KSC employees in the VAB on May 4, 1970. Seated at right are KSC Deputy Director Mike Ross and Kapryan.

Bottom right: Lovell, Swigert, and Haise during a seventy-minute evening news conference in Bldg. 2 at MSC on April 20, 1970. The conference was broadcast live nationally. "I began wondering if we'd ever get home again," said Lovell, who also laughingly reported that a number of airline stewardesses were in the audience—friends of Swigert, he claimed.

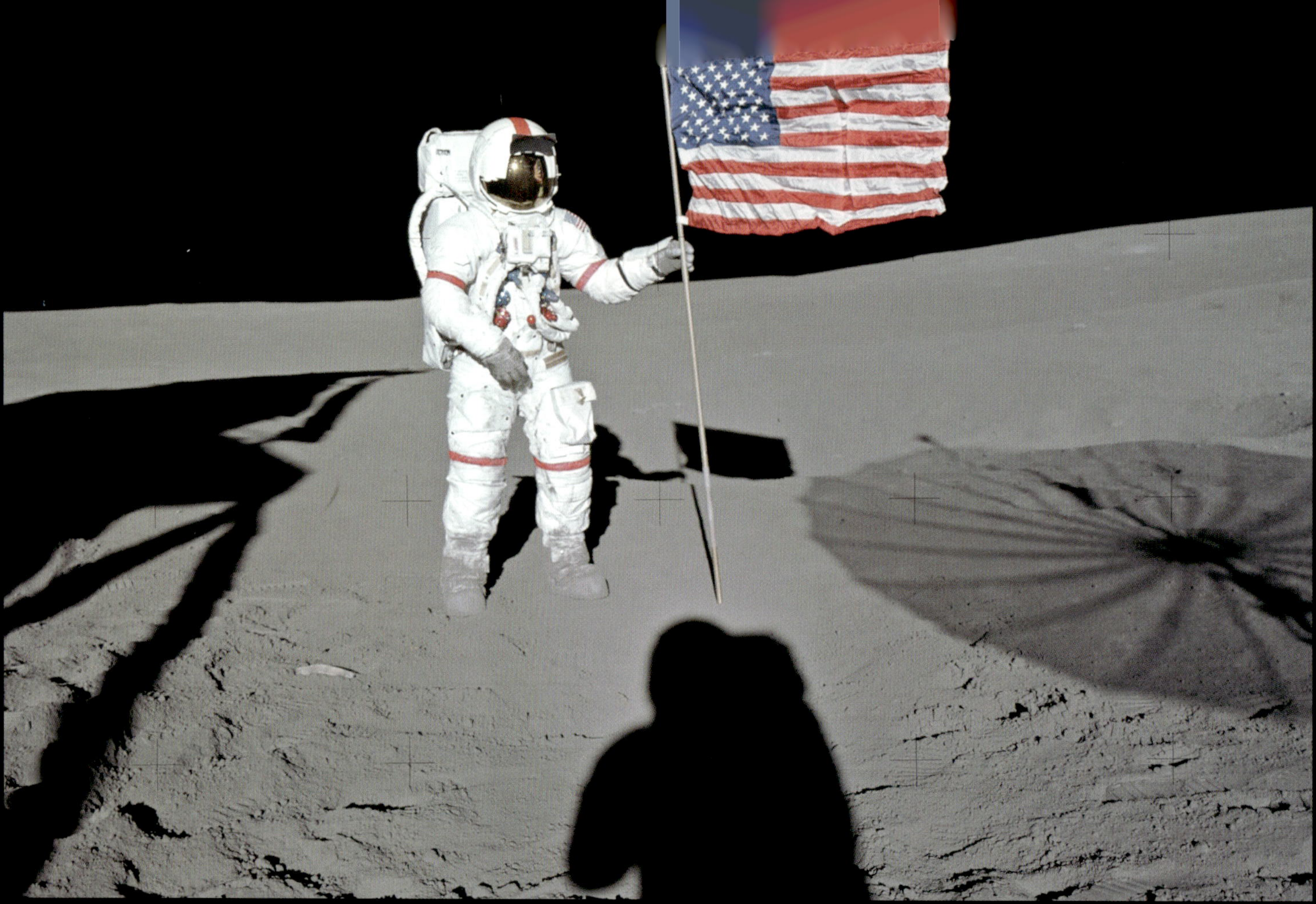

Alan Shepard, distinguished by the red markings on his arms, legs, and helmet, is photographed by Ed Mitchell on February 5. *Antares* casts the shadow on the left; the stand-alone S-band antenna's shadow is at right. Shepard had just turned the flag to face the 16mm data acquisition camera (DAC) on the Modular Equipment Transporter (MET). After both Shepard and Mitchell posed, Mitchell turned the flag so it again faced the TV camera.

CHAPTER 9

Apollo 14

January 31–February 9, 1971

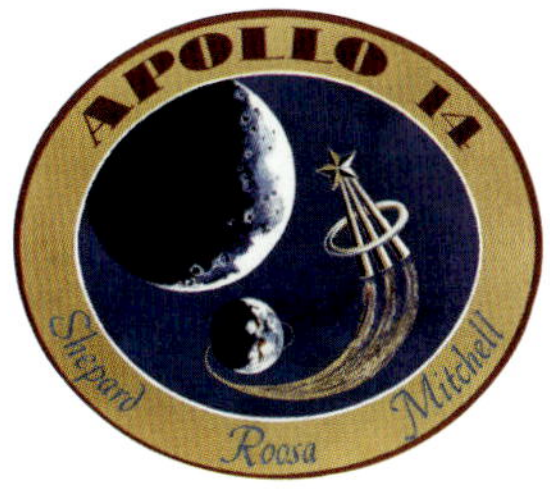

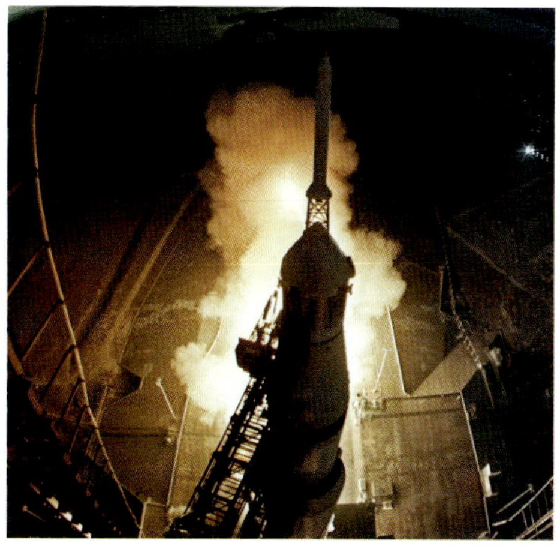

Launch: 4:03 p.m. (EST), January 31, 1971

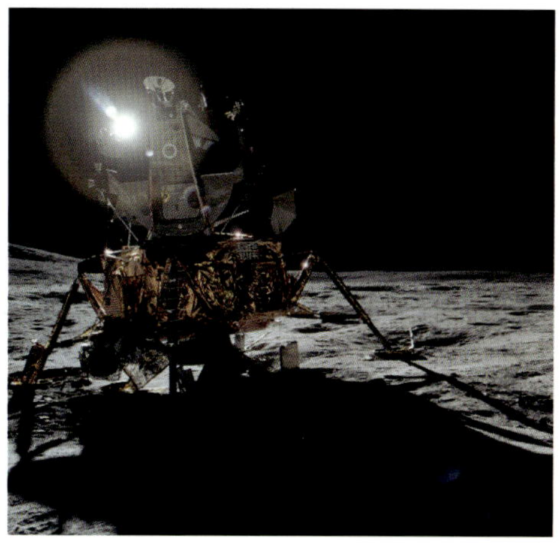

Lunar landing: 4:18 a.m. (EST), February 5.
Ascent: 1:48 p.m. (EST), February 6

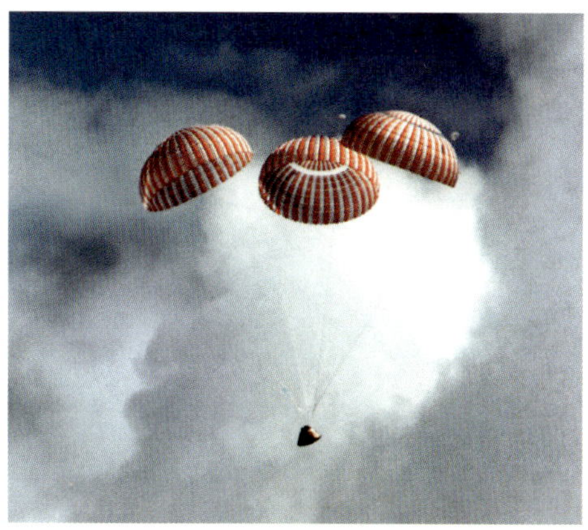

Splashdown: 4:05 p.m. (EST), February 9

In August 1969, CDR Alan Shepard and rookies CMP Stuart Roosa and LMP Edgar Mitchell were selected for what had been expected to be the fourth lunar landing, but became the third after Apollo 13's misfortune.

The Apollo 14 CDR would have been Gordon Cooper given Apollo's traditional crew assignments, but Shepard, finally cleared to fly after an operation to correct an inner ear problem, took the position for himself for what would have been the Apollo 13 crew. NASA managers, however, felt Shepard needed more training, and his crew was assigned to Apollo 14.

Because of Ken Mattingly's measles scare before Apollo 13, strict prelaunch crew-quarantine procedures were put in place. One NASA official claimed, "If any of the crew goes home during the quarantine, their children will be cleared out of the house twenty-four hours in advance, and guards will be placed on the doors to make sure no neighbors drop in."

A passing rain shower forced the first unplanned launch delay of Apollo—by just more than forty minutes—but on January 31, 1971, during a dreary Florida afternoon, the crewmen launched on their nine-day flight from LC-39A. Actors Bob Hope, Cary Grant, and Kirk Douglas were among the estimated five hundred thousand spectators.

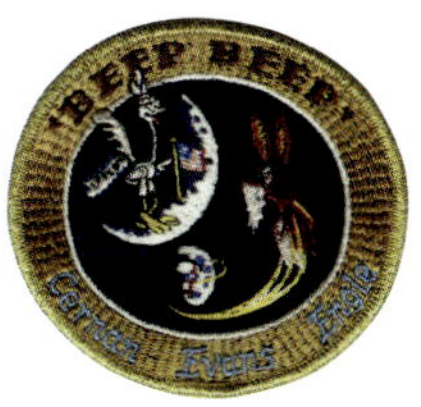

The members of the astronaut backup crew, Gene Cernan, Joe Engle, and Ron Evans, had created their own version of the mission patch. They had a number of them stashed aboard the CM and LM; these were first discovered during the outbound trip to the moon. One was even attached to Shepard's lunar backpack. (The backup crewmen also had their own "crew portrait" taken.)

Three days after launch, as *Antares* headed toward lunar landing, its radar altimeter failed to lock automatically onto the moon's surface, depriving the LM's navigation computer of critical information on the vehicle's altitude and ground speed. (The problem was later found to be an unintended consequence of a software patch.) After the astronauts cycled the landing radar circuit breaker, the unit successfully acquired a signal near eighteen thousand feet, just in the nick of time. Shepard manually landed the LM closer to its intended target than any of the other six moon landing missions.

Although Shepard called Apollo 13 "a successful mission," he knew there was a "certain stigma" attached to it. He wanted Apollo 14 to erase that, and Mitchell believed that Shepard would have continued to try to land without the radar, using the LM's inertial guidance system and visual cues. But a postflight review of the descent data showed that the inertial system alone would have been inadequate, and the astronauts probably would have been forced to abort the landing. Coming on the heels of Apollo 13, this could have potentially ended the moon landing program.

Shepard and Mitchell landed early on February 5, 1971, in the Fra Mauro formation, Apollo 13's target. It was the last of the three H-type missions (including that planned for Apollo 13).

The two spent about thirty-three hours on the moon, with roughly nine and a half hours spent outside on two lunar EVAs. They collected ninety-three pounds of rocks and conducted several experiments, including seismic studies.

They used a pull cart, officially known as the Modular Equipment Transporter (MET) but dubbed the "lunar rickshaw," to help carry experiments and rock samples. The Goodyear Tire & Rubber Company had supplied the two inflatable tires, touting them as "the first tires on the moon."

Shepard famously hit two golf balls on the surface with a makeshift club he had brought, and Mitchell threw a lunar scoop handle as a javelin. Video of these events exists, but photos do not, although one photo barely shows the golf ball and the club handle made from the tool designed for collecting the contingency sample.

Roosa remained in lunar orbit, conducting experiments, photographing the moon, and listening to the country music he had brought on a Sony audiocassette tape recorder.

Once on the lunar surface, Shepard and Mitchell only took a handful of good color shots. In fact, all of the photos taken during the second EVA were black and white. Shepard's first steps on the moon, however, were photographed in color by Mitchell from his LM window—the only time during the six landing missions that a CDR's initial steps were photographed with a Hasselblad.

At the end of an onboard telecast during the return trip on February 7, Shepard cited communication satellites as just one of the space program's contributions to global understanding, adding,

> We are reminded, however, as we look at that shimmering crescent tonight, which is the earth on our way back, that there still is fighting going on. The three of us all have acquaintances, friends, and even relatives in Vietnam. We are reminded that some of the people, some of the men who have gone to Vietnam have not returned, that are still being held there listed as missing in action or as prisoners of war, and it is our wish tonight that we can, in some way, contribute through our efforts to the space program to promote a better understanding of peace throughout the world and help to rectify these situations which still exist. And with that thought, for Ed, and Stu, and myself, I will say good night to you from Apollo 14.

It was the only mention of the war from a spacecraft during the Apollo program.

The trio splashed down in the Pacific Ocean on February 9. President Richard Nixon was in London doing an interview; the reporter said that the president appeared not to want to be bothered by the mission's return, and only reluctantly stopped long enough to place a congratulatory call to the astronauts.

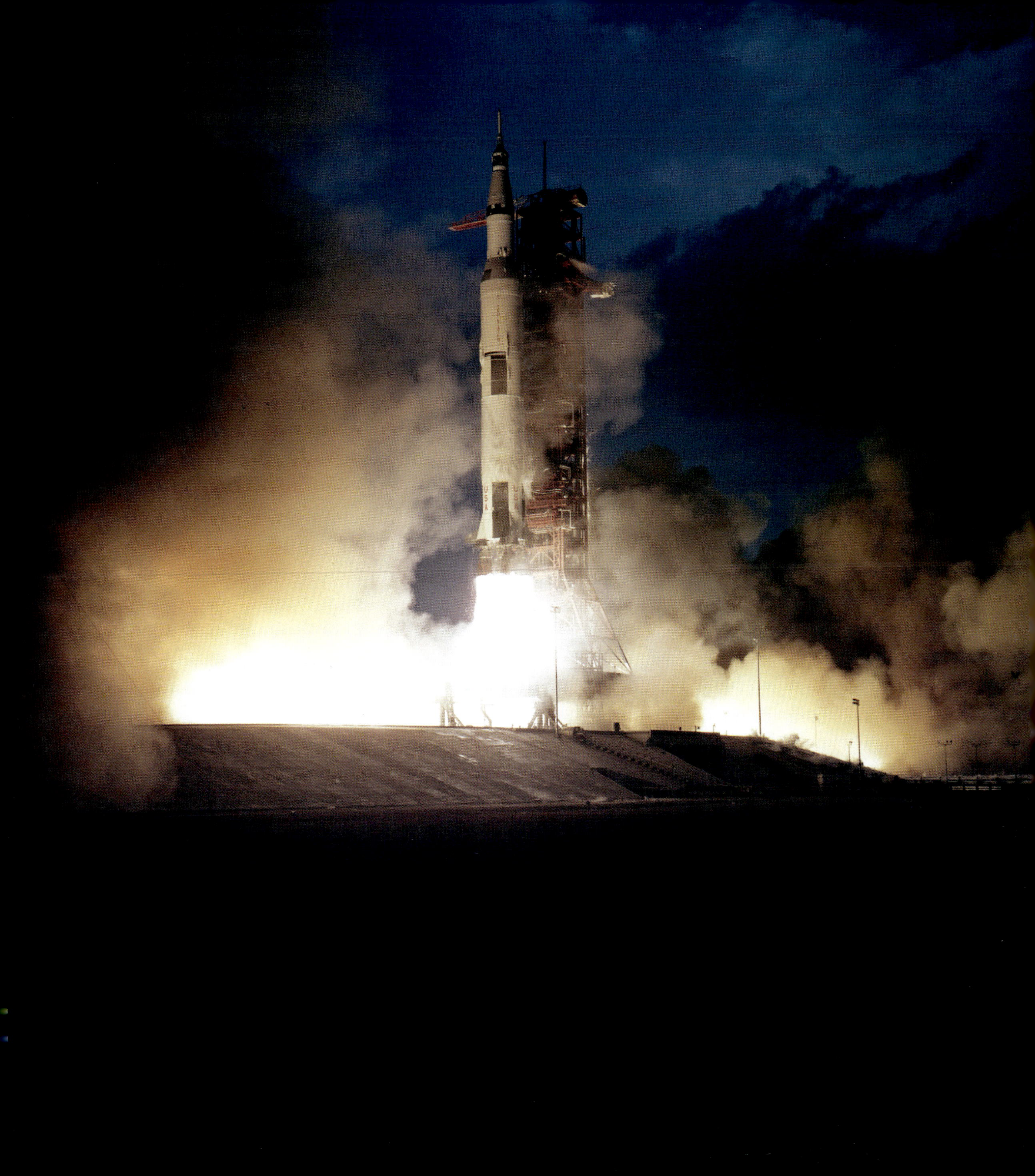

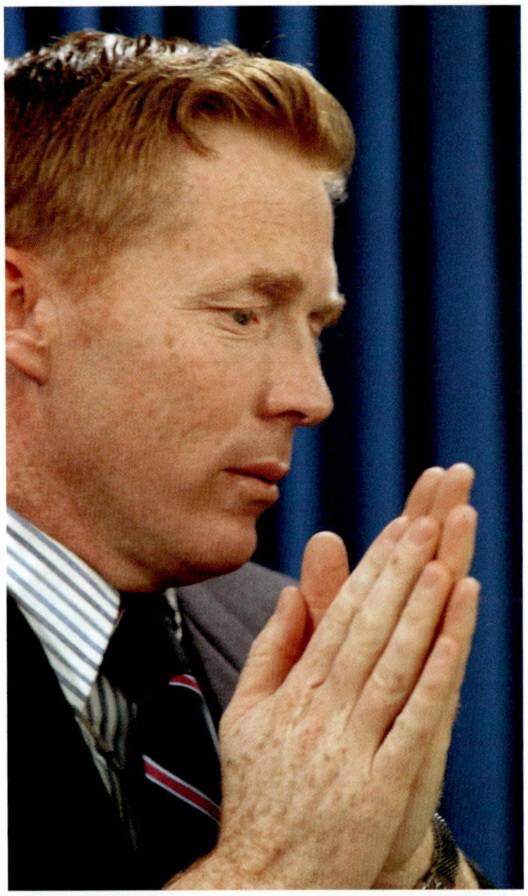

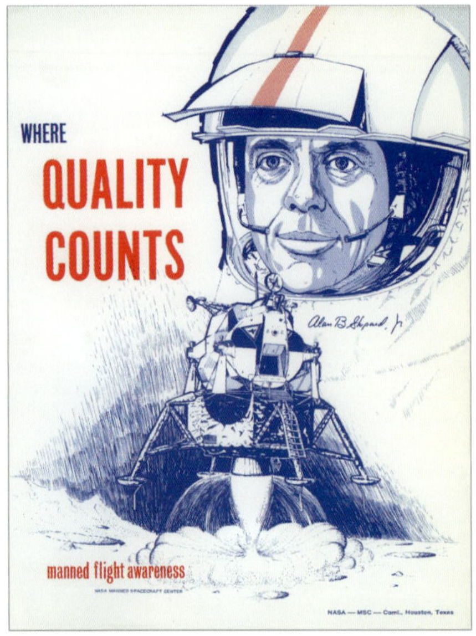

Top left: CMP Stu Roosa, thirty-five, had been a smoke jumper with the US Forest Service before joining the USAF in 1953. A test pilot at California's Edwards AFB when NASA selected him in 1966, Roosa was also an astronaut support crewman for Apollo 9.

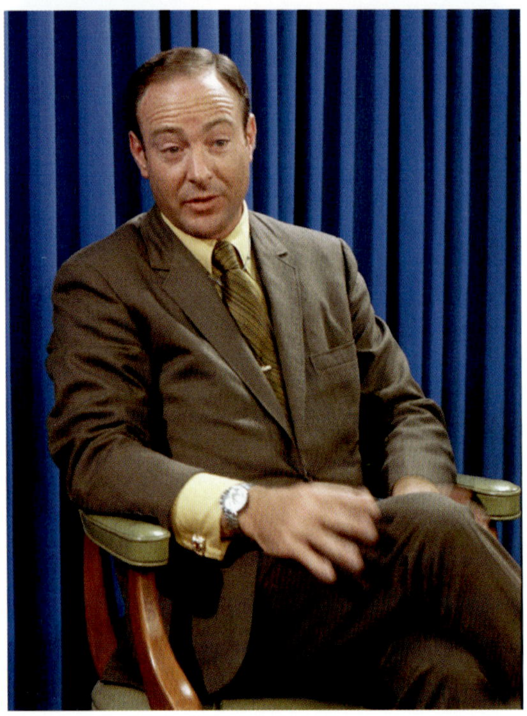

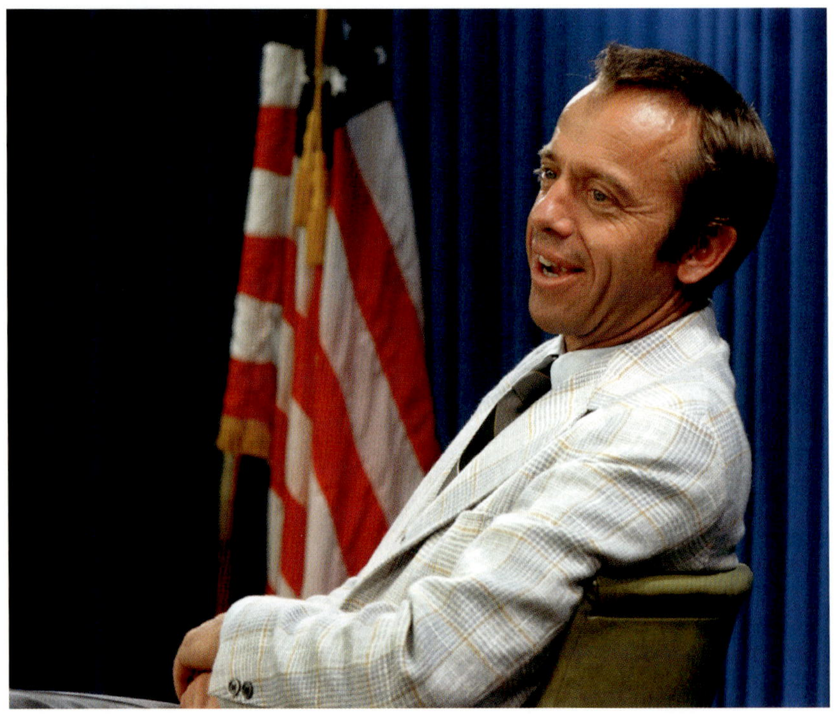

Bottom left: LMP Ed Mitchell, forty, had been a US Navy pilot. He earned a PhD in aeronautics and astronautics from MIT. Also selected by NASA in 1966, Mitchell served as backup LMP for Apollo 10. These images are from a crew news conference at MSC on July 10, 1970, when Shepard and Mitchell expressed concerns about the impact that significant layoffs at KSC in recent months might have on workmanship. Shepard remarked that some employees might be thinking, "Well, I won't be working here tomorrow. Why should I worry about that little screw I just dropped down there?" NASA's budget, on a steep decline since 1966, would continue to drop through 1975.

Bottom right: CDR Al Shepard, forty-seven, had been American's first man in space in 1961, but he was diagnosed with an inner ear disorder in 1964. This kept him off flight status until he had corrective surgery in 1969. He spent the intervening years as chief of the Astronaut Office at MSC.

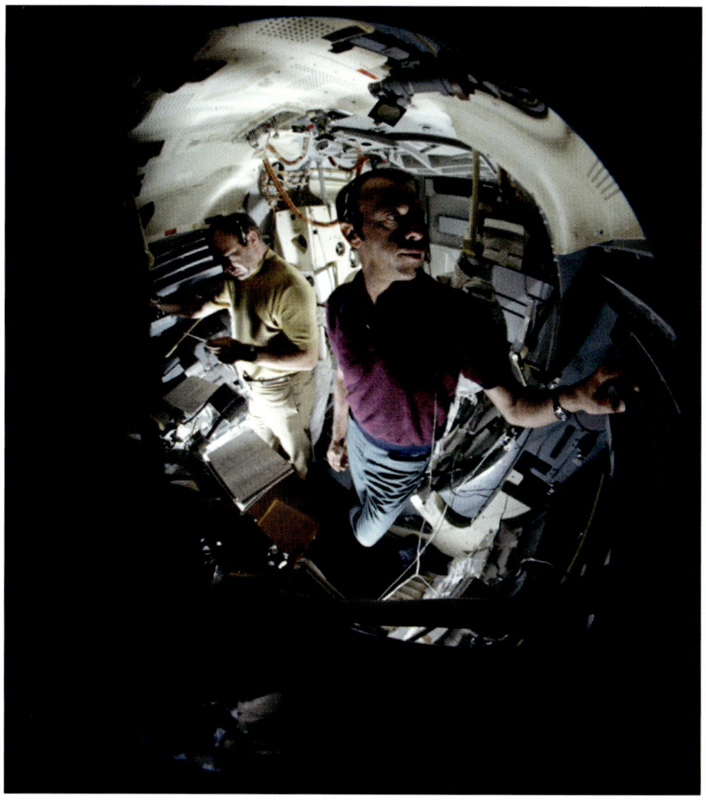

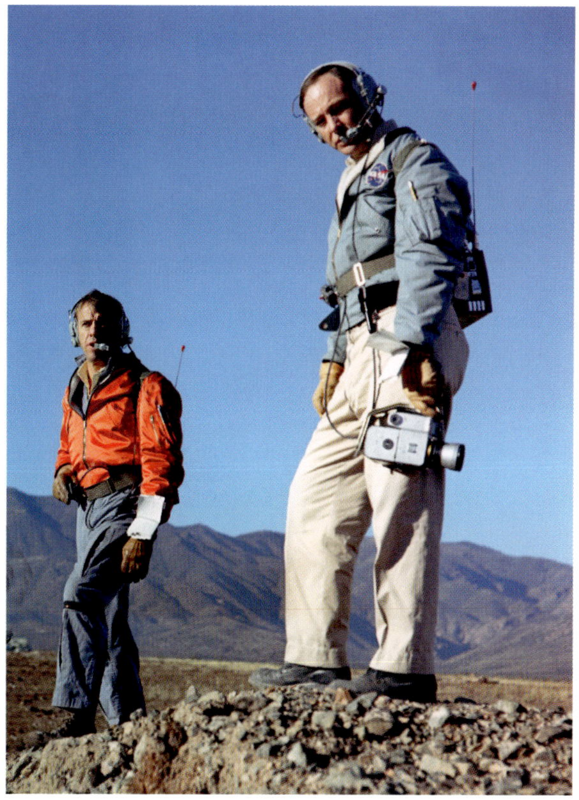

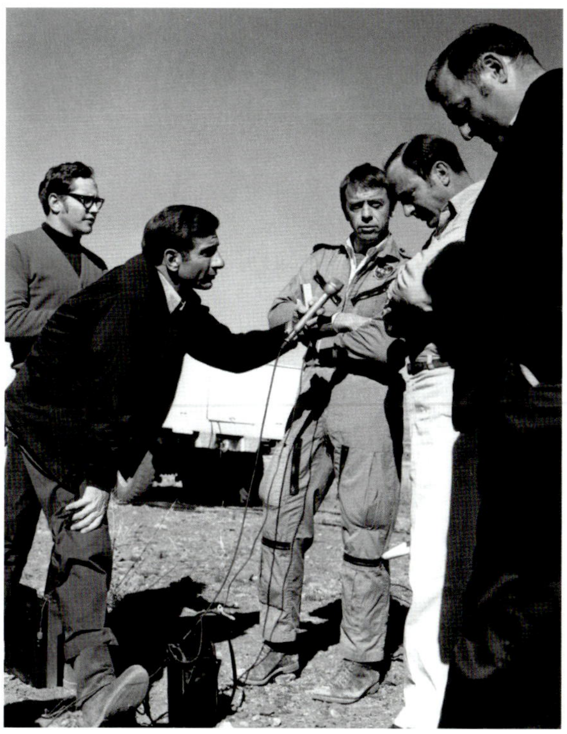

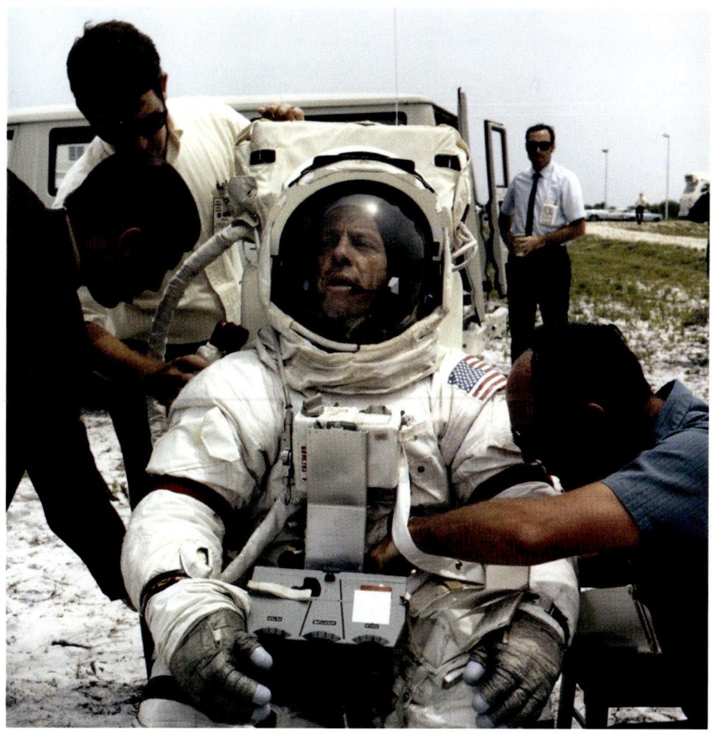

Top left: A fisheye lens captures Mitchell (*left*) and Shepard in the Apollo Lunar Module Mission Simulator in the Flight Crew Training Building on July 15, 1970. Such simulators at KSC and MSC were built by the Singer-Link Company. Link, purchased in 1968 by the Singer Corporation of sewing machine fame, had been making aircraft flight simulators since 1927.

Top right: Shepard and Mitchell during geology field training on November 16, 1970, near Cottonwood, Arizona. Mitchell carries a 70mm Hasselblad EDC. The checklists on the men's left wrists were similar to those they would wear on the lunar surface.

Bottom left: ABC News correspondent Jules Bergman interviews Shepard and Mitchell on February 13, 1970, during geology field training in the Pinacate Peaks, a group of volcanic peaks and cinder cones in the Mexican state of Sonora, south of Arizona. At right is MSC Public Affairs Officer Milt Reim.

Bottom right: Hamilton-Standard suit tech Danny Schaewitz (*left*) and NASA suit tech Byron Smith (*right*) tend to Shepard during EVA training at KSC in July 1970. Smith had been a USAF technician; he was assigned to NASA after the cancellation of the USAF's Manned Orbiting Laboratory (MOL) program two years earlier.

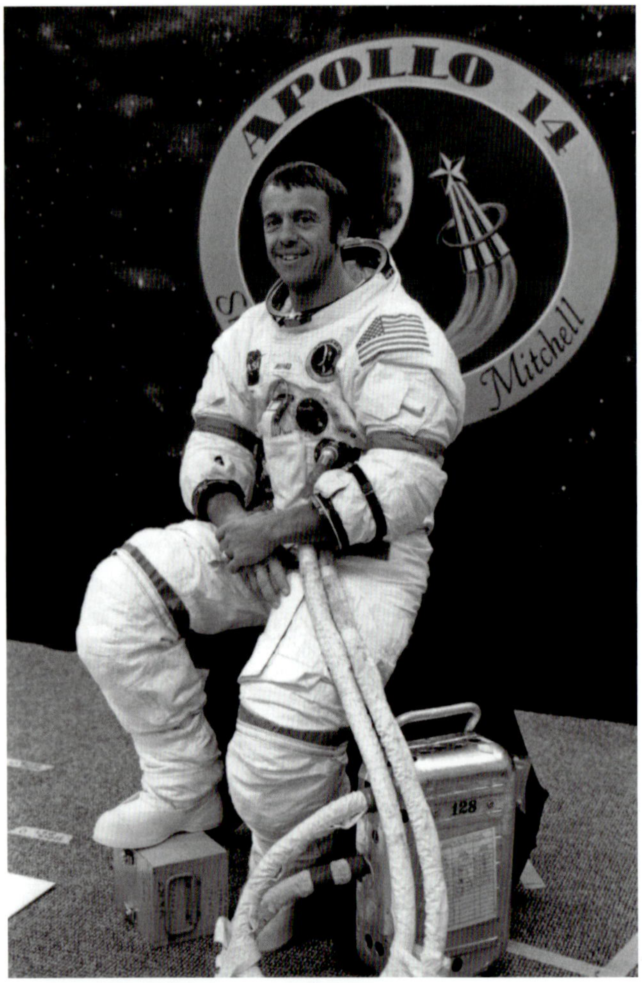
APOLLO 14
Mitchell
128

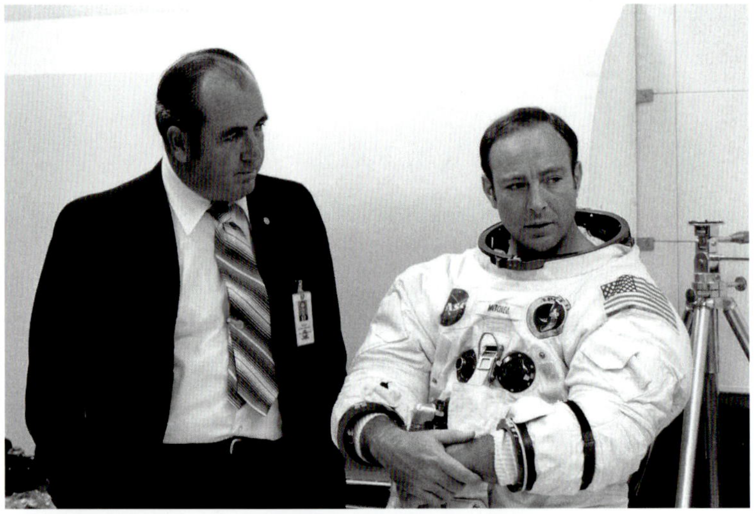

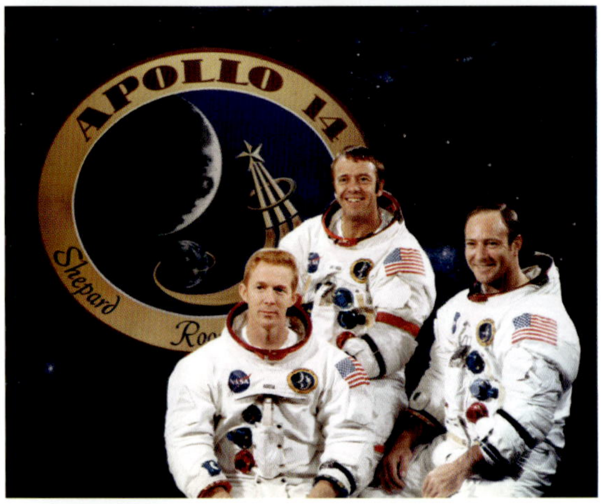
APOLLO 14
Shepard

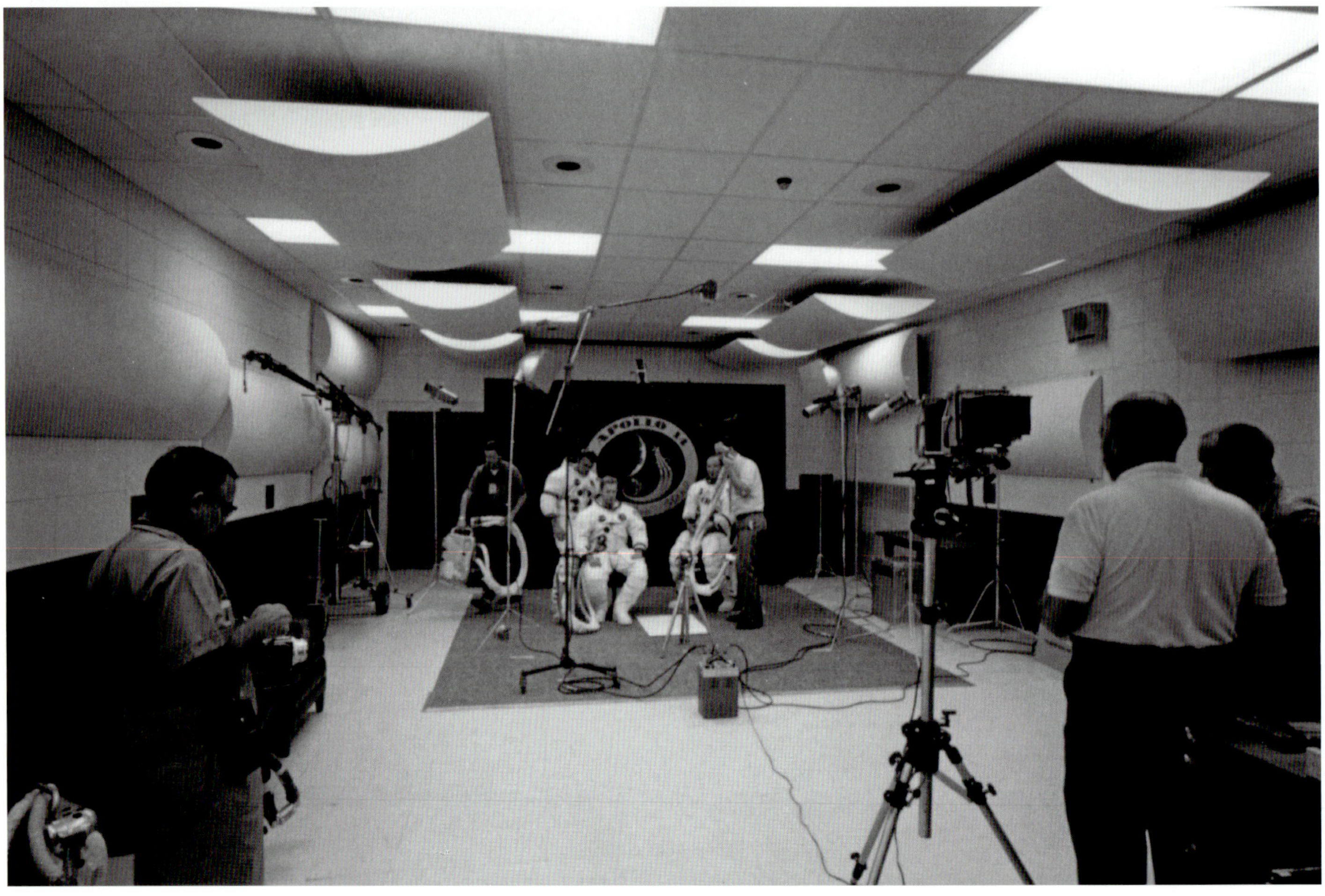

another giant leap
APOLLO 14

Above: Bumper sticker

Opposite page, top left: The portable ventilator at Shepard's feet provided welcome cooling during the photo shoot.

Opposite page, top right: KSC Security Chief Charles Buckley with Mitchell during the October 14 photo shoot. A former Massachusetts state trooper, Buckley had joined the Atomic Energy Commission Security Force at Los Alamos, New Mexico, in 1947. He later had security responsibility for shipments of nuclear material to the South Pacific. He joined NASA at the Cape in 1960, and retired in 1981.

Opposite page, bottom: Photo and suit technicians pose the crew on October 14, 1970, in the MSOB. Suit tech Al Rochford (*left rear*) moves Shepard's suit ventilator. The simple crew portraits begun during Gemini became far more elaborate for the moon missions—they featured dramatic backdrops consistent with the magnitude of the missions.

Right: Shepard with the Modular Equipment Transporter (MET) during training at KSC on December 8, 1970, with an armadillo nearby. A pull cart for carrying equipment and lunar samples made of aluminum tubing, it was dubbed the "lunar rickshaw." Designed after the Apollo 12 astronauts struggled to move equipment significant distances to and from their LM, it was also planned for Apollo 15 until the Lunar Roving Vehicle was included on that mission.

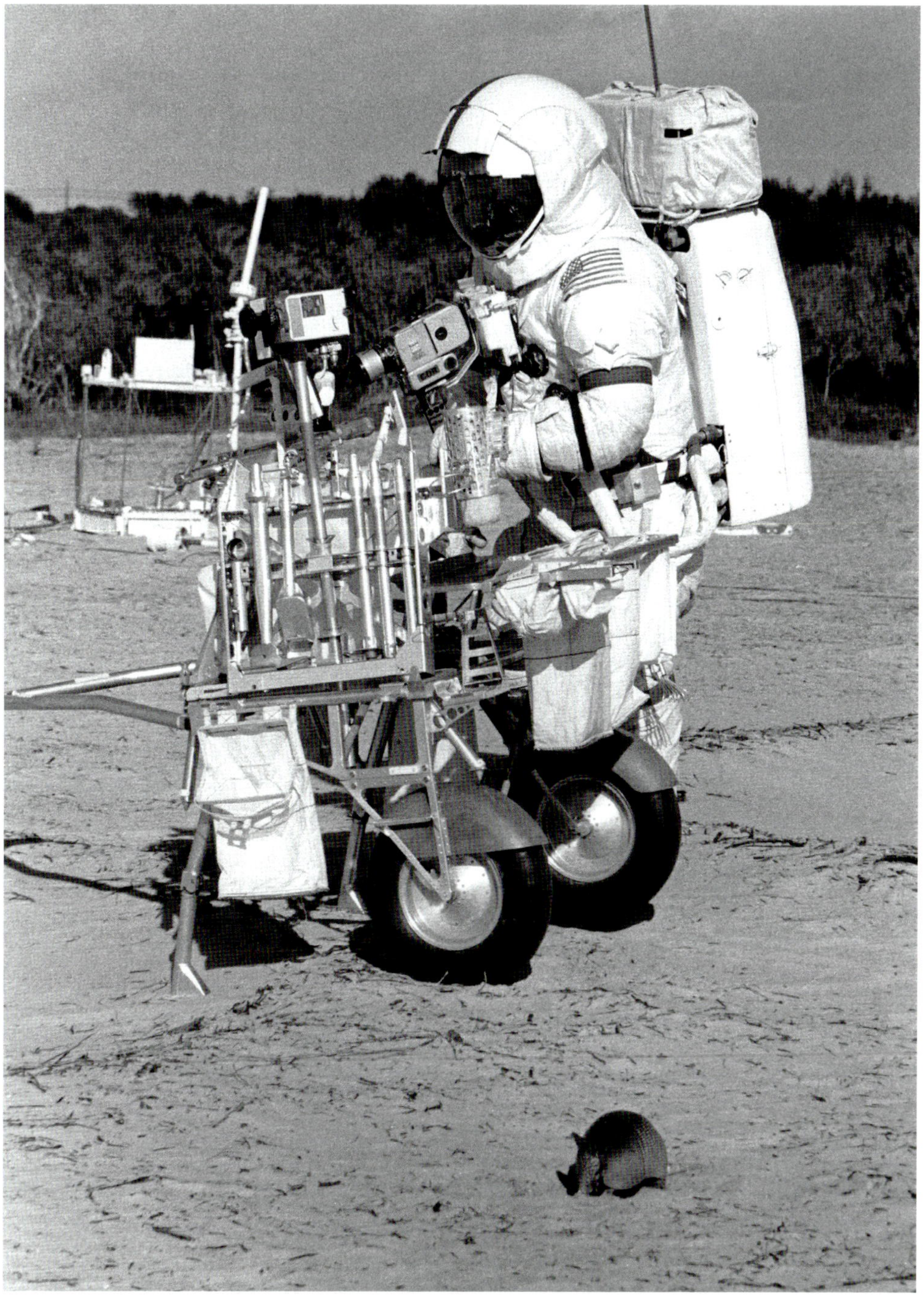

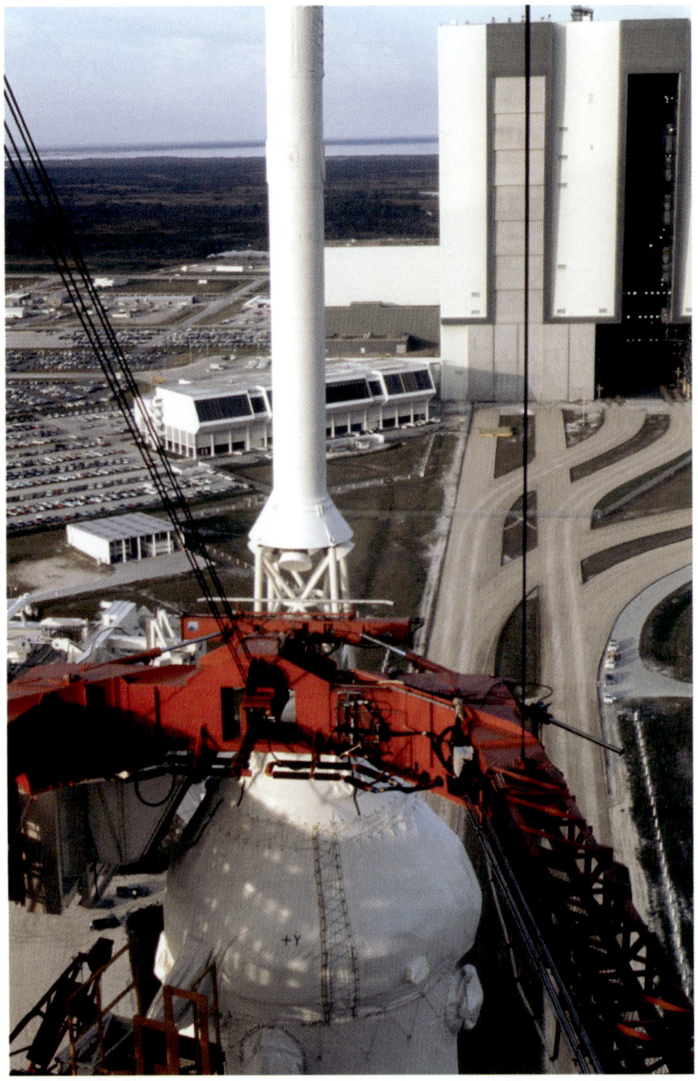

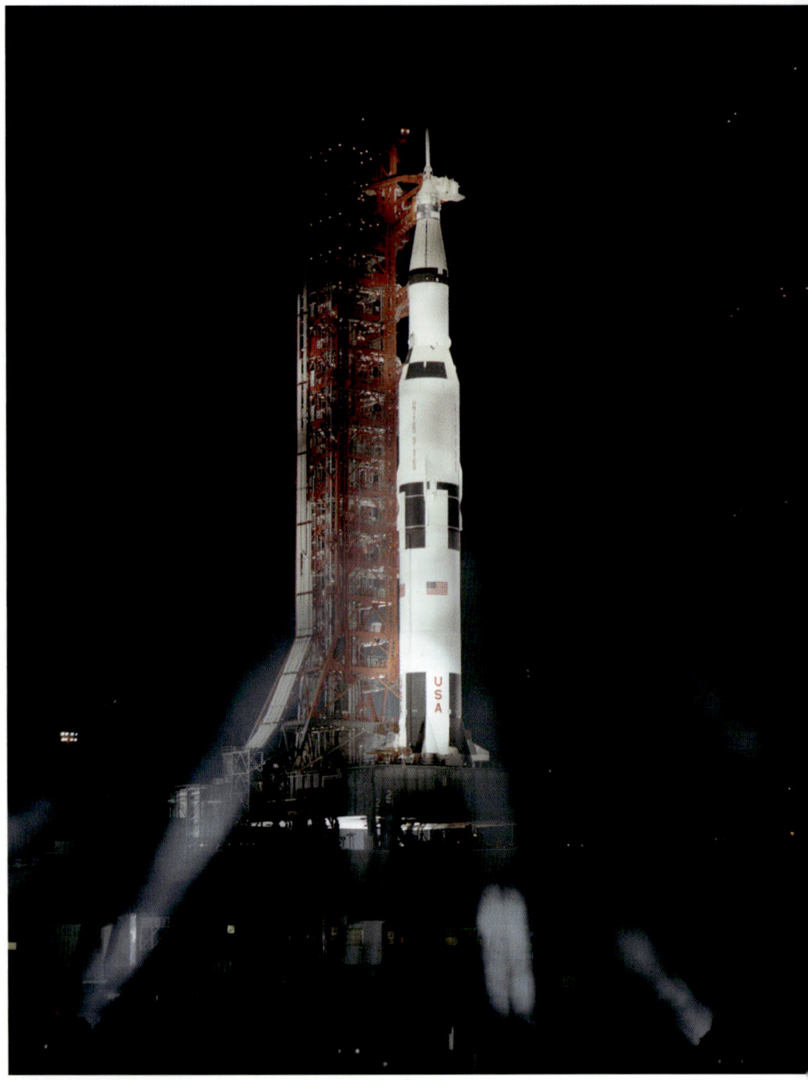

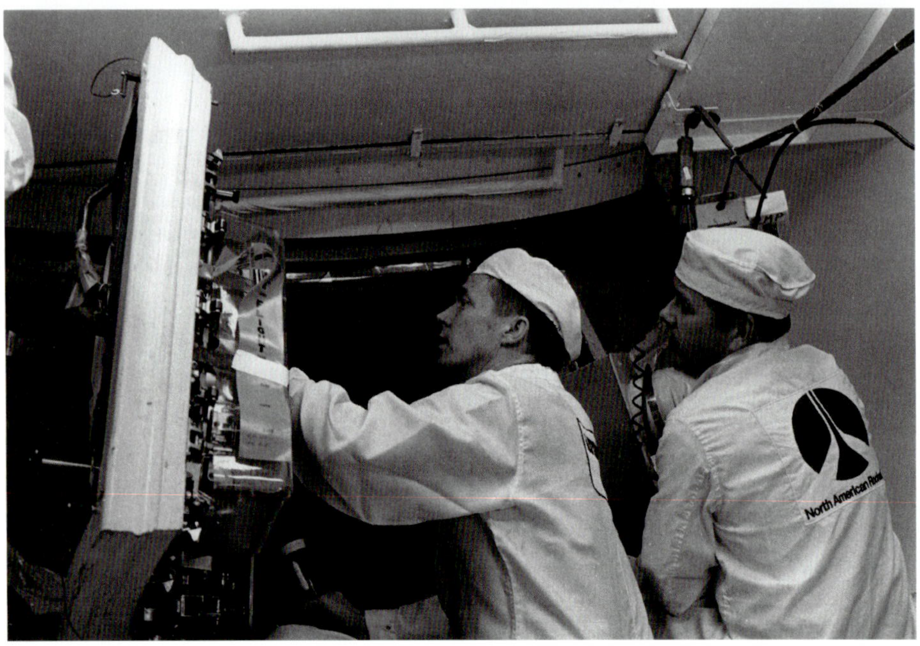

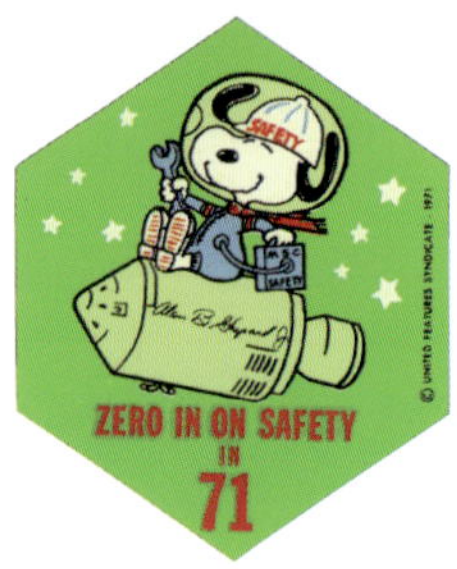

Top left: The LES tops the spacecraft with its three solid-fueled rocket motors. Here, the CSM is under wraps as the Saturn V is moved to Pad 39A on November 9, 1970. The low, black-shuttered LCC and the VAB (*right*) are in the background. (photo by Jacques Tiziou)

Above inset: One in a series of stickers given to Apollo workers as part of a safety-awareness program prompted by the Apollo 13 accident. The program began after Apollo 14 and ran through 1973.

Top right: During the CDDT, Apollo 14's Saturn V is bathed in xenon floodlights at Pad 39A on January 19, 1971. The CDDT, with the flight crew aboard for the final count, lasted 101 hours.

Above: In the white room, Roosa (*left*) inspects the Apollo 14 CM hatch with an NAR technician on December 3, 1970.

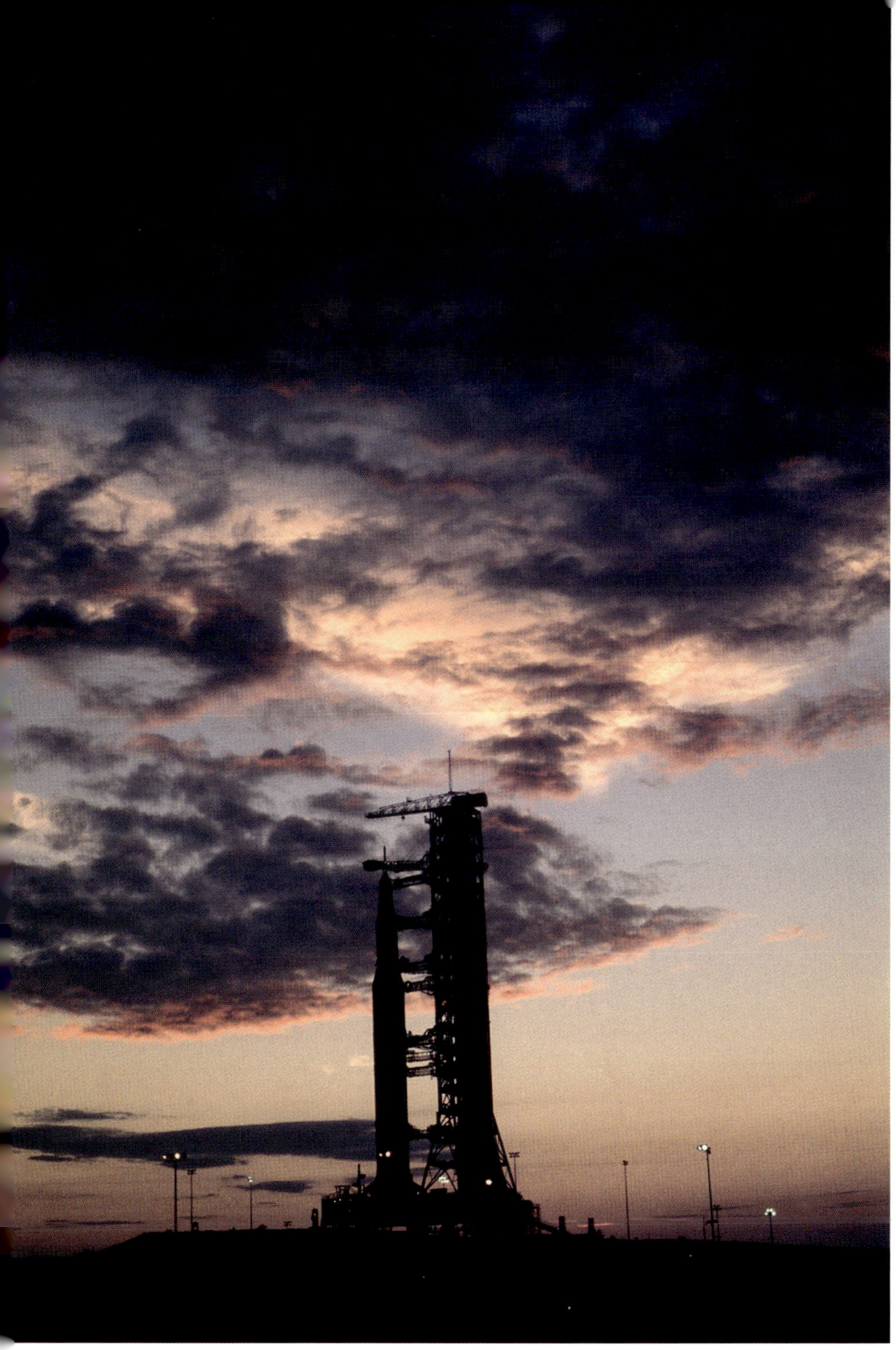

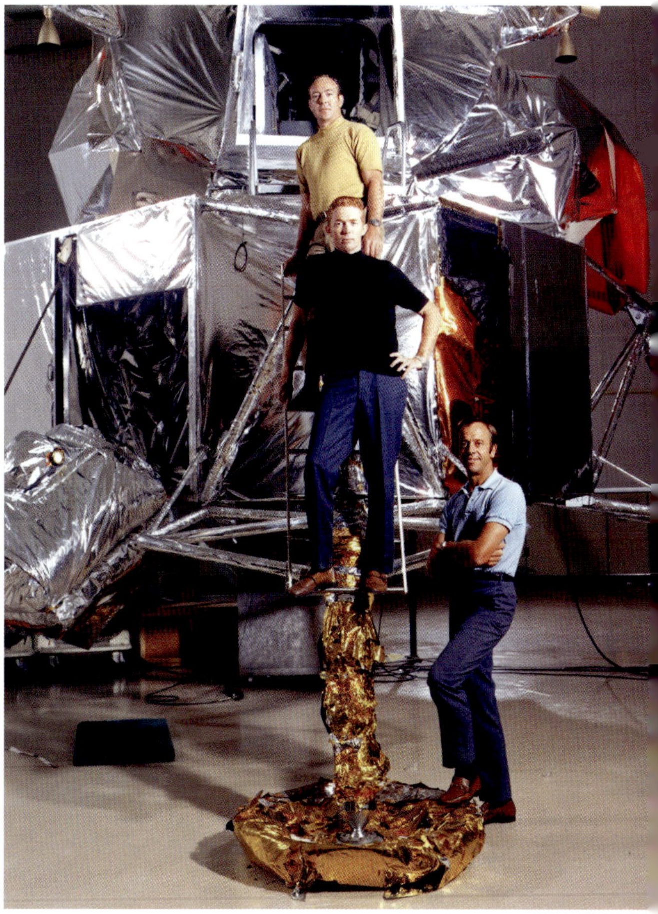

Top left: Apollo 14's Saturn V at Pad 39A during the CDDT, silhouetted at sunset on January 19, 1971. "We're now ready to proceed with our next flight to the moon," declared Rocco Petrone, KSC launch operations director, at the conclusion of the test. (photo by Jacques Tiziou)

Top right, top to bottom: Mitchell, Roosa, and Shepard pose with the full-scale LM mock-up in the Flight Crew Training Building at KSC on September 21, 1970.

Bottom left: The VAB as seen from Pad 39A at sunset on January 30, 1971, the evening before launch. Cochran Cove is in the foreground. (photo by Jacques Tiziou)

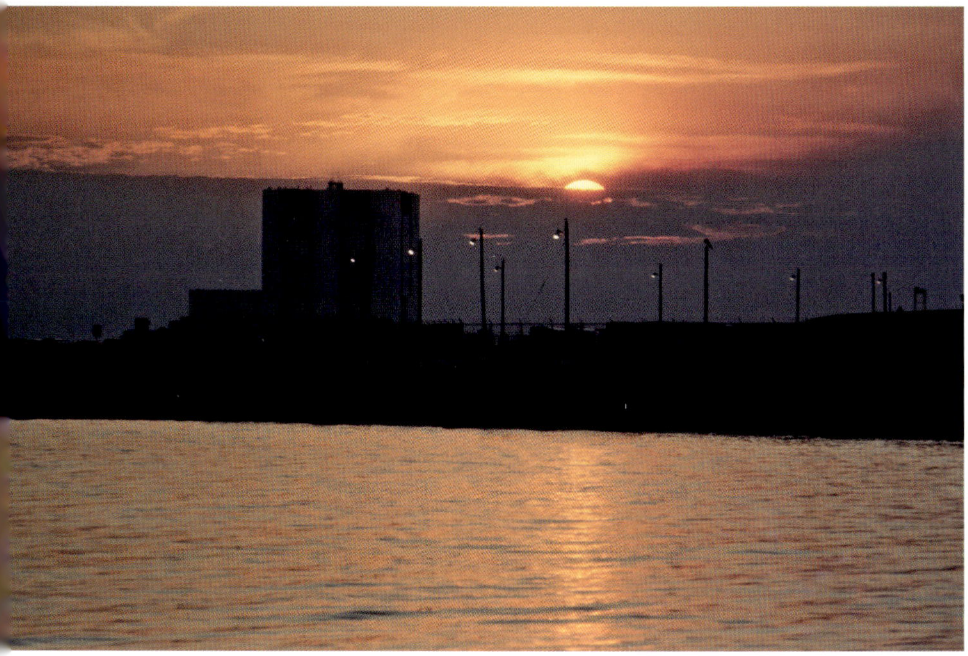

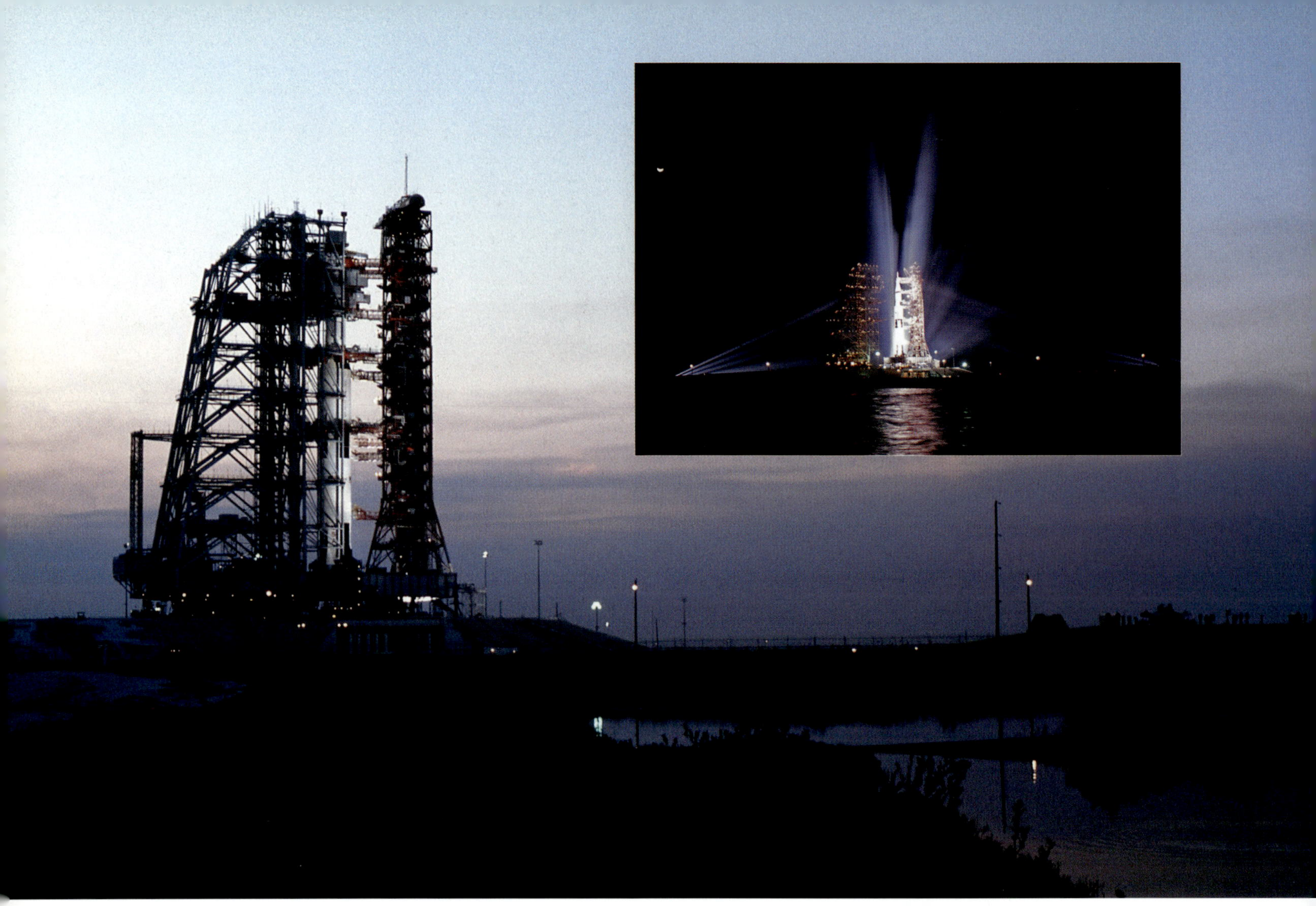

Above: At sunset on January 30, 1971, the evening before launch, the MSS (*left*) surrounds Apollo 14's Saturn V at Pad 39A. The 410-foot-tall MSS, originally intended primarily for ordinance installation, had five work platforms (the lower two were vertically adjustable) that provided access to the Saturn, and a base that contained several compartments. (photo by Jacques Tiziou)

Above inset: The MSS begins to move away, with the pad illuminated by xenon floodlights later that night. The moon is at left. (photo by Jacques Tiziou)

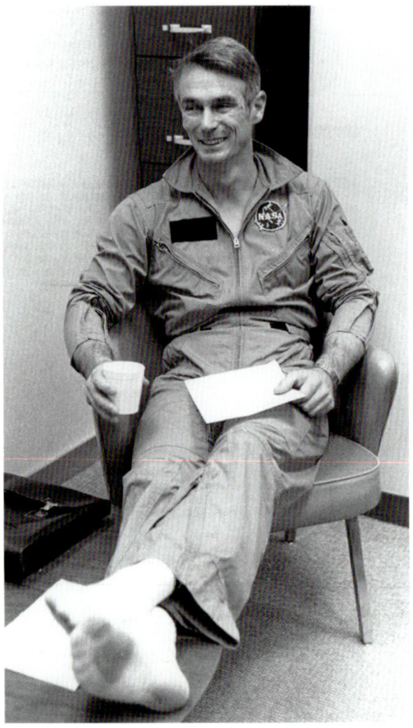

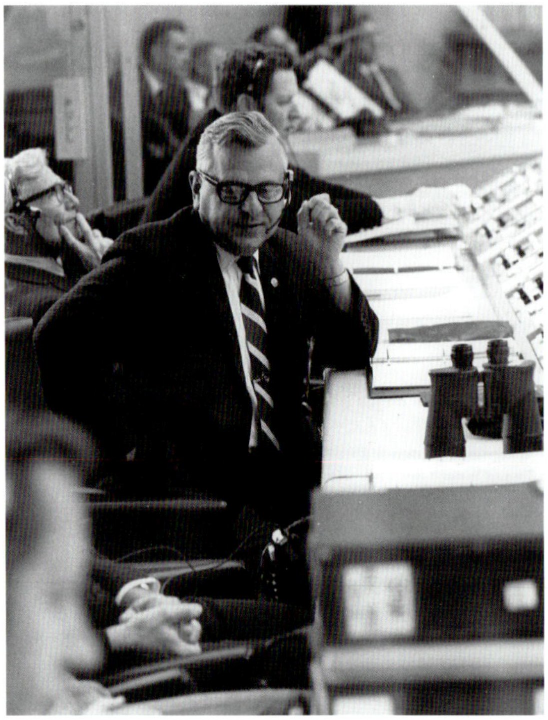

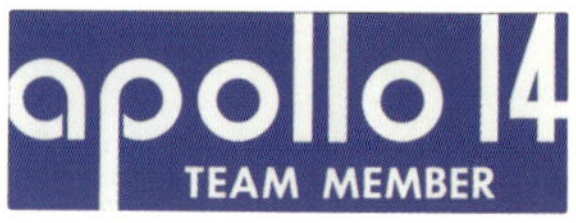

Above: Posters and bumper stickers from NASA's MFA program were meant both to instill pride and to remind workers of the need for vigilance in safety.

Bottom left: Apollo 14 backup CDR Gene Cernan at the astronaut crew quarters, after he was pulled from the Indian River following a helicopter crash on January 23, 1971. The craft burst into flames and sank before he managed to escape. Cernan admitted in his 2000 autobiography that he had been flying too low, to show off for nearby boaters.

Bottom right: Apollo 14 Launch Director Walt Kapryan in Firing Room 2 during the CDDT on January 19, 1971. Behind him are Hans Gruene (*left*), director of launch vehicle operations, and Dick Smith, who would become KSC's third director in 1979.

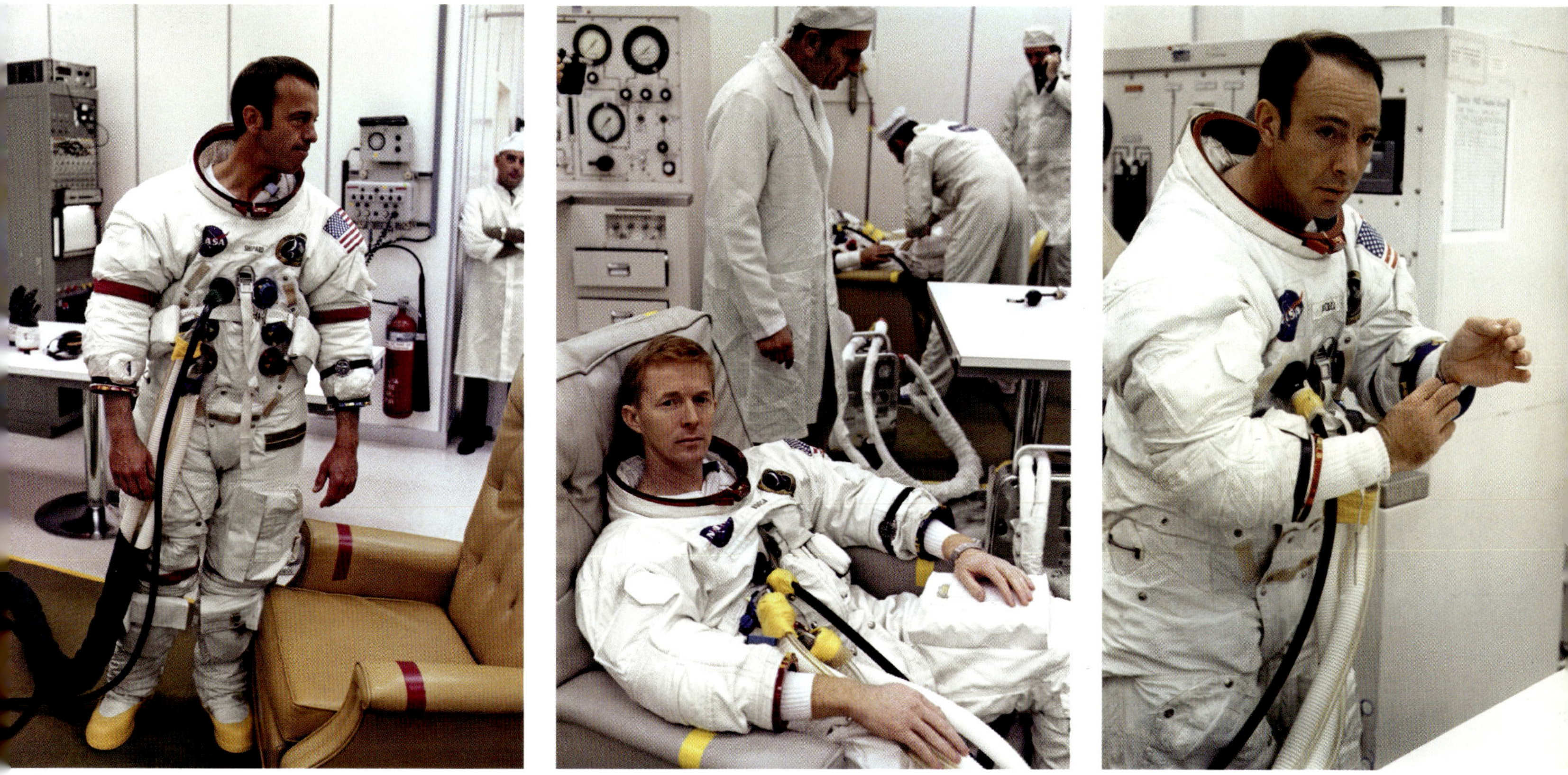

Above: The flight crew suits up on the morning of January 31, 1971, in the MSOB. As CDR, Shepard (*left*) wears red bands to aid in crewman identification in lunar surface photos. The bands were added to the Apollo 13 suits after there was confusion in identifying the Apollo 12 crewmen in lunar surface photos. Director of Flight Crew Operations Deke Slayton stands behind CMP Roosa (*center*). LMP Mitchell (*right*) would later claim to have conducted extrasensory perception experiments with friends on Earth during the mission.

Below: The transfer van heads for Pad 39A along the Saturn Causeway. The launch vehicle is visible in the distance at right. A band of cumulus clouds with showers developed over the northern MILA about thirty minutes before the scheduled launch time. The rain and the threat of lightning required a hold of forty minutes and two seconds at T-minus eight minutes, until the showers had moved a sufficient distance from the pad.

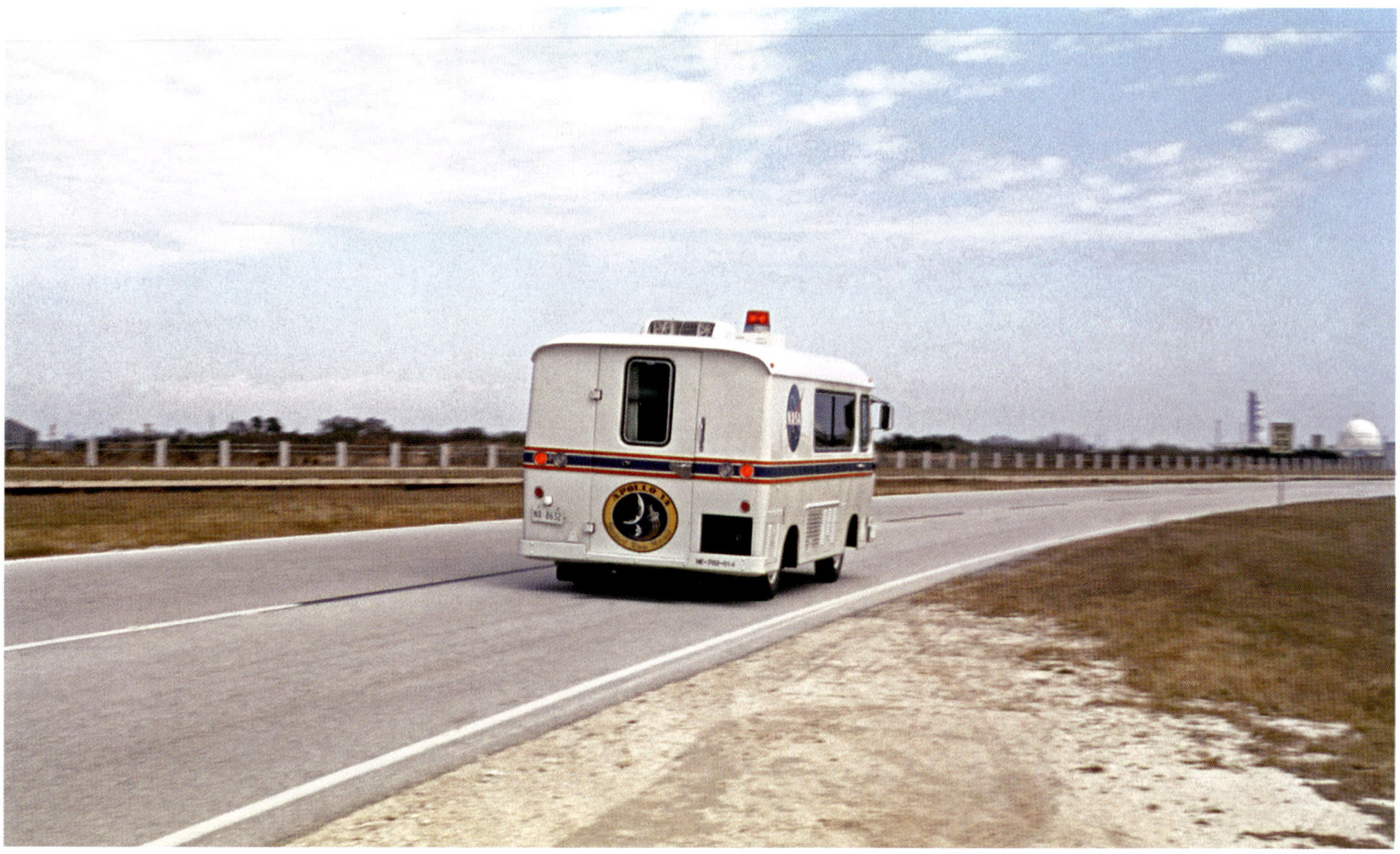

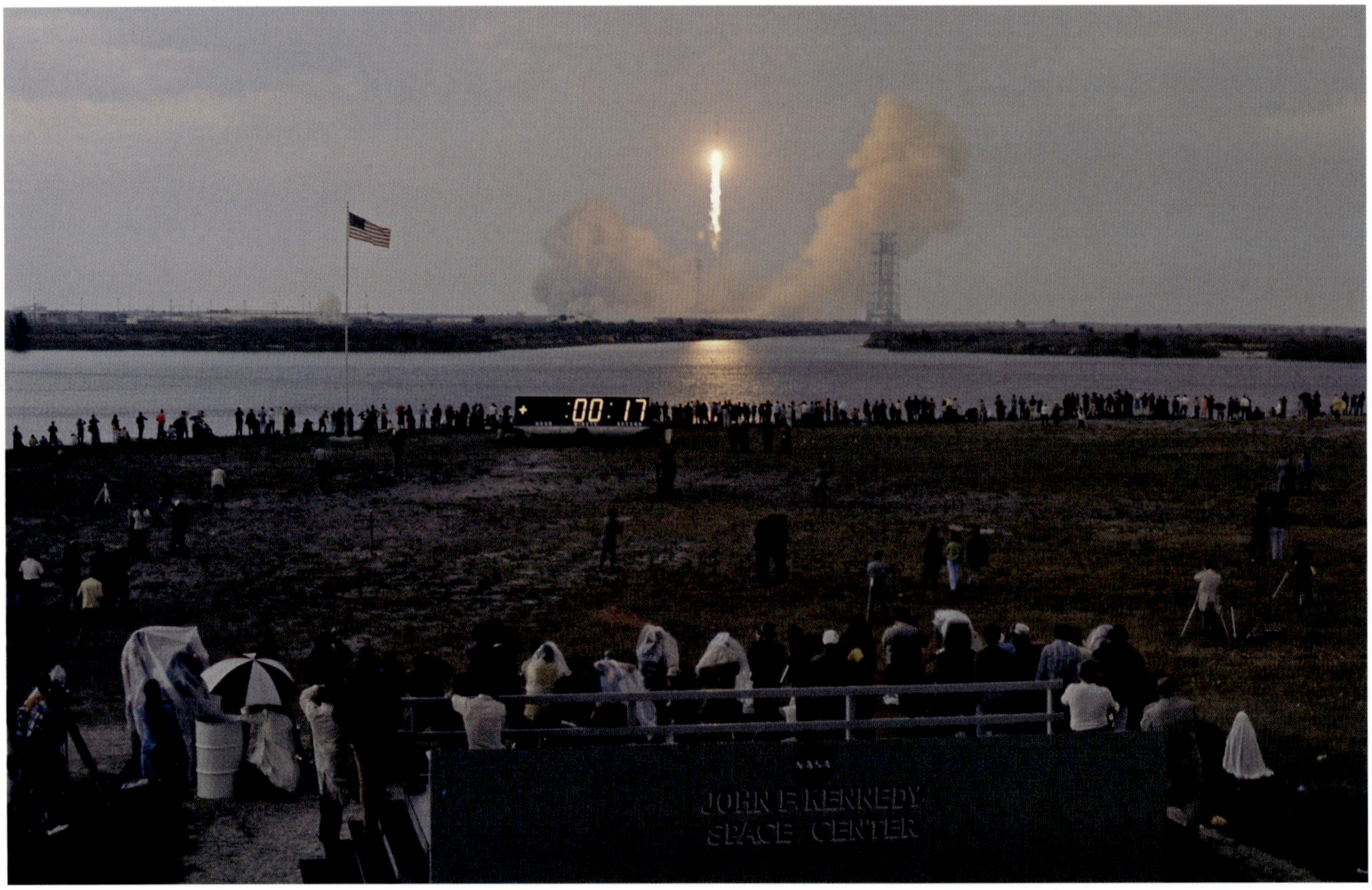

Journalists crowd the shoreline along the Turning Basin as Apollo 14 roars skyward, seventeen seconds into its nine-day mission. This view is from the LC-39 Press Site grandstand, three miles south-southwest of Pad A; the platform in the foreground was used for prelaunch and postlaunch news conferences. Light rain just prior to launch had forced the first unplanned countdown hold in the Apollo program. The temperature was seventy-one degrees Fahrenheit, with 86 percent relative humidity. The vehicle passed through a low cloud deck on its way to a parking orbit lasting two hours and twenty minutes, before translunar injection. (photo by Jacques Tiziou)

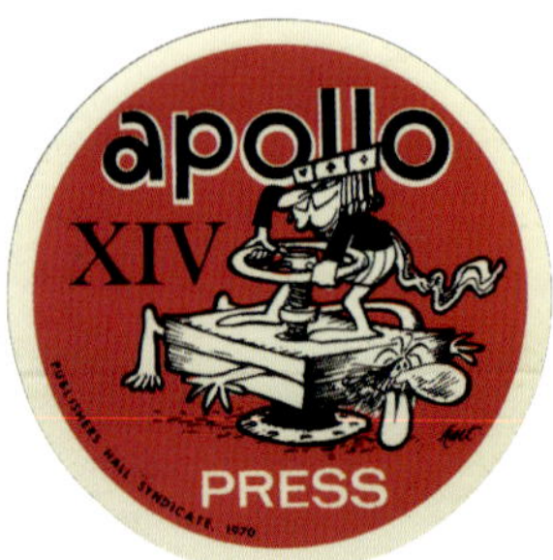

Cartoonist Johnny Hart created several emblems during Apollo. Here he used characters from his *Wizard of Id* comic strip.

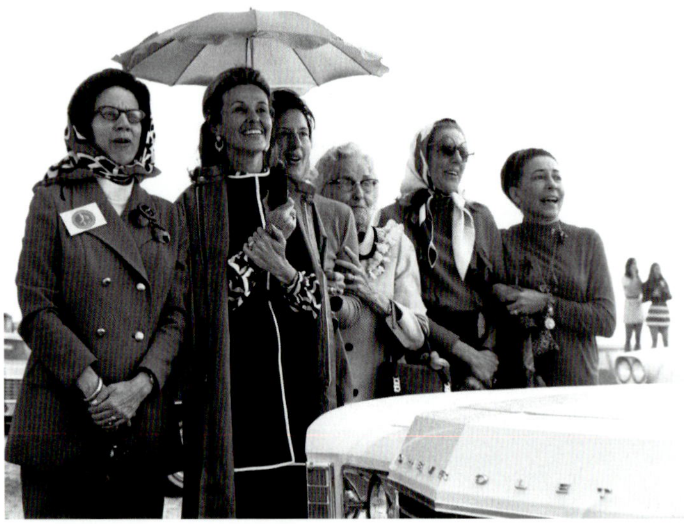

From the VIP area just north of the VAB, Shepard's wife, Louise (*second from left*), watches the launch with friends and family. *Left to right*: Loraine Meyer, a friend from Houston; Shepard; Doral Abbott, a school friend; Judy Brewer, Louise's mother; Jo Schirra; and Marjorie Slayton.

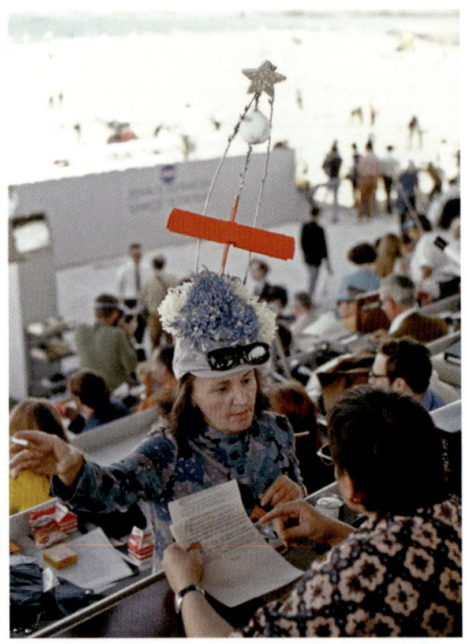

Reporter Mary Bubb—in the LC-39 Press Site grandstand on launch morning—had been covering the space program since 1957 for Fairchild Publications. She was known for creating a different elaborate hat for each Apollo launch. (photo by Jacques Tiziou)

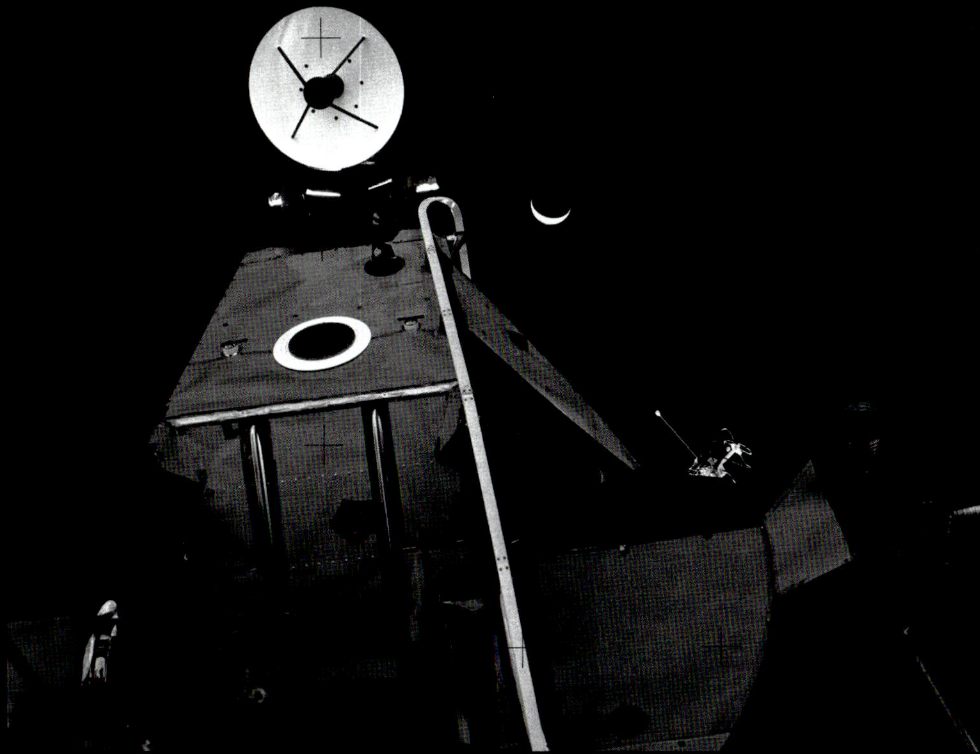

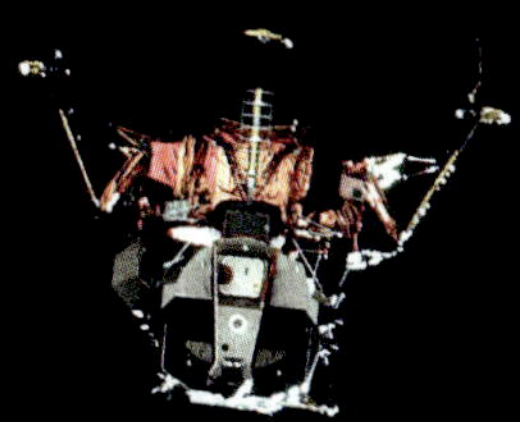

Top-to-bottom sequence: In this sequence, LM *Antares* (photographed by Roosa from CM *Kitty Hawk*) descends from orbit to the Fra Mauro highlands. Because of a faulty cabin switch, Mitchell was forced to manually reprogram the LM's software during the twelve-and-a-half-minute descent, to prevent a potential automatic abort.

Top right: Crescent Earth above the ascent stage of *Antares,* in a photograph taken by Shepard at the end of EVA 2. The photo also captured Venus as a tiny white dot, just to the right of the rendezvous radar dish.

Bottom right: Close-up of the descent-stage engine bell, showing a "sweep" pattern caused by the engine's exhaust during landing. The shield that protects the landing radar from the exhaust is to the left of the bell. No crater is evident because the engine was throttled far down before landing, and the lunar soil is compact below its surface dust layer. The landing was the closest of any Apollo to its target—just 165 feet away.

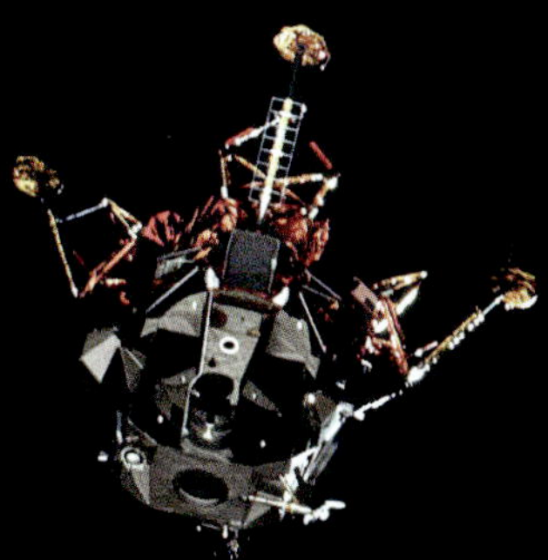

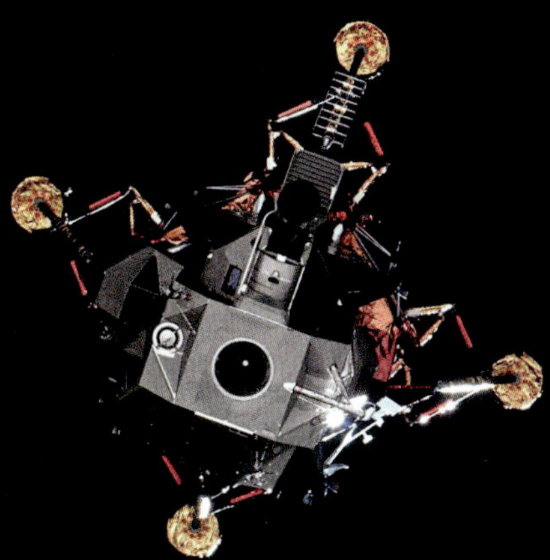

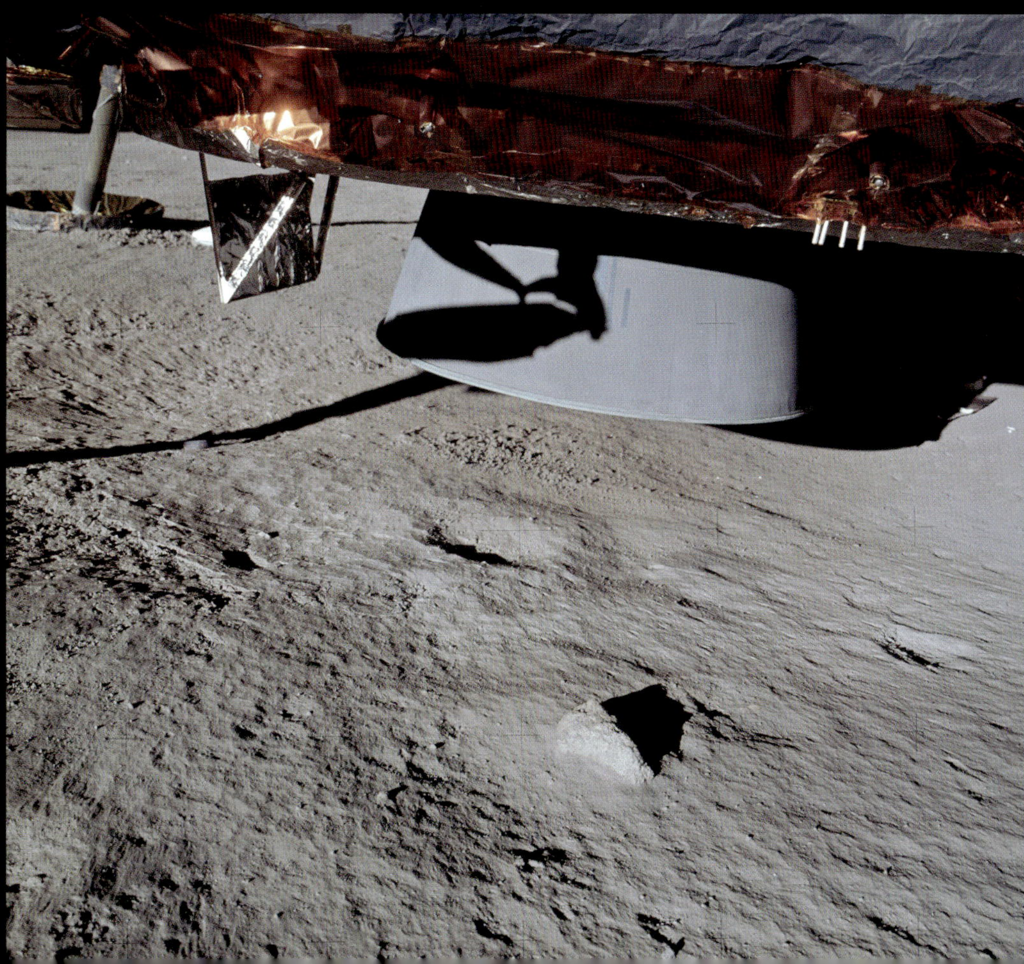

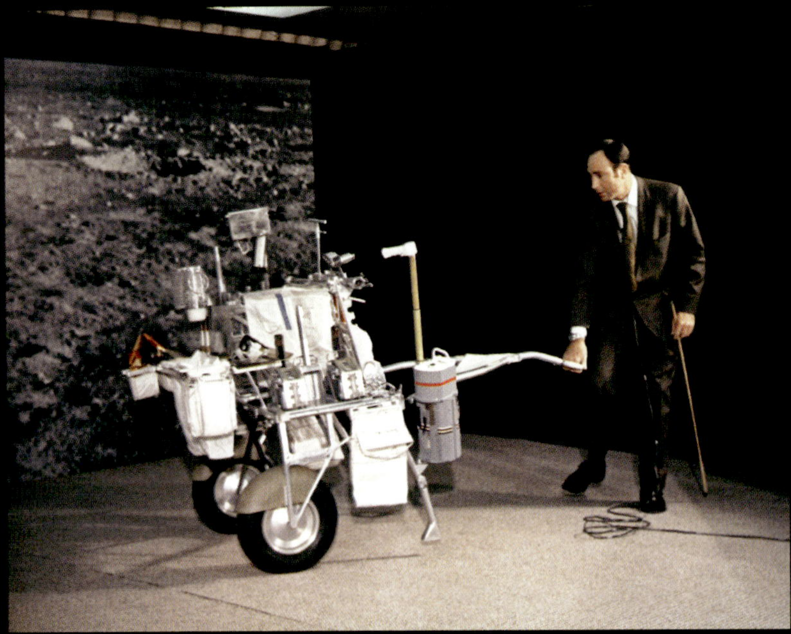

Left: Mitchell demonstrates the MET using a full-fidelity mock-up during a news conference at NASA HQ in Washington, DC, on October 16, 1970. The gray close-up stereo camera with its pole and white handgrip is mounted at right. Other equipment included two 70mm Hasselblad cameras, a 16mm DAC, film magazines, sample weigh bags, a dispenser for sample bags, a trenching tool, a worktable, and the lunar portable magnetometer. The MET rode on Goodyear XLT tires pressurized with nitrogen gas.

Below: Mitchell casts a shadow on February 6, 1971, during the second of two lunar EVAs. The wheeled MET is parked at upper left. In the foreground is a gnomon (Greek for "indicator") with color and grayscale charts, used to calibrate photos during processing. The shadow cast by the gnomon's vertical rod indicated sun angle and, therefore, direction. The rod length and the painted scale provided a reference for estimating the sizes of nearby objects. Shades of gray on the rod ranged in reflectivity from 5 to 35 percent, and a color scale helped determined rock and soil colors by comparison. Here, Mitchell is photographing the sample rock near the front leg of the gnomon. *Antares* can be seen just over the horizon at left.

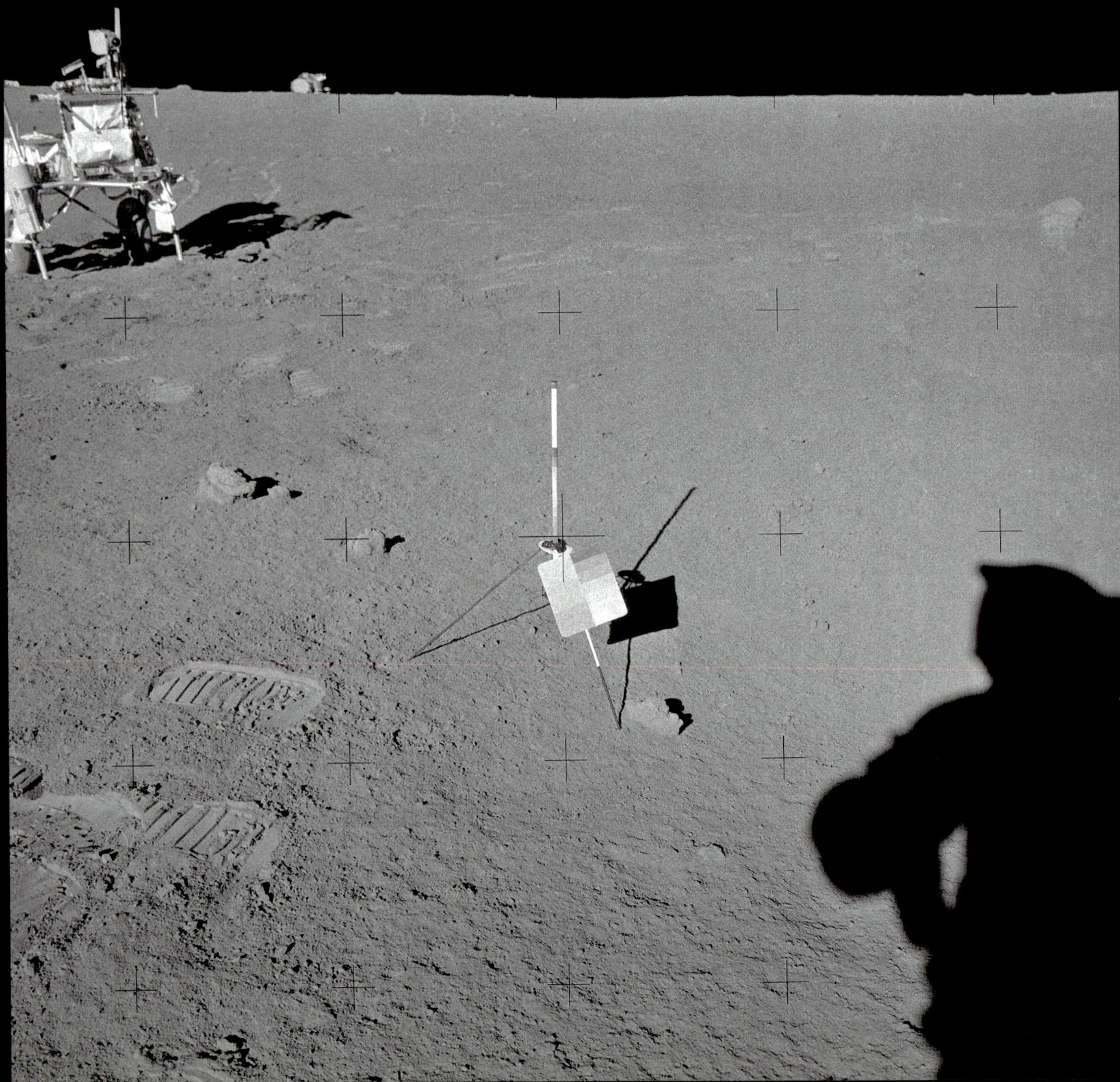

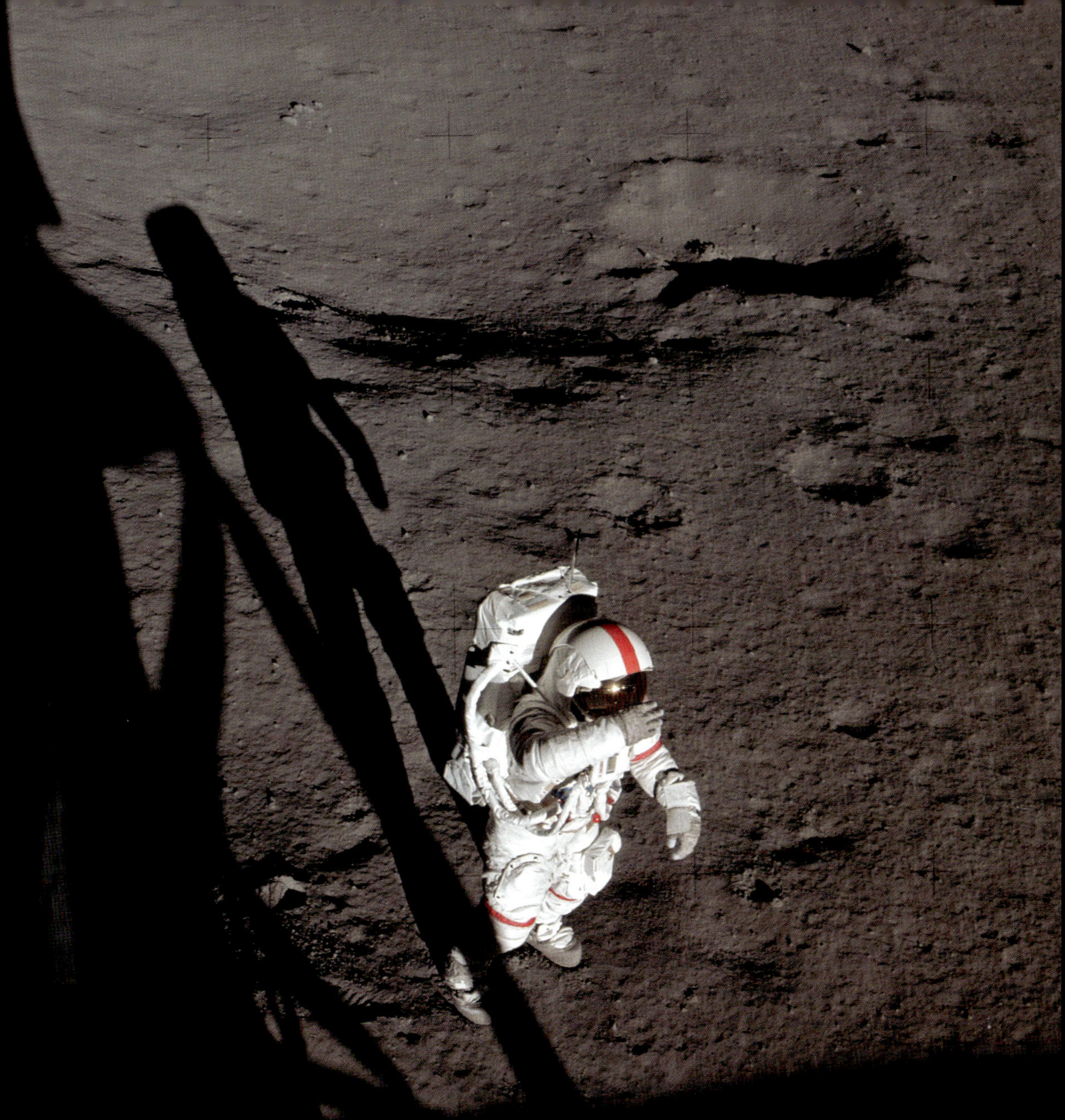

Shepard, on the north side of the LM, shades his eyes as he looks east toward Cone crater on February 5. Mitchell took this photo from his window shortly before he joined Shepard on the surface, about five hours after landing. "It certainly is a stark place here at Fra Mauro," said Shepard. "I think it's made all the more stark by the fact that the sky is completely black."

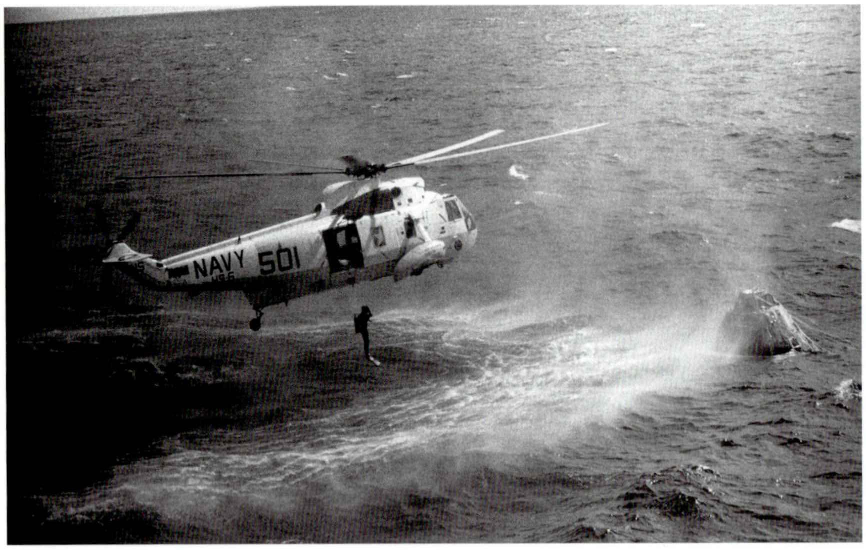

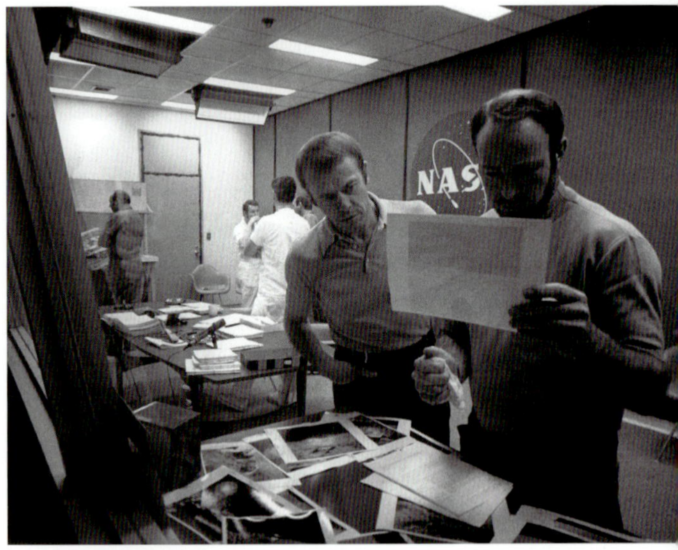

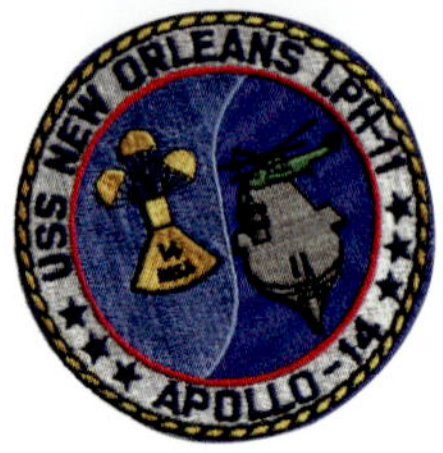

Top left: Robert R. Rohrbach, officer in charge and decontamination swimmer, jumps from a Sea King helicopter about five minutes after *Kitty Hawk* has splashed down in the Pacific. The UDT-11 swim team arrived before the mandatory ten-minute wait, during which the spacecraft's uprighting bags could be inflated. A diver had to cut shroud lines from one parachute, which had entangled the CM.

Top right: Shepard and Mitchell look over photographs from their mission on February 19, during the astronauts' fifteen-day confinement in the LRL. Scientists were already examining their returned samples elsewhere in the lab. The astronauts participated in debriefings from the table at left (with the microphone).

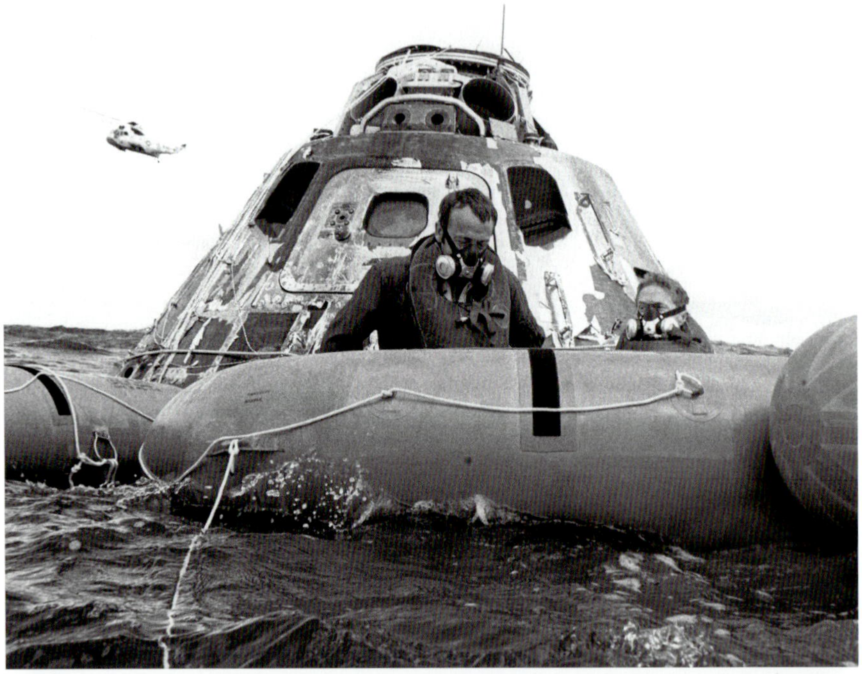

Center left: Shepard and Roosa wait to be lifted to a recovery helicopter. This was the last time a returning lunar crew wore respirators and went through decontamination and isolation.

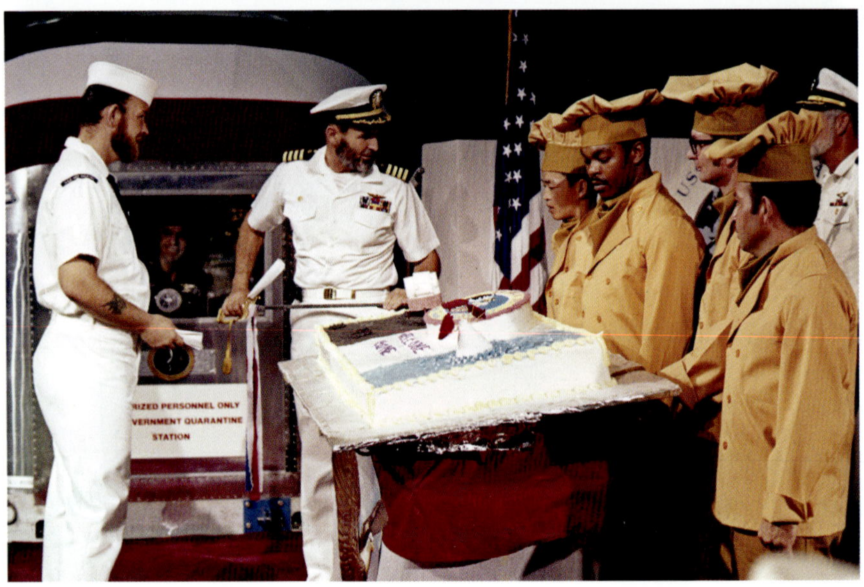

Bottom left: USS *New Orleans* Capt. Robert Moore cuts a cake honoring the crewmen, who watch from inside their MQF on February 9, 1971.

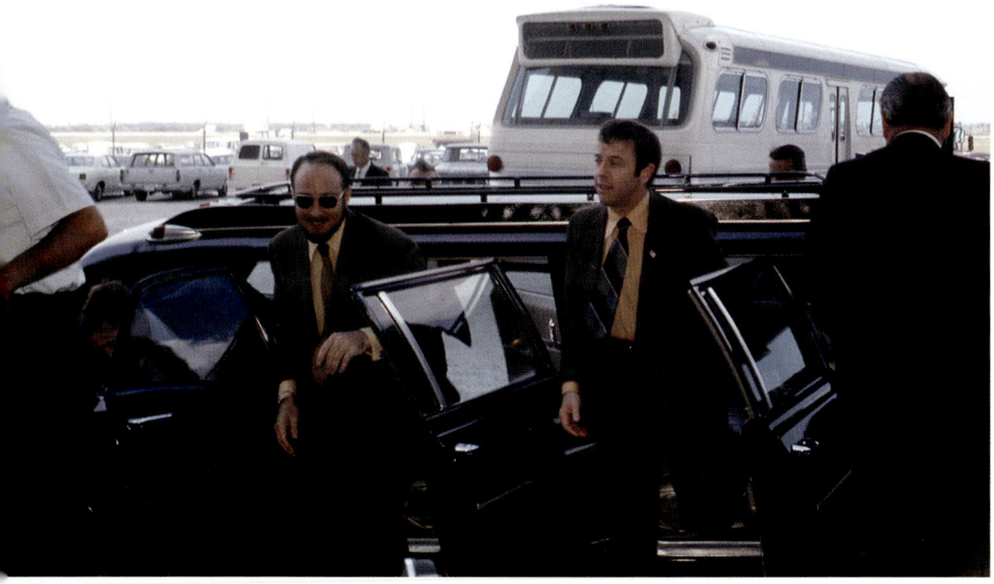

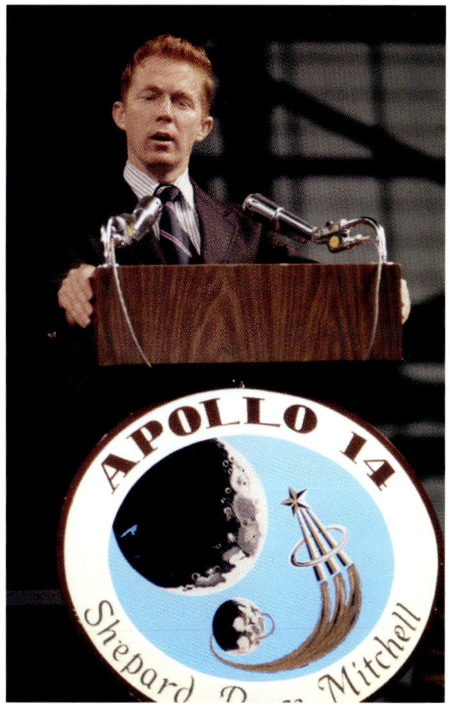

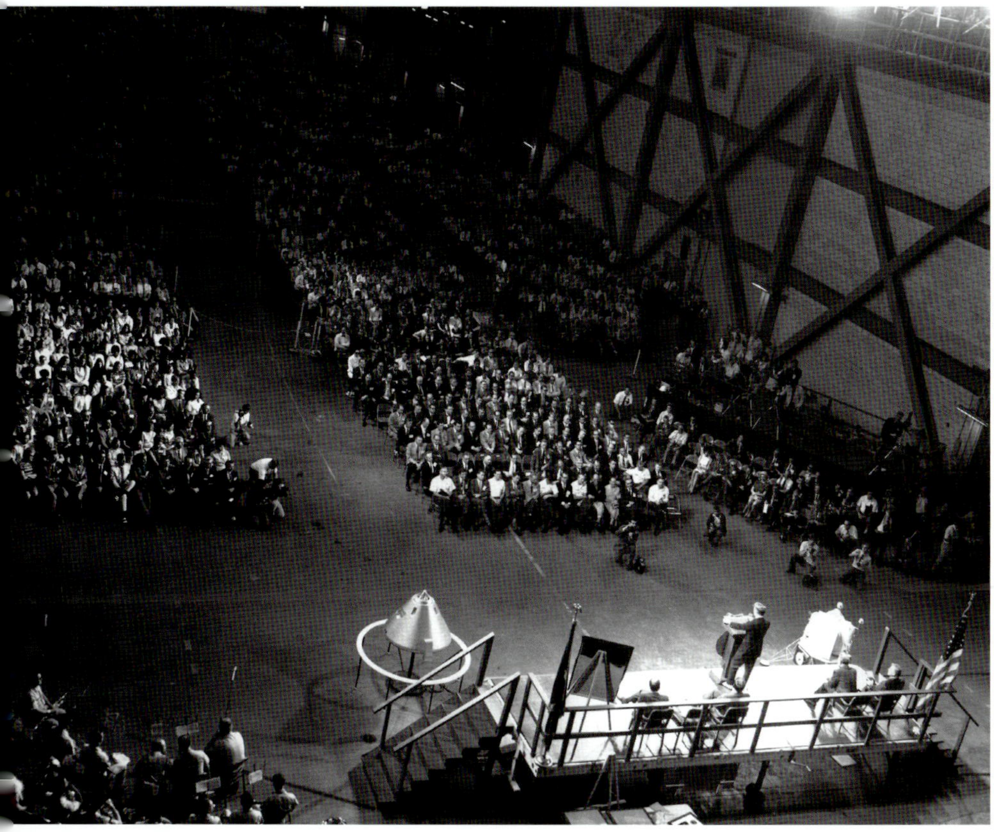

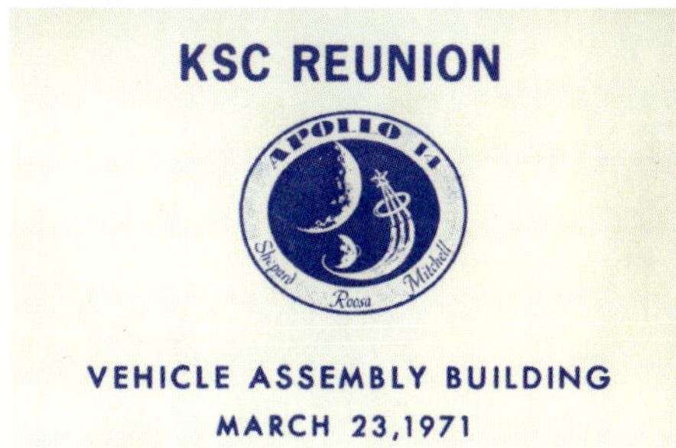

Top left: On March 23, 1971, six weeks after their mission, the astronauts arrive via limousine at the VAB, to greet and thank members of the KSC launch team—an event known as the KSC Reunion. KSC Security Chief Charles Buckley, nicknamed 007, is at right.

Top right: Roosa speaks at the crew reunion. He served as backup CMP for Apollo 16 and 17, and was then assigned to the space shuttle program until he retired in 1976. Roosa became owner and president of Gulf Coast Coors in 1981.

Center left: KSC workers in the VAB's transfer aisle listen to Roosa during the ceremony. The news-media platform is at right; a band is at lower left.

Bottom left: KSC Director Kurt Debus addresses NASA and contractor employees. Behind the stage is a dummy SLA, which was temporarily stacked atop the Saturn V boosters for Apollos 8 and 15 until the actual SLAs arrived. The SLA and fake CSM were later displayed at the KSC Visitor Center.

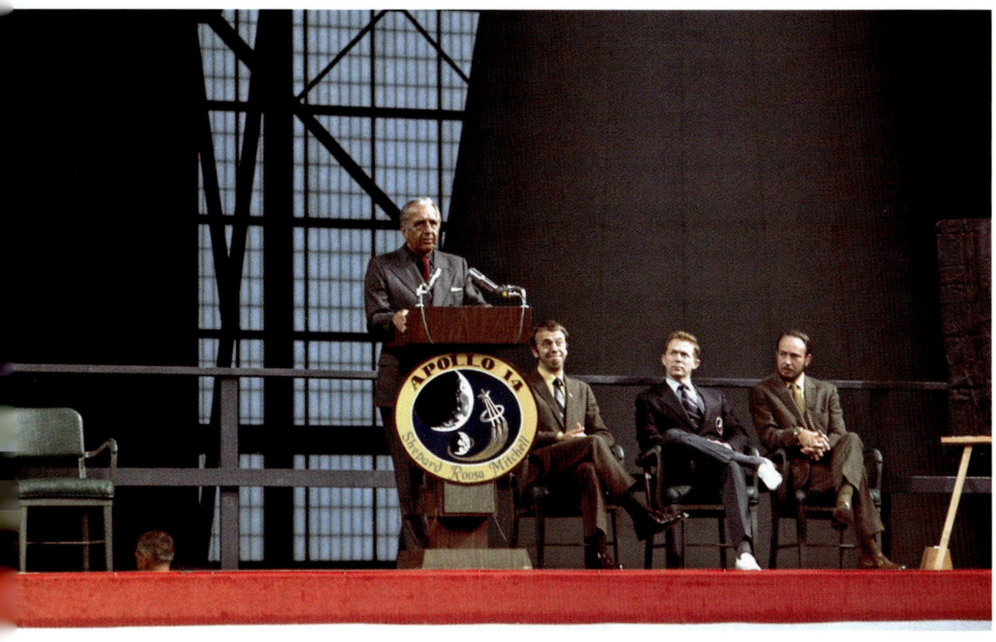

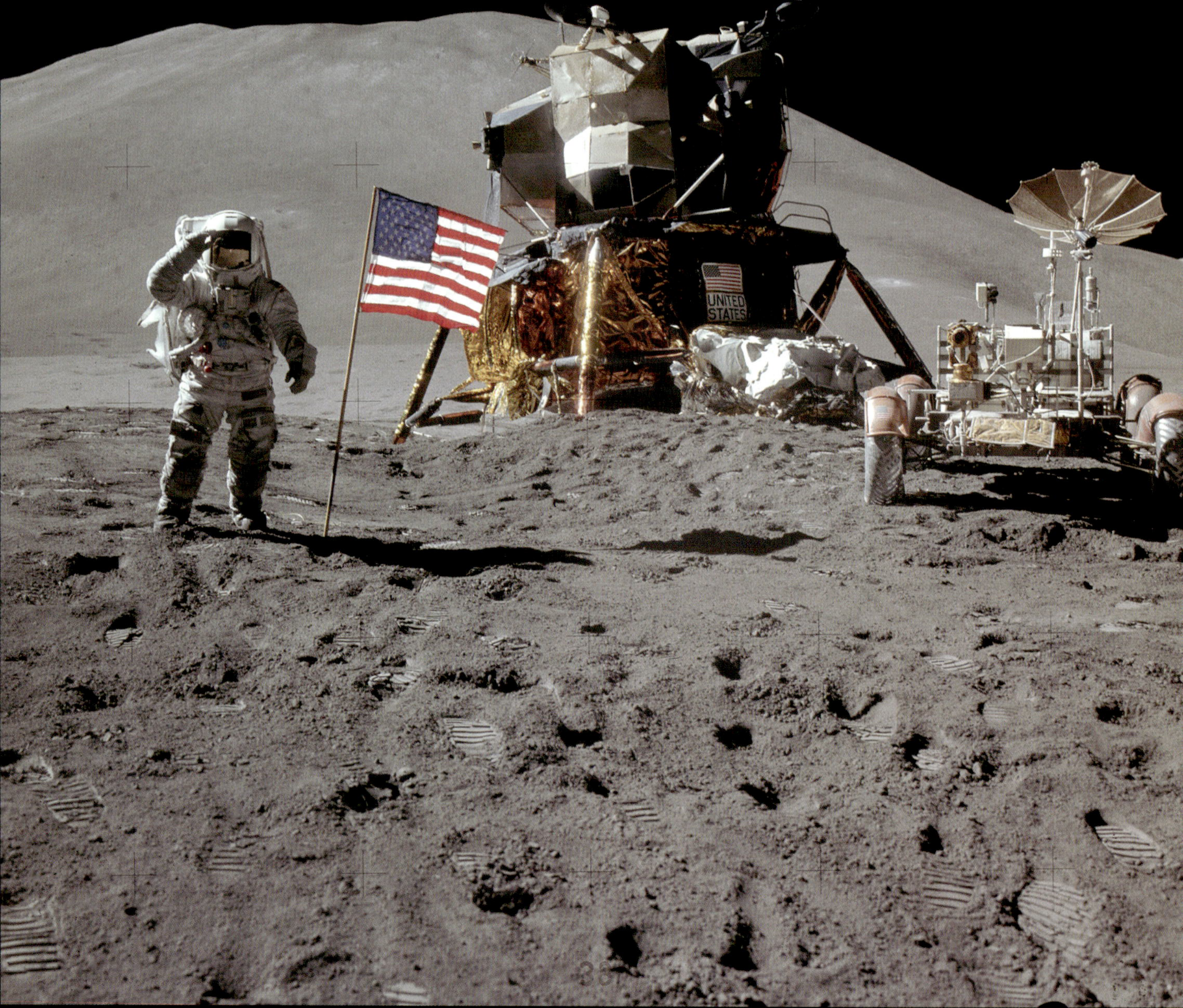

Few Apollo images have been reproduced more often than this color photo of James Irwin, the US flag, the Lunar Roving Vehicle (LRV, or "rover"), and LM *Falcon* during EVA 2, with Mount Hadley Delta in the background. Sunlight—which was reflected from mirrored tiles on the top of the TV camera—produced the bright rectangular pattern on the LRV's high-gain antenna. The US flags on the rover's fenders were made of Beta cloth.

CHAPTER 10

Apollo 15

July 26–August 7, 1971

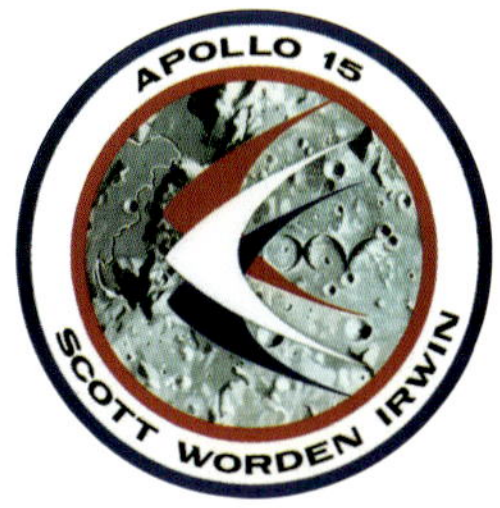

Before launch, Apollo 15 CDR David Scott called the flight "the most significant scientific expedition ever mounted." When it was completed, NASA called the twelve-day mission the most successful manned flight ever achieved. The first of the three J missions—with long-duration stays on the moon and a greater focus on science—Apollo 15 doubled the time and extended tenfold the range of lunar surface exploration.

The J missions were designed to change the role of the astronauts from test pilots to scientific explorers. Scott and LMP James Irwin used the first LRV to bring back a wider variety of rock and soil samples. CMP Alfred Worden conducted new experiments in lunar orbit, using the first Scientific Instrument Module (SIM)—mounted in the SM—to study the lunar surface and environment with cameras, spectrometers, and a laser altimeter. Worden also deployed the first subsatellite of Apollo into lunar orbit.

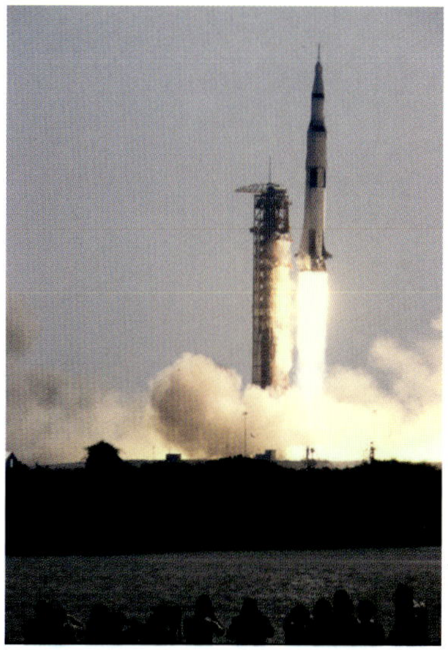

Launch: 9:34 a.m. (EDT), July 26, 1971. Photo from the LC-39 Press Site (photo by Steve Nolte)

At dawn on May 11, 1971, Apollo 15's Saturn V rolled out of the VAB to Pad 39A. Speaking at the ceremony was Wernher von Braun, by then NASA's deputy associate administrator for planning, who had attended only one other manned Saturn rollout (that of Apollo 11). Even considering that central Florida is the "lightning capital" of the United States, an abnormally high number of lightning strikes hit while Apollo 15 was at the pad. Five strikes on the MSS in June delayed several tests, and there were almost daily thunderstorms in July, leading up to launch.

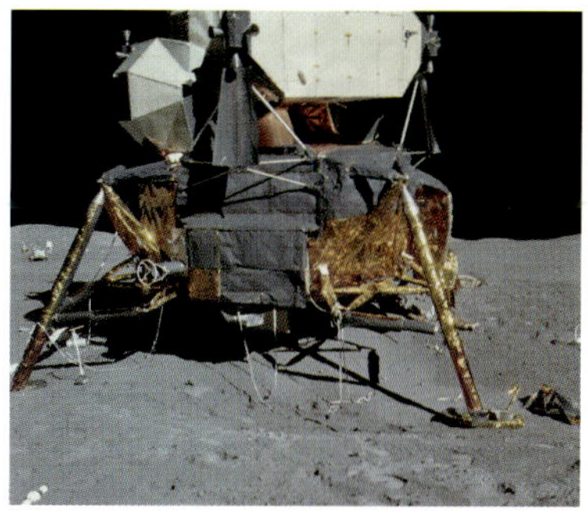

Lunar landing: 6:16 p.m. (EDT), July 30. Ascent: 1:11 p.m. (EDT), August 2

On launch day, July 26, officials expected somewhere between 1 and 1.5 million spectators, noted in the countdown commentary by Public Affairs Officer Jack King: "We are at now two minutes, fifty seconds and counting. We understand that there was an estimate that there are more than a million people in the area here to view the launch. The traffic has been heavy since two o'clock this morning. The beaches are packed and the roads are packed."

During their outbound trip, as the crewmen inspected the LM, they discovered glass fragments floating in the cabin; these had come from the broken cover of a tapemeter in the instrument panel. The men collected the glass with duct tape. Later, they noticed a drinking water system leak, which Scott quickly fixed by tightening a loose chlorine injection fitting.

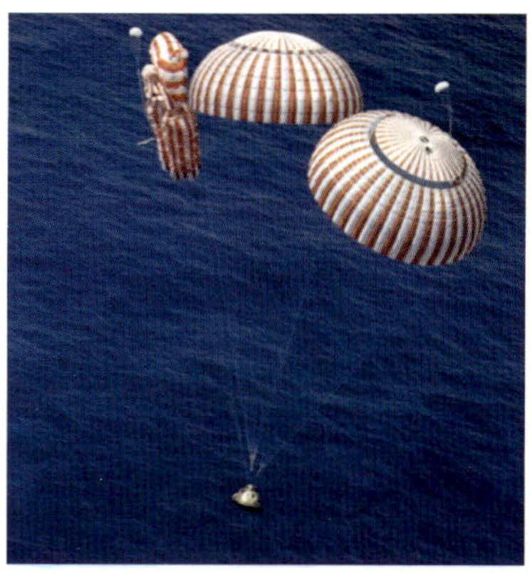

Splashdown: 4:47 p.m. (EDT), August 7. One main chute has failed to open.

The spacecraft entered lunar orbit on July 29. No images show the LM, *Falcon*, as it began its descent to the landing site. The mission was the first not to land in a lunar mare, instead landing near Hadley Rille in an area of

Mare Imbrium. After the landing on July 30, Scott poked his head out of the hatch on the top of the LM and took a number of photographs of the surrounding terrain.

Scott and Irwin spent three days on the moon and a total of eighteen and a half hours on lunar EVA. Relatively few color images exist of the crewmen on the surface, although the color photo of Irwin next to the LM with the US flag is one of the most widely used photos taken on the moon. Surface photography overall was quite good; however, Irwin did not photograph Scott's demonstration—using a feather from a falcon and his geology hammer—showing that two bodies of unequal weight fall at the same rate in a vacuum. Photographs were also not taken of the deployment of the LRV.

During EVA 2, the astronauts collected what would become the most famous lunar sample of the Apollo program, dubbed the "Genesis Rock" by the media. The sample was later shown to possibly be a piece of the moon's original crust, some 4.5 billion years old.

Before departing, Scott set up a memorial to the cosmonauts and astronauts known to have died in the pursuit of space exploration, with a plaque bearing their names and a small Fallen Astronaut figure. The rover's TV camera then captured the LM's ascent stage lifting off, as Worden played the official USAF song, popularly known as "The Wild Blue Yonder."

The next day, during the return trip to Earth, Worden climbed outside the spacecraft to retrieve exposed film from the SIM bay. It was the first such EVA in deep space. The Hasselblad images of this EVA are not good.

Following reentry on August 7, all three main parachutes inflated, but after dumping the RCS fuel, Worden noticed that one had deflated. Postflight analysis found that two or three of the six shroud lines on the failed parachute were missing, probably dissolved by the fuel.

In Trapani, Sicily, twenty inmates burned their cell mattresses when prison officials refused to let them watch television coverage of the splashdown.

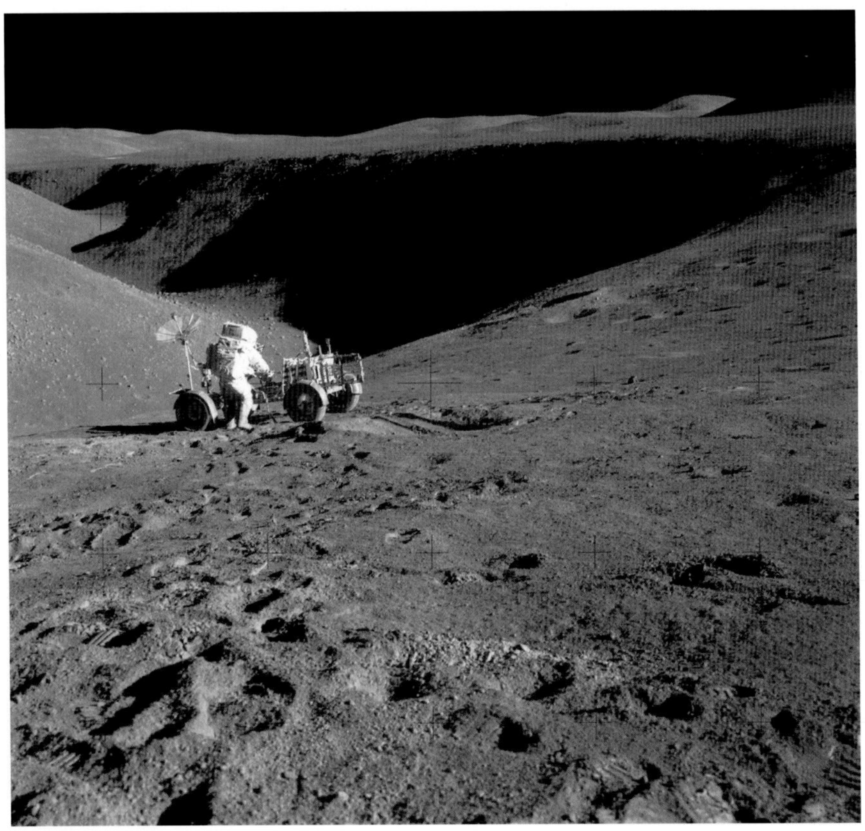

David Scott works at the LRV during EVA 1 on July 31, 1971. The view, up the rille, is toward the northwest.

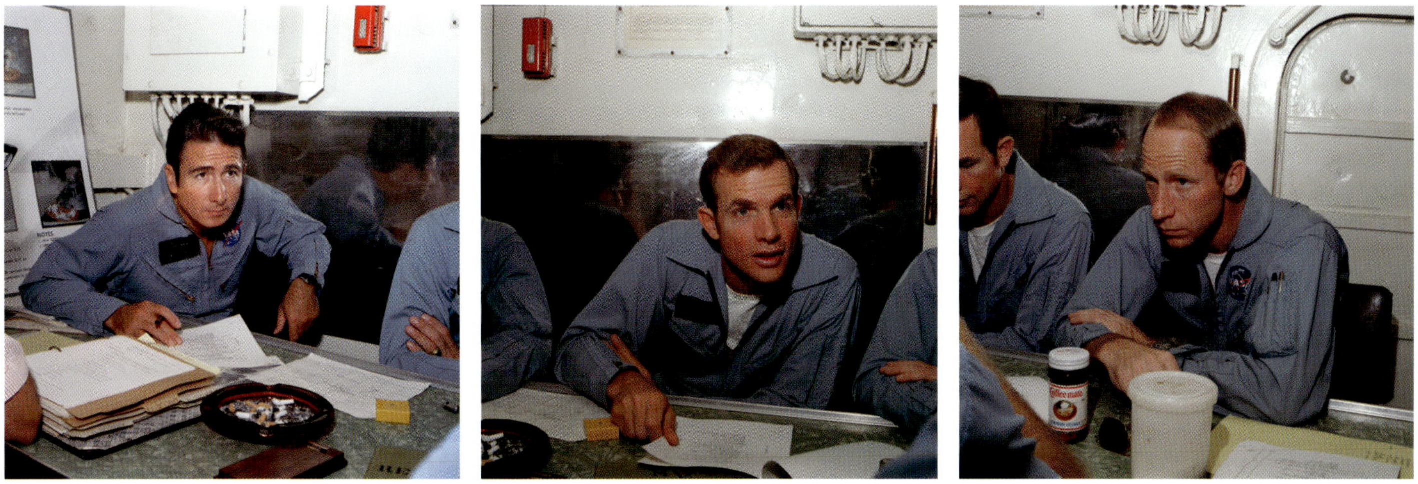

In May 1971, the Apollo 15 crew meets aboard the NASA recovery training craft MV *Retriever. Left to right*: Lt. Col. James B. Irwin, forty-one; Col. David R. Scott, thirty-nine; and Maj. Alfred M. Worden, thirty-nine. All were USAF officers and graduates of US military service academies. Irwin and Worden were also University of Michigan graduates; Scott had taken classes there. Smoking was not uncommon among astronauts of the era.

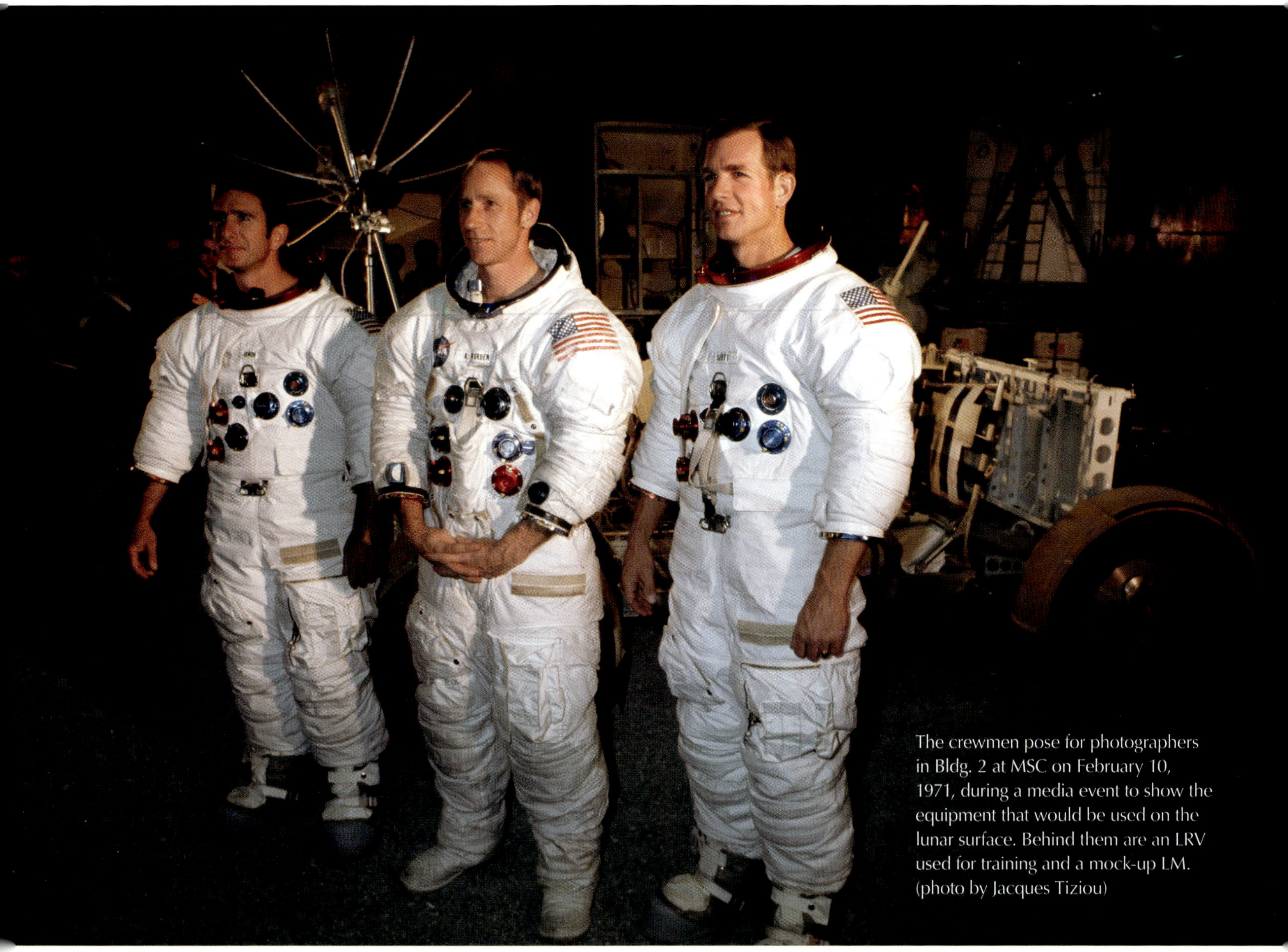

The crewmen pose for photographers in Bldg. 2 at MSC on February 10, 1971, during a media event to show the equipment that would be used on the lunar surface. Behind them are an LRV used for training and a mock-up LM. (photo by Jacques Tiziou)

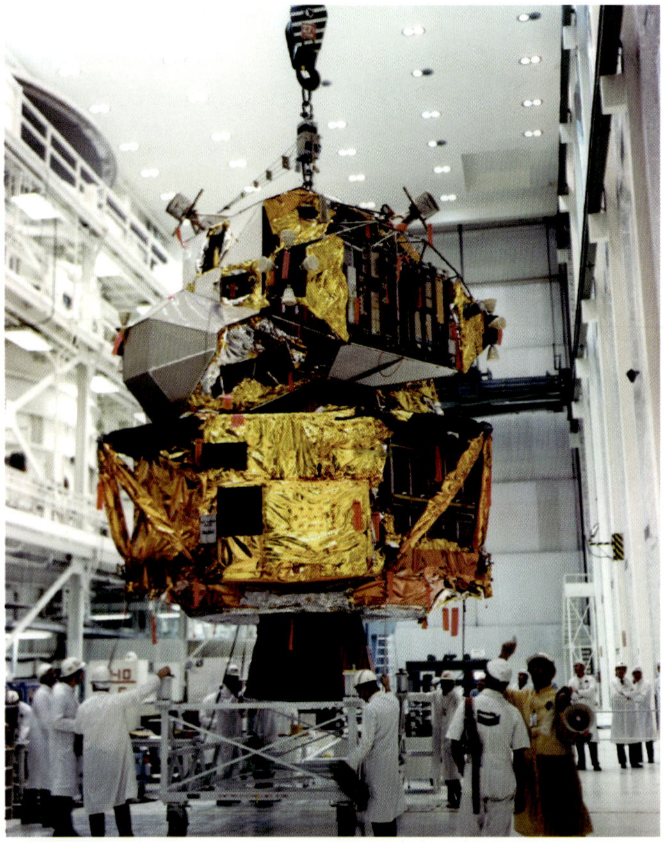

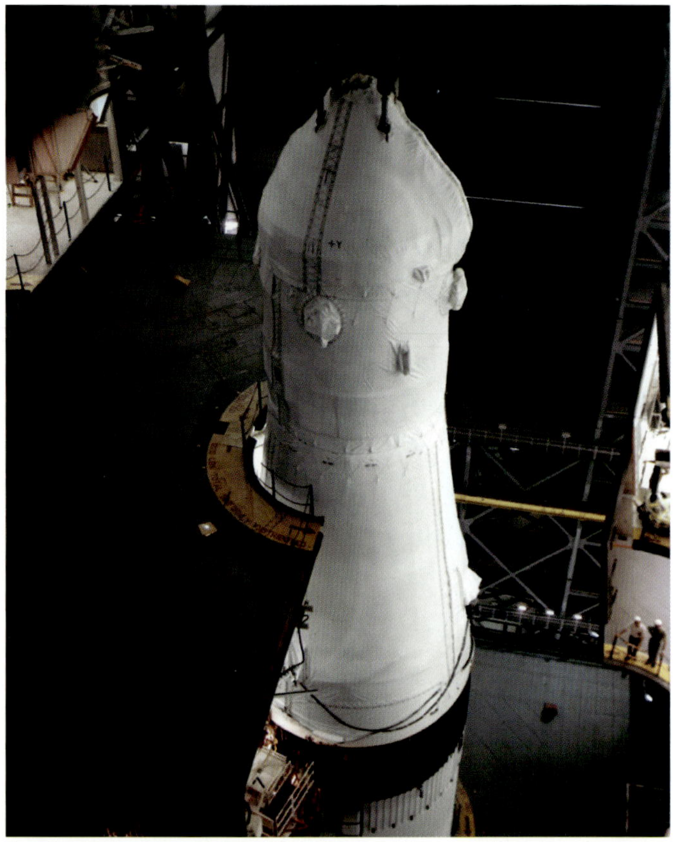

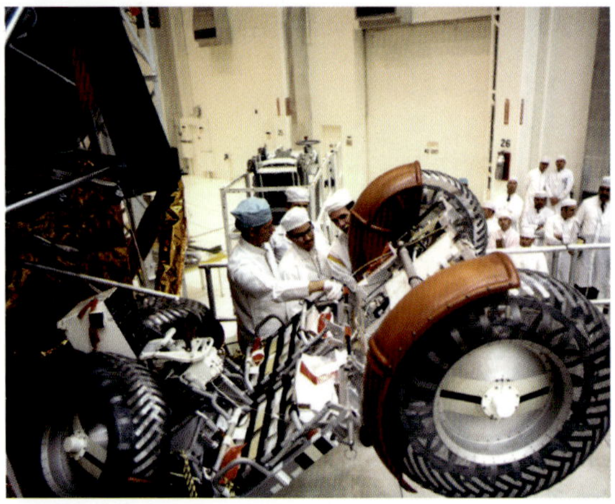

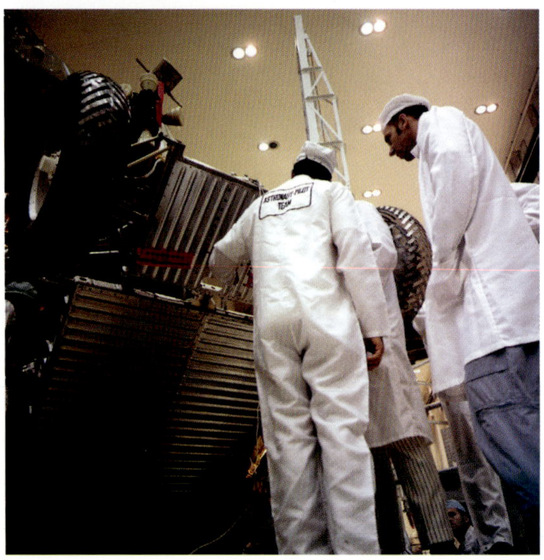

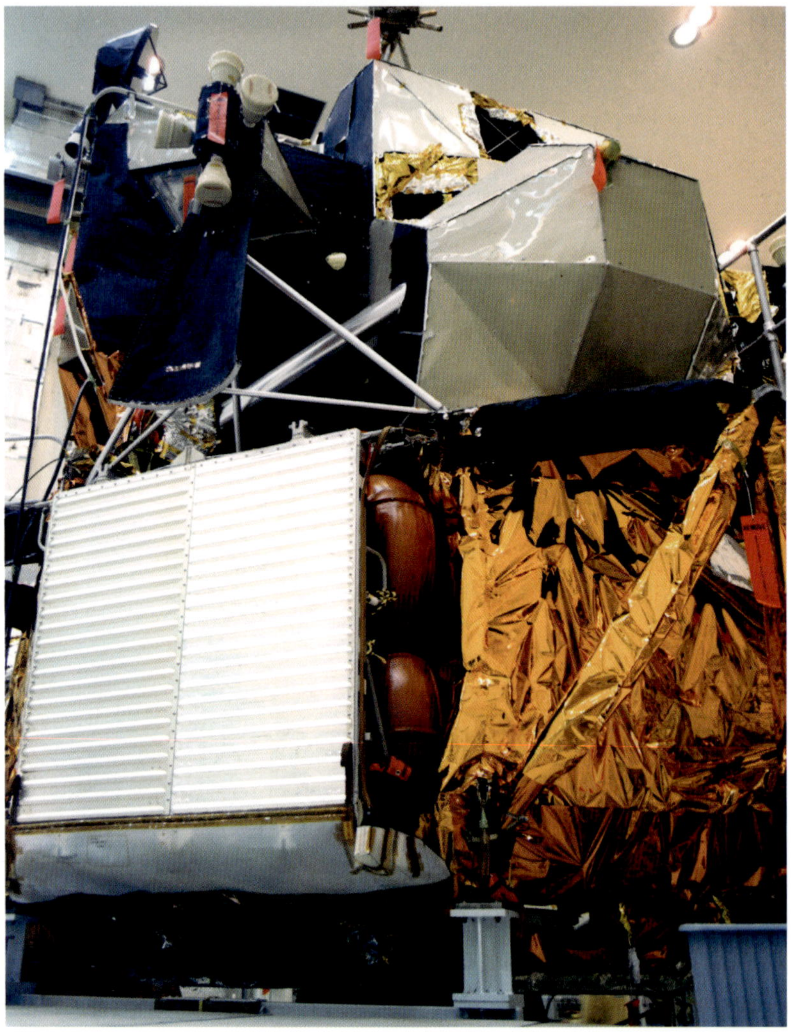

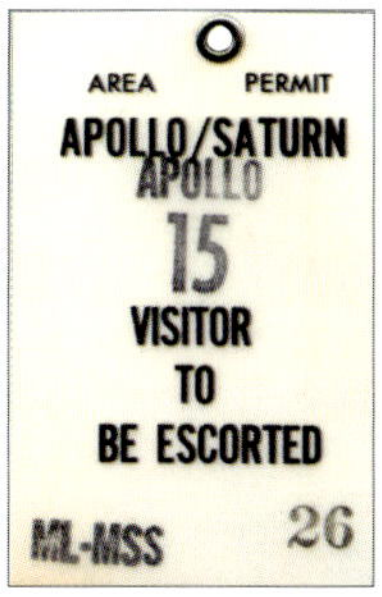

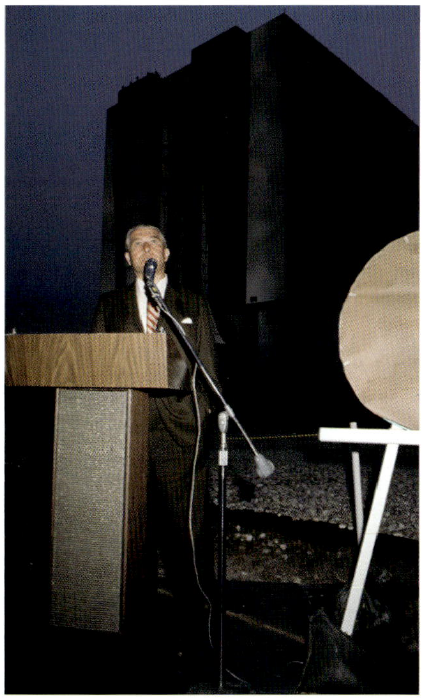

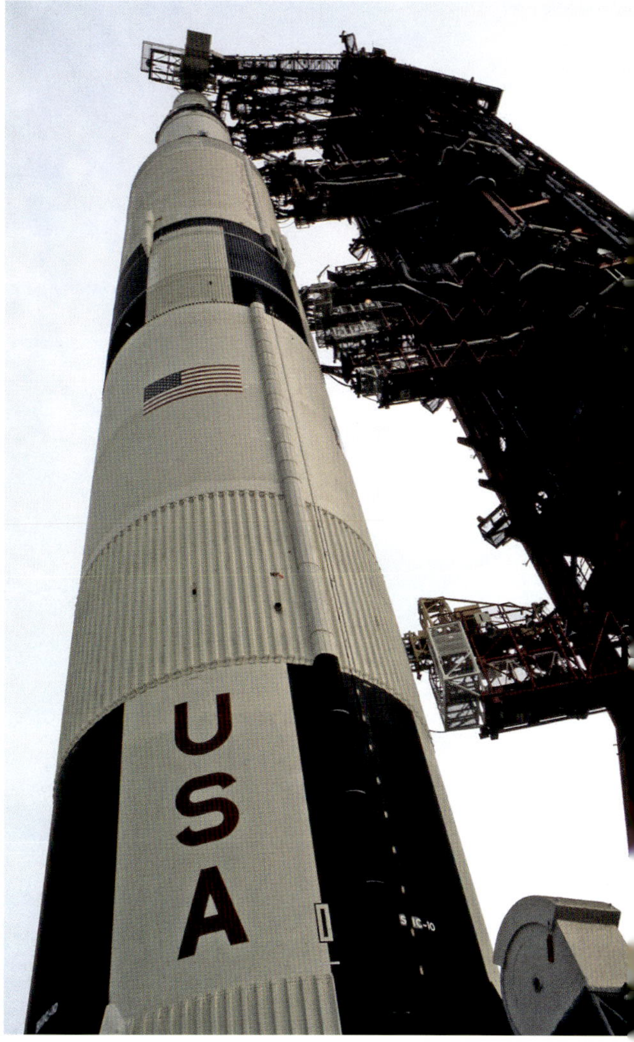

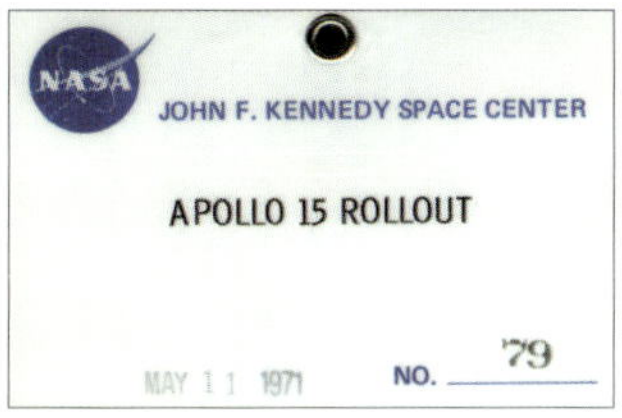

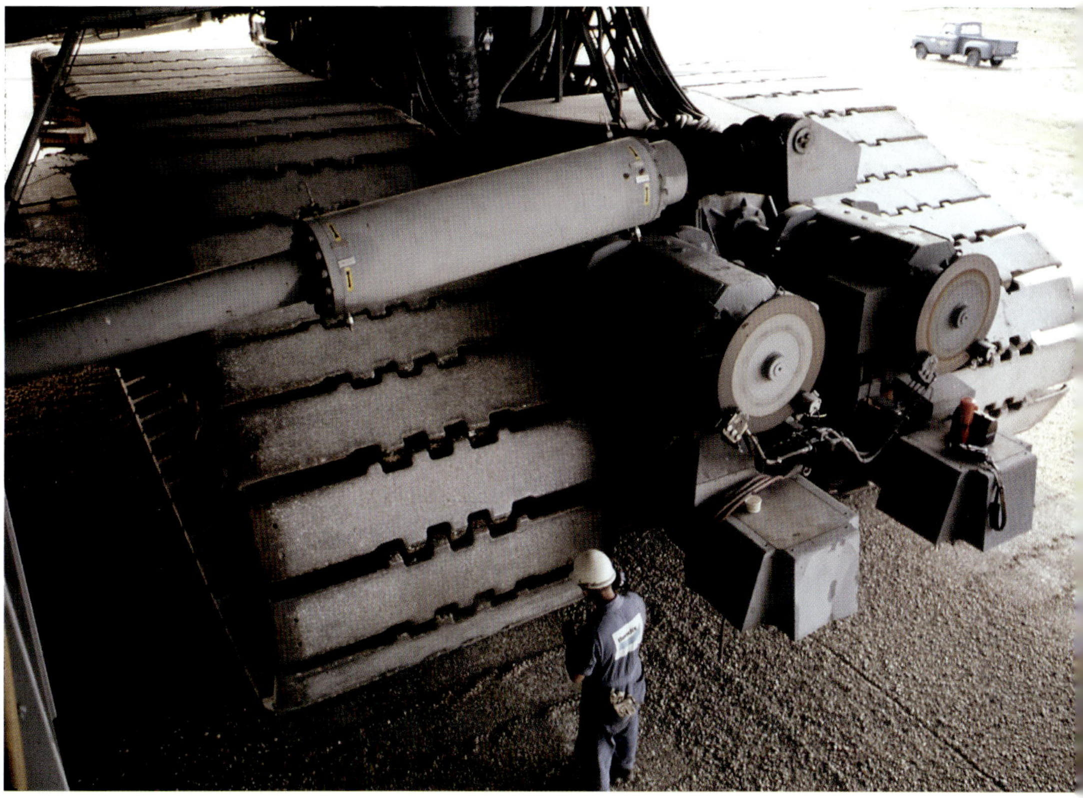

Opposite page, top left: The LM, minus its landing gear, is removed from Altitude Chamber R in the MSOB on April 23, 1971, after a chamber run with Scott and Irwin.

Opposite page, top right: In high bay 3 in the VAB, the combined CSM/SLA is mated to the Instrument Unit atop the Saturn's third stage on May 8. The crane attach points are where the LES would be mounted. The vehicle had been moved from high bay 1 in February.

Opposite page, center left: Scott (*third from left*) with NASA technicians as the LRV is loaded aboard the LM on May 3, 1971.

Opposite page, bottom left: Irwin (*left*) and astronaut Joe Engle inspect stowage.

Opposite page, bottom right: Only the LRV's brown fenders are visible after stowage is complete.

Top center: NASA's deputy associate administrator for planning, Wernher von Braun, addresses the news media at dawn on May 11, 1971 (ten weeks before launch), as Apollo 15 moves out of high bay 3 behind him. At right is the mission emblem, which the crew would soon unveil. (photo by Jacques Tiziou)

Top right: Apollo 15, with its Saturn V and LUT, moves toward Pad 39A. At lower right is one of three tail service mast assemblies, which instantly disconnected the first stage from ground support lines at launch. (photo by Jacques Tiziou)

Bottom right: A Bendix technician monitors one of four pairs of tracks beneath the crawler-transporter as it moves to the pad, three miles away. The monster, which weighed 6 million pounds, could carry 12.6 million pounds, traveling at less than one mile per hour. Each tread shoe weighed one ton. Although the Marion Power Shovel Company of Marion, Ohio, built the two vehicles, the Bendix Corporation's Launch Support Division supplied their crews. (photo by Jacques Tiziou)

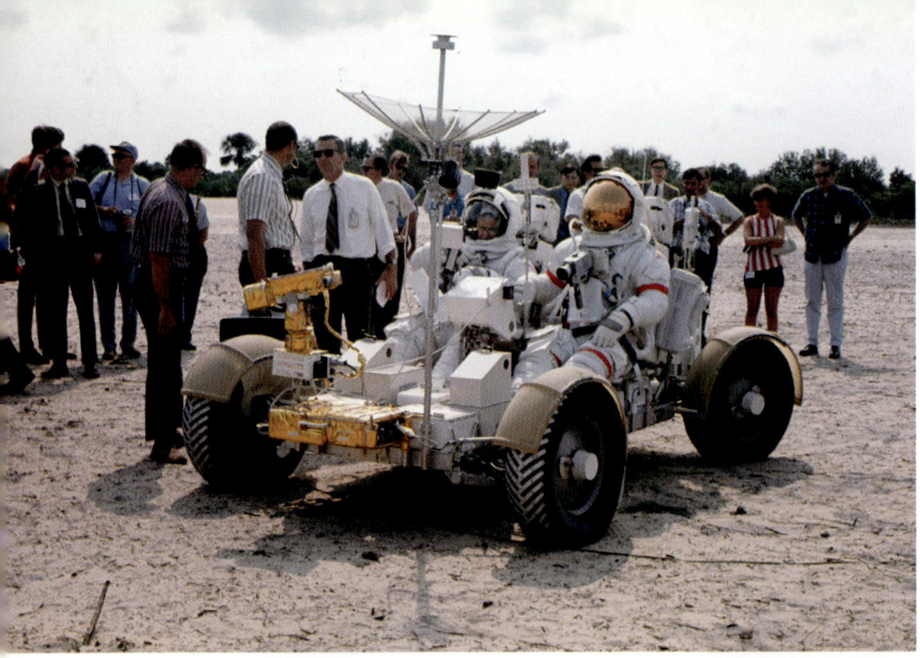

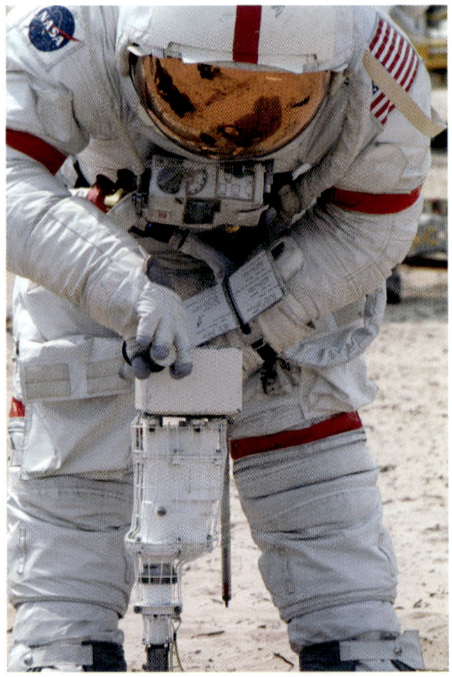

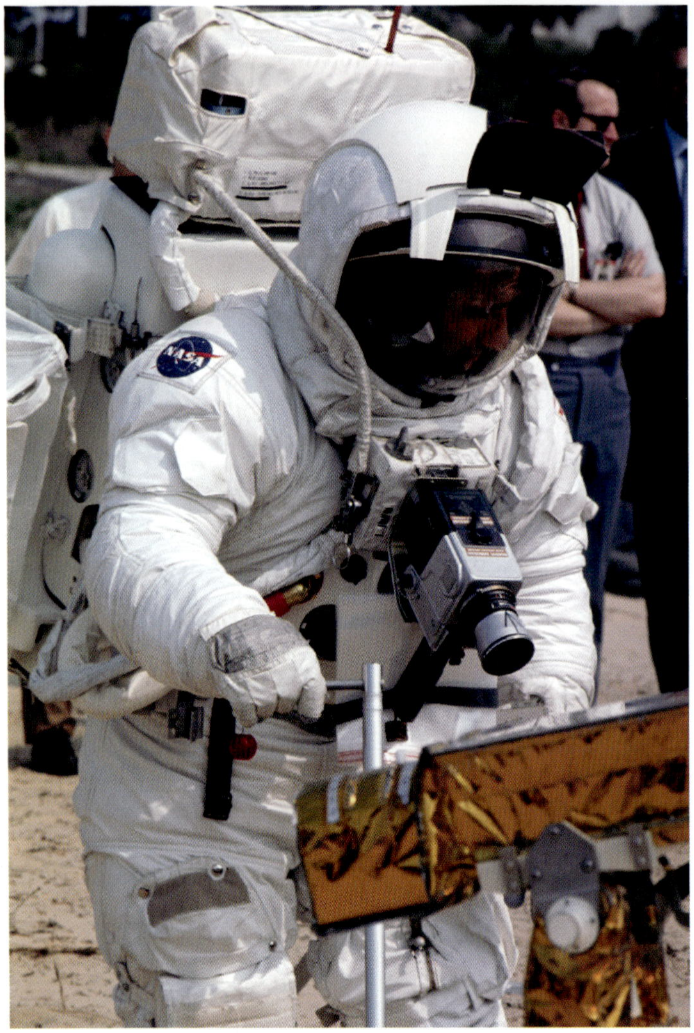

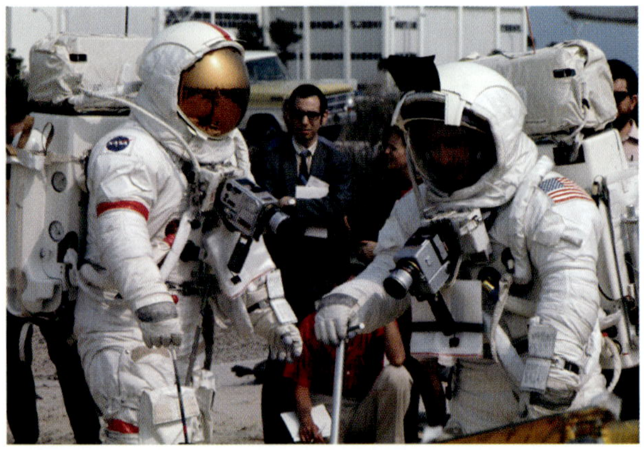

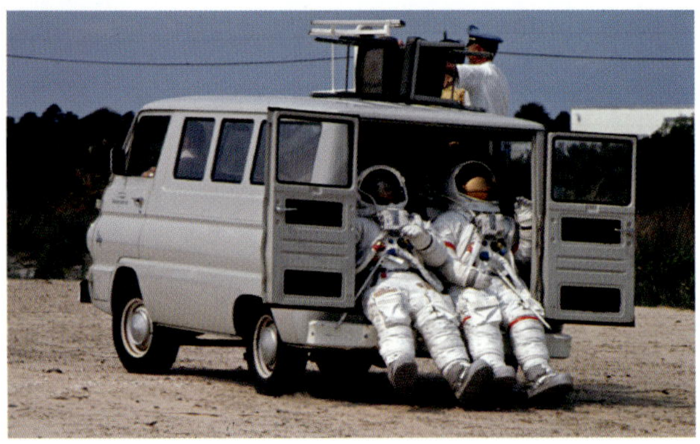

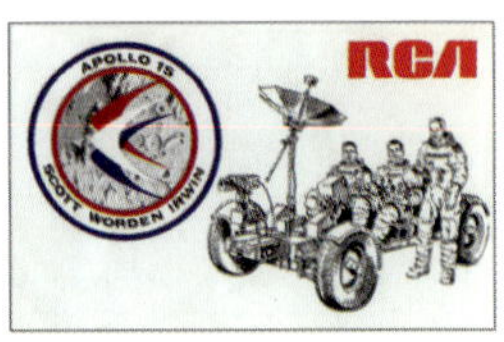

Direct left: RCA built the LRV's color TV camera.

Top left: Irwin and Scott drive the LRV one-gee trainer near the MSOB on the morning of May 11, for a photo opportunity with the media. Freelance photographer Jim Long is in a blue-checkered shirt immediately behind the rover. (photo by Jacques Tiziou)

Top right: Scott practices using the lunar drill. (photo by Jacques Tiziou)

Center left: Irwin, with the gold Mylar–covered LRV television camera in the foreground. (photo by Jacques Tiziou)

Center right: Scott (*left*) and Irwin with 70mm Hasselblad film EDCs. The MSOB is in the background. (photo by Jacques Tiziou)

Bottom right: The May 11 photo opportunity ends ignominiously for Irwin (*left*) and Scott, who return to the MSOB in a van after a battery-related failure on their LRV. (photo by Jacques Tiziou)

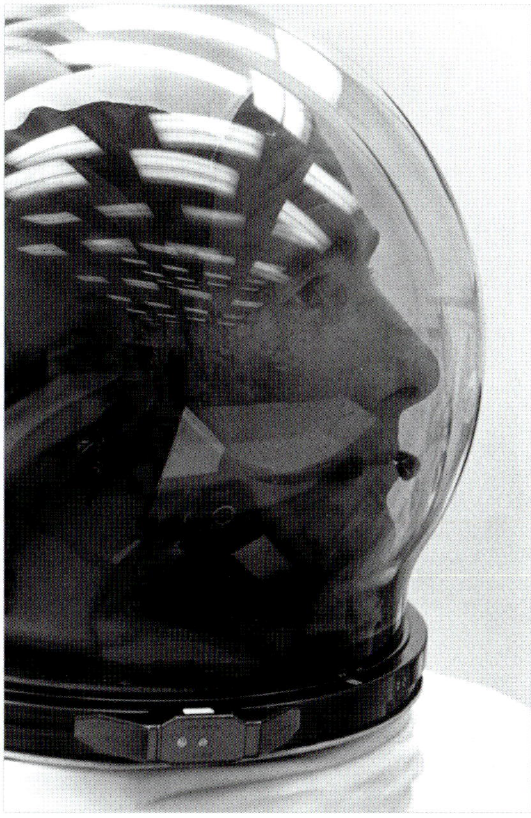

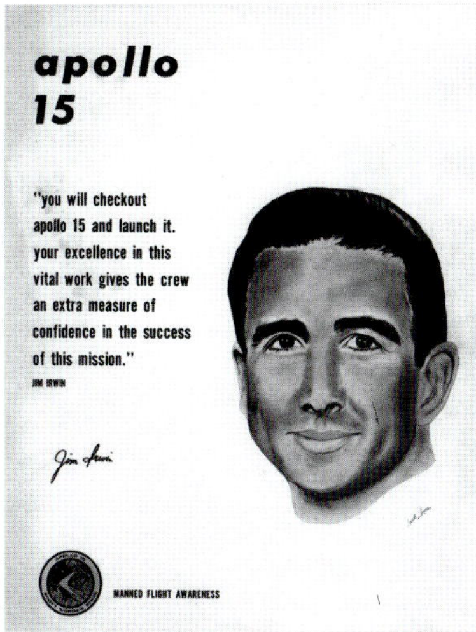

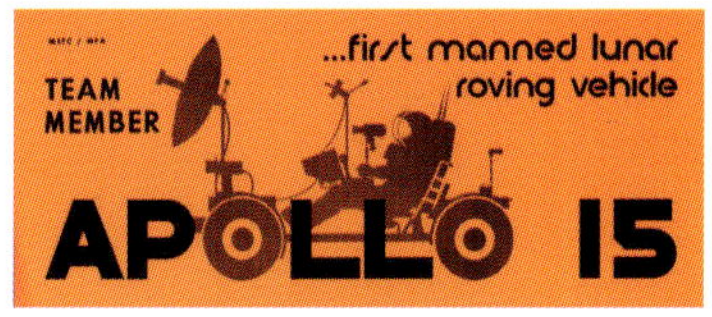

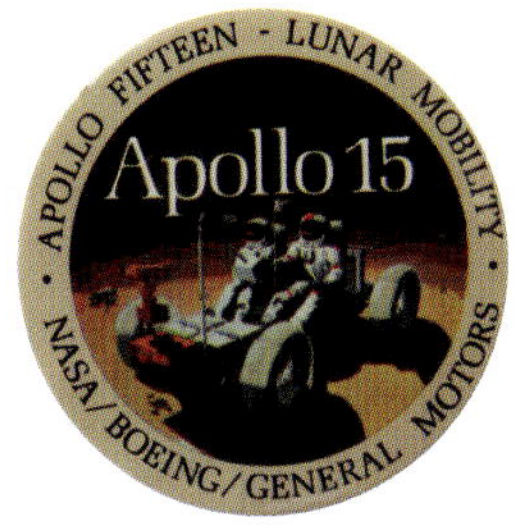

Contractor pin-back button

Above: LMP Jim Irwin trained as a USAF test pilot. In 1961, a student he was training crashed their plane. Irwin suffered two broken legs, a fractured jaw, and a concussion that temporarily wiped out part of his memory. He was rejected by NASA twice before being accepted as an astronaut in 1966. A devout Christian, Irwin would lead several expeditions in search of Noah's Ark between 1973 and 1982.

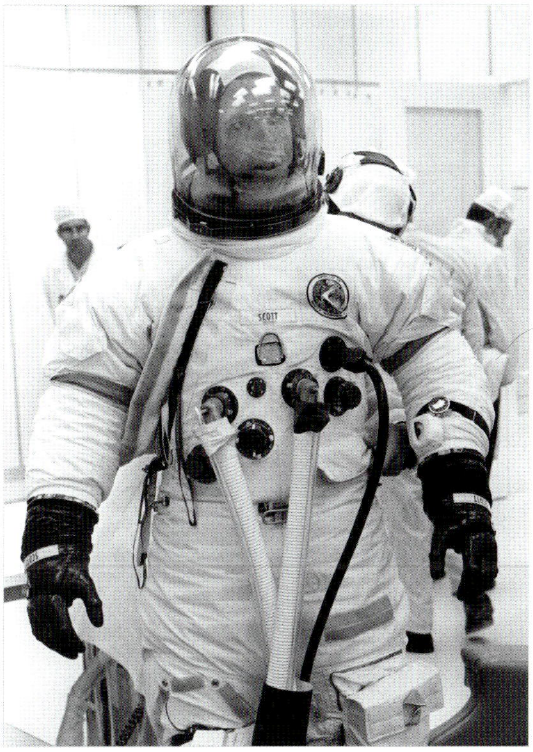

Direct left: CDR Dave Scott's space suit is pressurized on the morning of launch. An Omega Speedmaster Professional watch is strapped to his left arm with a black Velcro band. The crew wore A7LB suits with modified joints from the A7L, allowing greater mobility when riding on the LRV.

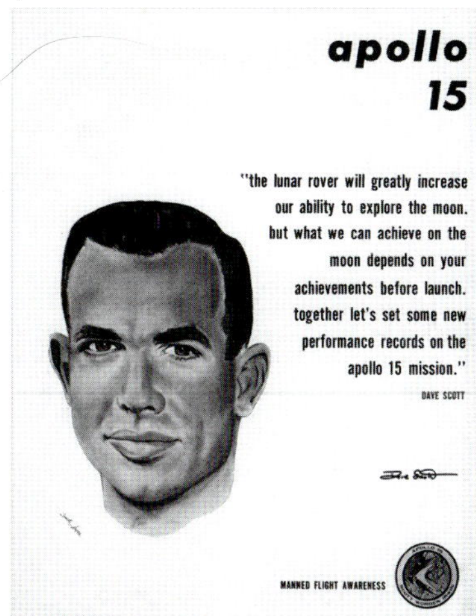

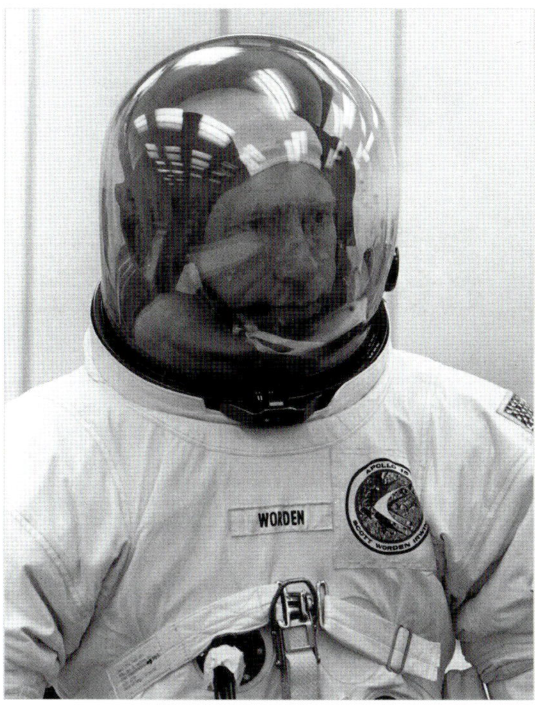

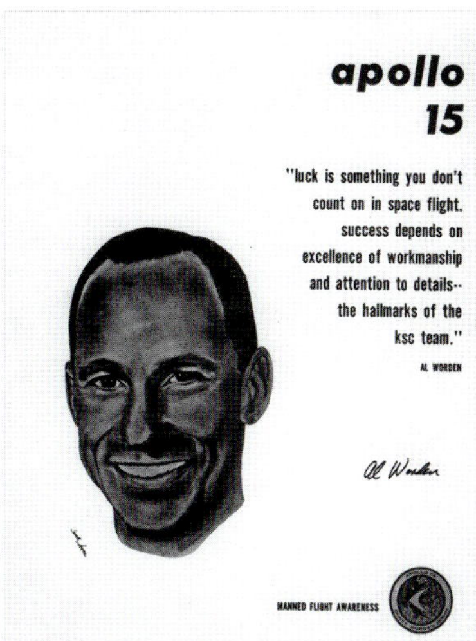

Far left: CMP Al Worden was the first divorced astronaut in space; his marriage had ended in 1968. Like Irwin, he was making his first space flight. All three crewmen would forfeit additional flights after being reprimanded for taking stamped envelopes on the mission. They had planned to sell the envelopes for profit.

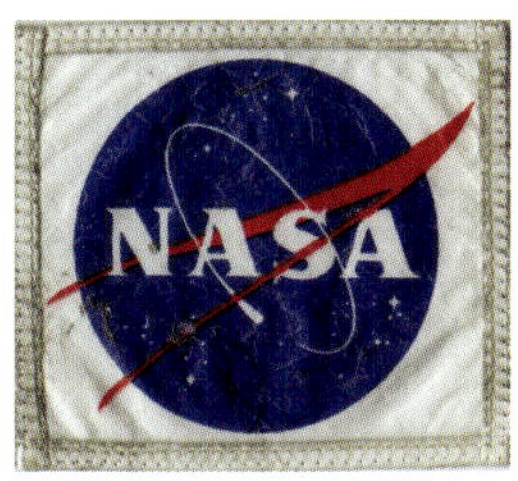

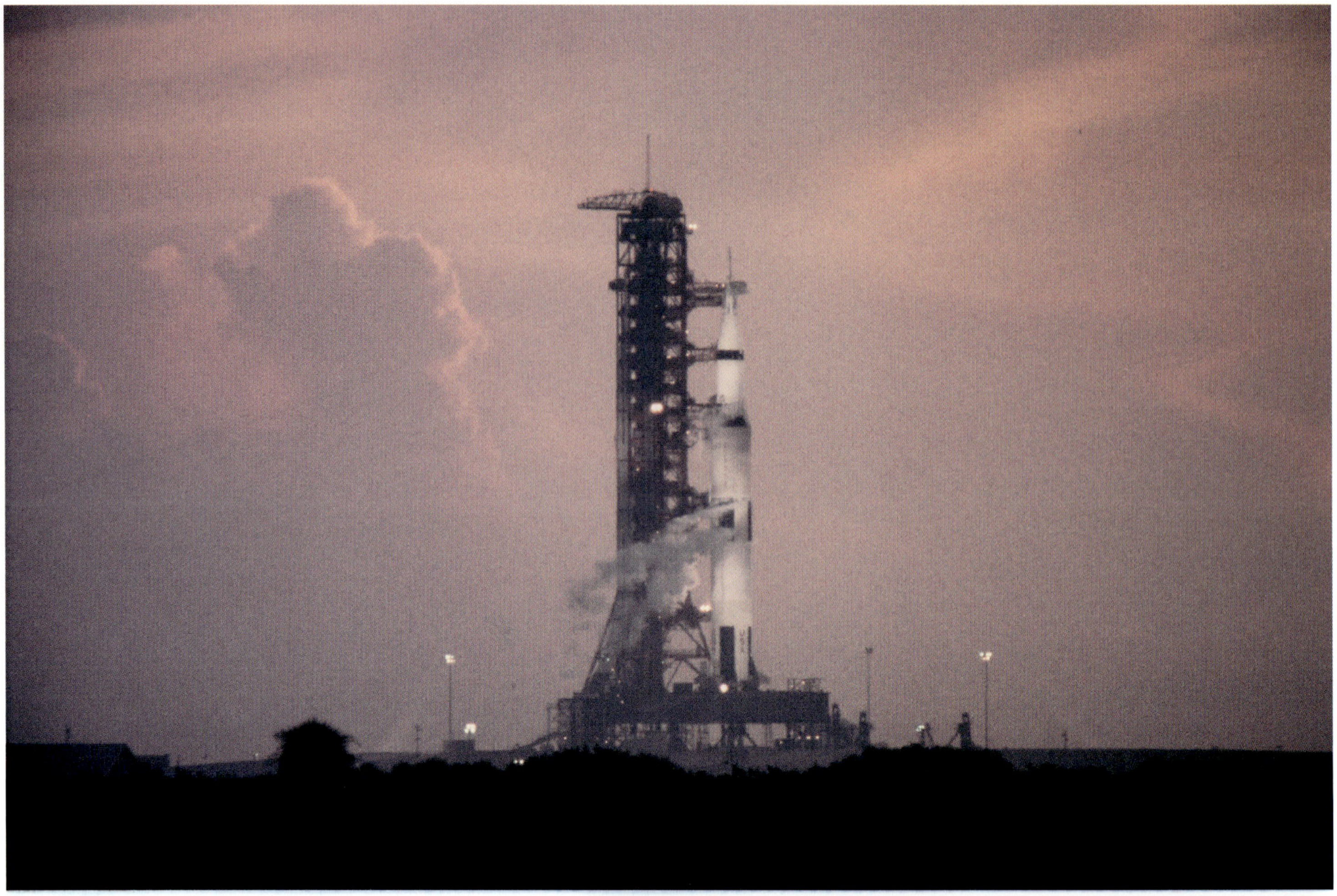

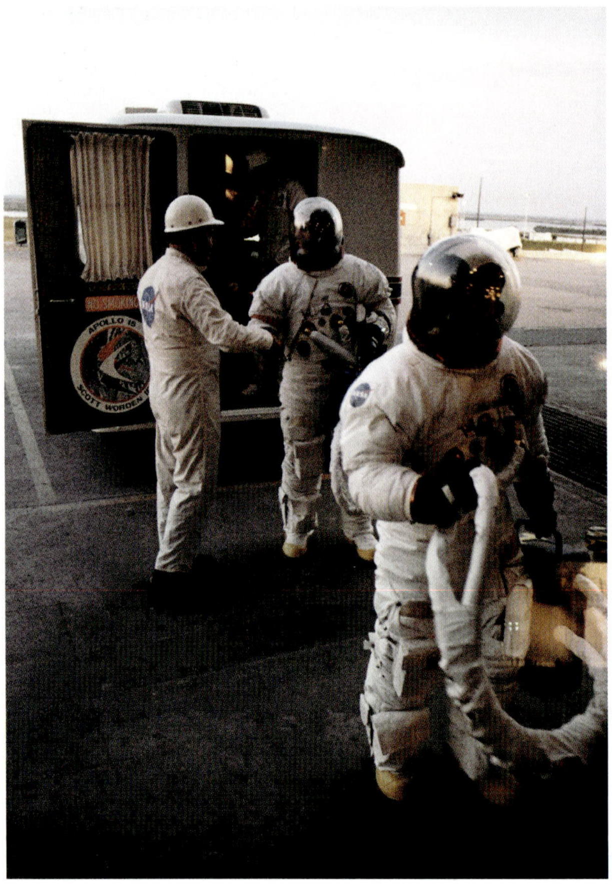

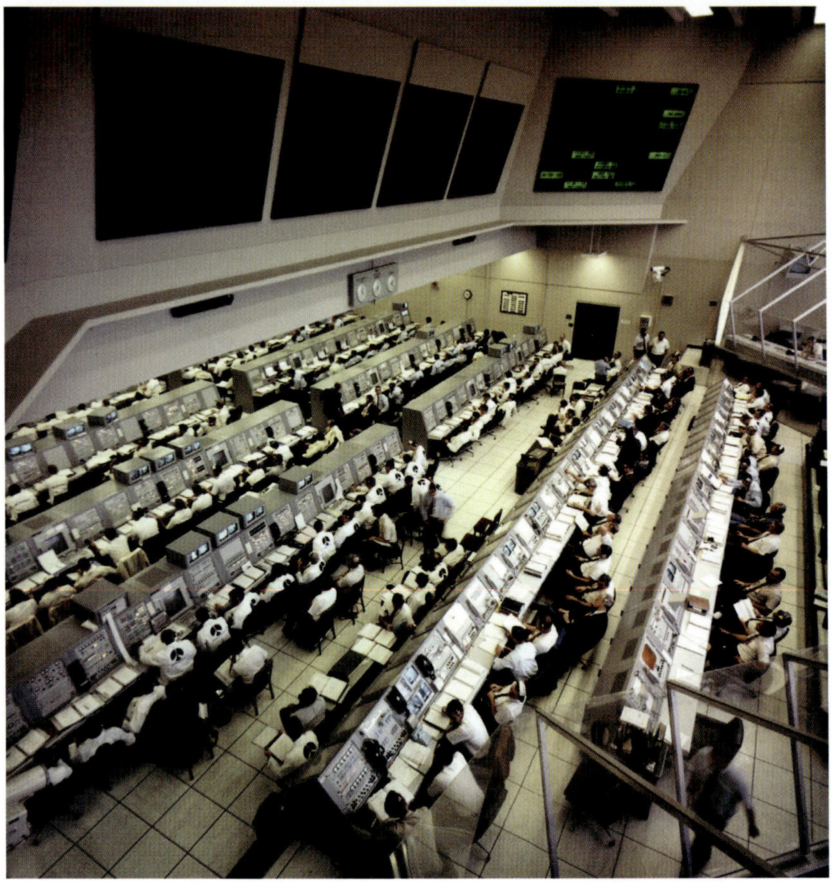

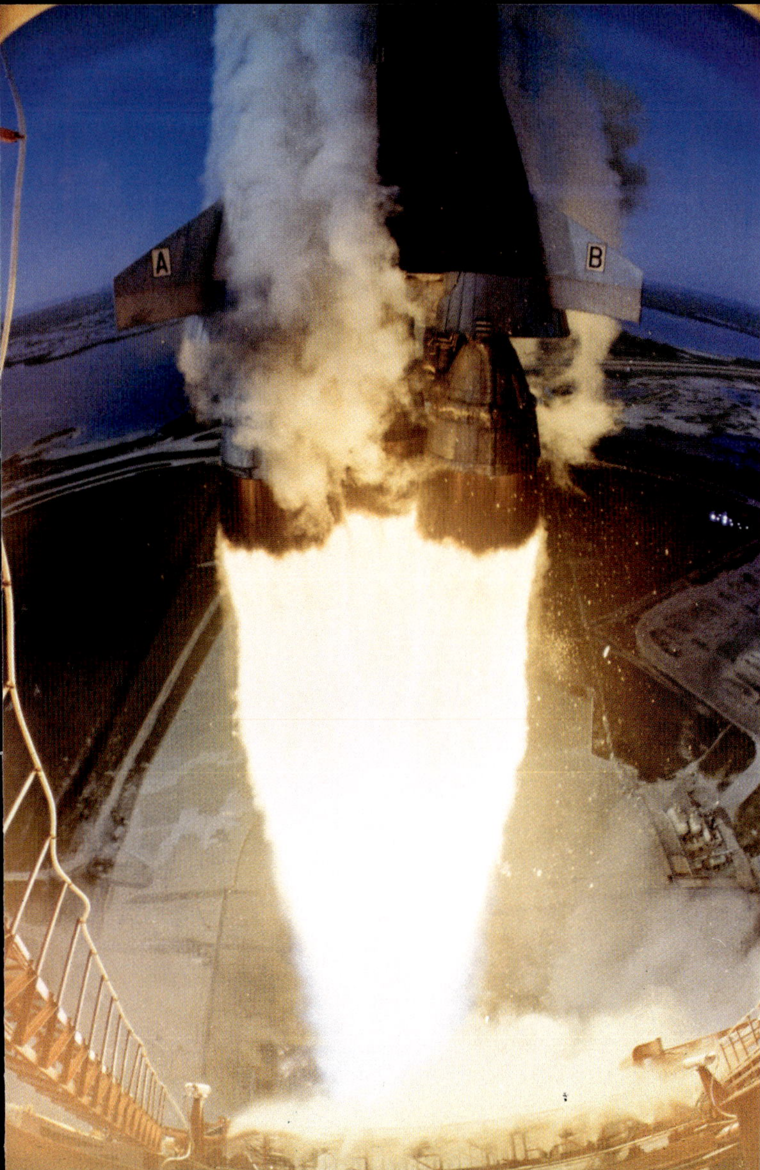

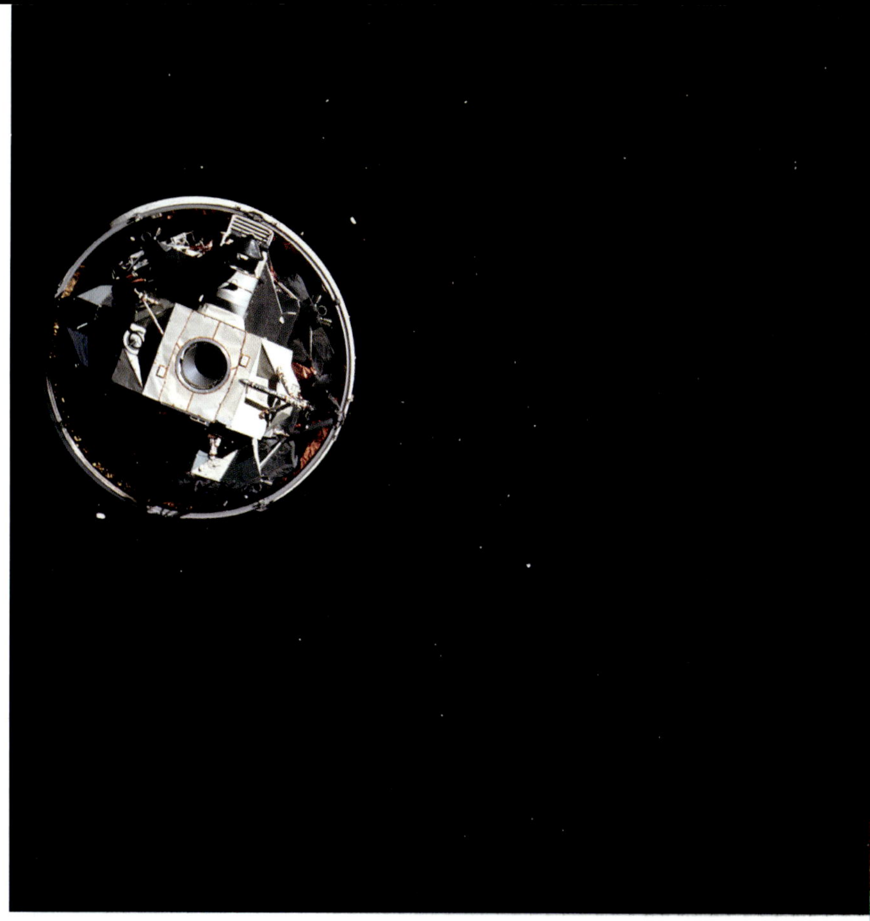

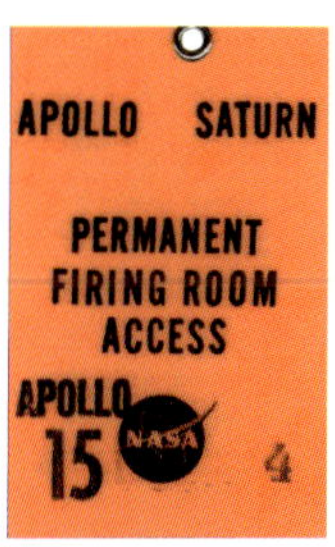

Opposite page, top: Sunrise on launch morning greets Apollo 15 at Pad 39A. The Saturn vents oxygen vapors. (photo by Tom Usciak)

Opposite page, bottom left: Irwin leads the way as the crew arrives at the pad. Scott shakes hands with Charles Buckley; suit tech Joe Schmitt is inside the transfer van.

Opposite page, bottom right: The launch team in Firing Room 1 in the LCC on July 26, 1971

Top left: The main engines of the Saturn's first stage are captured by a camera mounted on the LUT at Pad 39A, as Apollo 15 heads into an Earth parking orbit.

Top right: The LM, still inside the base of the SLA atop the Saturn's third stage, is photographed from the CM during the transposition and docking maneuver, three and a half hours after launch. The white specks are ice.

Center left: Kurt Debus's firing-room badge

Center right, left to right: Astronaut office chief Deke Slayton, Apollo 15 support crew member Karl Henize, and back-up CDR Richard Gordon in MOCR 1 during lunar surface activity. Henize would fly on a Space Shuttle mission in 1985.

Bottom right: Eugene Kranz, lead White Team flight director, in MOCR 1 during lunar EVA activities.

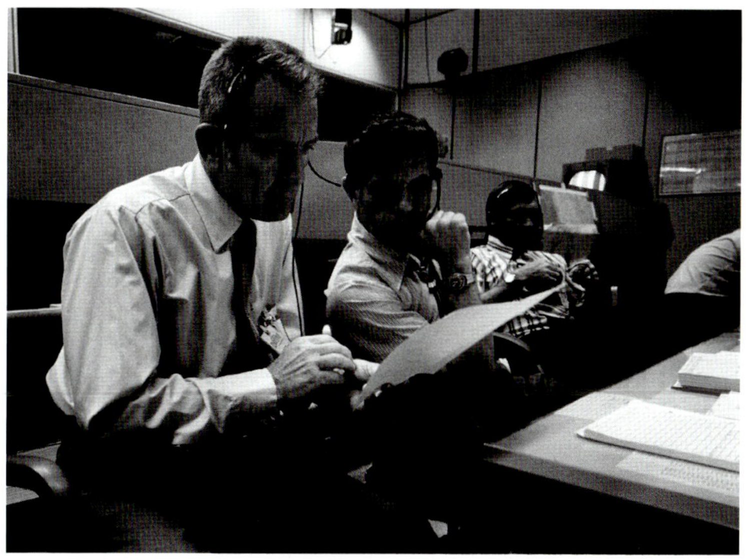

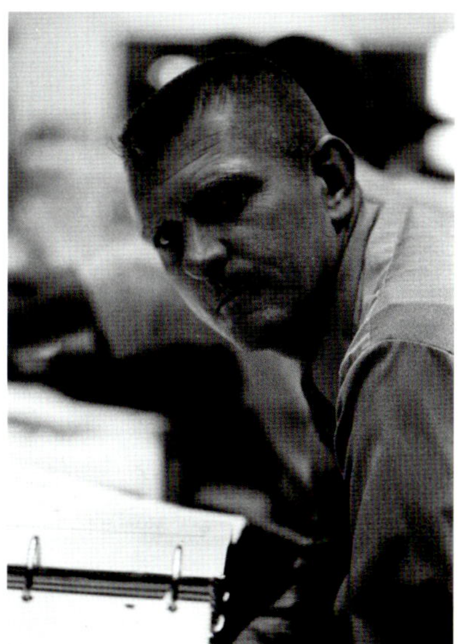

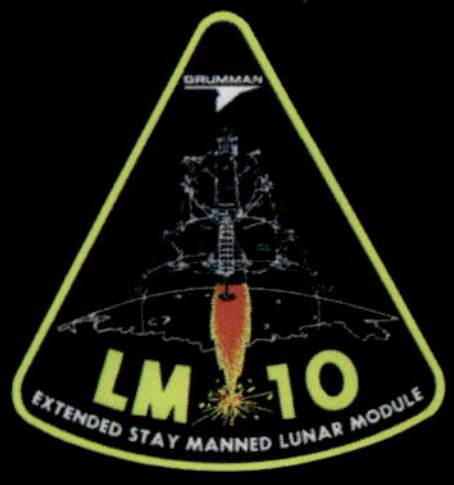

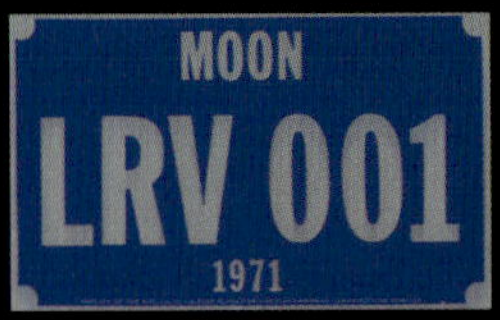

Replica of the LRV's "license plate," built by Boeing. Although the LRV did not bear such a plate, Scott carried several small versions on his lunar EVAs.

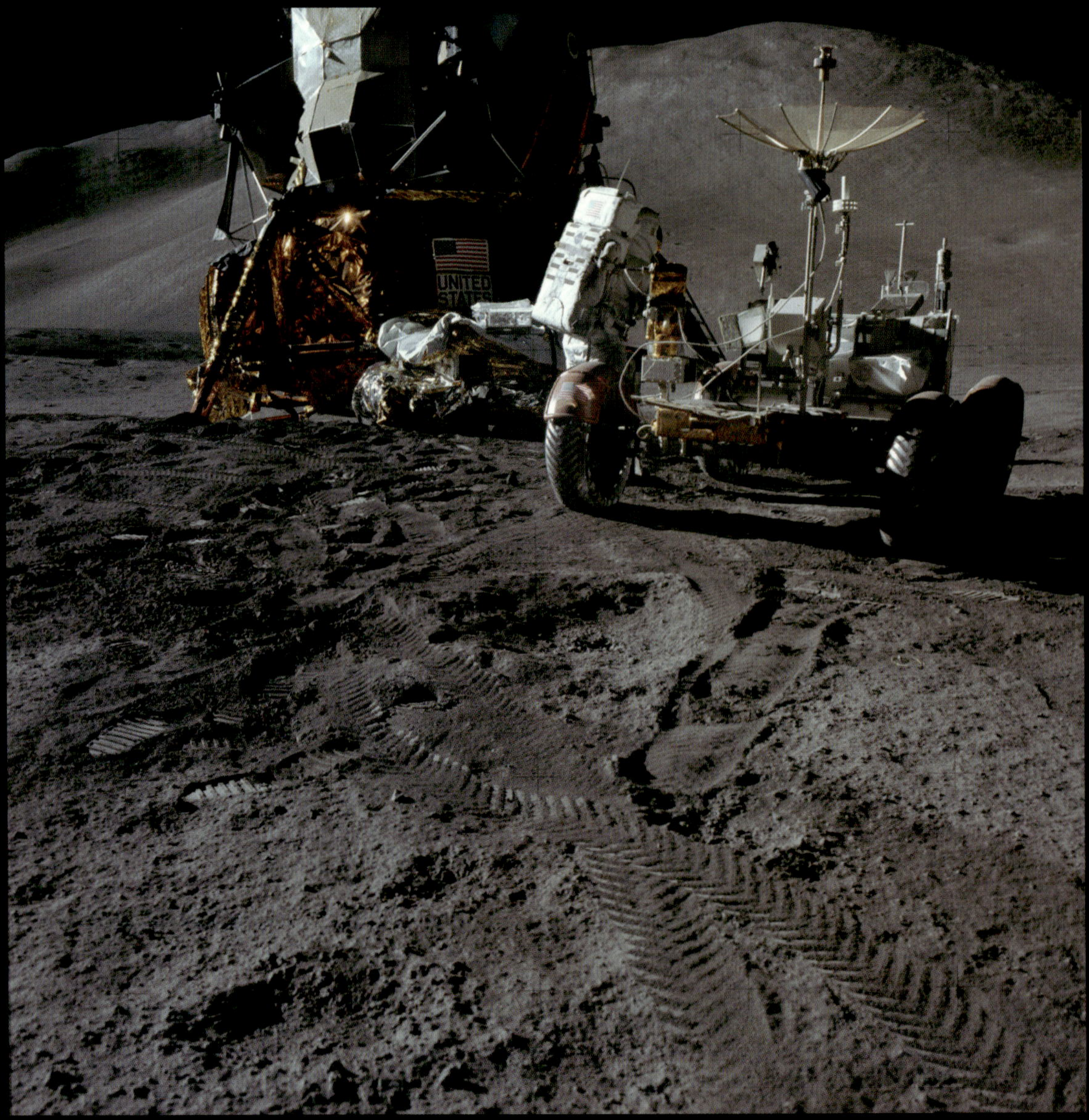

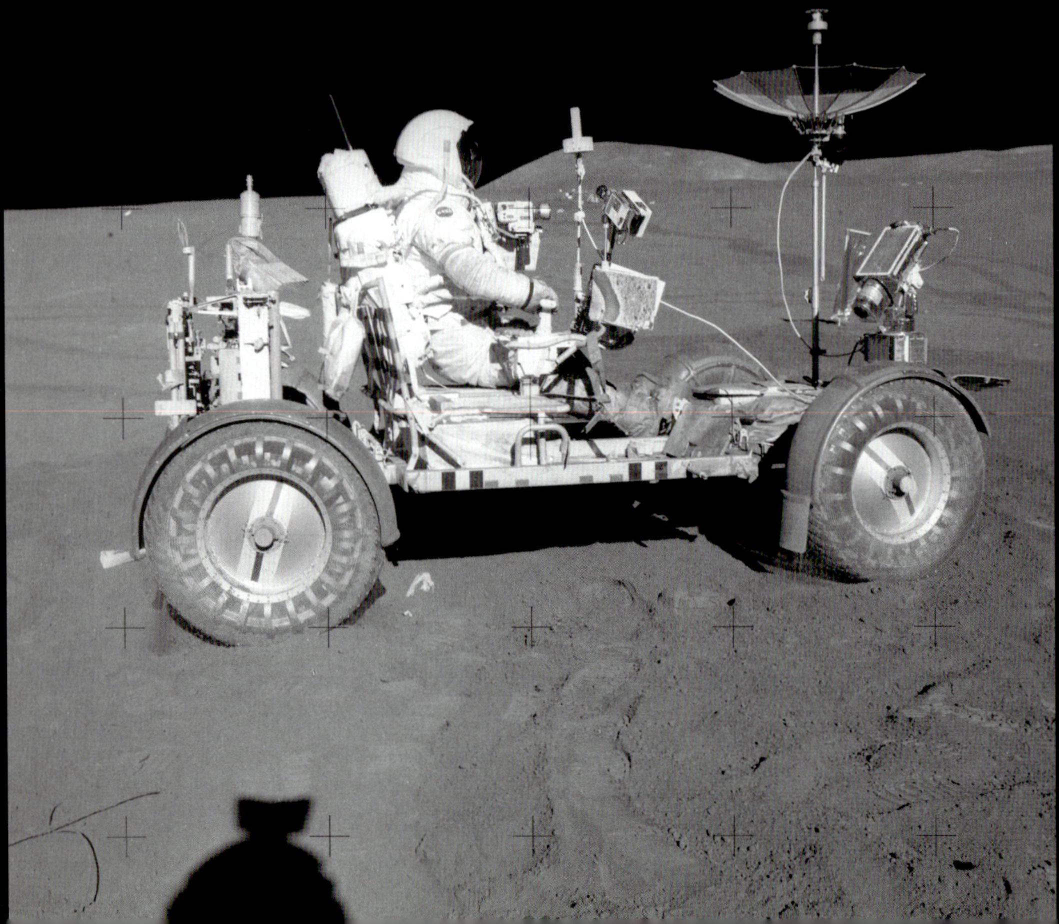

Above: Scott's photo of *Falcon* and the LRV shows the tracks made when he backed up the rover to put it closer to the MESA at left. The rock box is visible on the MESA table, as Irwin works on his side of the LRV. On the rover, the sighting scope on the back of the high-gain antenna, the 16mm DAC, and the penetrometer can be seen. Mount Hadley Delta is in the background on the right, and Silver Spur is to the left of the ascent stage. If the LRV was parked for extended periods, it had to be in the correct orientation to the sun, or critical components could overheat.

Left: Irwin's photo of Scott aboard the LRV at the beginning of the EVA 2 traverse, showing dust on the wheels. The SWC is visible beyond the TV camera. Scott wears a Hasselblad camera. The traverse maps are mounted on Irwin's handhold, and the Buddy Secondary Life Support System bag hangs from the rear of Irwin's seat.

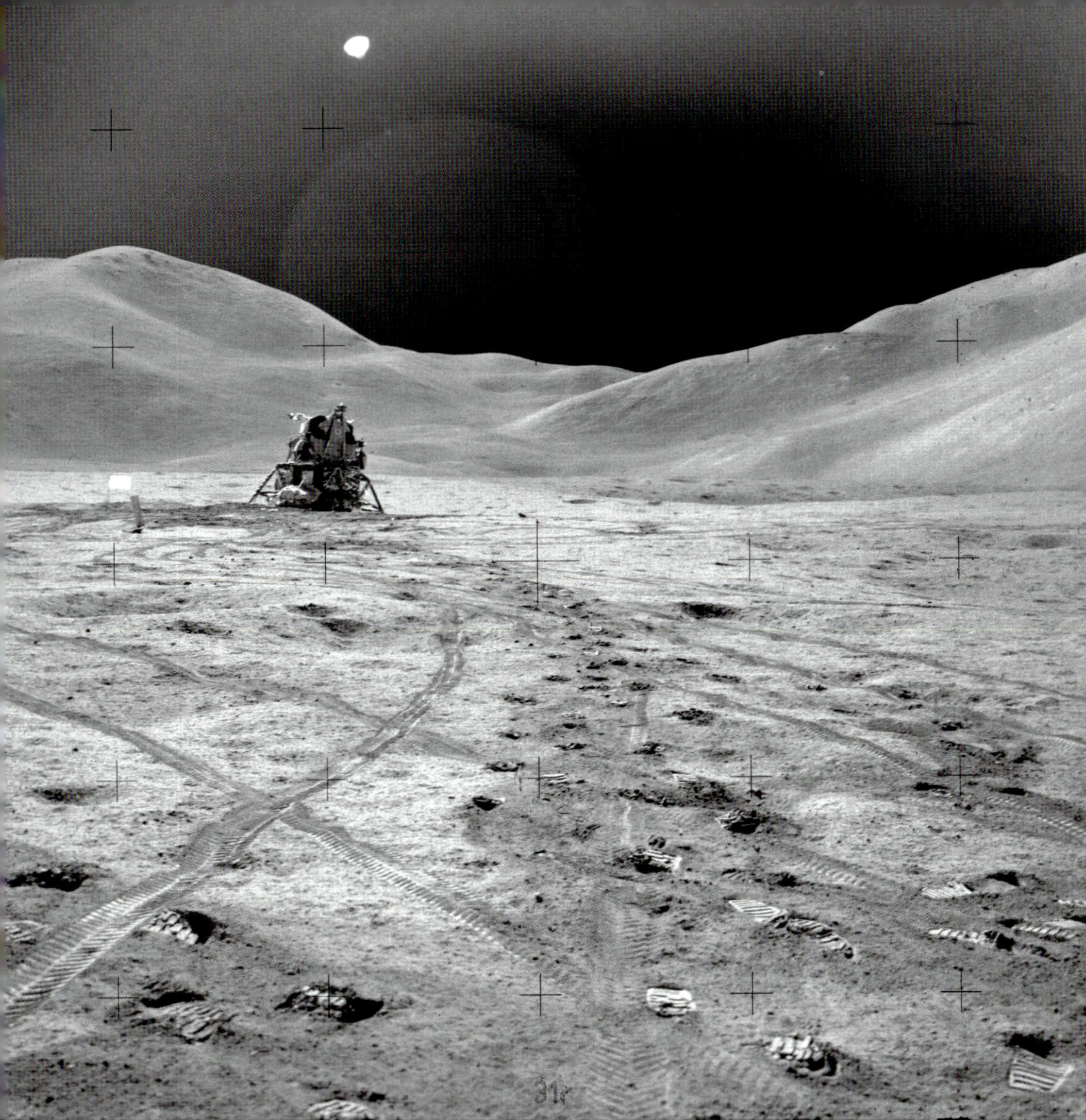

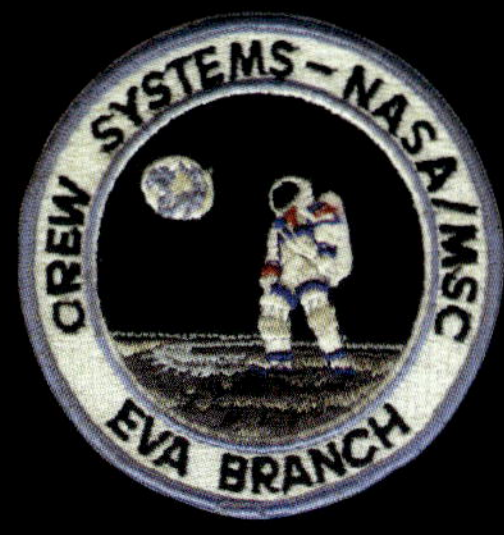

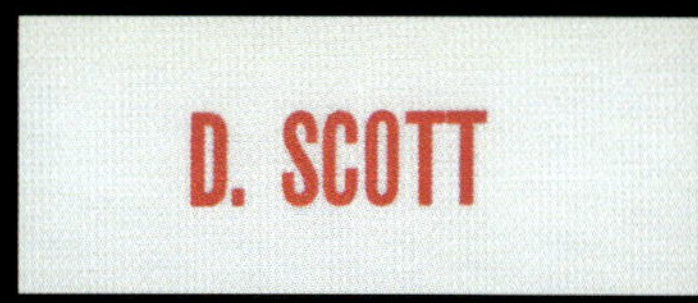

Red-lettered Beta cloth name tags were used on the EVA crewmen's EMUs.

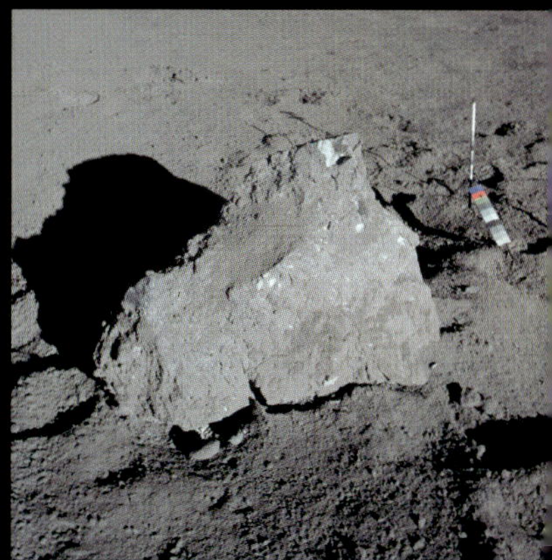

Above: *Falcon* shows its eleven-degree tilt, after landing with its rear footpad on the edge of a twenty-foot crater. The Apennine Mountains are at left and Mount Hadley Delta is at right in this photo by Irwin looking southeast, near the beginning of EVA 3 on August 2. The SWCE is to the right of the US flag at left; at top (white spot) is a reflection in the camera lens. LRV tracks and boot prints from previous trips fill the foreground. The astronauts named the small crater to the right of the LM "Last."

Center right: A boulder at Station 2 shows where Scott has removed a sample at top. The boulder may be ejecta from the impact that created the Aristillus and Autolycus Craters to the north; it would have then bounced onto the hillside. Note the color scale on the gnomon.

Bottom right: Irwin's photo of the LRV at Geology Station 2, during EVA 1 on July 31

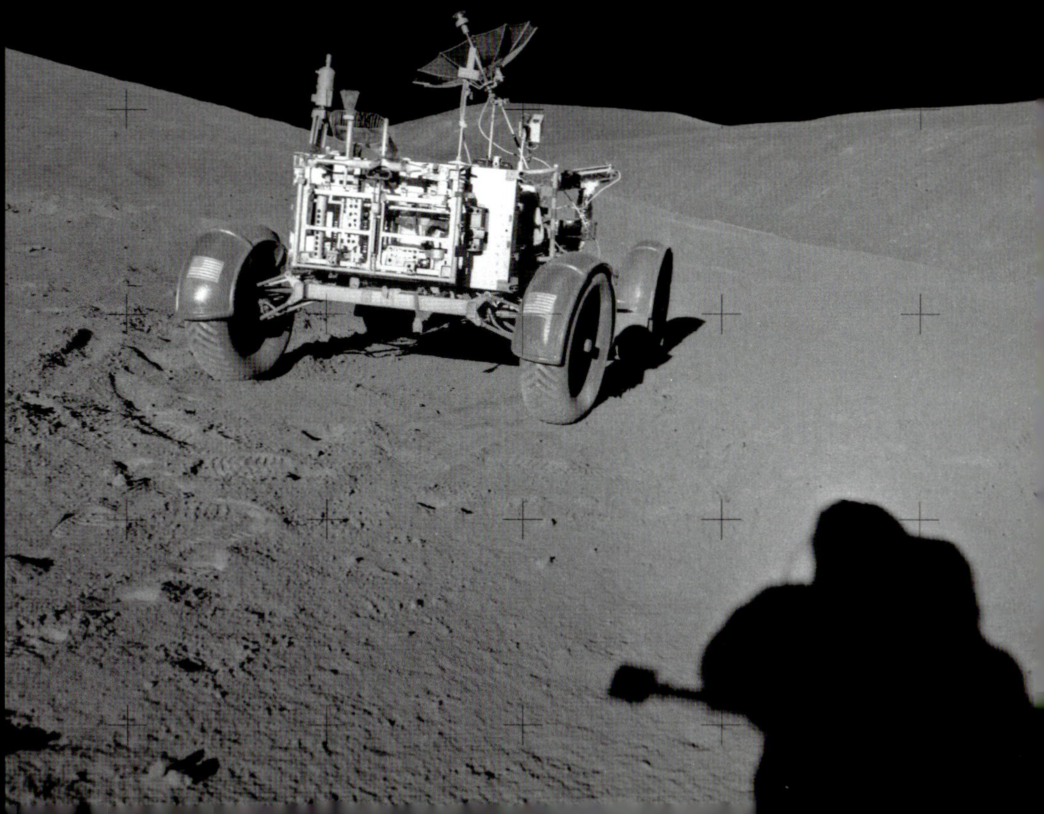

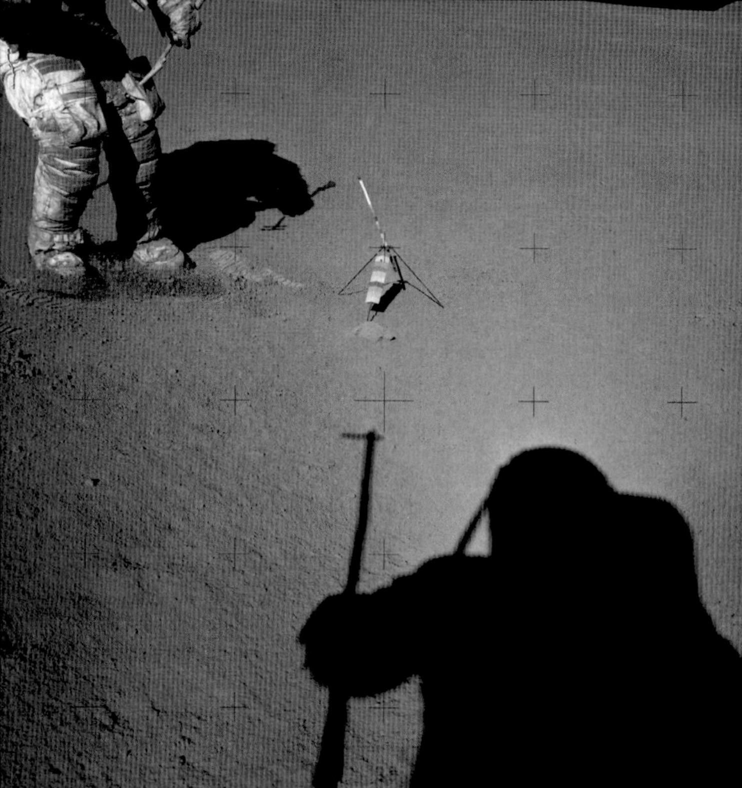

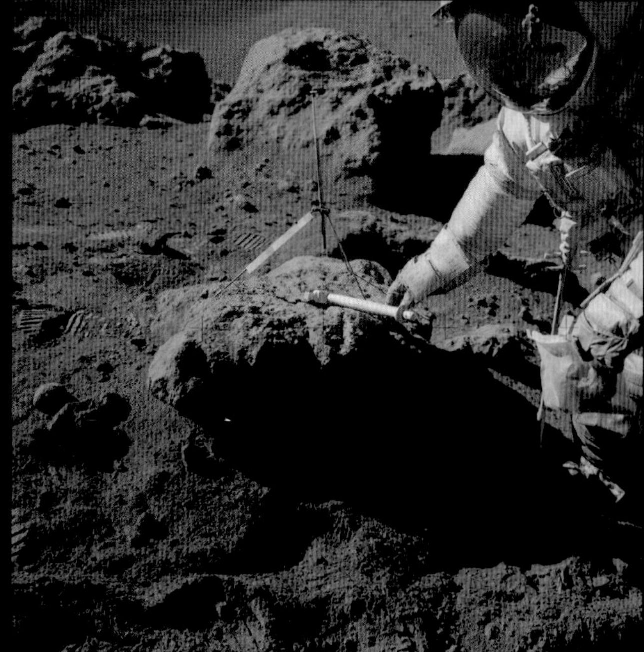

Top left: As he backs away, Scott kicks up some dust after positioning the gnomon at the third Station 6 sample area during EVA 2.

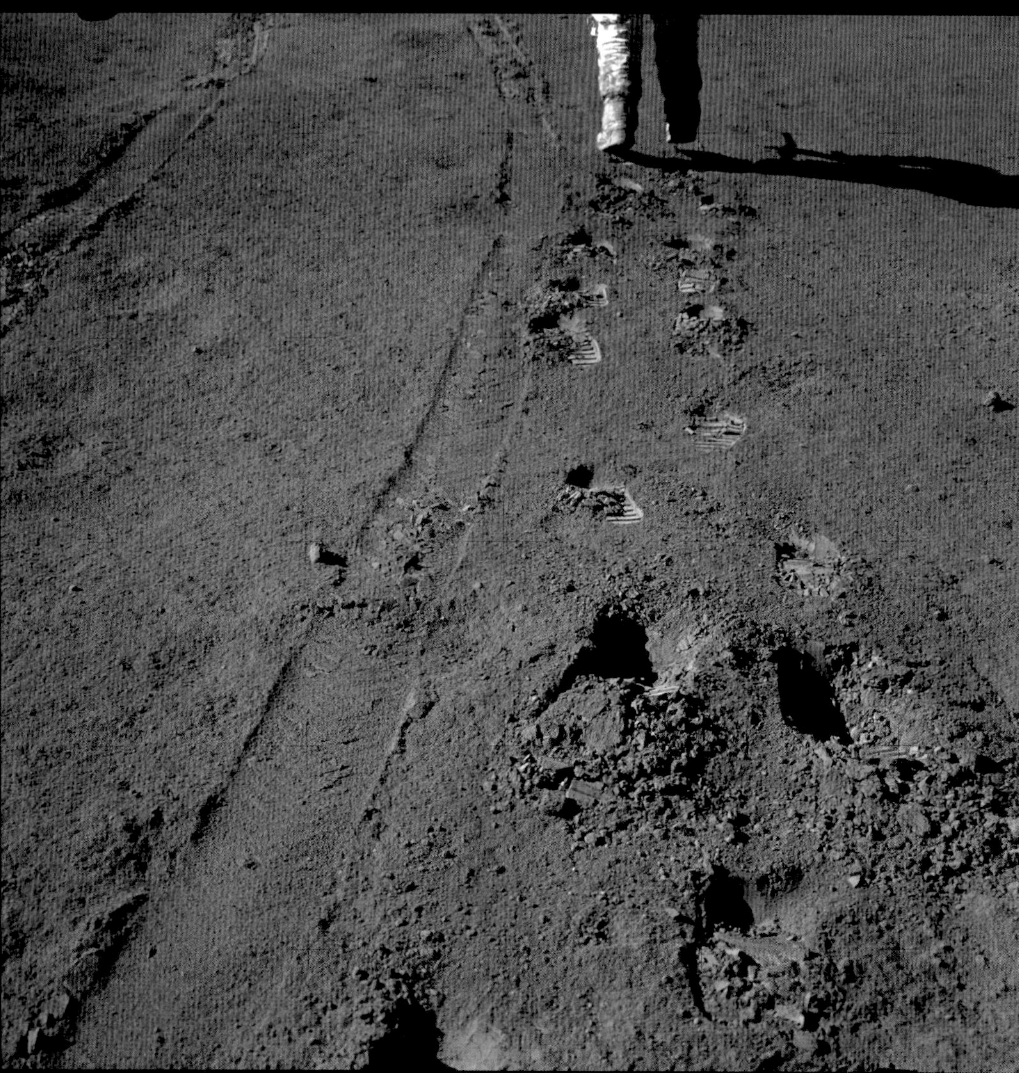

Top right: Scott reaches for a hammer on a boulder during EVA 3. A checklist on his right wrist identifies his drilling assignments for all three EVAs. He holds a sample bag—with a rock visible—in his left hand.

Left: During EVA 2, Irwin leaves boot prints as he climbs back toward the LRV from the sample site in the Station 6 crater. The rover's tracks can be seen in this photo by Scott.

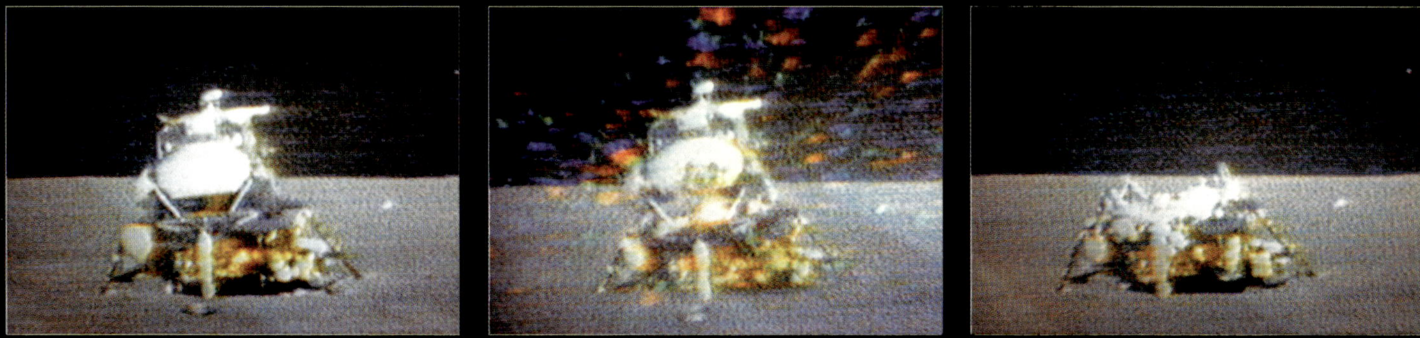

Above: This sequence of still video frames from the LRV's TV camera, controlled from the MOCR, shows *Falcon*'s ascent stage lifting off at 1:11 p.m. (EST) on August 2. The LRV was parked about one hundred yards east of the LM. Its camera's tilt mechanism malfunctioned, preventing a longer view of an ascent. (A longer view was captured on Apollo 16.)

Bottom left: The CSM, photographed from an LM window during rendezvous on August 2, 1971. The SIM bay is visible on the right side of the SM. This was the first time a SIM flew on Apollo; it proved troublesome during installation at KSC.

Bottom right: Worden is hidden by the SM as he makes his way to the SIM bay to retrieve two film cassettes on August 5. The twenty-minute EVA during the return trip was the first in deep space. Handholds can be seen on either side of the bay.

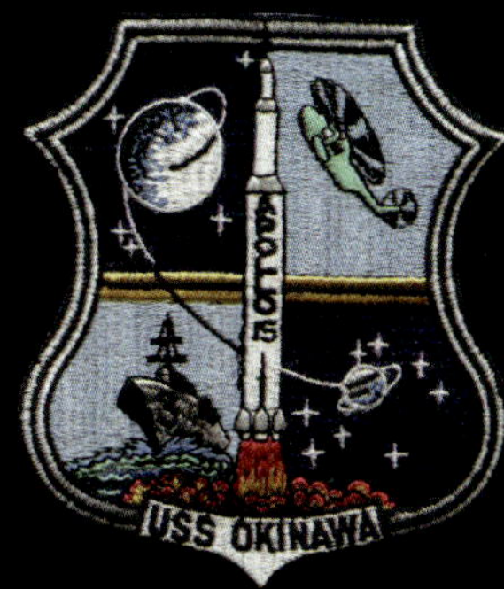

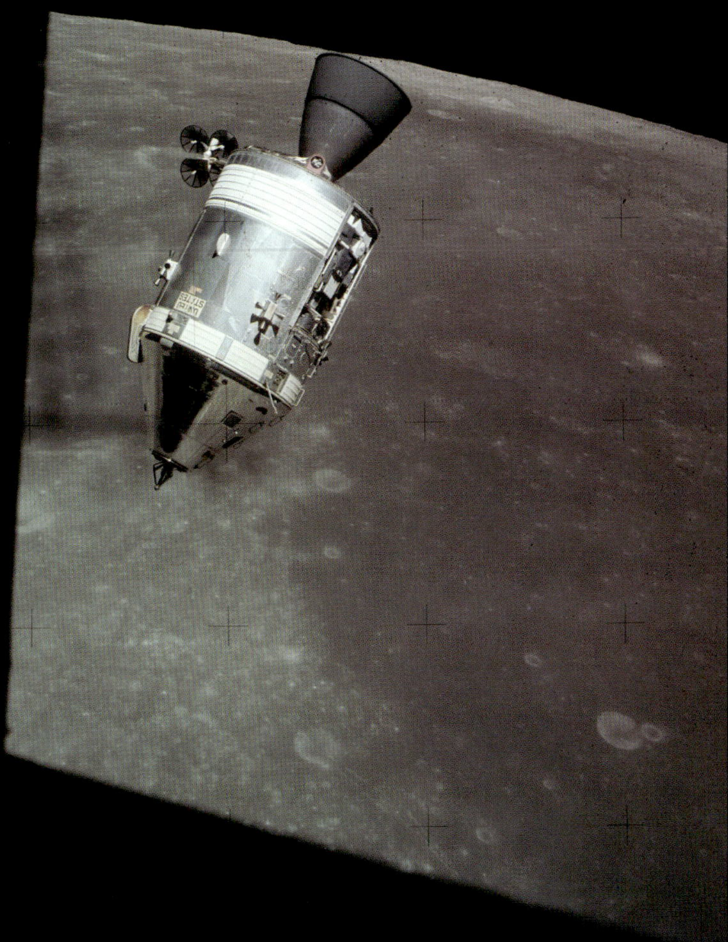

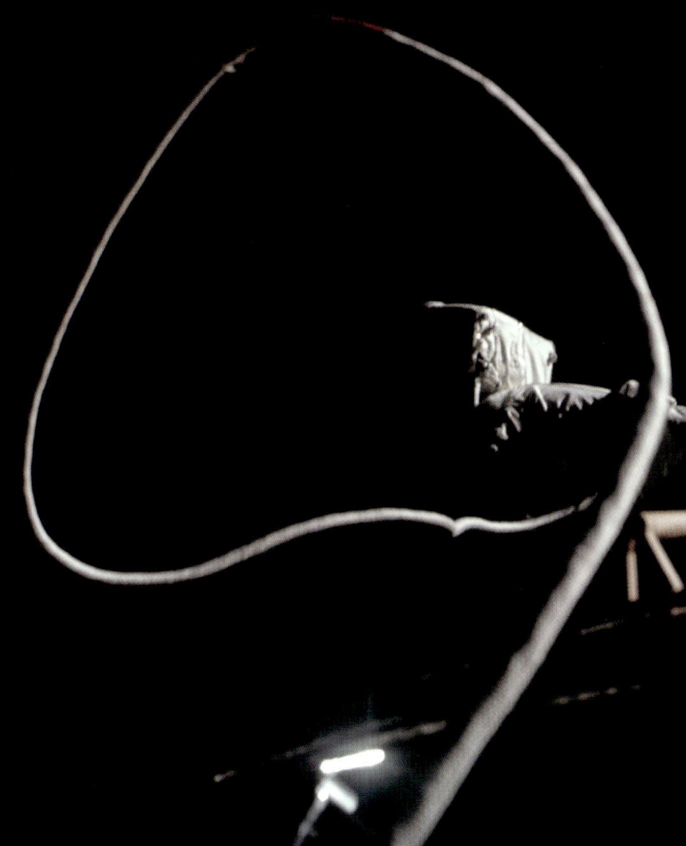

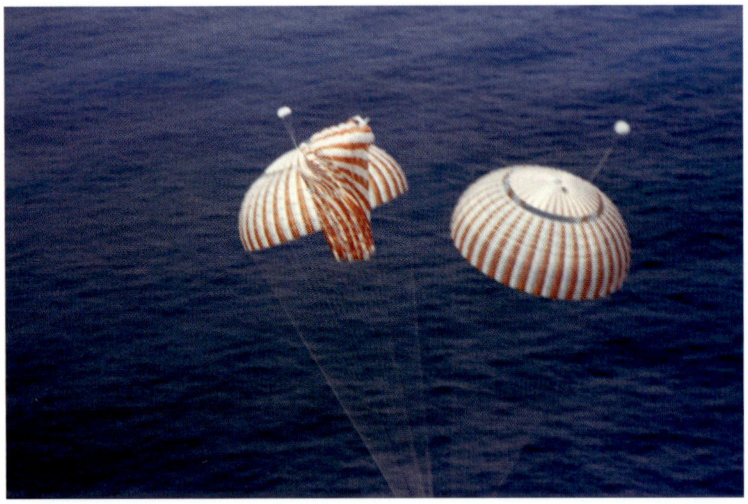

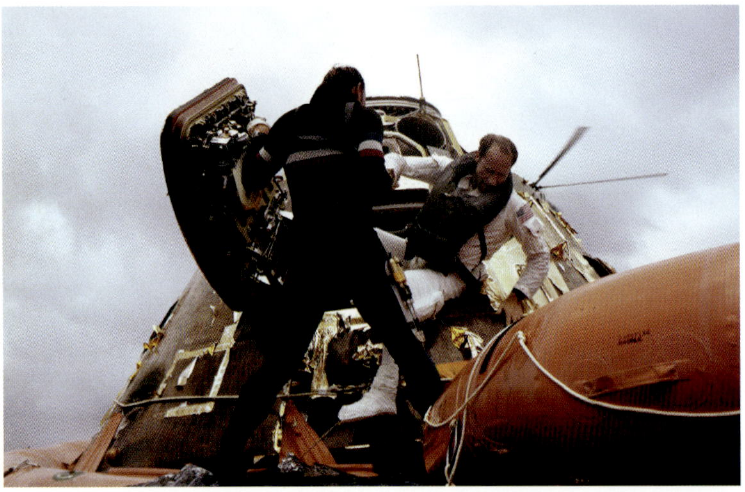

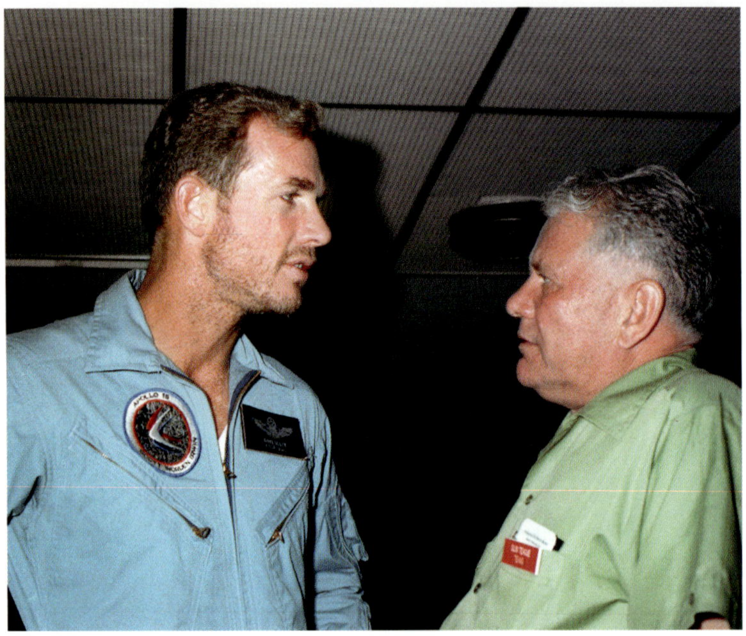
APOLLO 15

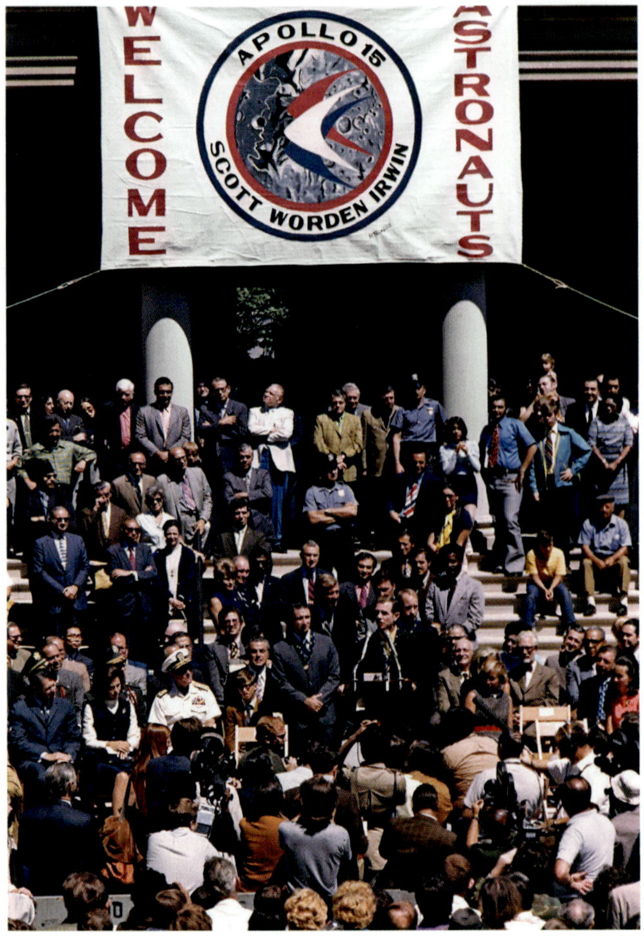
WELCOME
APOLLO 15
SCOTT WORDEN IRWIN
ASTRONAUTS

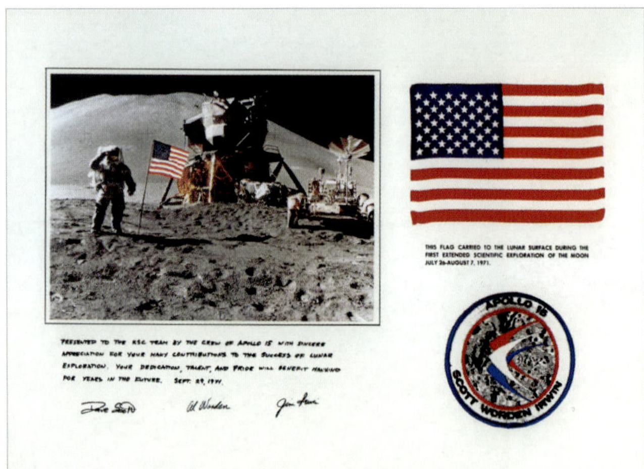
THIS FLAG CARRIED TO THE LUNAR SURFACE DURING THE FIRST EXTENDED SCIENTIFIC EXPLORATION OF THE MOON JULY 26-AUGUST 7, 1971.
PRESENTED TO THE KSC TEAM BY THE CREW OF APOLLO 15 WITH SINCERE APPRECIATION FOR YOUR MANY CONTRIBUTIONS TO THE SUCCESS OF LUNAR EXPLORATION. YOUR DEDICATION, TALENT, AND PRIDE WILL BENEFIT MANKIND FOR YEARS IN THE FUTURE. SEPT. 29, 1971.
APOLLO 15
SCOTT WORDEN IRWIN

Opposite page, top left: One main parachute deflated, likely because its nylon lines had burned during a standard RCS fuel dump shortly before splashdown.

Opposite page, center left: A navy frogman helps Worden egress from CM *Endeavour* after splashdown in the Pacific Ocean on the afternoon of August 7, 1971. The landing took place less than seven miles from the prime recovery ship, USS *Okinawa*, an amphibious assault ship.

Opposite page, bottom left: Scott speaks with Rep. Olin Teague (D-Texas) aboard *Okinawa* on August 7. Teague chaired the Manned Space Flight Subcommittee of the House Committee on Science and Astronautics, later becoming the committee's chairman. He is in the white jacket on the top step in the City Hall photo at top right.

Opposite page, top right: Scott speaks on the steps of New York City Hall on August 24, after Mayor John Lindsay (*between Scott, at the lecturn, and Irwin*) presented each crewman with the New York City Gold Medal of Honor following a motorcade down Fifth Avenue. The astronauts gave Lindsay a US flag, which they had carried on their mission.

Opposite page, bottom right: Printed certificate—with reproductions of the flown flag and crew patch—presented by the astronauts to the KSC workforce

Top right: Scott, Worden, and Irwin hold launch photos, which they received during the crew reunion ceremony in the VAB on September 29. The mission emblem on the lectern came from the transfer-van door.

Bottom right: Scott, Irwin and Worden discuss the mission on NBC-TV's *Today* show with space correspondent Jim Hartz and host Joe Garagiola in New York on August 24. Their first joint TV interview was the previous day on the PBS program *Newsroom* produced by KERA-TV in Dallas, hosted by Jim Lehrer.

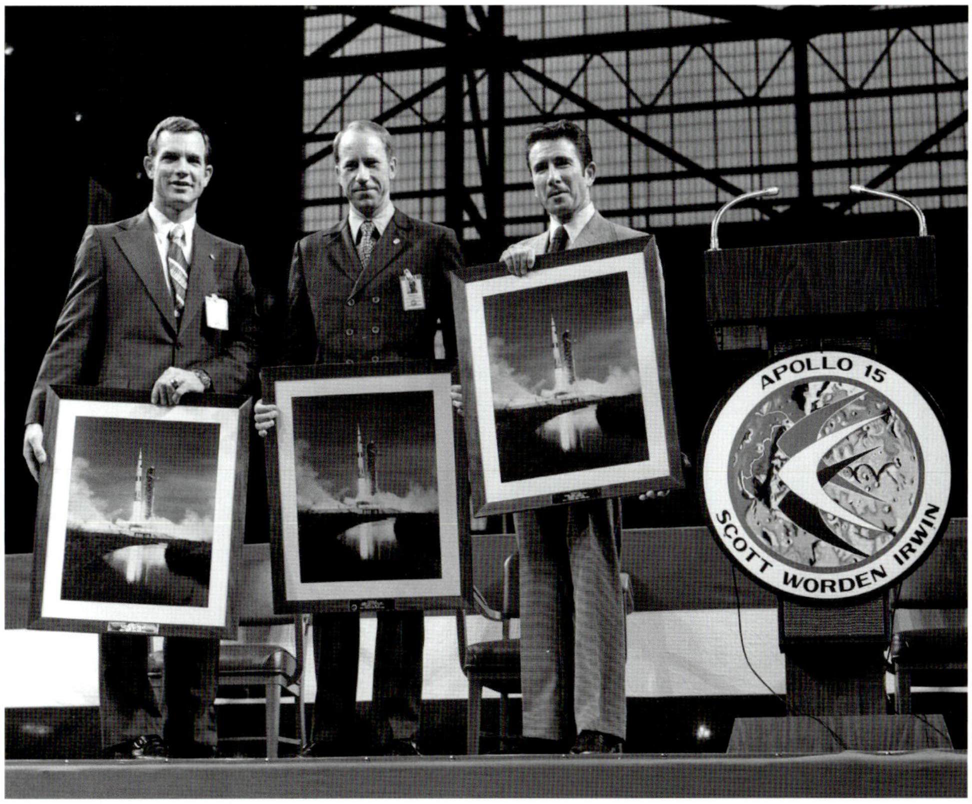

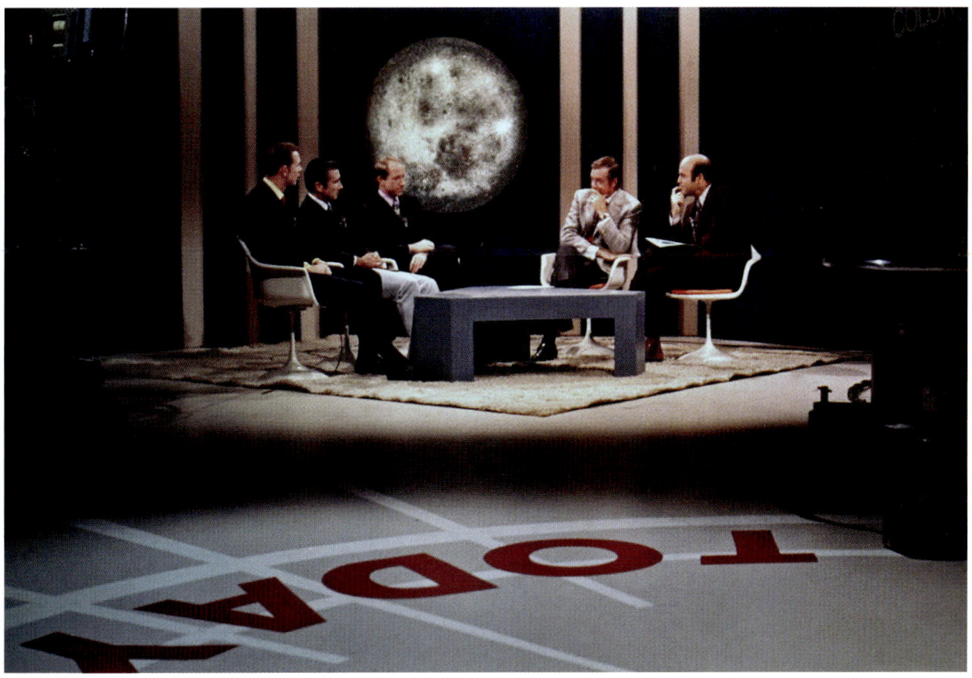

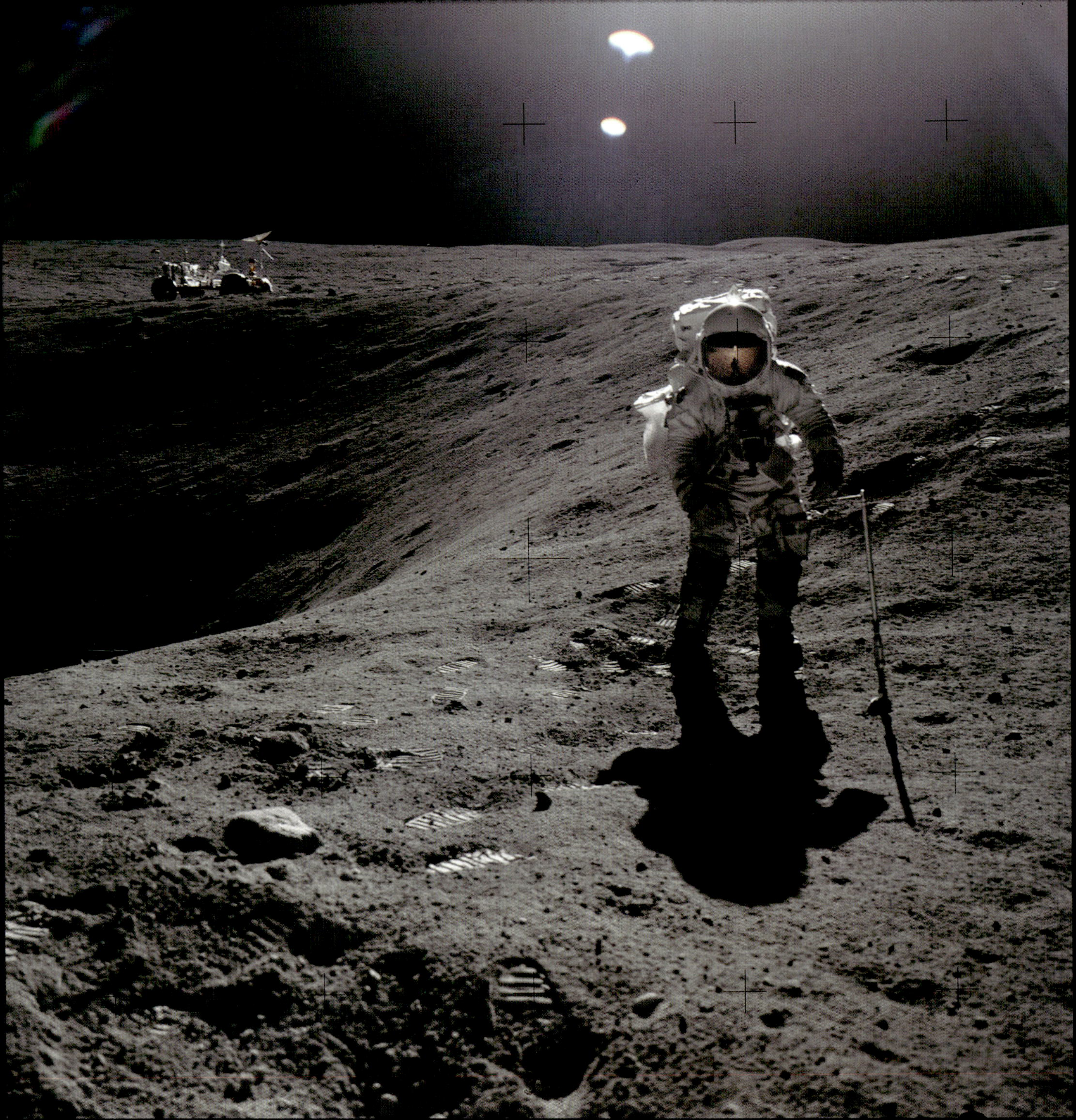

John Young photographs Charles Duke at Station 1 (with Plum Crater at left) during EVA 1 on April 21, 1972. Duke had planted the scoop to provide scale in photos he took

CHAPTER 11

Apollo 16

April 16–27, 1972

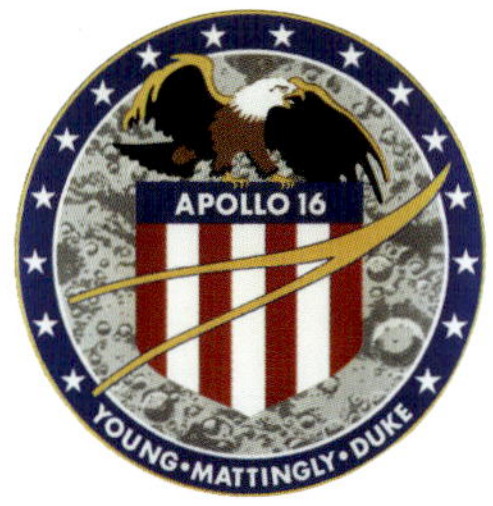

The fifth moon landing mission was the first to explore a lunar highlands area. Apollo 16 also featured the program's second LRV. The mission returned 208 pounds of lunar samples—including a 25-pound rock, the largest of the Apollo program.

Despite a malfunction in the CM that almost aborted the lunar landing, CDR John Young and LMP Charles Duke landed the LM *Orion* successfully in the Descartes Highlands on April 21, 1972. They spent nearly three days on the lunar surface.

CMP Ken Mattingly orbited above, piloting CM *Casper* and supervising the program's second SIM. He also released a subsatellite to study magnetic fields and solar particles. Like Duke, he was a first-time space flyer, as he had been bumped from the Apollo 13 crew after being exposed to measles by Duke in 1970.

Launch was planned for March 17, 1972. Some ten thousand people attended the rollout of Apollo 16 to Pad 39A on December 13, 1971. Mattingly was the only member of the crew present. On January 27, however, the Apollo/Saturn was moved back to the VAB after a problem with a fuel tank supplying the CM's thrusters, the first time a Saturn was returned to its hangar. After repairs, the entire crew attended the second rollout, on February 9, 1972, a rainy day.

Launch was rescheduled for April 16; one million spectators were expected in Brevard County, Florida. (Following Walt Disney World's opening the previous October, KSC tourism was up 40 percent over 1971.) Clear, bright blue skies at launch allowed the best photos and general viewing of any Saturn V launch.

Guests included Vice President Spiro Agnew; President Richard Nixon's daughter Julie and her husband, David, son of former president Dwight D. Eisenhower; and Jordan's King Hussein. Actor John Wayne, who was invited, was a fan of Duke; he sent him an autographed photo inscribed, "From one Duke to another." The photo was presented to Duke when the crew returned to KSC in May.

The outbound leg of the flight was uneventful, with the exception of some thermal paint peeling off *Orion*, later attributed to a poor paint job. Once in lunar orbit, however, on April 20, a problem was discovered in a backup system of *Casper*'s main engine, the SPS. It raised concerns about firing the engine to adjust the spacecraft's lunar orbit, and nearly caused the moon landing to be aborted.

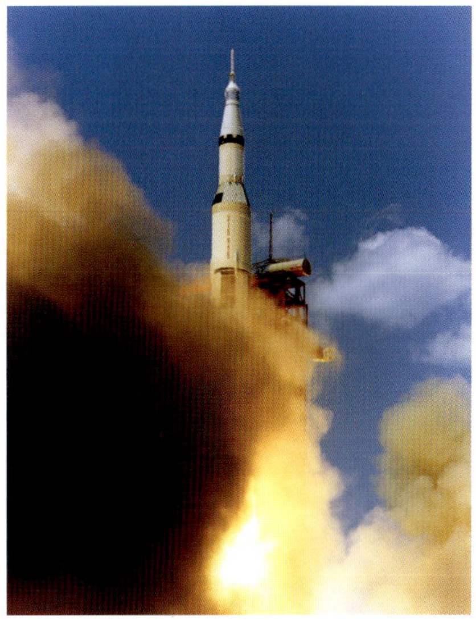

Launch: 12:54 p.m. (EST), April 16, 1972

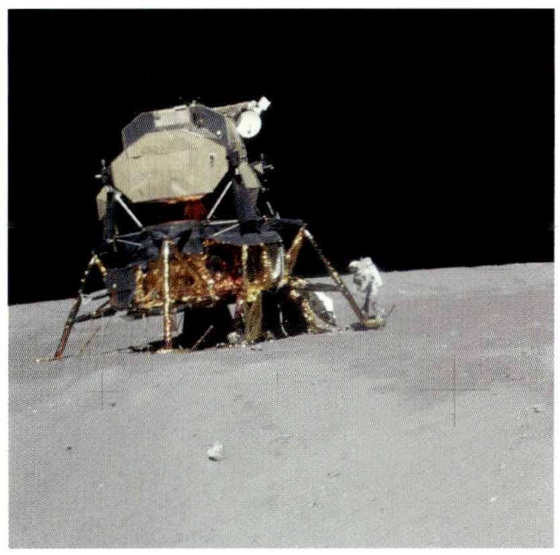

Lunar landing: 9:23 p.m. (EST), April 20.
Ascent: 8:25 p.m. (EST), April 23

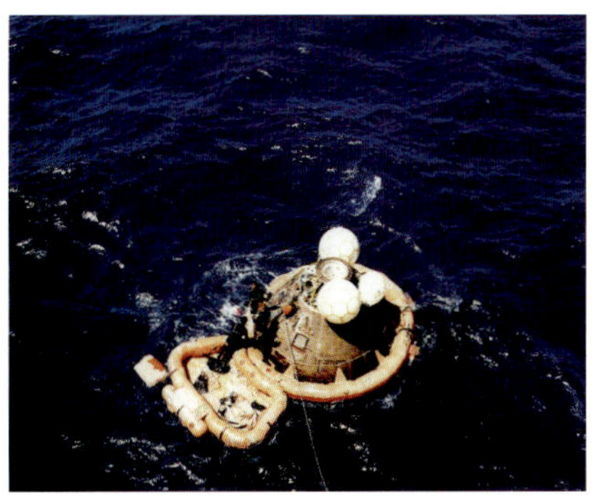

Splashdown: 2:45 p.m. (EST), April 27

After a delayed first landing attempt, the malfunction was determined to present acceptable risk, and Young and Duke (who had already undocked *Orion* when the problem was identified) were cleared to land. During the descent to the lunar surface, Duke received a "helmet full" of orange juice when a valve on his in-suit drink bag began to leak; it was not the last time the fruit drink would become an annoyance during the mission. The landing delay meant there was no live television of Young's first steps on the surface, as the communications network was not in synch.

The two set up their ALSEP. This included an experiment to measure heat flow between two probes, which the men were to insert into holes drilled in the surface. Young accidentally got one foot tangled up in the cable to one of the probes, detaching it and rendering the experiment useless.

Young used a far-UV camera/spectrograph to photograph Earth and the stars—the first astronomical observations from the moon. Dust filtered into one of the Hasselblad cameras on the lunar surface; it is evident in many of the images.

The astronauts also conducted performance tests with the LRV. Young reached a top speed of eleven miles an hour, which stands as the record for any wheeled vehicle on the moon (listed in the *Guinness Book of Records*). Duke commented, "Man, I'll tell ya—Indy's never seen a driver like this."

Duke referred to the LRV's television camera as "the big eye," as Mission Control would often remotely pan and tilt it to watch what the crewmen were doing. After the mission, Young said that Duke was such a talker on the moon that he limited him to "five thousand words a mile" while they were aboard the rover.

In the most famous faux pas transmission of Apollo, Young commented on his indigestion, caused by drinking potassium-enriched orange juice (to prevent an electrolyte deficiency identified in the crew of Apollo 15). He was unaware that his microphone was still transmitting after a conversation with Mission Control. "I mean, I haven't eaten this much citrus fruit in twenty years!" he said. "And I'll tell you one thing, in another twelve fucking days, I ain't never eating any more."

Few good images exist of Mattingly's EVA on the return trip, during which he retrieved film from the SIM in the SM bay. He later said that during his spacewalk, he had such a tight hold on the handrails that his fingerprints were still on them.

Apollo 16 was originally scheduled for splashdown on April 28, but the mission was shortened by a day (by reducing the time in lunar orbit) because of the concerns about the CM's SPS. As Duke recalled, "The more you waited up there—if you did have a problem—the less time you had to think of something brilliant to fix it. They got a little nervous and brought us home a day early, I think, just to make sure we could have some ample time to fix any problems."

After recovering the CM in the central Pacific Ocean on April 27, the aircraft carrier USS *Ticonderoga* delivered the spacecraft to the North Island Naval Air Station in California on May 5. Three days later, equipment being used to empty the CM's residual RCS fuel exploded in a hangar, blowing a hole in the roof and shattering windows. Forty-six people were hospitalized for observation; most suffered from inhalation of toxic fumes. One technician suffered a fractured kneecap when a cart overturned, and the CM had a three-inch gash.

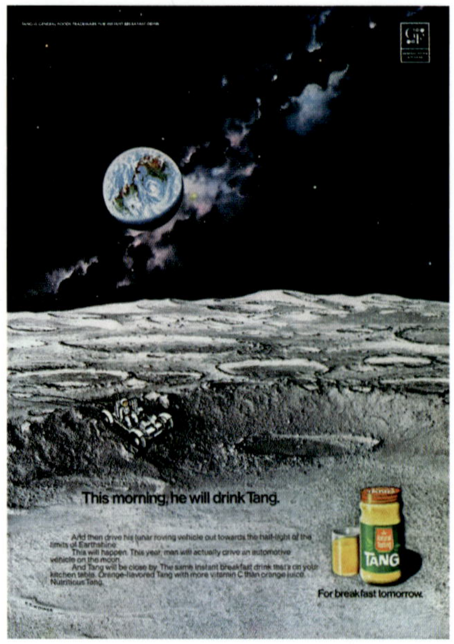

Opposite page: Young is photographed by Duke collecting lunar samples near North Ray Crater during EVA 3 at the Descartes landing site. He uses the lunar surface rake and a set of tongs. The LRV is parked in the field of large boulders in the background.

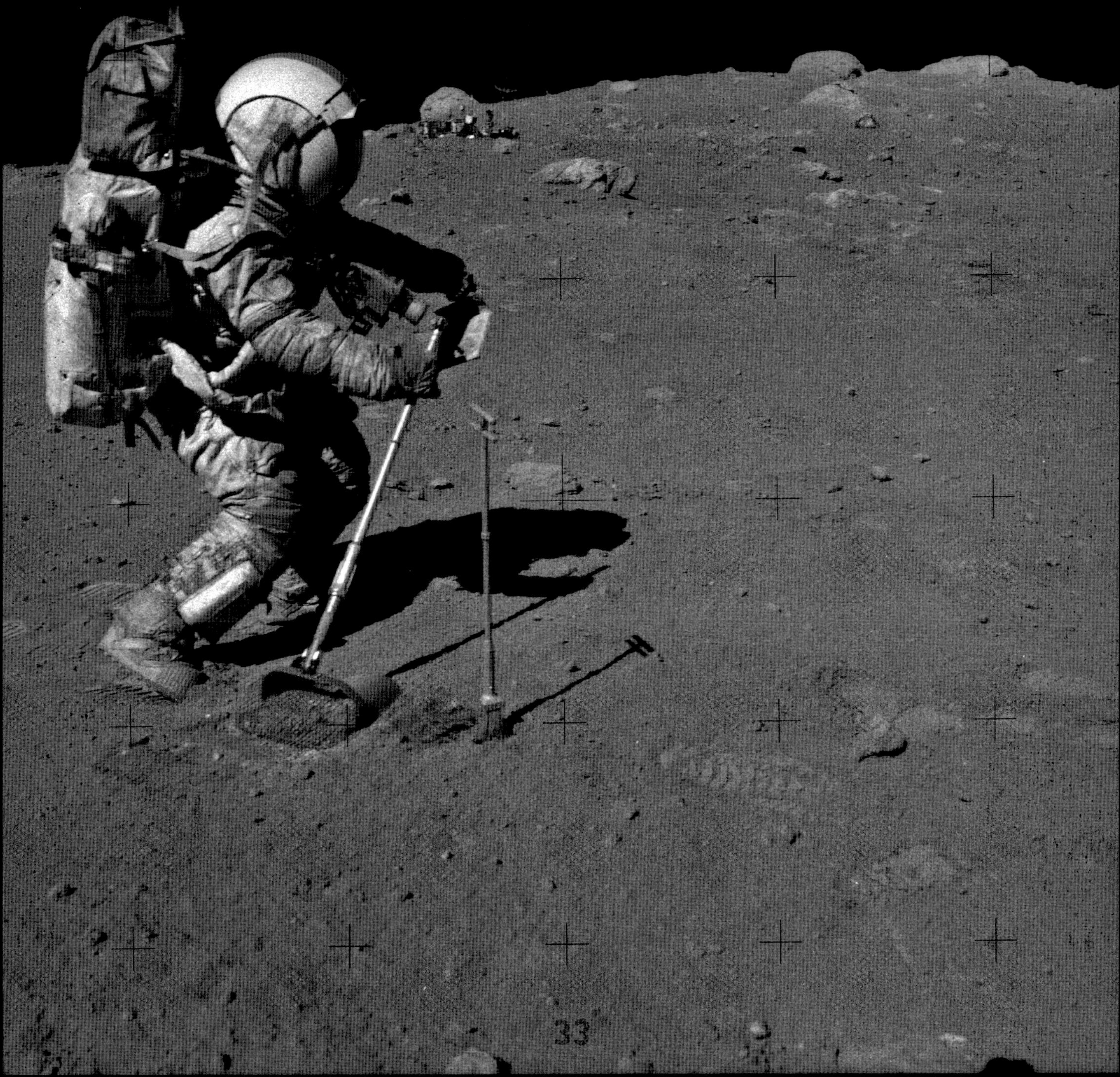

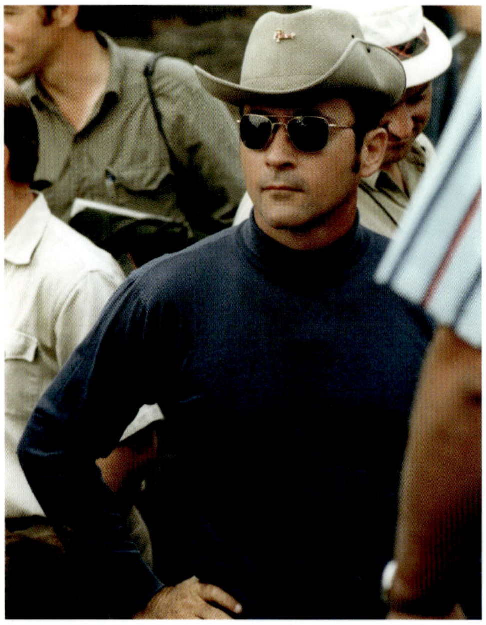

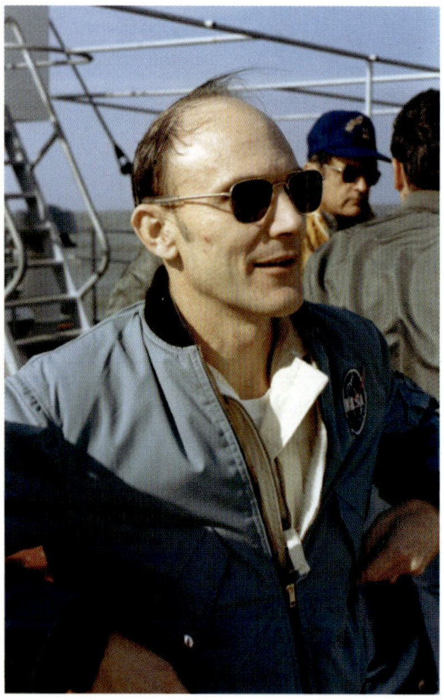

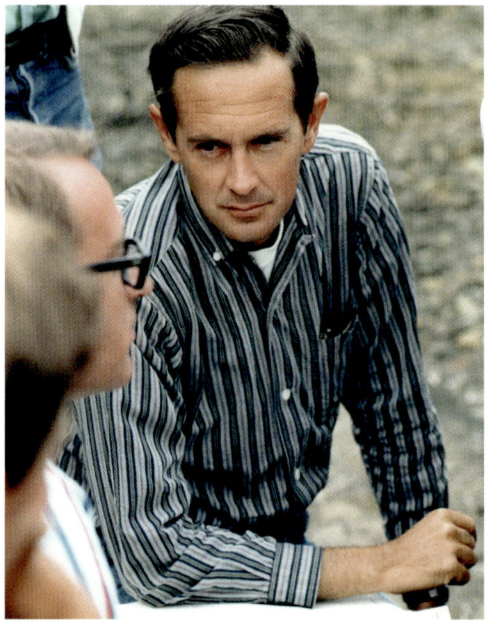

Above: Apollo 16 CDR John Young (*left*) and LMP Charles M. Duke (*right*) during geology training in July 1971, at Bishop Tuff, a volcanic plateau with craters near California's border with Nevada. Young had flown on two Gemini missions and served as CMP on Apollo 10. CMP Thomas K. (Ken) Mattingly (*center*, aboard MV *Retriever* in February 1971) and Duke were rookies. Duke had inadvertently exposed the Apollo 13 crew to the German measles in 1970, leading to Mattingly being bumped to Apollo 16

Below: Young, Mattingly, and Duke at the final preflight crew news conference, in MSC's Bldg. 2 on March 17, 1972, their original launch date. Duke's shamrock name tag references St. Patrick's Day. (photo by Jacques Tiziou)

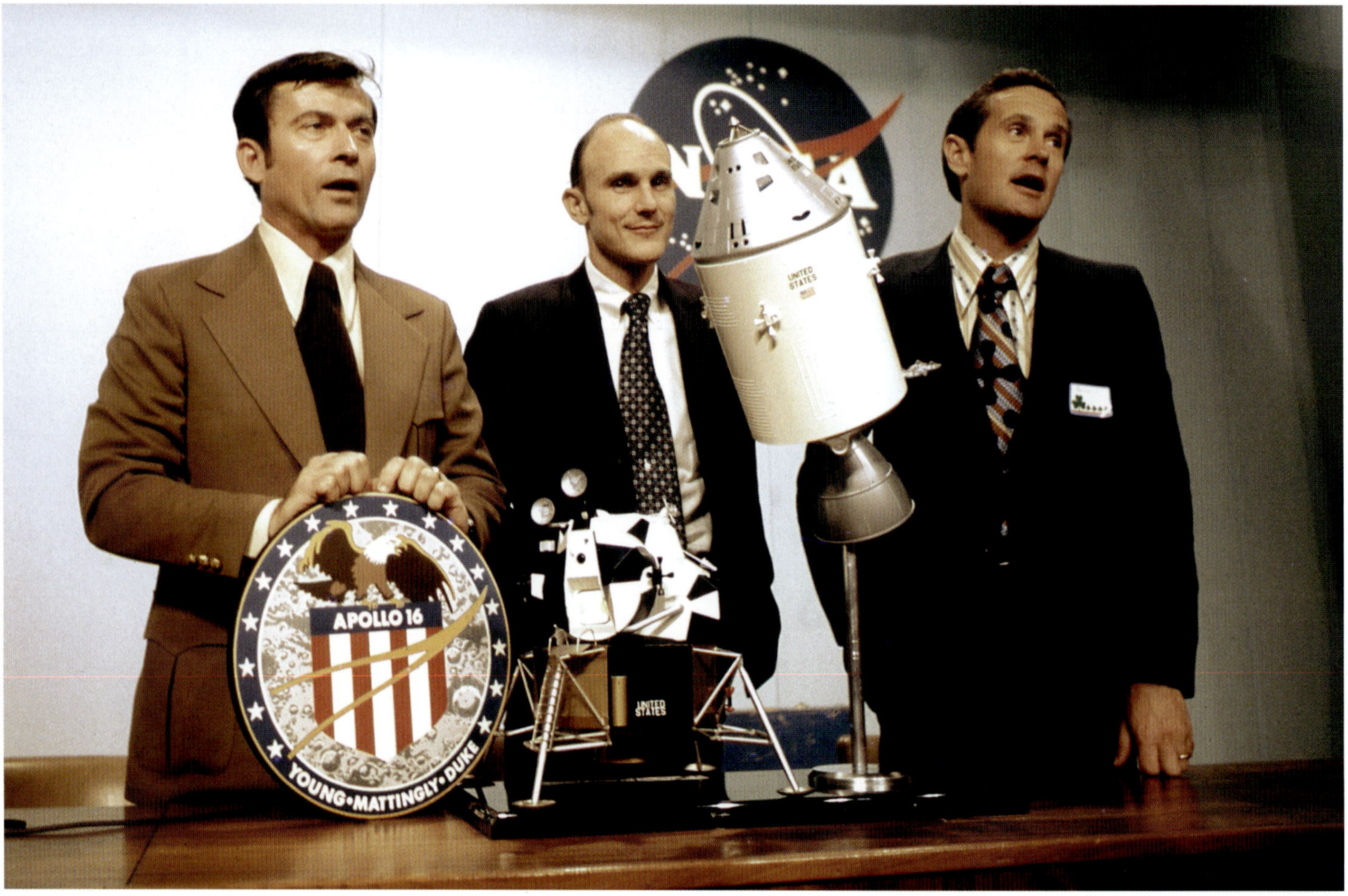

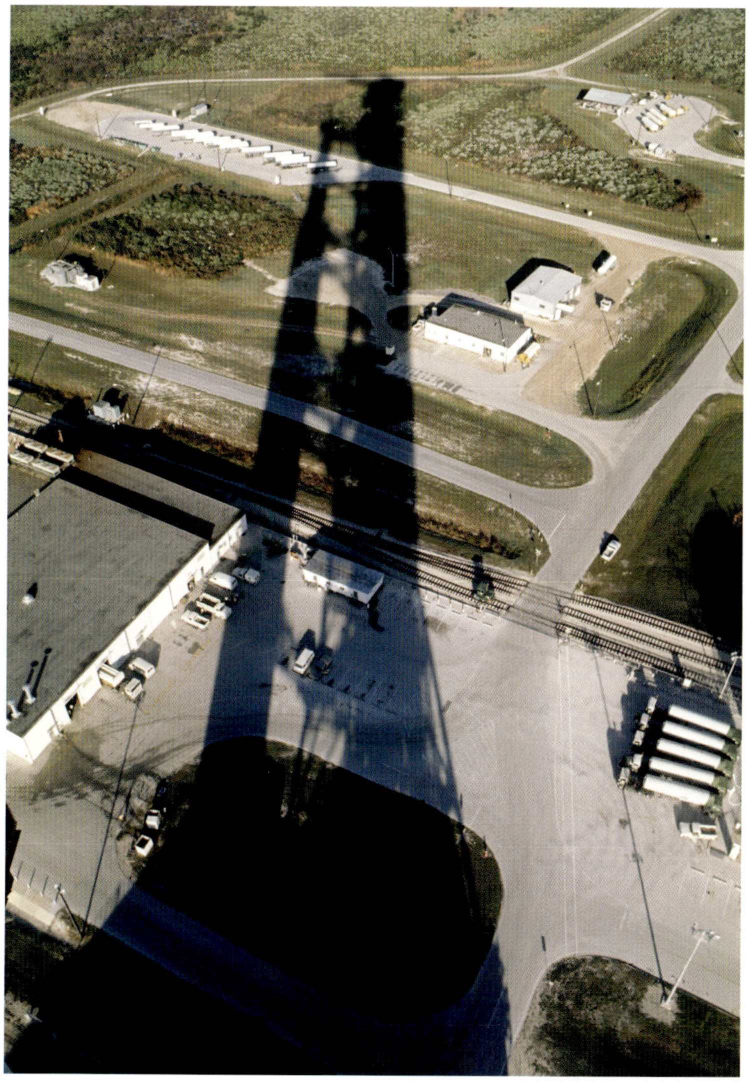

Above: The 363-foot Saturn V and its LUT cast a long morning shadow during the vehicle's first rollout, on December 13, 1971. (photo by Jacques Tiziou)

Below: The aft end of the Saturn V's first stage on ML-3 during its first rollout, on December 13. One hold-down arm is visible at center right; two tail service masts are far left and right.

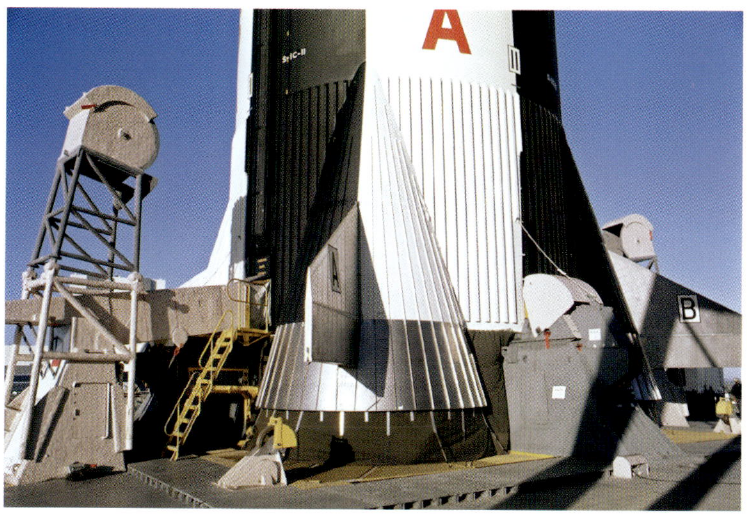

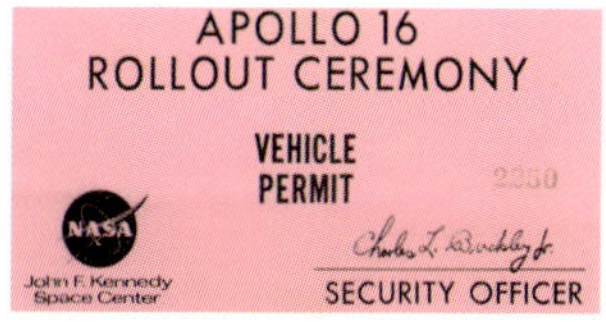

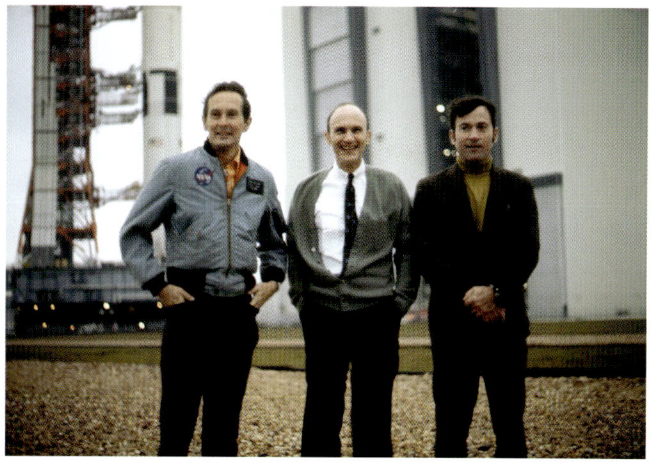

Above: Duke, Mattingly, and Young at the second rollout of their launch vehicle on February 9, 1972—a chilly, overcast day at KSC. Almost two weeks earlier, problems with a thruster system on the CM had forced the Saturn to be returned to the VAB for repairs. (photo by Jacques Tiziou)

Below: A crawler-transporter slowly moves Apollo 16's Saturn V along the crawlerway to the pad, on the morning of December 13. (photo by Jacques Tiziou)

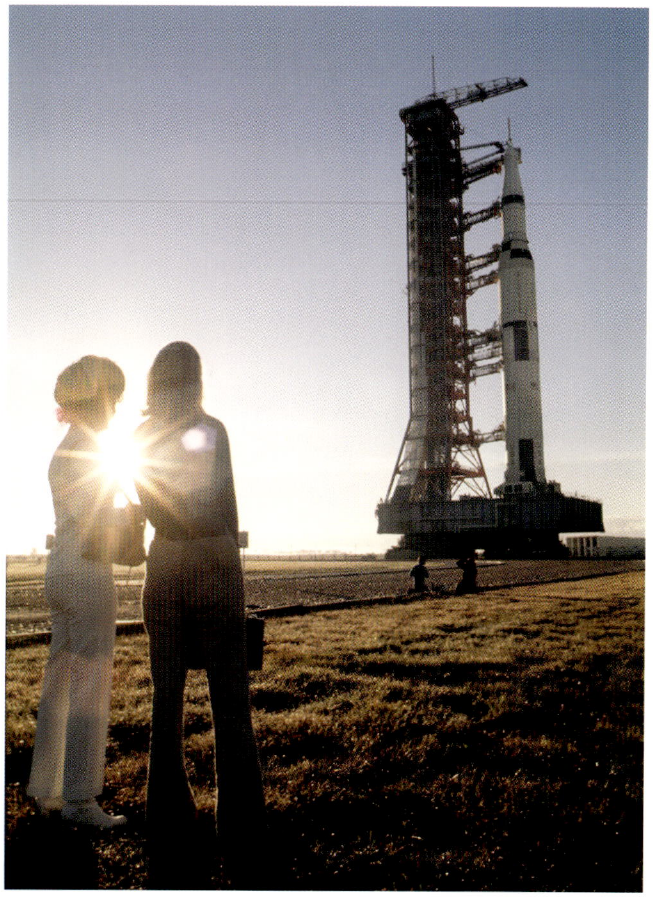

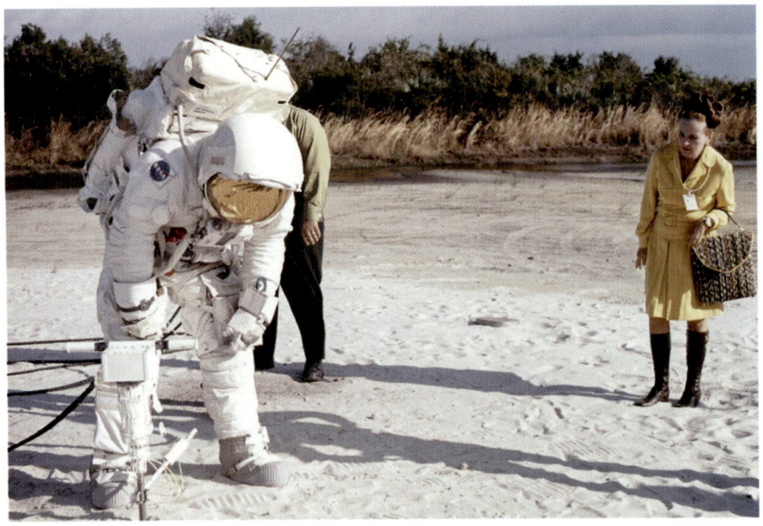

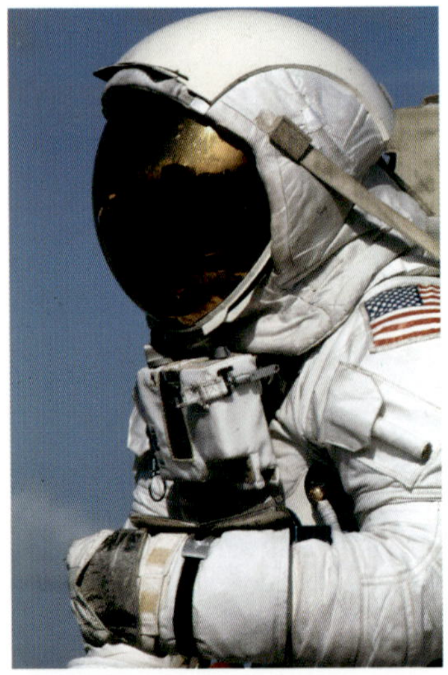

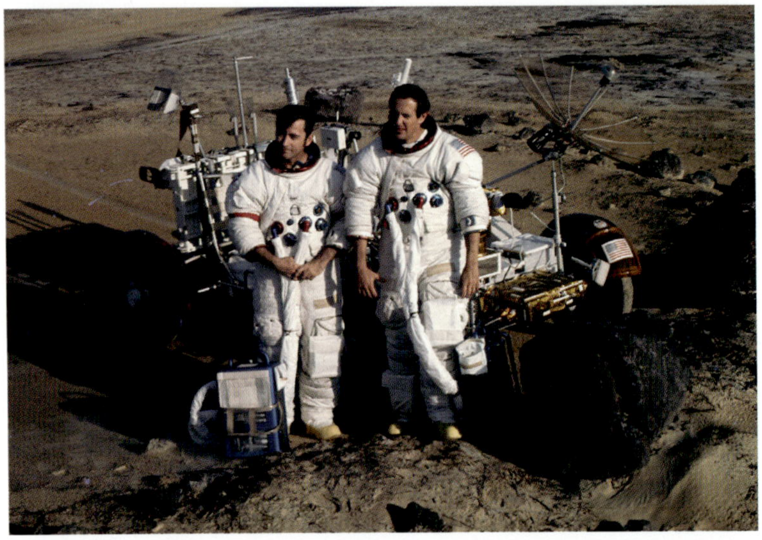

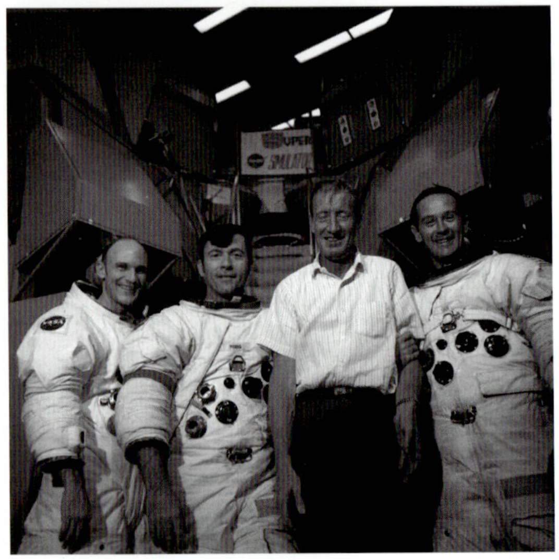

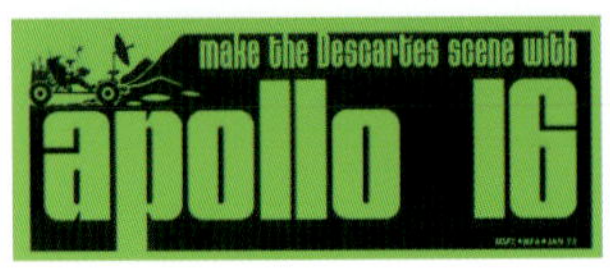

Direct left: MFA bumper sticker

Top left: Fairchild reporter Mary Bubb, holding a cigarette, watches Duke use the lunar surface drill during EVA training at KSC on December 22, 1971. (photo by Jacques Tiziou)

Top right: Duke on December 22, 1971, wearing his LMP checklist on his left wrist. The remote control unit for his PLSS is on his chest. (photo by Jacques Tiziou)

Center left: Young and Duke pose with the LRV trainer at the "rockpile," west of the Flight Crew Training Building at KSC, on March 2, 1972. (photo by Jacques Tiziou)

Center right: Mattingly, Young, and Duke with NASA suit tech Clyde Teague on December 3, 1971. They are in front of the CM simulator in the Flight Crew Training Building.

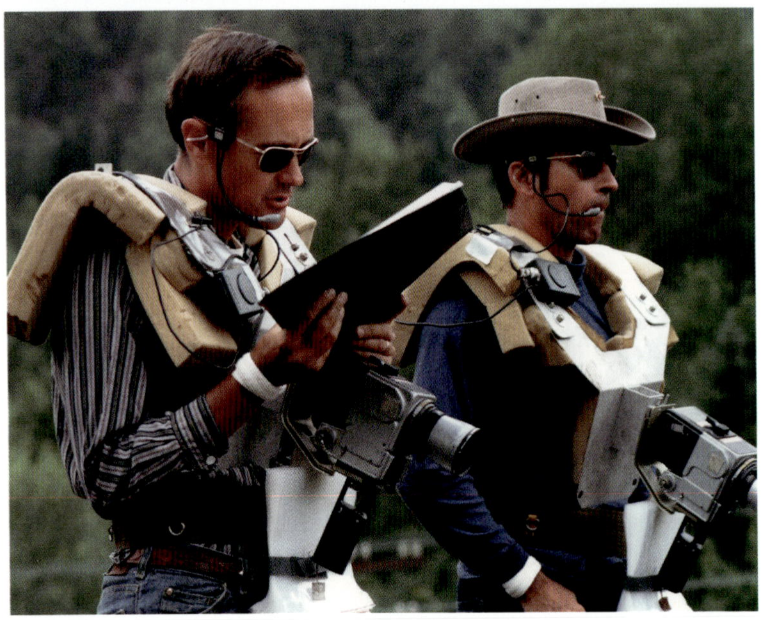

Above: Duke and Young wear yokes with Hasselblad EDC 70mm cameras during training at Bishop Tuff, an igneous formation in east-central California, in July 1971. Because the cameras were worn on the chest during lunar EVAs and had no viewfinders, the crewmen had to practice in order to master their use.

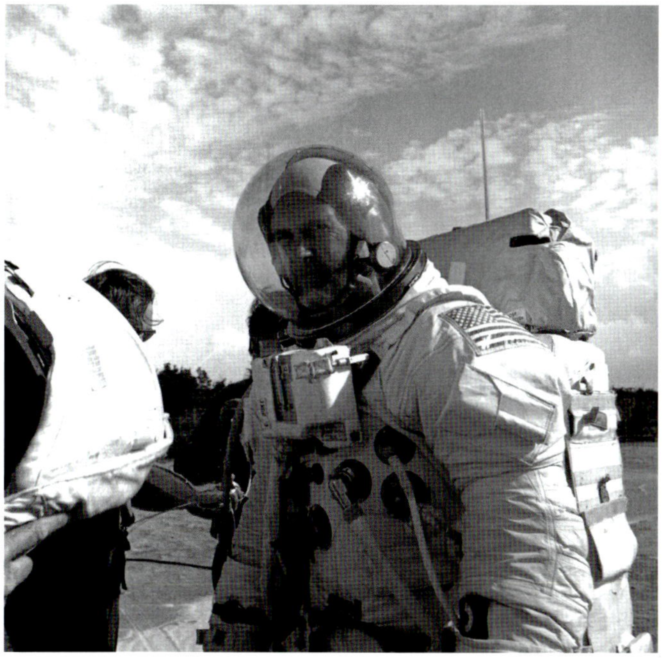

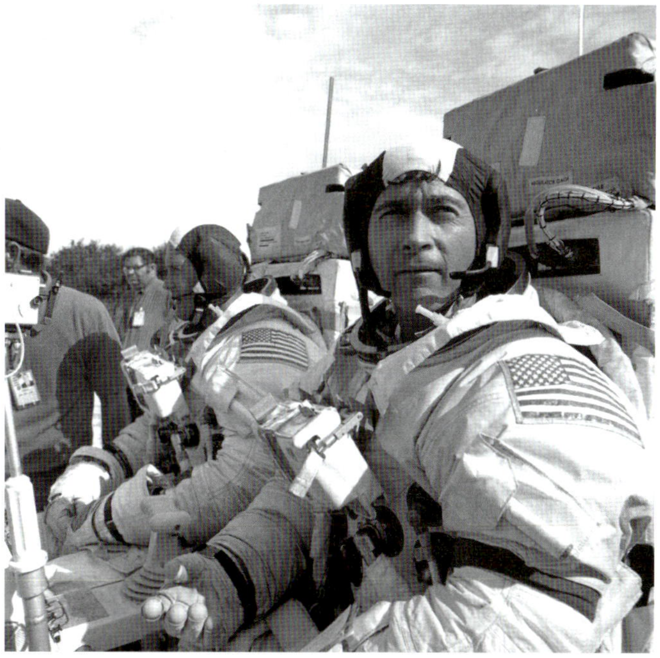

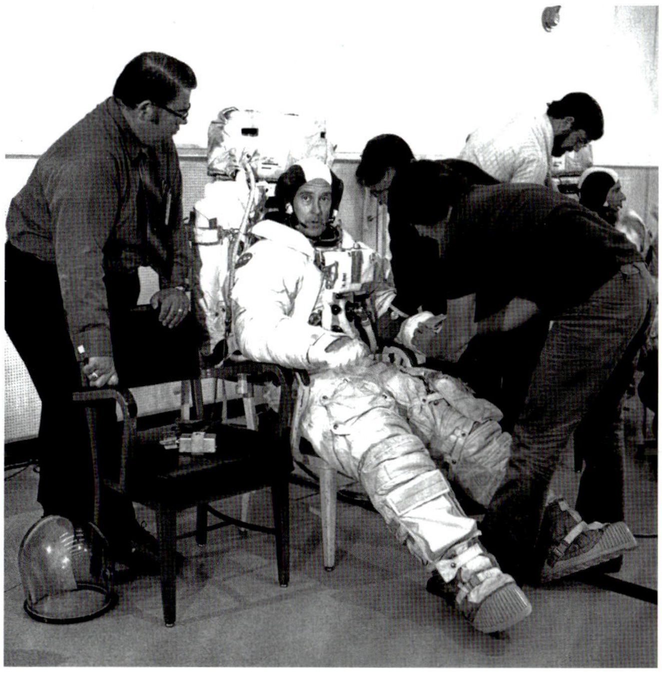

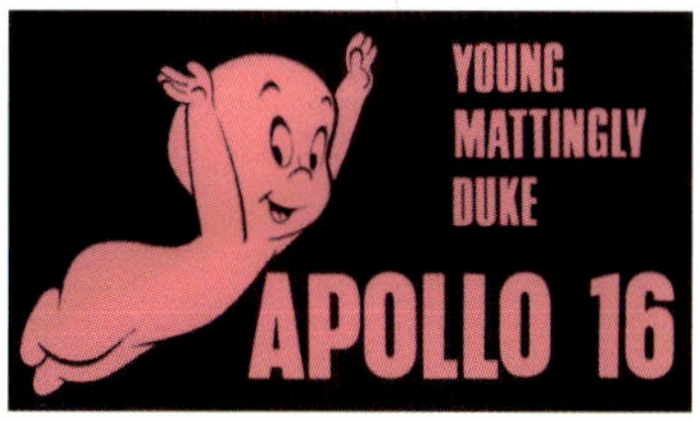

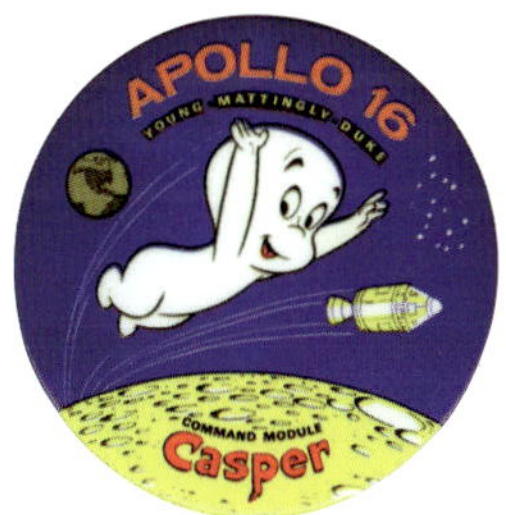

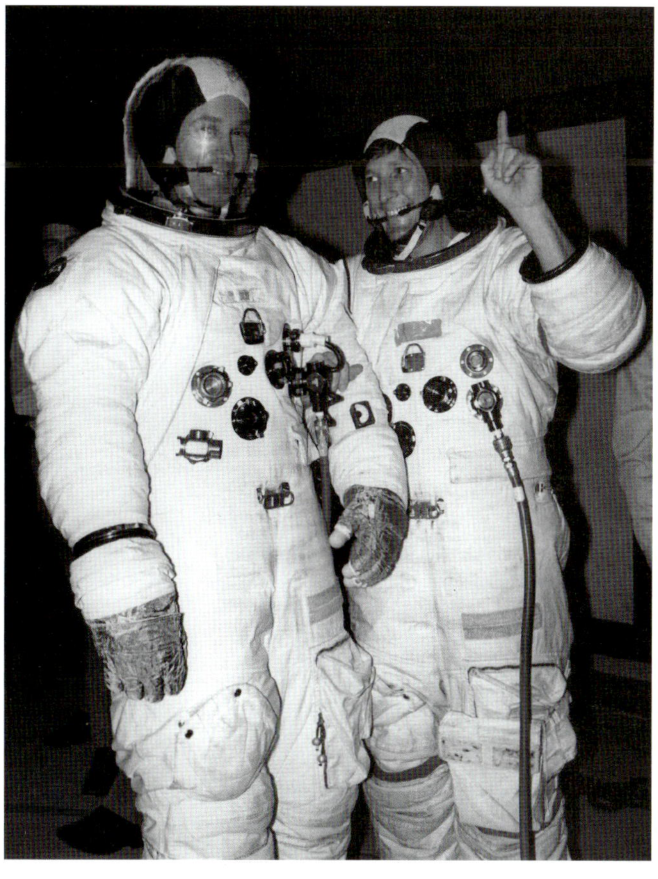

Top left: A steely-eyed Duke in the Florida sun

Top right: Duke (*left*) and Young, outfitted in their EMUs, on their LRV one-gee trainer at KSC on December 22, 1971.

Center right: The Apollo 16 CM's radio call sign was *Casper*—named for the Harvey Comics character—because previous Apollo moonwalkers had appeared ghostly in live television transmissions.

Bottom left: Suit technicians work with Duke (*center*) and Young in the Flight Crew Training Building on December 22, 1971. Duke wears his PLSS backpack. His helmet is on the floor at left.

Bottom right: Duke (*left*) with NASA suit tech Troy Stewart in an undated photo. Stewart would later help suit Space Shuttle crews.

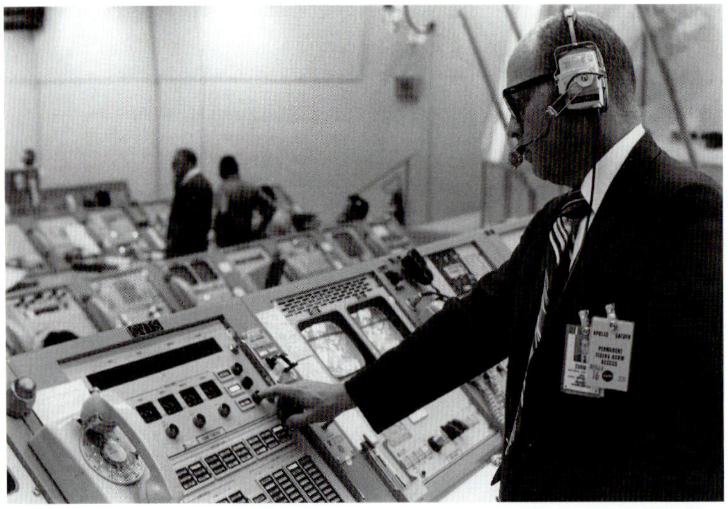

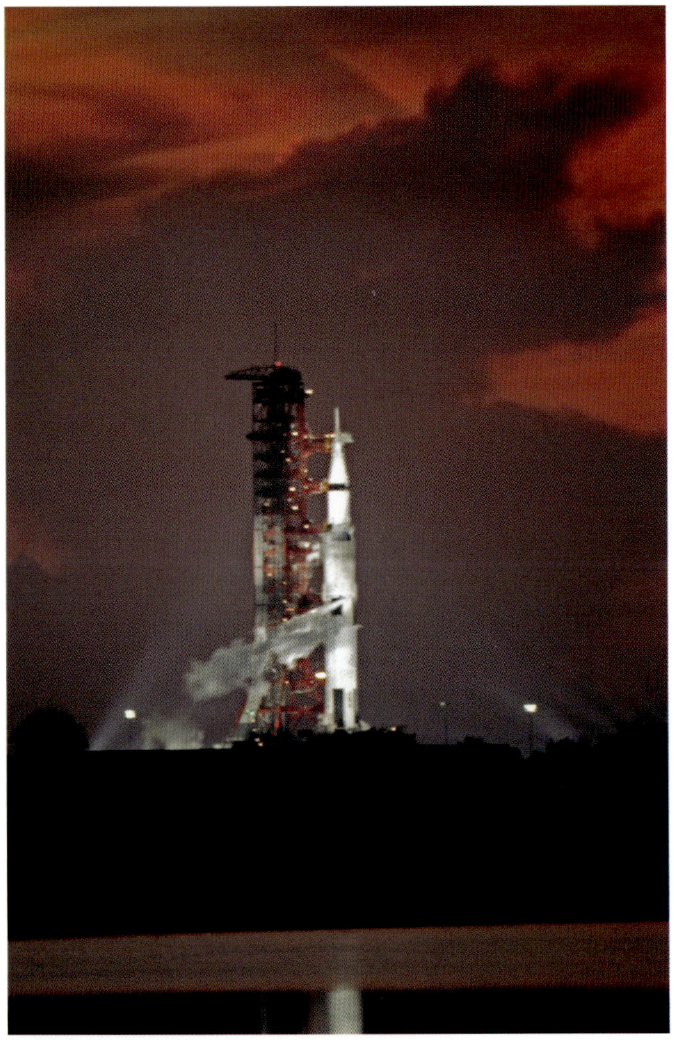

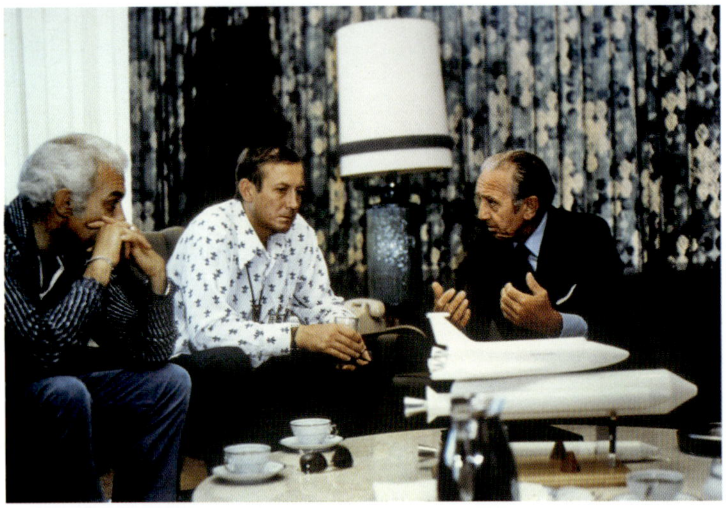

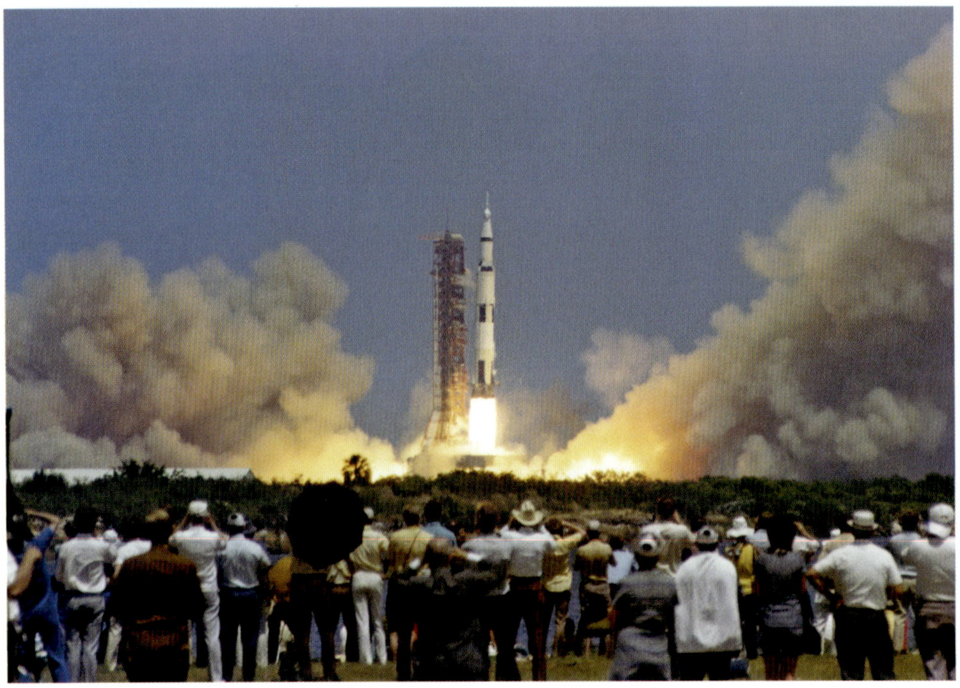

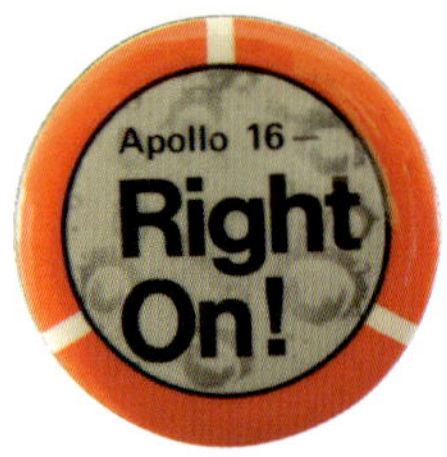

Top left: NASA Test Supervisor Gordon Turner starts the Apollo 16 countdown in Firing Room 1 on April 10, 1972.

Above: Button from IBM, contractor for the Saturn V's Instrument Unit

Top right: Sunrise brings a dramatic sky to Pad 39A on launch morning, as the Saturn's LOX tanks vent. (photo by John de Bry)

Center left: Yevgeny Yevtushenko (*center*), the Soviet Union's foremost poet and the first Russian to attend a manned US launch, meets with KSC Director Kurt Debus at KSC HQ the day before launch. A space shuttle model is in the foreground. The Space Shuttle program had been approved by the White House on January 5.

Above: Apollo 16 as seen from the LC-39 Press Site, seconds after liftoff at 12:54 p.m. (EST) on April 16. Spectators sweltered in 88-degree-Fahrenheit temperatures, with 44 percent humidity—the hottest of any Apollo launch. (photo by Tom Usciak)

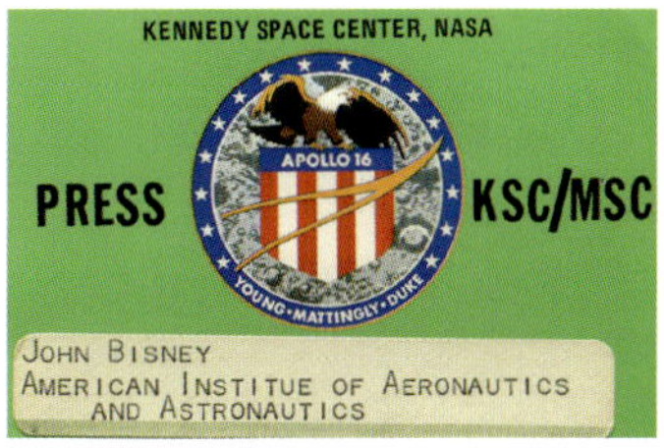

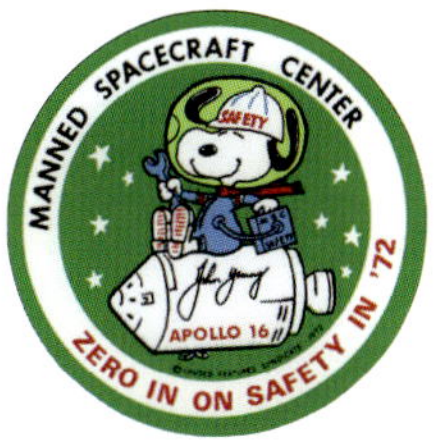

MFA sticker

Top center: Mattingly, thirty-six, was selected as a NASA astronaut in 1966. Like Duke, Mattingly embarked on his first space flight with Apollo 16. He would command the fourth and fifteenth Space Shuttle missions, and he retired from the US Navy as a rear admiral. All photos are from launch morning.

Top right: Duke, thirty-six, had already achieved some level of fame as the Mission Control CAPCOM who acknowledged the Apollo 11 landing with a flustered "Roger, twank . . . *Tranquility*, we copy you on the ground. You got a bunch of guys about to turn blue. We're breathing again. Thanks a lot!"

Center right: Young, forty-one, was NASA's most experienced astronaut. He had piloted the first Gemini mission, commanded Gemini X, and, as CMP on Apollo 10, spent two and a half days in lunar orbit. Director of Flight Crew Operations Deke Slayton stands behind Mattingly (*center*) at left.

Bottom left, left to right: Young, Duke, and Mattingly exit the MSOB on April 16. At the door of the astronaut transfer van is KSC Security Chief Charles Buckley. Between Young and Duke is William D. Killen, a member of the Astronaut Rescue Team. (photo by Jacques Tiziou)

Bottom right: The photo at left was taken moments before by French photojournalist Jacques Tiziou, seen in this photo at far right, standing on a stepladder.

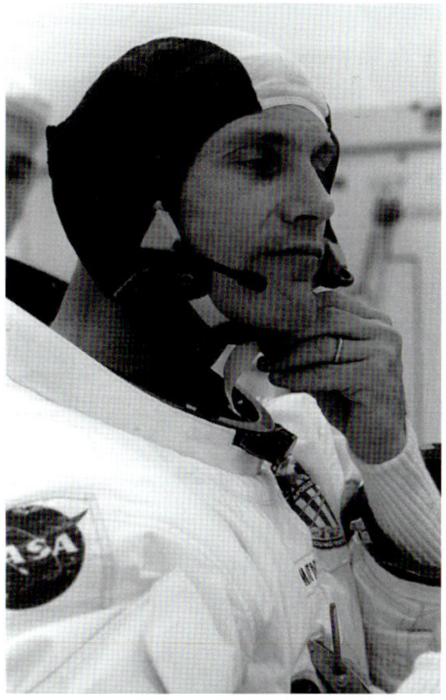

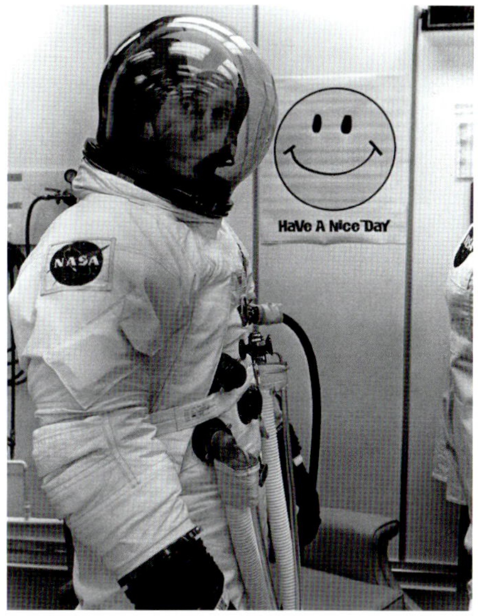

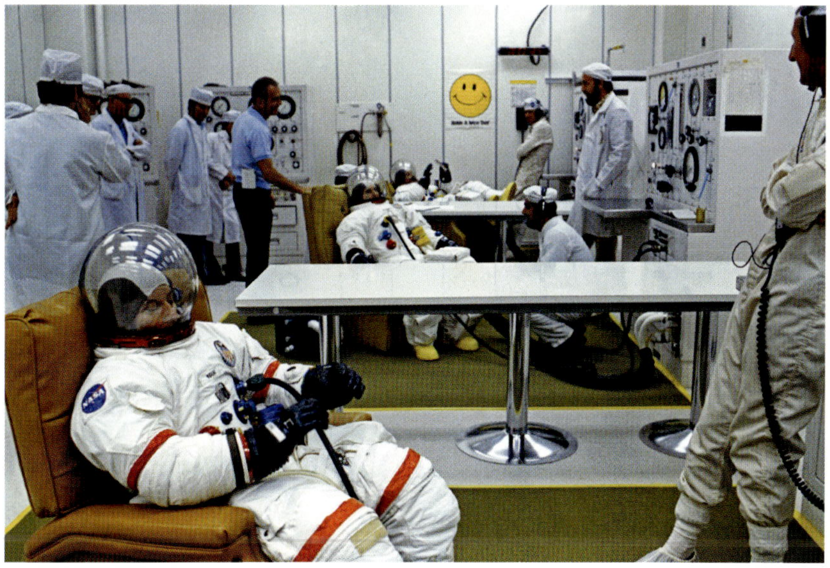

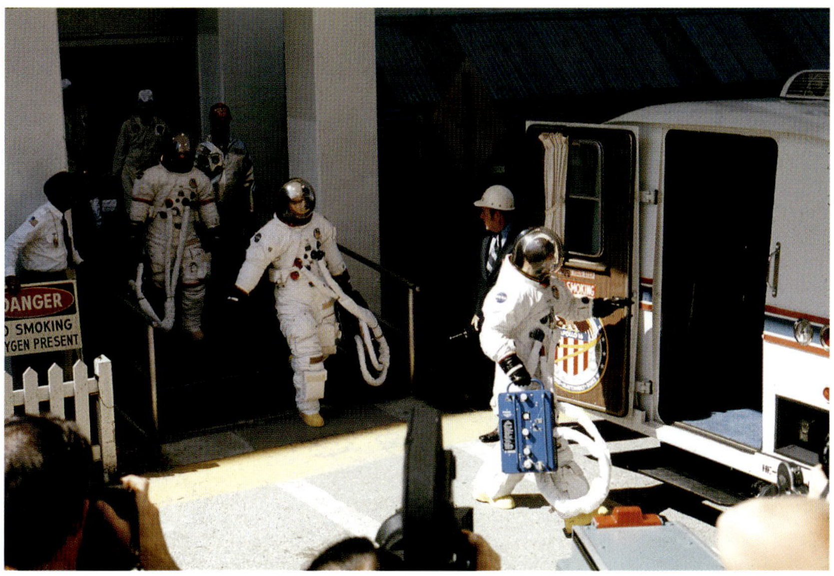

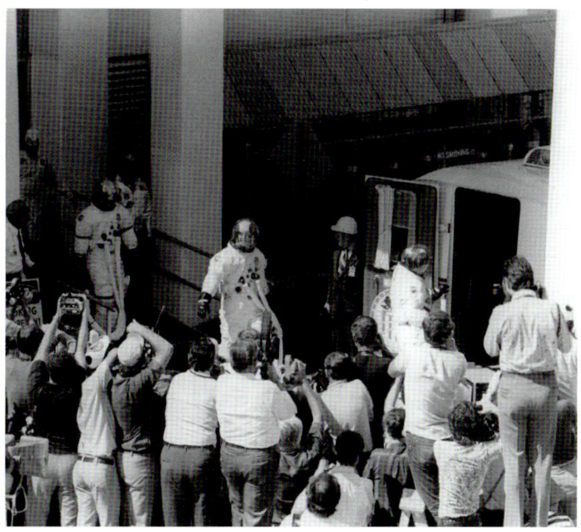

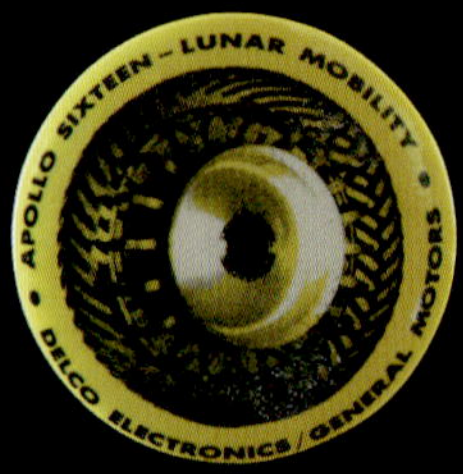

Left: Boeing's major subcontractor for the rover was GM's Delco Electronics Division.

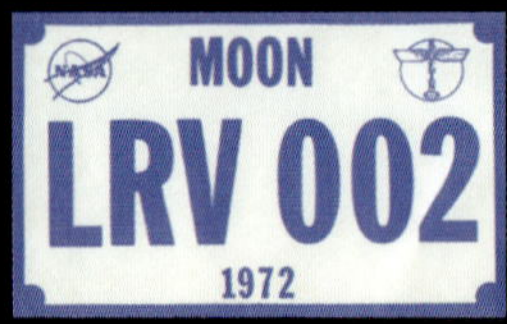

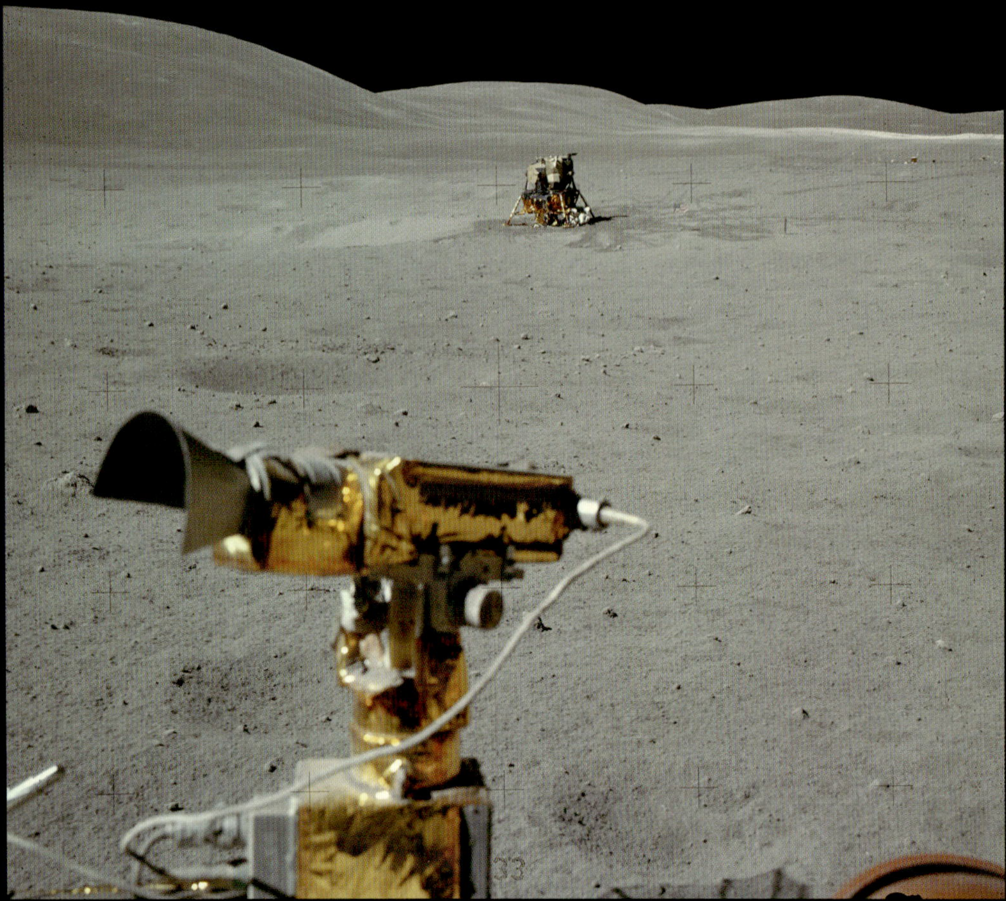

Above: A photo of *Orion* from the rover during the traverse from Station 13 to 10 Prime. In the foreground is the rover's color television camera, which could be operated remotely by Mission Control in two axes: pan and tilt. Photo is from EVA 3 on April 23, 1972.

Bottom left: Young aligns the rover's high-gain antenna to establish communications with Earth. The rake basket is at upper left, and the magnetometer tripod is to the right of it, with a red-orange ribbon cable. Photo is from EVA 3 on April 23, 1972.

Bottom center: The first of two pictures Duke took during EVA 3 to show how dirty Young's suit had become. Young is working near Duke's rover seat. Photo is from EVA 3 on April 23, 1972.

Bottom right: Duke beside *Orion* with the gold UV Camera-Spectrograph, which provided the first astronomical observations made from the moon. The astronauts brought back its exposed film for analysis. Photo is from EVA 3 on April 23, 1972.

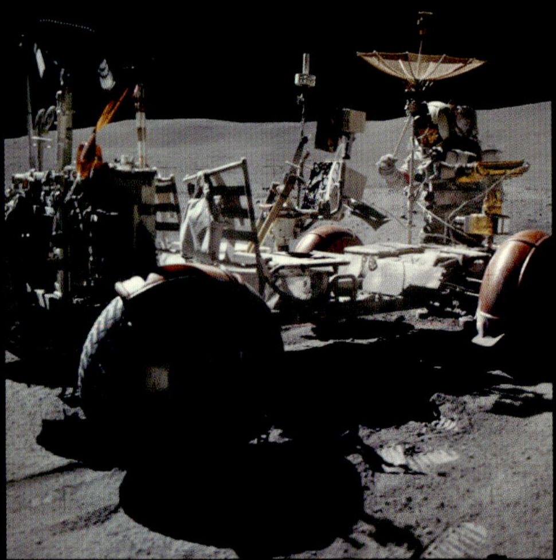

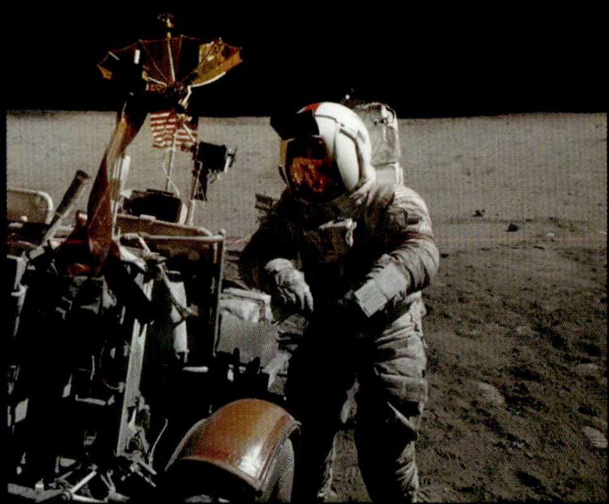

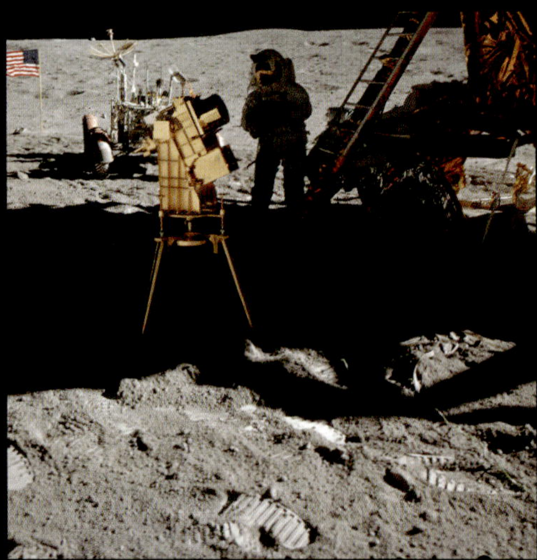

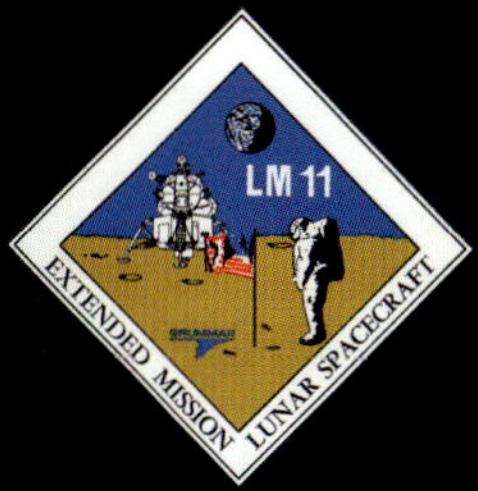

Right: Duke's photo of the ALSEP site, less than one hundred yards southwest of the LM, shows the undulation of the terrain. The PSE (covered in a Mylar shroud in the foreground) was designed to detect moonquakes—either naturally or artificially created—to help study the moon's interior. The seismometers deployed at the Apollo 12, 14, 15, and 16 landing sites radioed data to Earth until September 1977.

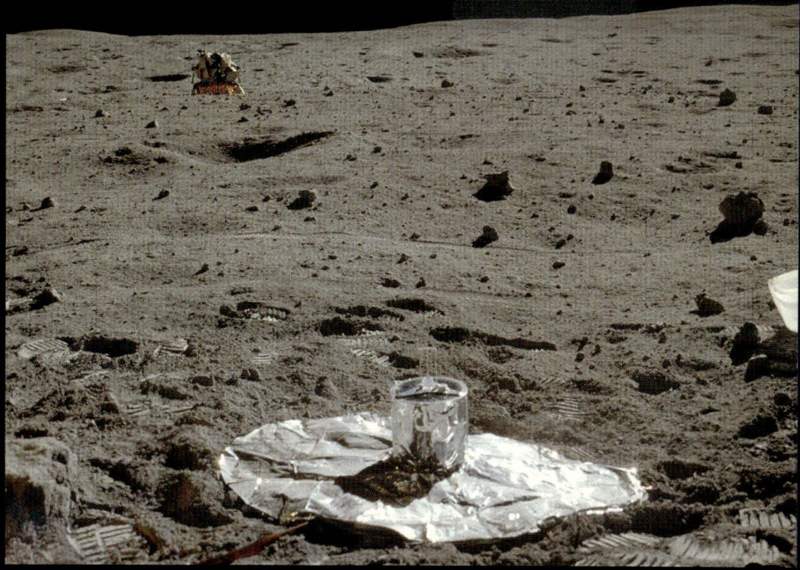

Below: Young works at the LRV during EVA 3 on April 23, with *Orion* at center. The MESA is deployed to the left of the ladder (the horizontal rungs of which are just below the hatch), and the UV camera is set up to the right. Some of the LRV deployment hardware can be seen near the UV camera.

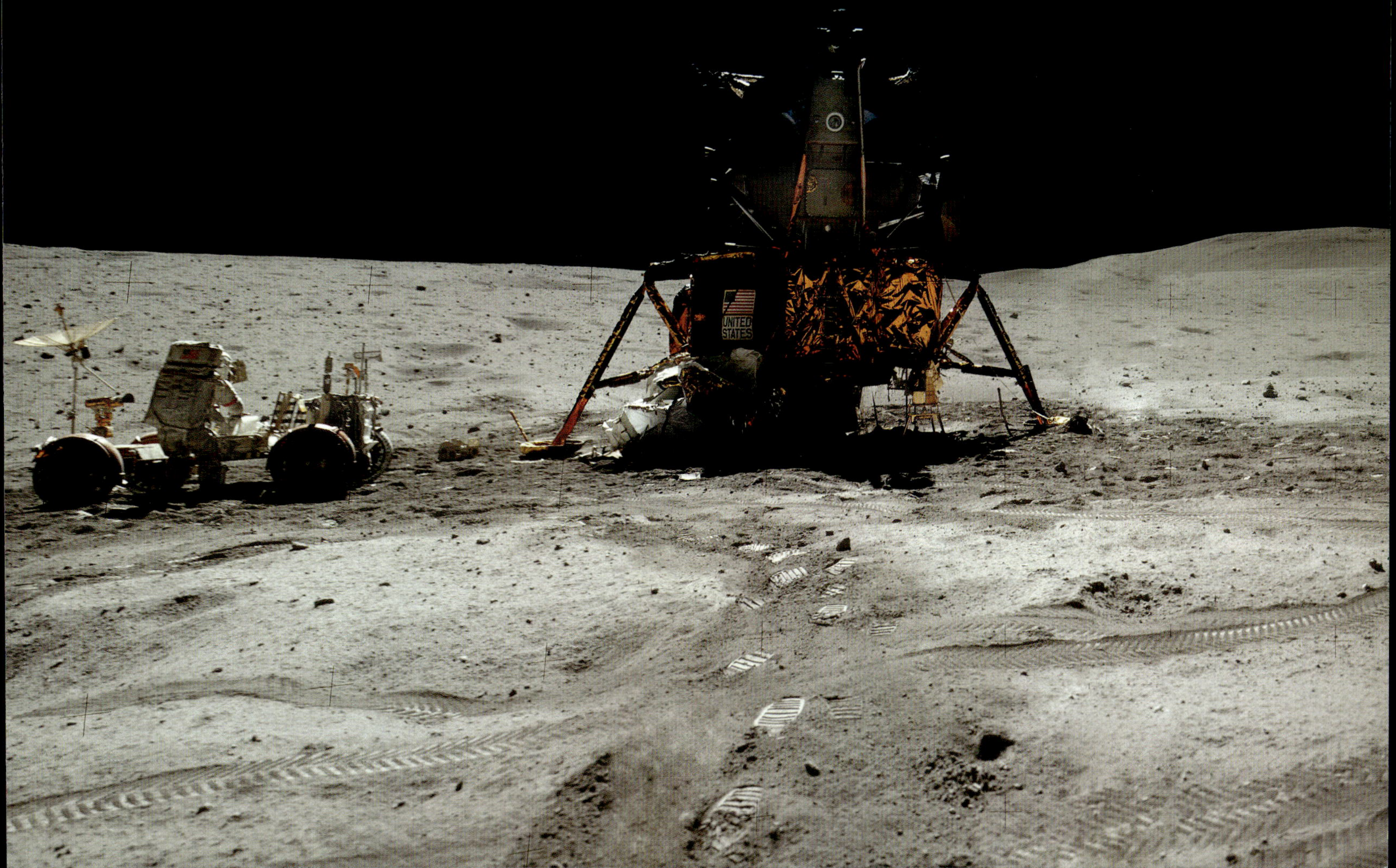

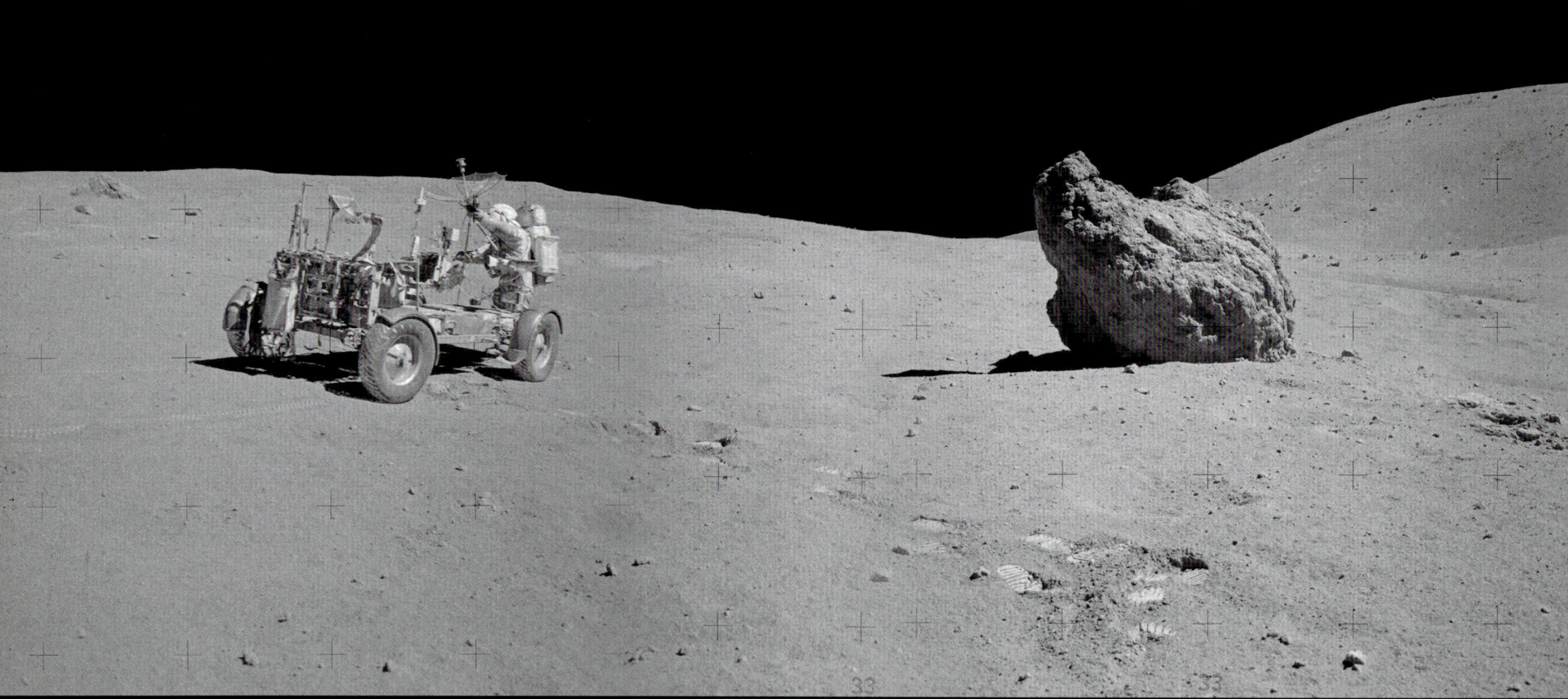

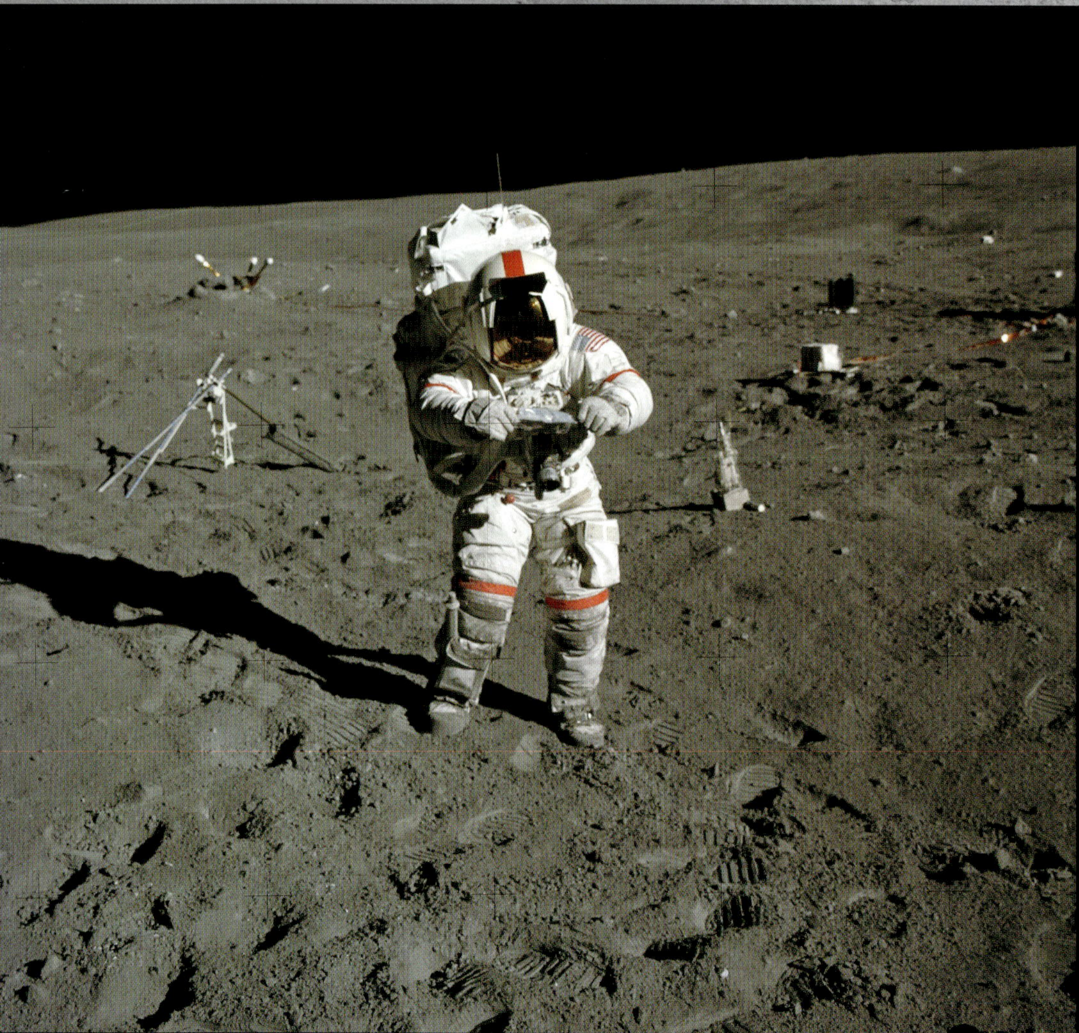

Above: During EVA 3, Young aims the LRV's high-gain antenna to allow live S-band TV from Station 13. Shadow Rock is at right. The rover, on a slight slope in this view looking north, has a large sample-collection bag hanging from the left rear. Duke collected a sample of permanently shadowed soil from under the overhang on Shadow Rock's left side. Station 13, 2.3 miles north of *Orion*, was the last before Station 11, the northernmost stop.

Left: Young bags a sample from a small crater behind the LRV at the ALSEP site, during EVA 1 on April 21, 1972. The lunar surface drill, upside down, is just behind and to the left of Young, with the drill's rack and bore stems at left. The three-sensor lunar portable magnetometer is beyond the rack at upper left. The dark object in the right background is the Radioisotope Thermoelectric Generator (RTG), where Young drilled an eight-foot hole for the heat-flow experiment below the RTG. He then put the RTG out of commission by accidentally catching the gold cable to the central station (*just out of frame to the right*) on his boot and pulling it loose.

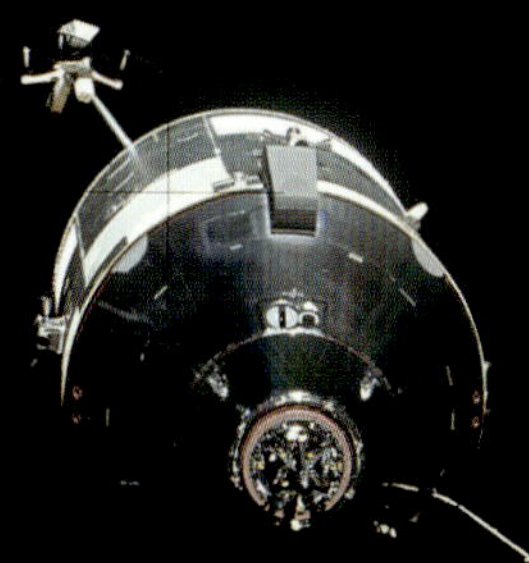

Top left: CM *Casper*, seen from *Orion* during station-keeping on April 20. The steerable antenna at upper left used Apollo's new "Unified S-Band System," built by the Collins Radio Company of Cedar Rapids, Iowa. First tested on Apollo 7, the system combined voice, telemetry, television, and tracking into a single radio signal, to streamline communications at lunar distances.

Right, top to bottom: *Orion*'s ascent stage as it prepares to rendezvous with *Casper*, about two hours after lunar liftoff. Mattingly took this sequence of photos.

Bottom left: At the end of EVA 3 on April 23, Duke is at the front of the LRV, which has been parked near *Orion*. Three gold Mylar–lined dust covers have been raised to expose black radiators, used to cool the rover's two batteries and the heat sinks for its drive control electronics and signal processing unit.

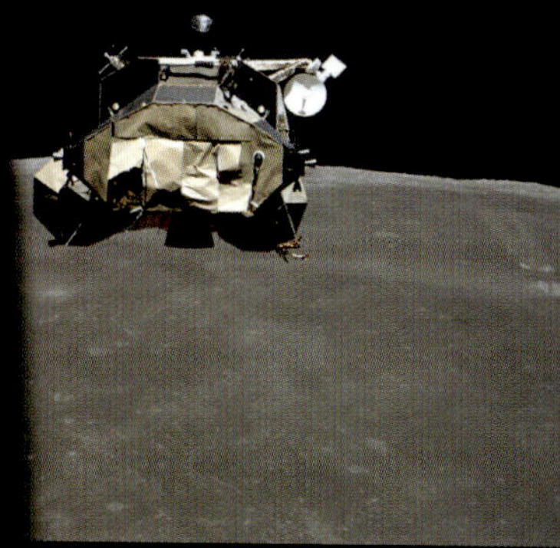

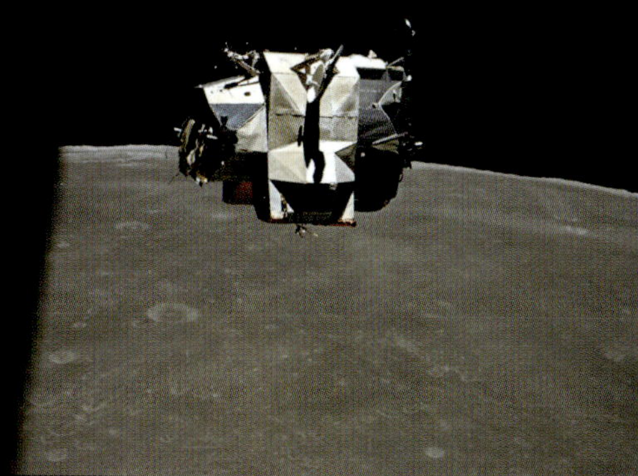

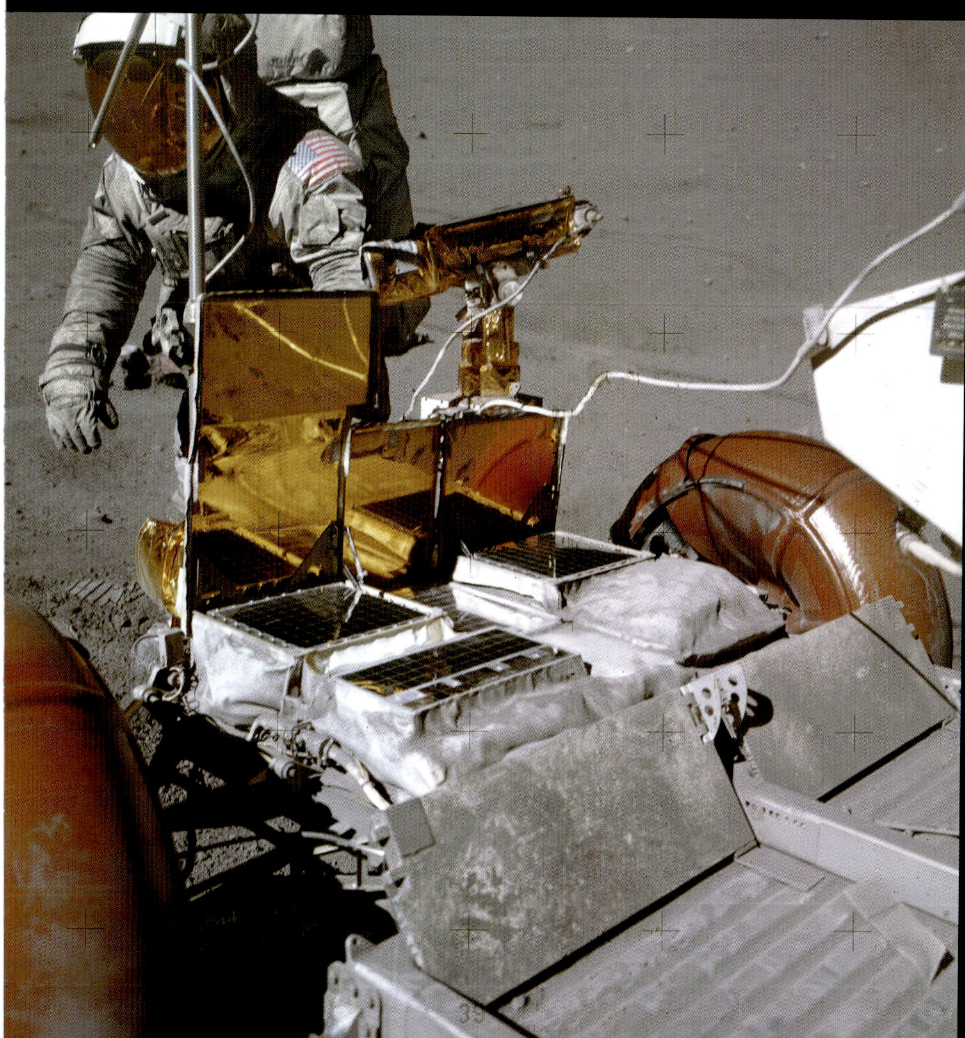

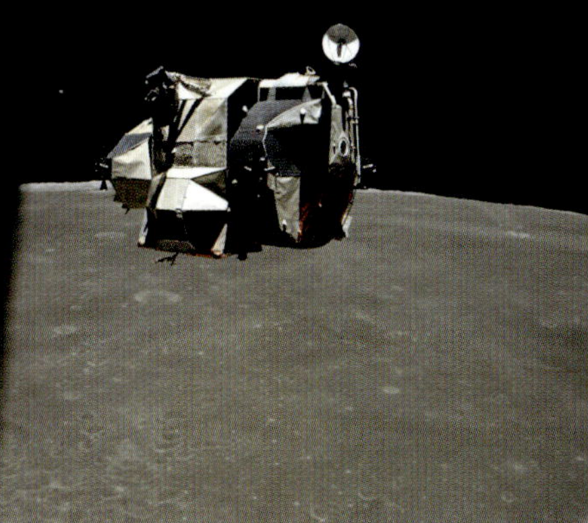

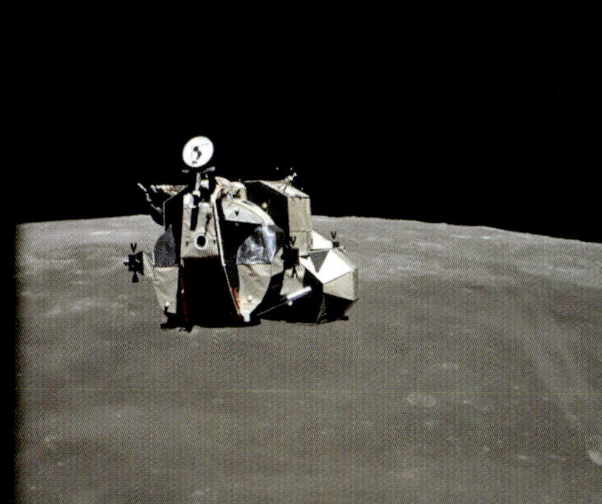

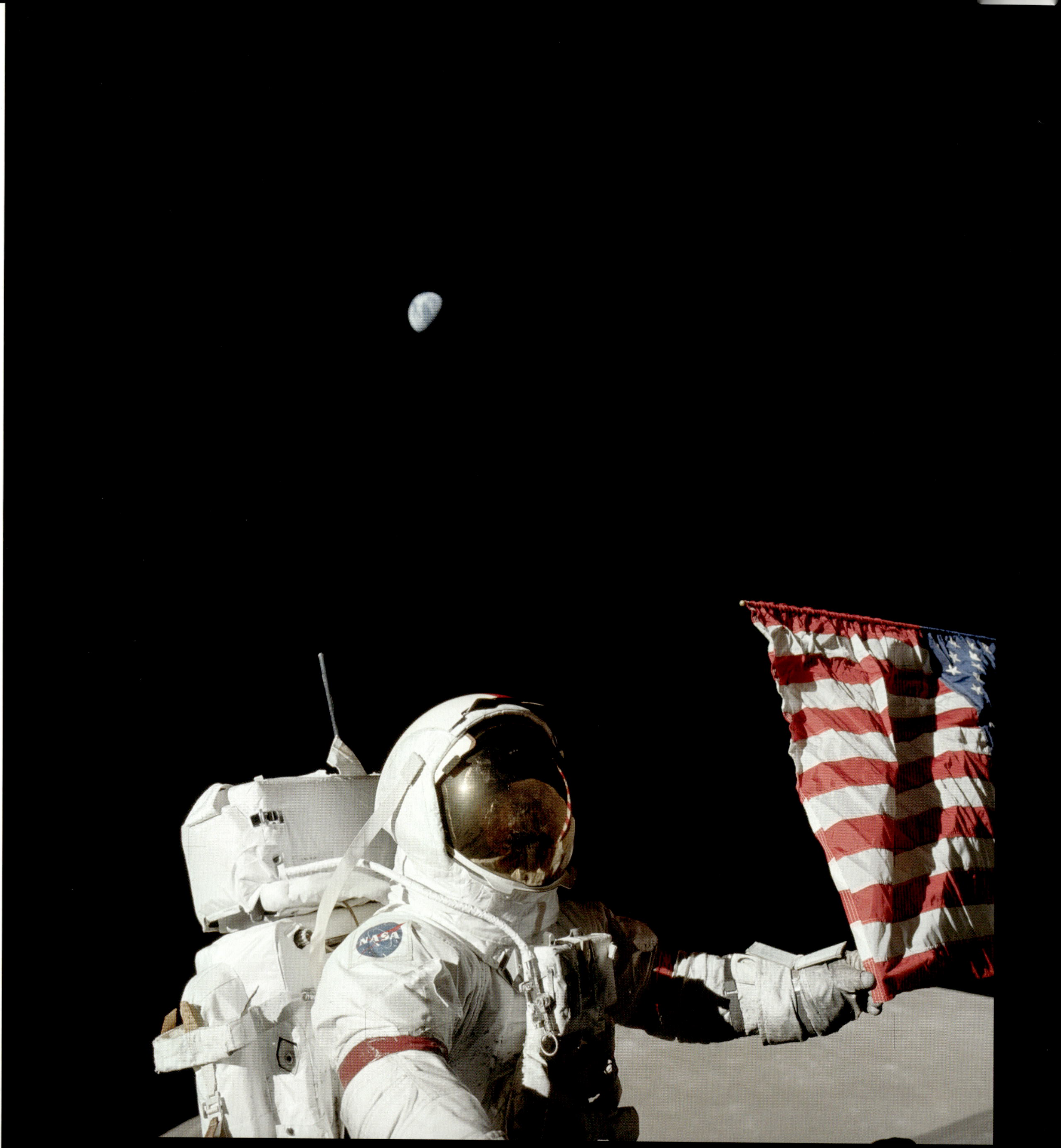

With Earth above him, Gene Cernan holds the American flag on December 11, 1972, during the first of three lunar EVAs. His backpack contains his PLSS topped by an oxygen purge system, where his communication antenna is mounted. His checklist booklet is open on top of his glove.

CHAPTER 12

Apollo 17

December 7–19, 1972

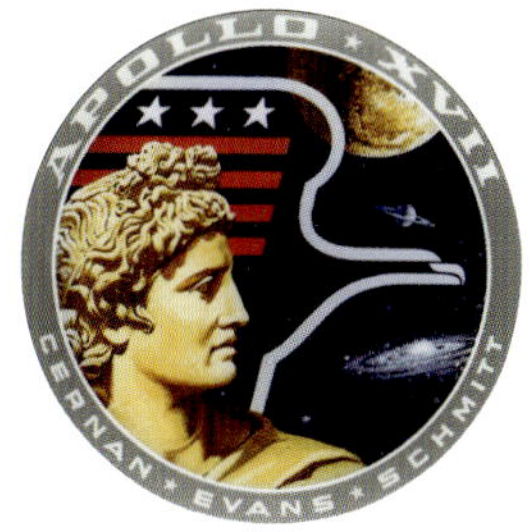

In September 1970, NASA canceled the last two remaining Apollo missions of the three originally planned beyond Apollo 17, making 17 the final lunar landing of the program. The crew was composed of CDR Gene Cernan—making his third space flight and second trip to the moon—and two rookies: CMP Ron Evans and LMP Harrison "Jack" Schmitt, a member of NASA's 1965 group of scientist-astronauts. A geologist, Schmitt likely would have flown on Apollo 18; its cancellation prompted NASA to add him to this crew, bumping former test pilot Joe Engle. Cernan reluctantly accepted the substitution.

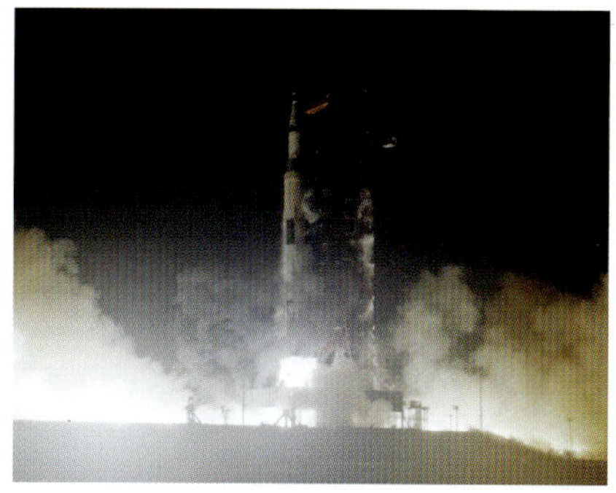

Launch: 12:33 a.m. (EST), December 7, 1972

The alignment of Earth and the moon for a December 1972 launch target meant that Apollo 17 would lift off from Florida at night. NASA issued a fact sheet with advice on how best to photograph the launch. With just thirty seconds to go on the evening of December 6, launch was delayed by two hours and forty minutes because of a glitch in the automatic countdown sequencer. It was the only equipment delay for any manned Apollo launch. At 12:33 a.m. (EST) on December 7, the Saturn V roared to life, briefly but dramatically bathing KSC in an eerie yellow light as the booster climbed into the night sky. An estimated 750,000 spectators were on hand.

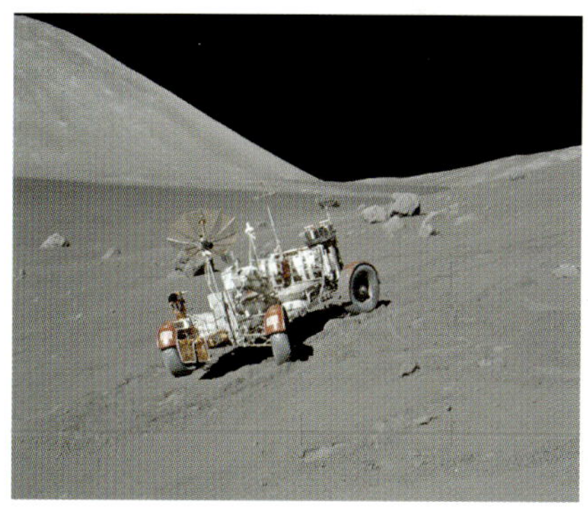

Lunar landing: 2:55 p.m. (EST), December 11. Ascent: 5:55 p.m. (EST), December 14

On December 10, the crew reached lunar orbit. The next day, as Evans remained behind in CM *America*, Cernan and Schmitt landed LM *Challenger* in a valley in the Montes Taurus (mountain range), south of the Littrow Crater in the Sea of Serenity. As an ancient lunar highland, the site was chosen because of its potential to offer clues about the moon's formation.

For the next seventy-five hours, Cernan and Schmitt conducted the longest—and, in many ways, most productive—lunar exploration of the Apollo program. Late on December 11, they set up Apollo's third LRV and conducted the first of three seven-hour EVAs. They also deployed the ALSEP, conducted other surface experiments, and collected samples.

Just before this first outing, Cernan accidentally pulled off the rover's right-rear fender extension with a hammer. He tried to tape it back on, but the duct tape eventually failed because of lunar dust. Before their second EVA the next day, Cernan attached a makeshift fender section, made by taping together four lunar maps. He successfully clamped it to the rover. Schmitt sampled an area of unusual orange soil during the outing; the soil was later believed to have come from an ancient volcanic eruption. They also drove the rover more than 4.5 miles away from the LM, the farthest of any lunar EVA.

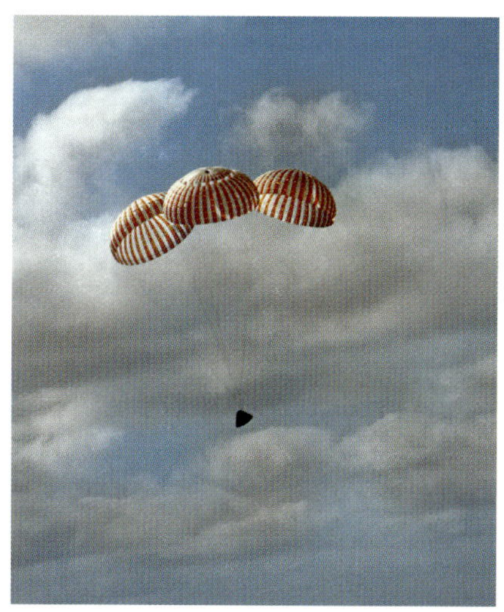

Splashdown: 2:25 p.m. (EST), December 19

The public, however, was not tuning in. The three television networks had cut back coverage of the Apollo 16 moonwalks, airing less than a third of that flight's time spent in EVA. NASA approached the Public Broadcasting Service about carrying all of Apollo 17's lunar surface activities, but PBS declined, citing the transmission and production costs involved. CBS and ABC aired about six hours each, while NBC carried just more than ten.

While Cernan and Schmitt were exploring the lunar surface, Evans conducted various scientific tests aboard *America*, using three new SIM experiments: an infrared radiometer, a UV spectrometer, and a lunar sounder.

During their final expedition on the lunar surface, on December 13, Cernan and Schmitt collected samples of basalts, rocks, and soil; they also unveiled a plaque on the landing gear of *Challenger*, commemorating all of the Apollo landings. They collected a total of 243 pounds of lunar samples during the mission.

Cernan was the last man on the moon. He entered *Challenger* shortly after midnight (EST) on December 14, after carefully positioning the rover so that its TV camera could transmit the liftoff of *Challenger*'s ascent stage to Earth, with Houston controlling it. The camera successfully tilted up to follow the launch for the first time.

After docking with *America*, *Challenger*'s ascent stage was sent back to the lunar surface. Its impact was recorded by the four Apollo 17 geophones and by the ALSEPs at the Apollo 12, 14, 15, and 16 landing sites.

Apollo 17 then spent a full day in lunar orbit for some final experiments, before heading back toward Earth. On December 17, Evans left *America* during a one-hour deep space EVA to recover several film cassettes from the SM; the SM, like that of the previous two missions, carried a SIM.

Apollo 17 splashed down in the Pacific Ocean at 2:25 p.m. (EST), 348 miles west of the islands of Samoa and 4 miles from the recovery ship, USS *Ticonderoga*. A recovery helicopter retrieved Cernan, Evans, and Schmitt, who were safely aboard the aircraft carrier fifty-two minutes after landing.

Below: Ron Evans, Harrison "Jack" Schmitt, and Gene Cernan with an LRV trainer near Pad 39A, during a photo session on August 28, 1972. The top of the Saturn V's LUT can be seen between Evans and Schmitt. (photo by John de Bry)

Opposite page, top left: Cernan, thirty-eight, would be making his third space flight as Apollo 17 CDR, and his second to the moon. (photo by Jacques Tiziou)

Opposite page, top right: When the crew was officially named on August 13, 1971, NASA announced that geologist Schmitt, thirty-seven, would be the flight's LMP, rather than former X-15 test pilot Joe Engle (who, with Cernan and Evans, had been on the backup crew for Apollo 14). (photo by Jacques Tiziou)

Opposite page, center right: Evans, thirty-nine, was an F-8 fighter pilot during the Vietnam War aboard USS *Ticonderoga*—which would be Apollo 17's recovery ship—when he was chosen as a NASA astronaut in 1966. Like Schmitt, Evans had his first and only space flight on Apollo 17. (photo by Jacques Tiziou)

Opposite page, bottom: The Apollo 17 prime crewmen pose for photographs following their preflight news conference at MSC's Bldg. 2 on November 10, 1972. Space artist Robert McCall designed the mission emblem. (photo by Jacques Tiziou)

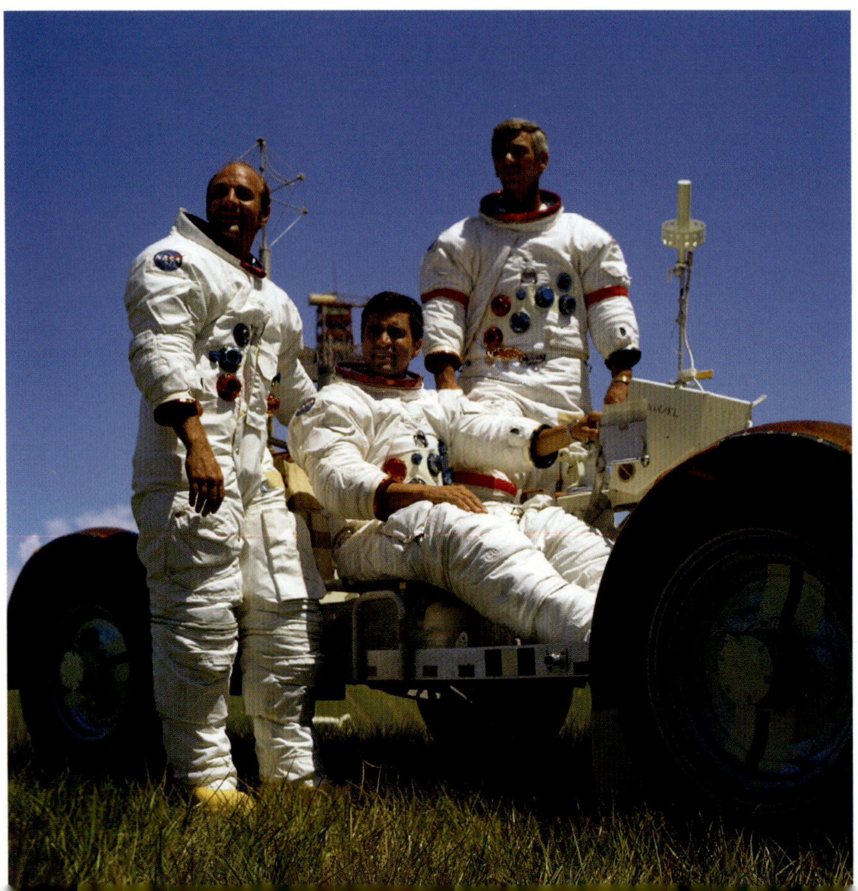

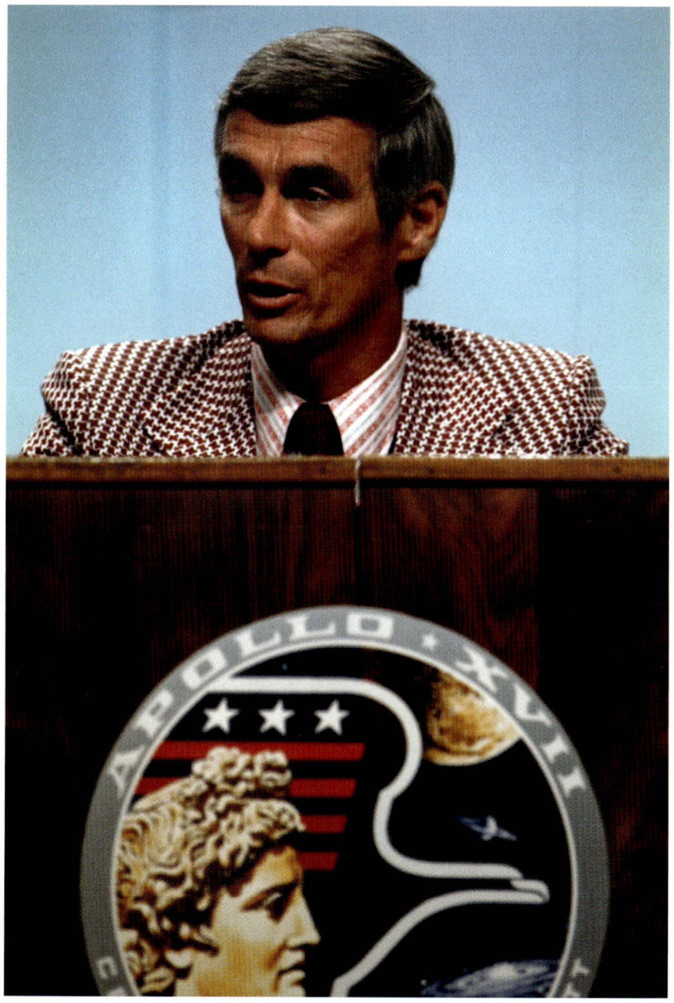
APOLLO · XVII

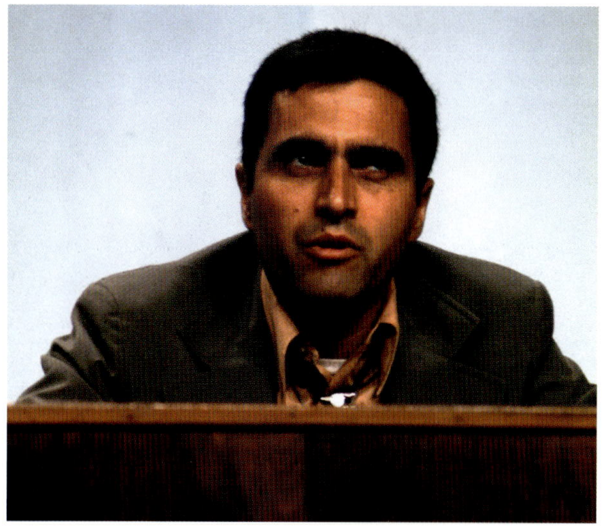

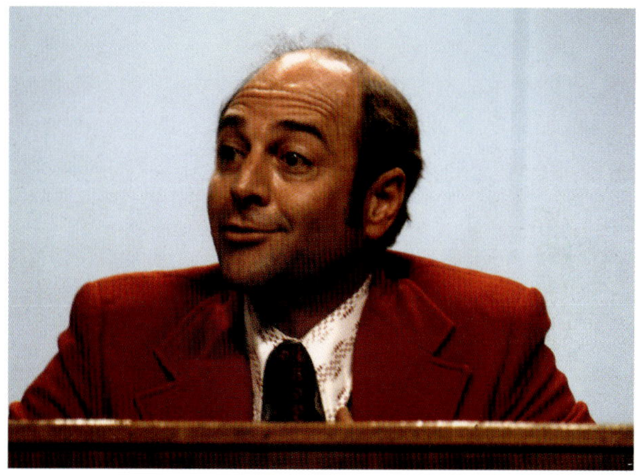

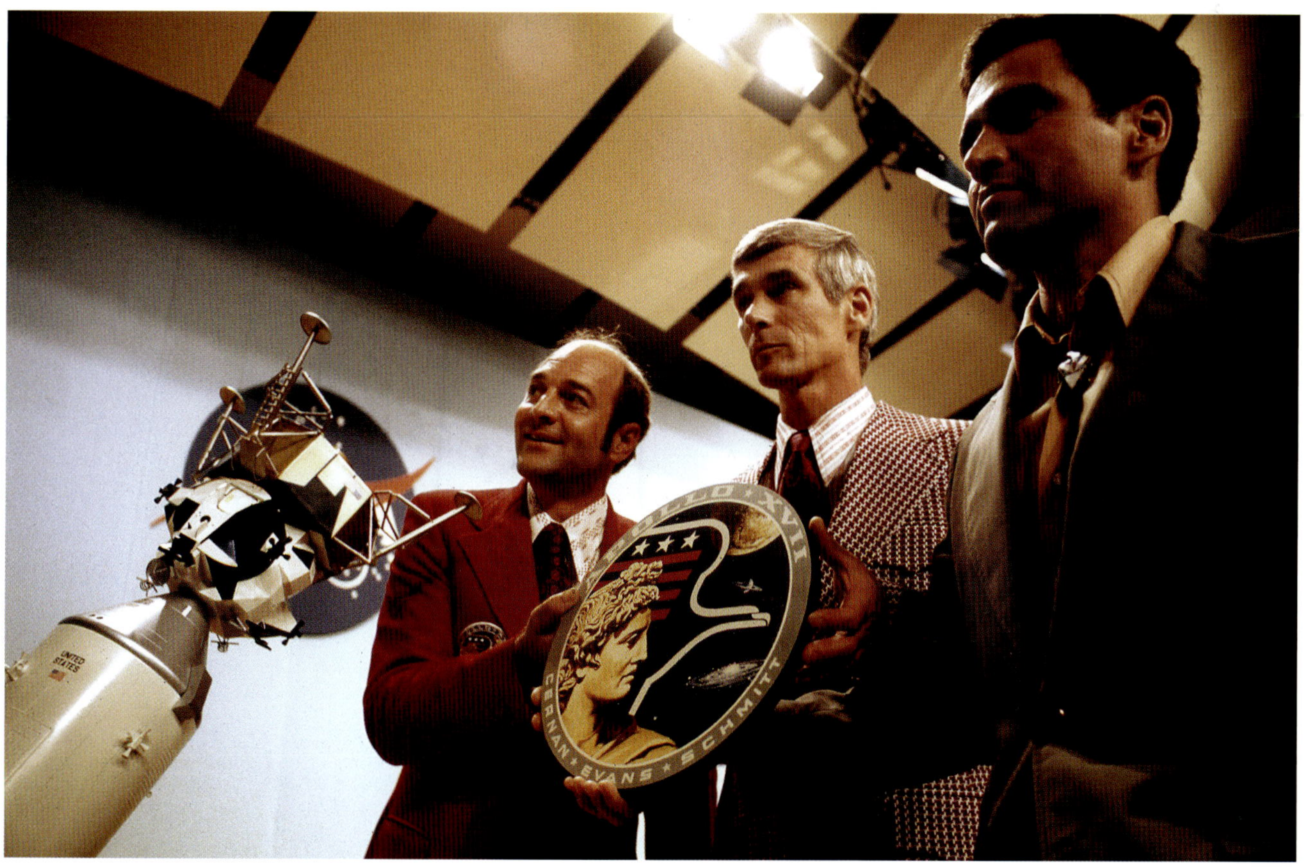
APOLLO XVII
CERNAN · EVANS · SCHMITT

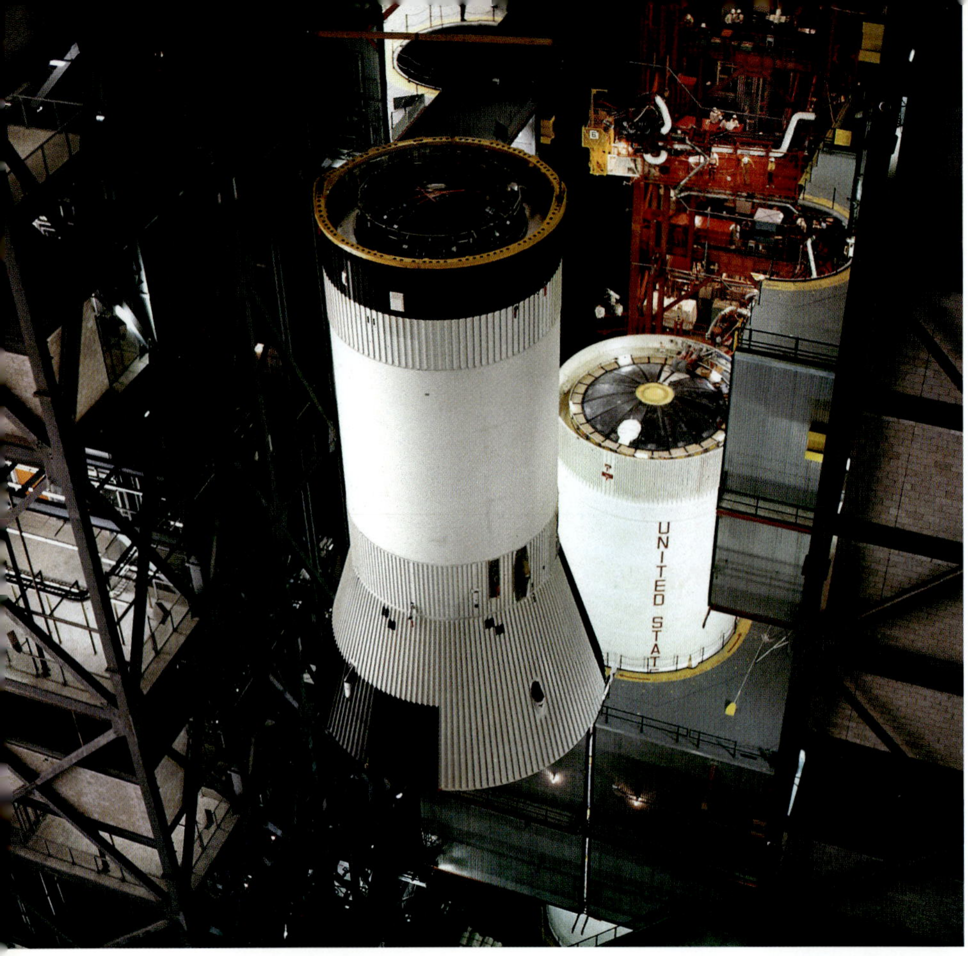

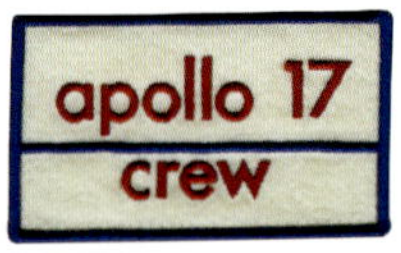

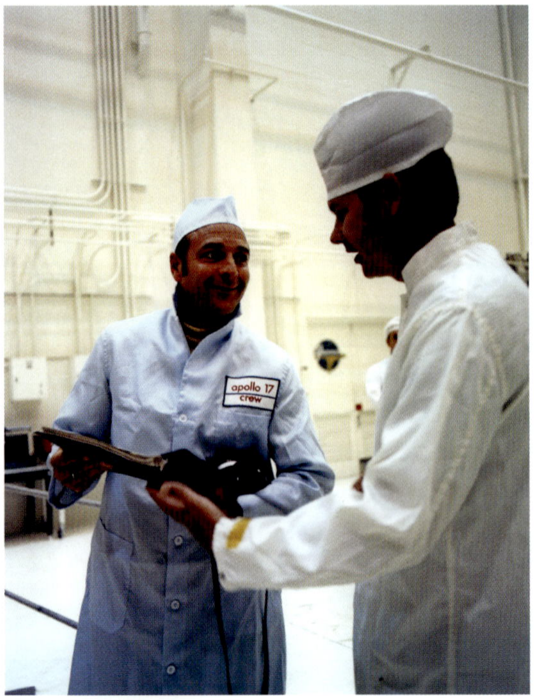

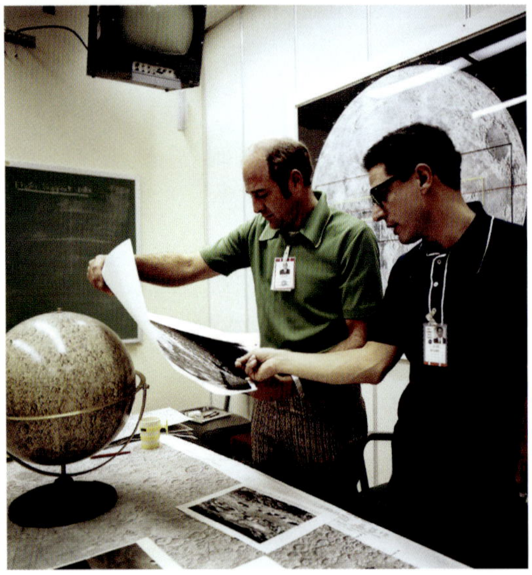

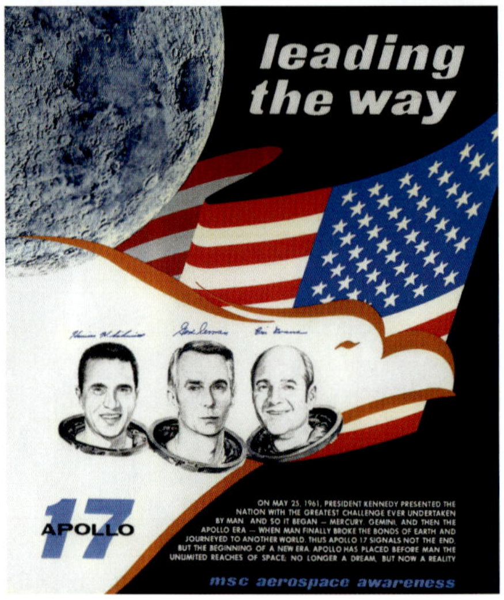

Top left: The eighty-foot S-IVB of Apollo 17's Saturn V is raised in the VAB's high bay 1 on June 20, 1972, as workers watch from swing arm 6 on the LUT (*top right*). The S-IVB would be mated with the Saturn's S-II stage behind it, surrounded by work platforms.

Top right: Evans (*left*) visits the MSOB on August 15, 1972, to check on the CSM, which is being readied for a move to the VAB a week later. The Grumman logo is on the wall in the background.

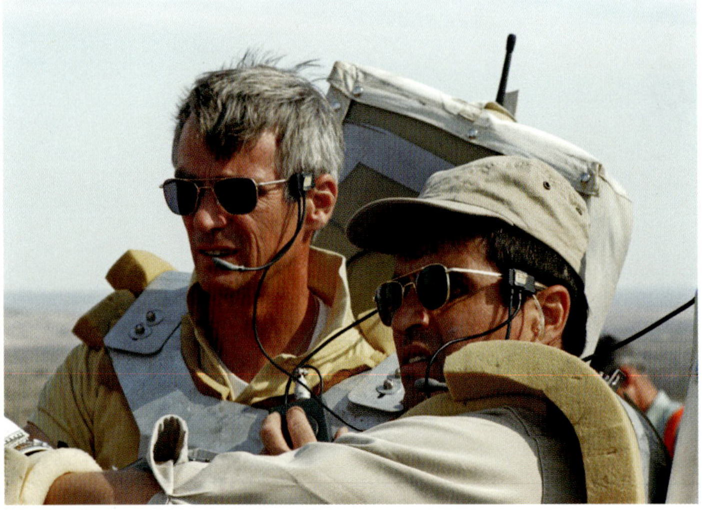

Center left: Evans (*left*) studies lunar photographs and charts on October 11, 1972, in the MSOB crew quarters with Farouk El-Baz, an Egyptian American geologist. El-Baz was the supervisor of Lunar Science Planning for Bellcomm, an AT&T division that analyzed mission planning for NASA.

Bottom left: Cernan (*left*) and Schmitt wear mock-up lunar backpacks during geology training near Sudbury, Ontario, Canada, on July 14, 1972. Apollo astronauts visited Sudbury, the site of the second-largest known impact crater on Earth, to study shatter cones—rare rock formations that only form in the bedrock beneath meteorite impact craters.

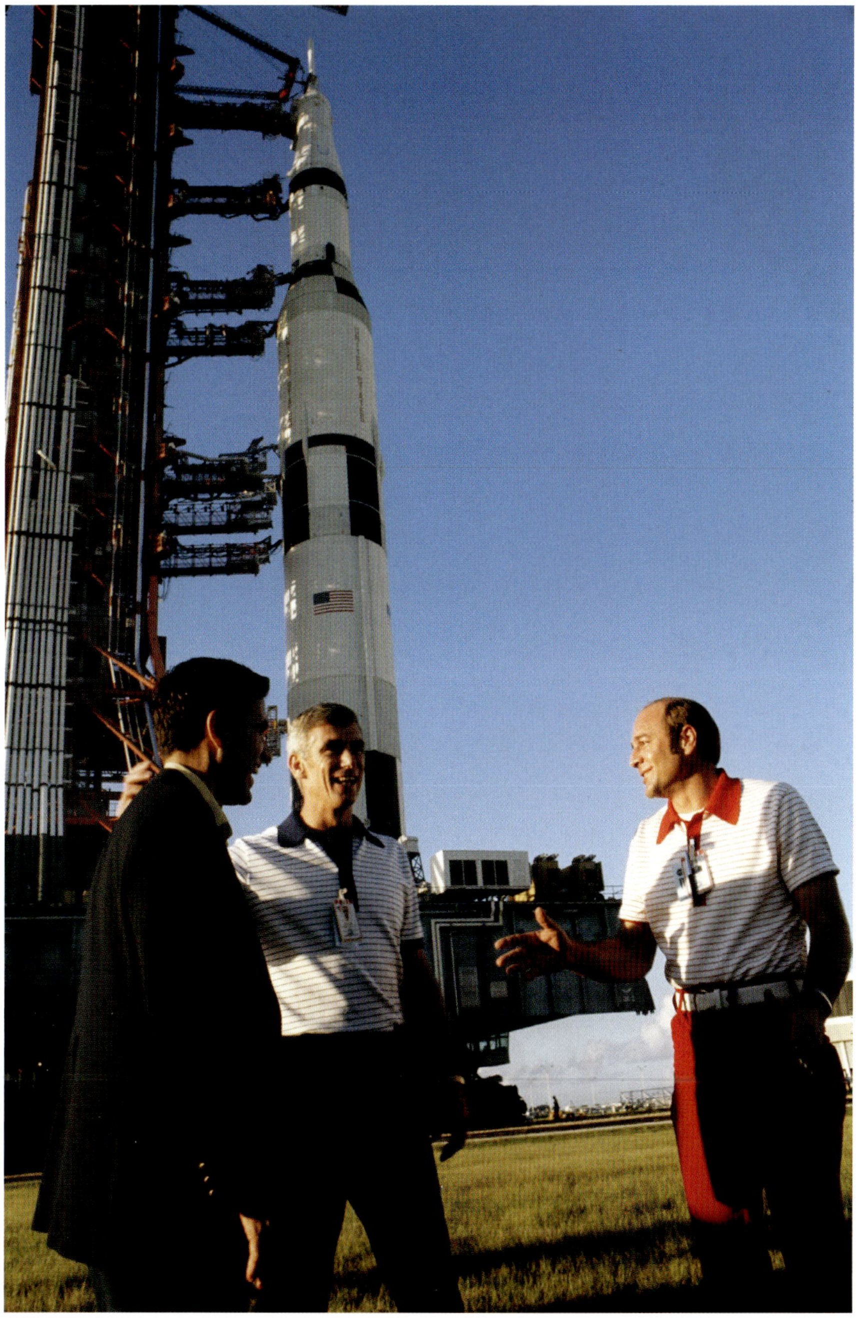

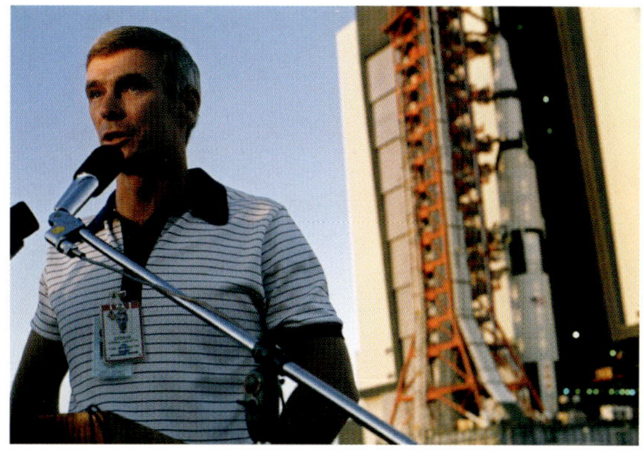

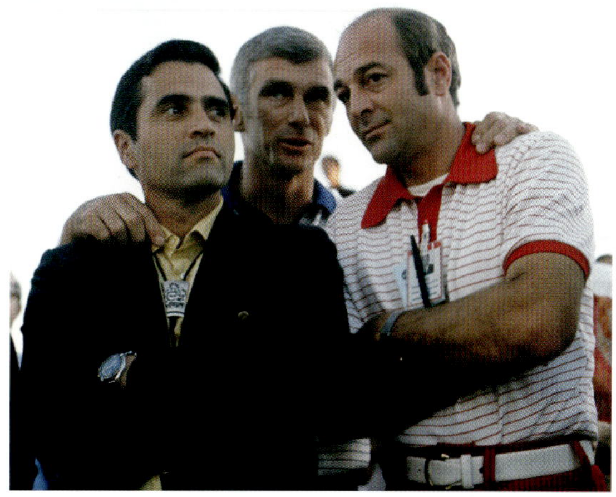

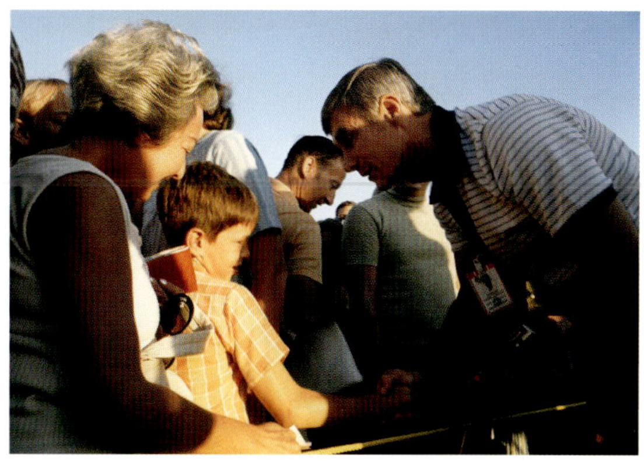

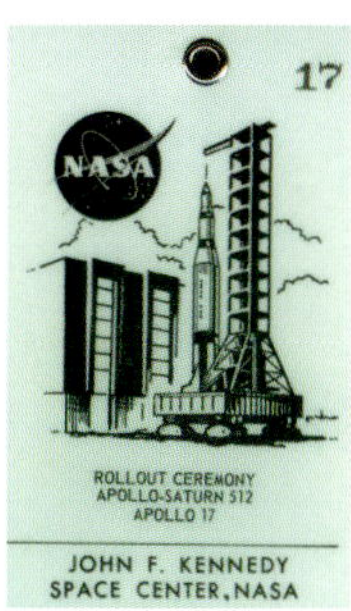

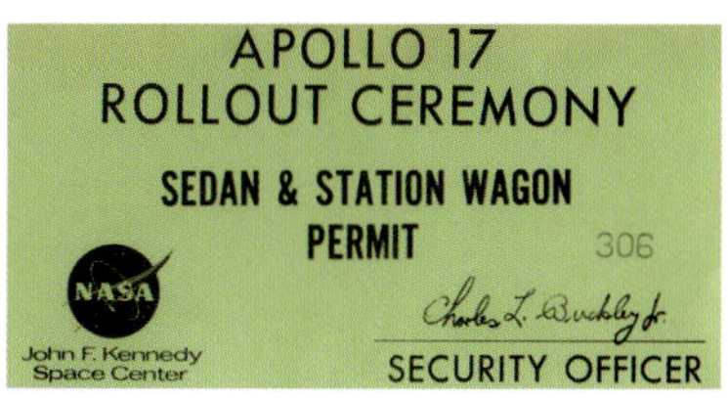

Top left: Schmitt, Cernan, and Evans climbed aboard the gray crawler-transporter behind them during the Saturn's rollout to Pad 39A, to take photos with the crawler-transporter's operators. After the launch vehicle reached the pad six hours later, the crewmen changed into pressure suits and posed with the LRV one-gee trainer for the photo on page 180.

Top right: Shortly after 7:00 a.m. (EDT) on August 28, 1972, Cernan calls Apollo 17, the final manned lunar mission, "not the end but the beginning of a continuing quest for knowledge." The flight's Saturn V slowly emerges from the VAB.

Center right: Schmitt, a New Mexico native, often wore bolo ties. He received his doctorate in geology from Harvard University and would later serve as a one-term US senator from his home state.

Bottom right: Cernan greets one of the nearly five thousand spectators who attended the rollout of the last Saturn V to carry astronauts.

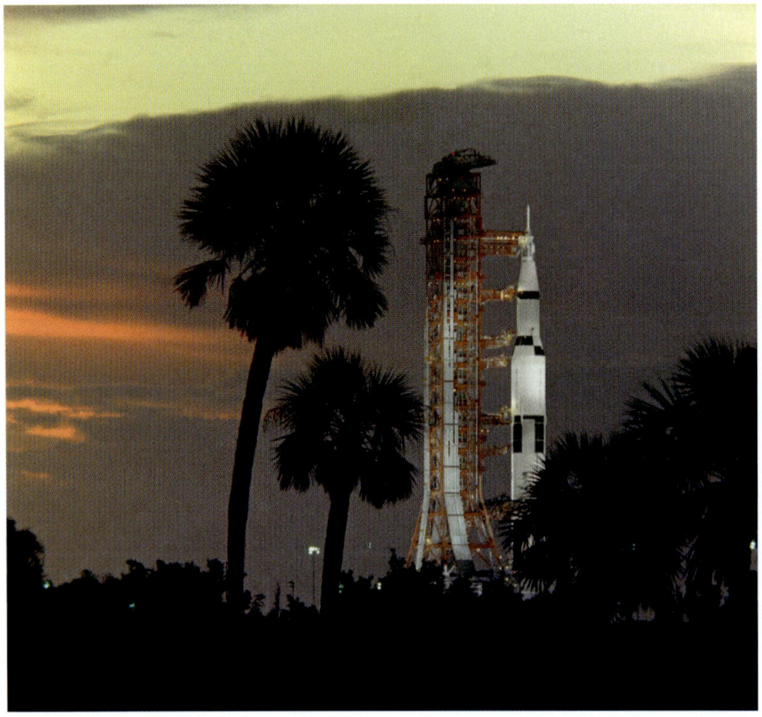

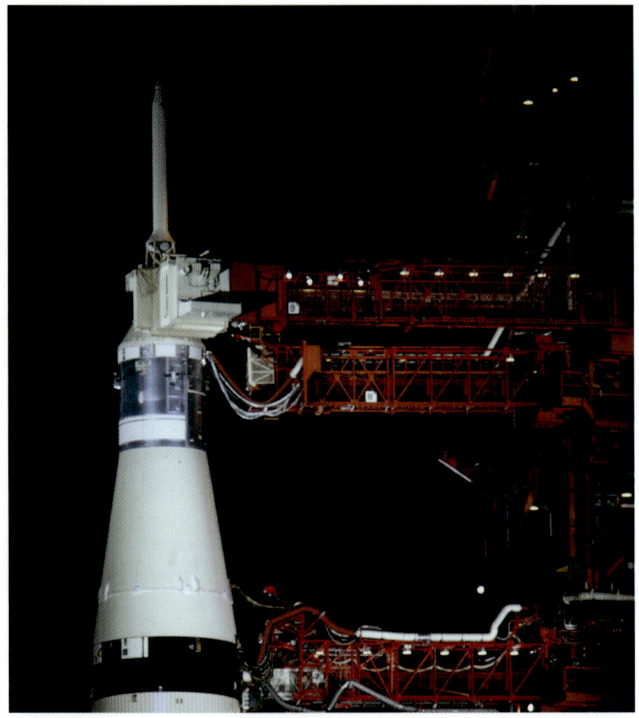

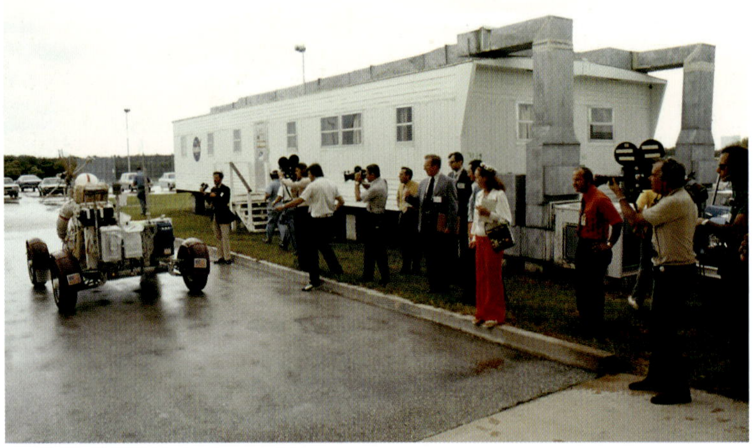

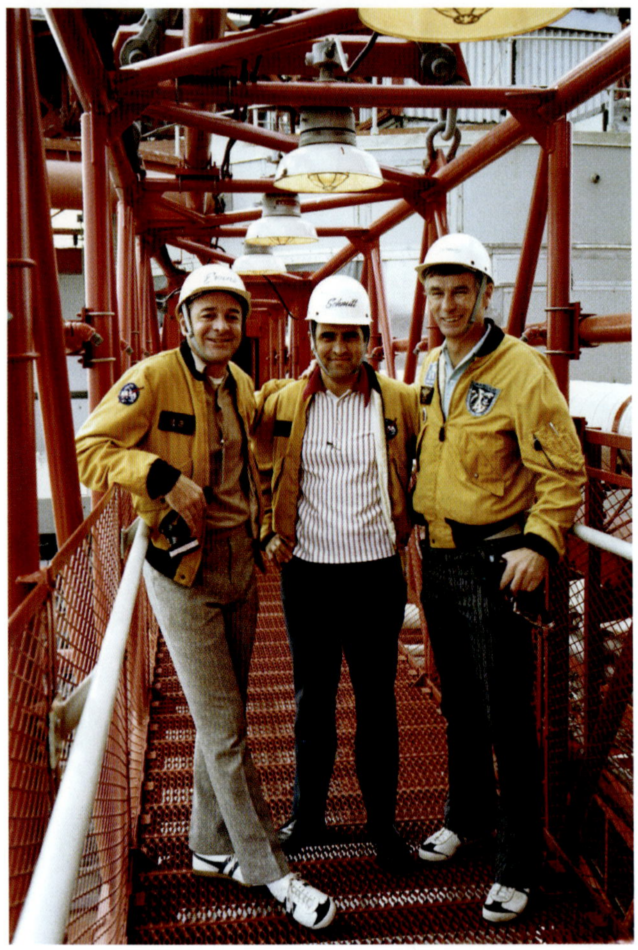

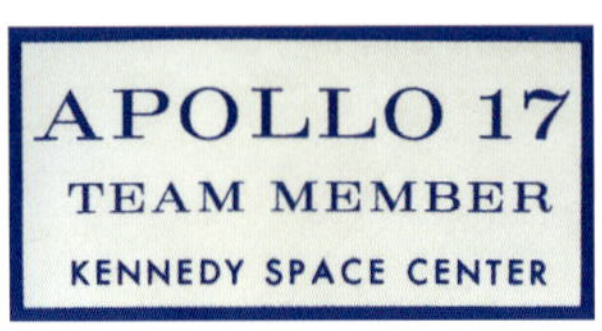

Top left: Sunrise at Pad 39A on launch morning, December 6, 1972. Many varieties of palm trees flourish in the 140,000-acre Merritt Island National Wildlife Refuge, an overlay of KSC established in 1963. The refuge is home to more than five hundred wildlife species, including Florida panthers, bald eagles, alligators, West Indian manatees, and wild boars. (photo by John de Bry)

Top right: Early on launch day, a row of lights illuminates the top swing arm at the LUT's 320-foot level, with the white room at left against Apollo 17's CM. Aluminum paint covers the spacecraft's SM. (photo by Steve Nolte)

Center left: Cernan drives the LRV trainer on October 11, 1972, from the MSOB toward the outdoor simulated lunar training area known as the "rockpile." Journalists include NBC's Roy Neal (in the blue jacket) and Mary Bubb, of Fairchild Publications (*right of Neal*). To her right, in the red shirt, is Ed Harrison, NASA public affairs officer. The Gemini suiting trailer, relocated from LC-16, is in the background. (photo by Jacques Tiziou)

Bottom right: During a tour of Pad 39A on December 4, 1972, Evans, Schmitt, and Cernan pose on the swing arm that they will cross on launch day to board Apollo 17. All three carry respirators.

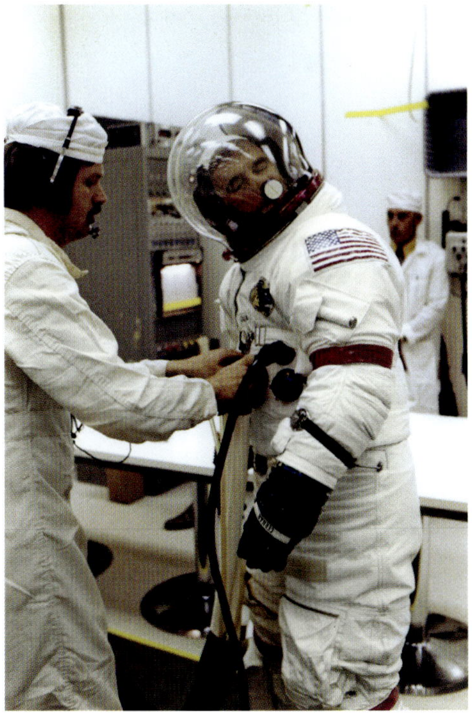

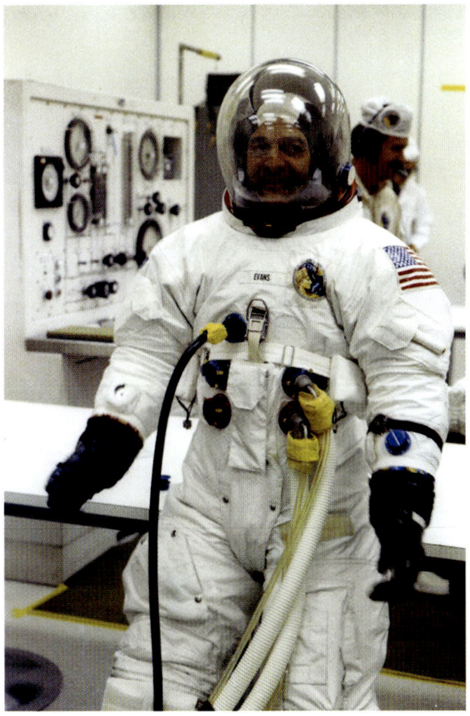

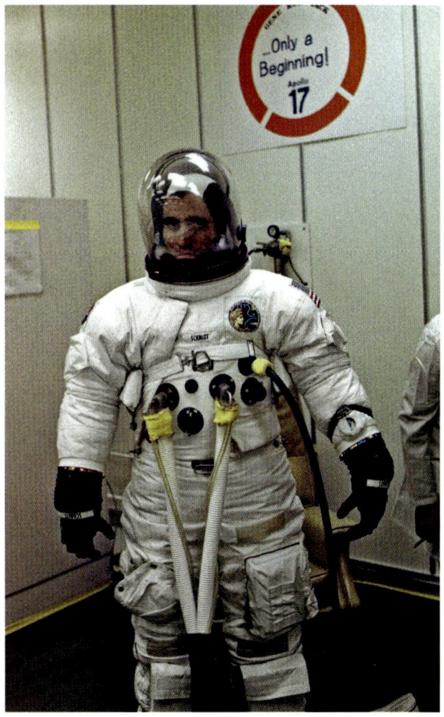

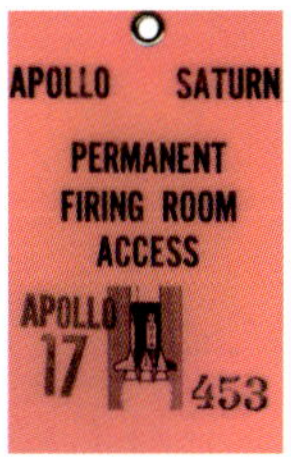

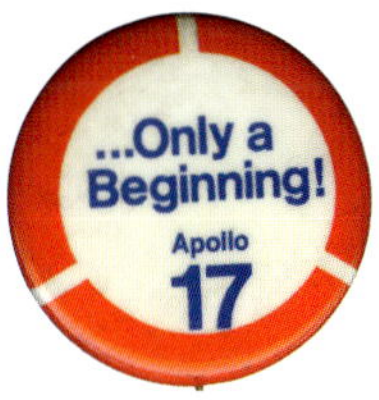

IBM button

Top left: During the CDDT on November 21, NASA suit tech Al Rochford connects an oxygen hose to Cernan's suit in the MSOB's suiting room. Cernan rubs his nose on the helmet's drinking port, which could also be used for consuming rehydrated food.

Top center: Evans during a suit leak and back pressure resistance check in the suiting room on the evening of December 6, with Rochford behind him. The black cable carried communications and biomedical data.

Top right: Schmitt on December 6. He wears an A7LB-model space suit, made by ILC. Oxygen entered the suit through the red inlet connector and left through the blue outlet connector. On his left wrist, Schmitt wears an Omega Speedmaster Professional watch, the only watch certified for space flight by NASA at the time.

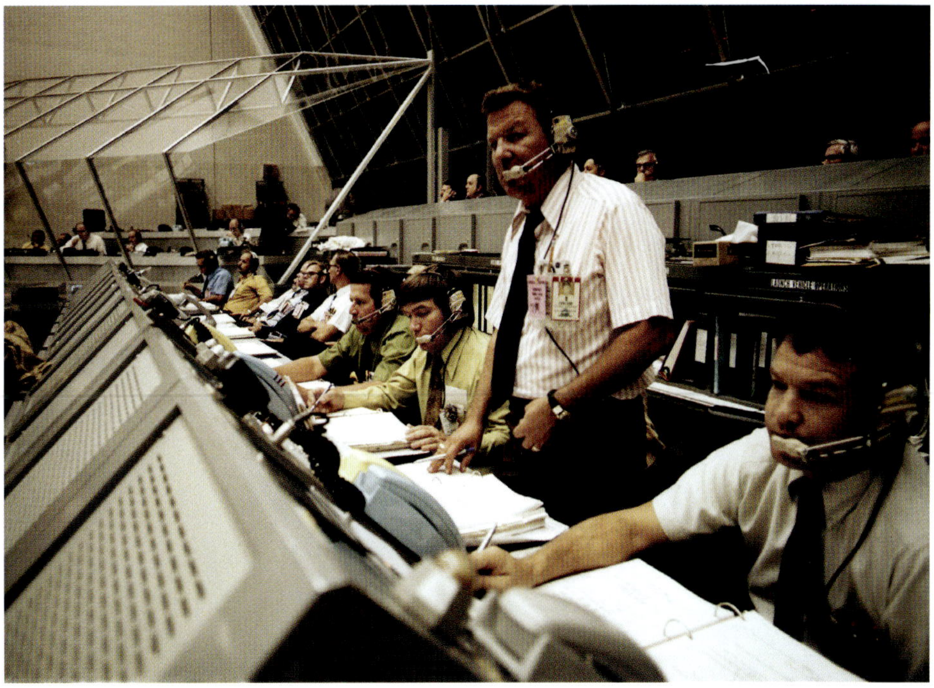

Above: Norm Carlson, launch vehicle test conductor, stands at his console in Firing Room 1 during the CDDT on November 21, 1972. At right is test conductor Gary Powers; to Carlson's right are John "Tip" Talone (*next to Carlson*) and Bill Schick (in green), chief test supervisor. The glassed-in area at left was the operations management room. Carlson and Schick would continue as test managers into the Space Shuttle program, while Talone took on other KSC management roles.

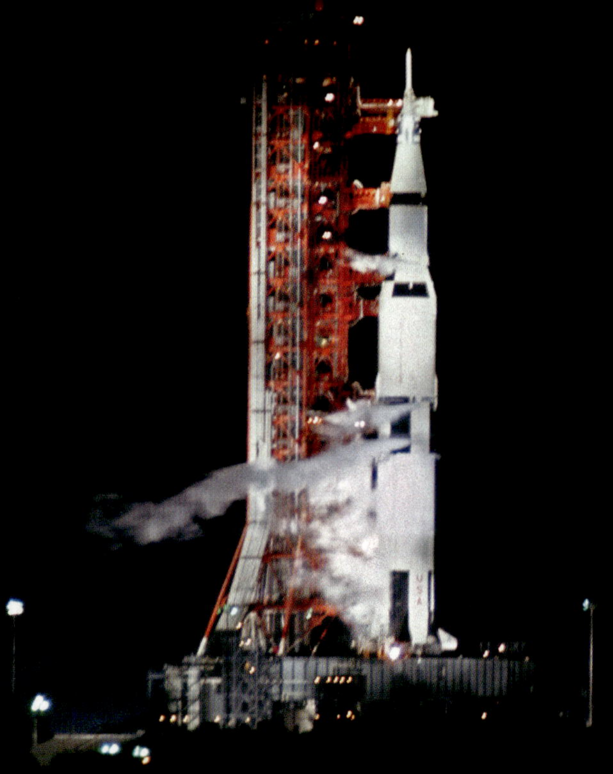

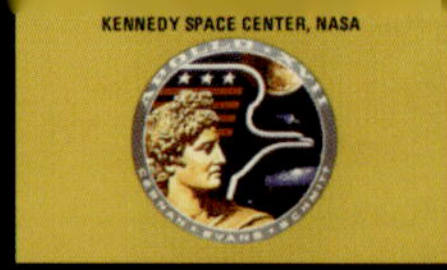

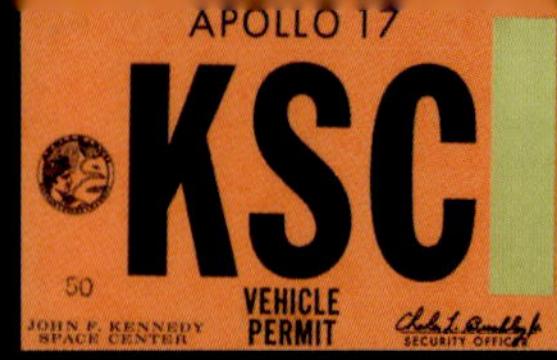

Top left: Excess oxygen vapor vents from the Saturn's stages on December 6, as the countdown enters an unplanned hold just thirty seconds before the booster's expected launch. An automatic cutoff had been ordered by the launch sequencer, forcing a launch delay of one hour and forty minutes. (photo by John de Bry)

Top right: The reflective strip at right was used for increased visibility at the night launch's security checkpoints.

Bottom right: Apollo 17, seconds after main engine ignition. This photo was taken with a camera using a wide-angle fisheye lens, mounted near the 320-foot level of the LUT. One of the nine swing arms is at bottom.

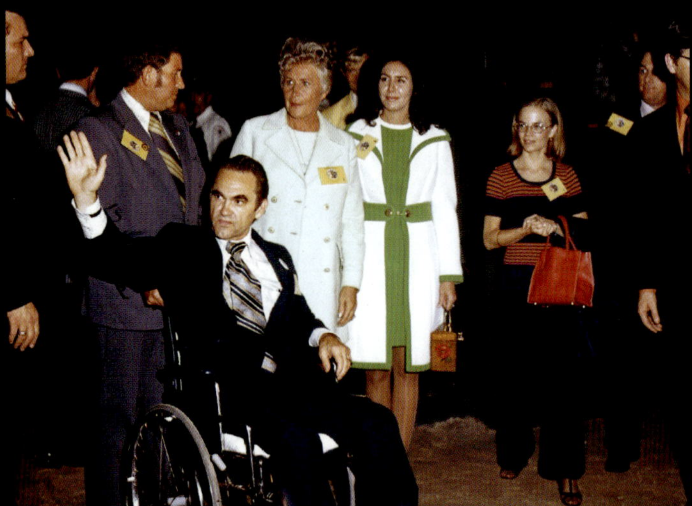

Above: Governor George Wallace of Alabama was among those at the VIP viewing stand near the VAB. He had been left partially paralyzed during an assassination attempt less than seven months earlier. Gay Debus is second from left; Wallace's wife, Cornelia, is in white. Other guests included Vice President Spiro Agnew; actors Connie Stevens, Eva Gabor, and Hugh O'Brian; singer Frank Sinatra; and comedian Jonathan Winters.

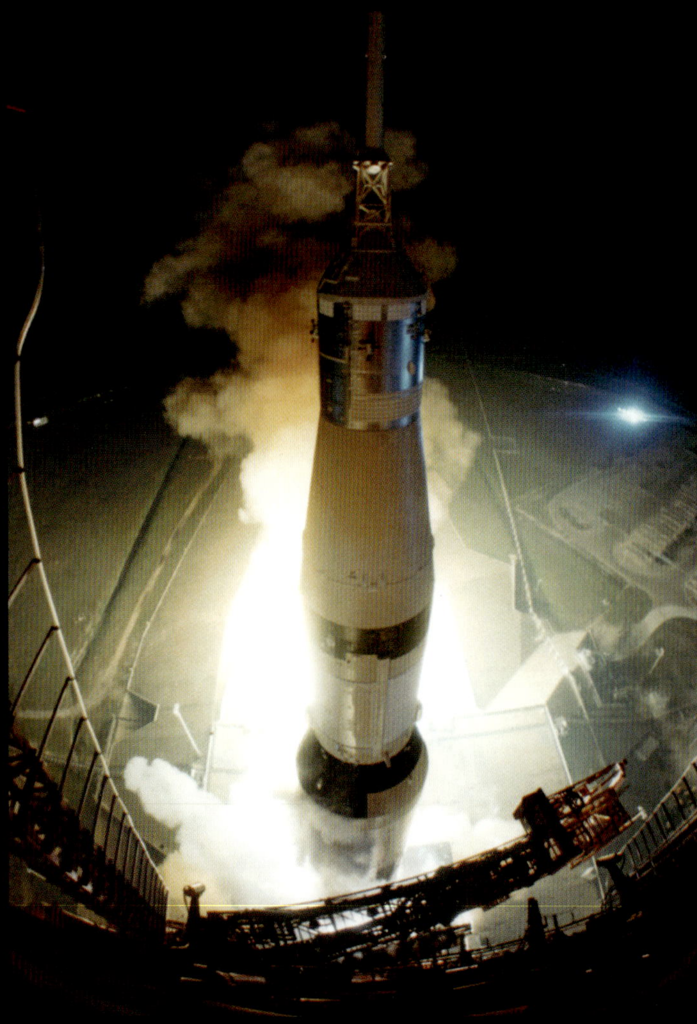

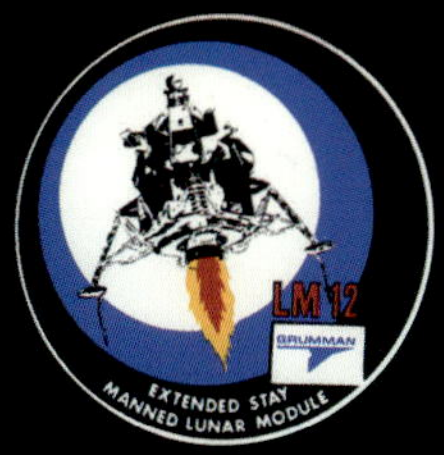

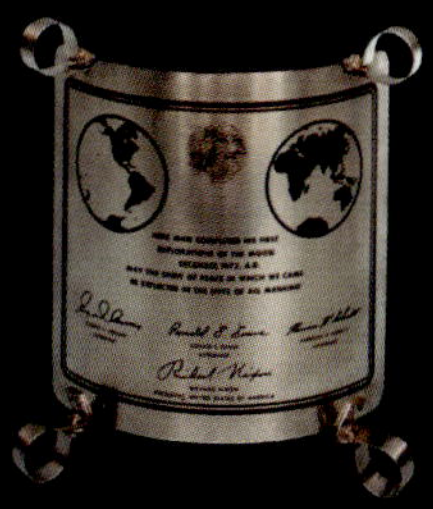

Top right: Much of *Challenger*'s descent stage is covered with gold Mylar thermal blankets, which were hand-crinkled to minimize the heat transfer between layers. At left, a white blanket protects the MESA pallet, containing lunar surface tools, sample containers, and film and TV cameras. Rover tracks can be seen in this photo, taken on December 11.

Bottom left: On December 13, Cernan, his gloves gray with lunar dust, holds the LM's ladder after uncovering the mission's stainless steel commemorative plaque (*center*), mounted on the primary strut of the forward landing gear. Each LM carried such a plaque; Apollo 17's was the only one to include a lunar disc showing all of the Apollo landing sites.

Bottom right: Cernan's photo of *Challenger* looking south (during the third and final EVA, on December 13) shows Earth, which is to the right of the black radar antenna used to rendezvous with CM *America*. The sticklike EVA communications antenna is at top left, with a VHF communications antenna next to it. A conical S-band inflight communication antenna protrudes from the ascent stage at far left. Two of the LM's four RCS quad thrusters are also visible at left and center.

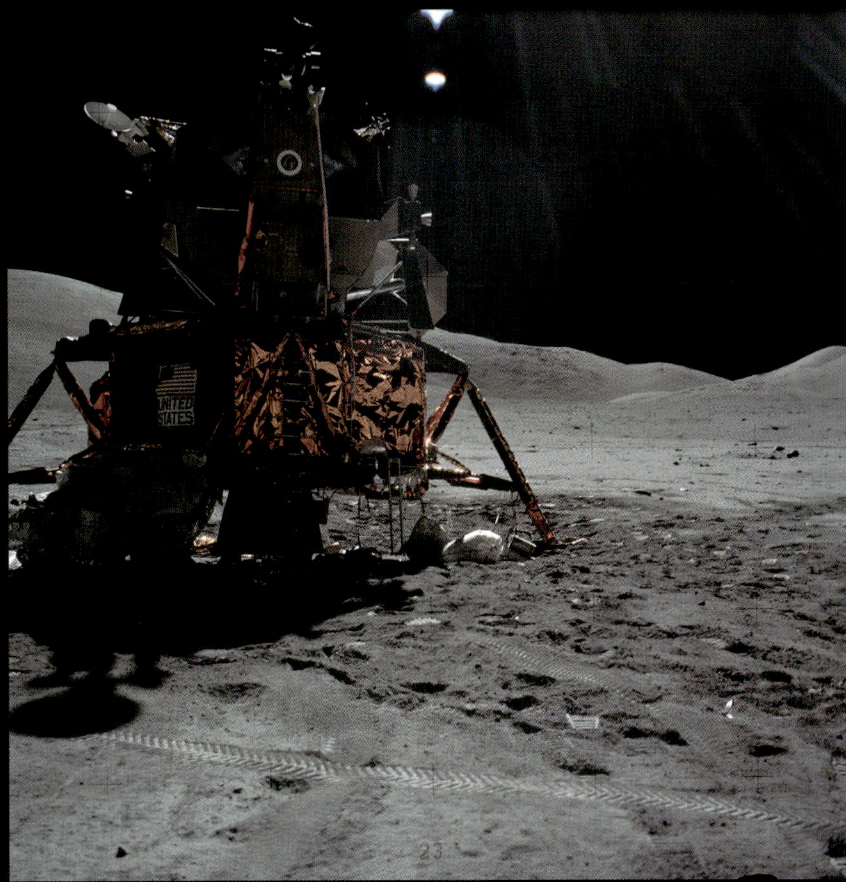

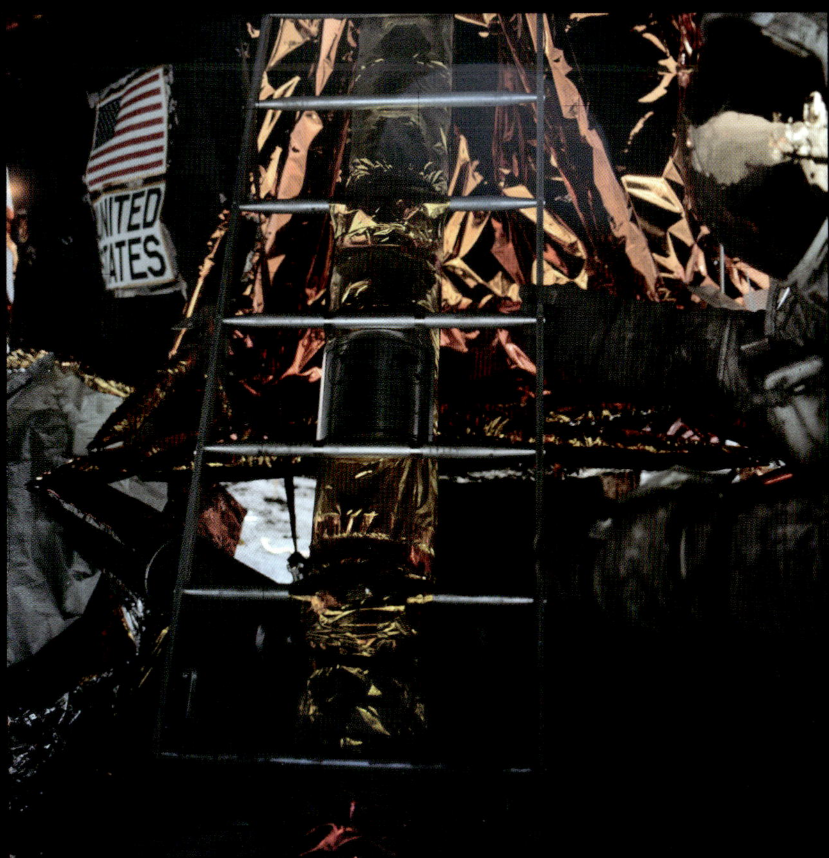

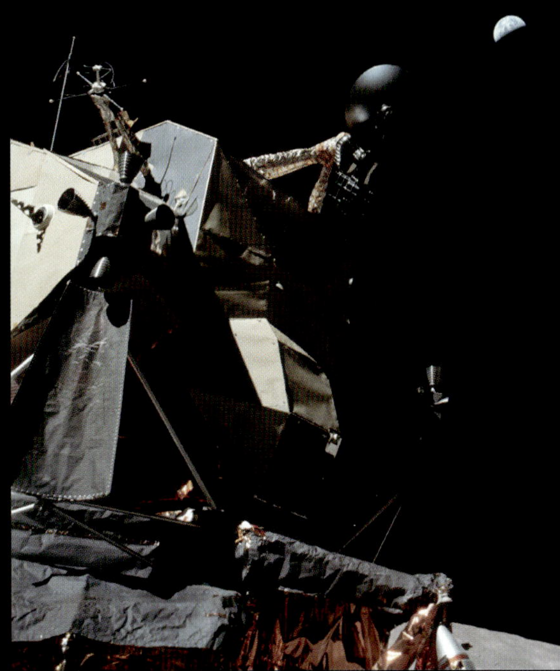

Top left: In Schmitt's photo, Cernan hammers a double core tube into the surface during EVA 3.

Top right: Schmitt works at the rover at the end of EVA 3, on December 13. The black object at upper right is the sighting scope for the rover's high-gain antenna, used to communicate directly with Earth. The EVA crewmen had to realign it each time they moved the rover to a new location. At left is the low-gain antenna.

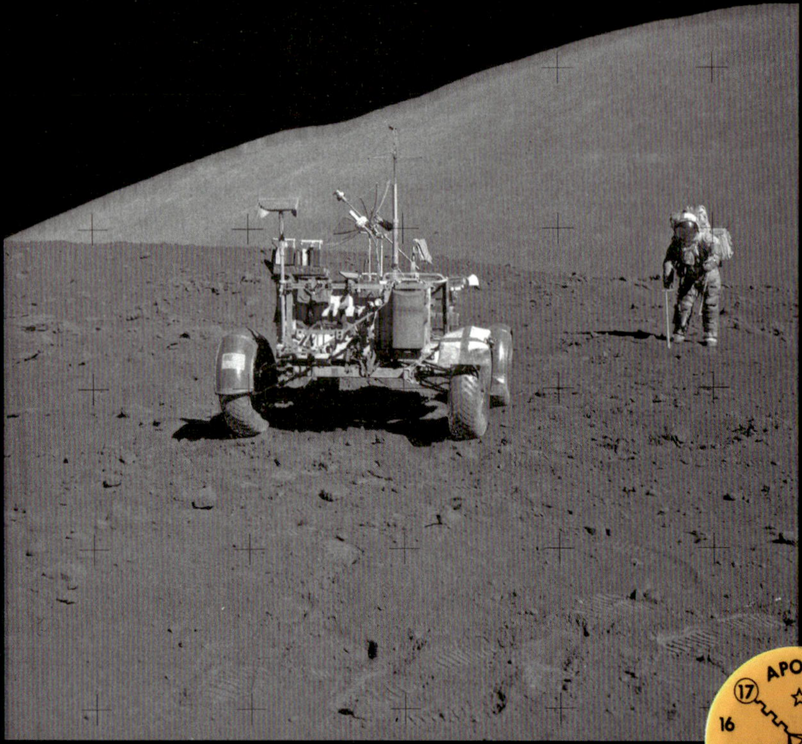

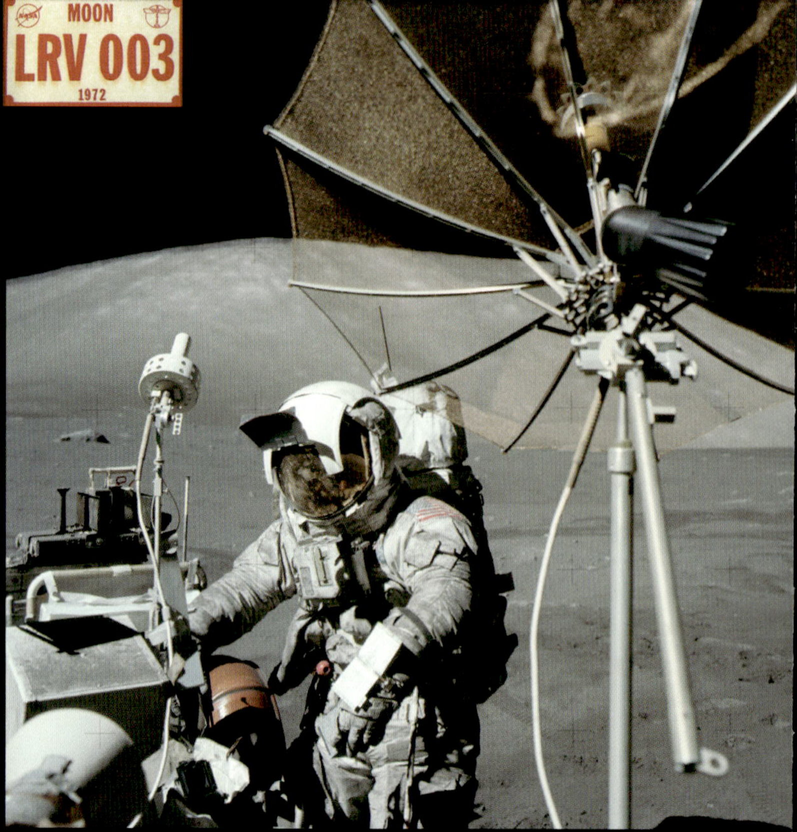

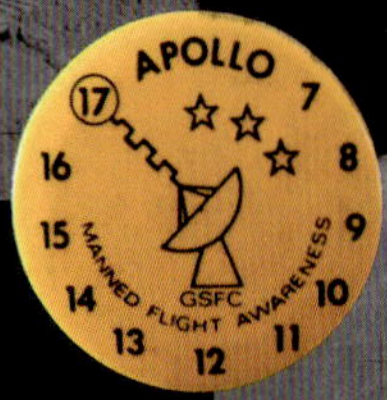

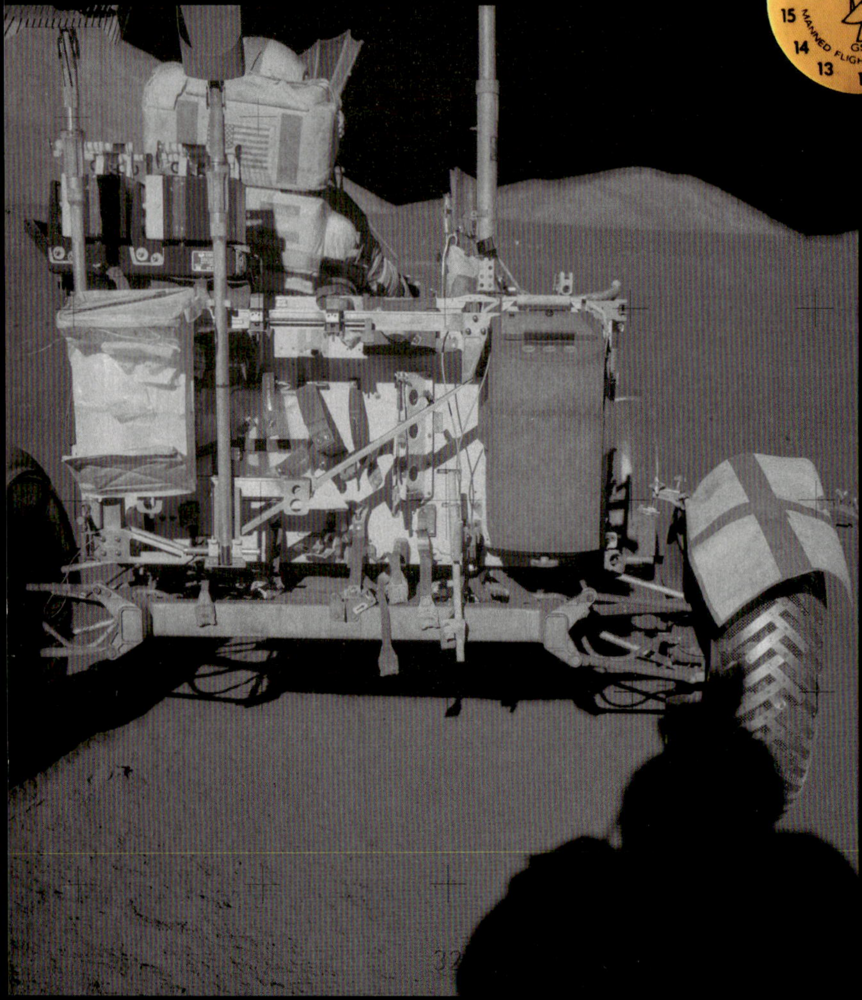

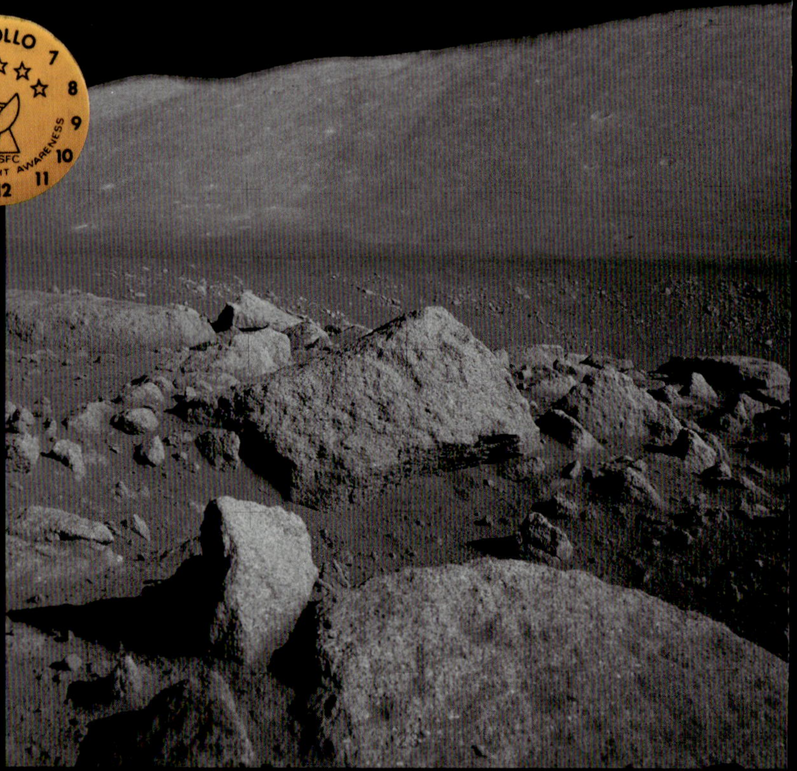

Bottom left: Schmitt's photo, taken December 12, shows the astronauts' makeshift repair to the LRV's right-rear fender, which Cernan had accidentally broken off the previous day. They used duct tape to fasten together four lunar charts, and then attached them with two clamps borrowed from the LM to block lunar dust kicked up by the rover's wheel.

Bottom right: Photo taken by Cernan during EVA 2 on December 12, showing an elevated area known as the North Massif (*background*) and the west wall of Camelot Crater (*closer to the viewer, mid-frame*).

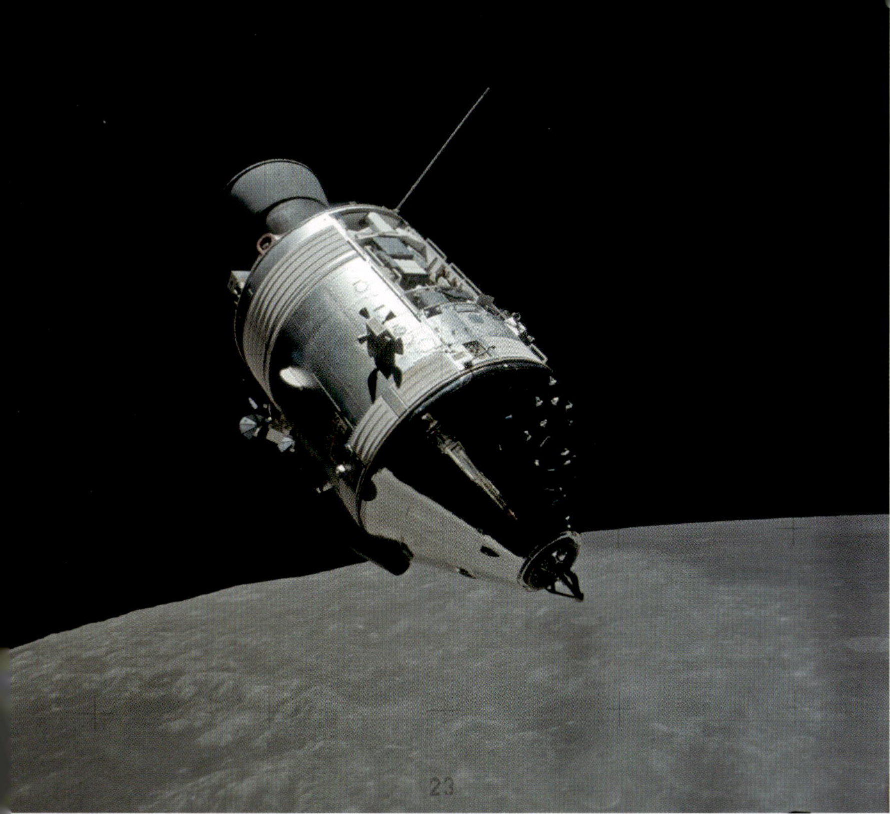

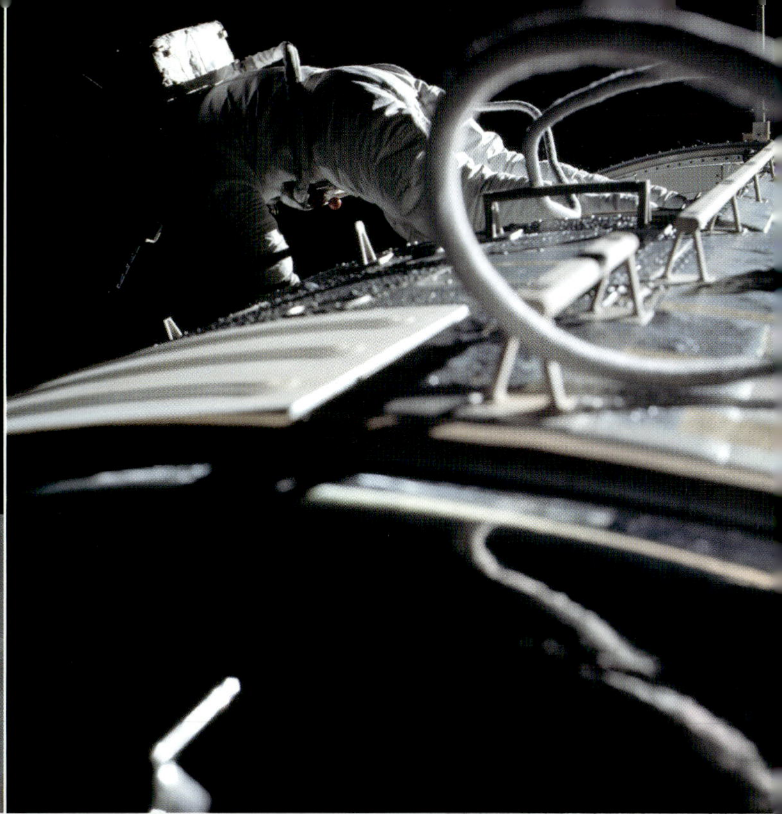

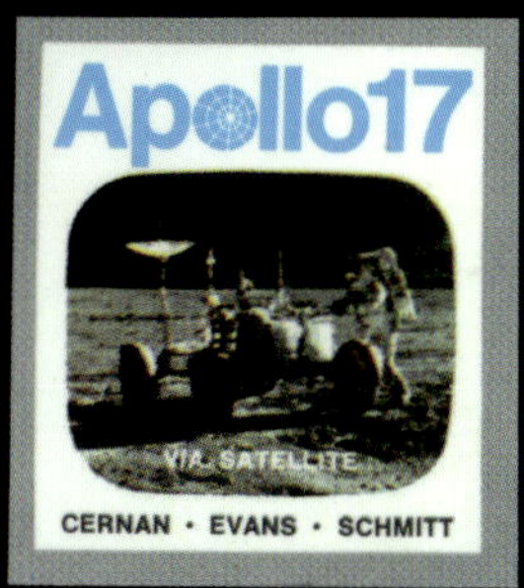

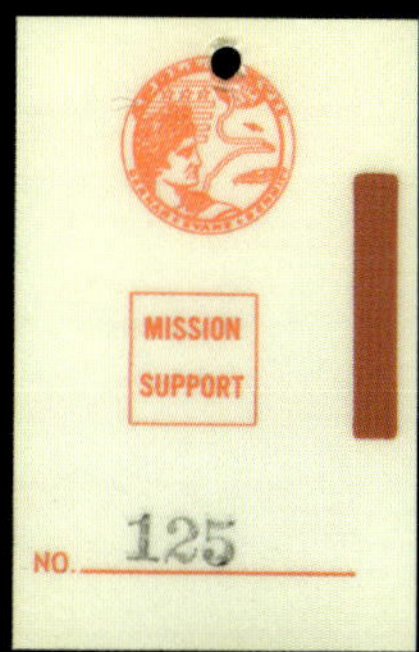

Top left: CM *America* (seen from *Challenger*), seventy miles above the moon prior to docking on December 14. The SIM bay, with an extended spectrometer boom, is on the right side.

Top right: Evans retrieves film cassettes from the SIM bay on December 17, as Apollo 17 heads toward Earth. Extensive handrails on the SM worked well for his EVA.

Center left: COMSAT (the Communications Satellite Corporation) provided ground stations that relayed Apollo TV from overseas antennas back to the United States.

Center right: MSC badge

Bottom: A crescent Earth is the backdrop in this photo of Evans and a loop of his tether, during his deep space EVA on December 17. He wears Cernan's red-striped lunar EVA visor assembly. Evans spent just over an hour outside the spacecraft, which was still 184,000 miles from Earth.

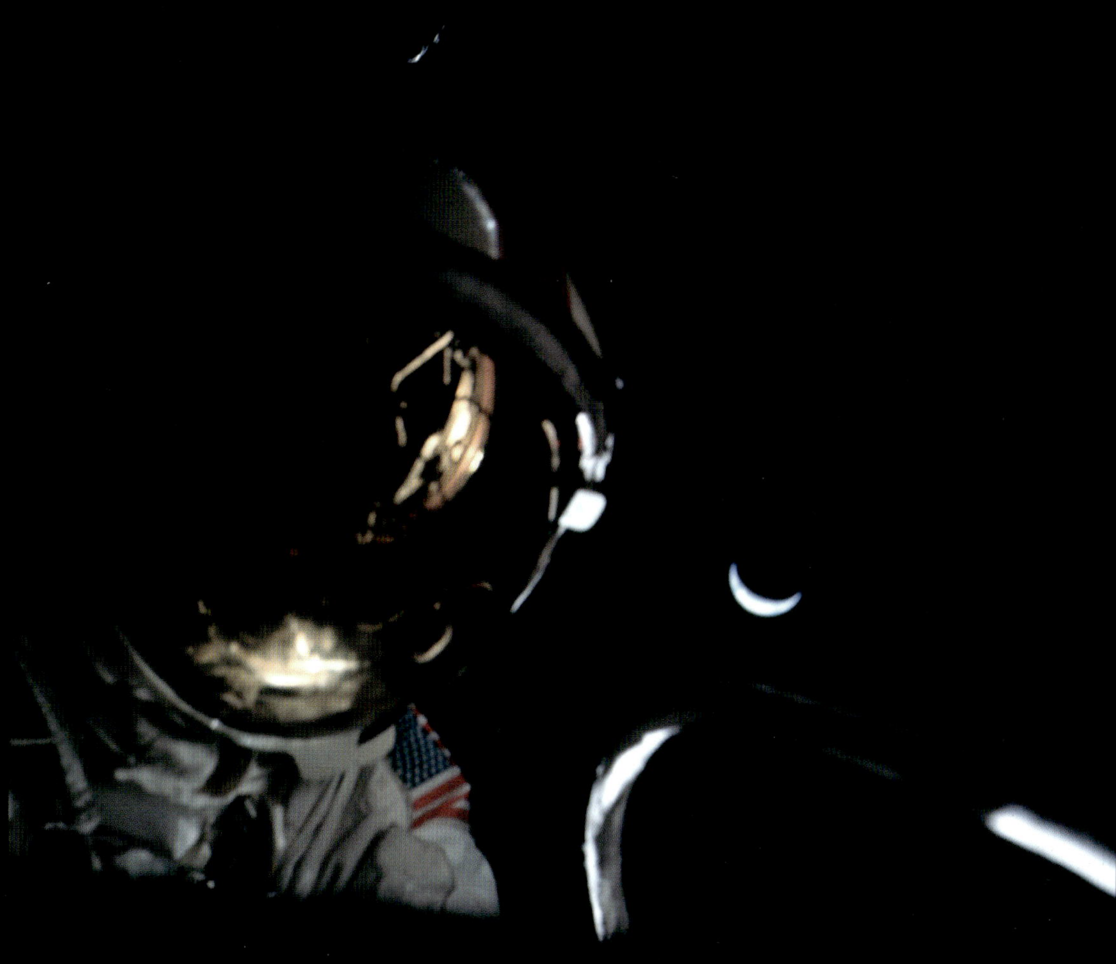

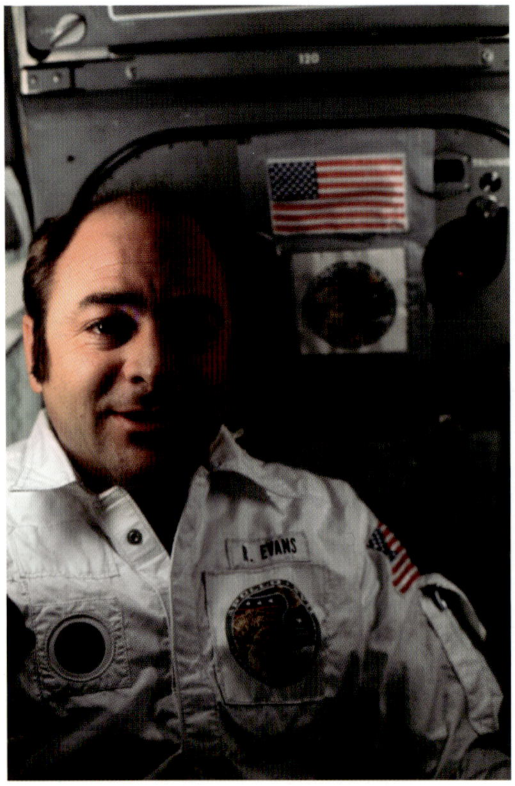

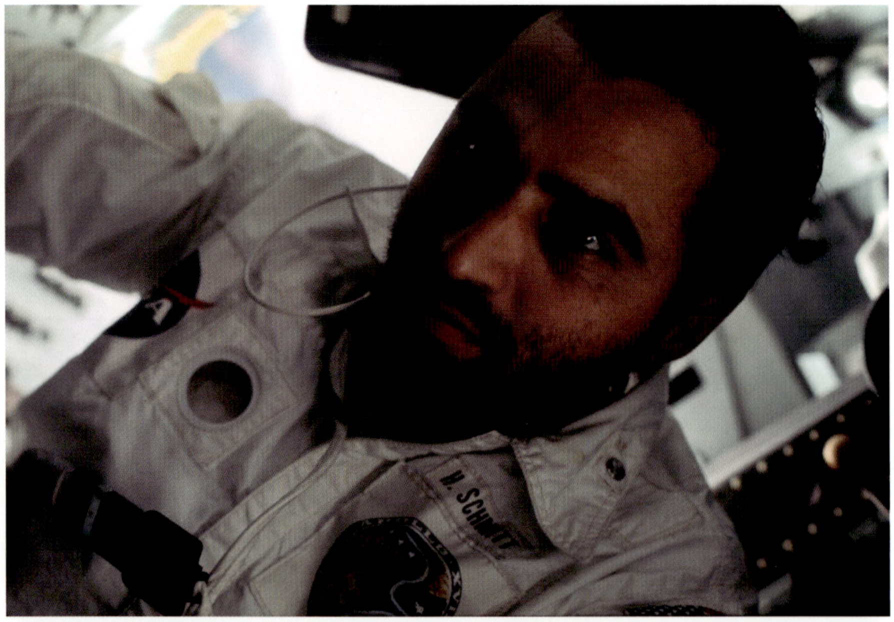

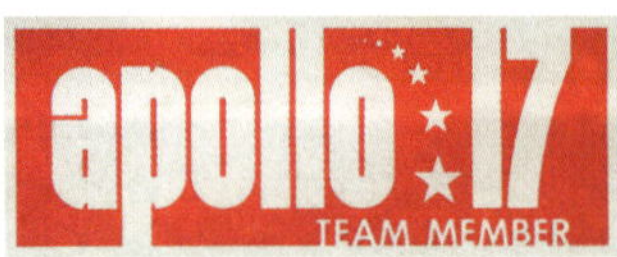

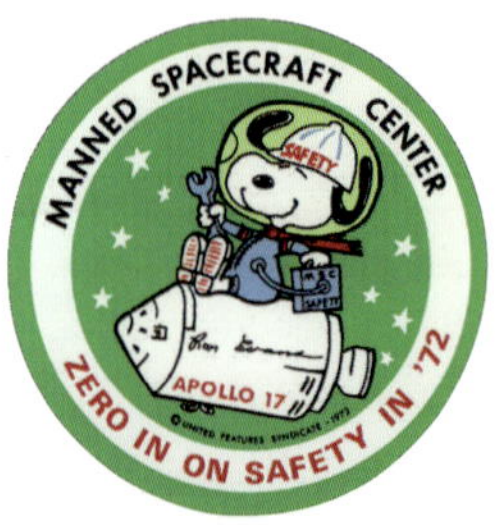

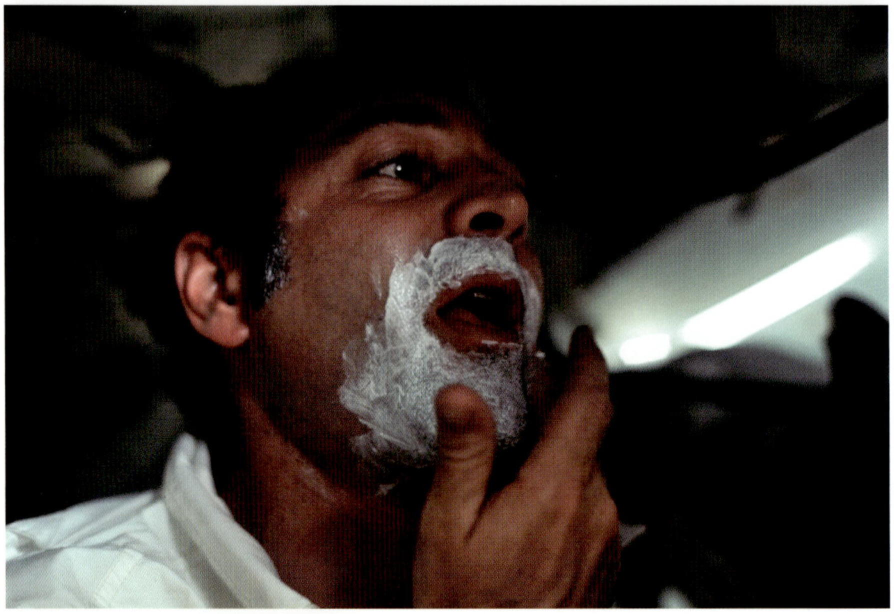

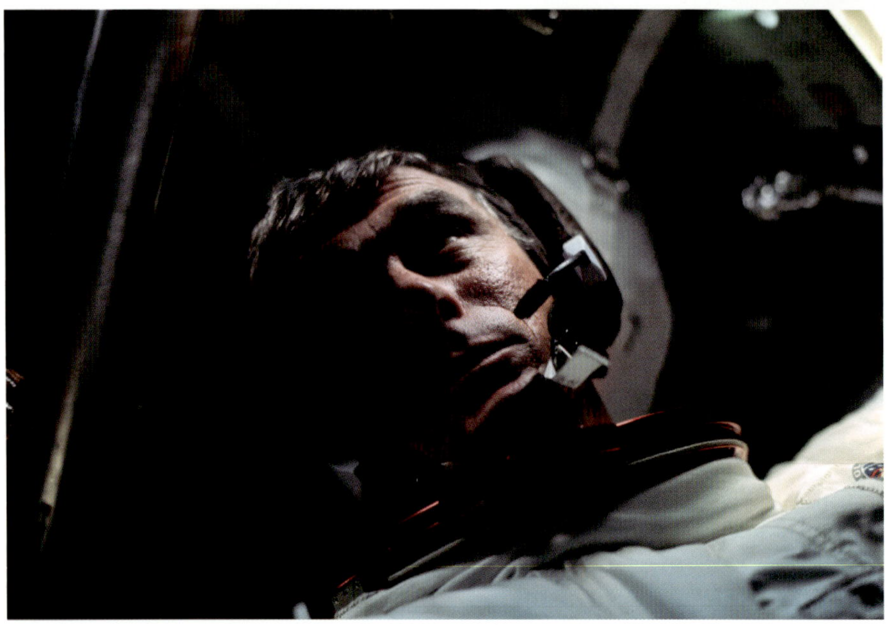

Top left: Evans in front of *America*'s optics panel, displaying a US flag and mission emblem on Beta cloth

Top right: Schmitt during the trans-Earth (return) portion of the mission

Bottom right: Evans applies brushless shaving cream from a tube. Apollo astronauts were issued Gillette Techmatic safety razors, but reported that they became easily clogged without the use of running water.

Bottom left: Cernan after returning to the CM from the LM

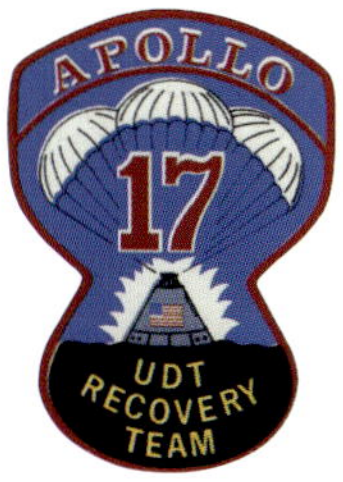

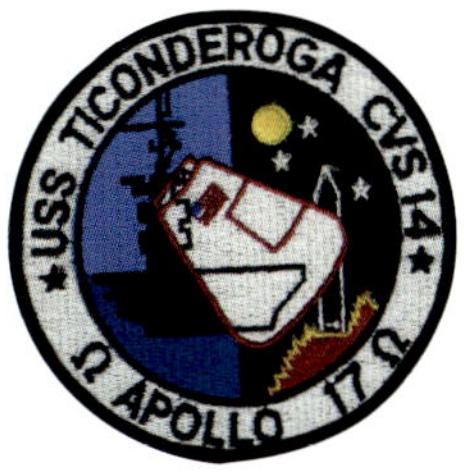

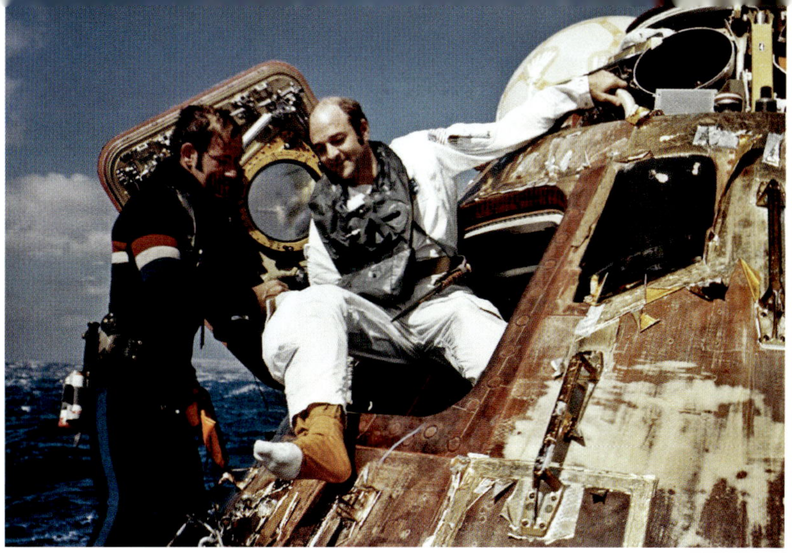

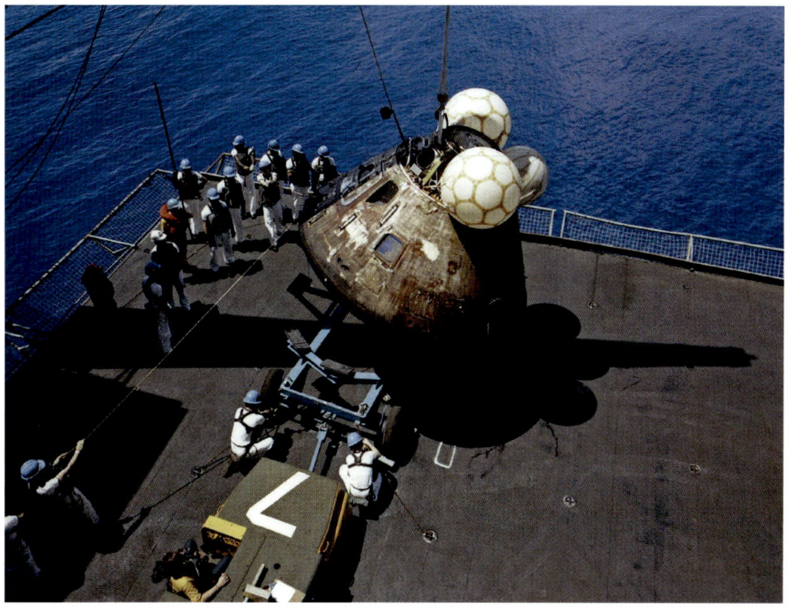

Top right: A diver with UDT-17 helps Evans egress from *America.* Foil covering the spacecraft had been mostly peeled away by the heat of atmospheric reentry. The large circular opening at top contained a drogue parachute mortar.

Second right: A crane lowers *America* onto a special carrier on the *Ticonderoga*'s deck, two hours after splashdown. The spacecraft's three uprighting flotation bags remain inflated.

Third right: On December 19, Schmitt chats with twelve-year-old Kevin Steen during dinner in the *Ticonderoga* officers' dining room. Steen, invited because he was believed to have terminal cancer, later went into remission. NASA's Donald Stullken is left of Schmitt.

Bottom right: US Navy Rear Adm. John Butts (*left*)—commander of Task Force 130, the navy support unit for the splashdown—and Capt. Norman K. Green flank the Apollo 17 crewmen as they prepare to cut sheet cakes baked in their honor, using gold-plated survival knives.

Bottom left: Cernan hugs NAR Pad Leader Guenter Wendt in the VAB's transfer aisle during the February 20, 1973, KSC Reunion.

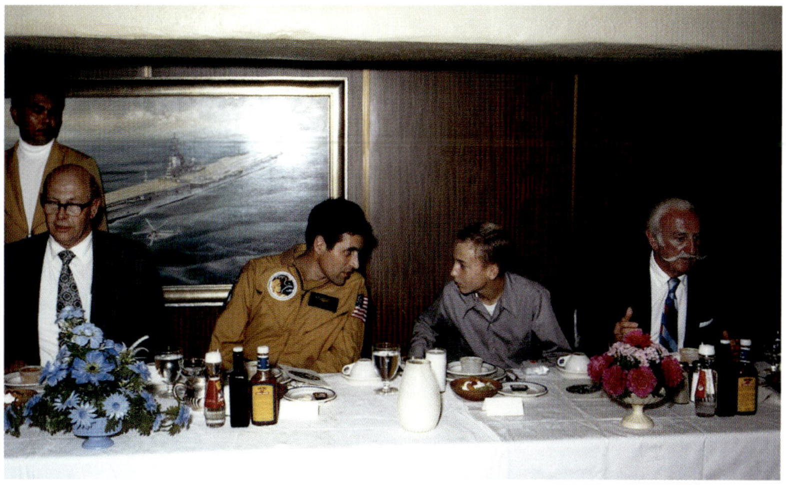

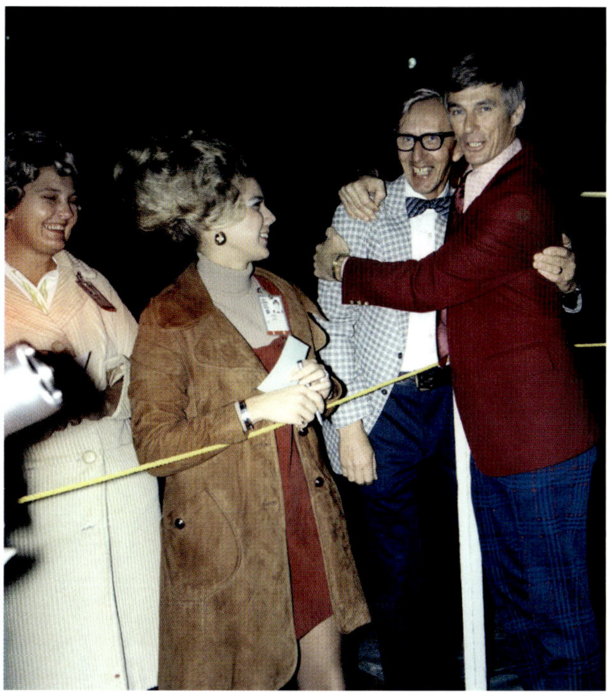

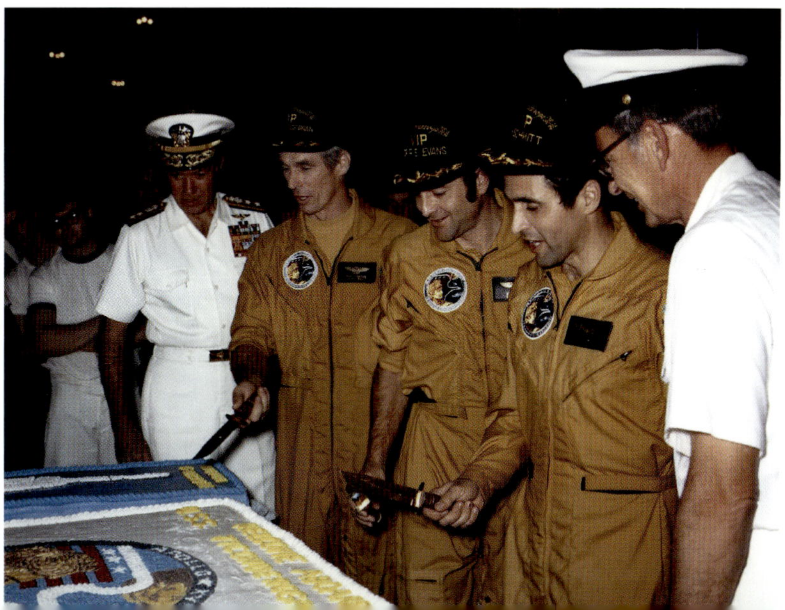

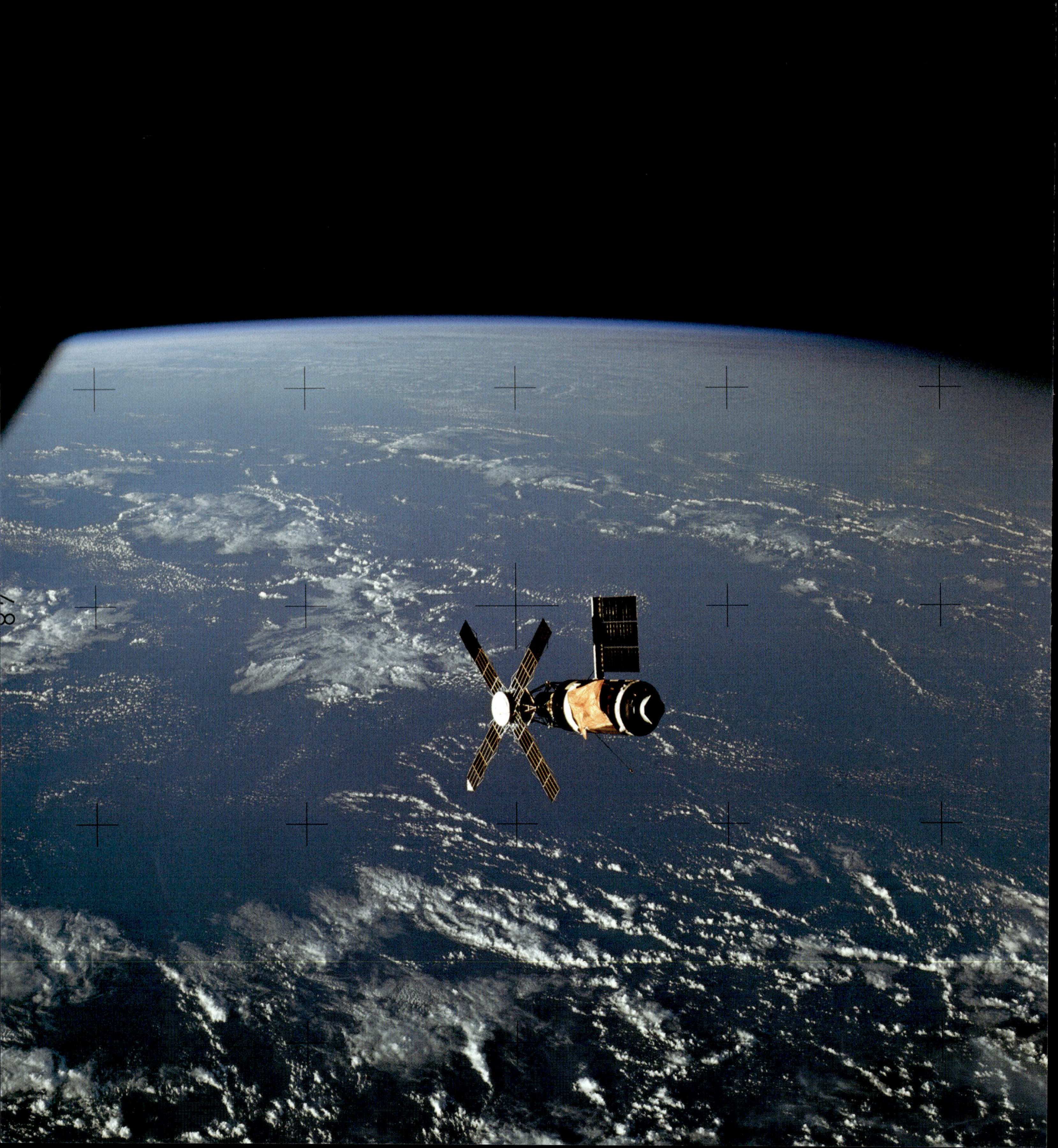

The workshop's single primary solar panel is at top right; the X-shaped solar panels on the Apollo Telescope Mount (ATM) are at left; and the mount's white circular solar shield is at center.

CHAPTER 13

Skylab 1 and 2

May 25–June 22, 1973

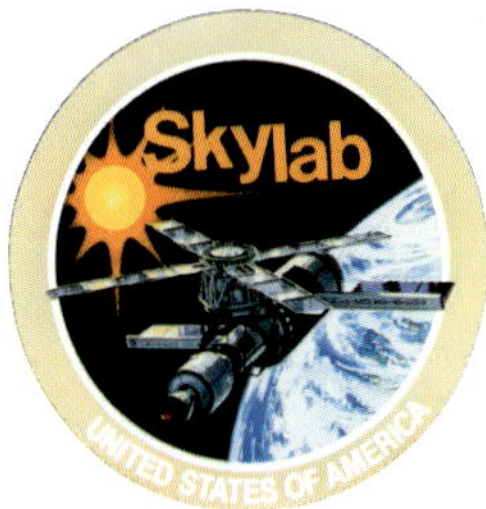

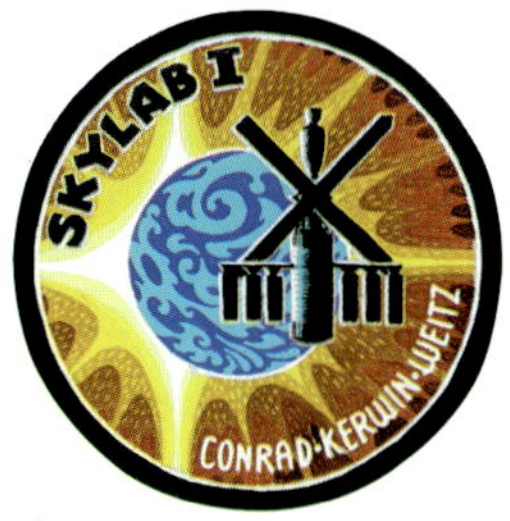

In the early 1960s, NASA and its contractors began studying various proposals for turning a spent rocket stage into a large manned orbiting laboratory. In 1965, NASA created the Apollo Applications Program to develop the concept. Budget cuts in 1969, and the resulting cancellation of the three final planned lunar landing missions, gave the space agency an opportunity to repurpose the surplus hardware for a program called Skylab.

The plan was to send three Apollo spacecraft to dock with an OWS—a converted Saturn V S-IVB stage—for missions of one, two, and three months. The primary goals were to demonstrate that astronauts could survive in space for long periods of time and to do extended useful work. Experiments would focus on solar and stellar astronomy, biomedical data–gathering, and studies of Earth resources.

The unmanned workshop, designated Skylab 1, was launched on May 14, 1973, using the last Saturn V booster to fly. The first crew of three men was to launch the following day. But sixty-three seconds after Skylab 1's liftoff, a meteoroid shield—also designed to shade the workshop from the sun's heat—deployed prematurely. It was torn from the space station by atmospheric drag, exposing the mounting for one of the workshop's two main solar array wings. The wing partially released. The exhaust plume of the Saturn's second-stage retro-rockets then hit the wing and literally blew it off. Finally, a strap from the meteoroid-shield debris wrapped around the remaining solar array wing, preventing its full deployment as planned in orbit.

The impact on the workshop was significant—with no meteoroid shield, it began to heat up, and the solar-wing damage meant that it could only generate less than half of the electrical power needed. Launch of the first manned mission, Skylab 2, was postponed. Within ten days, however, engineers at MSFC and elsewhere had devised ways for the first crew to make repairs.

On May 25, CDR Pete Conrad and two rookies, pilot (PLT) Paul Weitz and science pilot (SPT) Joe Kerwin, were launched into orbit, with tools and equipment to be used to cool the workshop by freeing the stuck solar wing and replacing the meteoroid shield. They rode atop a Saturn IB launched from Pad 39B, where a Saturn V ML had been modified to accommodate the shorter booster.

After rendezvousing with the workshop that afternoon, Weitz tried unsuccessfully to free the surviving solar array, reaching out from the CM's hatch

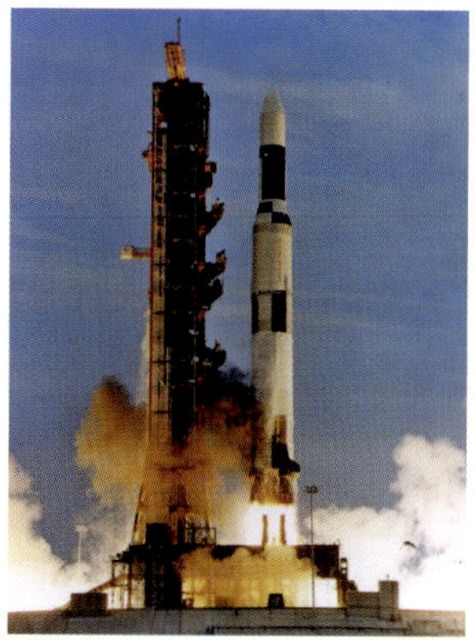

Skylab 1 launch: 1:30 p.m. (EDT), May 14, 1973

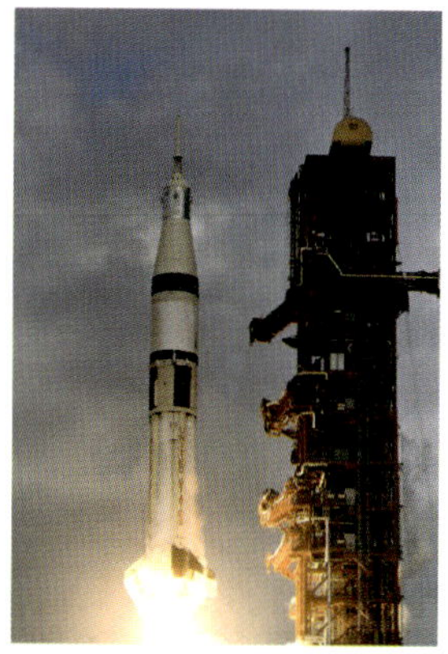

Skylab 2 launch: 9:00 a.m. (EDT), May 25, 1973

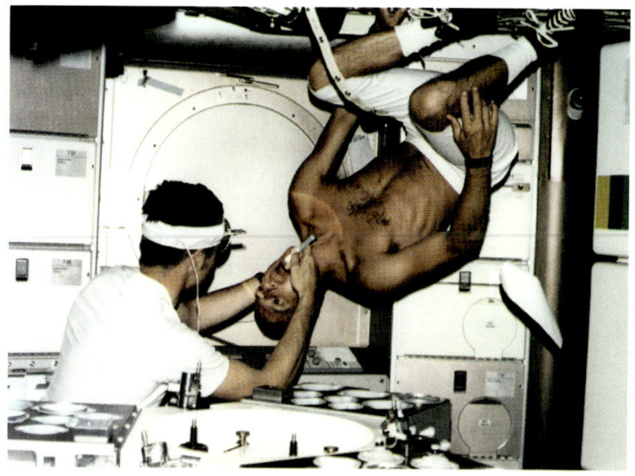

On orbit: twenty-eight days

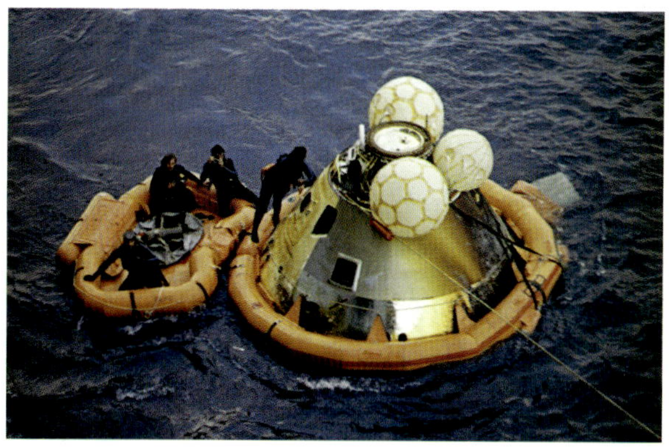

Splashdown: 9:49 a.m. (EDT), June 22, 1973

with tools on the ends of long poles while Kerwin held his legs. They docked with the workshop that night, and the next day they deployed a collapsible sunshield "parasol" with telescoping rods through a small airlock (intended to be used for various experiments) in the side of the workshop. The replacement sunshade allowed temperatures inside to drop low enough for the crew's comfort. Two weeks later, during a second EVA, Conrad and Kerwin used bolt cutters taped to a pole to snip an aluminum strap that had pinned the solar wing closed.

During their twenty-eight-day stay, the crewmen made further repairs to the workshop, conducted experiments, and gathered solar and Earth science data. The astronauts doubled Gemini VII's US space duration record, setting a new world record for the longest manned space flight. Conrad set the record for the most time in space by one astronaut.

The mission ended on June 22, 1973, when Skylab 2 splashed down in the Pacific Ocean just more than five miles from the recovery ship, USS *Ticonderoga*.

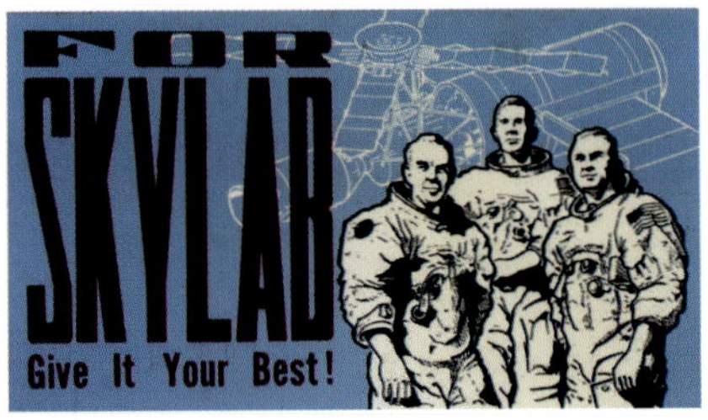

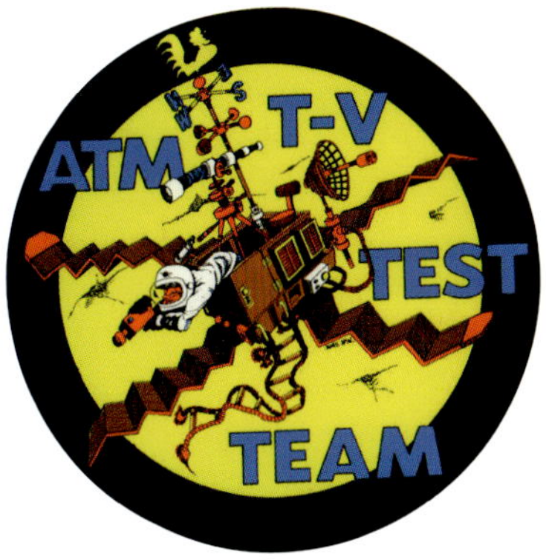

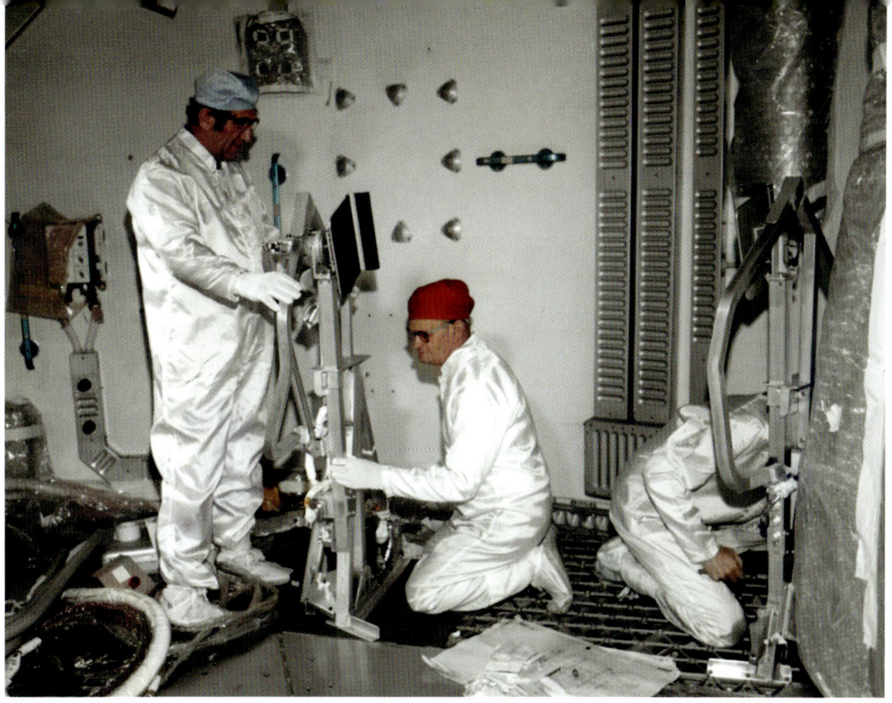

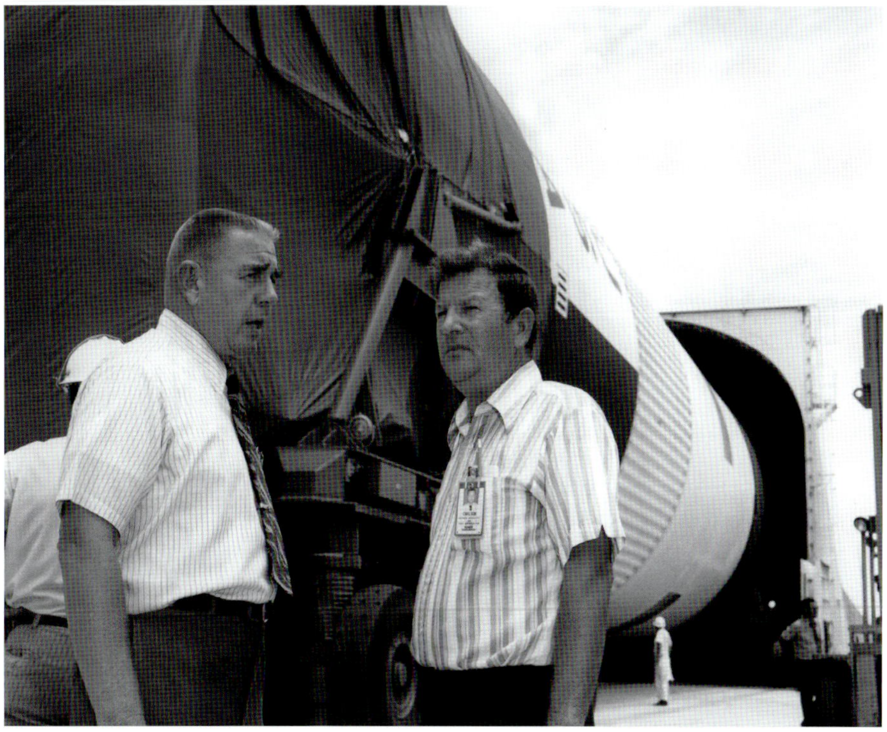

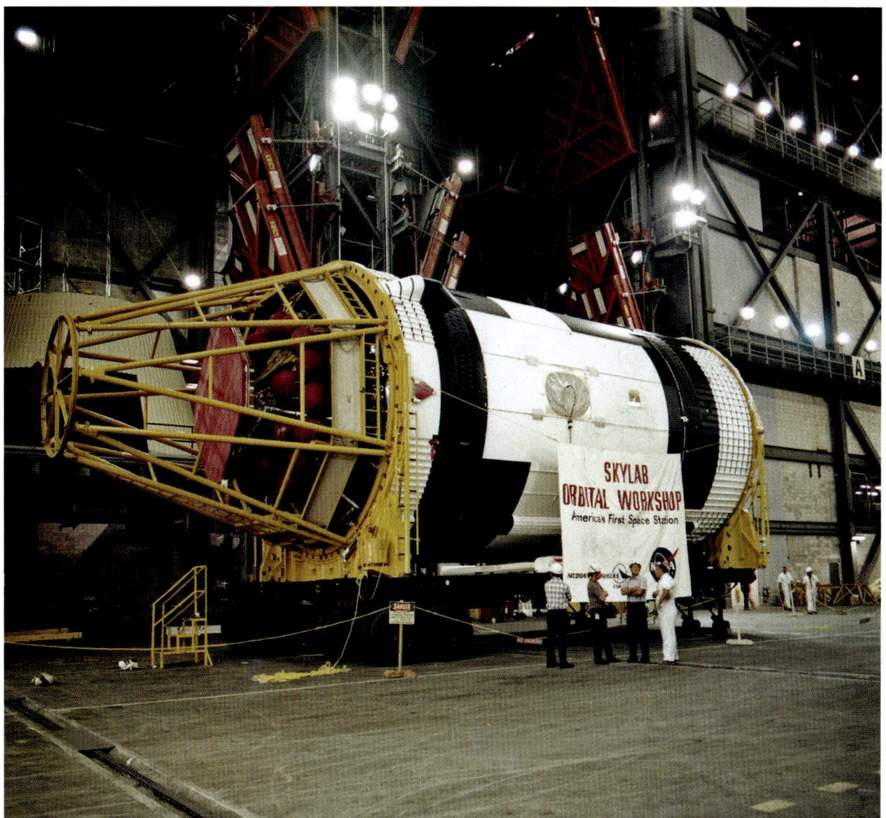

Top left: Test team emblem for ATM TV cameras, used to observe the sun

Top right: Workers at the McDonnell Douglas plant in Huntington Beach, California, install the end supports for the USAF's M509 experiment in the OWS's forward compartment in January 1972. The experiment involved testing a nitrogen jet–powered maneuvering backpack—known as the Automatically Stabilized Maneuvering Unit (ASMU)—and a handheld unit inside the workshop. Ribbed wiring conduits are on the wall at right.

Center right: Robert Hock, KSC Apollo/Skylab program manager (*left*), and Norm Carlson, launch vehicle test conductor, converse as the first stage of Saturn V No. 13 is off-loaded from the barge *Orion* at LC-39 on July 26, 1972. The five-day voyage to reach KSC traversed a thousand miles, from the Michoud Assembly Facility in New Orleans, Louisiana, through the Gulf of Mexico and up Florida's Atlantic coast.

Bottom right: The OWS, inside its converted Saturn V third stage, is parked on its transporter in the VAB's transfer aisle on October 2, 1972, after delivery to KSC. The red spheres at the left (aft) end would later contain nitrogen gas for attitude control thrusters; the flat refrigerator-system radiator is to the left of the spheres. The yellow support rings—at forward and aft—were used for hoisting, while the yellow cage-like structure at left is the dummy interstage protective support kit. Its inverted A-frame legs hinged outward to provide a support stand when the stage was moved to a vertical position, before it was hoisted to be mated with the Saturn's second stage.

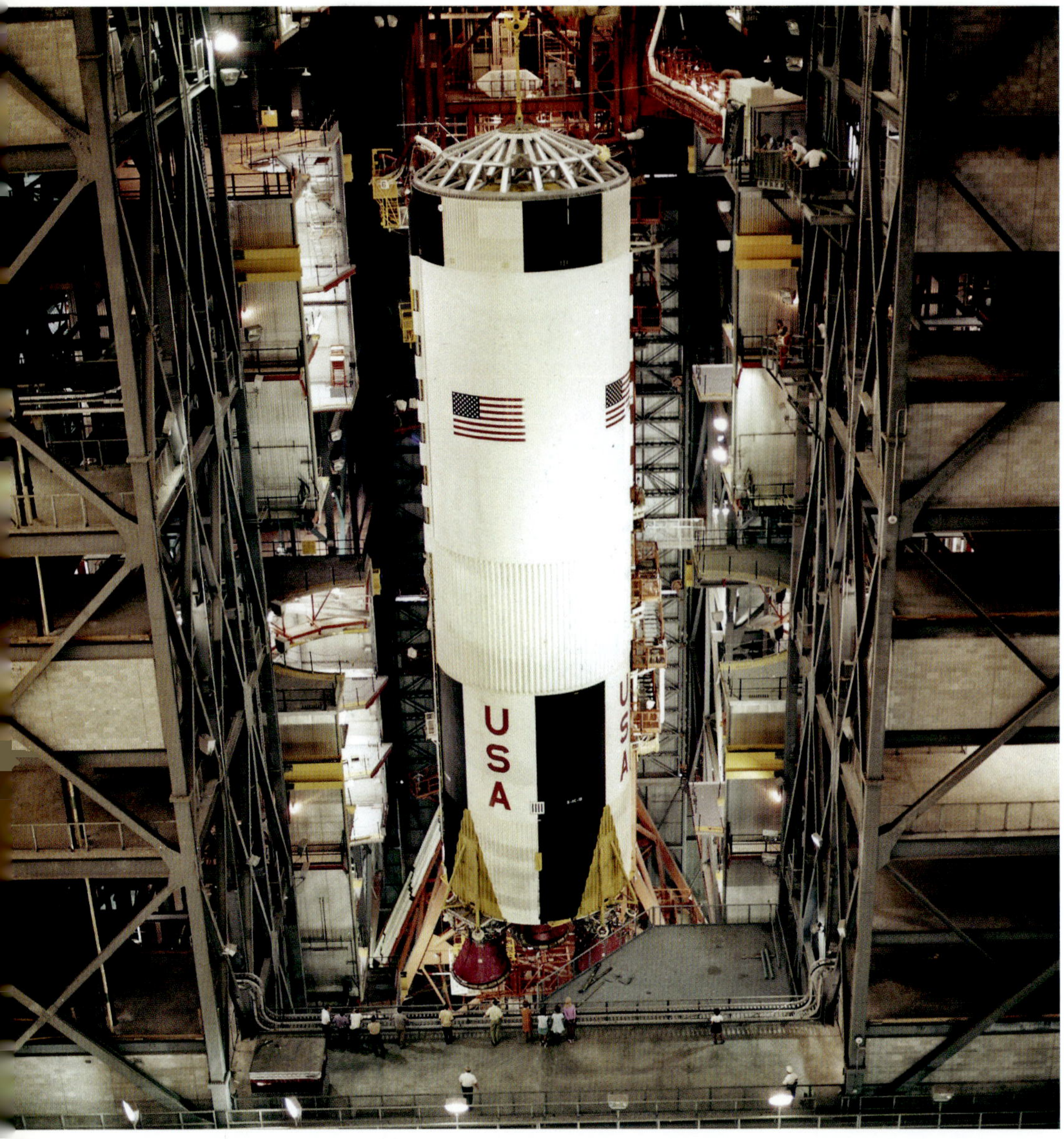

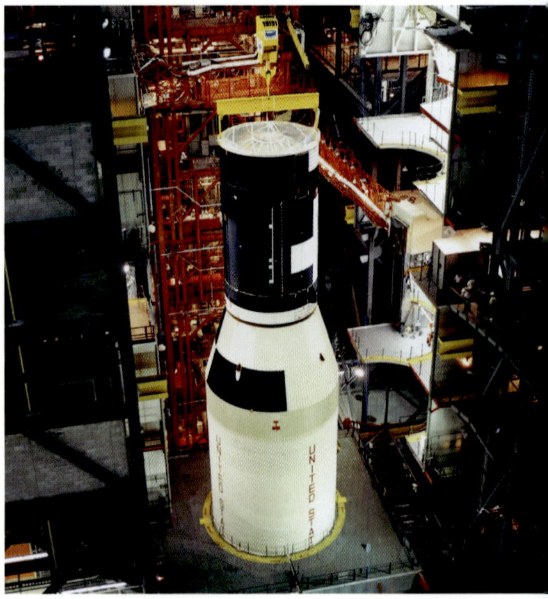

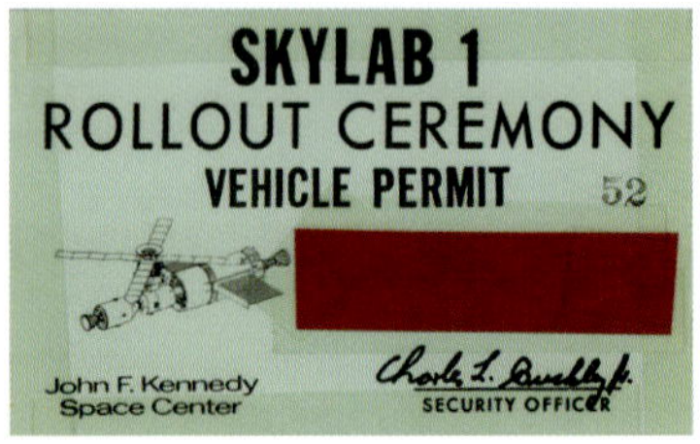

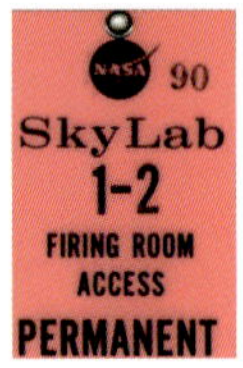

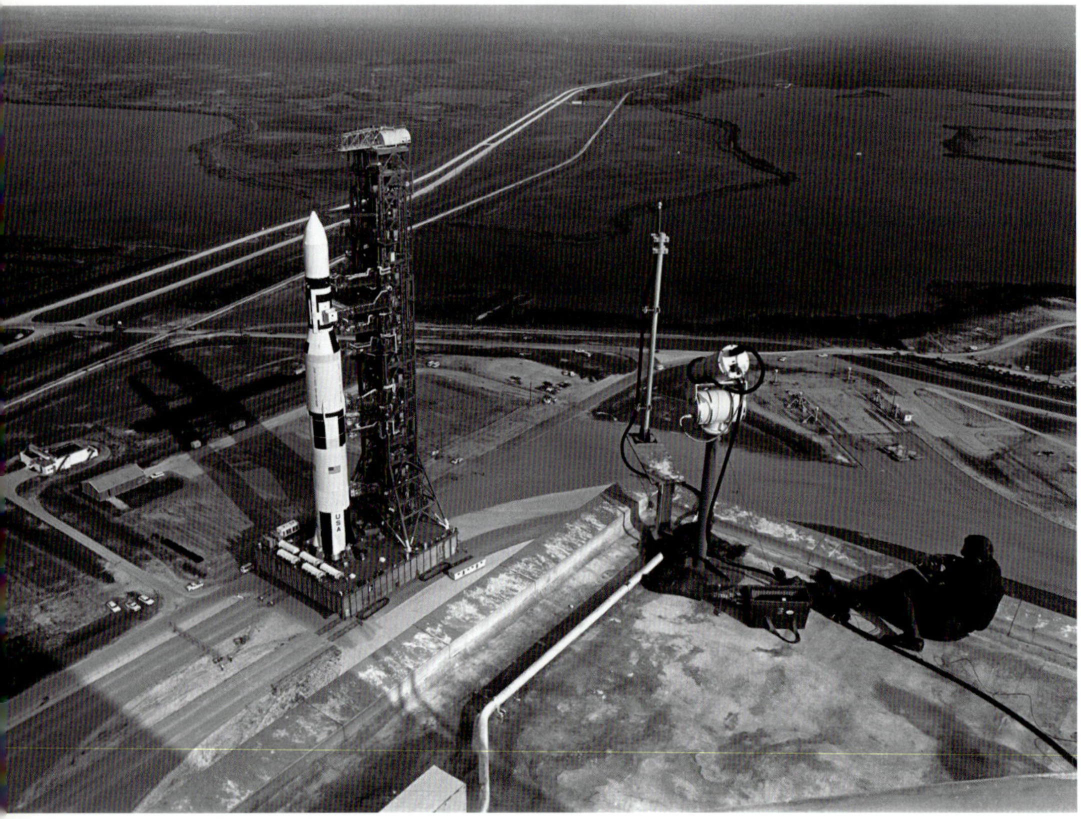

Top left: On August 2, 1972, KSC workers watch as S-IC-13, the first stage of the Skylab workshop's Saturn V, is lowered into place by the overhead crane onto ML-2 in the VAB's high bay 2. The fairings for its four outboard F-1 engines have not yet been attached.

Top right: The crane lowers the OWS onto the conical interstage on top of the Saturn's second stage—surrounded by a work platform—in high bay 2 on October 4, 1972. The red LUT is behind the stages. Apollo 17 occupied high bay 1 at the time.

Bottom left: A photographer positions himself near the northwest corner of the VAB roof, 525 feet up, during the Skylab 1 rollout on April 16, 1973. An ML parking area is just to the right of the closed-circuit television camera and lightning rod. The Kennedy Parkway stretches north, from left center to right top. (photo by Jacques Tiziou)

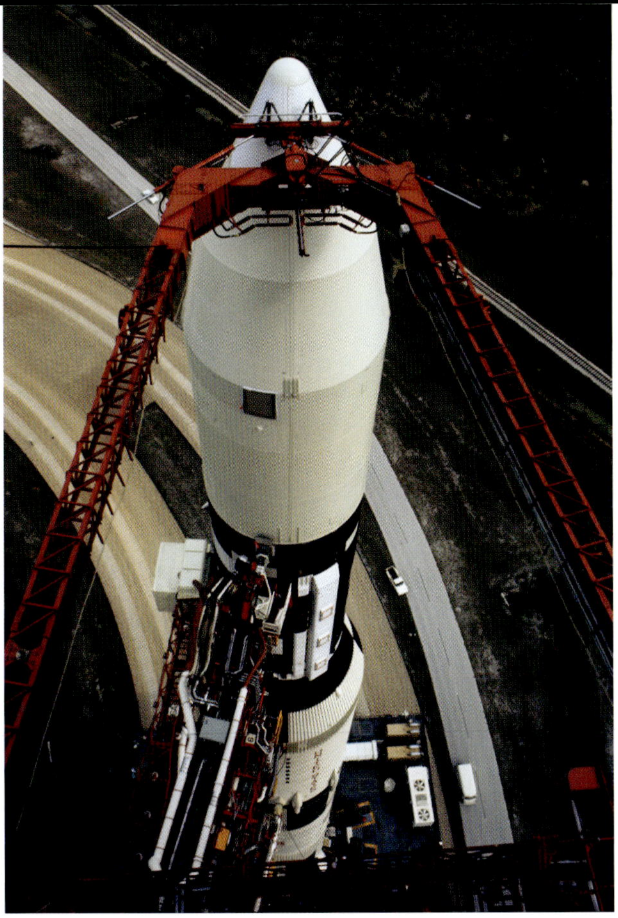

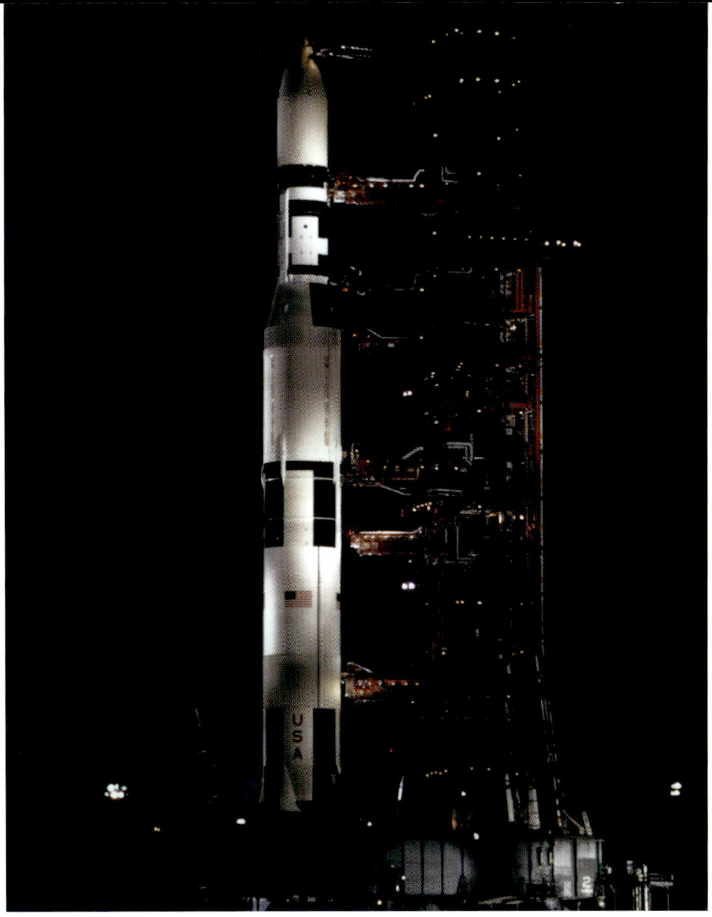

Top left: View from the LUT of the fifty-six-foot-tall payload shroud, during the Skylab 1 rollout on April 16, 1973. The damper arms extending to the shroud mitigated movement caused by winds. The swing arms and umbilical connections to the OWS and second stage are below. The uppermost swing arm, which had been used to access the Apollo CMs for manned launches, was removed for this launch. (photo by Jacques Tiziou)

Top right: Skylab 1 and its LUT and ML are illuminated by floodlights at Pad 39A on May 13, 1973, the night before launch. (photo by Tom Usciak)

Below: The final Saturn V to be launched comes to life at 1:30 p.m. (EDT) on May 14, in this photo looking west along Pintail Creek. The workshop's meteoroid shield prematurely deployed just over one minute later, setting off a chain reaction of problems. Atmospheric drag ripped away the shield, releasing one of the two solar panel wings. The exhaust plume of the Saturn's second stage retro-rockets then struck the wing and blew it off, while debris jammed the remaining wing. This would be the final launch from Pad A until the first Space Shuttle mission in 1981.

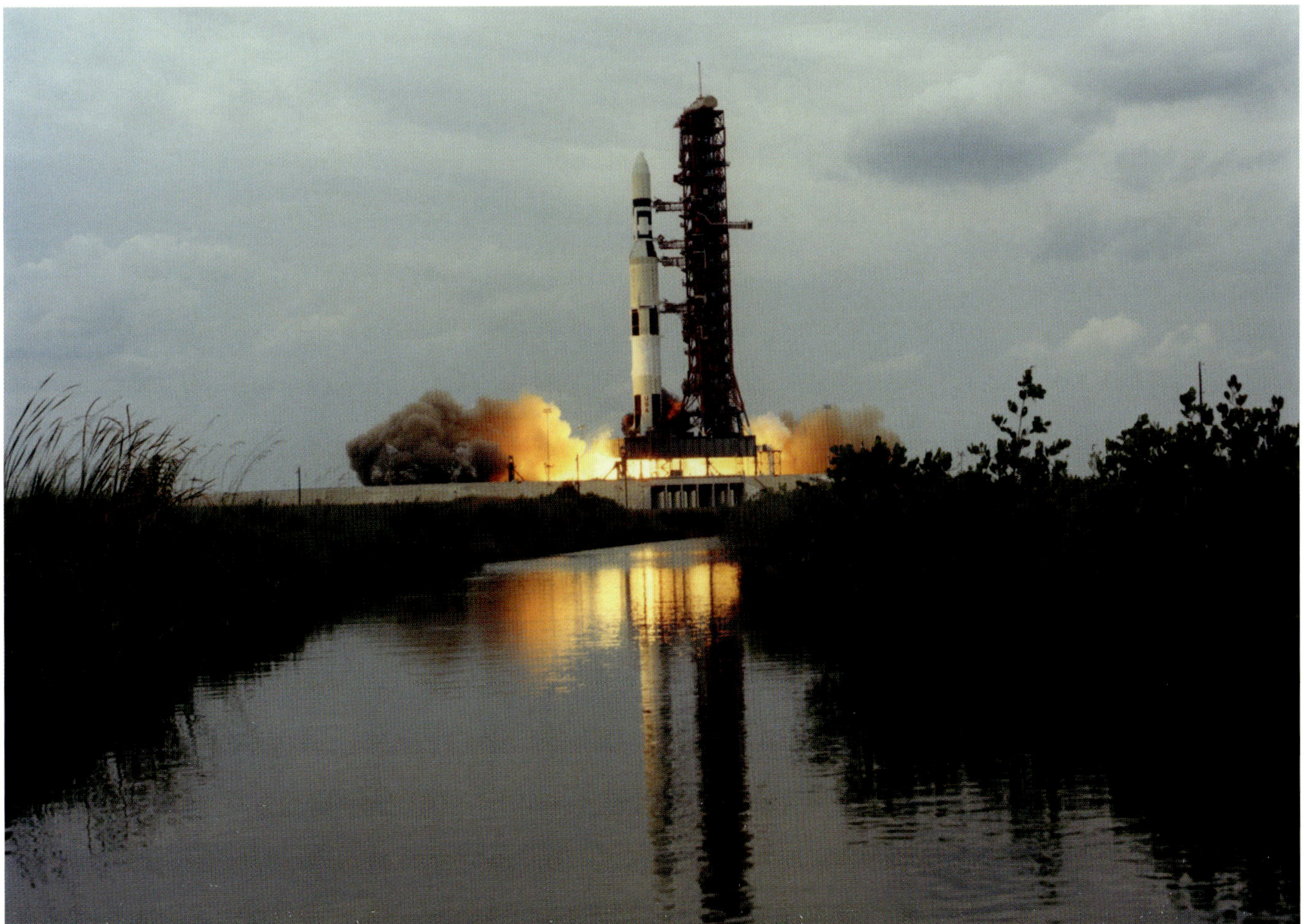

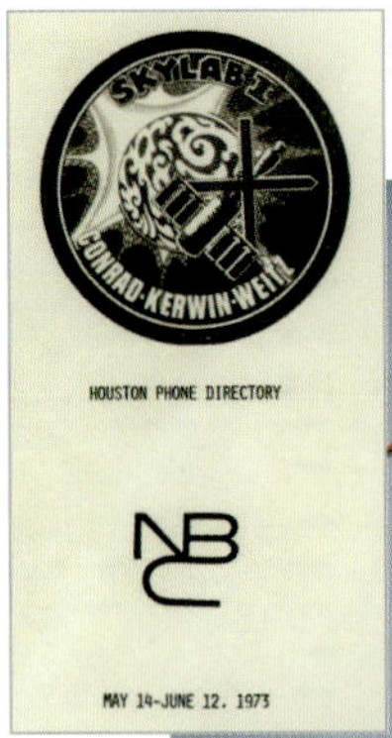

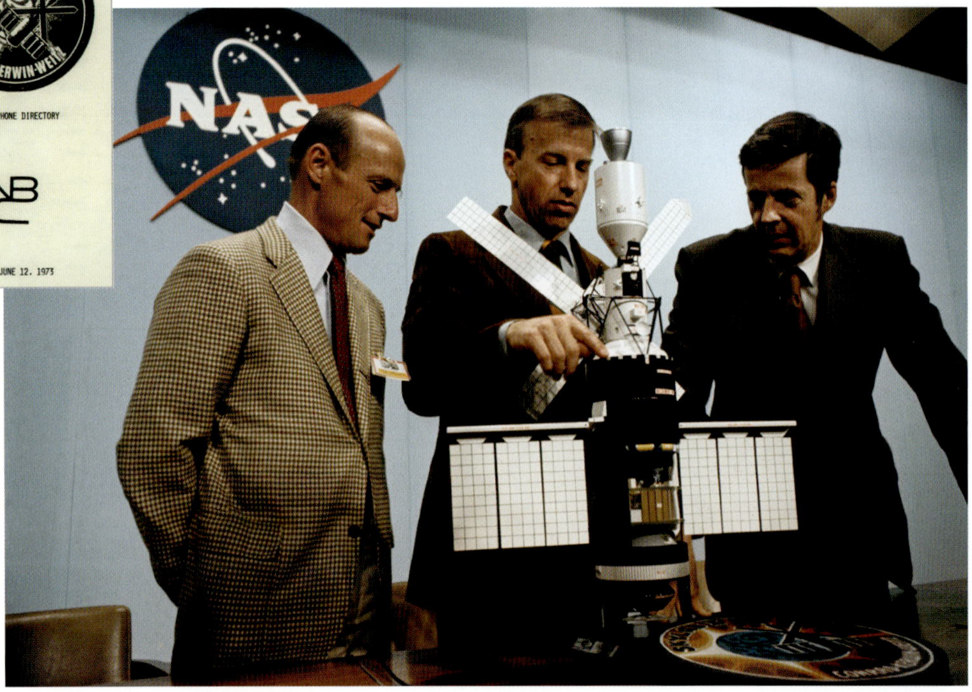

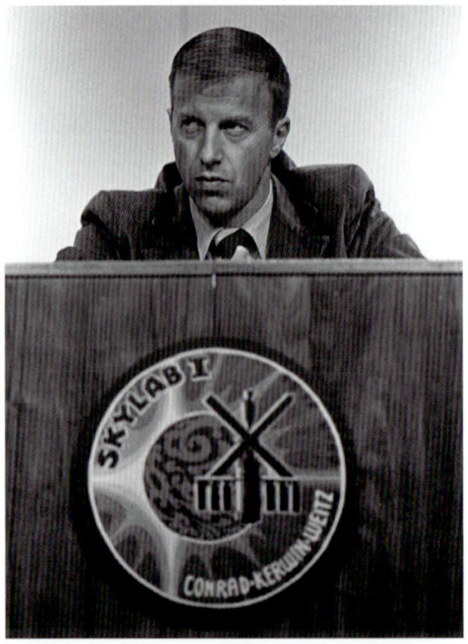

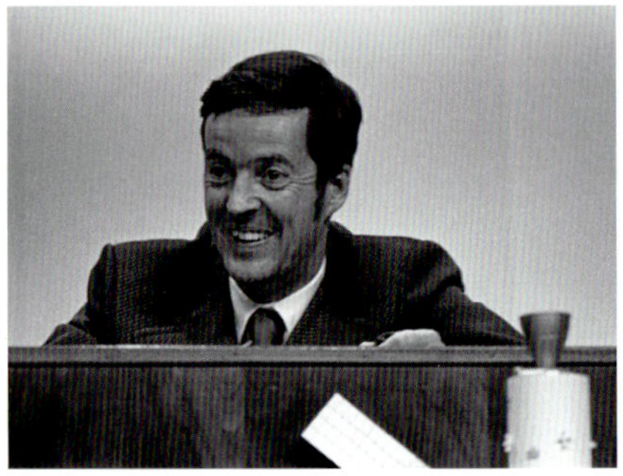

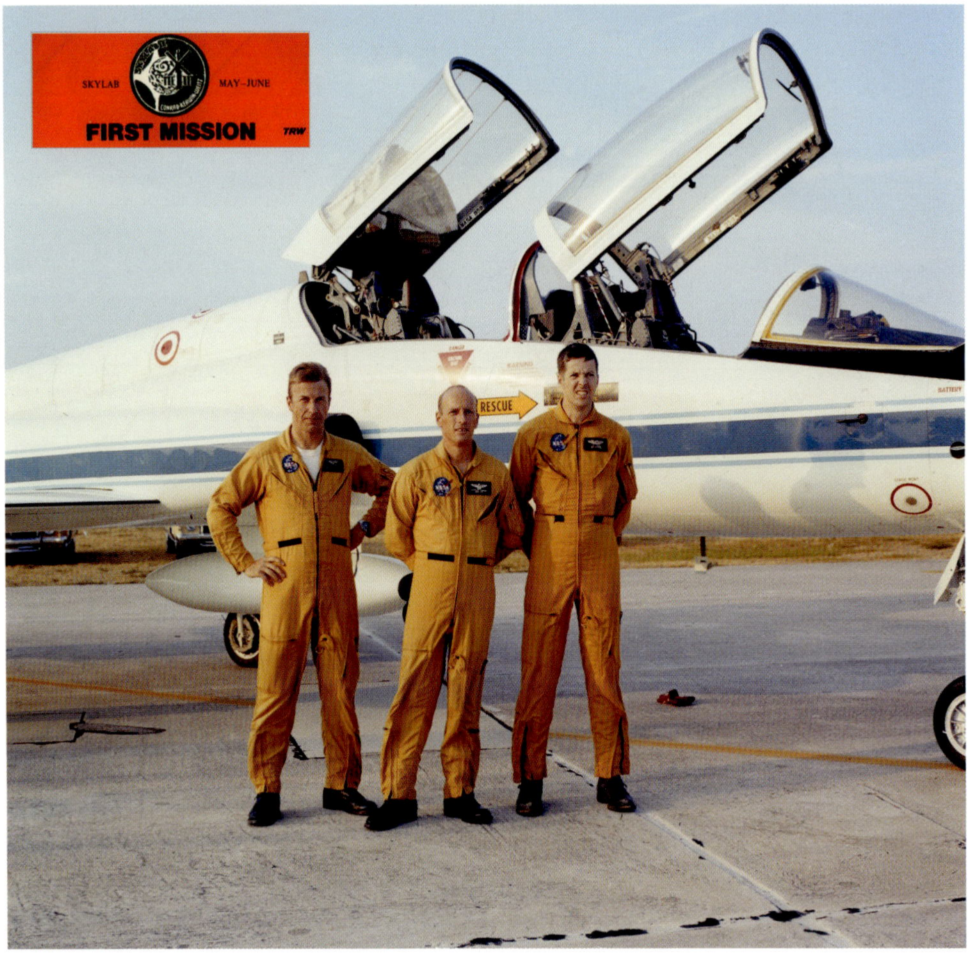

Above: Weitz, Conrad, and Kerwin stand next to a T-38 training jet at Patrick AFB on May 3, 1973. They have arrived from Houston to participate in the CDDT. The USAF began using the Northrop aircraft in 1961 as a jet trainer; NASA soon adopted it as an effective and critical part of astronaut training, used throughout the Space Shuttle program. (photo by Jacques Tiziou)

Top left: Pete Conrad, Weitz, and Kerwin, an all–US Navy crew, pose with a Skylab model built by MSFC, following their pre-flight news conference in the Bldg. 2 auditorium at Houston's Johnson Space Center (JSC) on April 17, 1973. MSC had been renamed JSC that February in honor of the late former president. On April 4, the space agency officially set their launch date as May 15. All three Skylab crews were named on January 19, 1972.

Top right: US Navy Cmdr. Paul J. Weitz, forty-one, was a rookie selected as an astronaut in 1966. He would make his second and final space flight as CDR of STS-6 in 1983. (photo by Jacques Tiziou)

Bottom right: US Navy Cmdr. Joseph P. Kerwin, forty-two, was a Medical Corps doctor who had been chosen as an astronaut in 1965. He was CAPCOM for the Apollo 13 mission. Skylab 2 would be his only space flight. (photo by Jacques Tiziou)

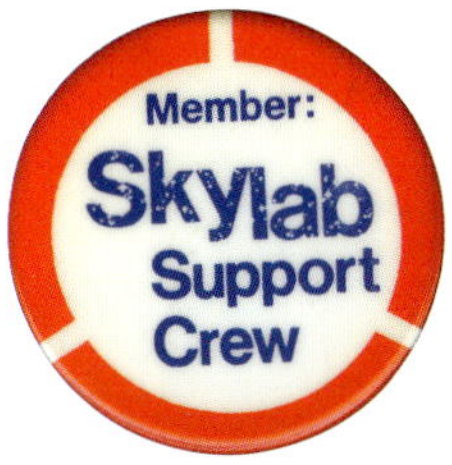

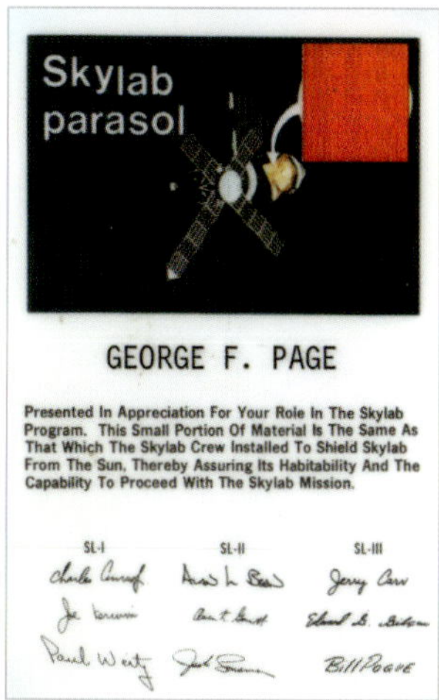

Direct right: Commemorative card—with a sample of the parasol material—given to Skylab ground personnel. George Page was chief of the KSC Spacecraft Operations Division.

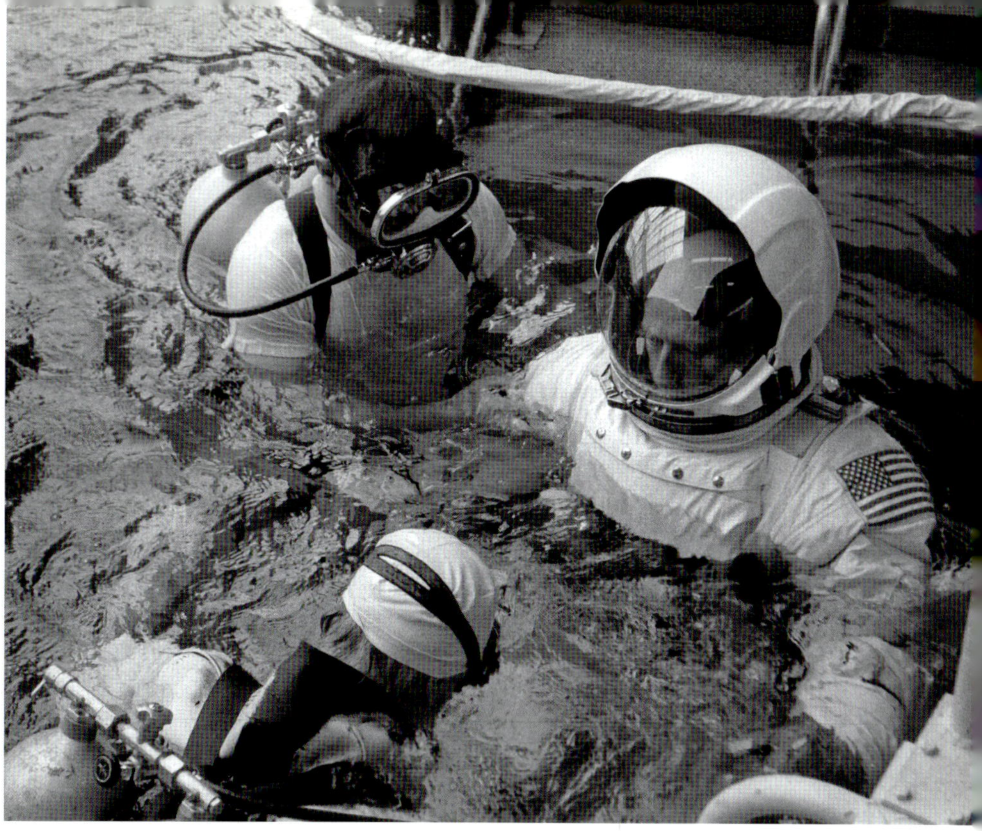

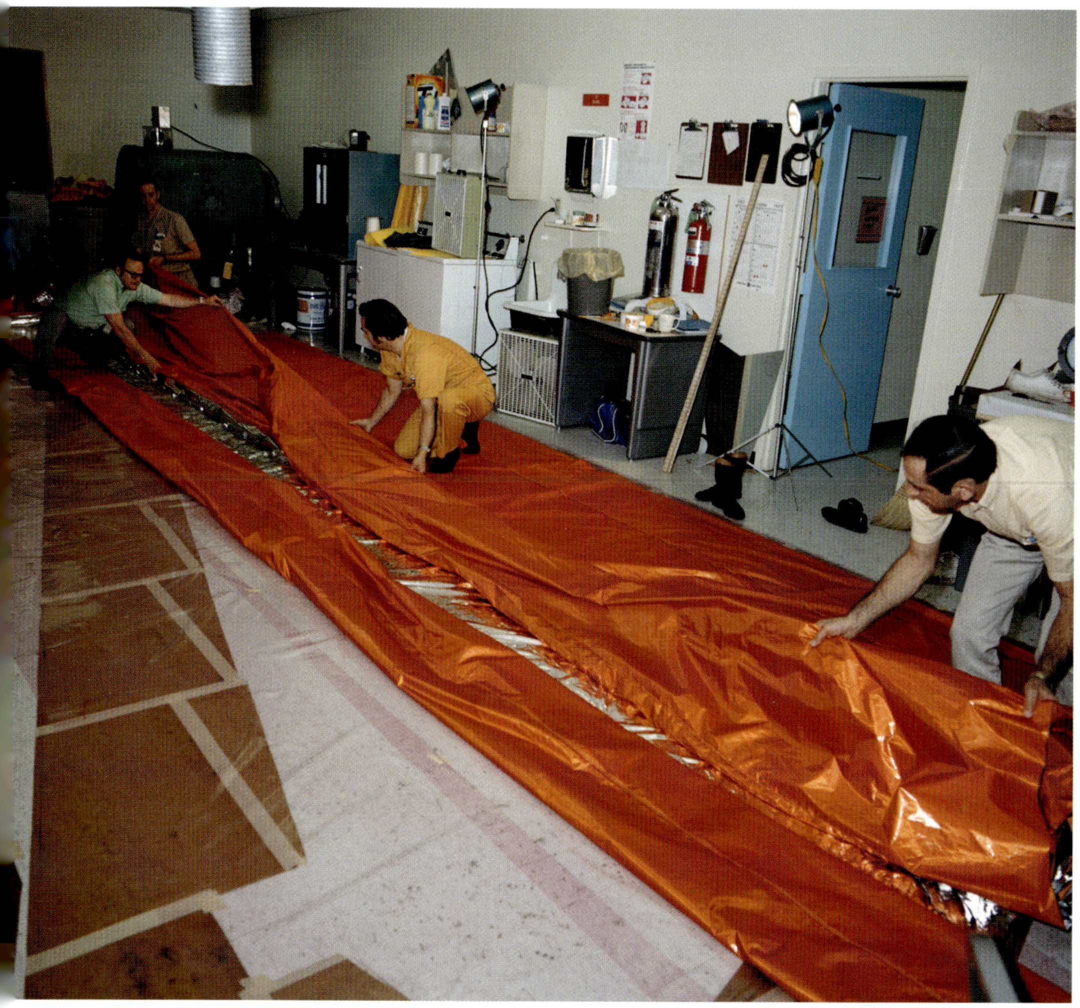

Above: Technicians fold the parasol, which was twenty-four by twenty-two feet, in the Technical Services shop in JSC's Bldg. 10 on May 18, 1973. Four telescoping extension poles were subsequently added. G. T. Schjeldahl Company of Northfield, Minnesota, created the sunshade, which was made of aluminized Mylar and laminated to orange nylon ripstop fabric. Schjeldahl had manufactured the first passive communications satellite, *Echo 1*, in 1960. Seamstresses from ILC sewed the parasol; it was designed, fabricated, and tested in seven days.

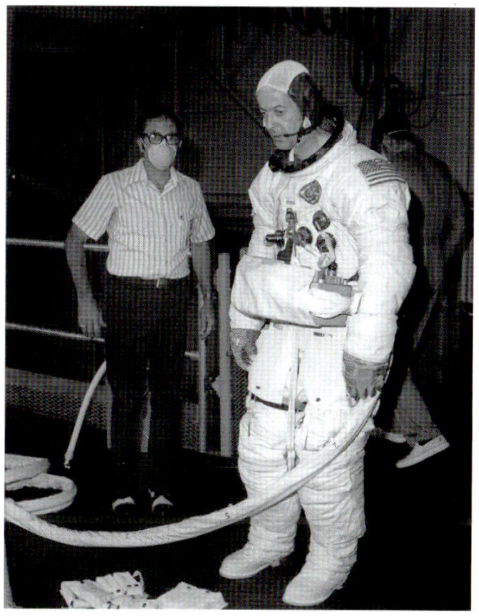

Top right: Divers assist Conrad in the Neutral Buoyancy Simulator at MSFC on May 22, 1973, while he practices EVA procedures to resolve the damaged workshop's power and thermal problems.

Center right: Kerwin prepares for underwater EVA training in the Neutral Buoyancy Simulator on May 22, 1973. His tether is attached to an oxygen pressure control unit around his waist, designed especially for Skylab. All Skylab crews wore modified Apollo A7LB space suits made by ILC.

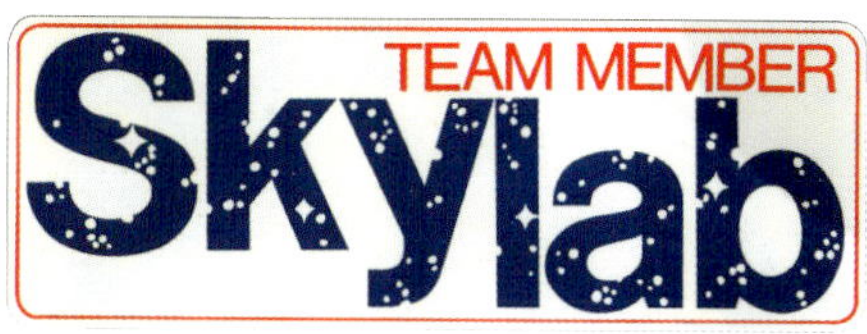

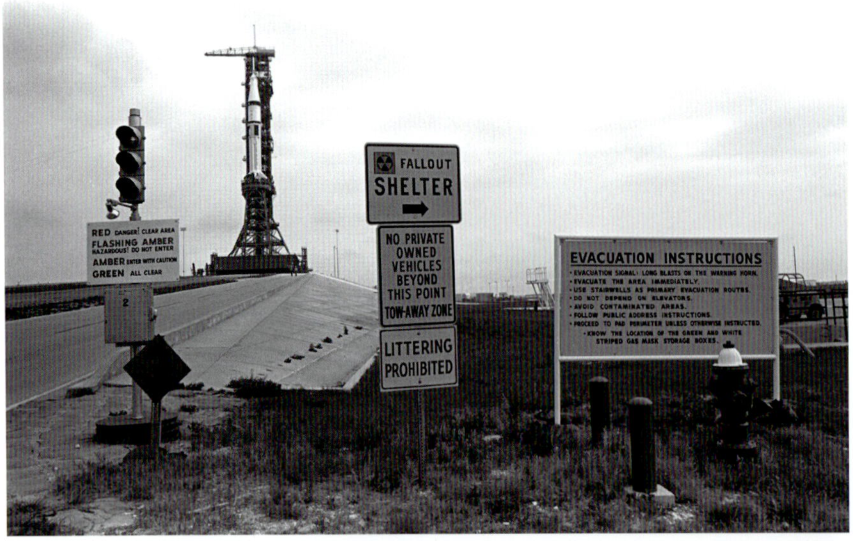

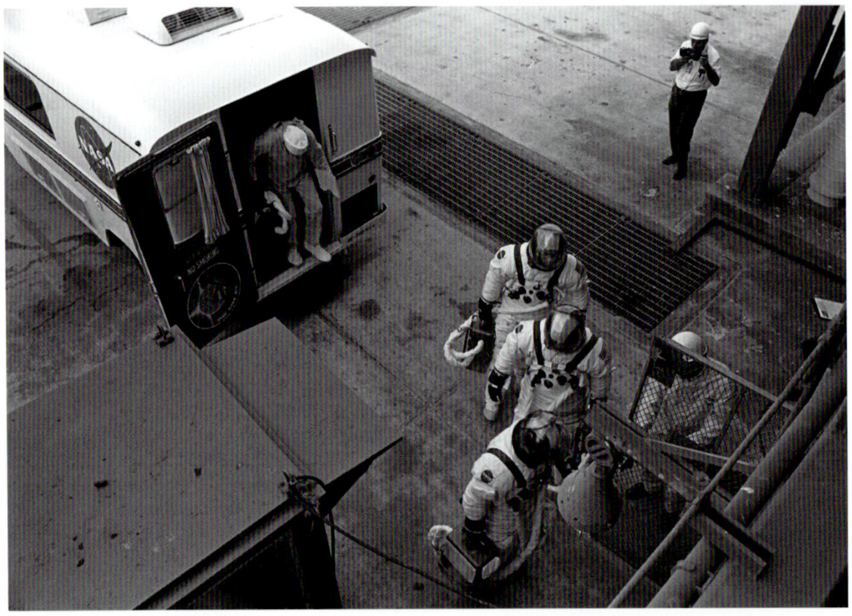

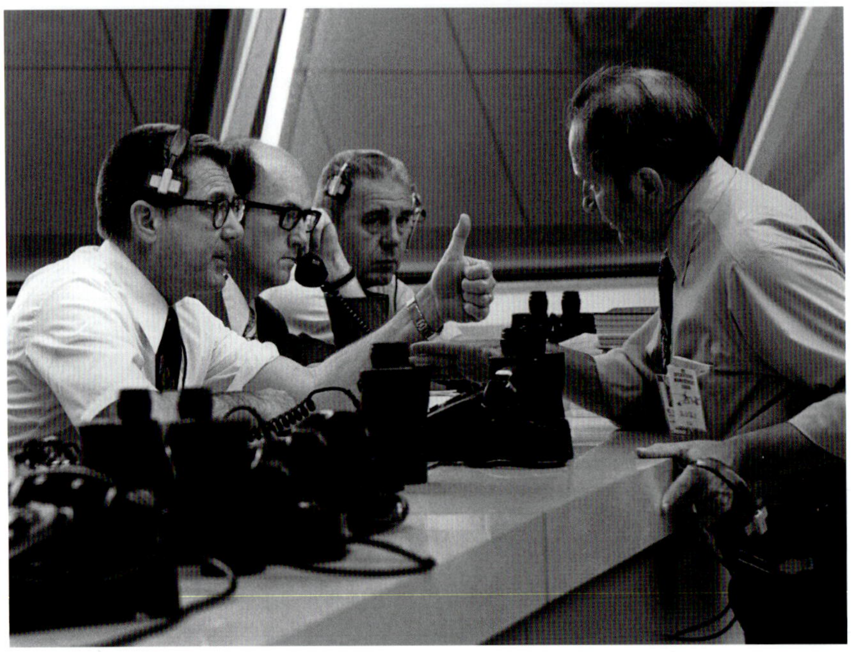

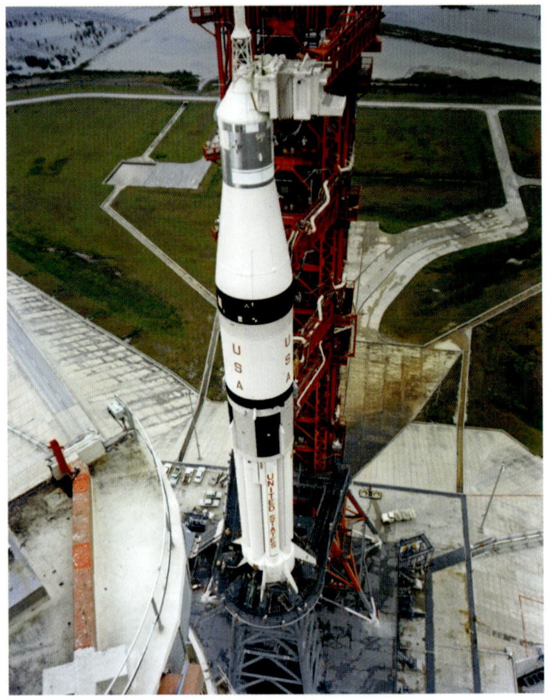

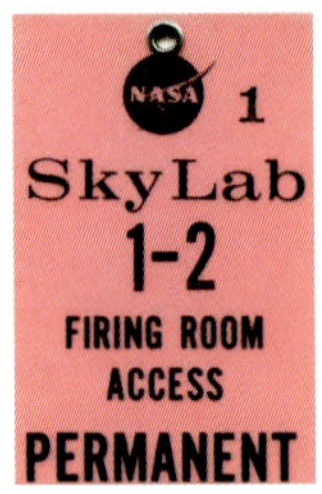

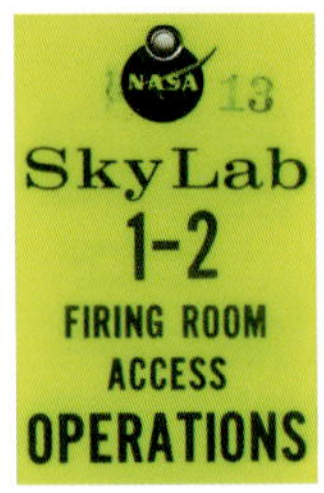

Top left: Skylab 2 at Pad 39B on May 4, 1973, on a 127-foot-tall pedestal known as the "milkstool." The milkstool was used to adapt the mission's Saturn IB booster to ML-1, designed for the taller Saturn V. The stack had first been moved to the pad in January for fueling checks; it was then returned to the VAB until a second rollout on February 26. (photo by Jacques Tiziou)

Top right: Skylab 2's 224-foot Saturn IB booster, seen from the MSS during MSS rollback on the afternoon before launch, May 24, 1973.

Center left: Conrad, Kerwin, and Weitz arrive at Pad 39B for the CDDT on May 4, 1973. Suit tech Joe Schmitt exits the transfer van carrying a spare suit ventilator. (photo by Jacques Tiziou)

Bottom left, left to right: Skylab Program Director William Schneider; test conductors James Harrington and Robert Hock; and Deputy Director of Launch Vehicle Operations Ike Rigell, in Firing Room 3 on launch morning, May 25, 1973.

Opposite page: Skylab 2, seconds after ignition at 9:00 a.m. (EDT) on May 25. The Florida Highway Patrol estimated that no more than thirty thousand spectators attended, the smallest crowd to see a manned US launch since Gemini. The crew's families watched the launch on TV from Houston.

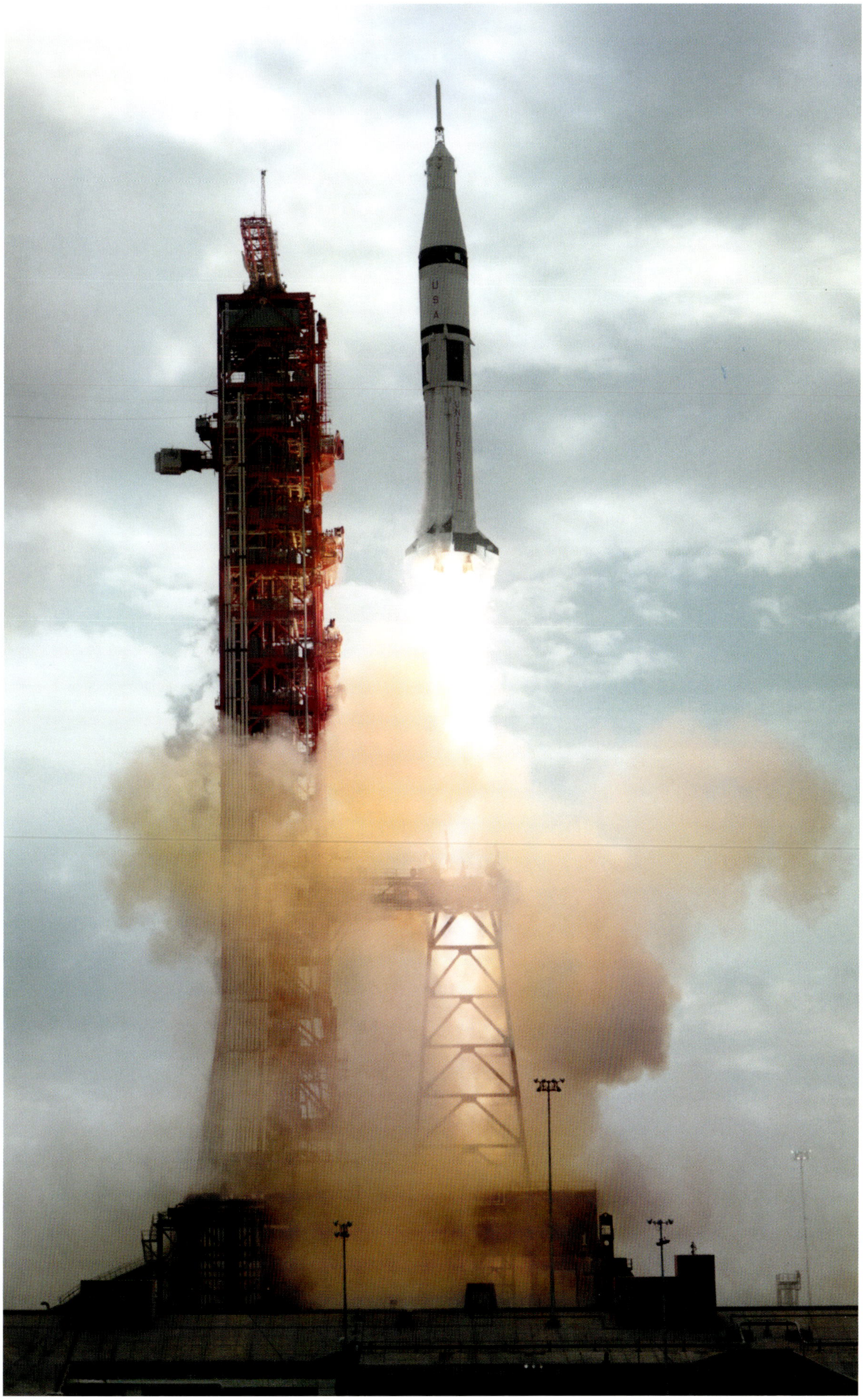
USA
UNITED STATES

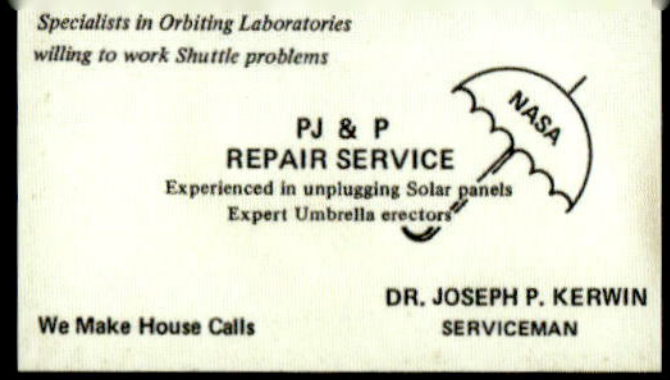

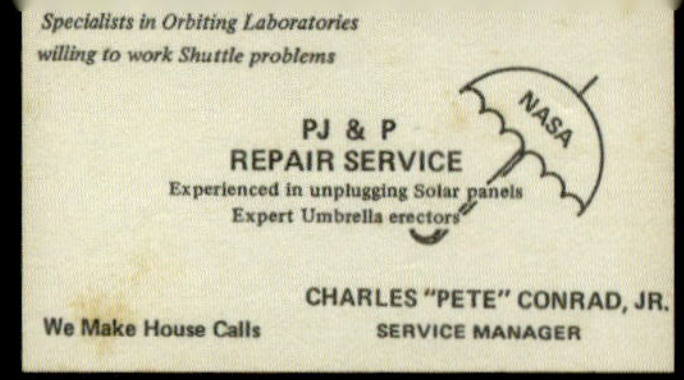

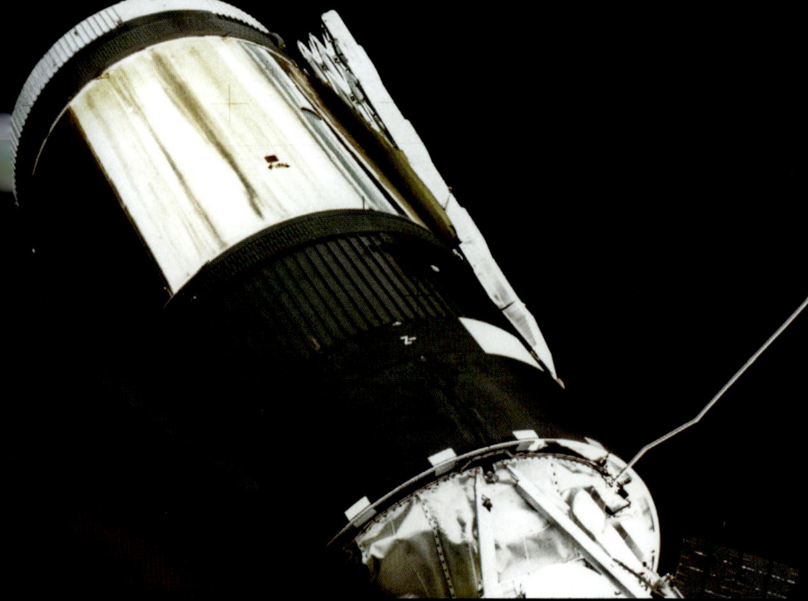

Top left: The Skylab 2 crew's view of the OWS during rendezvous on May 25, 1973. The craft's stuck solar panel wing is on the right, with a portion of a solar array wing on the ATM at bottom right. The CM's window edge is at left. At right is one of two discone antennas, used for telemetry, command, and voice. They used modified Gemini equipment and were on thirty-foot booms, intended to minimize interference from the solar arrays.

Bottom left: This rendezvous photo of the OWS shows the blackening of its goldized exterior skin. After its meteoroid shield was lost during launch, the OWS was exposed to the sun—and its three-hundred-degree-Fahrenheit temperatures—for ten days. Wires protrude from the missing solar wing mount in the center of the ribbed section.

Bottom right: Kerwin holds a pole that he and Conrad used to free the stuck solar wing during an EVA on June 7, 1973. A bolt cutter is taped to the other end. The unit around his waist is an astronaut life support assembly (ALSA), built by the AiResearch Division of the Garrett Corporation, based in Los Angeles, California. The ALSA regulated his oxygen supply from the tether at left.

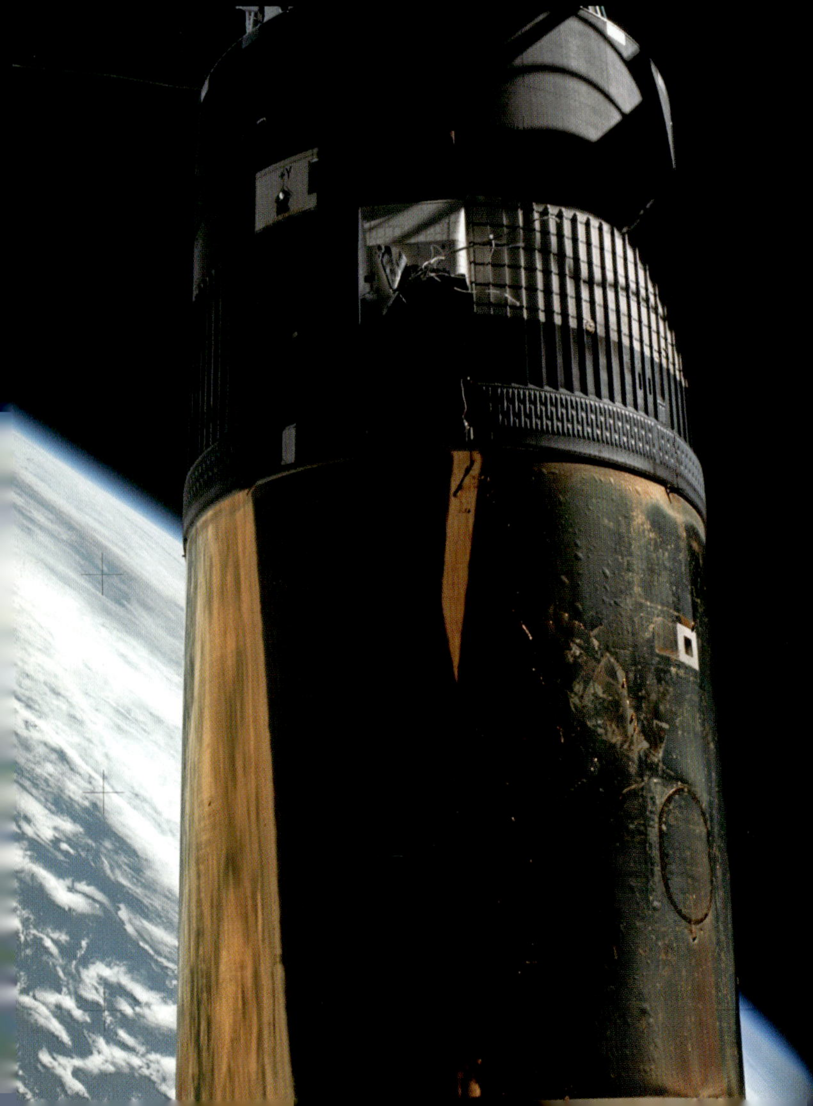

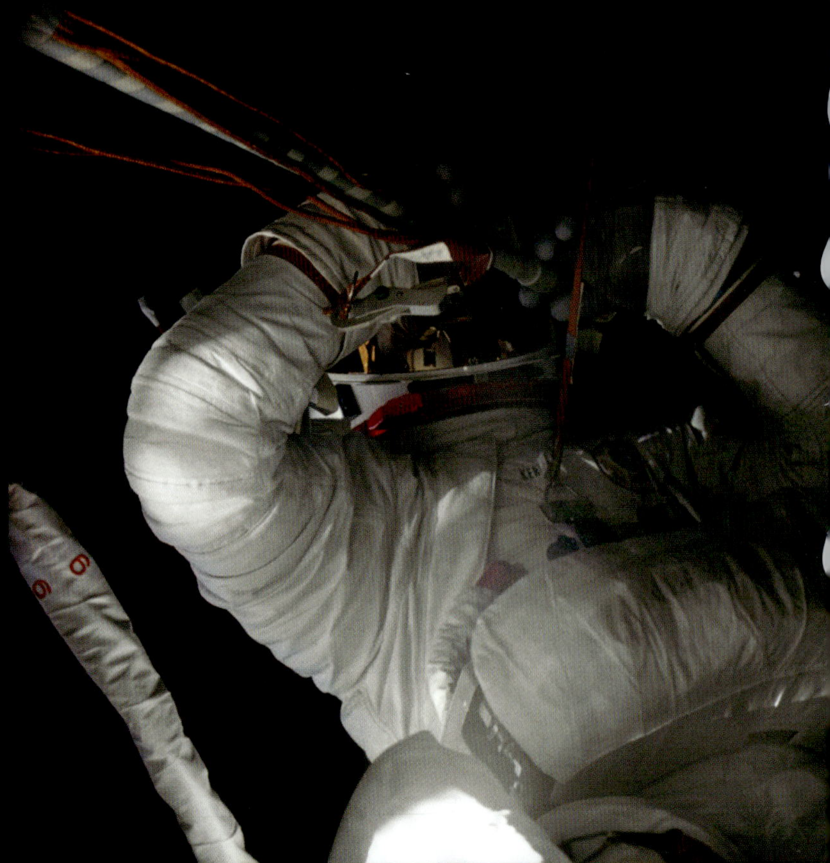

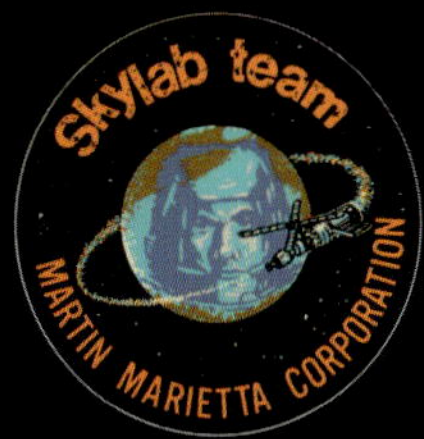

Top right: Kerwin with his EVA suit in the workshop's forward compartment. The suit's boots are in restraints on the blue portable panel mounted to the floor grid, used to assist in donning or doffing the suit, as well as during suit-drying operations (to remove moisture after use). The ASMU backpack, which was not tested until the next mission, is at top left.

Bottom right: Weitz (*left*) and Conrad were photographed by Kerwin in the workshop's Multiple Docking Adapter (MDA), which housed the control console for the ATM (*at right*); controls and sensors for Earth resources viewing; and other experiment facilities. They wear caps made from Beta cloth.

Bottom left: The OWS 270 miles above Earth, photographed during Skylab 2's final fly-around inspection prior to reentry and splashdown, on June 22. The workshop's single solar panel is at top right, with the orange solar shield below, rigged by the crew to replace the meteoroid shield lost during launch.

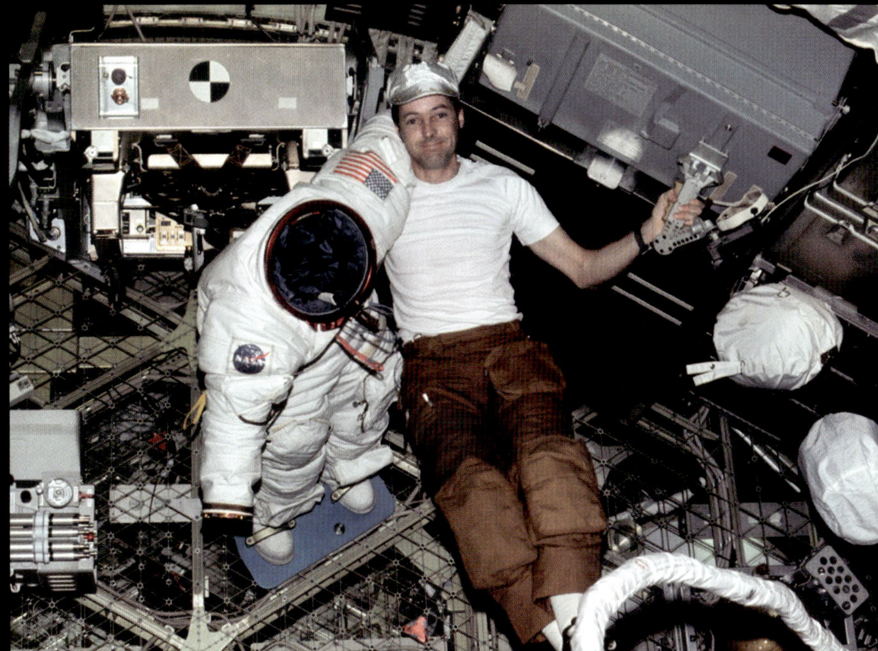

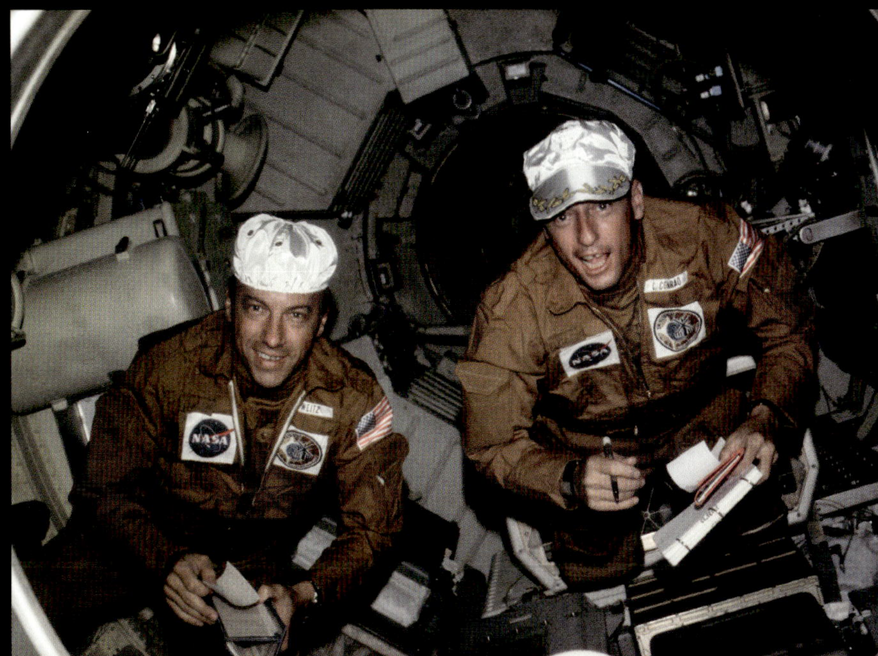

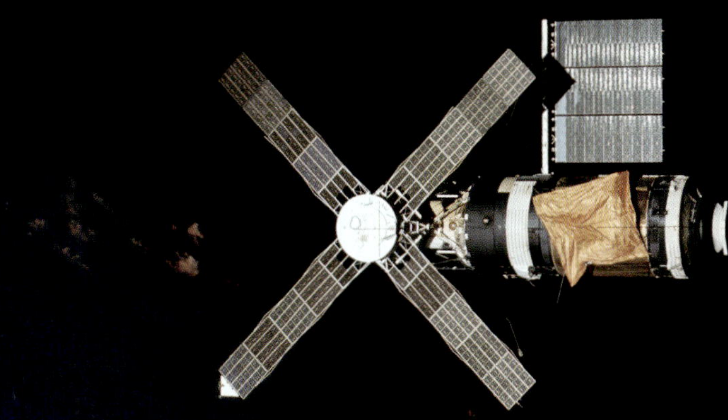

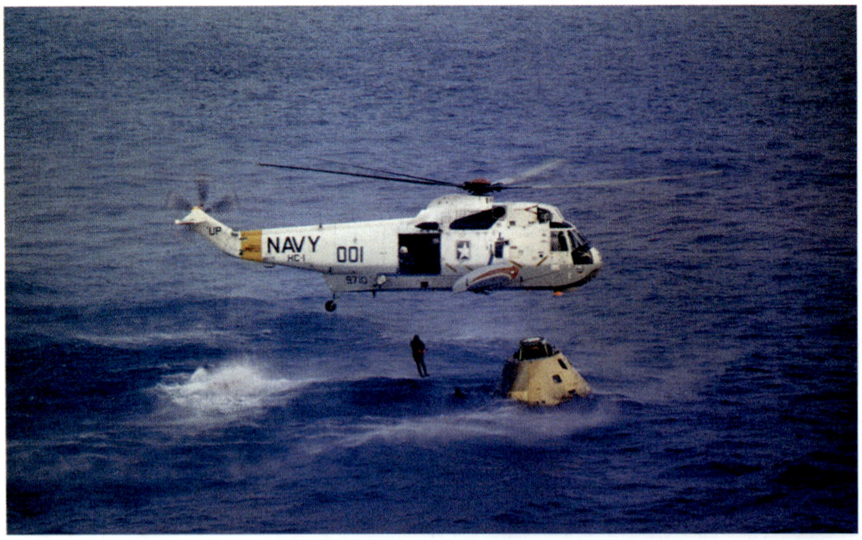

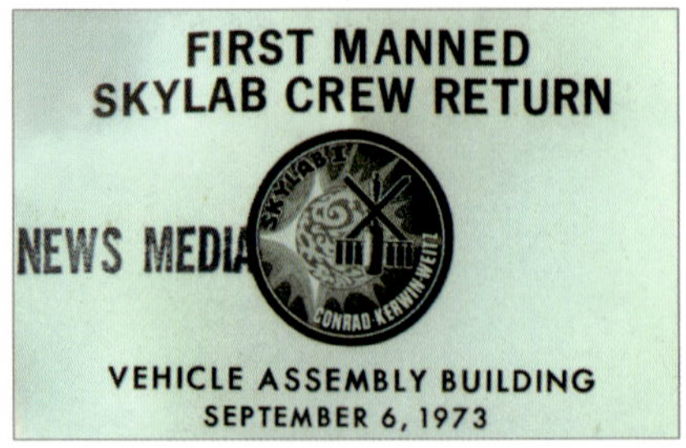

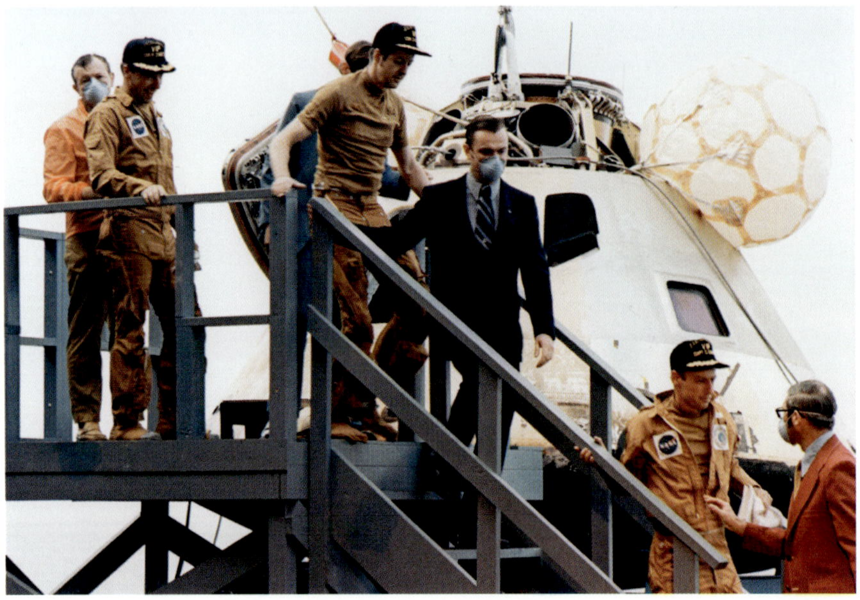

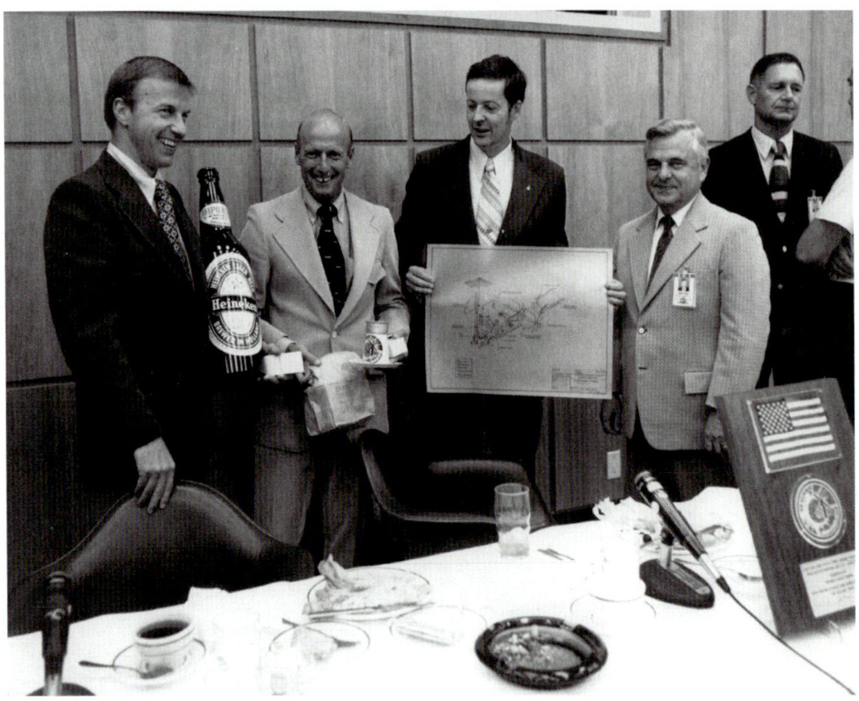

Above: Badge for the crew return ceremony at the VAB. The crowd of one thousand workers was the smallest to welcome home an astronaut crew.

Top left: A diver leaps from a Sea King helicopter, before attaching a flotation collar to Skylab 2. The spacecraft was recovered by USS *Ticonderoga* 830 miles southwest of San Diego on June 22, 1973.

Center left: NASA physicians assist Weitz, Kerwin, and Conrad after they exit from their CM aboard the *Ticonderoga*. Kerwin, wearing inflatable pants to prevent blood from pooling in his legs, was dizzy and had difficulty walking unaided.

Bottom left: After a luncheon at KSC HQ on September 6, 1973, the crew displays gag gifts: beer and cookies—which they had missed during their twenty-eight days in space—for Weitz and Conrad respectively, from reporter Mary Bubb; and, for Kerwin, an engineering drawing of a Rube Goldberg–like device, which would deploy the parasol sunshield and free the solar panel wing. At right are Launch Director Walter Kapryan and KSC Deputy Director Mike Ross. A plaque with a US flag and crew emblem is on the table. It had been carried on the mission. September 6 was the forty-second day in orbit for the Skylab 3 crew.

Opposite page: Weitz, Kerwin, and Conrad speak to the *Ticonderoga*'s crew as the aircraft carrier arrives in San Diego on June 24. They met with President Richard Nixon and Soviet leader Leonid Brezhnev at the helicopter pad at the "Western White House," Nixon's West Coast home in San Clemente, California, later that day, after the two leaders declared an end to the Cold War in a joint communiqué.

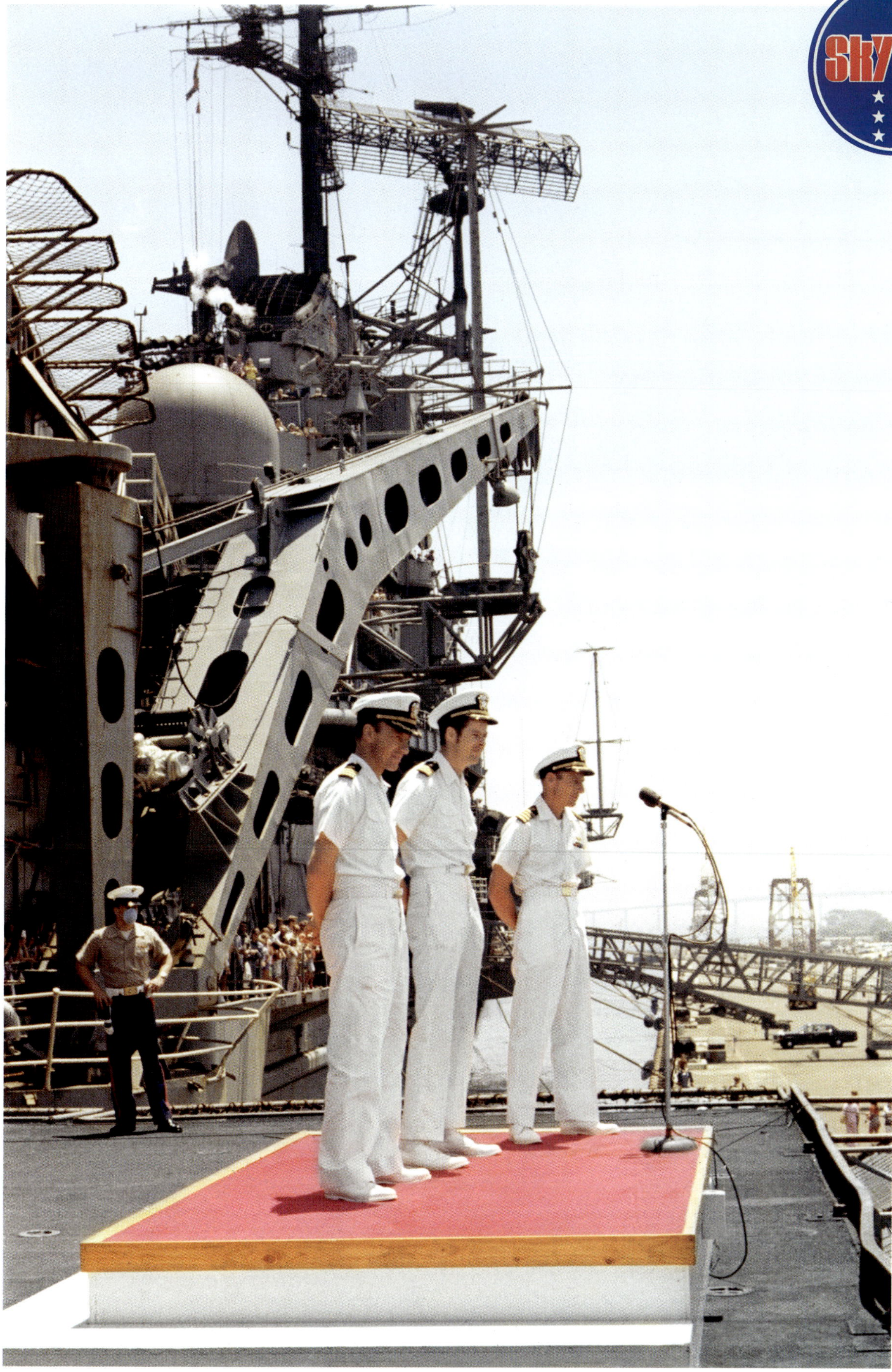

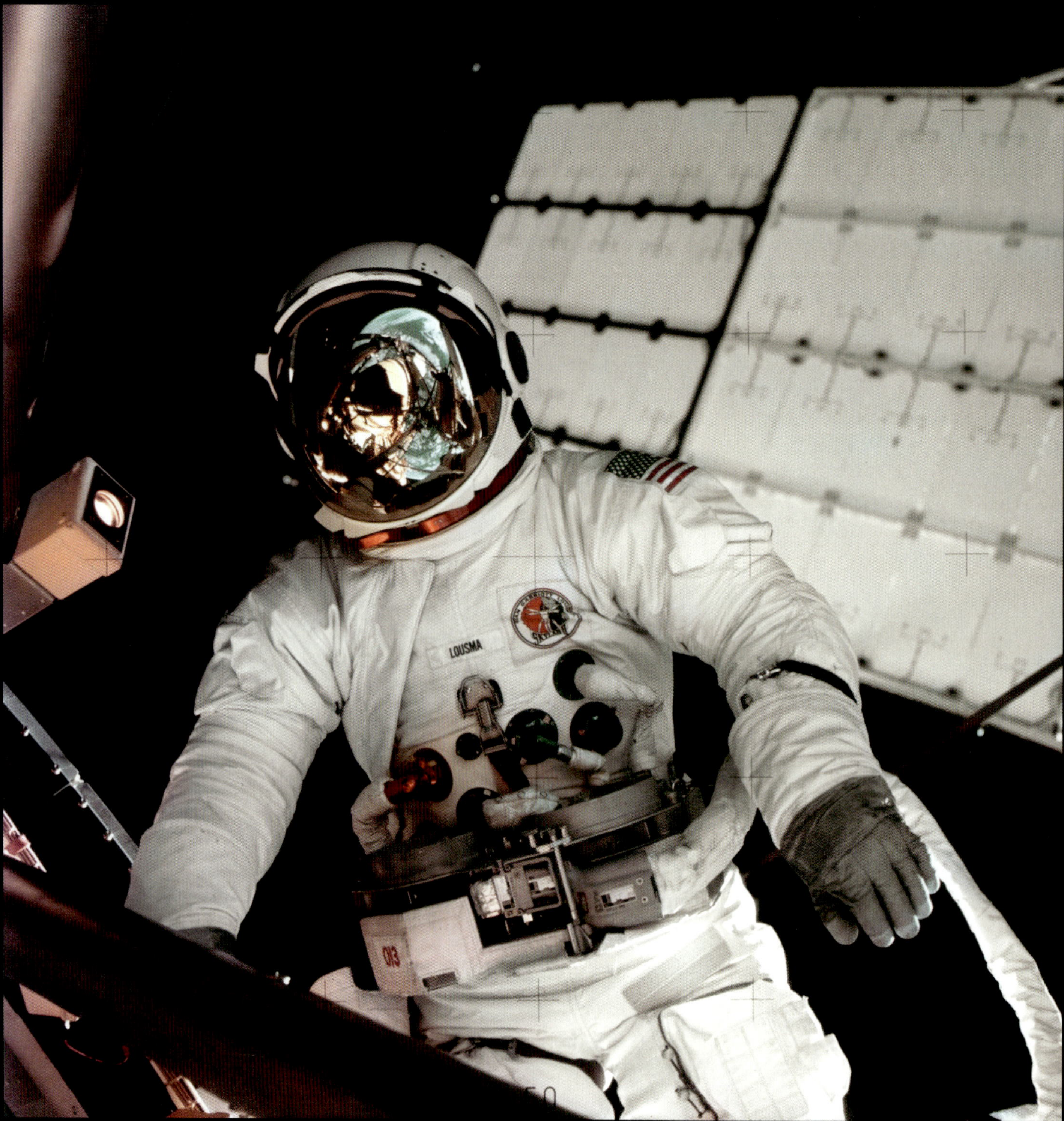

Jack Lousma—with an ATM solar array wing behind him and Earth and the ATM reflected in his visor—at a work station, during EVA 1 with Owen Garriott on August 6, 1973. His combined tether and oxygen line is attached to the pressure control unit around his waist. A twenty-watt EVA light is at left. The goals of the EVA were to replace the ATM's film and deploy a twin-pole sunshade, which would supplement the parasol the Skylab 2 crew had put in place. The August 6 EVA, the first of three, lasted six and a half hours. Garriott took this photo while standing in the airlock shroud.

CHAPTER 14

Skylab 3

July 28–September 25, 1973

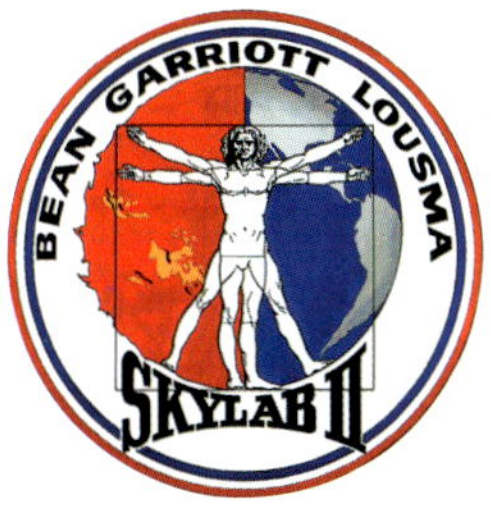

Launched just one day short of five weeks after Skylab 2's splashdown, the second manned Skylab mission more than doubled that flight's space endurance record—it lasted nearly two months. US Navy Capt. Alan Bean led the crew, which included civilian scientist-astronaut Owen Garriott (SPT) and Marine Col. Jack Lousma (PLT), both rookies.

Skylab 3 launched from Pad 39B atop a Saturn IB at 7:11 a.m. (EDT) on July 28, 1973, after what was called the smoothest countdown for a US manned flight. Three hours after launch, however, an oxidizer valve in one of the four SM RCS quads refused to close, forcing the crewmen to shut it down. They rendezvoused with the Skylab cluster about five hours later, and docked half an hour after that.

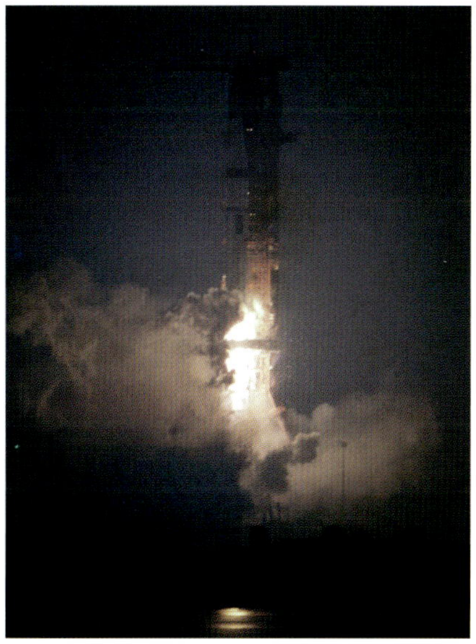

Launch: 7:11 a.m. (EDT), July 28, 1973 (photo by John de Bry)

Shortly after entering the OWS at about 4:30 p.m., all three astronauts began to experience motion sickness, something NASA still did not fully understand. Although some Apollo astronauts had initially experienced nausea, the first Skylab crew had not; neither had Bean during Apollo 12. An entire crew had never been hampered by the malady. The worst was over by July 31, and the astronauts felt much better the next day, but they had fallen behind on their activation timeline. They remained behind for the next week.

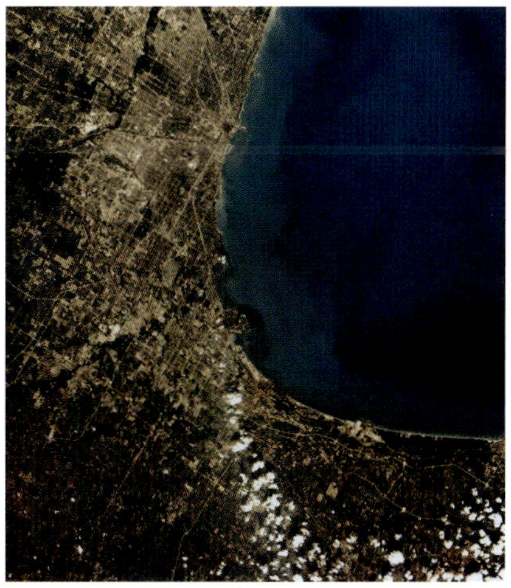

On orbit: Chicago, Illinois, on Lake Michigan

On August 2, a potentially serious problem cropped up: a second SM RCS quad was leaking oxidizer. With only two good quads left, the flight, according to mission rules, would typically have been terminated soon after; Mission Control did not rule this out as an option. But the second CM docking port on the OWS, along with another Skylab spacecraft and booster being prepared at KSC for the next mission, meant that a rescue flight was now a new option. Crews at KSC were put on a round-the-clock schedule, with September 10 as the launch target.

At the pad, storage lockers were removed from the rescue CM to make room for extra couches for the rescued crewmen. Concerns eased after engineers concluded that the two malfunctioning RCS quads did not share a common problem.

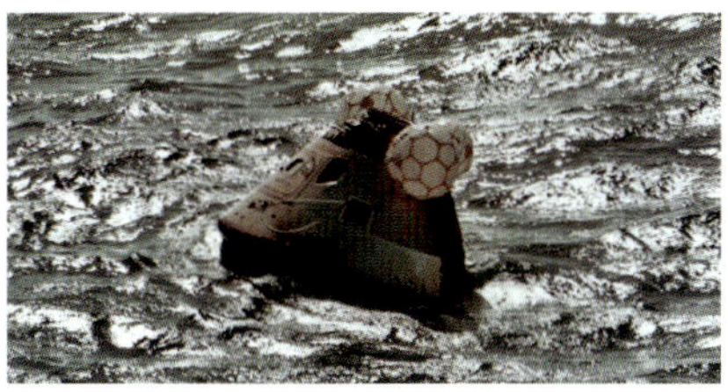

Splashdown: 3:19 p.m. (PDT), September 25, 1973

Dealing with the leak had delayed the first EVA, but on August 6, Garriott and Lousma went outside to deploy a large two-pole sunshield, since the nylon parasol installed by the first crew was eventually going to disintegrate from UV radiation. The twin-pole design was chosen over an improved parasol because it could be deployed over the first parasol; the OWS would thus not be left uncovered for even a few minutes. Each pole was assembled from eleven five-foot sections.

Garriott and Lousma also exchanged ATM film cassettes, deployed an experiment, and looked for signs of any problems, including with the SM's two faulty quads. The two astronauts spent six and a half hours outside—by far the longest EVA on record. Temperatures in the workshop quickly fell with the new sunshade in place.

The crewmen had run behind schedule for the first ten days, but after the EVA, they resolved to catch up, and do more if possible. Suddenly the tables were turned: Houston had previously given them more than they could handle, but now the flight planners had trouble finding enough tasks for these astronauts. Mission Control increased their daily workload from eight to twelve hours each during the third week, and by mission's end, the crewmen had surpassed their experiment goals by 50 percent.

Much of their astronomy work involved the ATM. On August 9, Garriott photographed a medium-sized solar flare, and the next day, he and Bean filmed the type of huge eruption that might happen every six to eight months. Two weeks of viewing culminated on August 21 with the discovery of a huge prominence on the sun's edge.

By August 14, the decision had been made to continue the flight. The rescue mission would revert to Skylab 4, although it would still be ready if serious problems cropped up. The previous day, and again on August 15 and 17, Bean tested an advanced Astronaut Maneuvering Unit (AMU) and a Hand-Held Maneuvering Unit (HHMU) for several hours. He also tested (and panned) a foot-controlled maneuvering unit on August 19.

Garriott and Lousma's second EVA, on August 24, lasted four and a half hours. Overheating had degraded at least five of the nine gyroscopes in the ATM, which provided data to control the station's attitude. Lousma, as the tallest crewman, was to run a twenty-three-foot cable from the ATM to a new package of six gyros they had brought. Since switching over the cable would mean no attitude data for several minutes, Bean, as CDR, decided to remain onboard, and Garriott went in his place. They also replaced the ATM's film and tended to experiments.

On August 27, the crew participated in the ceremony dedicating the Manned Spacecraft Center (MSC) as the Johnson Space Center. From orbit, they welcomed Lady Bird Johnson and her two daughters.

Flight planners had hoped the astronauts would wrap up their days by 9:00 p.m. (EST), leaving two hours free, but each man kept his own schedule. Bean stayed busy with experiments, rarely gazing out the window. On one of two air-to-ground radio channels, Lousma offered commentary on everything from the airlock's tight quarters to Garriott's barber skills, and put in frequent plugs for the Marines. He performed a routine with a fake set of barbells for TV viewers: after "straining" to lift the weights, he floated up with them over his head. Garriott had planned a series of science demonstrations; he illustrated the effects of weightlessness on magnets, water drops, and spinning objects.

Skylab 3's biological experiments studied the effects of microgravity on a variety of organisms—including the crewmen—although the day after launch, mice and gnats for two experiments died following a power failure. US high school students participated in many experiments during Skylab. One study that achieved some fame on the mission looked at spider-web formation in microgravity.

The third and final EVA, on September 22, lasted two hours and forty-one minutes and was undertaken by Bean and Garriott. They retrieved experiment packages and again replaced the ATM's film.

Skylab 3 splashed down at 3:19 p.m. (PDT) in the Pacific Ocean, 224 miles southwest of San Diego, California. USS *New Orleans* recovered the crew.

Top left: Garriott, Bean, and Lousma at the northwest corner of Pad 39B on July 9, 1973. Instead of conducting a separate CDDT, they spent several hours in the CM and reviewed their equipment. The CDDT was integrated into the final countdown for the first time.

Top right: Bean, Garriott, and Lousma pose for photographers with a 1/48-scale Skylab model after their preflight news conference in JSC's Bldg. 2 on June 30, 1973. (photo by Jacques Tiziou)

Bottom left: Bean, forty, was Skylab 3 CDR. It was his final space flight. He remained at NASA until 1981, when he left to devote himself to painting full-time.

Bottom center: Lousma (*partially visible at far left*), Bean (*left*), and Garriott (*right*) in Bldg. 220 at JSC on May 1, 1973, before training for ocean recovery with a helicopter-hoisted basket in case of high seas. The standard recovery plan, however, was to bring the CM aboard ship before crew egress. Garriott, forty-two, was a scientist-astronaut chosen for the corps in 1965.

Bottom right: Lousma, thirty-seven, was a Marine Corps attack and reconnaissance pilot before joining NASA in 1966. He was a support crewman for three Apollo flights. In 1984, he ran unsuccessfully for the US Senate from Michigan. Both space rookies, Lousma and Garriott would make Space Shuttle flights in the early 1980s.

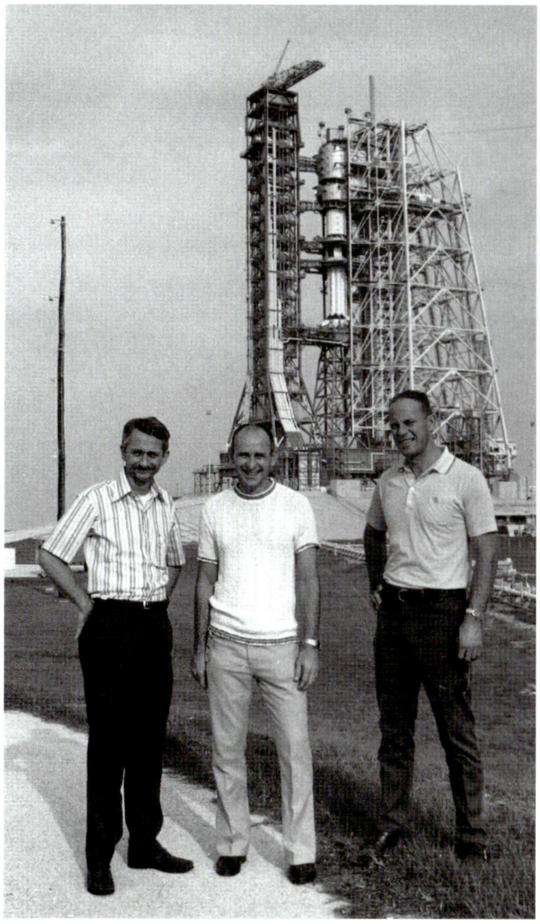

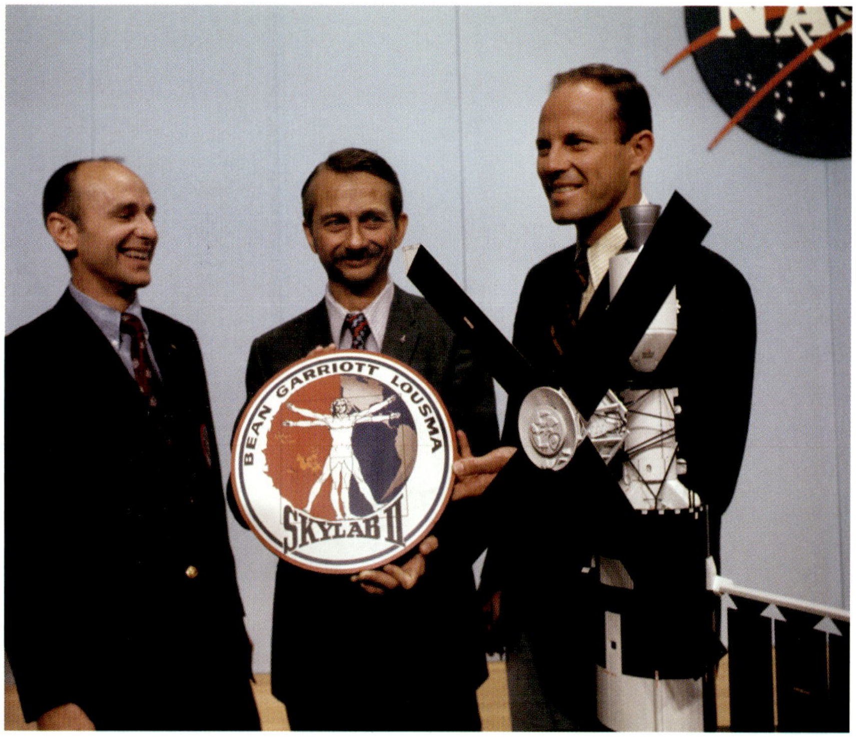
BEAN GARRIOTT LOUSMA
SKYLAB II

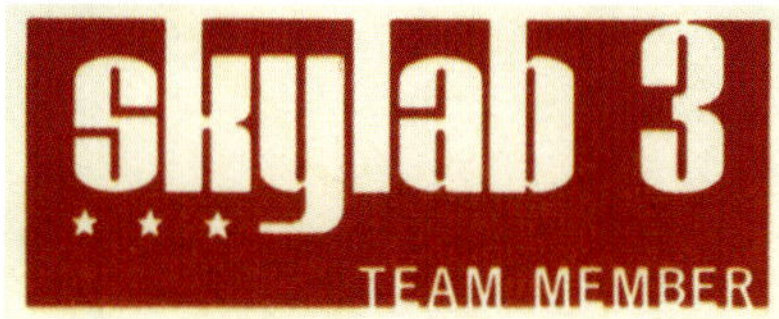
skylab 3
TEAM MEMBER

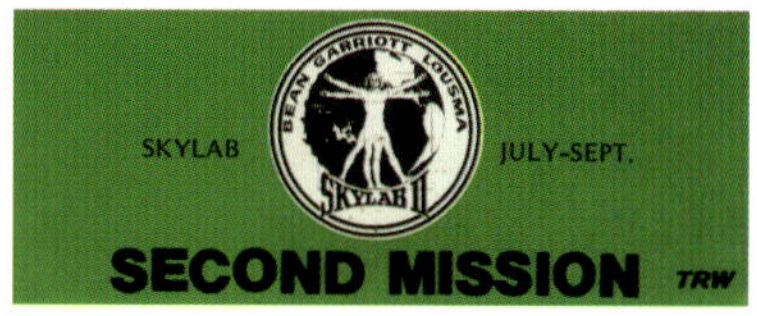
SKYLAB
JULY-SEPT.
SECOND MISSION
TRW

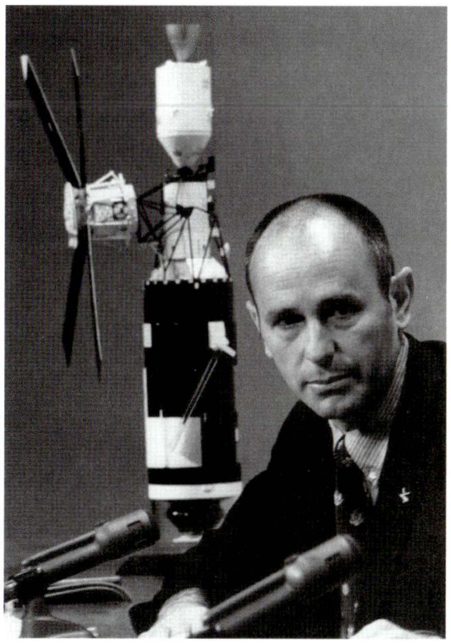

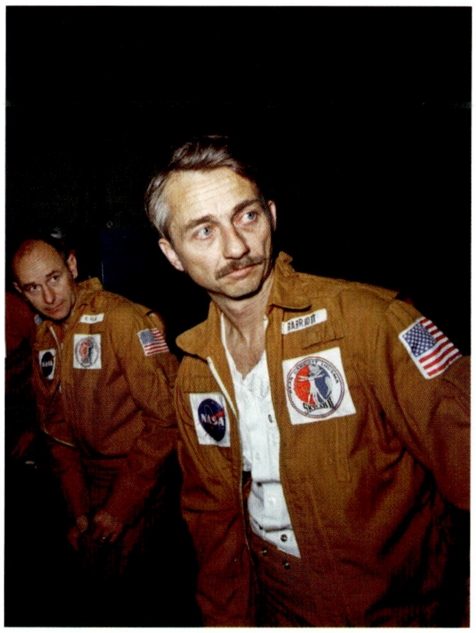

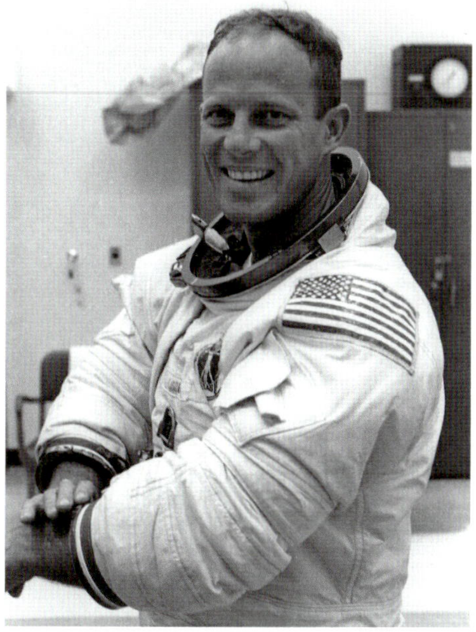

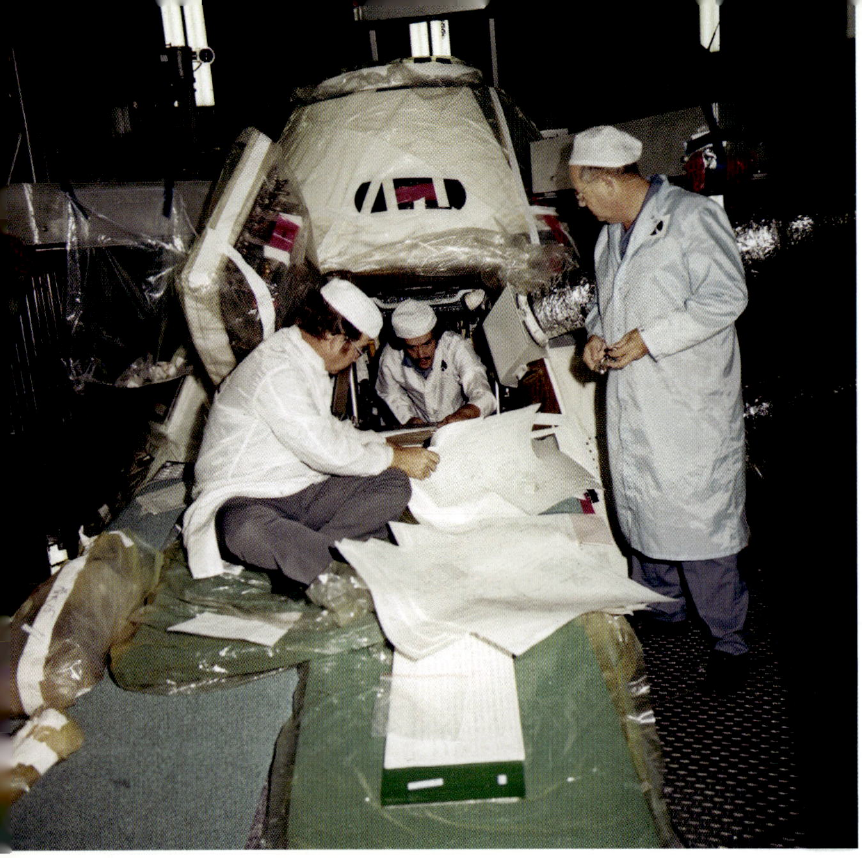

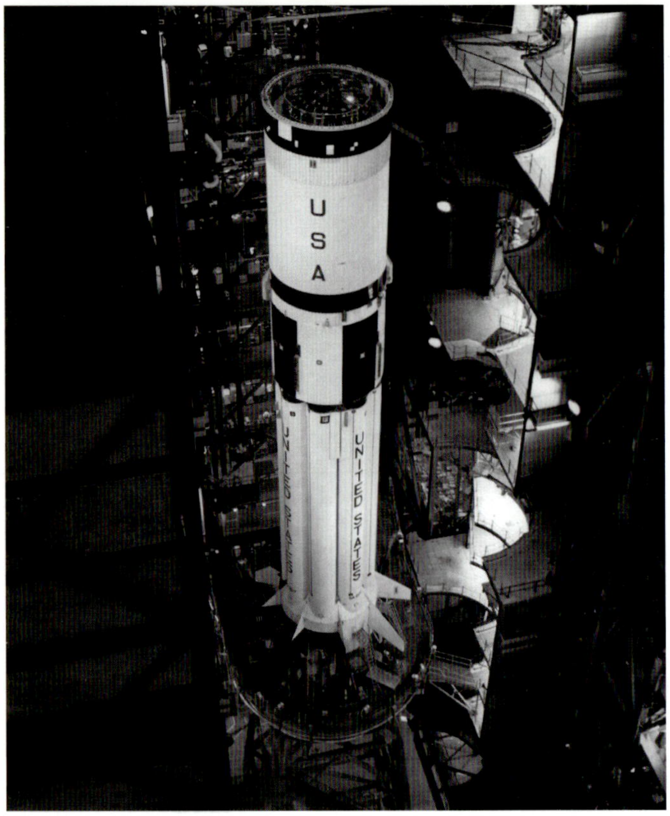

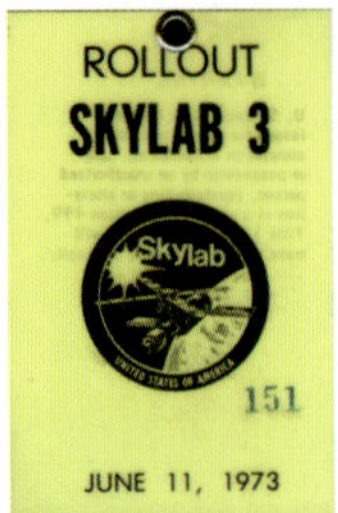

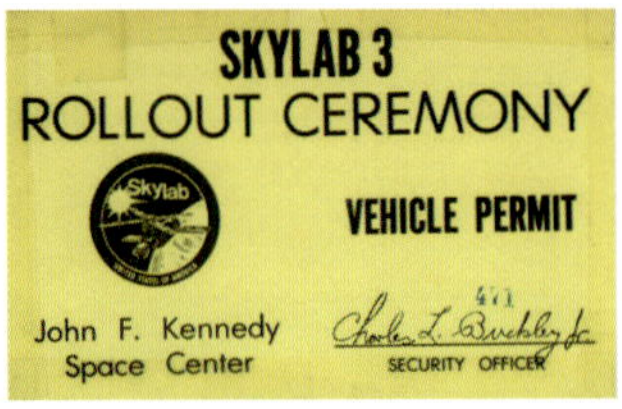

Top left: Rockwell International engineers refer to diagrams as they check the fit of a modification kit inside the Skylab 3 CM, itself inside a KSC altitude chamber. The kit added two more couches so that a rescue flight, manned by two astronauts, could rescue three orbiting crewmen in case of emergency. The kit was later installed in CM-119, stacked atop Saturn IB (AS-208), which was rolled to Pad 39B on December 3, 1973.

Top right: The S-IVB second stage is mated to the first stage, secured to the milkstool, in the VAB's high bay 1 on May 28, 1973.

Bottom left: Skylab 3 moves slowly toward Pad 39B on ML-1 on the morning of June 11, 1973. Several thousand KSC workers and guests were on hand. This view looks north from the MSS, parked near the fork in the crawlerway leading to Pad A. It was the eighteenth day in orbit for the Skylab 2 crew, who were engaged in Earth resources mapping and medical experiments.

Bottom right: Guenter Wendt (*left*) briefs Bean, Lousma, and Garriott on emergency pad escape procedures in the white room at Pad 39B on July 10, 1973. Because of increased spacecraft and launch vehicle reliability, the flight was the first manned Saturn mission without a separate CDDT.

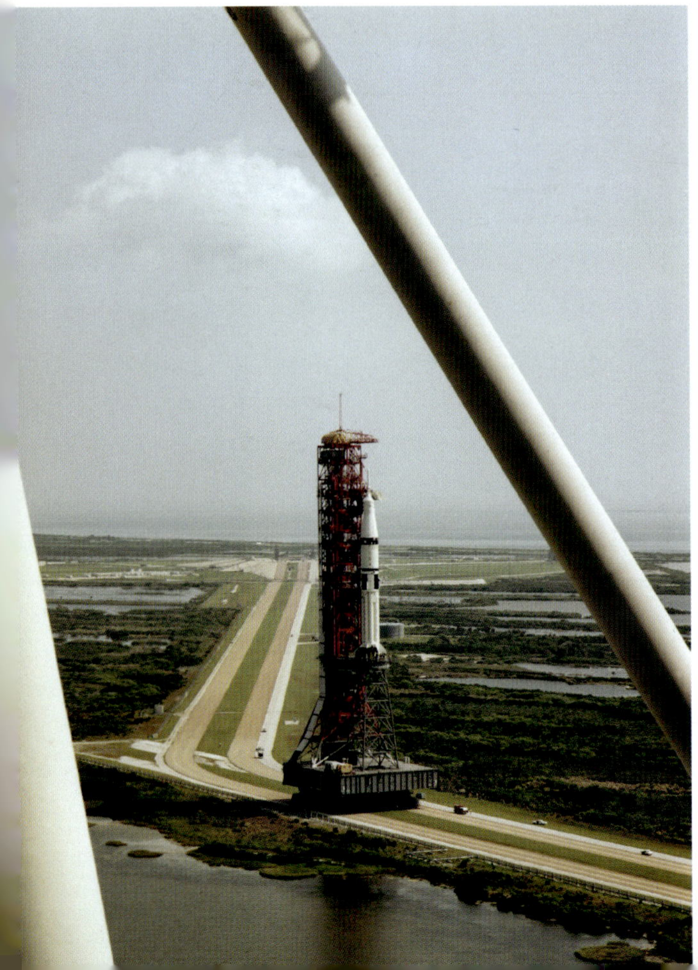

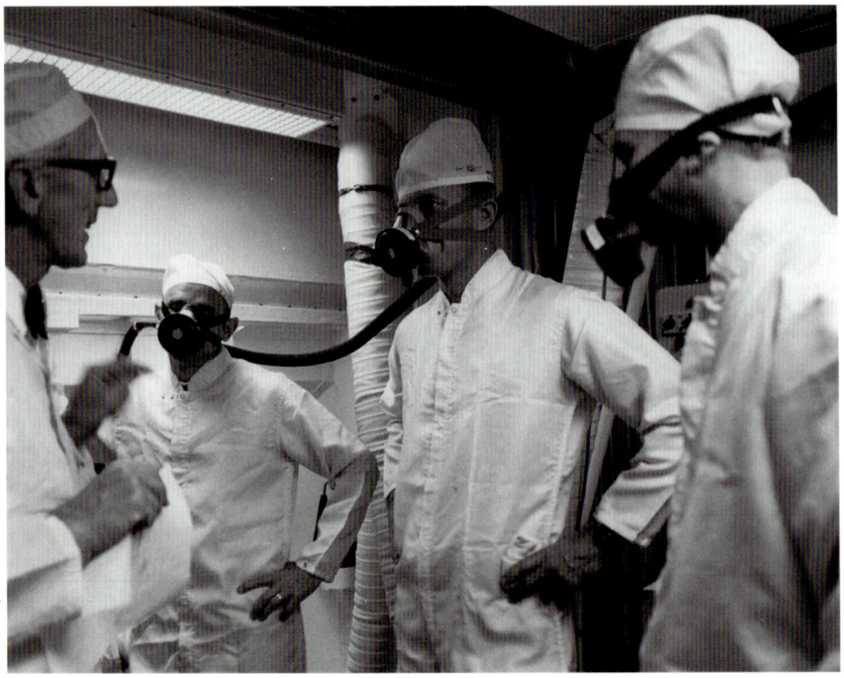

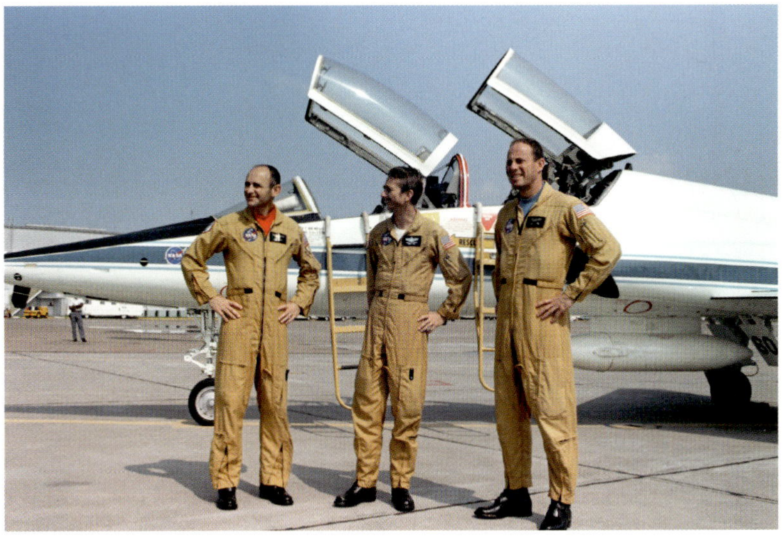

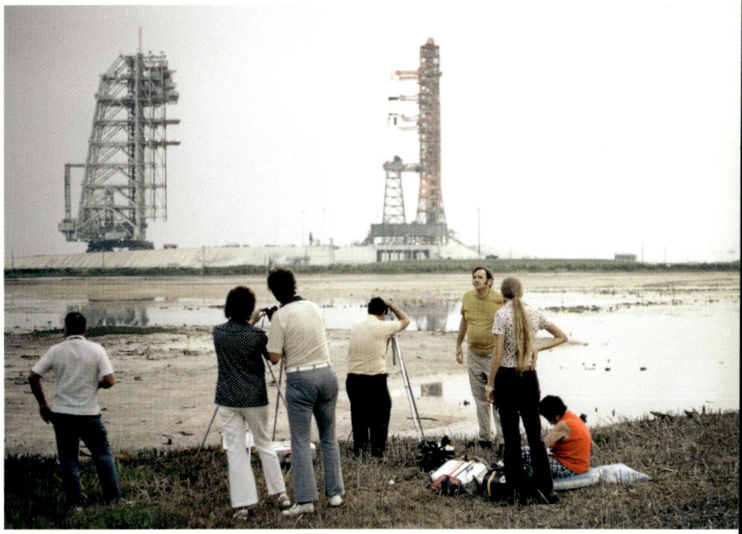

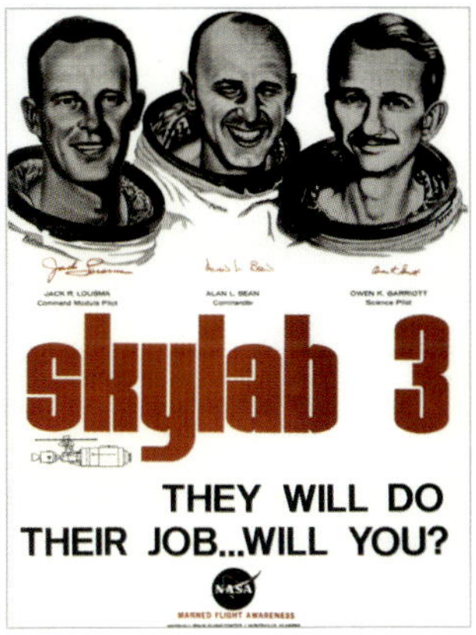

Top left: Bean, Garriott, and Lousma prepare to depart Ellington AFB for Patrick AFB on July 26, 1973. In the T-38 behind Lousma is Dr. Paul Buchanan, NASA flight surgeon. Buchanan flew with Garriott, and Bean with Lousma.

Top right: Jim Long (*fourth from left*) and other photographers shoot the Saturn booster at Pad 39B on July 27. The MSS, at left, has been rolled back. KSC photo escort Ray Yost is right of Long. (photo by Steve Nolte)

Center right: Fog envelops the mound at the LC-39 Press Site on launch morning, July 28. NASA and radio network trailers are at left, while the two-story CBS building is at far right, with the NBC trailers next to it. (photo by Tom Usciak)

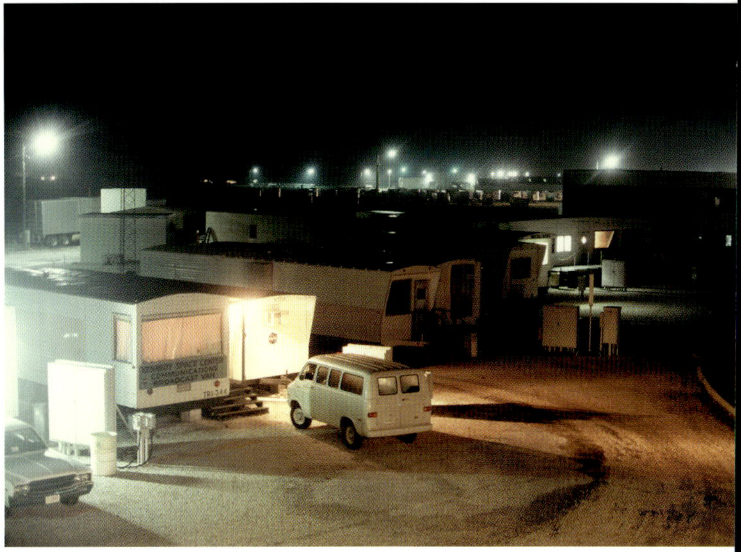

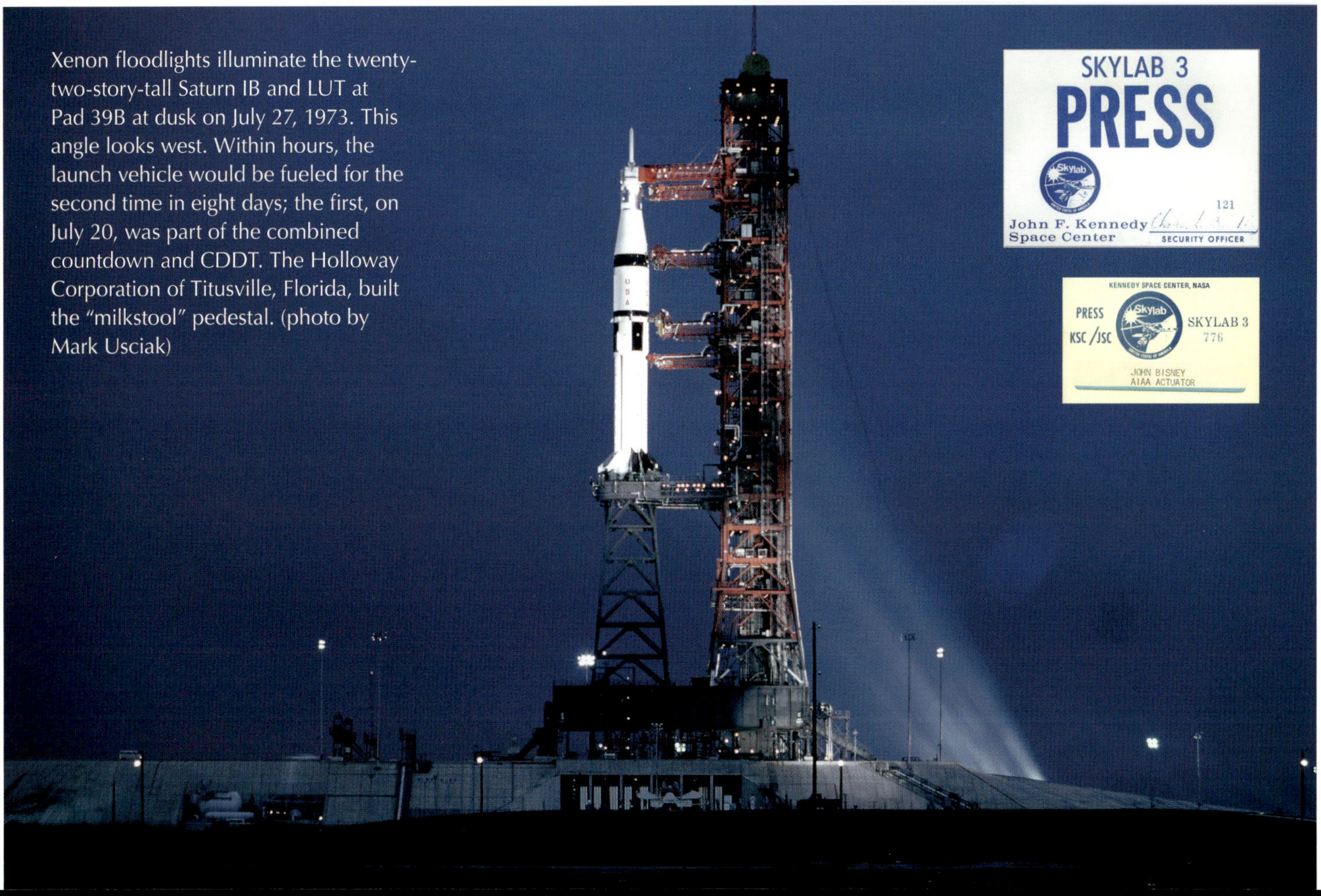

Xenon floodlights illuminate the twenty-two-story-tall Saturn IB and LUT at Pad 39B at dusk on July 27, 1973. This angle looks west. Within hours, the launch vehicle would be fueled for the second time in eight days; the first, on July 20, was part of the combined countdown and CDDT. The Holloway Corporation of Titusville, Florida, built the "milkstool" pedestal. (photo by Mark Usciak)

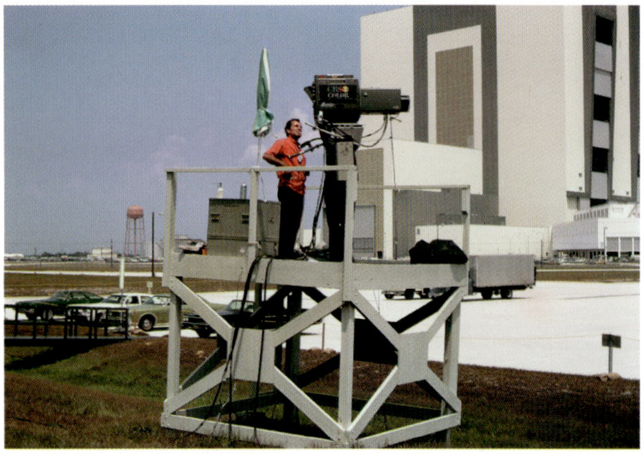

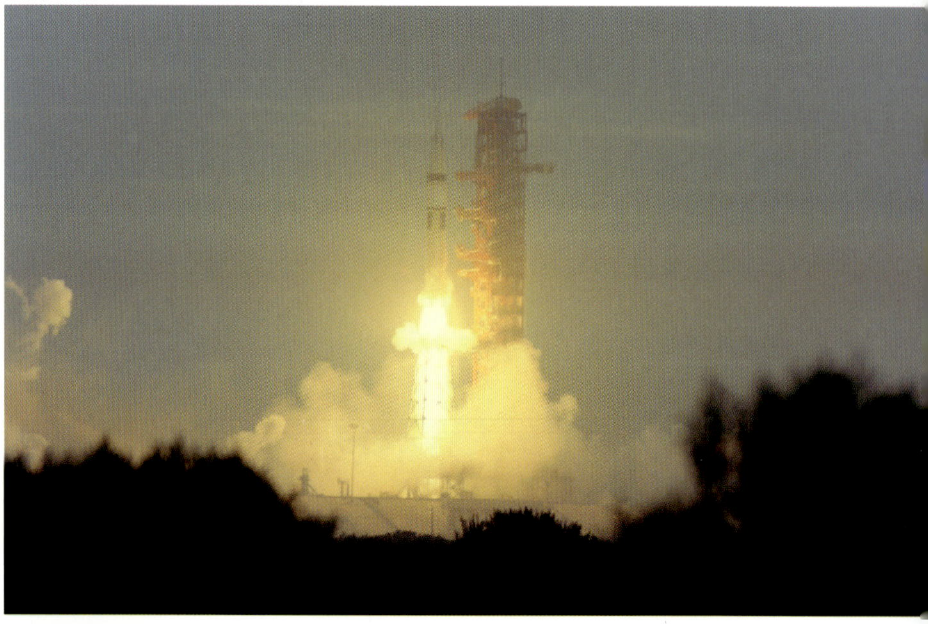

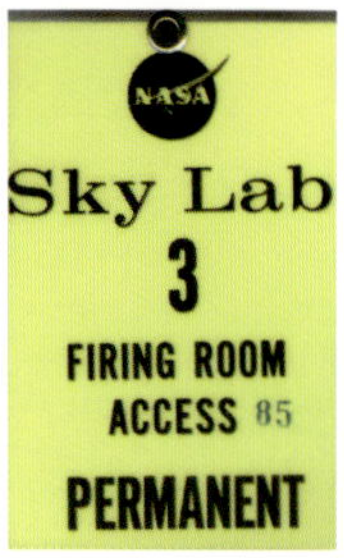

Top left: A CBS-TV cameraman at the LC-39 Press Site the day before launch with a Norelco PC-60 color camera, which had a specially built tilt head for tracking vertical launches. (photo by Steve Nolte)

Top right: Skylab 3 heads into Earth orbit on a foggy morning at 7:11 a.m. (EDT) on July 28, 1973, after a very smooth countdown. The CSM's initial orbit was one hundred miles below Skylab. Bean fired the SPS at 8:29 a.m. to boost the spacecraft's orbit to match the station's.

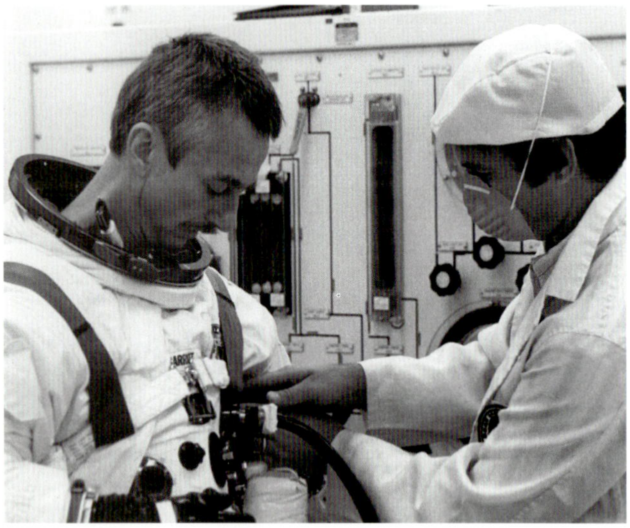

Second left: Garriott undergoes suit checks with suit tech Ron Woods in the MSOB on launch morning.

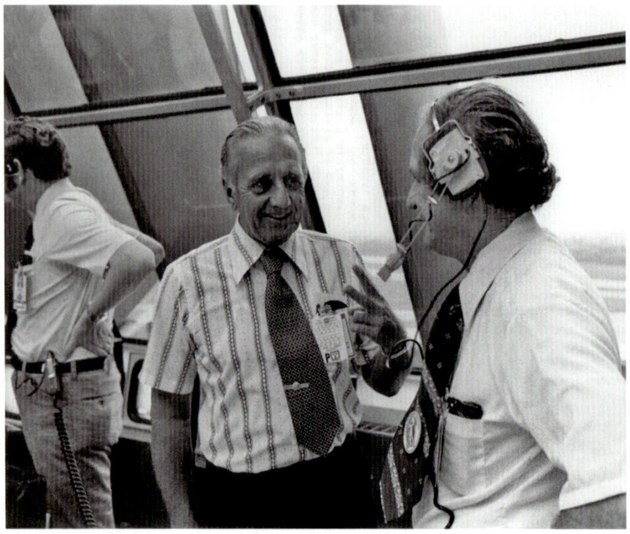

Third left: Kurt Debus and Ike Rigell, chief engineer for launch vehicle operations, in "management row" in Firing Room 3, after launch. MSFC's Dick Smith, the Saturn program manager, is at far left.

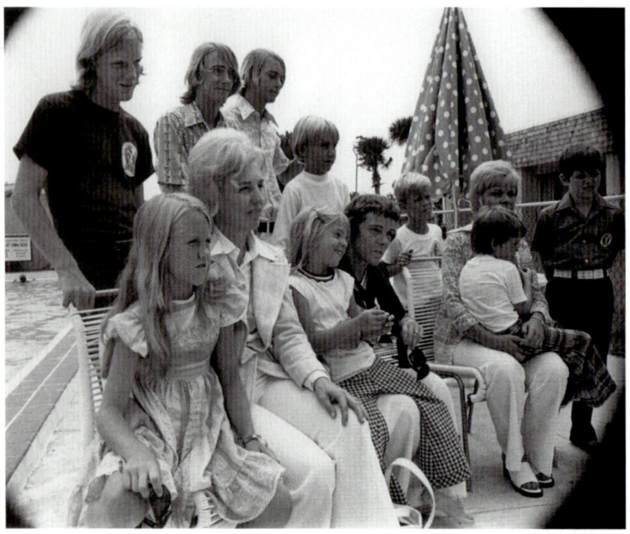

Bottom left: The crew's families meet with reporters poolside at the Cocoa Beach Holiday Inn after launch. *Front row, left to right*: Amy Sue Bean, Sue Bean, Linda Garriott, Helen Garriott, Mary Lousma, Gratia Lousma, and Timothy Lousma. *Back row, left to right:* Clay Bean, Randall Garriott, Robert Garriott, Richard Garriott, and Matthew Lousma.

Opposite page, top: The S-IVB second stage in Earth orbit above Italy, France, and the Mediterranean Sea, about three hours after launch. The SLA panels were not jettisoned from the Skylab S-IVB stages. This image was taken with a 70mm Hasselblad camera, using a 100mm lens and SO-368 medium-speed Ektachrome film. The "50" at the bottom of this and the following image is from a Reseau plate—a transparent piece of glass in the Hasselblad used to superimpose the camera's plate number onto photos.

Opposite page, bottom: The view from the CM during the fly-around inspection, before docking about five hours later. The primary docking port on the forward end of the MDA is visible below the ATM. The workshop's solar power wing, released by the first crew, is at lower left.

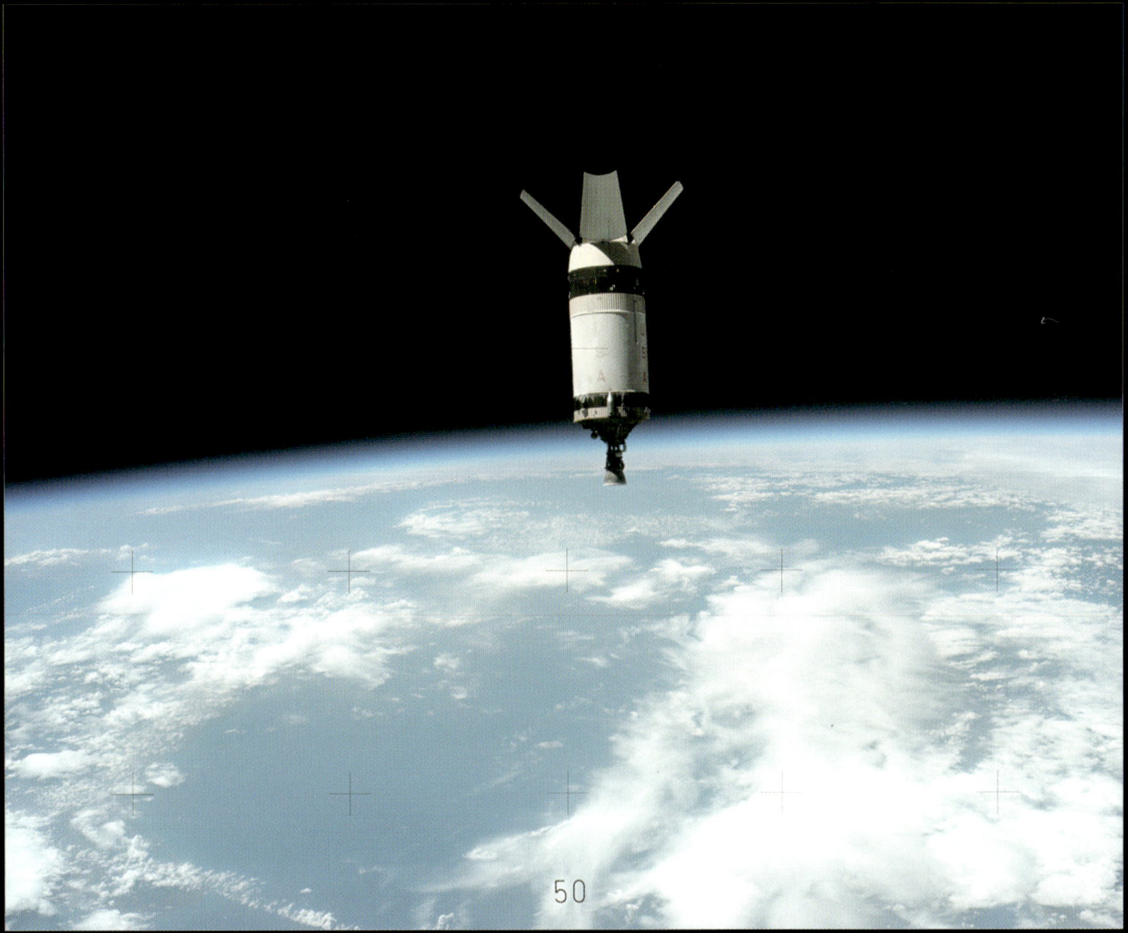
50

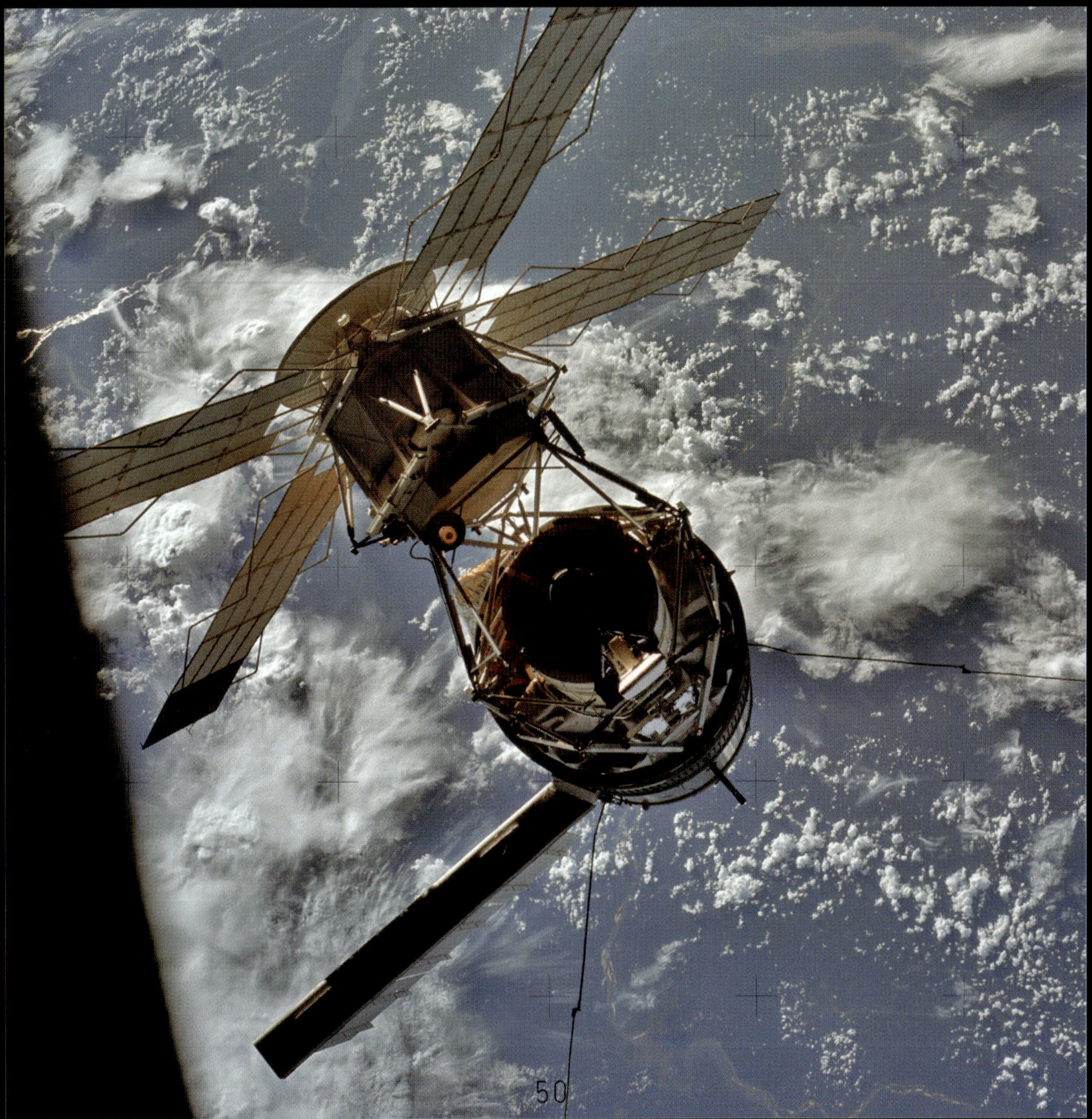
50

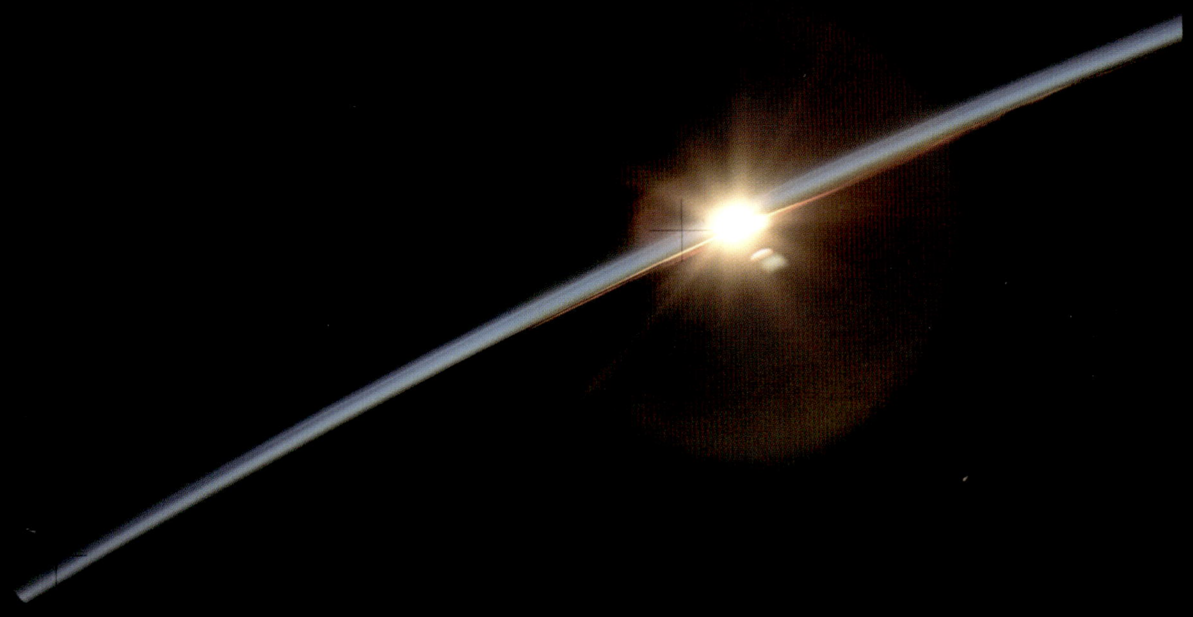

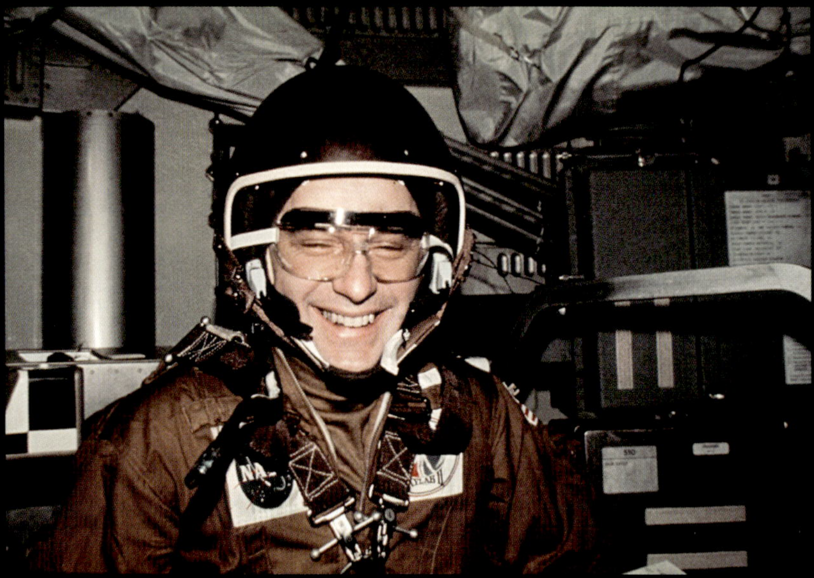

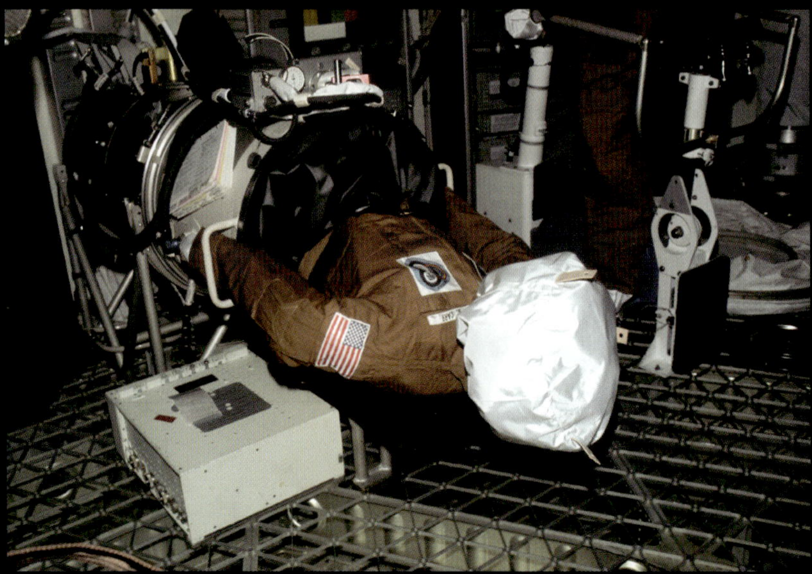

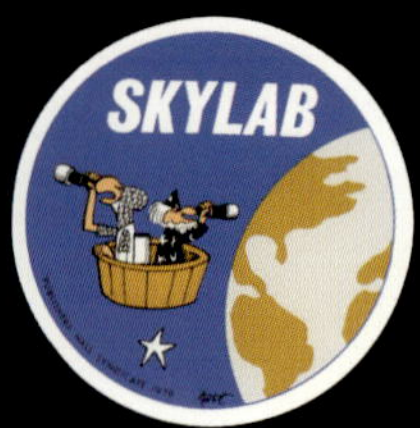

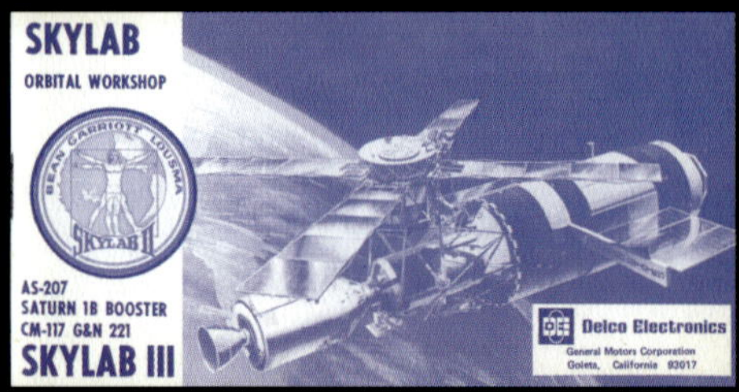

Top: Earth's atmosphere is a sliver as the sun sets behind the limb. The mesosphere appears dark blue, the stratosphere light blue, and the troposphere orange.

Center left: Bean prepares to test-fly the ASMU in the forward dome of the OWS. The nitrogen jet–powered system was the latest in a series of attempts, begun during Gemini, to perfect a practical astronaut jetpack. The jetpack finally came to fruition with the Manned Maneuvering Unit (MMU), used on three Space Shuttle flights in 1984.

Bottom left: Before departing, the crew left two dummies of the Skylab 4 astronauts, with name tags and crew insignia, posed in the experiments compartment of the OWS. At center, in the Lower Body Negative Pressure Device, is "G. Carr." The dummy for Bill Pogue, at right, is propped on the bicycle ergometer. Ed Gibson's was in the waste management compartment.

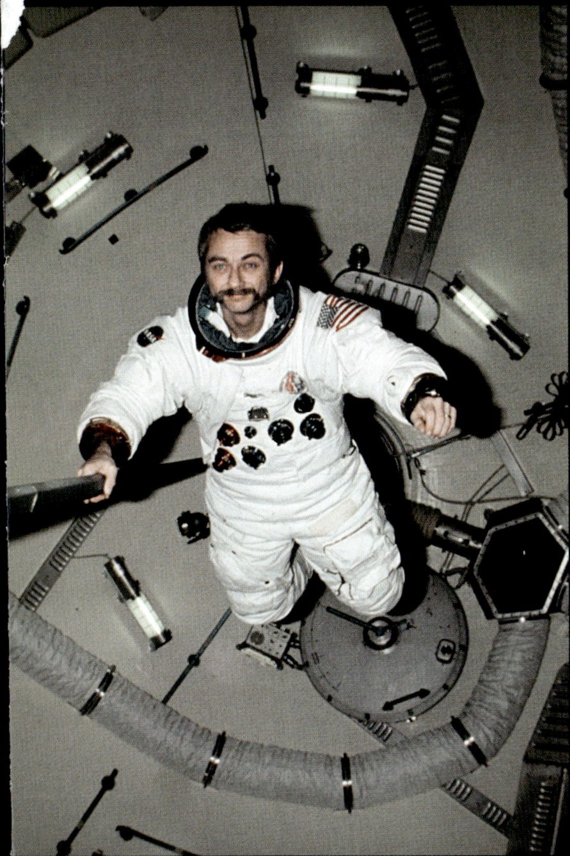
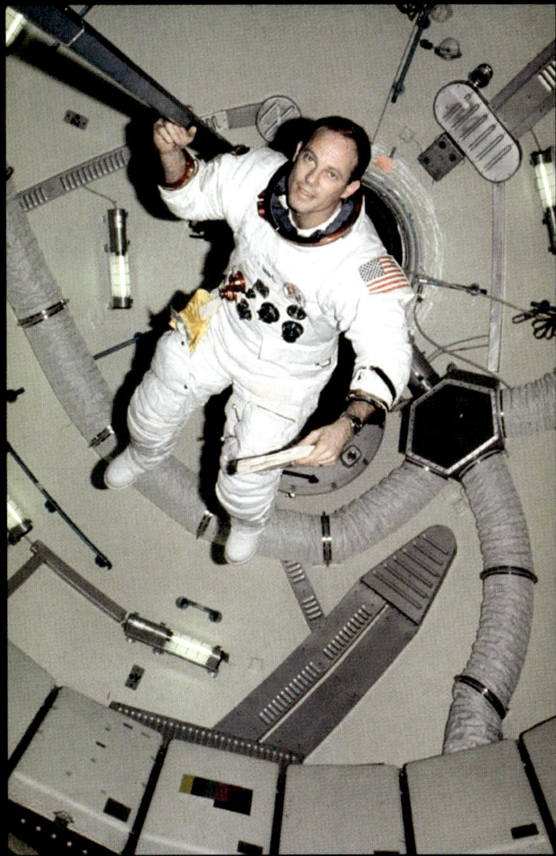
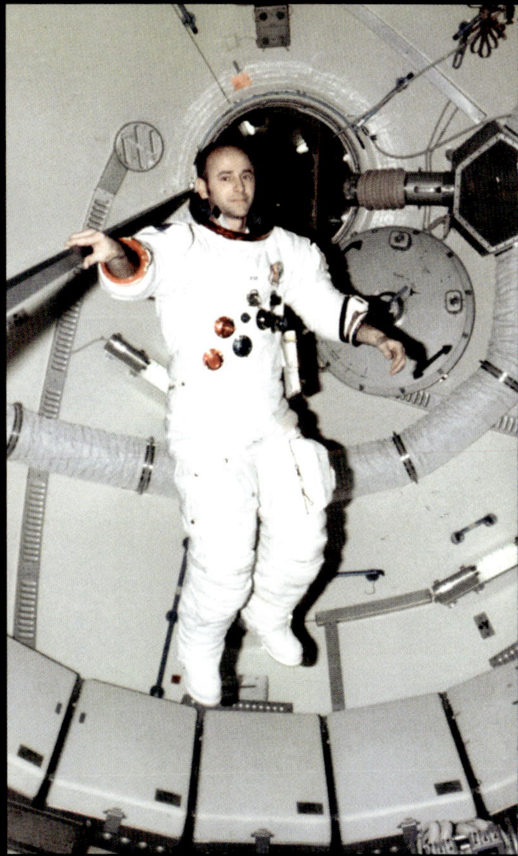

Above: Garriott (*left*), Lousma (*center*), and Bean float in the forward dome of the OWS before the first EVA, on September 6. Each crewman holds a translation pole, which they soon removed; only the first crew left it in place. The hatch to the airlock module is behind them, as are floodlights and air ducts for cabin ventilation. Several of the twenty-five lockers in the storage ring can be seen below Lousma and Bean. One locker door below Lousma has a color chart for photo calibration.

Right: Bean prepares a bag of garbage for disposal; Lousma, in the doorway to the sleep compartment, dons his liquid-cooled undergarment prior to an EVA. Skylab's trash dump was the modified unused LOX tank, "below deck." Behind Bean is the waste management compartment.

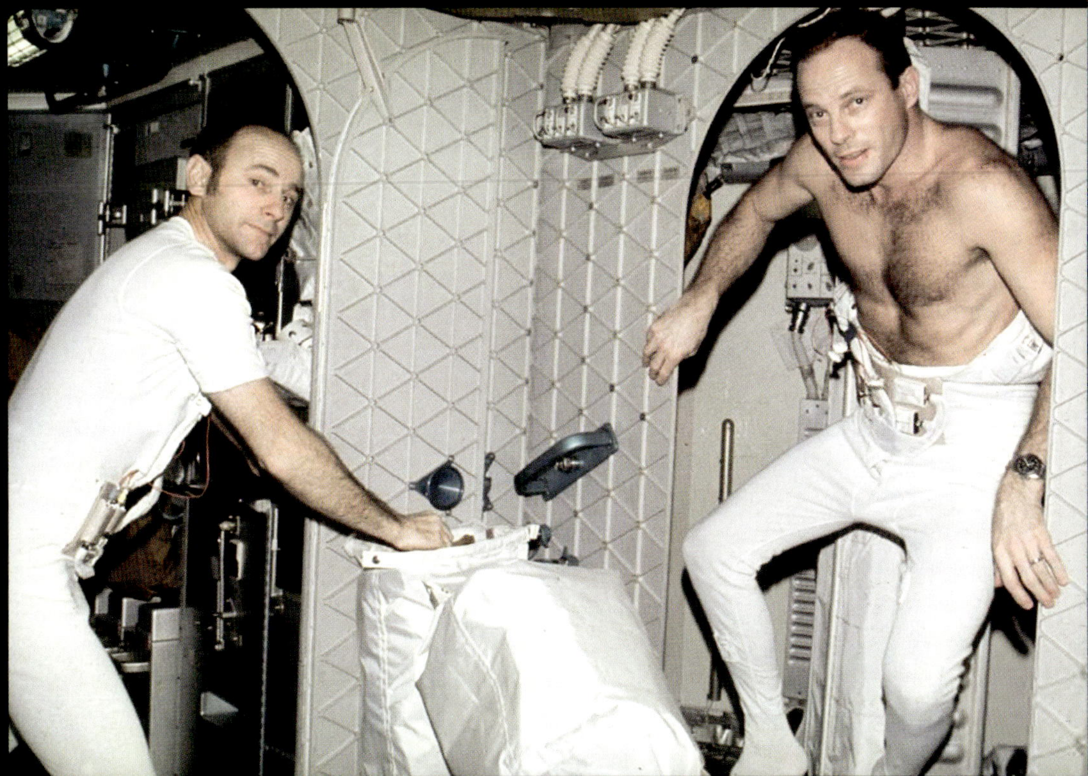

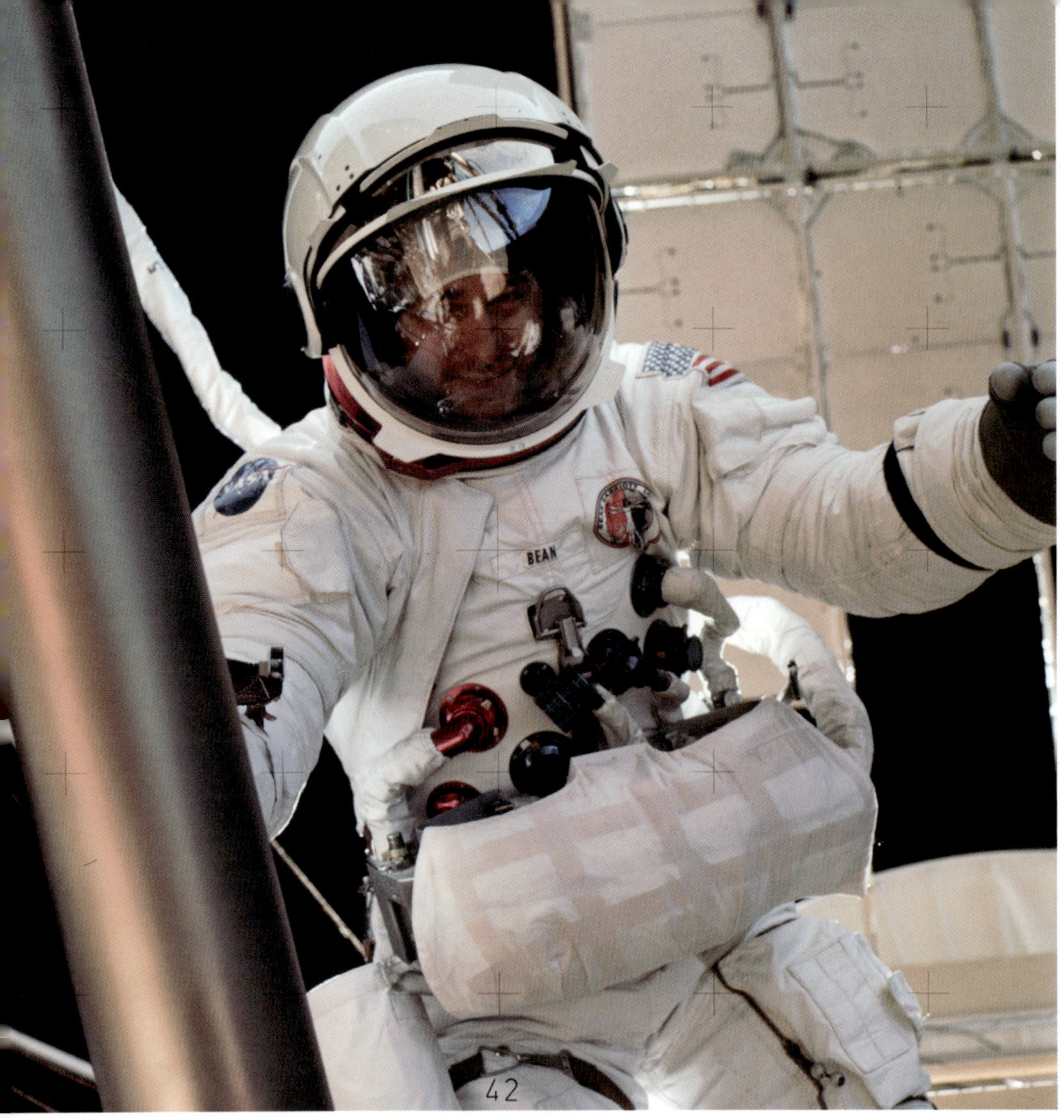

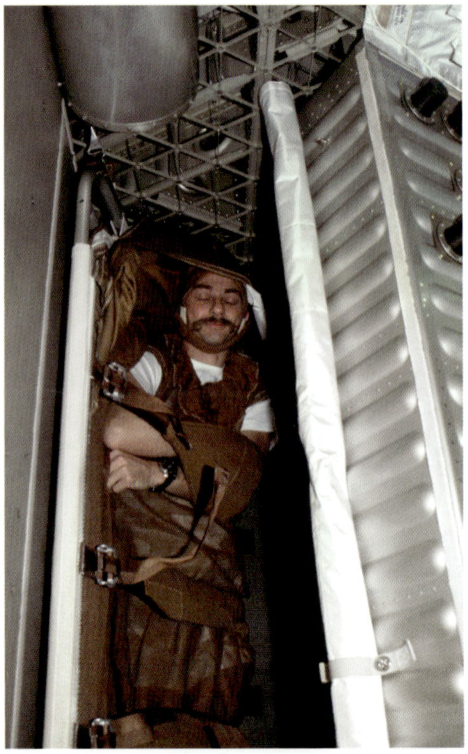

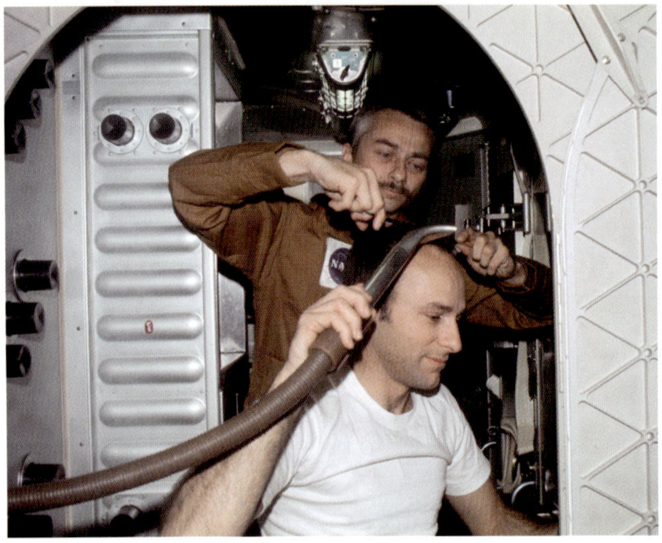

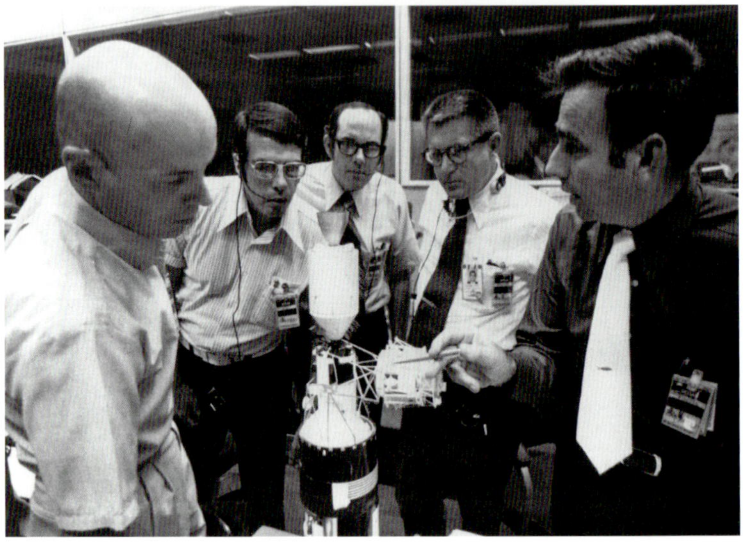

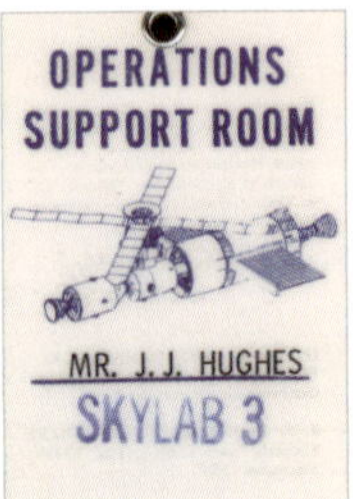

Top left: Bean, photographed by Garriott from the airlock shroud during the third EVA, on September 22, 1973. The goal was to retrieve the final set of six film cassettes from the ATM and a parasol sample for tests back on Earth. Bean remembered the egress from the airlock as being quite different from the exit from an LM on Apollo 12. The moon provided gravity and a solid surface; this time he floated out into nothingness. He later described the EVA as "much more like science fiction" than what he had experienced on Apollo.

Top right: Garriott in his sleep restraint, with the fabric door open. The restraint was attached to a removable frame in his sleep compartment, and a baffle had been installed above the open gridwork over his head to block light. Each crewman's sleep compartment included an intercom and lockers containing clothing and personal items.

Center left: Just inside the door to the waste management compartment, Garriott trims Bean's hair. Bean holds a vacuum hose to collect the clippings.

Bottom right: On August 6 in MOCR 1, Flight Director Milt Windler points to the location of the ATM's film magazines, before the first EVA to retrieve them that day. *Left to right*: CAPCOM and astronaut Story Musgrave; Robert Kain and Scott Millican, both with the Crew Procedures Division of the EVA Procedures Section; William Schneider, Skylab program director; and Windler.

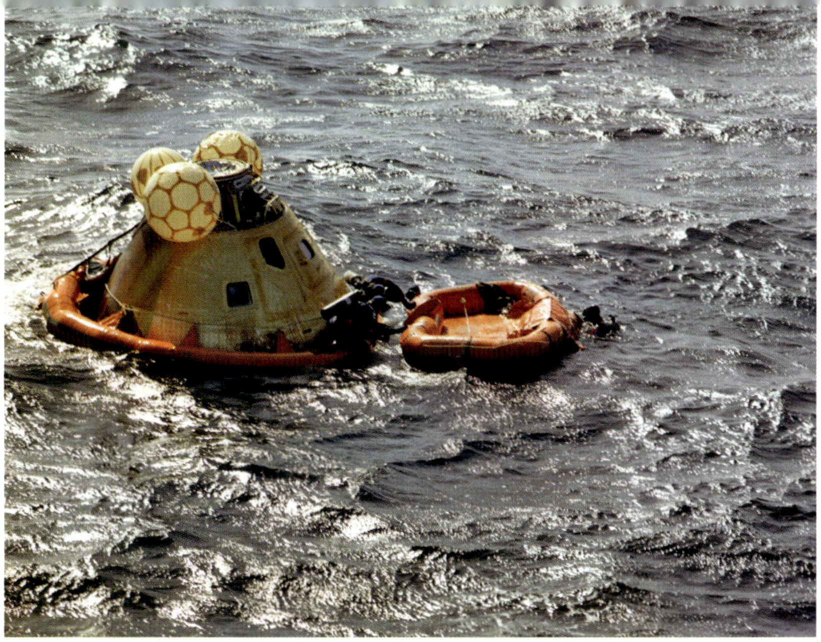

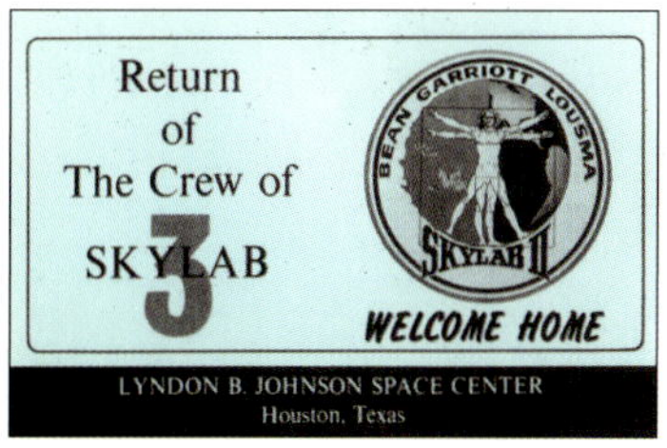

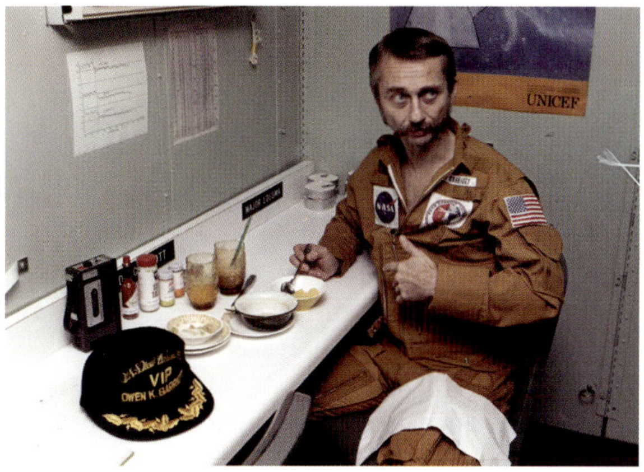

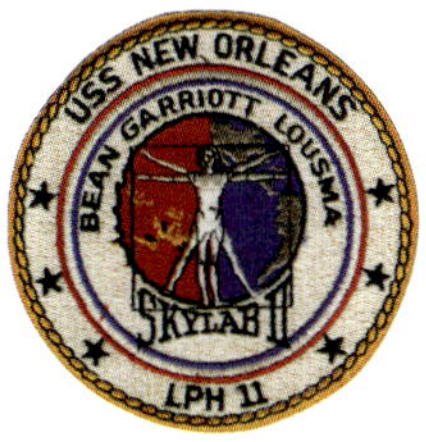

Top left: Navy divers secure a raft to the CM in case of emergency crew egress on September 25, 1973. Splashdown occurred 224 miles southwest of San Diego, California, at 3:19 p.m. (PDT). The CM had traveled 24 million miles during nearly sixty days in orbit. Landing was 6 miles from USS *New Orleans*, a helicopter carrier used for amphibious assault.

Second right: Garriott has a meal aboard ship, in the medical laboratory complex. The Skylab Medical Laboratory was made up of six modified army mobile medical labs, each devoted to a different physiological specialty. As the ship sailed to San Diego, the Skylab crew spent two days undergoing tests and eating a controlled diet.

Third right: Sue Bean, Helen Garriott and Gratia Lousma wait for their husbands to arrive at Ellington AFB on the afternoon of September 27 in a 1929 Ford Model A Standard Phaeton. It was Gratia Lousma's idea to borrow the car from Lousma's friend Leroy Moore, to symbolize the contrast between old and new times.

Below: Bean and the astronauts' wives wave as they prepare to drive off, with Lousma at the wheel. Behind the car is the VC-140B Lockheed JetStar, which flew the crewmen from San Diego.

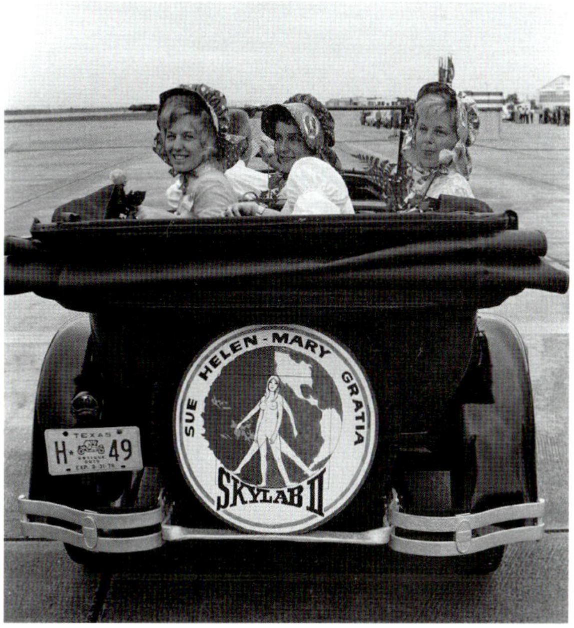

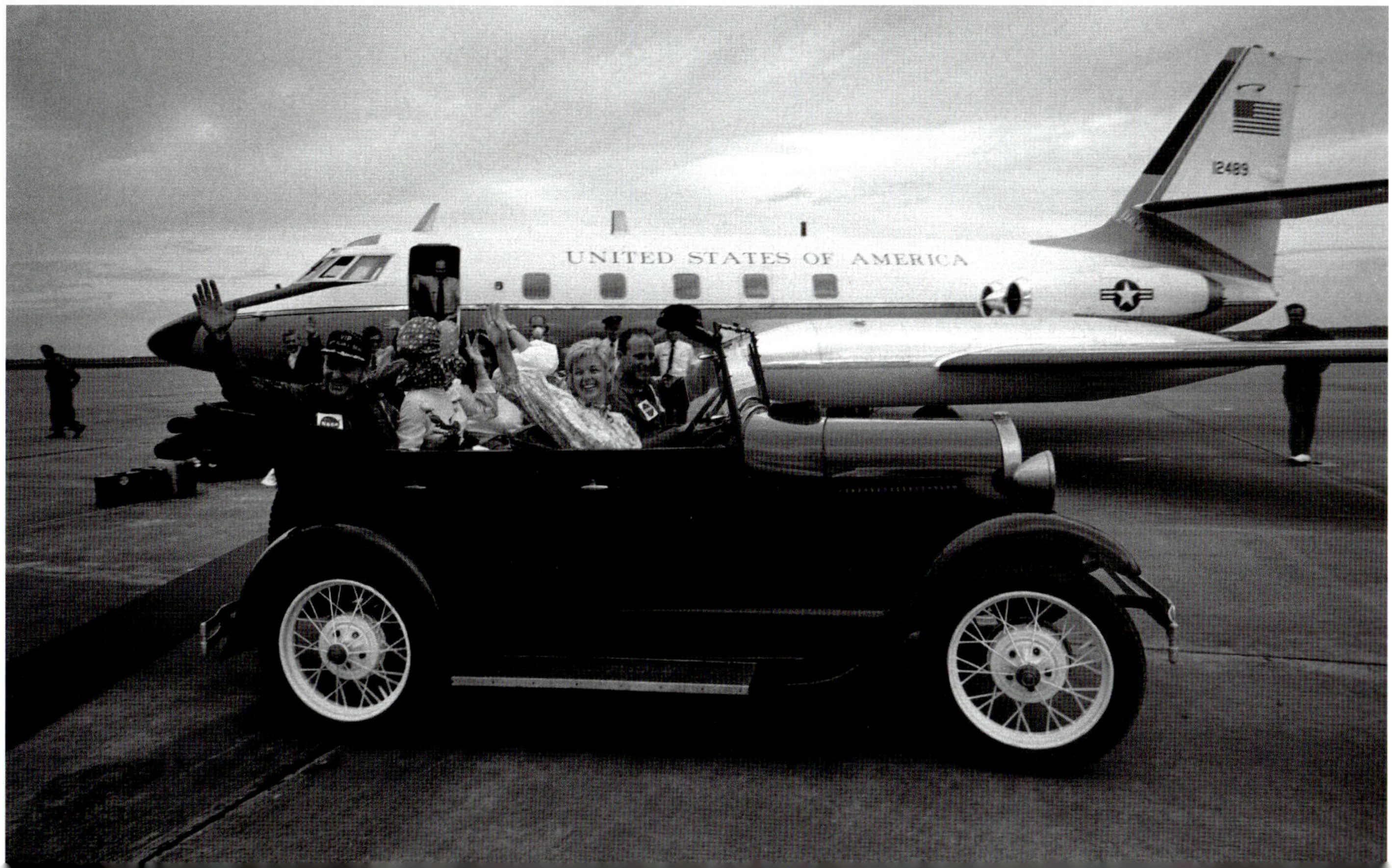

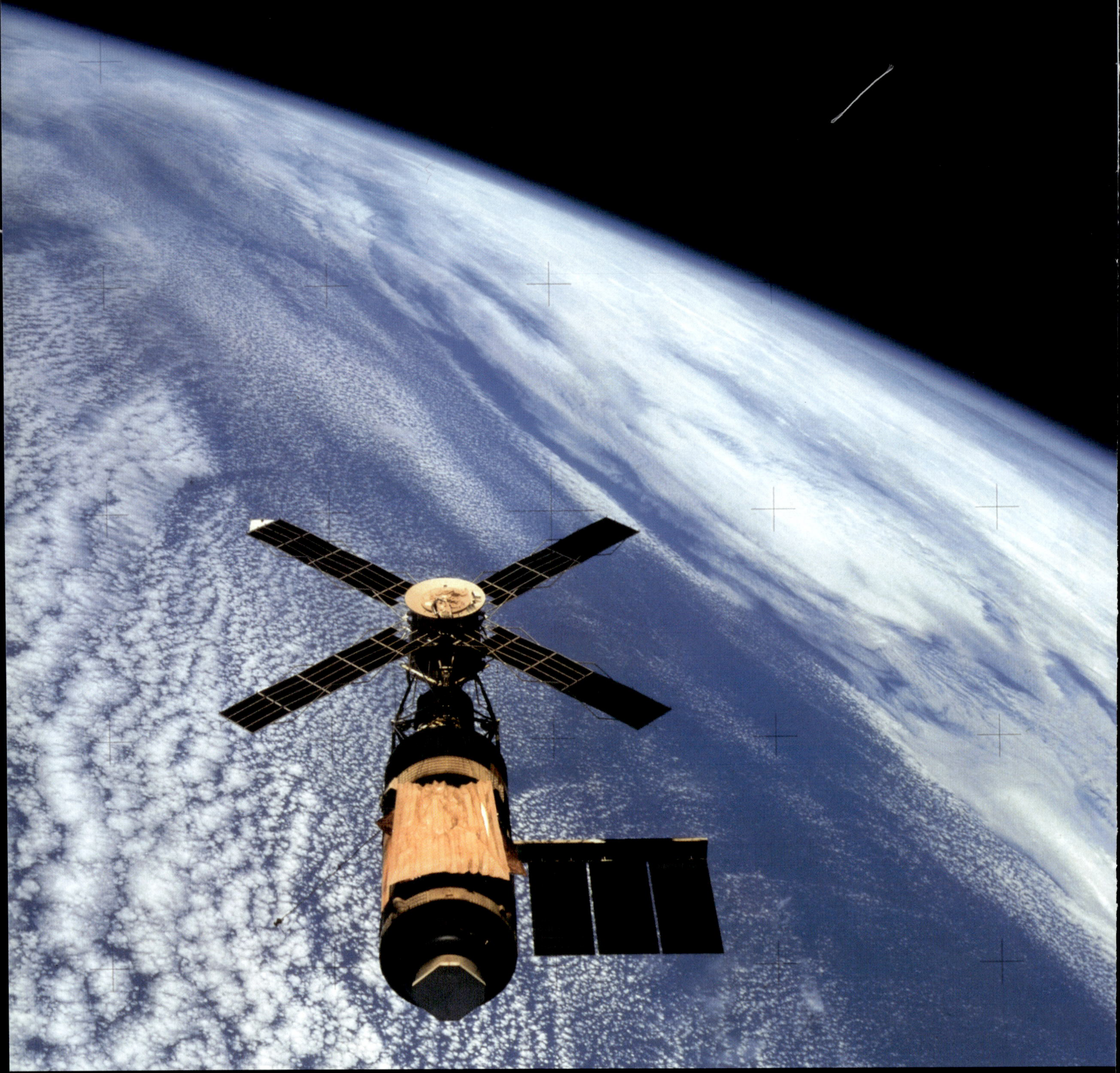

During the final fly-around before the crew leaves orbit, the Skylab cluster is photographed from the Skylab 4 CSM by Bill Pogue. The solar shield deployed by the second crew, made of aluminized Mylar film, covers much of the OWS. The crew boosted the workshop into a parking orbit about 275 miles high before undocking. The cluster reentered the atmosphere on July 11, 1979.

CHAPTER 15

Skylab 4

November 16, 1973–February 8, 1974

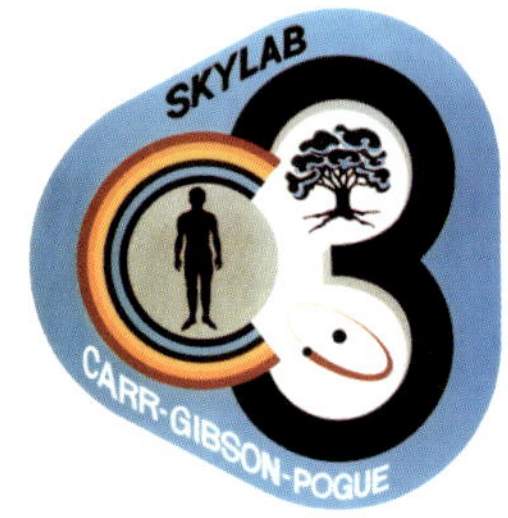

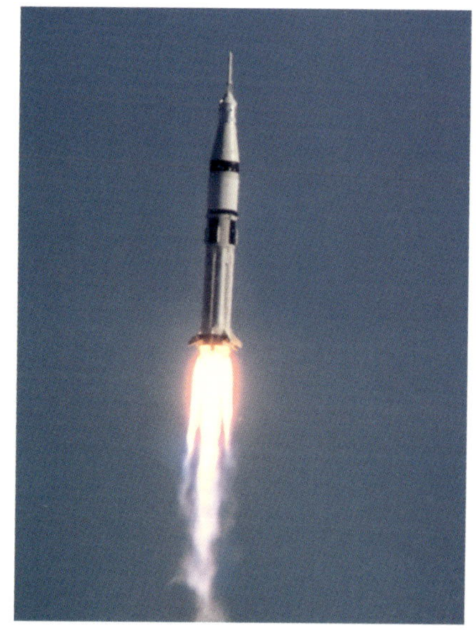

Launch: 9:01 a.m. (EST), November 16, 1973 (photo by Tom Usciak)

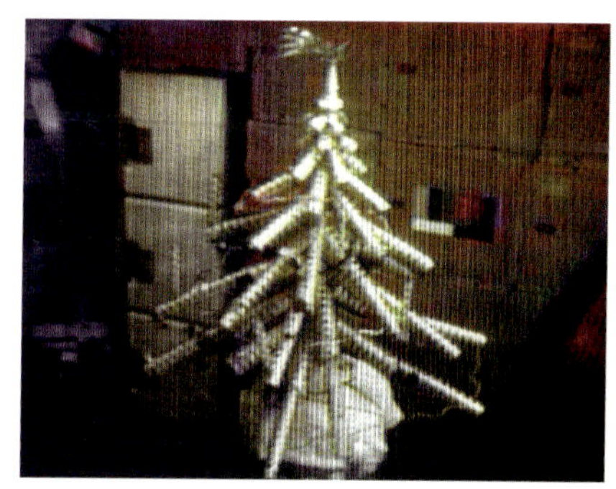

On orbit: TV image of a Christmas tree made of food containers, December 24

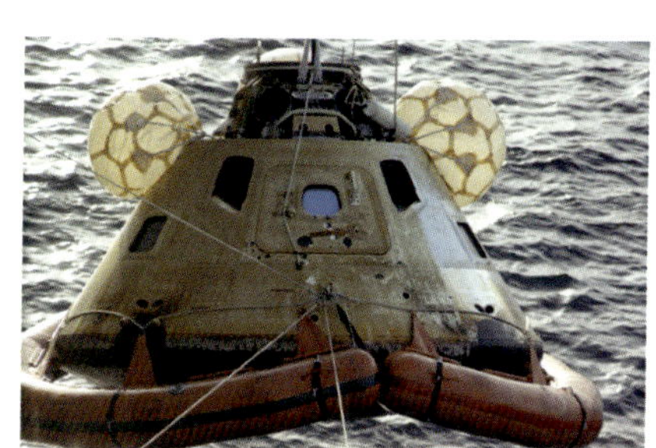

Splashdown: 8:17 a.m. (PDT), February 8, 1974

Skylab 4, the third and final manned visit to the OWS that wrapped up the Skylab program, lasted nearly three months and included four spacewalks totaling almost twenty-four hours. The all-rookie crew was led by CDR Gerald Carr, with William Pogue as PLT and Edward Gibson as SPT. Although a successful flight, the mission was affected beyond the first month by an admittedly awkward reluctance by the astronauts and Mission Control to openly discuss workload and scheduling issues.

It was booster problems, however, that delayed the launch. Chrysler had used high-strength aluminum alloys for the Saturn IB's first-stage tail-unit assembly—including its eight stabilizing and support fins, which engineers knew were subject to stress-induced corrosion cracking. On November 6, 1973, five days before the planned liftoff, an inspection after a fueling test four days earlier, which was intended to look for signs of stress corrosion, found fifteen tiny cracks in the fins. The cracks—around bolts in the rear spar fittings, where the fins were attached to the tail unit—were blamed on a combination of the salty ocean air around Complex 39, the stresses of fueling, and age. (The SA-208 booster was assembled in 1966.)

Replacement fins were flown in from NASA's Michoud, Louisiana, Saturn assembly facility, and all eight fins were replaced at Pad 39B from November 9 to 12. Reinforcement fittings were also installed. On November 12, technicians discovered larger but less worrisome cracks in seven of the eight support beams in the interstage ring, between the first and second stages. The beams were strengthened with aluminum bracing, and launch was set for November 16.

Skylab 4 got underway on schedule at 9:01 a.m. (EST), with docking at the Skylab cluster eight hours later. Because of the previous crew's early motion sickness, all three astronauts took medication; all three were nonetheless affected at first, with the motion sickness hitting Pogue particularly hard.

The crew's handling of the situation got the flight off to a rocky start with Mission Control. They spent the first night in the CM, and between tracking station coverage, Carr and Gibson debated what they should say during the evening status report, apparently forgetting that the CM's onboard tape recorder was running. They decided that Carr would report Pogue's nausea but not his vomiting, and he would report that Pogue had not been hungry.

The next morning, after a good night's sleep, they all felt better, and at

9:15 a.m. they entered the OWS—where they learned that the onboard recorder tapes had been routinely downloaded and transcribed, revealing Carr and Gibson's candid conversation. Mission Control called for a medical conference in midafternoon, and Astronaut Office Chief Alan Shepard later came on the line to give the crew a public reprimand. "I just wanted to tell you," he said, "that on the matter of your status reports, we think you made a fairly serious error in judgment here in the report of your condition." Carr replied, "Okay, Al, I agree with you. It was a dumb decision."

They got to work, but transferring the supplies they had brought, activating the OWS, and working with experiments proved overwhelming. Soon they were tired and behind schedule. Mission planners and experimenters only added to the workload. Planning for the new experiments was sometimes inadequate, and the crew would later report that they had never been trained for some of the experiments. Though the astronauts called their pace frantic, Houston, remembering the pace of the Skylab 3 crew, believed they were not working long or hard enough.

Six days into the mission, on Thanksgiving Day, Pogue and Gibson conducted the first EVA, which lasted six hours and thirty-four minutes. They replaced the ATM film and repaired a malfunctioning antenna on an Earth resources experiment package on the airlock truss, using a repair kit they had brought just for the job.

The next day, one of the gyroscopes in the Skylab cluster's attitude control system overheated; the issue was blamed on insufficient lubrication. It was shut down. The station used three large gyroscopes, sized so that any two of them could provide sufficient attitude control, with the third as a backup if one failed. Mission Control, however, called the problem more of a nuisance than anything and even developed a plan for the system to operate on only one gyroscope if necessary. Later in the mission, a second gyro showed similar problems, but temperature control and load-reduction procedures kept it operating (at reduced speed).

From December 18 to 26, the Soviet manned flight Soyuz 13 was on orbit. It was the first time the Soviets and the United States had manned crews in space at the same time, although the two spacecraft were never in the same orbital plane.

On Christmas Eve, the astronauts' families gathered in the MOCR 1 at JSC, and the crew sent back live video of a tree they had built from food containers.

On Christmas Day, Carr and Pogue began an EVA that would last for just more than seven hours, setting a duration record not broken until a 1985 Space Shuttle mission. In addition to reloading ATM film; pinning open a balky aperture door on the UV solar camera; and freeing a jammed filter wheel in the X-ray telescope, the astronauts took two cameras to photograph Comet Kohoutek. Czech astronomer Luboš Kohoutek had first spotted the comet on March 7, 1973; extensive observations of it were then added to Skylab 4's mission, increasing the crewmen's workload. The astronauts first saw it on December 13. Based on its unusual brightness, the comet had been predicted to become a spectacular sight in the night sky.

By December 25, the crewmen had no trouble keeping up with their work, but Houston's inflexible scheduling and approach to operations was still an extreme annoyance. The astronauts agreed that they needed to

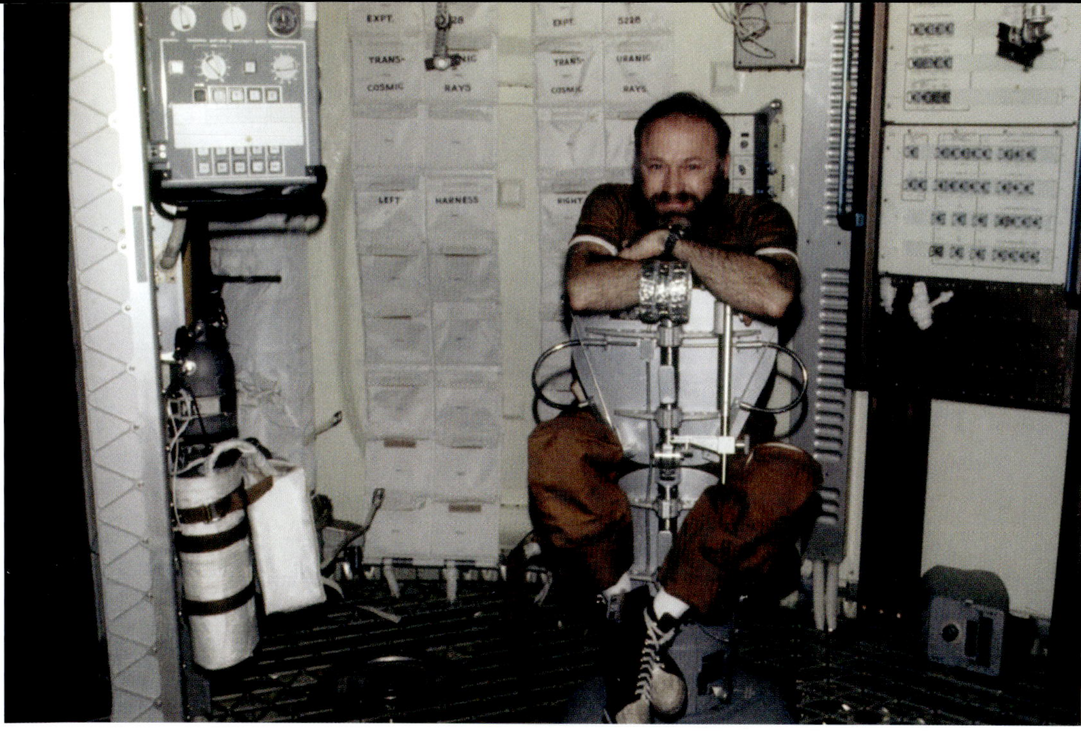

In late January, Gerald Carr clings to the Rotating Litter Chair in the experiments compartment as part of Experiment M131, Human Vestibular Function, which investigated space motion sickness. The chair could be used in a rotating or "tilt litter" mode, and could spin a seated subject up to thirty rpm. Equipment stowage pockets are on the wall behind Carr.

reach a better understanding with the flight planners. On December 28, at the mission's halfway point, Carr taped a six-minute request for a frank discussion with Mission Control.

During two hours of conversations with CAPCOM Richard Truly on December 30, the air was finally cleared. It had taken nearly seven weeks. Everyone was relieved to find that candid conversations could be held on the open communications channel without any repercussions. Carr, who was not prone to rocking the boat, later regretted having waited so long to broach the subject. The flight controllers acknowledged that they had not considered that this crew might have a different approach to their duties from the first two.

The next day, Carr and Gibson conducted the flight's third EVA, which was devoted to Kohoutek observations and photography and lasted three and a half hours. The comet had swept out from behind the sun on December 30, and then had dimmed; a week later, even the astronauts had trouble seeing it.

As was the case for the previous two missions, solar observations with the ATM were a central objective. The crew recorded about seventy-five thousand new X-ray, UV, and visible light images of the sun. On January 21, 1974, Gibson filmed the first recording from space of the birth of a solar flare to maturity.

Carr and Pogue operated the Earth photography and measurement experiments. One crewman photographed the classified USAF Area 51 facility in Nevada, despite specific instructions against doing so. NASA prevailed in a fight with the Pentagon over releasing the image with the rest of the photos, although the photograph went unnoticed until 2006.

On the morning of February 3, Carr and Gibson handled the mission's final EVA—lasting three hours and nineteen minutes—to retrieve the ATM film and some particle-collection experiments.

After eighty-four days, Skylab 4 ended with splashdown at 8:17 a.m. (PDT) on February 8, 1974. USS *New Orleans* recovered the crew and CM from the Pacific Ocean, 176 miles southwest of San Diego, California. Although it was the longest manned flight to date, the Soviets would easily break Skylab 4's record by 1978. They had been steadily gaining experience in long-duration manned flights aboard their space stations (first launched in 1971)—a capability they would build on for the next twenty years.

The Skylab cluster came to a fiery end in Earth's atmosphere during reentry on July 11, 1979, with some pieces landing in Australia.

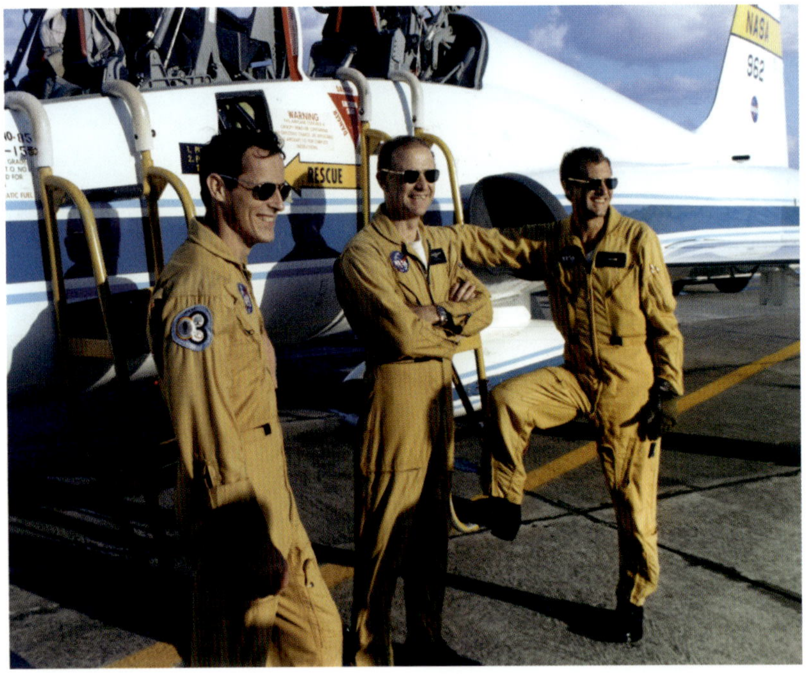

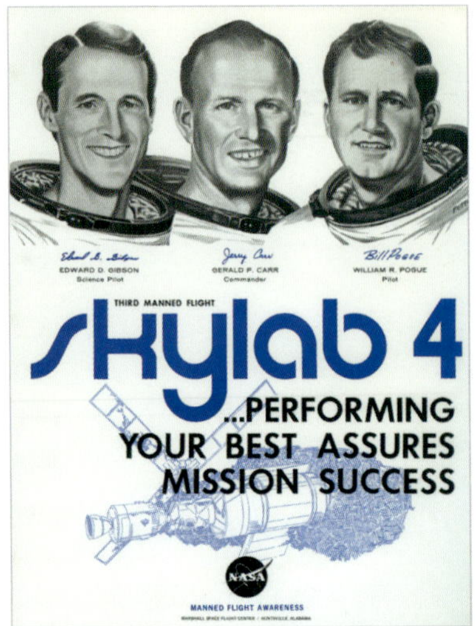

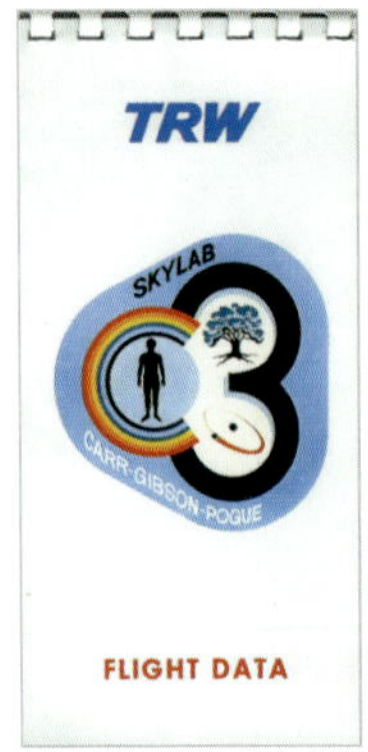

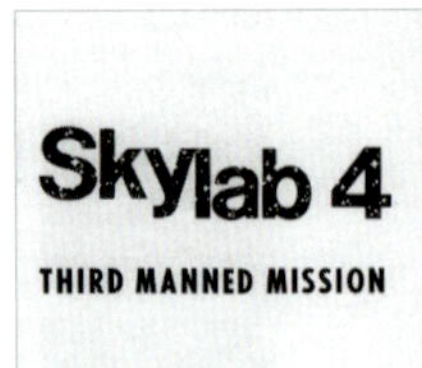

Top left: Gibson, Carr, and Pogue with a T-38 at Ellington AFB on November 13, 1973, before heading to Patrick AFB. That evening, NASA managers set November 16 for launch, after analysis showed that tiny cracks in the support structure between the Saturn's stages would not be a concern during flight.

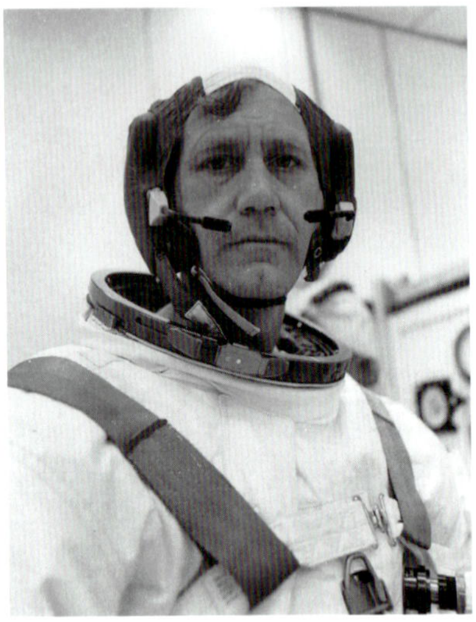

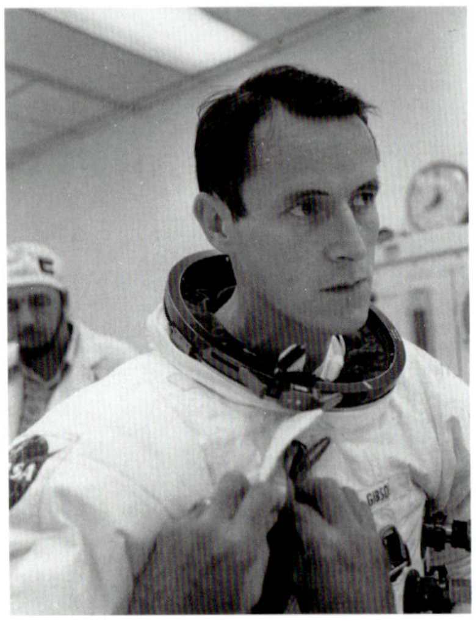

Center left: William R. Pogue, forty-three, suits up for an altitude chamber test at KSC on August 2, 1973. A USAF lieutenant colonel, Pogue had been a fighter-bomber pilot during the Korean War, and he was later a member of the Thunderbirds. He had also been a math professor at the Air Force Academy and a test pilot for the Royal Air Force. Like Carr, Pogue had been penciled in for the crew of the canceled Apollo 19. He would spend thirteen and a half hours outside the OWS on two EVAs. Skylab 4 was the only space flight for each of the three crewmen.

Center right: Edward G. Gibson, thirty-seven, Skylab 4 science pilot (SPT), also during suiting on August 2. The civilian Gibson earned his doctorate in engineering, with a minor in physics, from the California Institute of Technology in 1964. He had been a research scientist at Philco working on lasers before joining NASA in 1965, as part of the six-member scientist-astronaut class. He then earned his Air Force wings after a year of flight training at Williams AFB in Arizona.

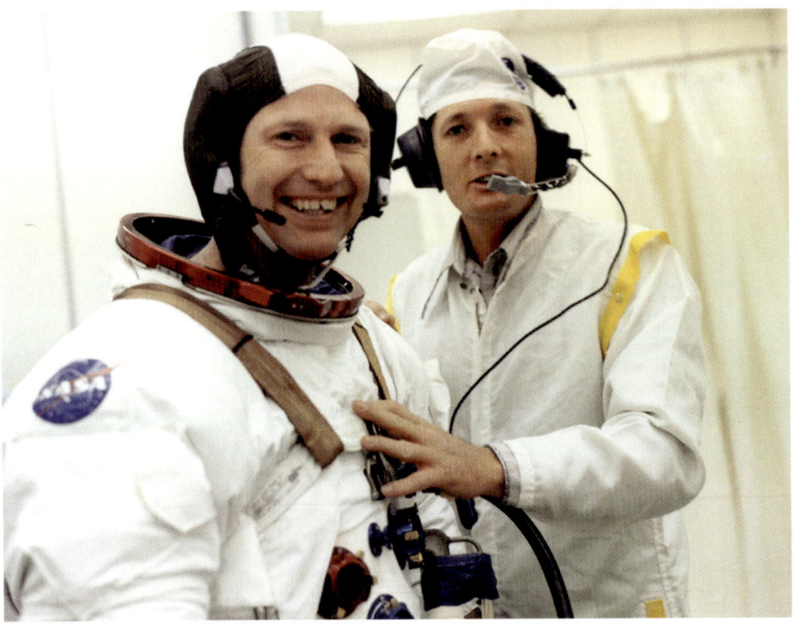

Bottom left: CDR Gerald P. Carr, forty-one, a Marine Corps lieutenant colonel, suits up on August 2 alongside suit tech Troy Stewart. Carr had been a carrier-based fighter pilot. He earned a bachelor of engineering degree in mechanical engineering from the University of Southern California in 1954, a bachelor of science degree in aeronautical engineering from the US Naval Postgraduate School in 1961, and a master of science degree in aeronautical engineering from Princeton University in 1962. Carr was the first rookie commander of a US space flight since 1966, when Neil Armstrong helmed Gemini VIII.

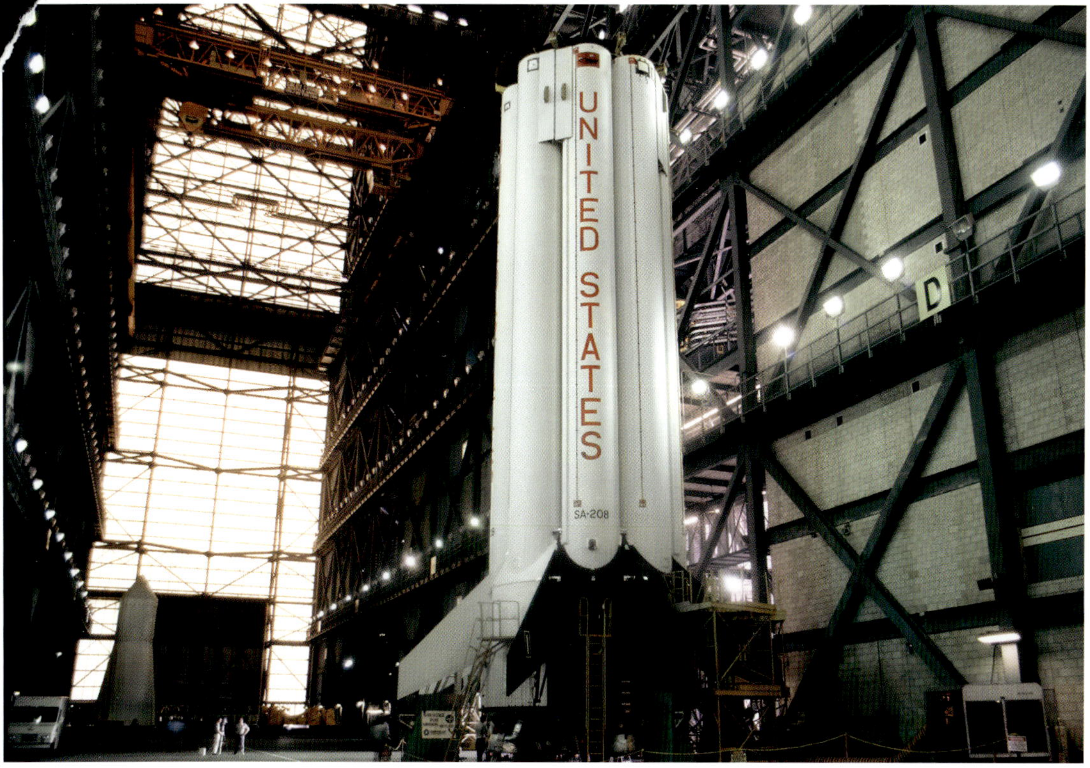

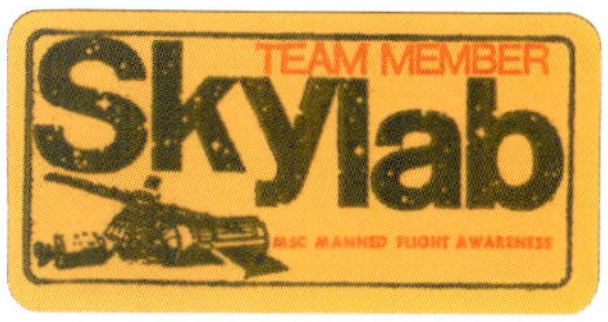

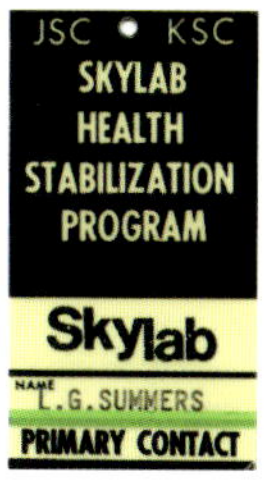

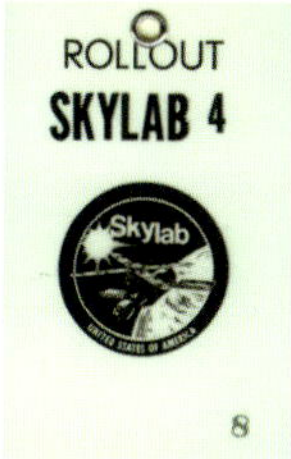

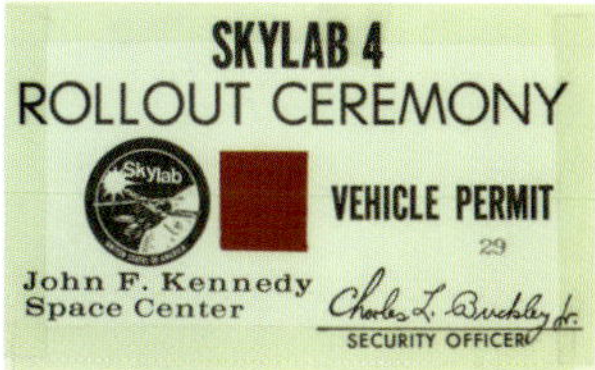

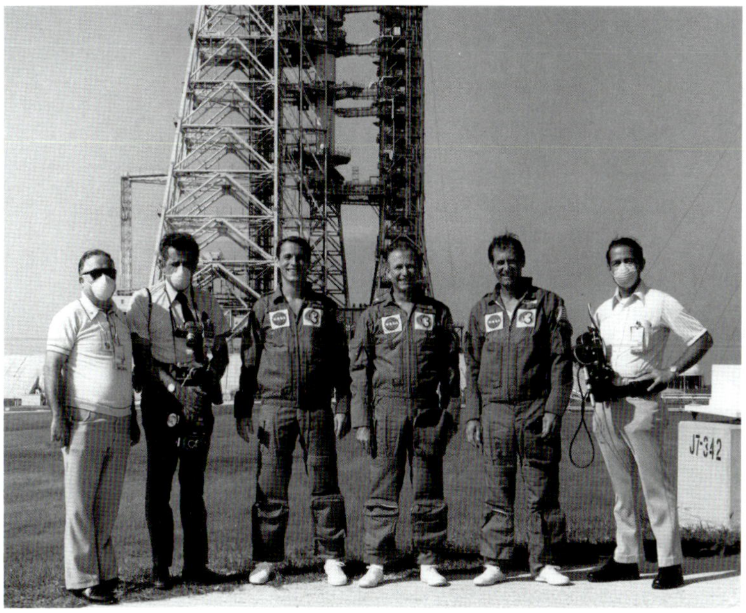

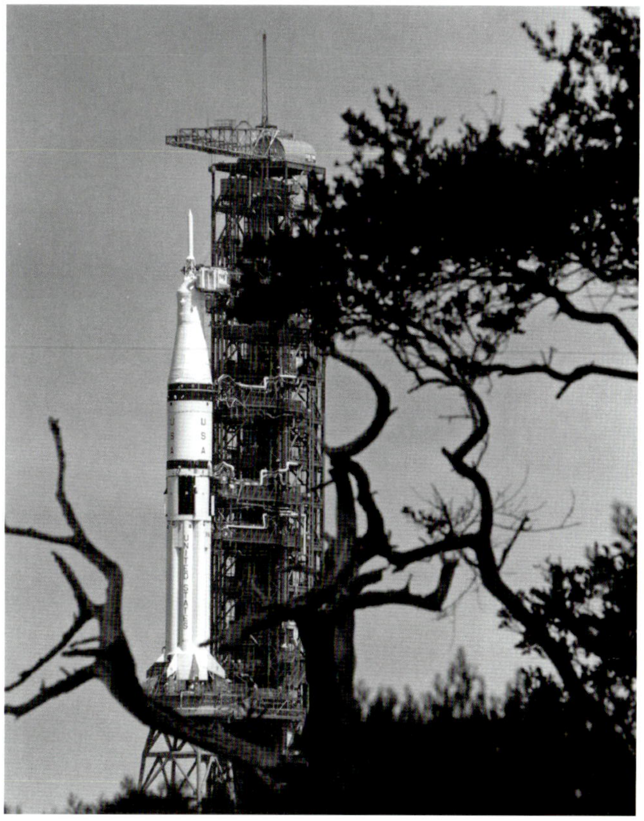

Top left: The S-IB first stage of Skylab 4's Saturn booster (SA-208) in the VAB's transfer aisle on July 27, 1973, during processing. On June 20, the stage had arrived at KSC by barge from the Michoud Assembly Facility near New Orleans, Louisiana. It had been static test–fired at MSFC in 1966 and then stored at Michoud until 1972. (photo by Steve Nolte)

Bottom left: The crew poses at Pad 39B with NASA photographers on November 8, 1973, two weeks into the astronauts' preflight quarantine. *Left to right*: Ed Thomas, Bill Taub, Gibson, Carr, Pogue, and Larry Summers (with an Arriflex S 16mm camera).

Above: The Skylab 4 stack crawls past Florida scrub oak trees to Pad 39B on ML-1 on August 15, 1973. Some 1,500 guests from surrounding communities were on hand to view the beginning of the rollout at 7:00 a.m. (EDT).

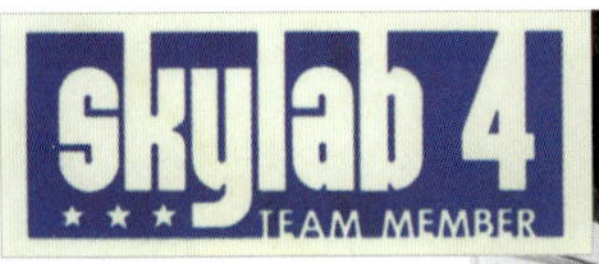

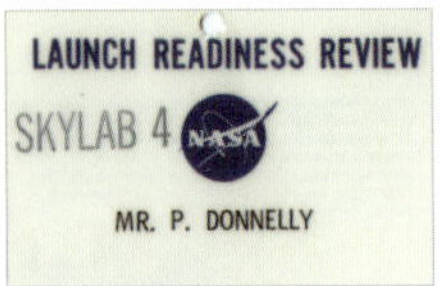

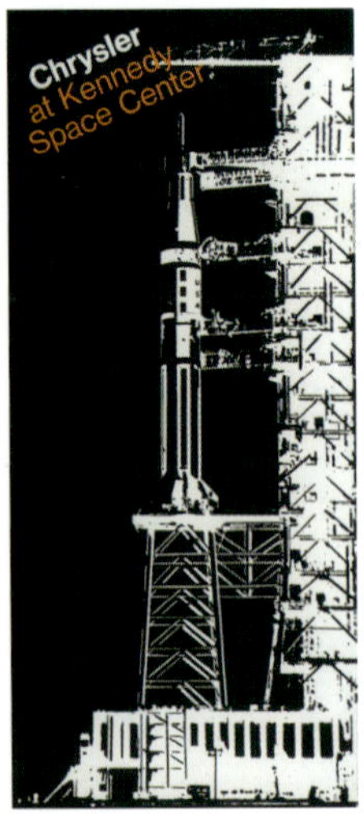

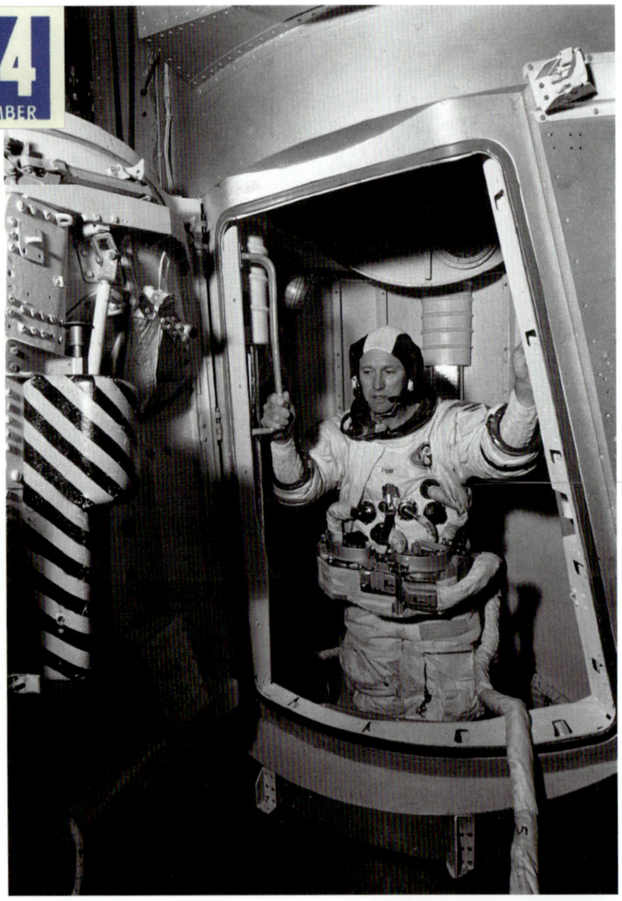

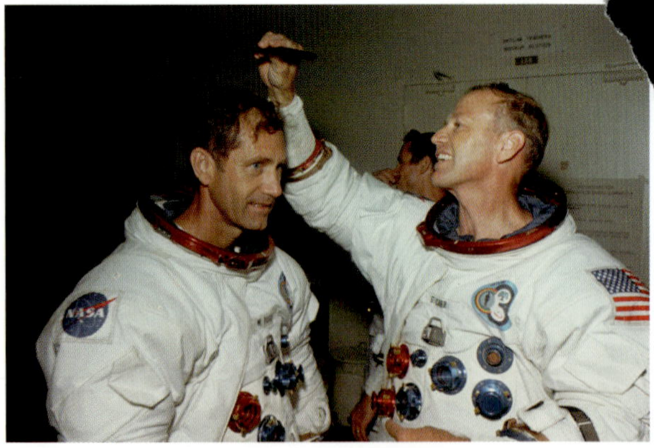

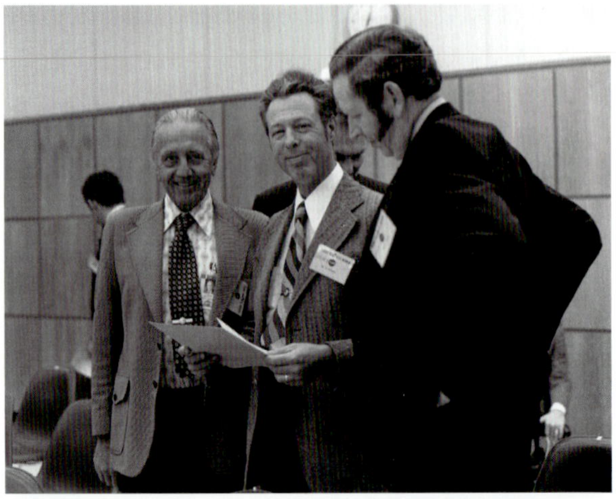

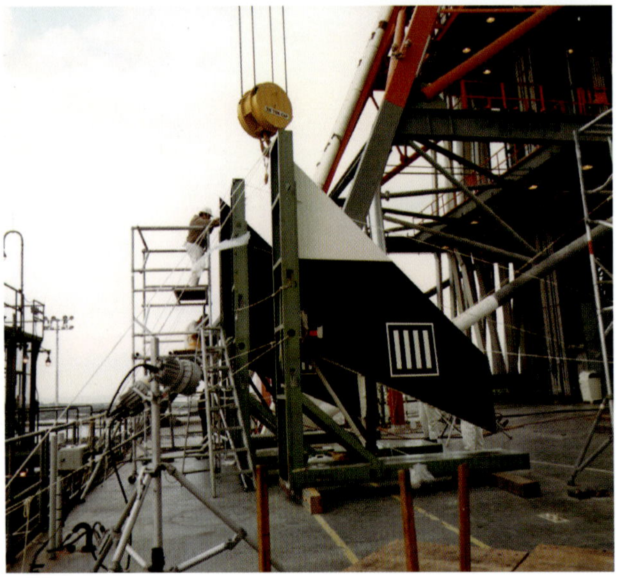

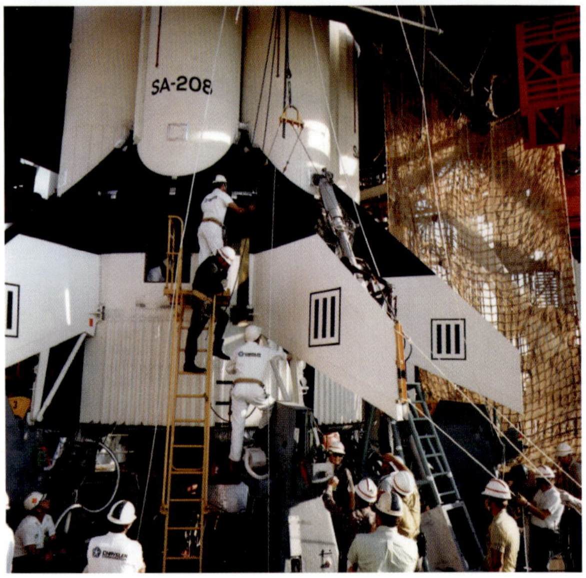

Top left: Carr at the airlock module hatch, a modified Gemini spacecraft hatch, in the Skylab trainer in Bldg. 5 at JSC on August 12, 1973.

Top right: Carr combs Pogue's hair before a photo session with a large Skylab model at JSC on August 10, 1973. The crew was preparing to conduct a simulated EVA using the Skylab trainer.

Bottom left: The booster's original fins, at Pad 39B on November 10, 1973. Members of the "Fin Team," who accomplished the change-out, were provided with VIP launch seats.

Center right: KSC Director Kurt Debus, Spacecraft Operations Chief Engineer George T. "Ted" Sasseen, and Dick Smith, Saturn program manager at MSFC, at the Flight Readiness Review in the MSOB on October 18, 1973.

Bottom right: The first of eight fins are removed from the Saturn IB's first stage at Pad 39B on November 9, 1973. Three days earlier, hairline cracks were discovered around bolts in the fittings that attached the fins to the booster's tail unit. Most were less than an inch long and were described as stress corrosion cracks.

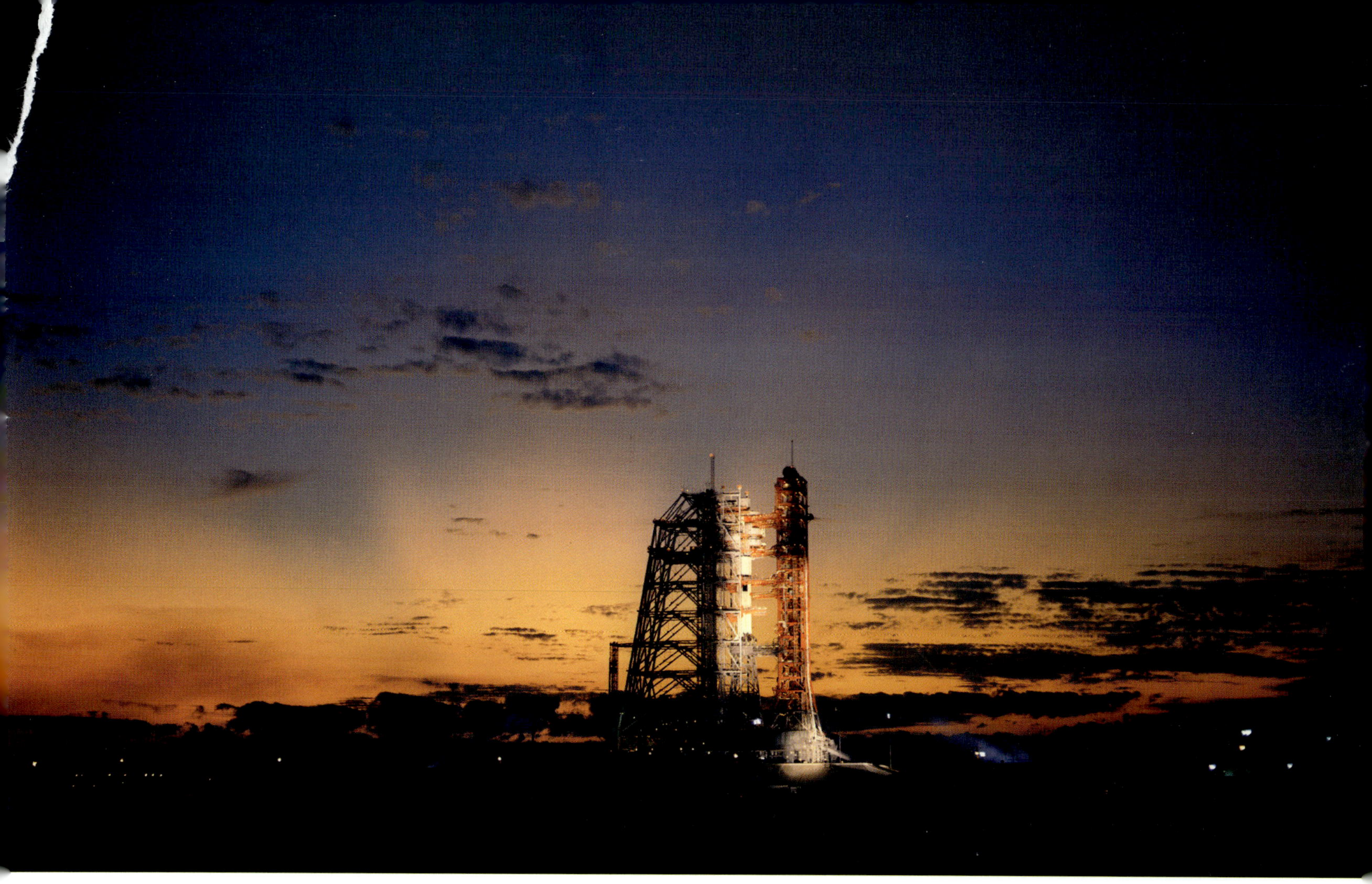

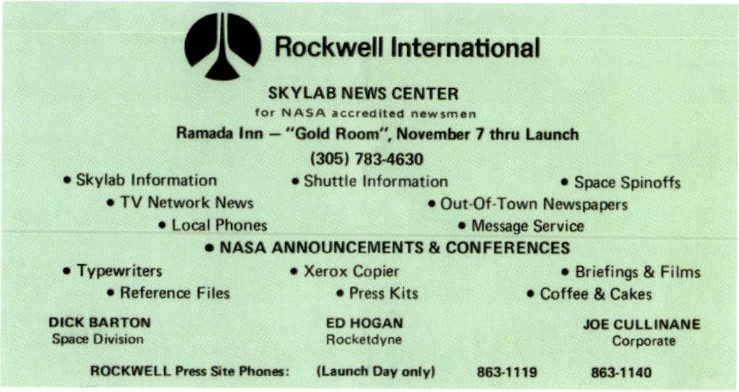

Rockwell International

SKYLAB NEWS CENTER

for NASA accredited newsmen

Ramada Inn – "Gold Room", November 7 thru Launch

(305) 783-4630

- Skylab Information
- Shuttle Information
- Space Spinoffs
- TV Network News
- Out-Of-Town Newspapers
- Local Phones
- Message Service
- NASA ANNOUNCEMENTS & CONFERENCES
- Typewriters
- Xerox Copier
- Briefings & Films
- Reference Files
- Press Kits
- Coffee & Cakes

DICK BARTON
Space Division

ED HOGAN
Rocketdyne

JOE CULLINANE
Corporate

ROCKWELL Press Site Phones: (Launch Day only) 863-1119 863-1140

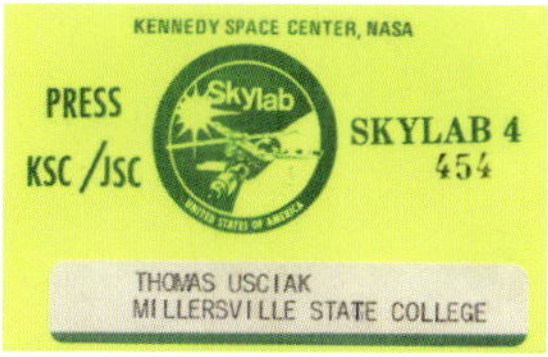

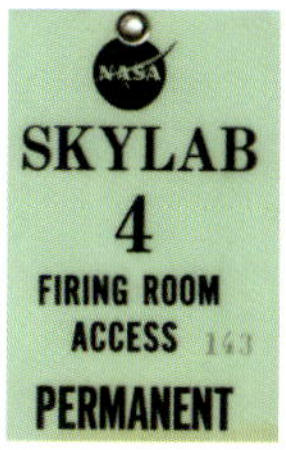

Top: Sunset at about 6:30 p.m. on November 15 at Pad 39B, with the MSS still in place. First-stage propellant loading would begin within hours. (photo by Mark Usciak)

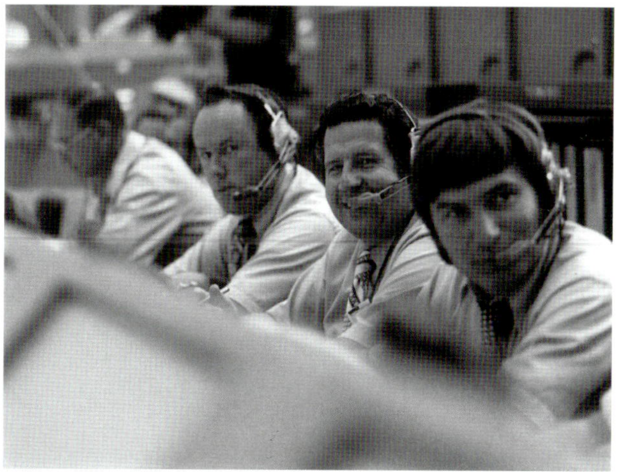

Center right, left to right: Chief Test Planning Officer Bob Moser, Test Supervisor Chuck Henschel, Chief Test Supervisor Bill Schick, and Assistant Launch Vehicle Test Conductor Tip Talone in Firing Room 3 on November 5, 1973, during the countdown toward the original launch date of November 10.

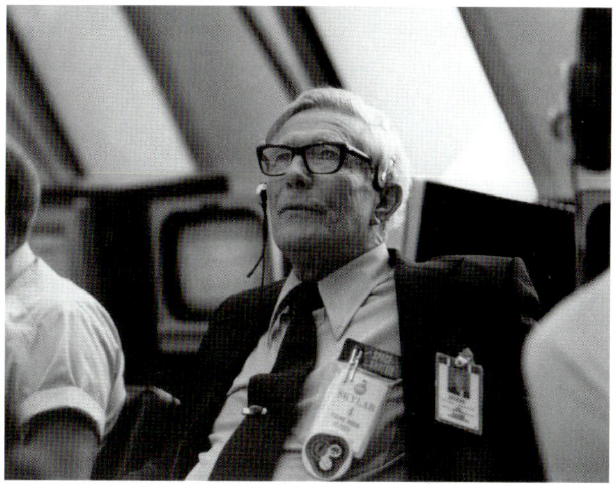

Bottom right: Director of Launch Vehicle Operations Hans Gruene in Firing Room 3 on November 5. Gruene, a German rocket engineer who was part of Operation Paperclip, had first come to the Cape in 1952 to help establish what would become the Eastern Test Range.

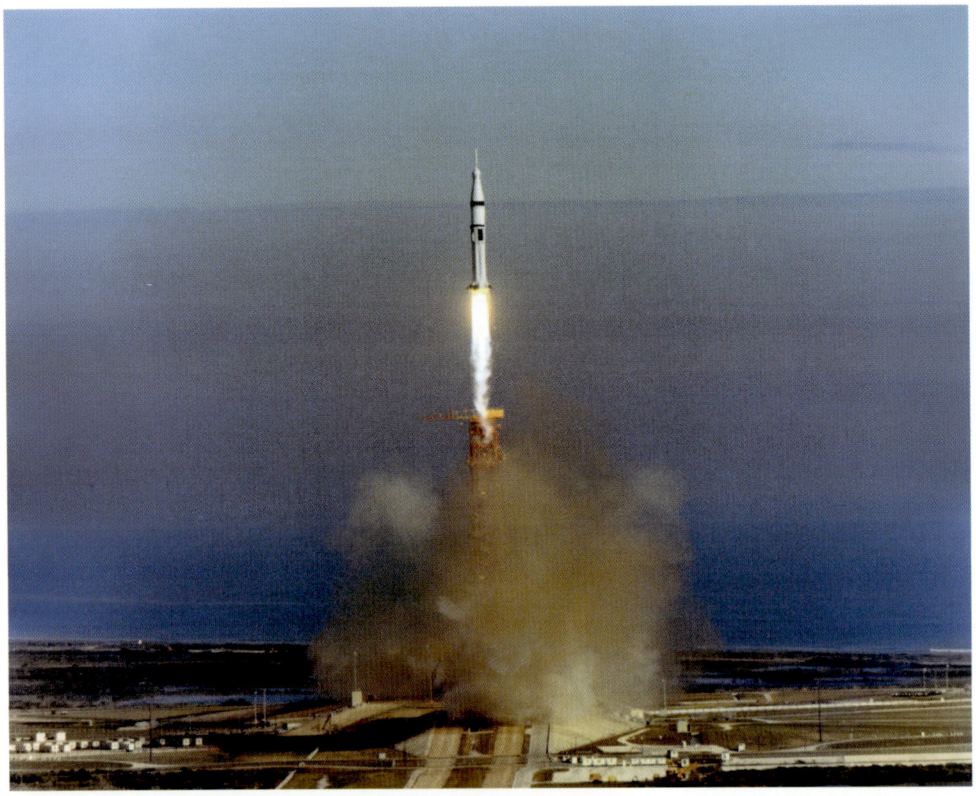

Left: Skylab 4 clears the LUT at 9:01 a.m. (EDT) on November 16 in this view looking north. NASA invited thirty thousand guests to the launch, and an additional ten thousand government and contractor personnel attended. Brevard County officials correctly forecast a low turnout and made no special provisions; public spectators were estimated in the tens of thousands.

Bottom left: Skylab 4 clearing the LUT, view from below

Bottom right: A flock of birds flies past the pad as, minutes after liftoff, water continues to drain from the LUT's swing arms and the launch pedestal.

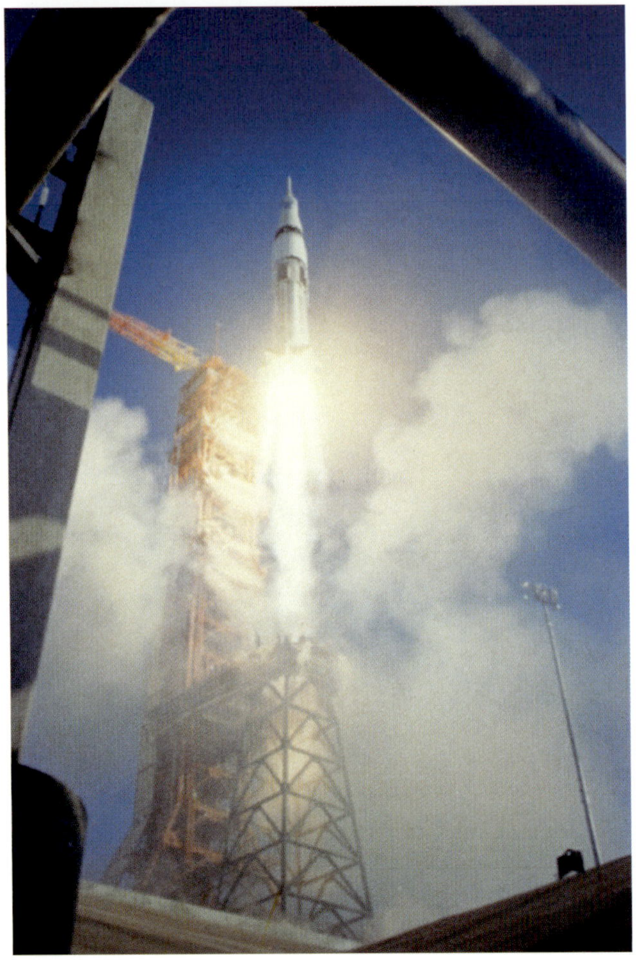

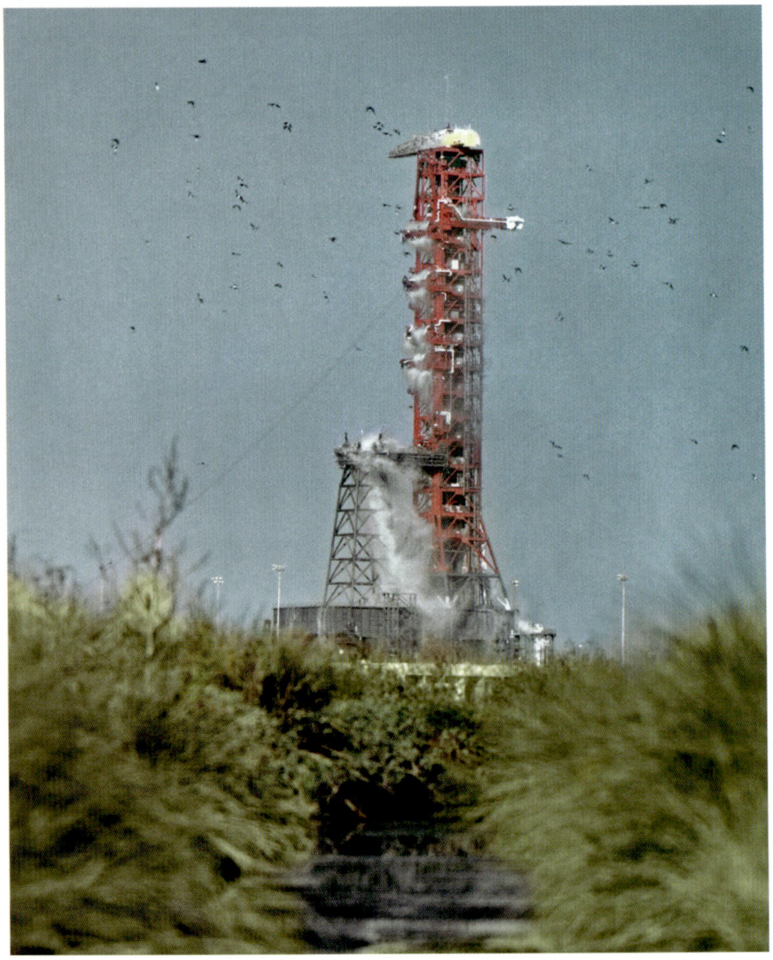

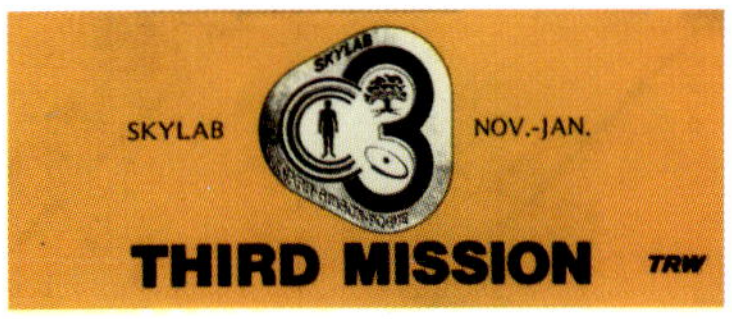

Top right: Gibson enters the upper dome area of the OWS through the hatch to the airlock module. Blue handrails and four general illumination fluorescent lights are mounted on the dome. The red stripe on the light at far right indicated it as an initial entry and emergency light. The air duct next to Gibson's arm at left provided environmental control in the airlock module.

Center right: In this gag photo, taken near the end of the mission in the experiments compartment, Pogue seeks revenge on the Experiment Support System, which centralized power, displays, and controls for four medical experiments. In the foreground at right is the control unit for the bicycle egronometer.

Bottom right: The Skylab 4 flight control team in MOCR 1. *Front row, left to right*: Tandy Bruce, instrumentation and communications officer; Dick Ramsell, organization and procedures officer; Chuck Lewis, flight director; Sy Liebergot, electrical generation and integrated lighting systems engineer; Ron Lerdal, guidance, navigation, and controls systems engineer; Pete Frank, chief, Flight Control Division; Pat O'Neill, ASCO; and Hank Hartsfield, CAPCOM. *Back row, left to right:* Bob Gordon, Public Affairs Office; Bob Heselmeyer, biomed experiments officer; Darrell Stamper, Apollo Telescope Mount; Tom Kaiser, flight activities officer; Dr. John Zieglschmid, flight surgeon; John Nelson, corollary experiments; and Earl Thompson, Earth Resources Experiment Package.

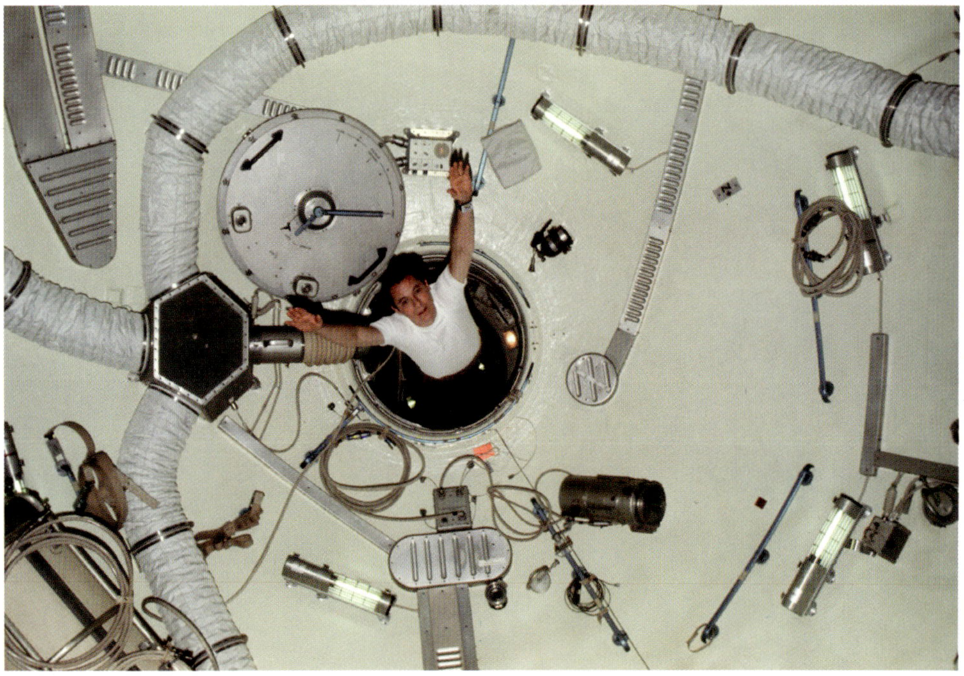

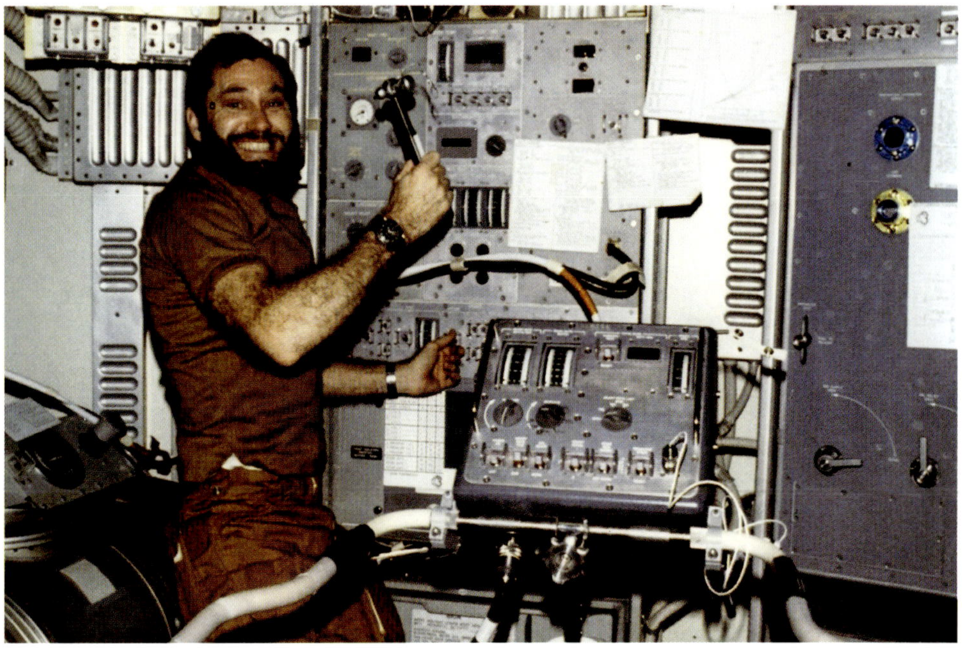

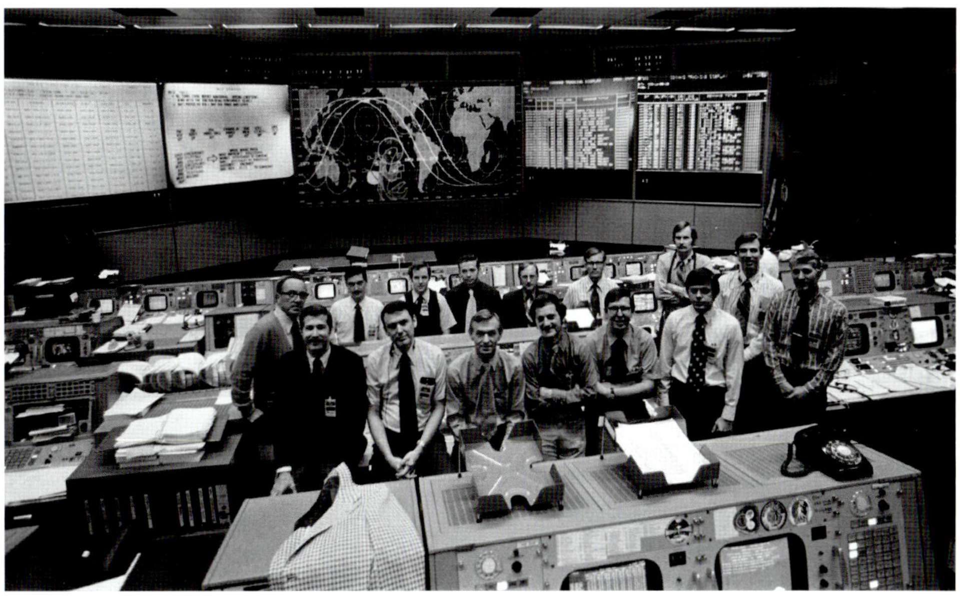

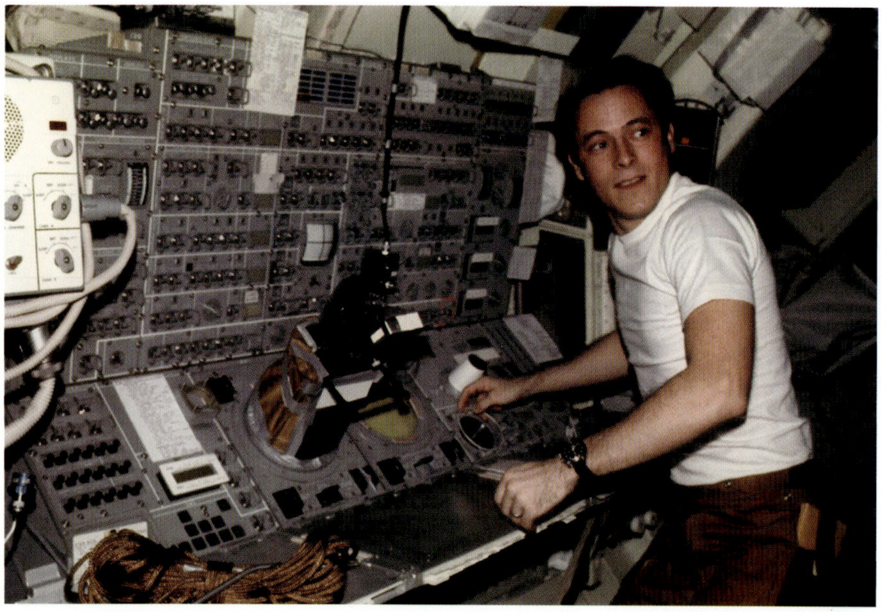

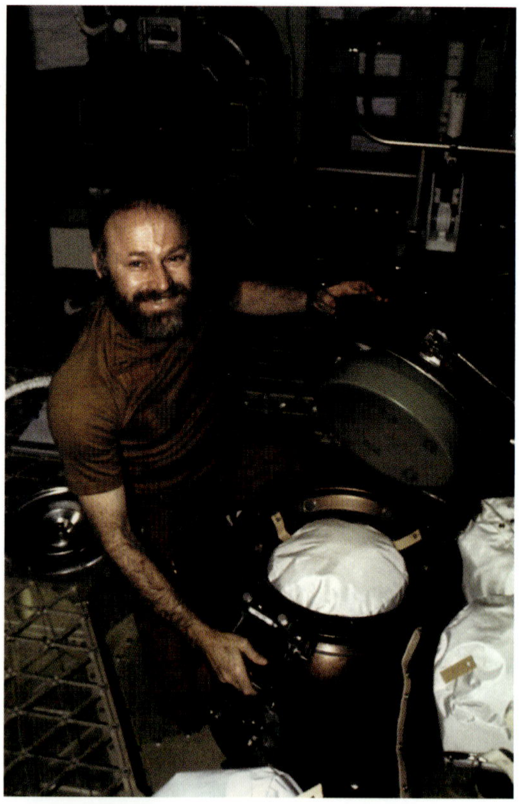

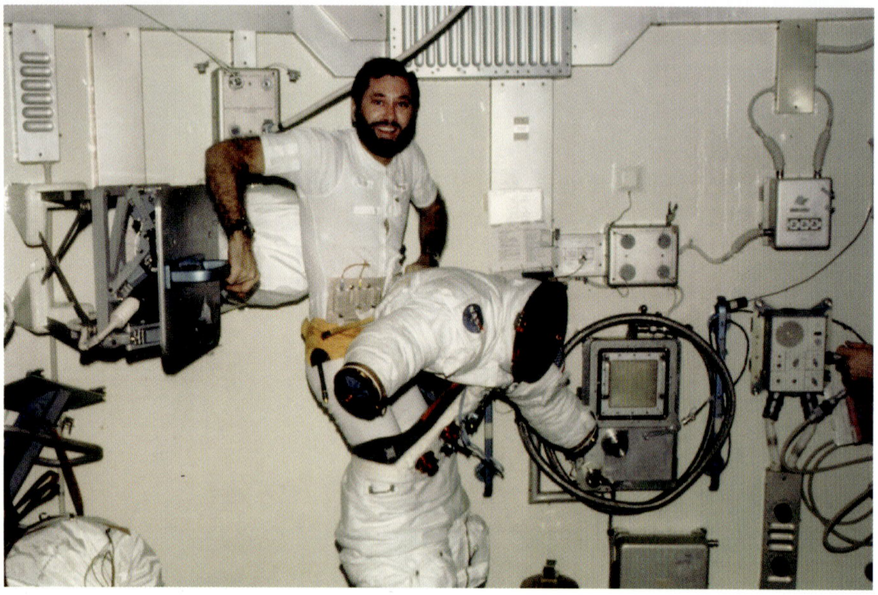

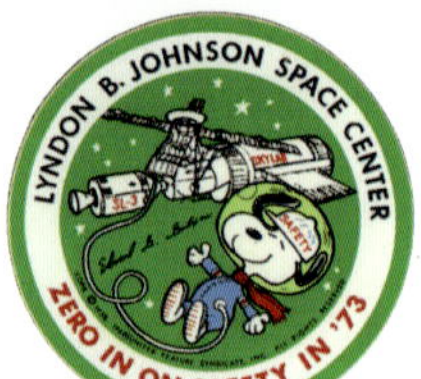

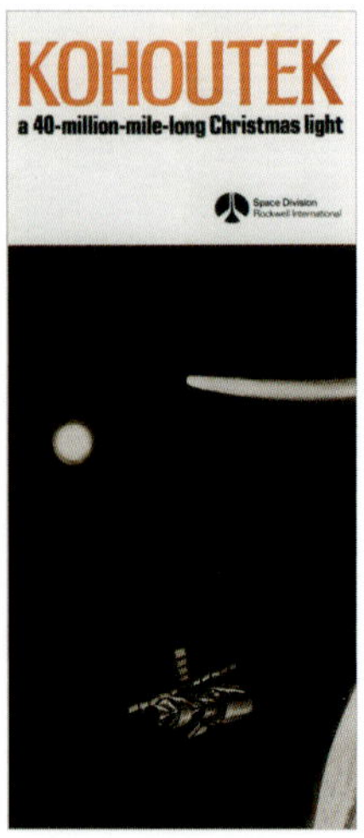

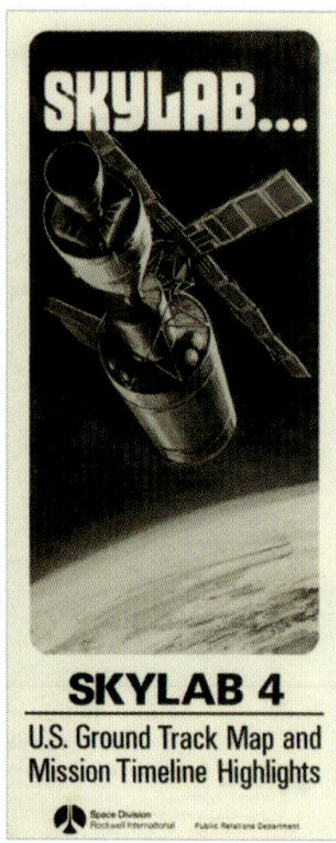

Top right: Carr closes the cover to the trash airlock in the center of the crew compartment floor. The brown tabs on the Beta cloth bags fit onto pins to hold each bag in place. The cover was closed and sealed and the compartment depressurized before a lever opened the outer door and a mechanism ejected the bag into the modified S-IVB oxygen tank below.

Center left: Pogue dons his pressure suit before his second EVA, on December 25, at the suit-drying station in the forward compartment. The gray hose is looped around the Scientific Airlock, used to expose experiment hardware to space. The workshop had two such airlocks, but the umbrella solar shield had been deployed through one, making it unusable.

Bottom left: Comet Kohoutek was first sighted on March 7, 1973, and was quickly added to Skylab's observational goals. Although predicted to be dramatic, it apparently partially disintegrated during its swing past the sun in December.

Top left: Gibson spent hours waiting for solar flares at the ATM Control and Display Console in the MDA. His hand is near the manual pointing control. The empty desk shows that the crew has packed up in the days before reentry. The previous crew brought the hood and new Polaroid SX-70 camera taped to one of the console's video displays. The camera supplemented the ATM's limited-magazine cameras—providing lesser-quality but still useful solar corona images—and let the crew compare solar features from one orbit to the next. The electronics in the console, made by Bendix in Teterboro, New Jersey, generated so much heat that the console had its own liquid cooling system.

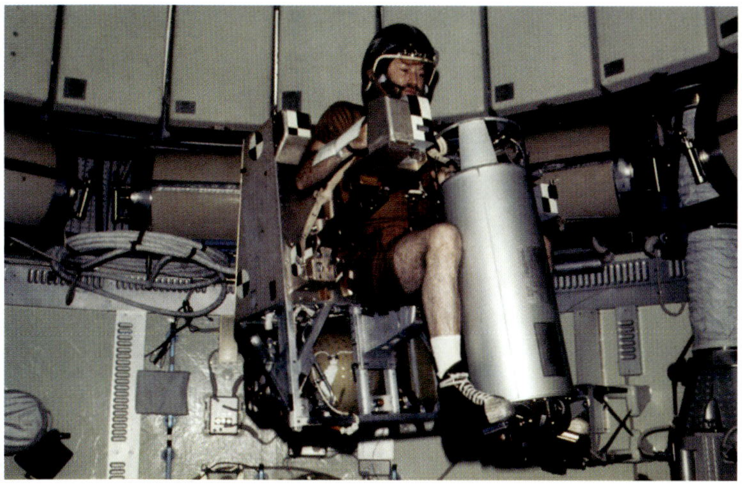

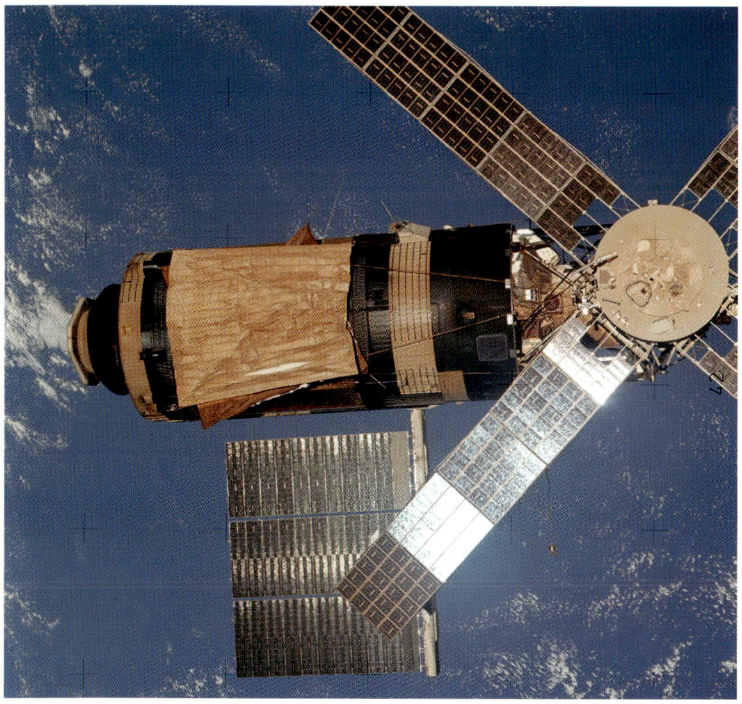

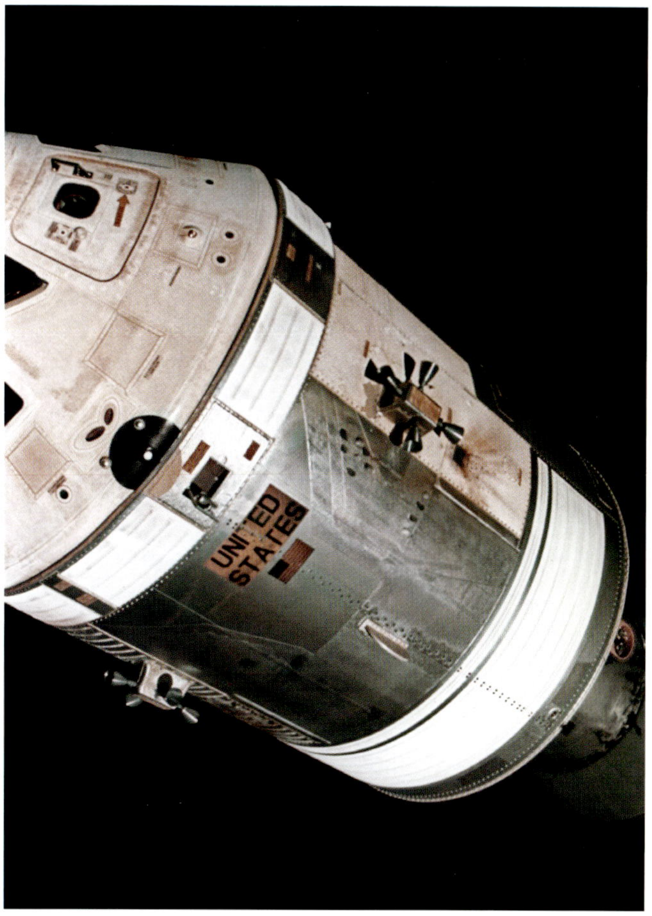

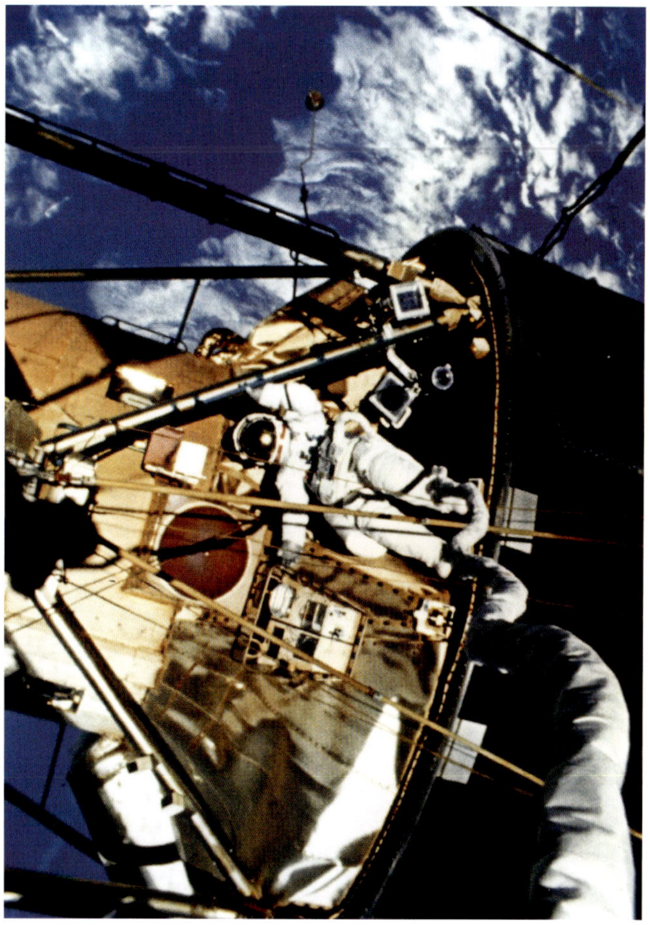

Top left: Pogue's photo shows Carr flying Experiment M509, the AMU, in the forward dome of the OWS on December 15, 1973. He holds a canister between his legs to evaluate how the unit performed with an object that moved the center of gravity. The CAPCOM was Bruce McCandless, who would go on to fly the first untethered MMU during a 1984 Space Shuttle mission.

Top right: The CSM docked with the OWS, as photographed by Carr during the final Skylab EVA on February 3, 1974, four days before splashdown. During the EVA, which lasted five hours and twenty minutes, Carr and Gibson retrieved material samples and the ATM film canisters, and tested a clothesline-like device for moving equipment.

Bottom left: The OWS, photographed from the Skylab 4 CM during the final fly-around inspection on February 8, 1974. The ATM's solar arrays provided about half of the OWS's electrical power.

Bottom right: Carr's umbilical stretches toward Gibson in this February 3 EVA photo by Carr. Gibson, his feet near the top edge of the Fixed Airlock Shroud, has egressed from the EVA hatch. He holds a lower ATM strut. The ATM is behind him and the MDA is above his head at left. A small discone antenna extends on a boom at top center.

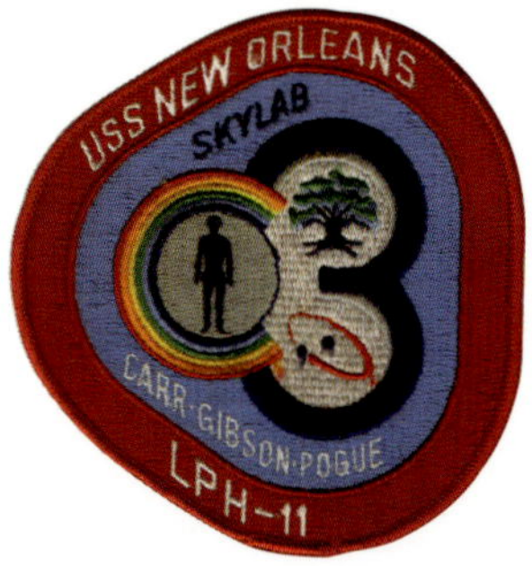

Top left: Divers wait with Skylab 4 shortly after splashdown in the Pacific Ocean, 176 miles southwest of San Diego, California, at 8:17 a.m. (PDT) on February 8, 1974. They are three and a half miles from USS *New Orleans*. It was the first manned splashdown not to be televised live by the three TV networks since Gemini VI, which had the first live television splashdown coverage (previously such broadcasts had not been possible).

Top right: Gibson prepares to egress from the CM aboard the *New Orleans*. "I never knew I weighed so much," he remarked. All three crewmen were deemed to be in good health, and all managed to walk a short distance.

Center left: President Richard Nixon presents Pogue with the NASA Distinguished Service Medal before a speech to some five thousand JSC employees and visitors. All three crewmen were thus honored. Nixon also toured displays on the upcoming ASTP mission.

Center right: Carr waves as the crewmen prepare to head to Houston after a brief dockside welcoming ceremony in San Diego on February 10, 1974. Three thousand people attended the ceremony.

Bottom right: Carr addresses KSC workers in the VAB on April 19, 1974, during the ceremony for the Skylab 4 crew return and KSC Skylab awards. Behind him are NASA Associate Administrator for Manned Space Flight Dale Myers, NASA Deputy Administrator George Low, and NASA Administrator James Fletcher. (photo by John de Bry)

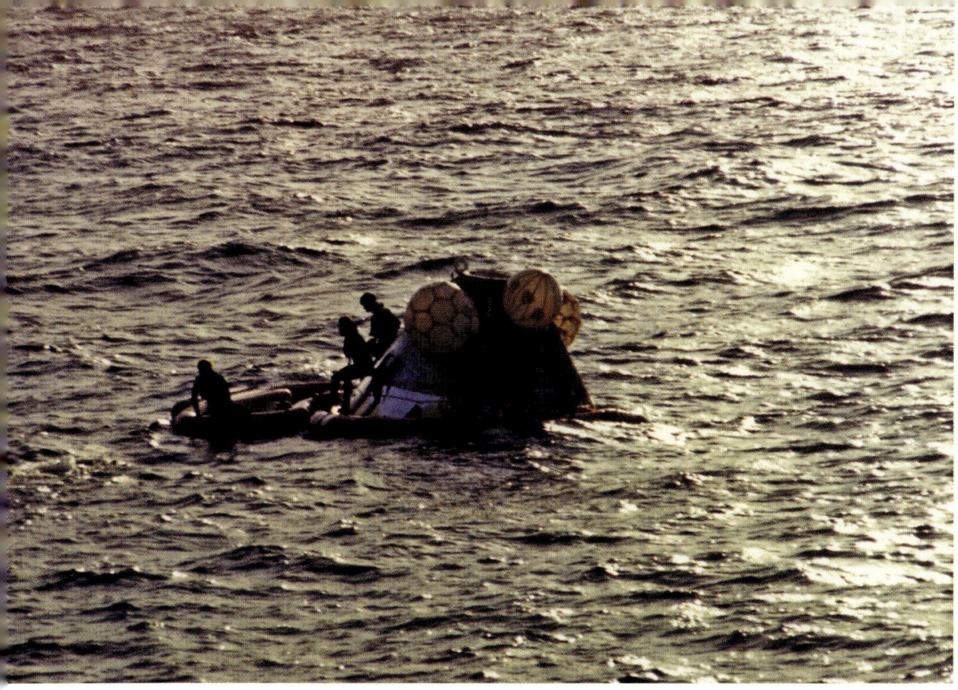

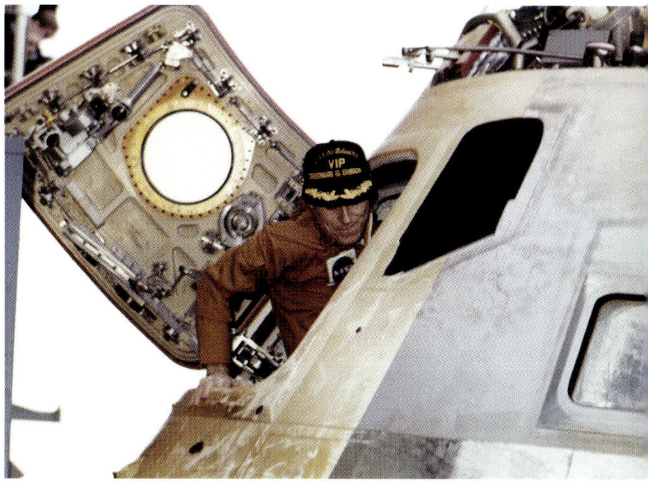

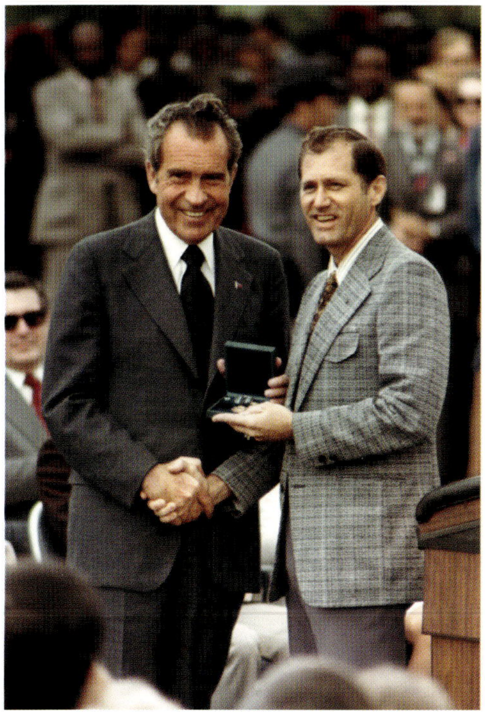

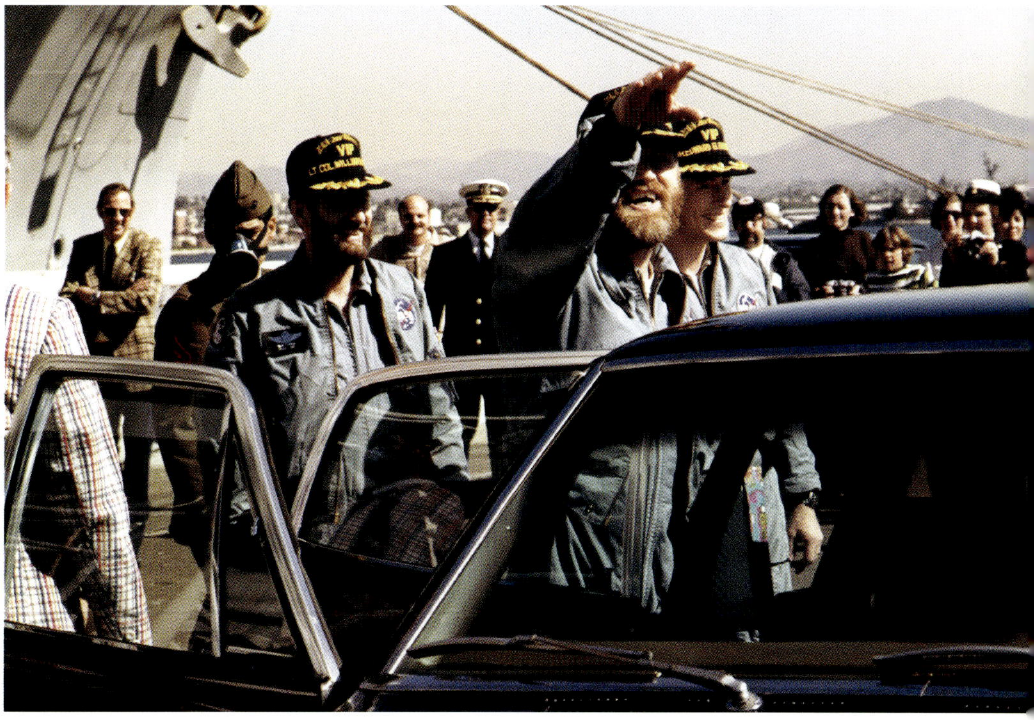

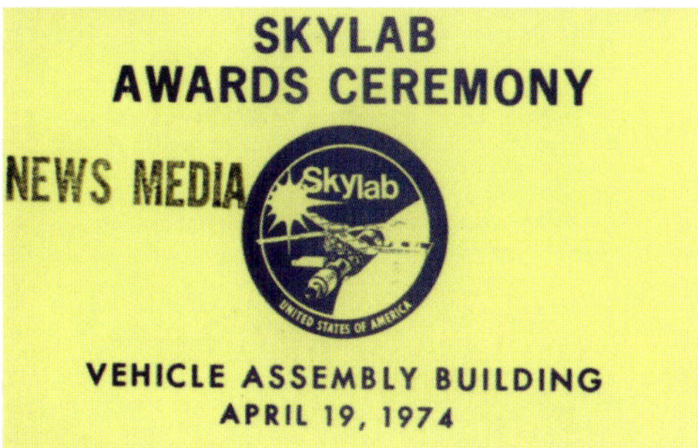
SKYLAB
AWARDS CEREMONY
NEWS MEDIA
Skylab
UNITED STATES OF AMERICA
VEHICLE ASSEMBLY BUILDING
APRIL 19, 1974

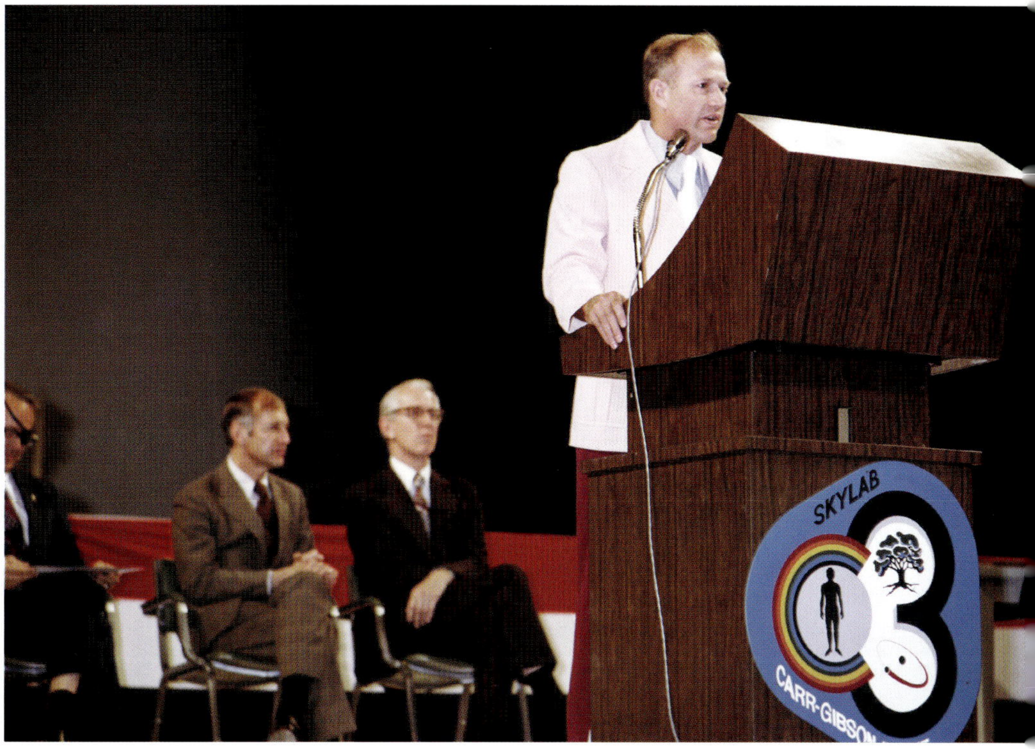
SKYLAB

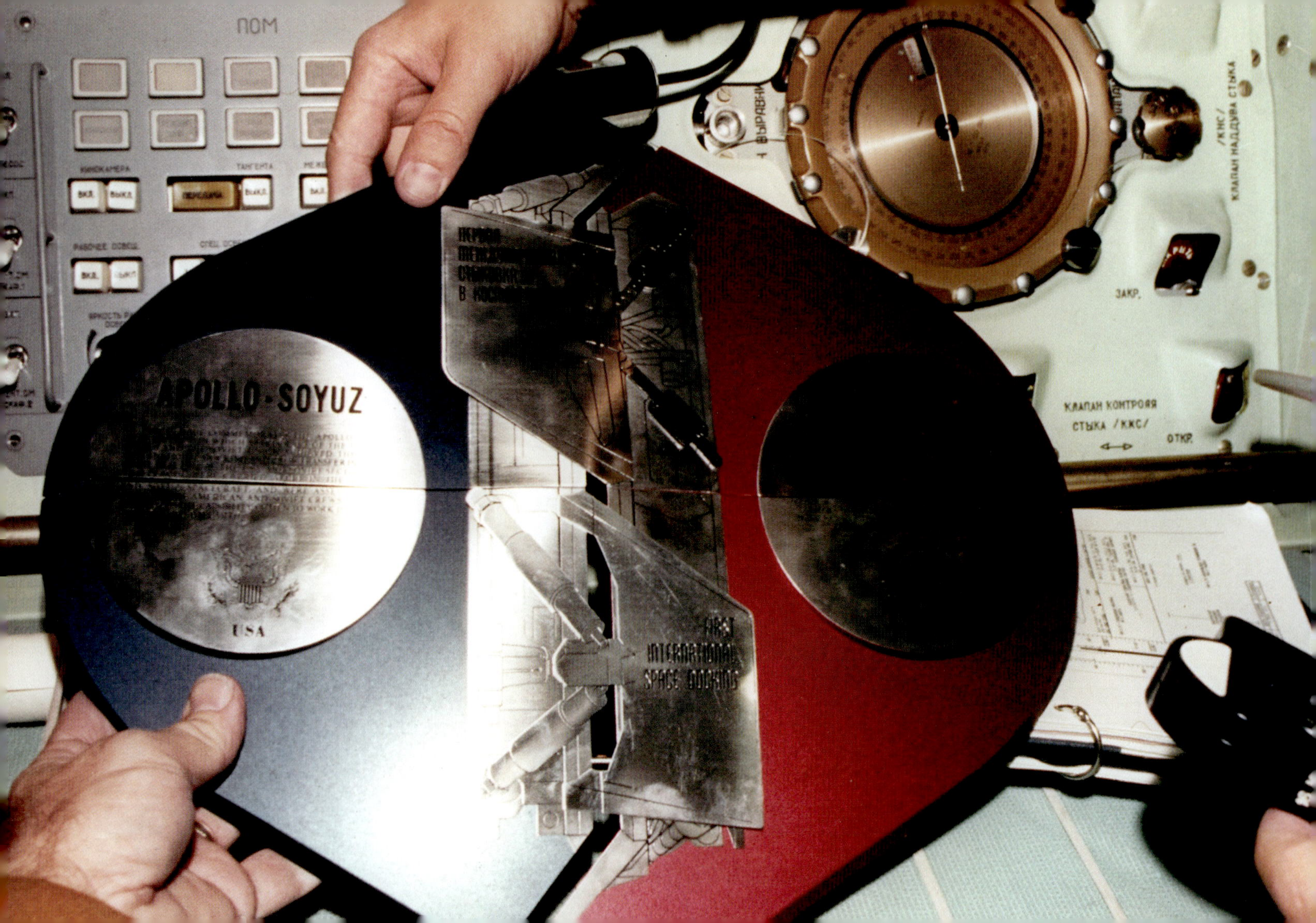

ПОМ
КИНОКАМЕРА
ТАНГЕНТА
APOLLO-SOYUZ
USA
FIRST
INTERNATIONAL
SPACE DOCKING
КЛАПАН КОНТРОЛЯ
СТЫКА /КЖС/
ЗАКР.
ОТКР.
/КЖС/
КЛАПАН НАДДУВА СТЫКА

CHAPTER 16

Apollo-Soyuz Test Project

July 15–24, 1975

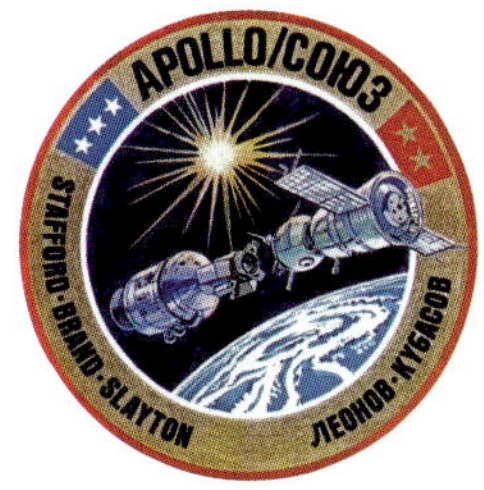

Seventeen months after the final Skylab mission, the last three Apollo-era crewmen spent nine days in Earth orbit in July 1975, including two days docked with a manned Soviet Soyuz spacecraft—the result of a 1972 diplomatic agreement between the United States and the USSR. The flight, called the Apollo-Soyuz Test Project (ASTP), was the first international manned space mission. It briefly united the world's two space powers in the aftermath of the Space Race.

Space officials in both countries were open to cooperation following the Apollo 11 moon landing. One obvious area of collaboration was in the development of compatible docking systems for space-rescue capabilities. Those 1970 discussions led to the formal 1972 summit agreement.

The US mission would use CSM-111, the last of the H-mission Apollo CSMs, which was available because its original flight had been canceled in 1970 due to budget cuts. The CDR would be Tom Stafford—now a brigadier general—who would be accompanied by two civilians: Vance Brand as CMP and Deke Slayton, who had been chosen for Mercury but was grounded due to atrial fibrillation, as docking module pilot (DMP).

The Soviets would use a transitional version of their second-generation, two-man Soyuz spacecraft, carrying Air Force Col. Alexei Leonov, forty-one, and civilian flight engineer Valeri Kubasov, forty.

The United States built a special docking module (DM)—with a Soyuz docking mechanism on one side and a US docking port on the other—which would be launched with the Apollo spacecraft on a Saturn IB booster. The DM would also serve as an airlock and transfer facility.

On July 15, 1975, Soyuz 19 was launched from Kazakhstan's Baikonur Cosmodrome at 8:20 a.m. (EDT). For the first time, the Soviet people—and the world—watched a Russian launch on live TV. The broadcast drew a huge audience in the USSR. Western reporters were not allowed at the launch site, but the US ambassador and his wife were.

Apollo was launched more than seven hours later, at 3:50 p.m. (EDT). It was the ninth and final Saturn IB launch. Among the guests was the Soviet ambassador to the United States—the first time a Russian official attended an American launch. Other VIPs at LC-39 included singer John Denver, Wernher von Braun, presidential daughter Susan Ford, Senator John Glenn, and Jim Lovell. Local officials had estimated that other than that of Apollo 17, the liftoff would have the largest crowd of spectators since Apollo 11, but fewer than one million people attended.

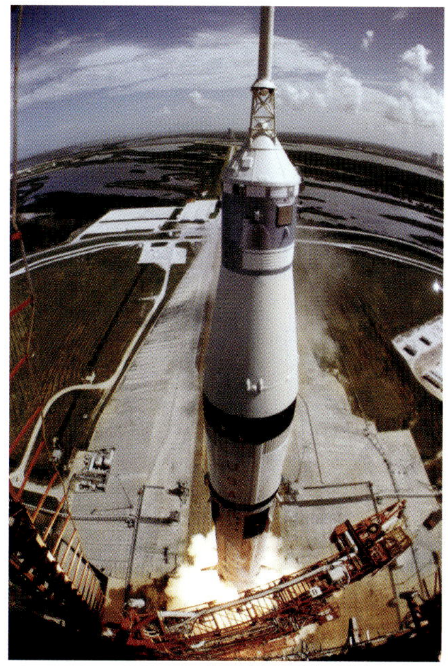

Launch: 3:50 p.m. (EDT), July 15, 1975

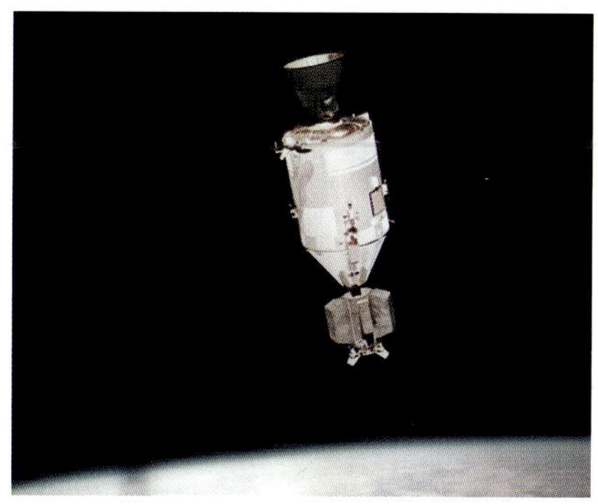

Docking: 12:09 p.m. (EDT), July 17

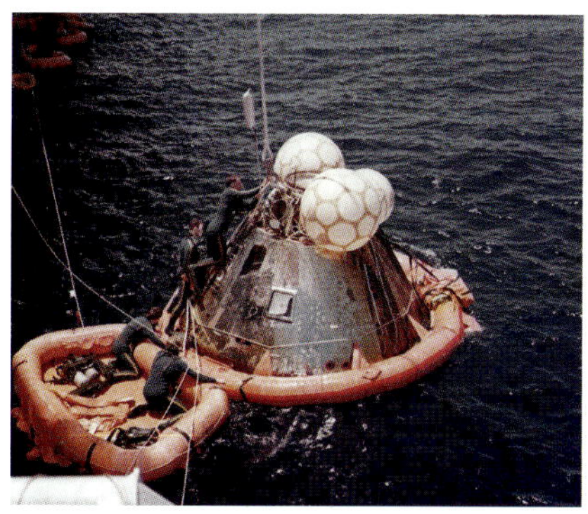

Splashdown: 11:18 a.m. (HST), July 24, 1975

The astronauts separated from the S-IVB stage at about 4:00 p.m., and then turned around and linked up with the DM, much as lunar missions had done with their LMs.

The crew spent the first full day getting settled in. Brand freed a balky docking probe assembly from the tunnel between the CSM and DM, and all three men conducted experiments and collected biomedical data.

The next morning, July 17, Apollo rendezvoused with Soyuz 19, docking at 12:09 p.m. in a 140-mile-high orbit. At 2:17 p.m., Stafford opened the hatch in the DM leading into the Soyuz cabin. The two commanders shook hands as their docked spacecraft passed over France.

During the next forty-two hours, the five crewmen visited each other's spacecraft, shared meals, and took part in a live thirty-minute TV news conference. They also spoke with US president Gerald Ford, heard a message from Soviet general secretary Leonid Brezhnev, and provided TV transmissions to their home countries. The crews conversed with their respective mission controllers in their native tongues, although each had studied the other's language.

A variety of commemorative items were exchanged or signed, including the International Aeronautical Federation certificate of docking; small American and Soviet flags; one UN flag; commemorative plaques; copies of the May 1972 Nixon-Kosygin Space Agreement, in English and Russian; American and Russian tree seeds; silver medallions (gifts to Leonov and Kubasov); and copies of papers by rocket pioneers Robert Goddard and Konstantin Tsiolkovsky.

On July 18, Slayton, Brand, and Kubasov assembled the two halves of a medallion commemorating the flight, and they exchanged the tree seeds. Stafford shook hands with the two Soviets, saying good-bye at about 3:50 p.m. He then moved back into the DM. The hatches were closed for the final time at 4:00 p.m.

The two spacecraft undocked the next day at 8:12 a.m. They briefly docked again as a test at 8:34 a.m., and undocked for the last time at 11:27 a.m. on July 19.

Soyuz 19 landed in the Soviet Union on July 21; Apollo remained in space for three more days, conducting experiments. Before releasing the DM on July 23, the astronauts took part in a half-hour TV news conference.

During reentry, with high noise levels in the cabin, Brand apparently did not hear Stafford call off one task for him to complete from the reentry checklist: flipping two switches, which would have automatically shut off the CM's RCS and triggered drogue parachute deployment. Instead, toxic nitrogen tetroxide fumes from the RCS oxidizer venting entered a cabin intake, as the spacecraft drew in outside air. Brand briefly lost consciousness. Stafford retrieved emergency oxygen masks, put one on Brand, and gave one to Slayton. The drogue parachute was deployed manually.

The last Apollo spacecraft to fly splashed down at 11:18 a.m. (HST) on July 24, 1975, in the Pacific Ocean. It was recovered by USS *New Orleans*, the only recovery ship assigned to the Pacific at that time. The three astronauts spent five days at a US Army hospital in Honolulu, but recovered fully.

With the Space Shuttle program still six years away, ASTP brought an end to the first era of US manned space flight, fourteen years after it had begun.

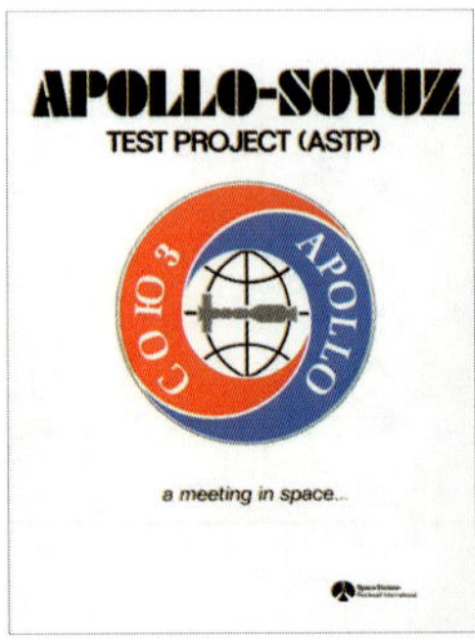

Left: The program emblem was mistakenly shown in the Soviet configuration on this Rockwell folder; the emblem was designed to be rotated to show "Apollo" on the left for the American use.

Opposite page, top left, left to right: Tom Stafford, Alan Bean (*behind Slayton*), Deke Slayton, Ron Evans, Vance Brand, and Alexei Leonov at a reception on November 20, 1973. This was the start of an eleven-day stay for the astronauts at the Gagarin Cosmonaut Training Center in Star City, near Moscow, for spacecraft familiarization. The dinner following the reception was held to welcome the visiting Americans. Bean was backup CDR; Evans was backup CMP.

Opposite page, bottom left: Brig. Gen. Stafford heads to a NASA T-38 jet at Patrick AFB for some proficiency flying on July 13, 1975, two days before launch. ASTP would be his fourth and final space flight. That November he became commander of the Air Force Flight Test Center at Edwards AFB in California. (photo by Jacques Tiziou)

Opposite page, top right: Slayton would finally make his first space flight as the ASTP DMP. He also became, at age fifty-one, the oldest person to fly in space (a record that was surpassed eventually by fellow Mercury astronaut John Glenn, who was seventy-seven during his 1998 Space Shuttle mission). This photo and the one below are also from July 13.

Opposite page, center right: Rookie Brand was CMP for ASTP. A former Marine Corps fighter pilot, he was a civilian experimental test pilot for the Lockheed Corporation from 1960 to 1966 before joining NASA. He was later CDR of three Space Shuttle flights.

Opposite page, bottom right: As Slayton looks on, Brand (*center*) and Stafford hold a model of the docked Apollo CSM and Soyuz 7K-TM spacecraft at MSC on January 30, 1973, the day they were named as the ASTP prime crew.

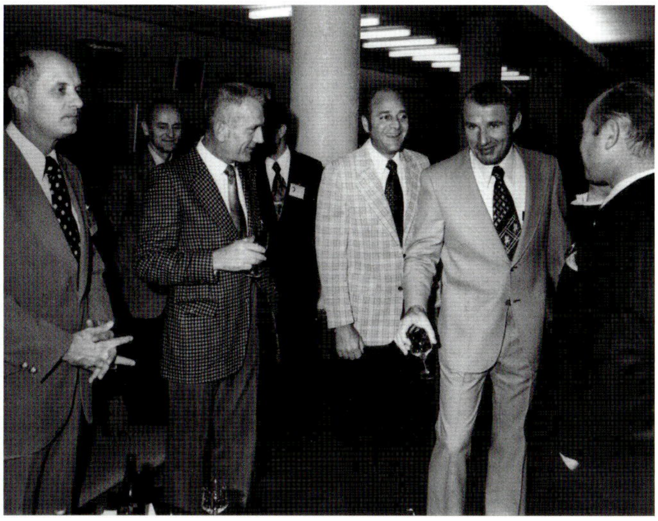

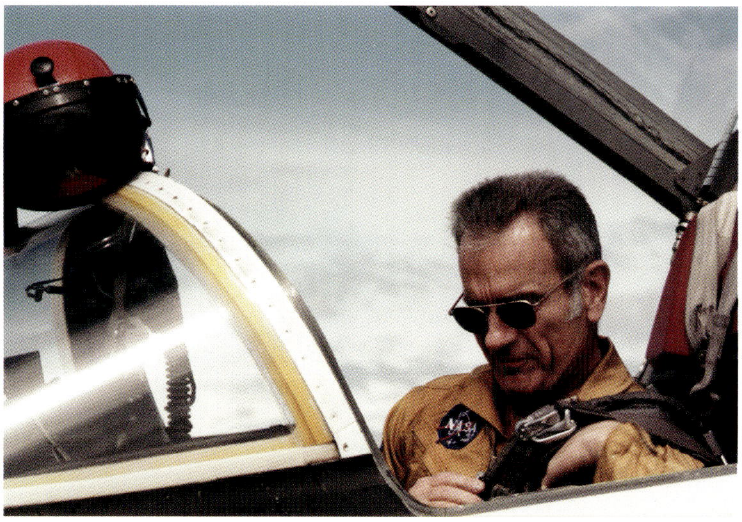

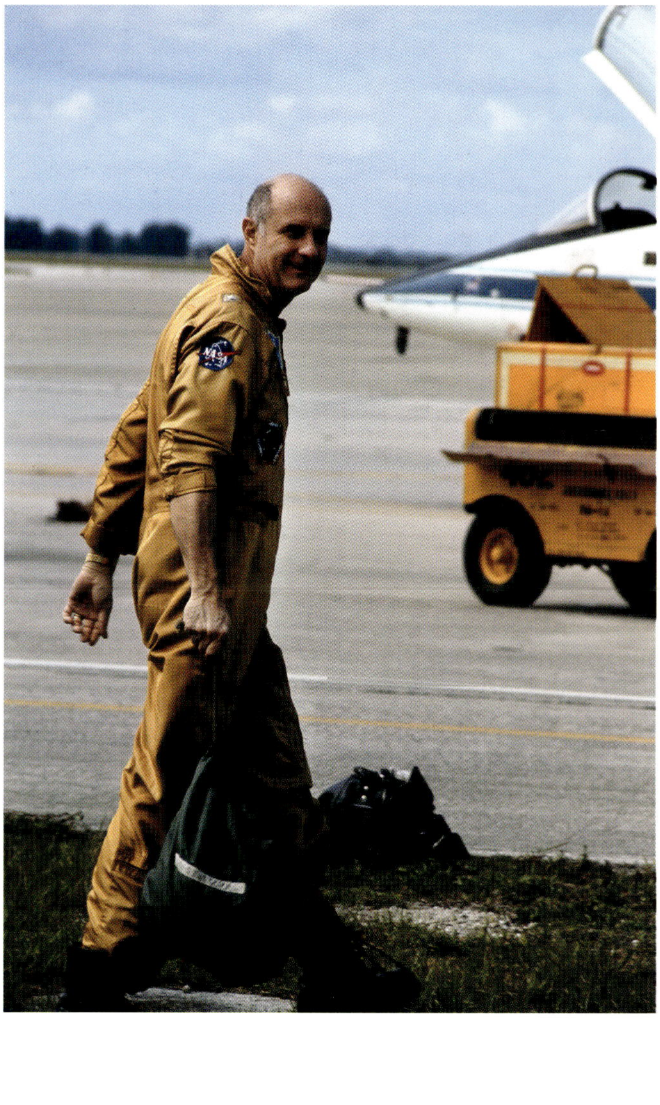

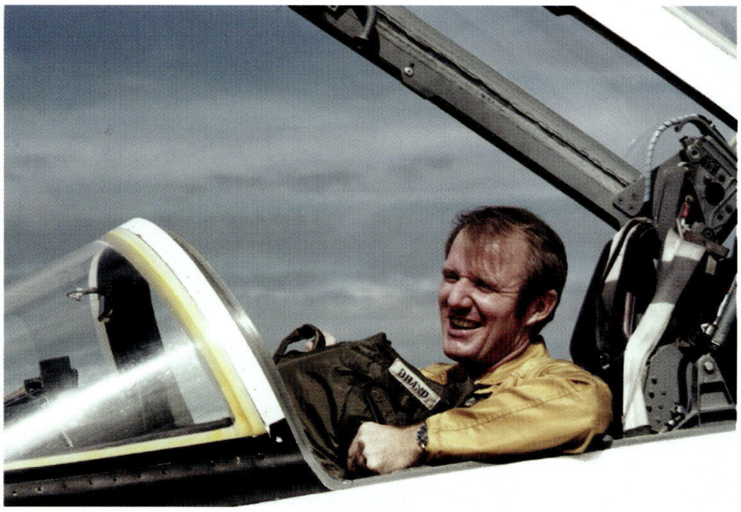

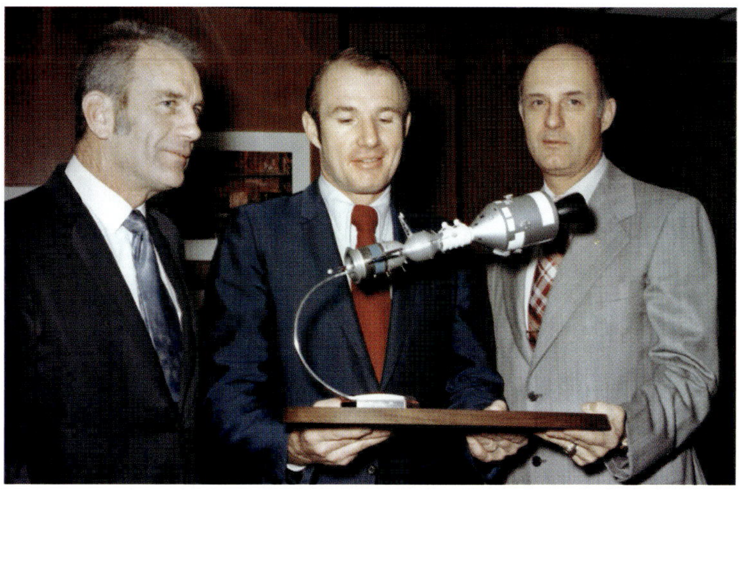

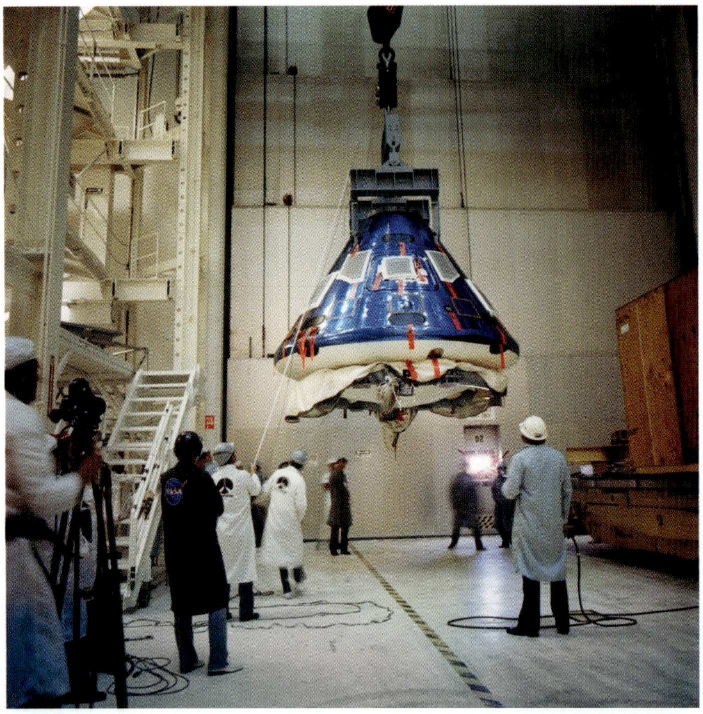

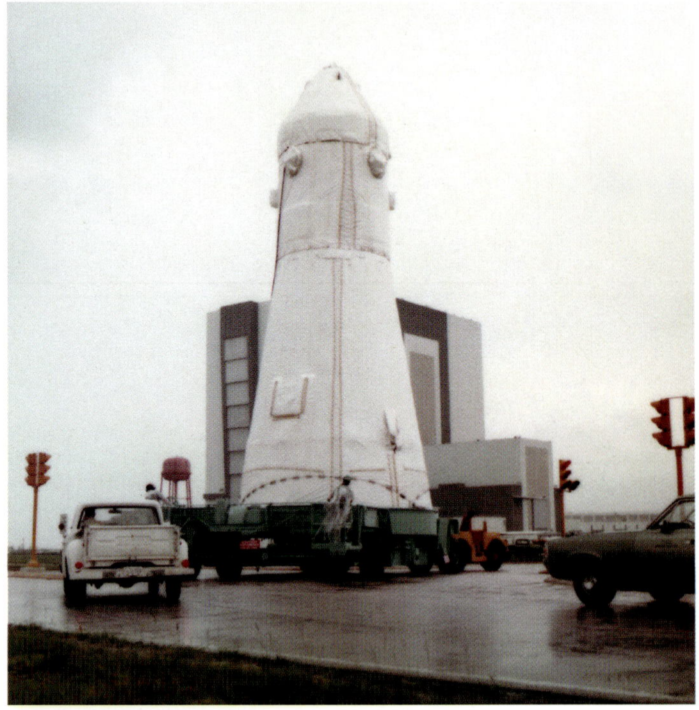

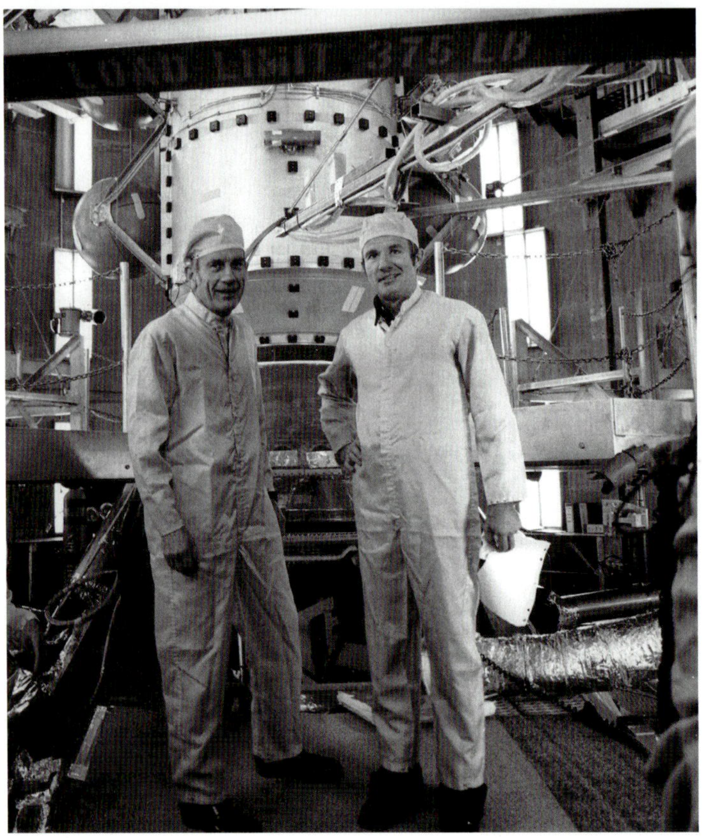

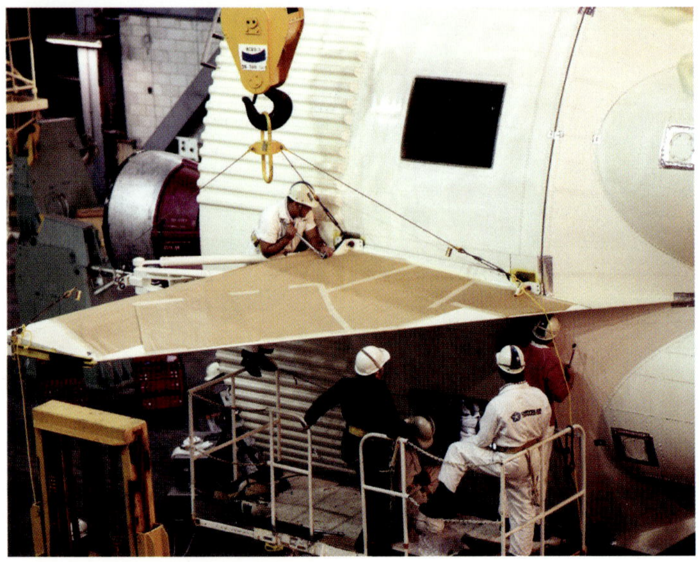

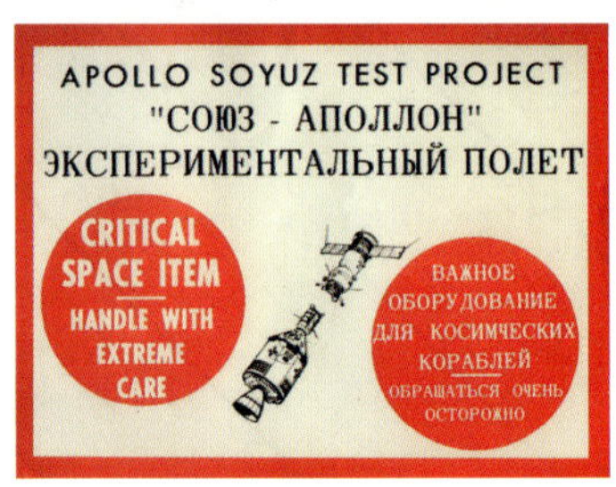

Bottom left: Slayton and Brand on a work platform. The top of the CM is behind them, docked to the DM (*above*) during fit checks inside Chamber L in the MSOB on January 27, 1975. The module's spherical oxygen and nitrogen gas storage tanks are to the left and right of the crewmen.

Top left: A Bendix crane operator (*right*) lowers the Apollo CM, which has been hoisted free of its shipping container for checkout, soon after arriving at the MSOB from Rockwell International in Downey, California, on September 8, 1974. Packing material is still attached to a white protective shell covering the heat shield.

Top right: The final CSM scheduled for flight, mated with an SLA carrying the DM, approaches the west side of the VAB on March 17, 1975—a rainy Florida morning—to be stacked atop its Saturn IB.

Center right: Chrysler technicians install one of eight fins to the tail-unit assembly of the S-IB first stage in the VAB transfer aisle on December 4, 1974. The aluminum fins provided aerodynamic stability during flight and supported the launch vehicle on the pad, using the hold-down fitting seen behind the worker at top left. One of the stage's eight Rocketdyne H-1 engines is visible above the fitting. All of the fins were replaced in March after stress corrosion cracks were discovered in two hold-down fittings.

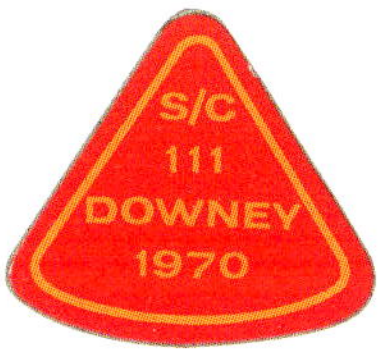

Top left: Soviet cosmonauts Valeri Kubasov (*front*), forty, and Leonov (*rear*), forty-one, meeting here with Slayton, were announced as the prime Soyuz crew on May 24, 1973, at the opening of the Paris Air Show. Leonov took the world's first spacewalk in 1965, beating Gemini IV's Ed White by more than two months. Both Russians were making their second space flight.

Top right, left to right: ASTP Program Director Chet Lee, Stafford, and Leonov on the Magic Kingdom's Main Street the same day. (photo by Jacques Tiziou)

Center left: Mickey Mouse shakes hands with Stafford, flanked by Leonov and Shatalov, before they ride RCA Space Mountain during a visit to the Magic Kingdom at Florida's Walt Disney World, on February 9, 1975. RCA, which sponsored Space Mountain, was also an Apollo contractor. The ride had opened three weeks earlier. (photo by Jacques Tiziou)

Center right: Jacques Tiziou created this unofficial support crew emblem for Americans Bo Bobko, Bob Overmyer, Bob Crippen, and Richard Truly, and Russians Johnny Dzhanibekov, Boris Andreyev, Yuri Romanenko, and Sasha Ivanchenkov. Space artist Robert McCall designed the official crew emblem.

Bottom: During a US visit, the American crew brought the cosmonauts to the thirty-first annual One Shot Antelope Hunt in Lander, Wyoming, on September 21, 1974. *Back row, left to right*: Slayton, Brand, Kubasov, two Shoshone Indians, cosmonaut training chief Gen. Vladimir Shatalov, Leonov, and Dave Scott. *Front row, left to right*: Evans, Stu Roosa, and Joe Allen. The Roosa-Scott-Allen team was the only one to score a perfect three hits. A miss by Shatalov cost his team—including Lenov and Kubasov—a tie.

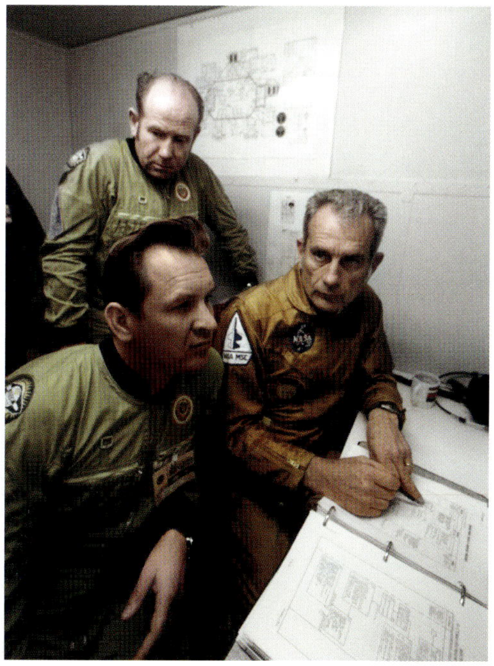

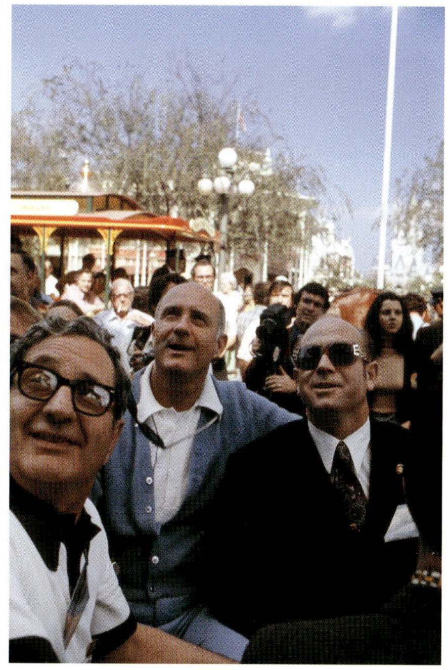

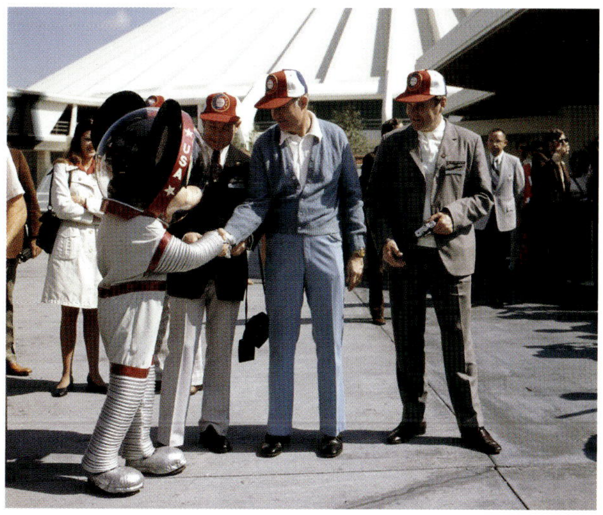

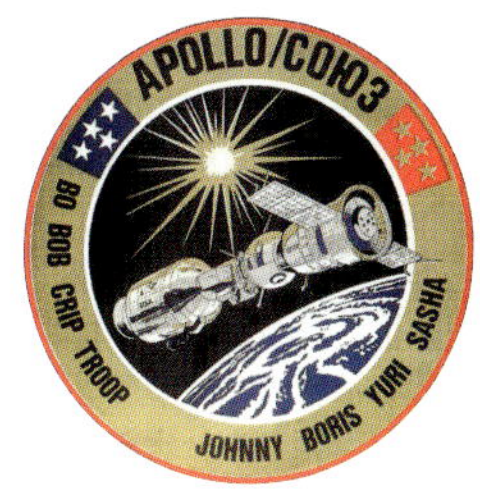

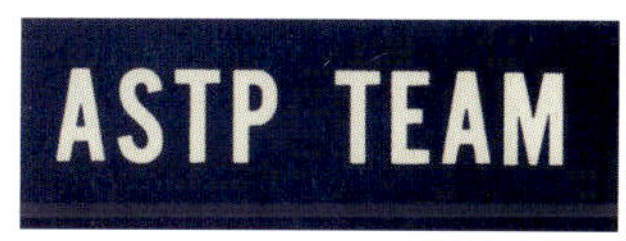

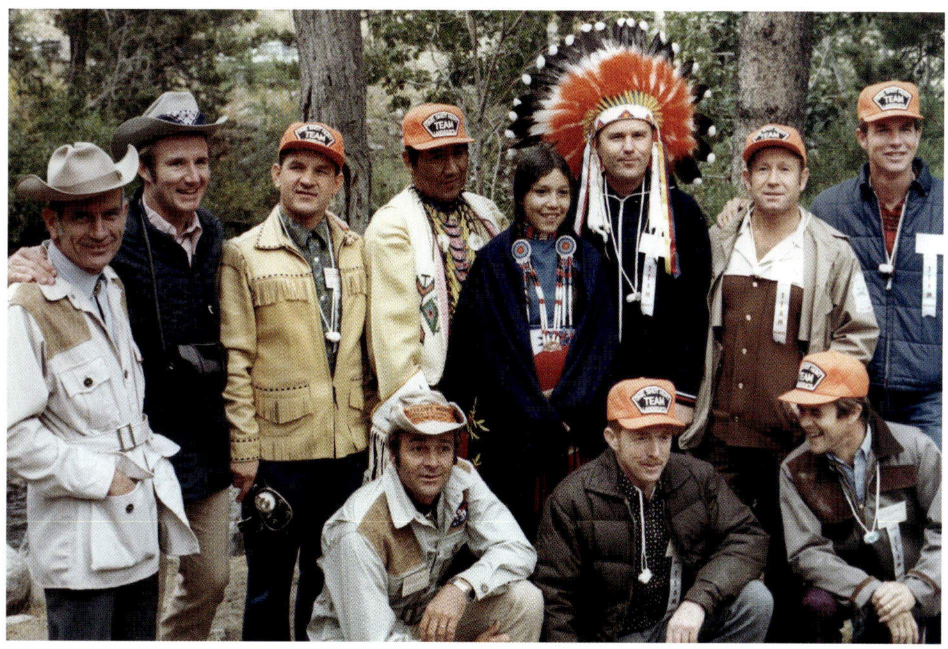

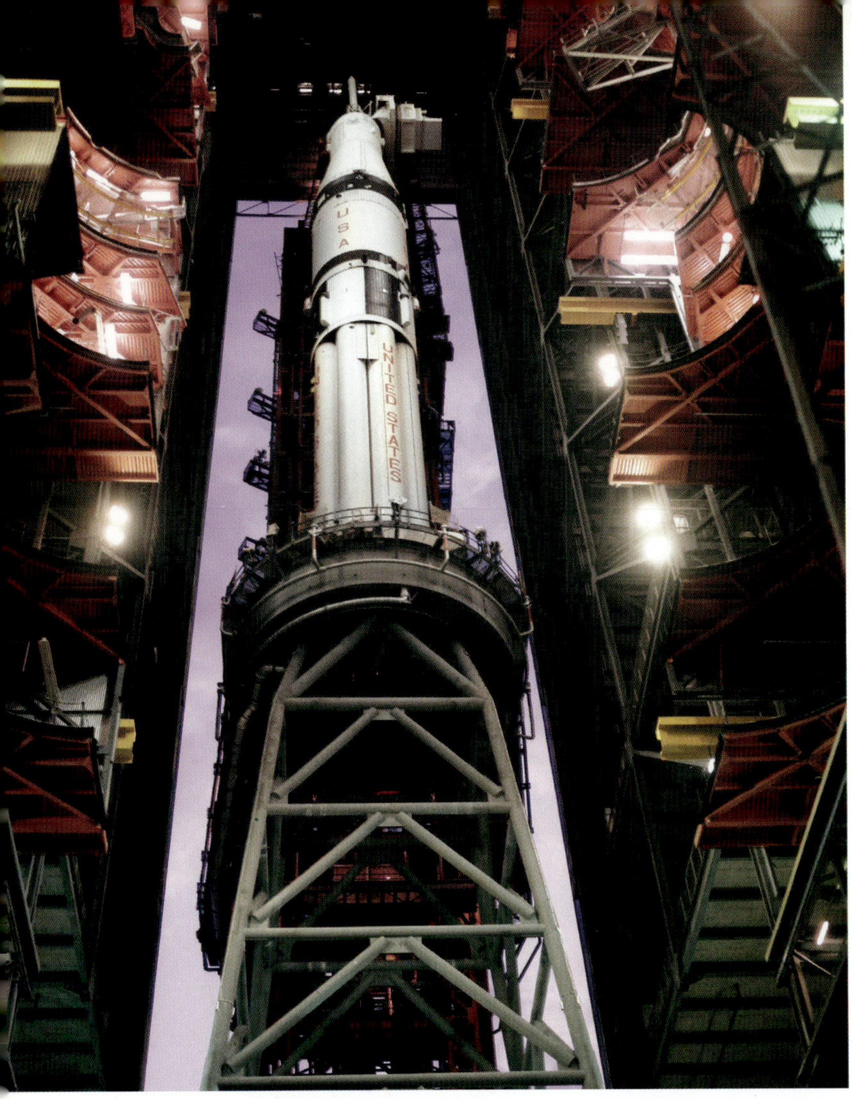

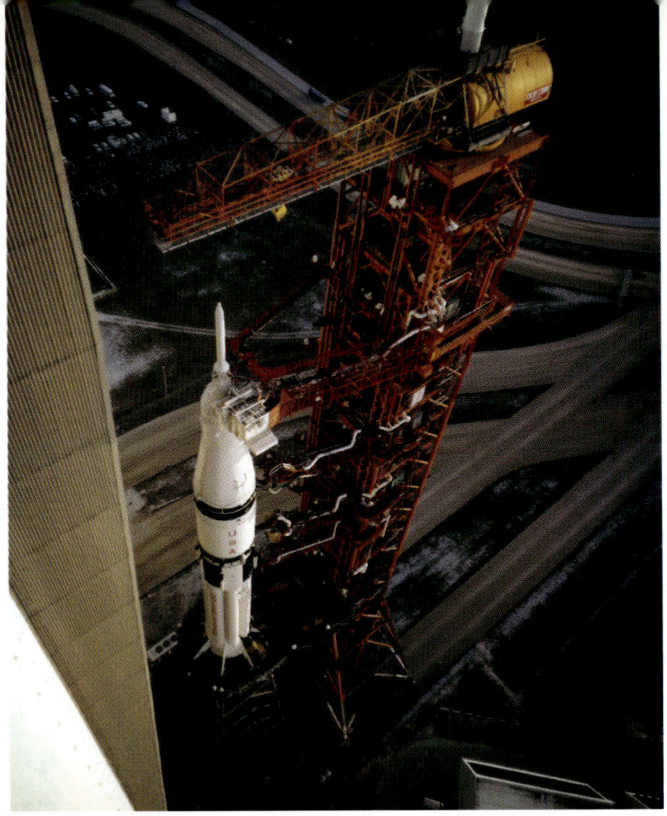

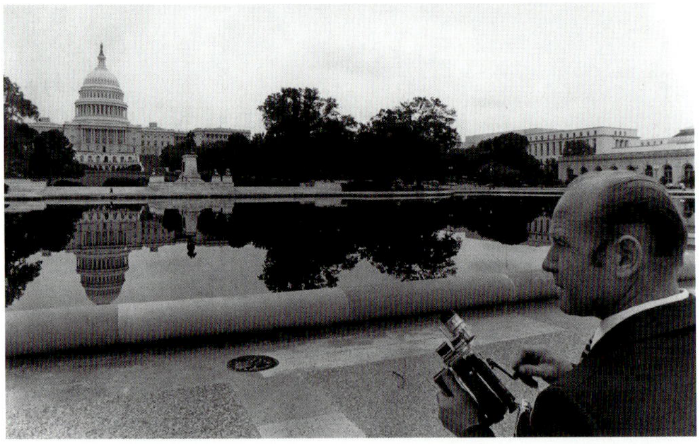

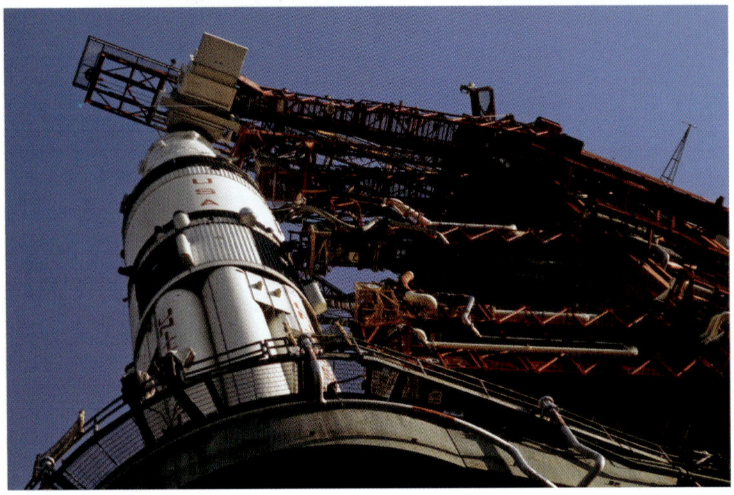

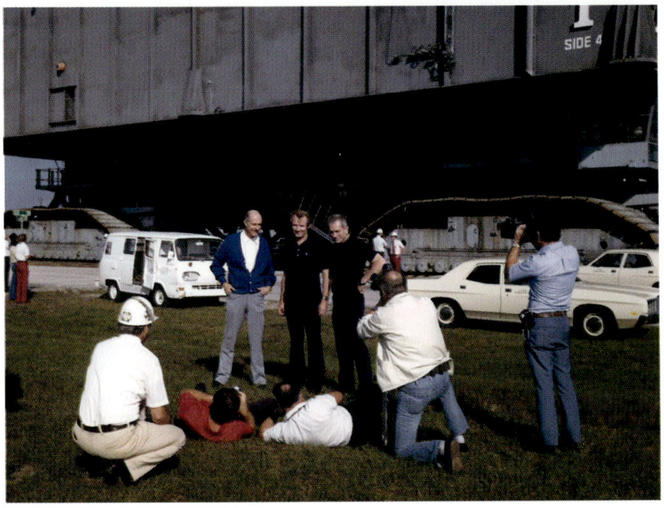

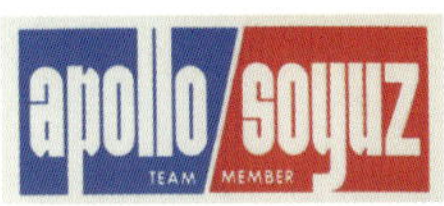

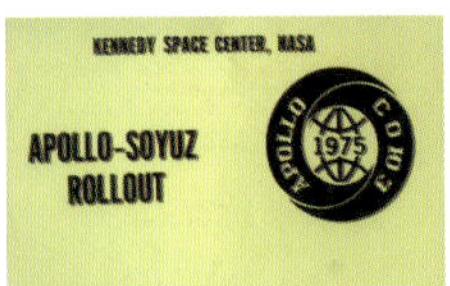

Top left: Framed by VAB work platforms, the final launch vehicle in the Saturn rocket family—SA-210—begins the trip to Pad 39B from high bay 1, at dawn on March 24, 1975. The first Saturn had been launched almost fourteen years earlier.

Top right: A twenty-five-ton hammerhead crane, designed by Colby Crane of Seattle, tops the LUT on rollout. This ML and ML-2, as well as the other two LUTs, were modified for Space Shuttle launches, which would begin in six years. The closed shutters of the Firing Room 1 windows in the LCC are at lower right.

Bottom left: "After working sixteen years in the space program, it is great to see it come out the door," said Slayton. KSC officials said that four thousand tourists were on hand for the Easter-week rollout. (photo by Jacques Tiziou)

Center right: Leonov takes in a view of the US Capitol Reflecting Pool in Washington, DC, on September 7, 1974, during a stop en route to JSC for three weeks of training. All five ASTP crewmen attended an annual policemen's "crab pick" picnic in Virginia that day with President Ford (who pardoned Richard Nixon the next day).

Bottom right: As the stack inches its way toward the pad, Stafford, Brand, and Slayton pose for cameramen along the crawlerway later that morning. NASA photographer Larry Summers is at right.

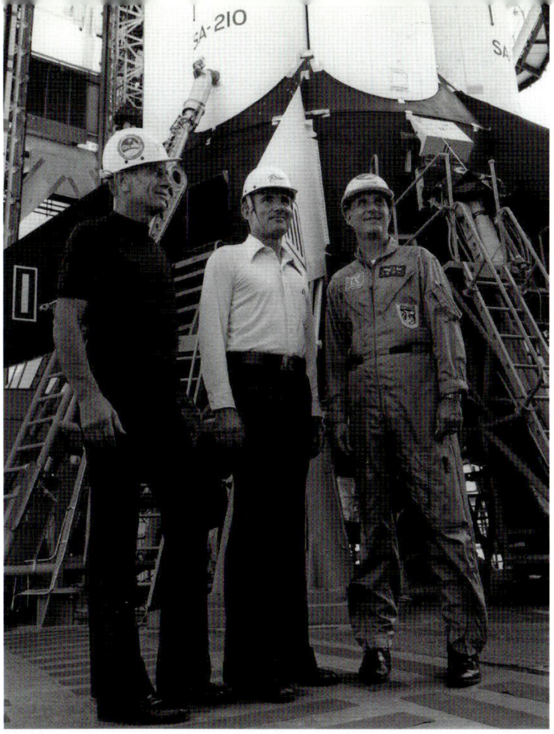

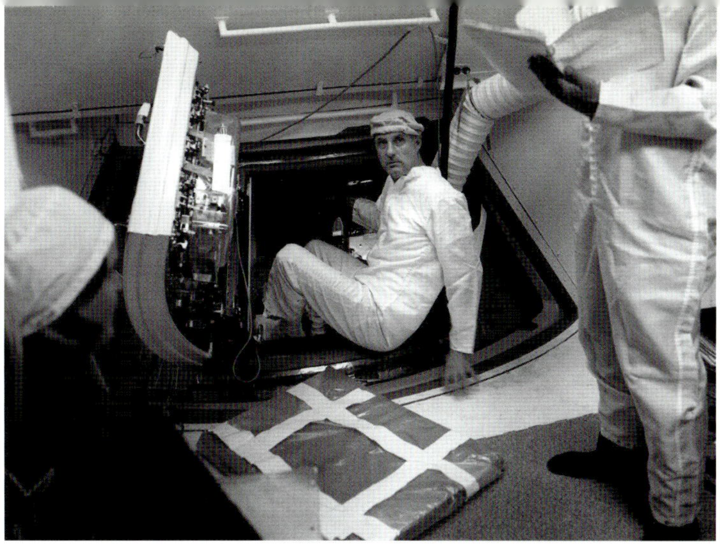

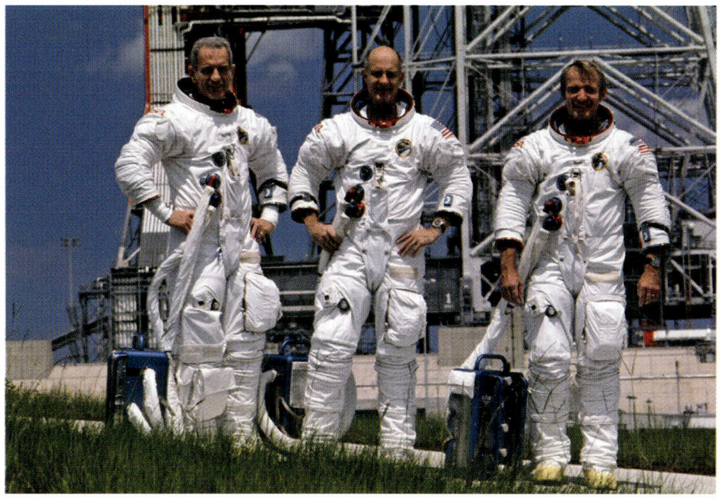

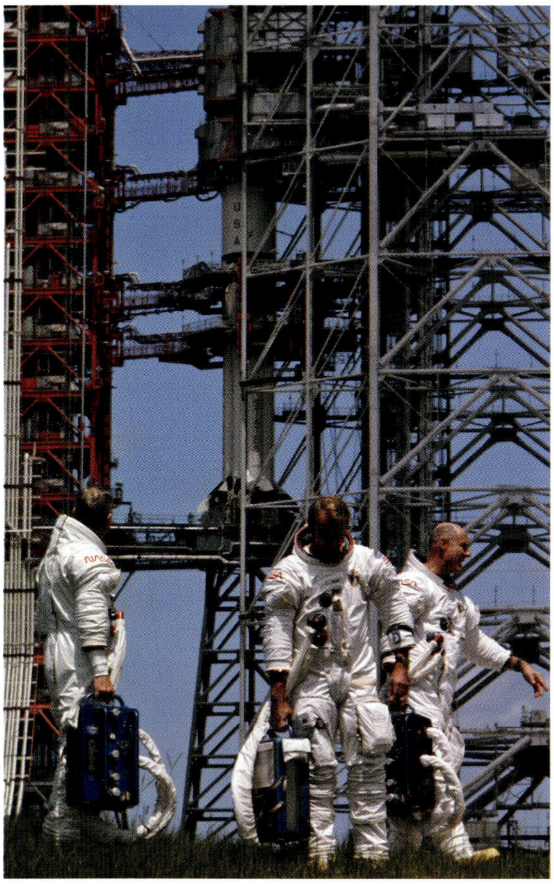

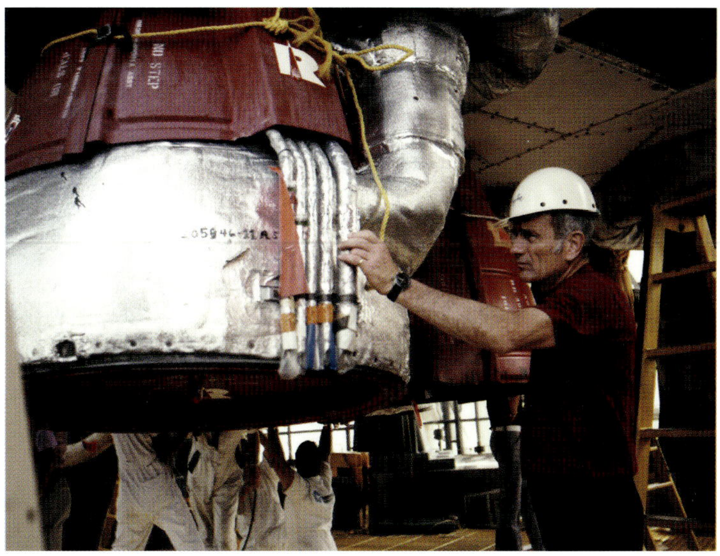

NASA

Direct left: NASA's new logotype, adopted that January, was used for the first time on ASTP suits, replacing the original agency insignia. The logotype was retired in 1993.

Top left: Slayton, Brand, and Stafford on ML-1 on June 10. The first stage's RP-1 fuel mast is behind Slayton; a short cable mast to the right of Stafford carried electrical cables and pneumatic service lines.

Top right: Stafford uses a handhold to enter the CM in the Pad 39B white room on June 10, 1975. He had been sitting on the protective pad in front of the hatch. KSC crew systems engineer Ann Montgomery (*left*) became a flow, logistics, and quality manager during the Shuttle program.

Bottom left and center right: Slayton, Stafford, and Brand on the west side of Pad 39B following the CDDT on July 3, 1975. Each wears a modified A7L pressure suit with an intravehicular cover layer, called an A7LB. (photos by John de Bry)

Bottom right: On June 10, 1975, Slayton inspects the turbopump lubricant drain on an inboard H-1 engine bell at the base of the Saturn IB. Chrysler technicians work nearby. The large duct above his hand directed exhaust from the engine's turbine into the engine bell.

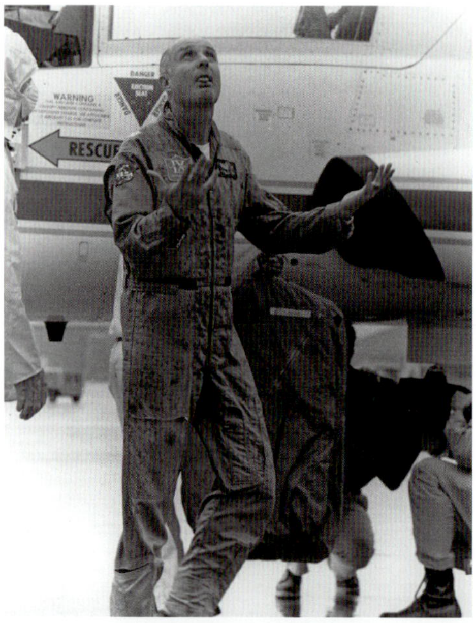

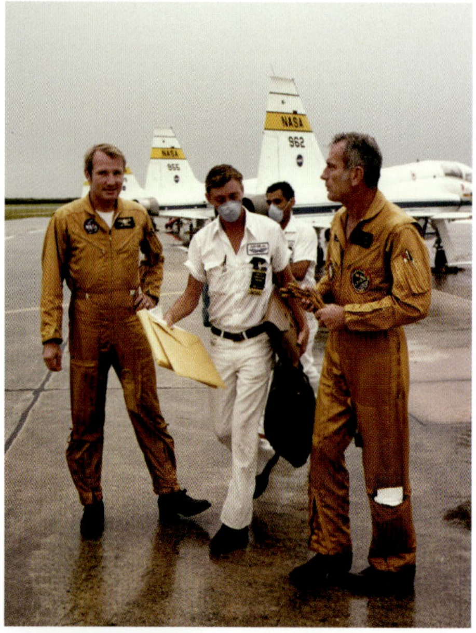

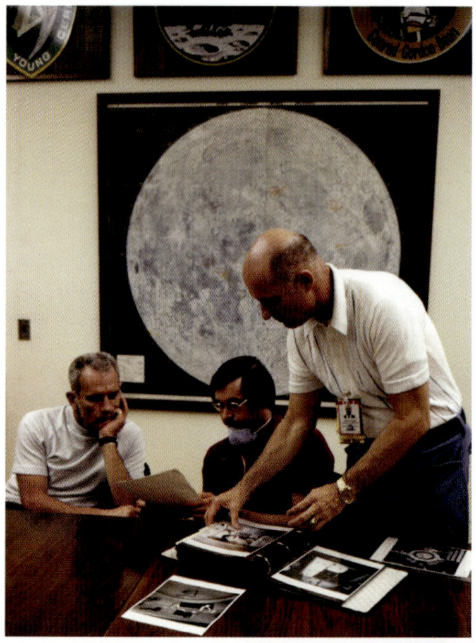

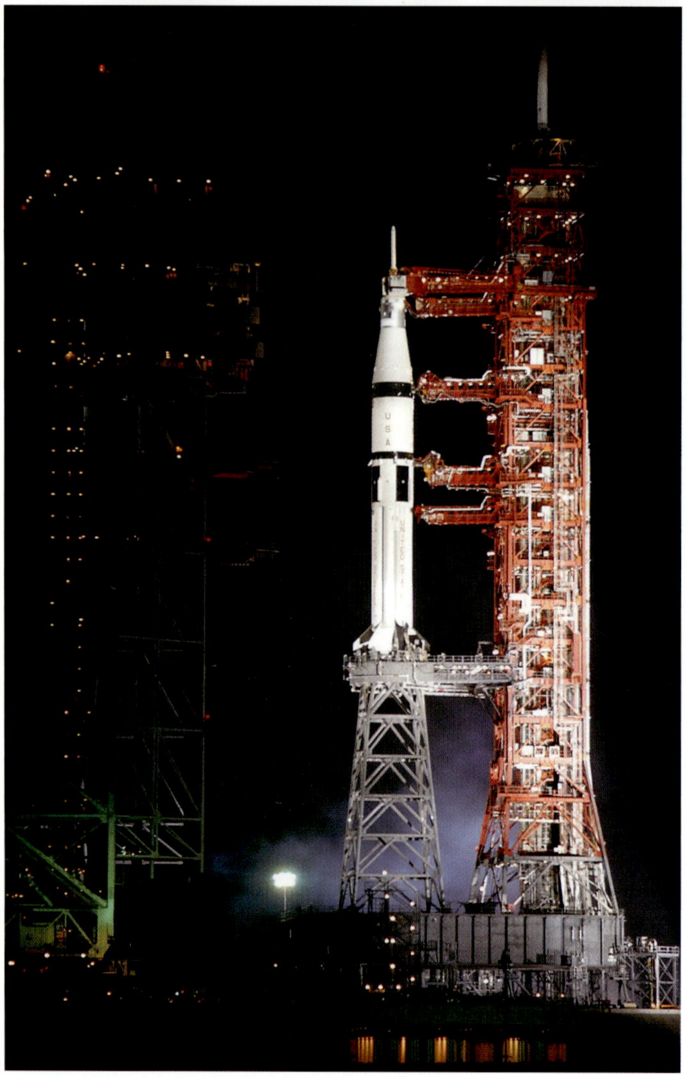

Above: The Saturn IB is bathed in xenon floodlights about nineteen hours before launch, after the MSS (*left*) has been rolled back. (photo by J. L. Pickering)

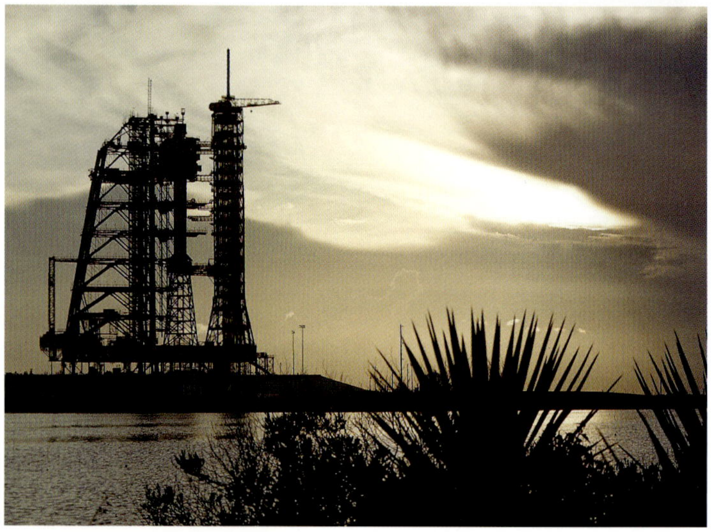

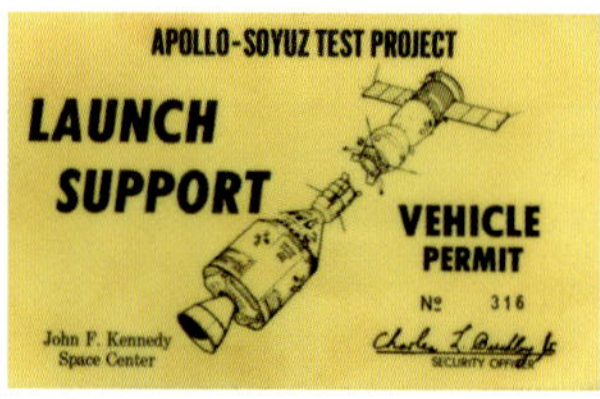

Top left: Stafford dodges raindrops after arriving for the launch at Patrick AFB on July 12, 1975. He came from Houston in a T-38.

Top center: The arrivals of Brand and Slayton were delayed by the afternoon rain. Daily thunderstorms in the days before the launch became a concern. Slayton wears the ASTP program patch in the Soviet configuration. Two members of the Patrick air services crew wear surgical masks; they had been prescreened to have close contact with the astronauts, who had entered a medical isolation program three weeks earlier.

Top right: Slayton (*left*) and Stafford (*right*) go over photos of flight hardware in the conference room of the MSOB crew quarters on the night before launch. They are with JSC Crew Station Engineer Robert "Terry" Neil, responsible for integrating all flight hardware with crew interfaces.

Center right: The sun sets for the last time on an Apollo/Saturn launch vehicle on July 14, 1975. This view of Pad 39B looks west. (photo by J. L. Pickering)

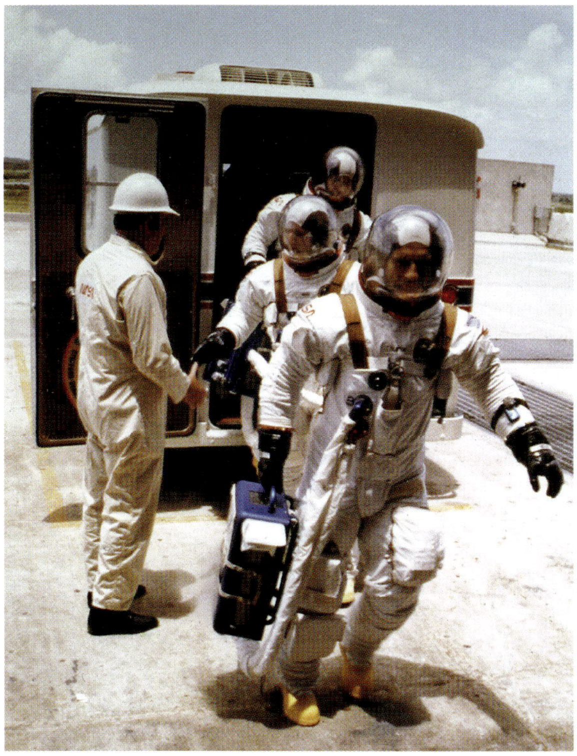

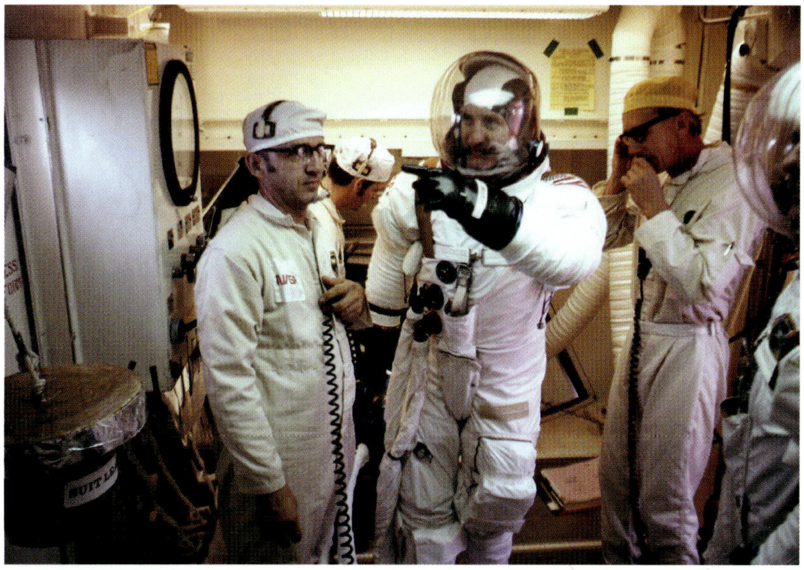

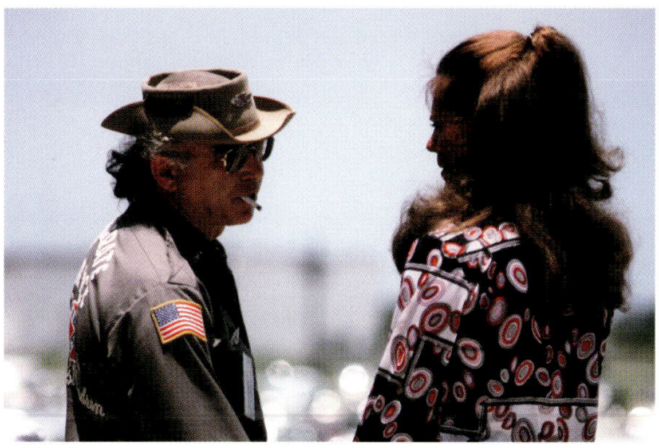

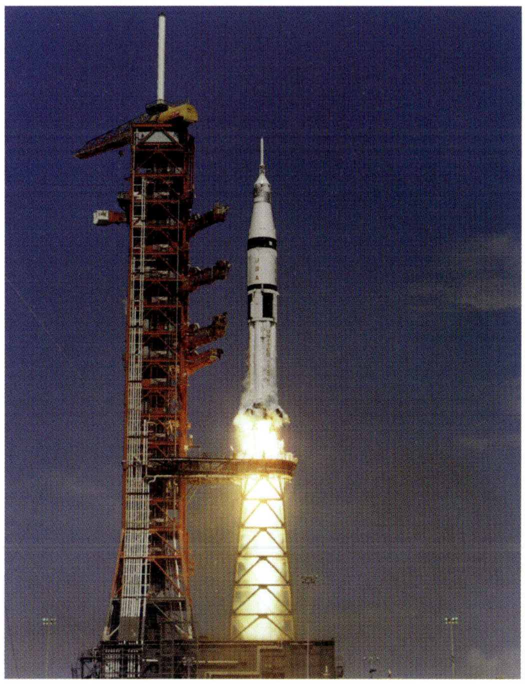

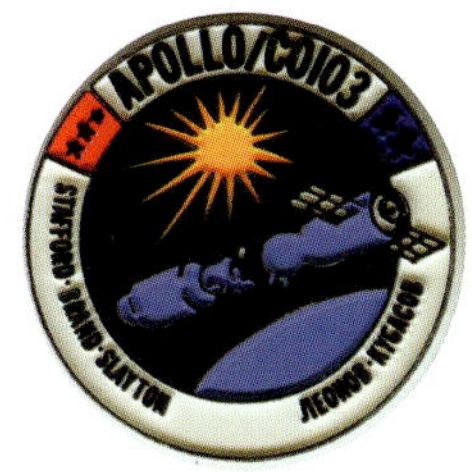

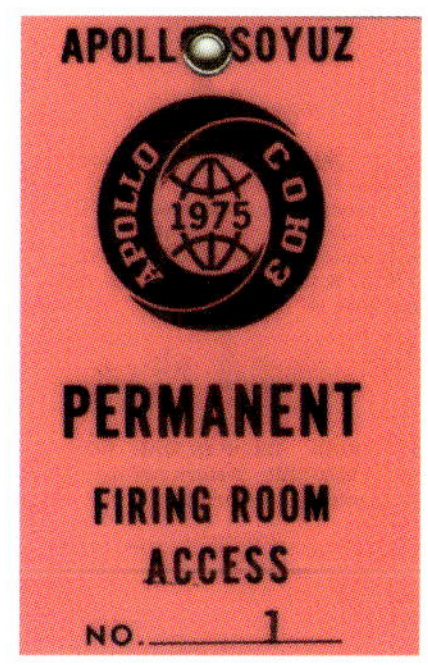

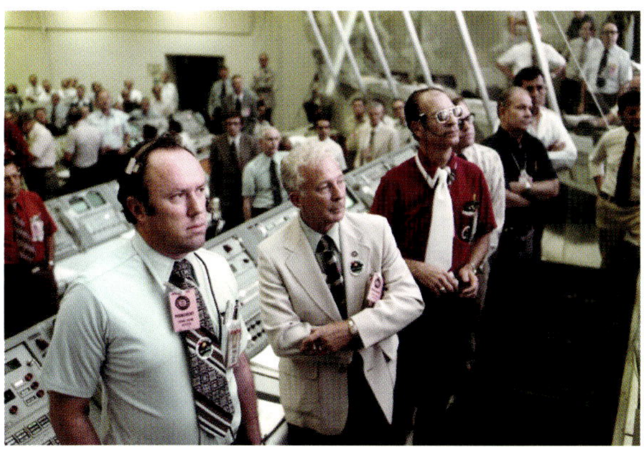

Top left: Slayton leads Brand (reaching for KSC Security Chief Charles Buckley's hand) and Stafford out of the crew transfer van at Pad 39B on launch morning.

Top right: Stafford (*center*) stands near the CM hatch in the white room on launch morning, with Pad Leader Guenter Wendt at right. ILC's Jim McNair is in the background; Brand is visible at far right.

Above: Liftoff at 3:50 p.m. (EDT) sends a column of flames down the milkstool—and sends the last Apollo CSM into an Earth orbit, about one hundred miles in altitude. The Soyuz had launched at 8:20 a.m.

Center left: Author Martin Caidin and an unidentified woman at the LC-39 Press Site before launch on July 15. Caidin had become well known for his 1972 novel *Cyborg*, the basis for the ABC-TV series *The Six Million Dollar Man*. (photo by Tom Usciak)

Bottom left, left to right: Test Supervisor Chuck Henschel, Launch Operations Manager Paul Donnelly, and Chief Test Planning Officer Bob Moser watch liftoff from Firing Room 3, which would be converted in 1993 for space shuttles. NASA's Wayne Stollard (*far right*) and Dale Carruthers (dark-blue shirt) are at right. Wernher von Braun, then an executive with Fairchild Space and Electronics, was among the seventy thousand invited guests outside.

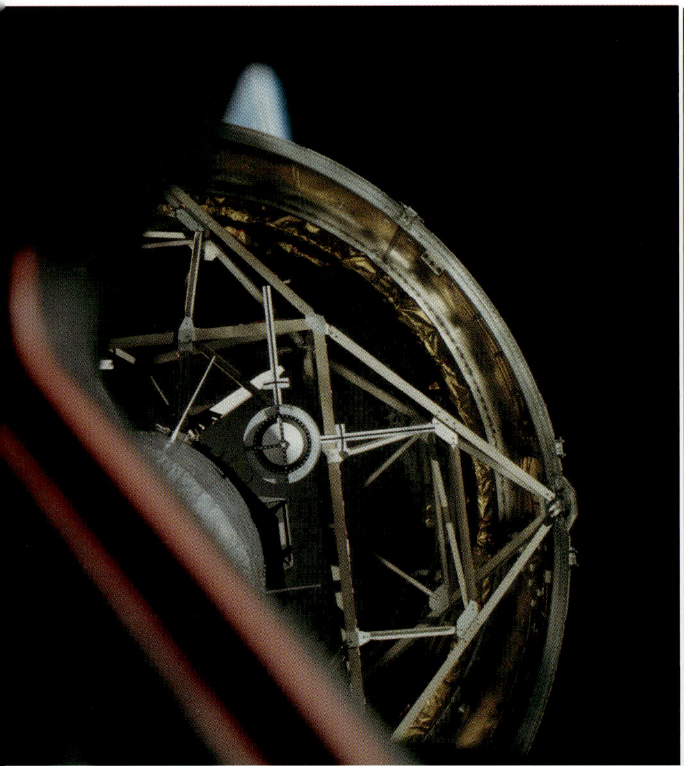

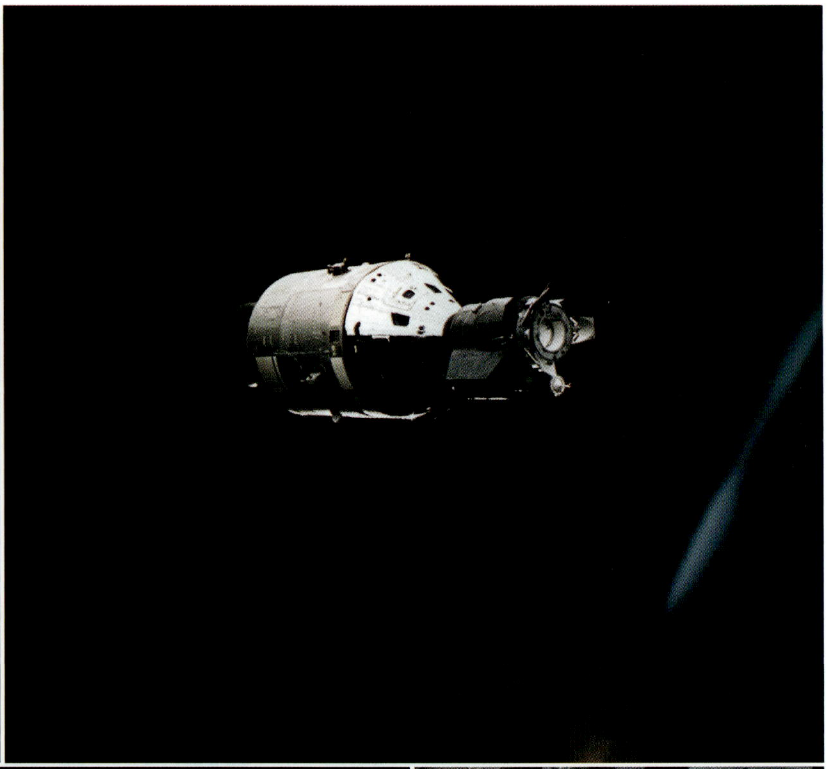

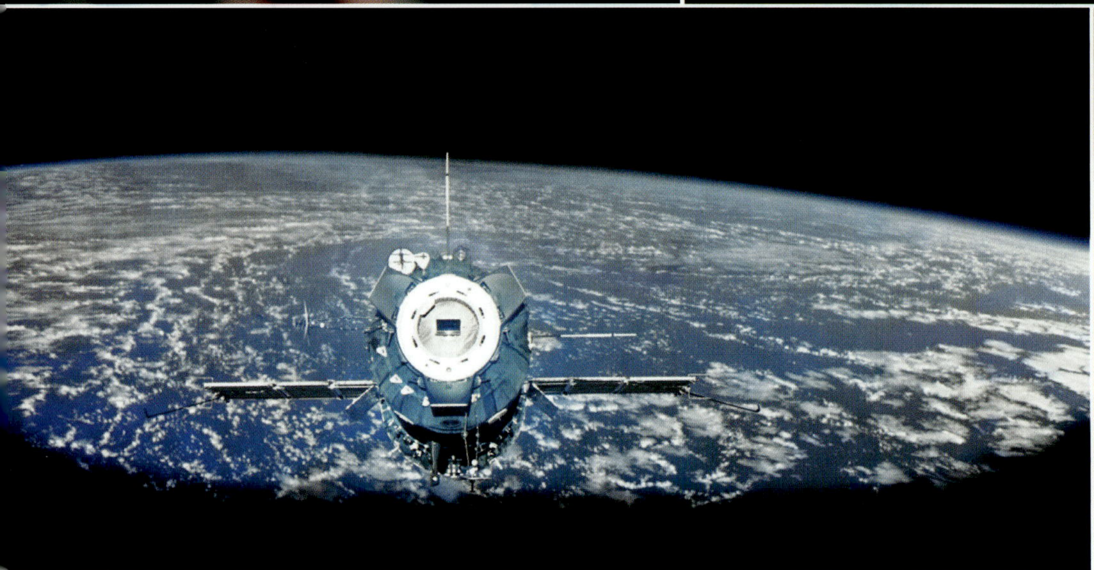

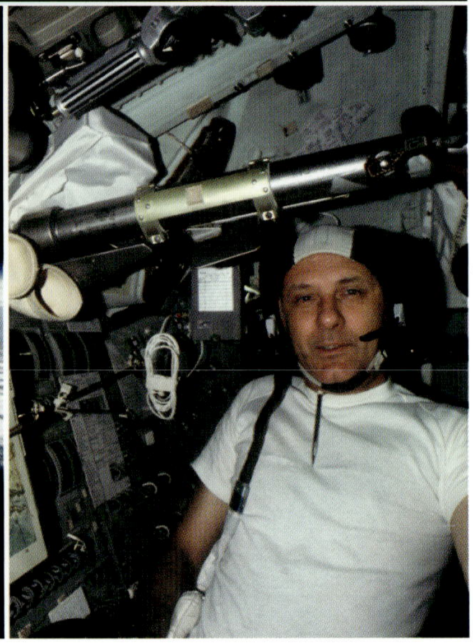

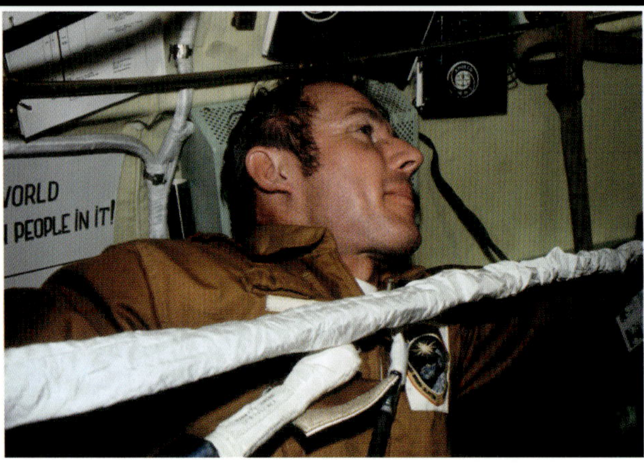

Above: Brand exercises his upper body with a bungee cord in the Soyuz cabin. At left is the edge of a quote from Shakespeare's *The Tempest*, which Brand later called "the Soyuz crew's way of trying to make us feel welcome in their ship."

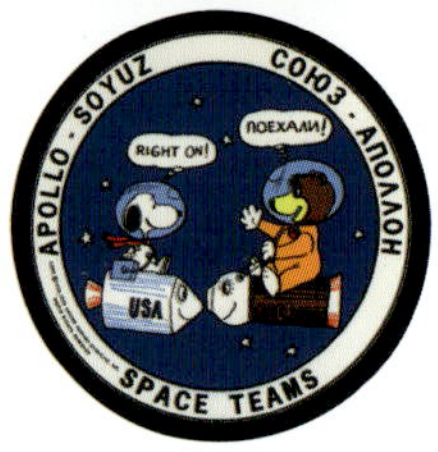

Direct left: "Let's go," says the Russian bear.

Top left: View from a CM rendezvous window as the spacecraft moves in to link up with the DM, mounted in the SLA above the S-IVB's fuel tank, on July 15, 1975.

Top right: The Apollo CSM with the DM attached, as seen from Soyuz. The two space agencies had agreed in advance to later swap inflight photos.

Center left: The docking port on the Soyuz 7K-TM spacecraft points toward the Apollo CSM, shortly before docking on July 17. The Soyuz featured improved solar panels for an extended mission, an APAS-75 docking mechanism to link up with the DM, and a modified environmental control system to lower the cabin pressure before docking with the American spacecraft.

Center right: Stafford, with the main CM instrument panel to his left. The hatch handle is at top left, and a couch landing attenuation strut is just above his head.

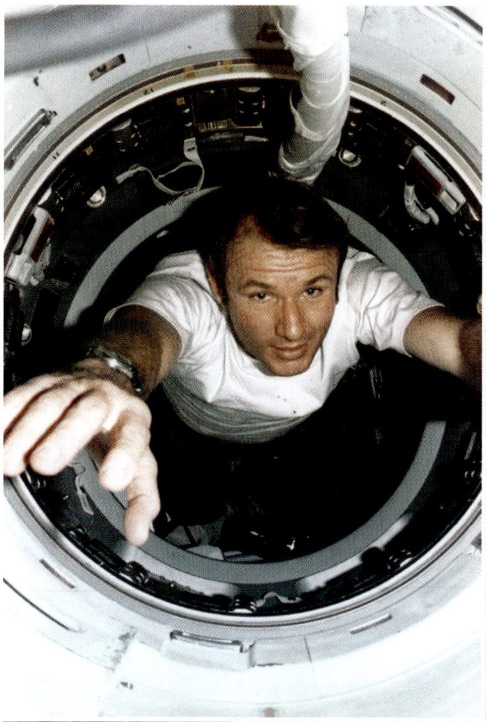

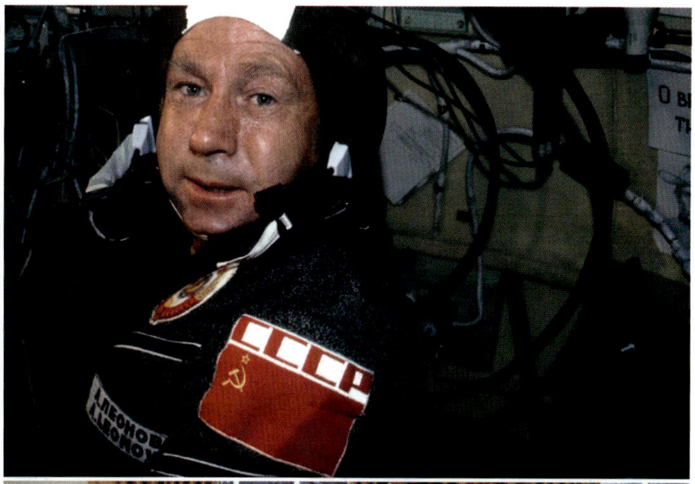

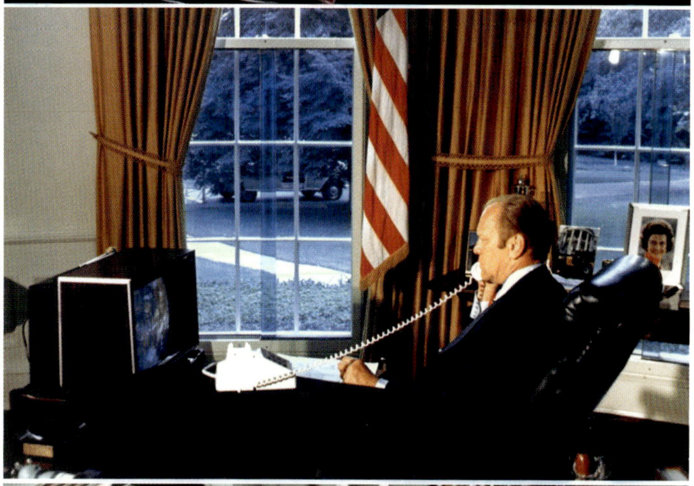

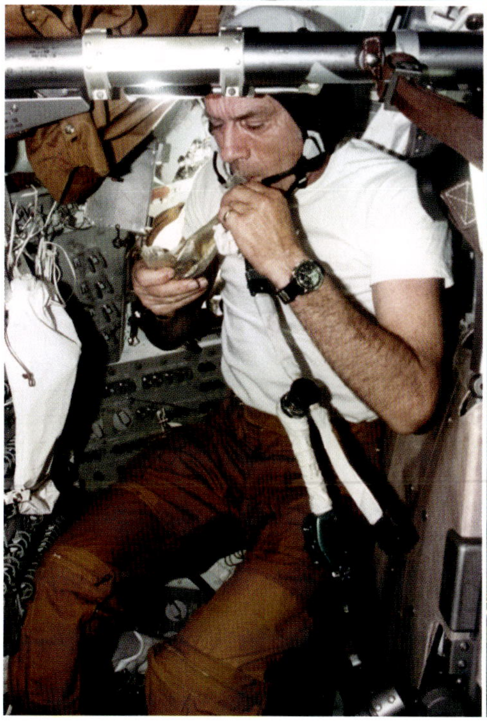

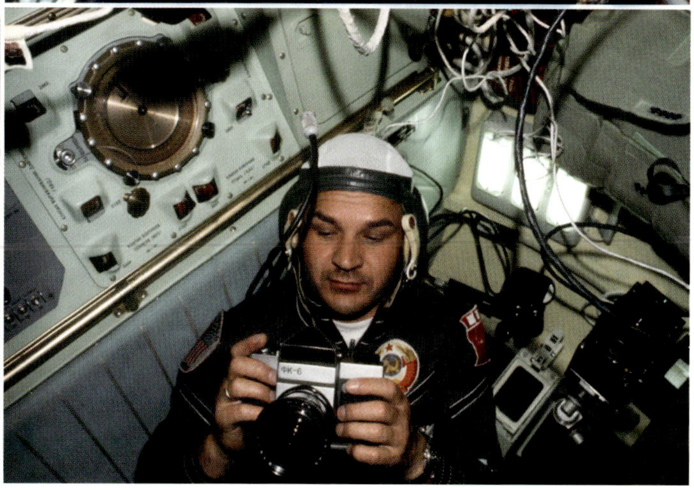

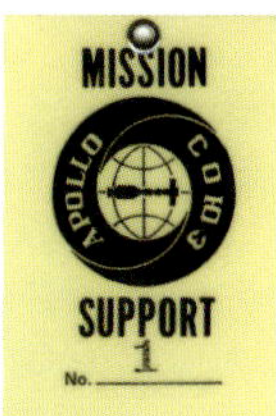

Top left: Brand enters the DM through the hatchway of the Apollo CM. The white cable above his head routed communications between the two spacecraft.

Top right: Leonov aboard Soyuz, wearing an American Snoopy cap. At right are the edge of a quote from *The Tempest* and one of three sketches of the US crew by Leonov. Because of flammability standards, NASA would not permit the Soviet crew aboard the CM in wool or cotton clothing. The United States offered to give the Soviets fabric to meet NASA standards, but the Soviets instead quickly produced a fabric of their own, which was equal or superior to Beta cloth.

Bottom left: Slayton takes a drink in the CM, with his jacket stashed at left above his head and his communications cable in the foreground. He informed Houston at one point that he and his crewmates were in the "food intake mode."

Center right: From the Oval Office at the White House, President Ford speaks with the crewmen—who are gathered in the Soyuz (except for Brand)—for about ten minutes on the afternoon of July 17. US TV and radio networks carried the conversation with Ford live. He questioned all five men and reminded the cosmonauts of a crab picnic they would be attending in Virginia in September. A Soviet space official had earlier read a short statement from General Secretary Brezhnev.

Bottom right: Kubasov holds a Soviet-made FK (Foto Kamera)-6, one of only four made as a prototype for the Kiev 6-C. It used medium-format 6x6cm 120 or 220 roll film, a film type introduced by Kodak in 1901.

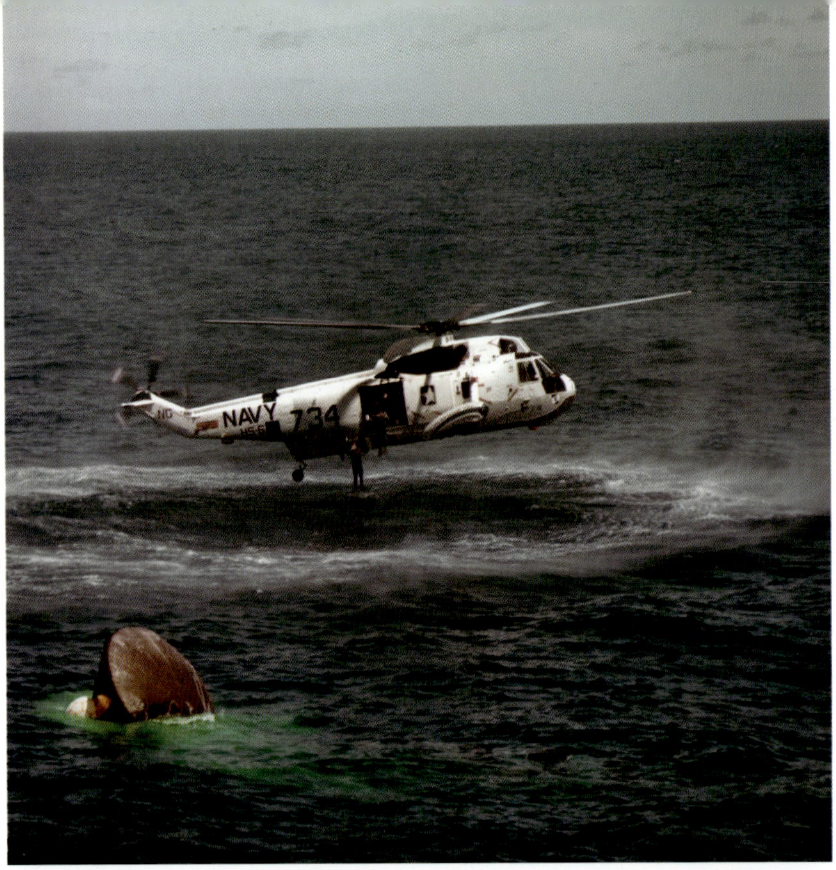

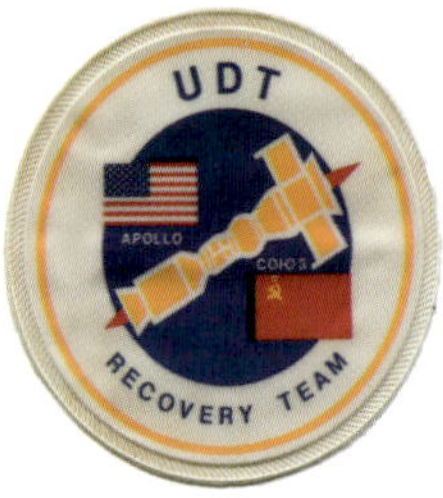

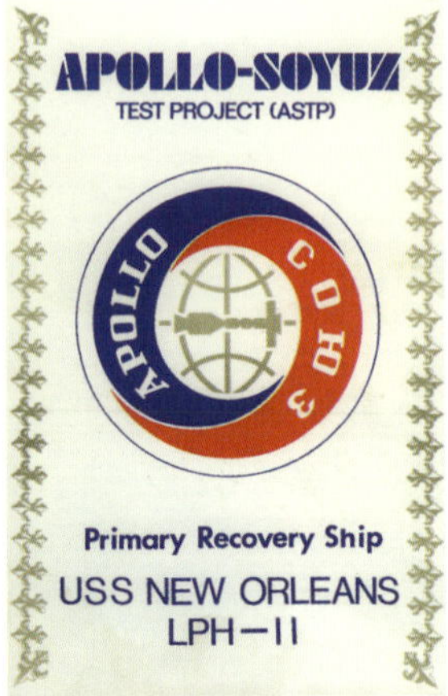

Top left: Divers from UDT-12 jump into the water from a Sea King helicopter from HS-6 (Helicopter Squadron), based aboard USS *New Orleans*, after splashdown at 11:40 a.m. (HST) on July 24. The CM is temporarily in the "Stable II" (apex down) position; inflating flotation bags would soon right it.

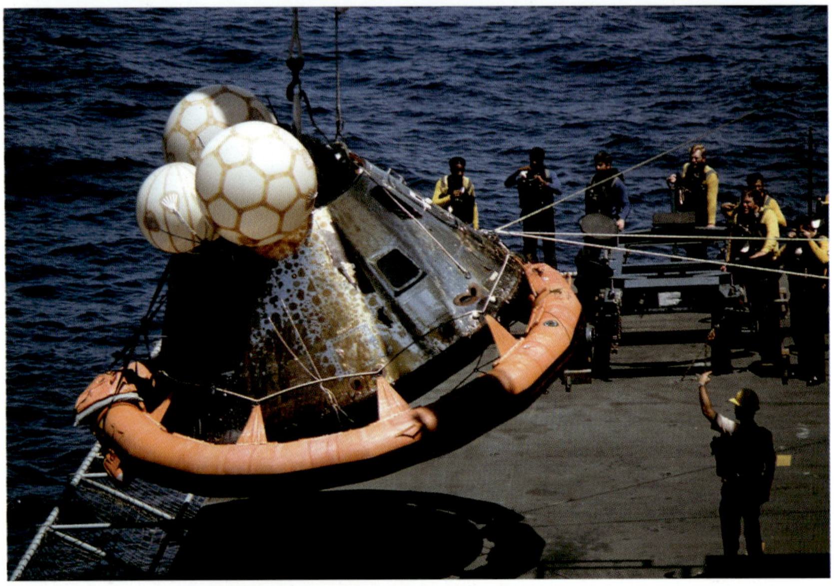

Center left: A crane brings the CM aboard the *New Orleans*, forty minutes after splashdown. Within three weeks, the KSC workforce would be cut by almost 17 percent. (photo by John de Bry)

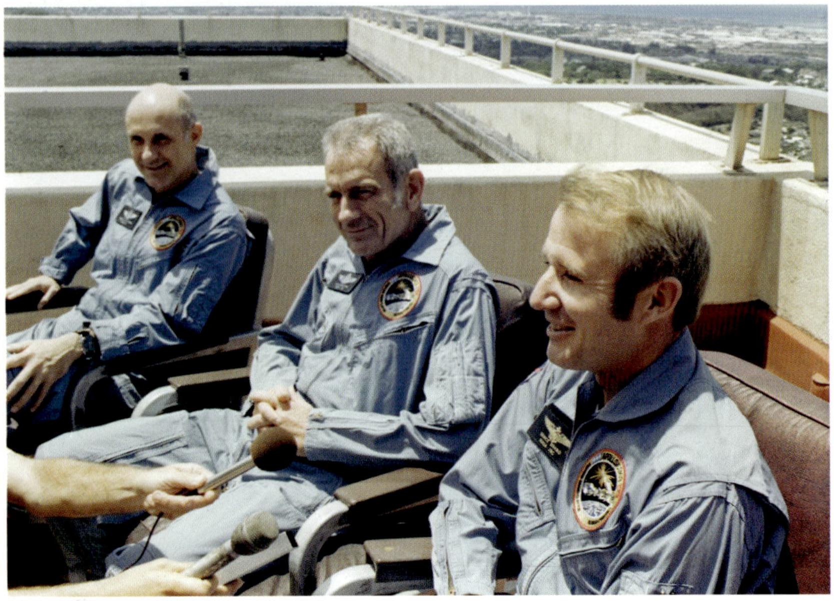

Bottom left: Stafford, Slayton, and Brand during a medical debriefing on the roof of Tripler Army Hospital in Honolulu, Hawaii, on July 26, 1975, just before leaving for Los Angeles. They had spent five days at the facility recuperating from breathing toxic fumes, which accidentally entered the CM just before splashdown. They then moved to cottages at the Kaneohe Marine Corps Station for another eight days of rest, with their families.

Abbreviations

AFB	Air Force Base
AFS	Air Force Station
ALSA	astronaut life support assembly
ALSEP	Apollo Lunar Surface Experiments Package
AMU	Astronaut Maneuvering Unit
ASMU	Automatically Stabilized Maneuvering Unit
ASTP	Apollo-Soyuz Test Project
ATM	Apollo Telescope Mount
BIG	biological isolation garment
CAPCOM	capsule communicator
CDDT	countdown demonstration test
CDR	commander
CM	command module
CMP	command module pilot
CSM	command/service module
DAC	data acquisition camera
DM	docking module
DMP	docking module pilot (ASTP)
EDC	electric data camera
EDT	Eastern Daylight Time
EMU	Extravehicular Mobility Unit
EST	Eastern Standard Time
EVA	extravehicular activity
FIDO	flight dynamics officer
HHMU	Hand-Held Maneuvering Unit
HQ	Headquarters
HST	Hawaii-Aleutian Standard Time
ILC	International Latex Corporation
JSC	Johnson Space Center
KSC	Kennedy Space Center
LC	Launch Complex
LCC	Launch Control Center
LEMS	Lunar Excursion Module Simulator
LES	Launch Escape System
LEVA	Lunar Extravehicular Visor Assembly

LLTV	Lunar Landing Training Vehicle
LM	lunar module
LMP	lunar module pilot
LOX	liquid oxygen
LRL	Lunar Receiving Laboratory
LRRR	Laser Ranging Retro Reflector
LRV	Lunar Roving Vehicle
LUT	Launch Umbilical Tower
MCC	Mission Control Center
MDA	Multiple Docking Adapter
MESA	Modularized Equipment Stowage Assembly
MET	Modular Equipment Transporter
MFA	Manned Flight Awareness
MILA	Merritt Island Launch Area
ML	Mobile Launcher
MMU	Manned Maneuvering Unit
MOCR	Mission Operations Control Room
MOL	Manned Orbiting Laboratory
MQF	Mobile Quarantine Facility
MSC	Manned Spacecraft Center
MSFC	Marshall Space Flight Center
MSOB	Manned Spacecraft Operations Building
MSS	Mobile Service Structure
NAA	North American Aviation
NAR	North American Rockwell
NASA	National Aeronautics and Space Administration
NASC	National Aeronautics and Space Council
OWS	Orbital Workshop
PDT	Pacific Daylight Time
PLSS	Portable Life Support System
PLT	pilot
PSE	Passive Seismic Experiment
PST	Pacific Standard Time
RCS	reaction control system
RTG	Radioisotope Thermoelectric Generator
SCE	Signal Conditioning Equipment
SIDE	Suprathermal Ion Detector Experiment
SIM	Scientific Instrument Module
SLA	Spacecraft Lunar Module Adapter
SM	service module
SPS	Service Propulsion System
SPT	science pilot (Skylab)
SWC	Solar Wind Collector
SWCE	Solar Wind Composition Experiment
TEI	trans-Earth insertion
UDT	Underwater Demolition Team
USAF	United States Air Force
UV	ultraviolet
VAB	Vehicle Assembly Building
VHF	very high frequency

Bibliography

Anderton, David A. *Apollo 17 at Taurus Littrow*. Washington, DC: Office of Public Affairs, National Aeronautics and Space Administration, 1973.

Apollo Mission Commentary Transcripts. Houston, TX: Public Affairs Office, Johnson Space Center, 1967–1972.

Apollo Operations Handbook Extravehicular Mobility Unit. Revision V. Houston, TX: Crew Systems Division, Manned Spacecraft Center, National Aeronautics and Space Administration, 1971.

Apollo Recovery Operational Procedures Manual. Revision C. Houston, TX: Landing and Recovery Division, Manned Spacecraft Center, National Aeronautics and Space Administration, 1971.

Apollo Spacecraft Familiarization. Downey, CA: North American Rockwell, 1967.

Apollo Spacecraft News Reference. Downey, CA: Space Division, North American Rockwell Corporation, 1969.

Apollo Spacecraft News Reference. Bethpage, NY: Grumman Aerospace Corporation, 1970.

Barton, Elmer E. Interview by Rebecca Wright. Houston, TX: Johnson Space Center Oral History Project, April 12, 2000.

Belew, Leland F., ed. *Skylab, Our First Space Station*. Washington, DC: Scientific Technical Information Office, National Aeronautics and Space Administration, 1977.

Benson, Charles D. *Moonport: A History of Apollo Launch Facilities and Operations*. Washington, DC: Scientific and Technical Information Office, National Aeronautics and Space Administration, 1978.

Blair, Don. *Splashdown! NASA and the Navy*. Paducah, KY: Turner Publishing Company, 2004.

Brooks, Courtney G., and Ivan D. Ertel. *The Apollo Spacecraft: A Chronology*. Vol. III. Washington, DC: National Aeronautics and Space Administration, 1976.

Butler, Sue. Interview by Dr. Patrick Moore. Indian Harbour Beach, FL: KSC Oral History Project, Kennedy Space Center, June 25, 2002.

"Cape Canaveral Air Force Station, Launch Complex 39, Altitude Chambers." Atlanta, GA: Historic American Engineering Record, National Park Service, Southeast Region, Department of the Interior, 2009.

Donnelly, Paul C. Interview by Dr. Orville Butler. Indian Harbour Beach, FL: KSC Oral History Project, August 23, 2003.

Ertel, Ivan D., and Mary Louise Morse. *The Apollo Spacecraft: A Chronology*. Vol. I. Washington, DC: National Aeronautics and Space Administration, 1969.

Ertel, Ivan D., and Roland W. Newkirk. *The Apollo Spacecraft: A Chronology*. Vol. IV. Washington, DC: National Aeronautics and Space Administration, 1978.

Ezell, Edward Clinton, and Linda Newman Ezell. *The Partnership: A History of the Apollo-Soyuz Test Project*. Washington, DC: National Aeronautics and Space Administration, 1978.

Koons, Wayne E. Interview by Rebecca Wright. Houston, TX: Johnson Space Center Oral History Project, October 14, 2004.

Kozloski, Lillian D. *US Space Gear: Outfitting the Astronaut*. Washington, DC: Smithsonian Institution Press, 1994.

Larsen, Curtis E. "NASA Experience with Pogo in Human Spaceflight Vehicles." Houston, TX: NASA Engineering and Safety Center, Johnson Space Center, 2008.

Lunney, Glynn S. Interview by Carol Butler. Houston, TX: Johnson Space Center Oral History Project, January 28, 1999.

McAllister, Fred A. *Apollo Experience Report: Crew Provisions and*

Equipment Subsystem. Washington, DC: National Aeronautics and Space Administration, 1972.

Morse, A. R. *MSFC Skylab Apollo Telescope Mount*. Huntsville, AL: George C. Marshall Space Flight Center, National Aeronautics and Space Administration, 1974.

Morse, Mary Louise, and Jean Kernahan Bays. *The Apollo Spacecraft: A Chronology*. Vol. II. Washington, DC: National Aeronautics and Space Administration, 1973.

MSFC Skylab Orbital Workshop. Vols. 1–5. Huntsville, AL: George C. Marshall Space Flight Center, 1974.

Orloff, Richard W. *Apollo by the Numbers: A Statistical Reference*. Washington, DC: National Aeronautics and Space Administration, 2000.

Press Reference Book: Gemini Spacecraft Number Twelve. St. Louis, MO: External Relations Division, McDonnell Company, 1966.

Rigell, Ike. Interview by Dr. Henry Dethloff and Dr. Lee Samples. Indian Harbour Beach, FL: KSC Oral History Project, Kennedy Space Center, June 18, 2001.

Saturn V Flight Manual—SA 507. Huntsville, AL: George C. Marshall Space Flight Center, National Aeronautics and Space Administration, 1969.

Saturn V News Reference. National Aeronautics and Space Administration, The Boeing Company, Douglas Aircraft Company, International Business Machines Corporation, Rocketdyne Division of North American Aviation, Inc., Space Division of North American Aviation, Inc., 1967.

Sherman, Madeline W., ed. *TRW Space Log: 1957–1982*. Vol. 19. Redondo Beach, CA: TRW, 1983.

Simmons, Gene. *On the Moon with Apollo 15*. Washington, DC: National Aeronautics and Space Administration, 1971.

———. *On the Moon with Apollo 16*. Washington, DC: National Aeronautics and Space Administration, 1972.

Skylab: A Chronology. Washington, DC: National Aeronautics and Space Administration, 1977.

Skylab News Reference. Washington, DC: Office of Public Affairs, National Aeronautics and Space Administration, 1973.

Skylab Saturn IB Flight Manual. Huntsville, AL: George C. Marshall Space Flight Center, 1972.

Stonesifer, John C. Interview by Kevin M. Rusnak. Houston, TX: Johnson Space Center Oral History Project, April 16, 2002.

USS Wasp (CV/CVS/CVA-18). Paducah, KY: Turner Publishing Co., 1999.

The Photographers

The number of photographers documenting the US manned space effort expanded in the early to late 1960s. As had been the case during the Mercury and Gemini programs, these photographers were primarily government contractors at KSC and federal employees at MSC and JSC. As interest in the moon landing program grew, however, the number of private photographers accredited through NASA as news media also increased; these private photographers took a significant number of the photos in this book.

Most of the media photographs are by French brothers Jacques and Michel Tiziou. Jacques, a retired radio, television, and print reporter, was born in Montélimar, France, in 1939. After earning a degree as an aerospace engineer in Paris in 1962, he began covering space topics for a variety of French outlets, including *Paris Match*, *Figaro*, and other dailies and weeklies; photo agencies; and French TV.

Jacques hired his younger brother, Michel, a former French Navy photographer, to be his KSC correspondent in 1968. They covered Apollos 8 through 11 from a one-bedroom apartment in Cape Canaveral. After Apollo 11, Michel started a successful photo company in France. He died in 1994.

After Apollo 13, Jacques moved to Cocoa Beach until the end of the Skylab program. He then relocated to Washington, DC, and started the Jacques Tiziou News Service, covering the White House, politics, and aerospace.

Another pair of brothers, Tom Usciak and his younger brother, Mark, were captivated by the US space program while growing up in Lancaster, Pennsylvania. Tom first traveled to KSC for the Apollo 15 launch, which he covered for his college. Mark joined him for Apollo 16. At age fourteen Mark may have been the youngest person to receive media accreditation during Apollo. NASA policy at the time stated that no one under the age of sixteen was permitted at the LC-39 Press Site.

The brothers went into photography as a career and covered the Skylab, ASTP, and Space Shuttle programs, traveling to NASA centers in Florida, Alabama, Texas, California, and New Mexico. Team members Steve Nolte and Bob Preston designed and built weatherproof housings and sound triggers for their remote cameras near the LC-39 pads.

Between them, Tom and Mark Usciak attended more than seventy-five shuttle launches and landings. Many magazines, including *Air & Space*, *Ad Astra*, and *Air Force Magazine*; websites; and newspapers have published their photos, one of which was also the cover of the James Oberg book *The New Race for Space*. Their work has won awards in the Aviation Week and Space Technology Annual Photography Contest. Today, Tom owns a digital printing lab and Mark works as a photographer and digital imaging specialist for a commercial studio in the Lancaster area.

John de Bry was born in Paris in 1943 and is the nephew of the famed opera diva Lily Pons (perhaps the greatest coloratura soprano of the twentieth century). He began covering the space program as an assistant to National Geographic Society photographer Otis Imboden in 1968, shooting the Apollo 8, 9, 10, and 11 missions. After he was hired by James L. Long Associates, he covered ASTP's splashdown aboard USS *New Orleans*. De Bry married Nancy Higgins, daughter of NASA Chief Engineer of Cryogenics Gerald Higgins, the co-designer of the LC-39 crawler-transporter.

Opposite page, top left: Photographers Larry Summers (*far left*) and Bill Taub (*second from left*) in Firing Room 3 in the LC-39 LCC on May 18, 1969, after the Apollo 10 launch. Al Bransford films behind them. Both Summers and Bransford hold 16mm 3-lens Arriflex S turret cameras.

Top center: NASA photographer "Pat" Patnesky aboard MV *Retriever* with the Apollo 1 crew during recovery training, off the Texas coast, on October 27, 1966

Top right: Photographer Bob Special at LC-34 for the Apollo 1 spacecraft arrival on January 6, 1967

Center left: John de Bry at the LC-39 Press Site before the Skylab 1 launch on May 14, 1973, with Pad A at left (de Bry collection)

Center middle: Jacques (*left*) and Michel Tiziou at Pad 39A on July 15, 1969, the day before the Apollo 11 launch

Center right: Technicolor photographer Alex Bosmeny films the Apollo 10 rollout along the crawlerway at LC-39 on March 11, 1969. (photo by Jacques Tiziou)

Bottom left: George Nevin at KSC in the mid-1970s. Nevin uses a Steadicam stabilization harness with a 16mm Eclair camera and an Angenieux zoom lens.

Bottom right: NASA photographer Ed Thomas (*wearing glasses*) confers with Apollo 16 LMP Charlie Duke, to his left, during EVA training at KSC on March 2, 1972. To Thomas's right is KSC Public Affairs Chief Chuck Hollinshead. NASA Public Affairs Officer Ed Harrison is at far right. (photo by Jacques Tiziou)

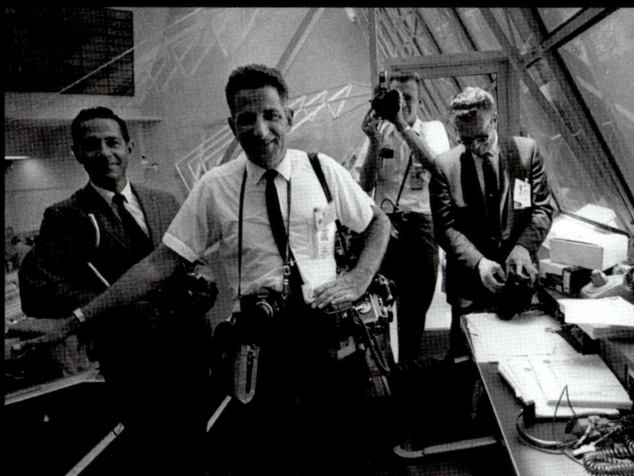
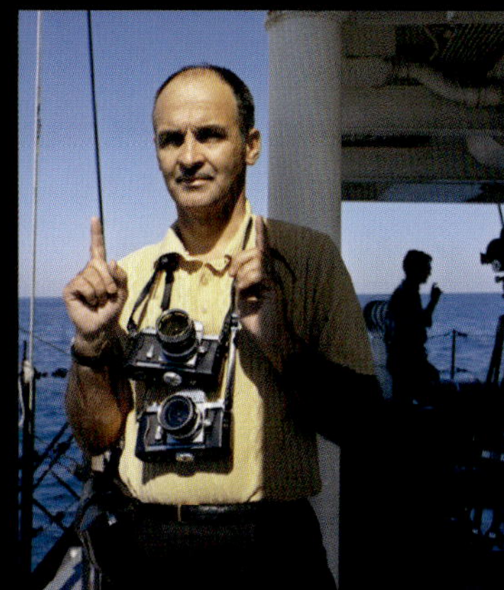
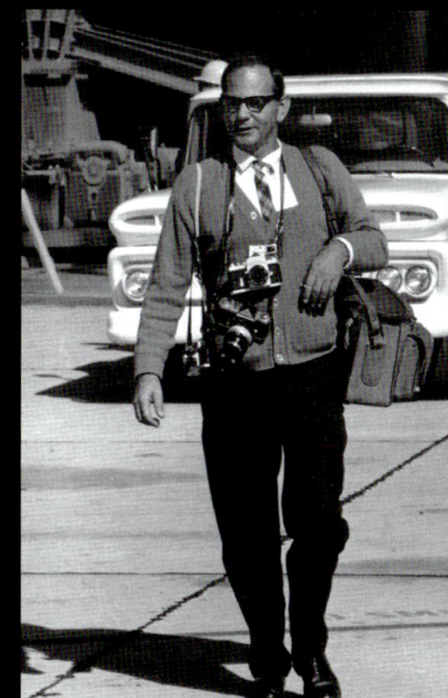
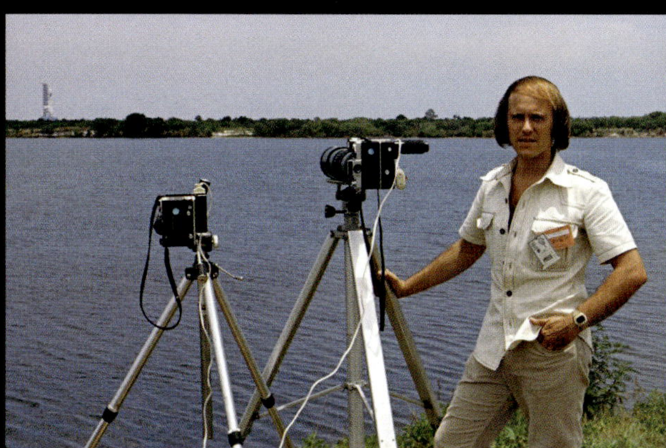
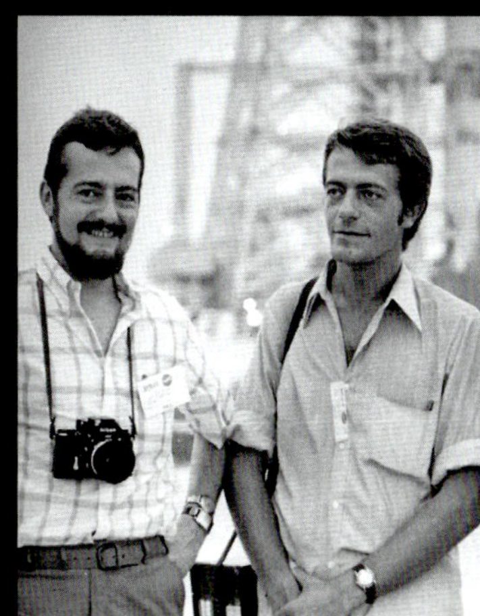
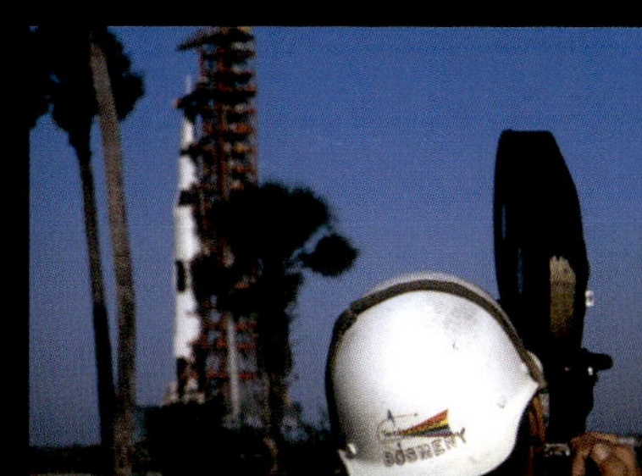
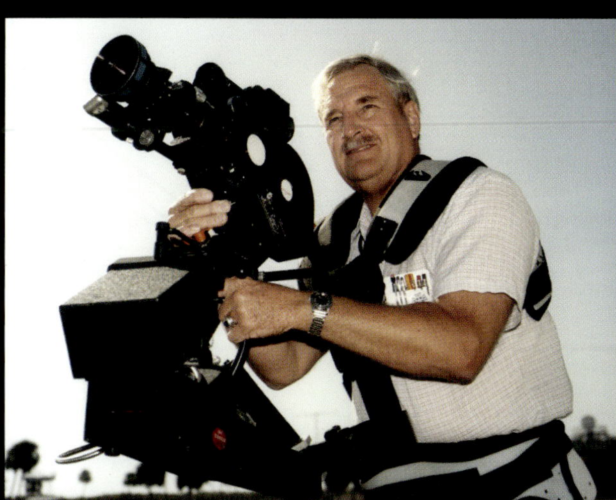
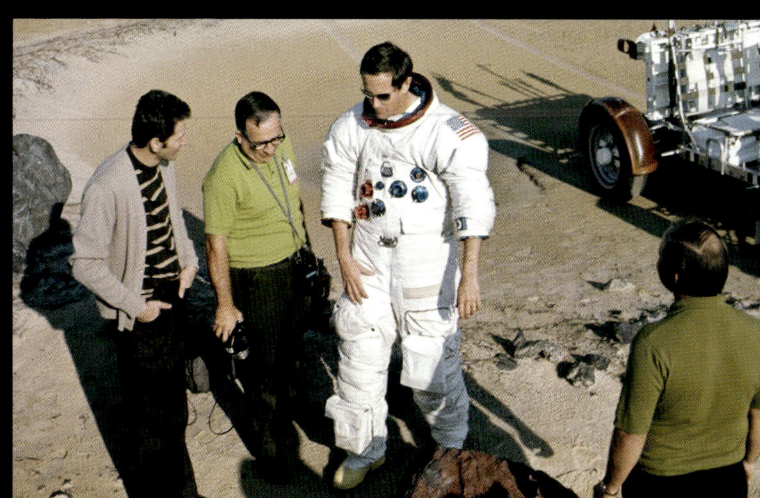

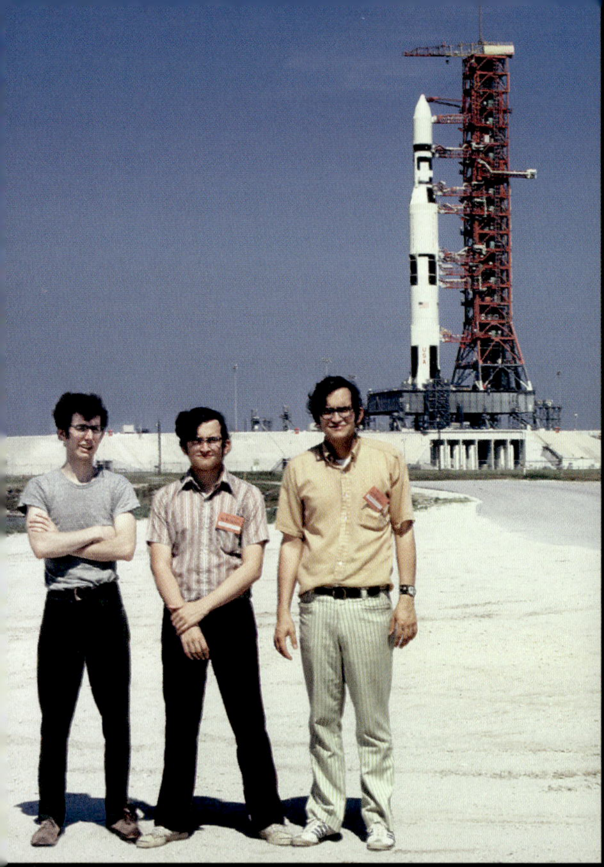

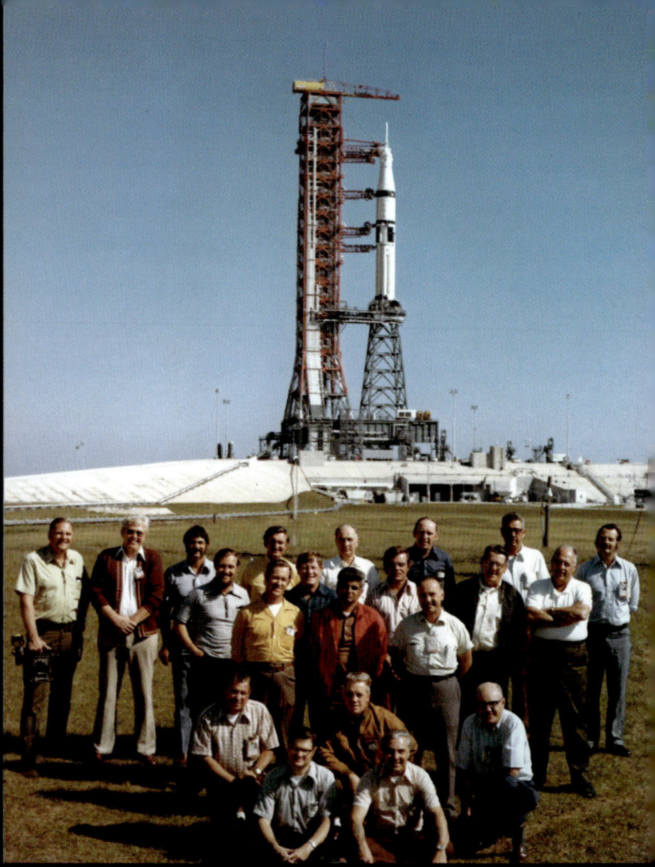

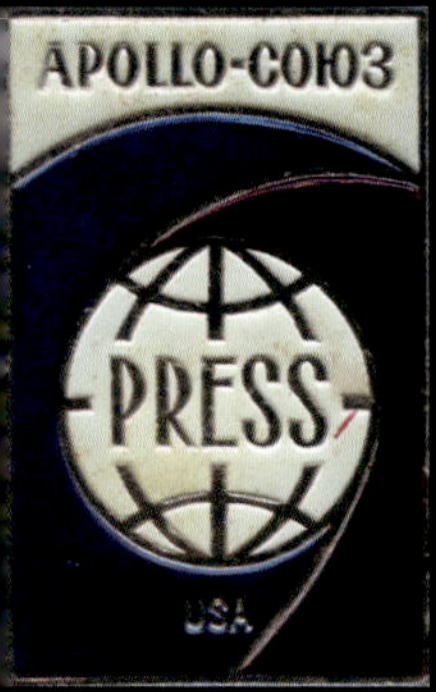

Top left: Steve Nolte (*left*) with Mark and Tom Usciak at Pad 39A on May 12, 1973, two days before the Skylab 1 launch (Usciak collection)

Top right: The KSC photo team at Pad 39B, after the Skylab rescue vehicle was rolled out on December 3, 1973

Bottom right: J. L. Pickering at Pad 39B before the ASTP launch on July 14, 1975 (Pickering collection)

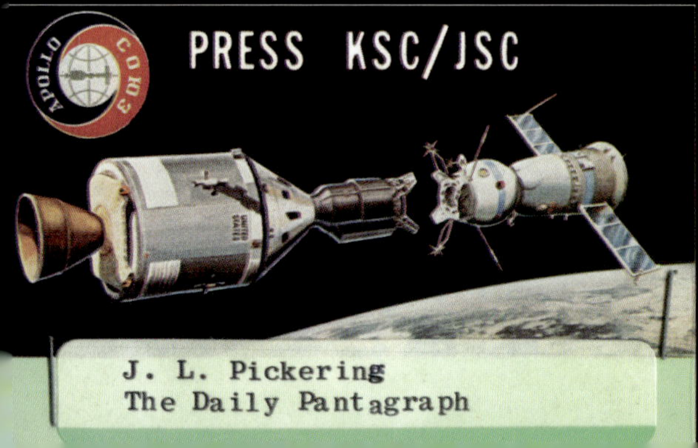

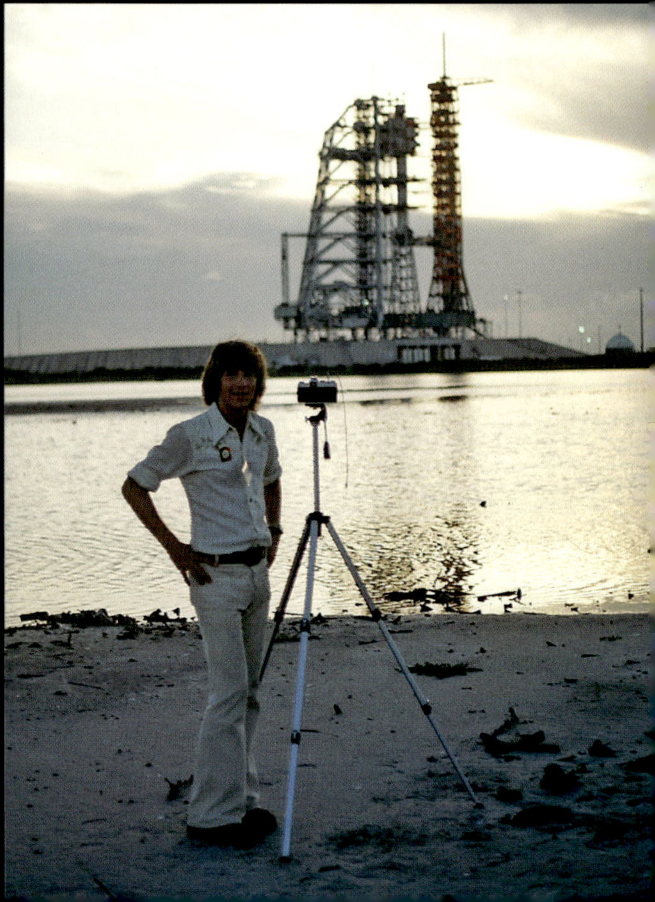